The Evolution and History of Human Populations in South Asia

Vertebrate Paleobiology and Paleoanthropology Series

Edited by

Eric Delson

Vertebrate Paleontology, American Museum of Natural History,
New York, NY 10024, USA
delson@amnh.org

Ross D. E. MacPhee

Vertebrate Zoology, American Museum of Natural History,
New York, NY 10024, USA
macphee@amnh.org

Focal topics for volumes in the series will include systematic paleontology of all vertebrates (from agnathans to humans), phylogeny reconstruction, functional morphology, Paleolithic archaeology, taphonomy, geochronology, historical biogeography, and biostratigraphy. Other fields (e.g., paleoclimatology, paleoecology, ancient DNA, total organismal community structure) may be considered if the volume theme emphasizes paleobiology (or archaeology). Fields such as modeling of physical processes, genetic methodology, nonvertebrates or neontology are out of our scope.

Volumes in the series may either be monographic treatments (including unpublished but fully revised dissertations) or edited collections, especially those focusing on problem-oriented issues, with multidisciplinary coverage where possible.

Published and forthcoming titles in this series are listed at the end of this volume.

The Evolution and History of Human Populations in South Asia

Inter-disciplinary Studies in Archaeology, Biological Anthropology, Linguistics and Genetics

Edited by

Michael D. Petraglia

Leverhulme Centre for Human Evolutionary Studies,
University of Cambridge, England

Bridget Allchin

Ancient India & Iran Trust, Cambridge, England

 Springer

A C.I.P. Catalogue record for this book is available from the Library of Congress.

ISBN 978-1-4020-5561-4 (HB)

ISBN 978-1-4020-5562-1 (eBook)

Published by Springer,
P.O. Box 17, 3300 AA Dordrecht, The Netherlands.

www.springer.com

Cover illustration: Sanyasula Cave, Kurnool District, India. Photo by Michael Petraglia

Printed on acid-free paper

Table of Contents

Preface

The two editors of this volume have spent considerable portions of their careers investigating the prehistory of the Indian subcontinent, though our research trajectories have different histories. One of us (Bridget Allchin) has been involved in study of South Asian prehistory for fifty years, conducted through various institutions, including the University of Cape Town, the Institute of Archaeology, London, the University of Cambridge, and the Ancient India and Iran Trust. On the other hand, the senior editor began his archaeological work in India a little over fifteen years ago, mostly under the affiliation of the Smithsonian Institution, in Washington, D.C. It was not until 2001 that the editors of this volume first met, as a consequence of Petraglia's appointment at the University of Cambridge. Our presence in Cambridge provided us with the opportunity to discuss our mutual interests in South Asian prehistory and to discuss plans for contributing to the future growth of archaeological investigations in the subcontinent. Our professional organizations, the Ancient India and Iran Trust, and the Leverhulme Centre for Human Evolutionary Studies, provided us with the institutional settings to facilitate our research plans. The Ancient India and Iran Trust, an independent charitable educational center dedicated to ancient culture and language, has facilitated the research of numerous South Asian scholars since the 1970s. The Leverhulme Centre for Human Evolutionary Studies is a university-affiliated center devoted to global studies of hominin evolution and culture change, with particular interests in expanding archaeological research projects in the subcontinent. Our mutual interests are to encourage research on human evolution and cultural diversity in South Asia and to promote the international exchange of scholars and students.

One of our first, collective projects was the organization of the joint conference, "South Asia at the Crossroads: Biological, Archaeological, and Linguistic Approaches to Cultural Diversity". On the 26th and 27th of November of 2004, two full days of talks were given at Brooklands House, Cambridge. Our goal was to reconsider what we knew about South Asian prehistory and begin to develop new ways of examining human evolutionary processes and cultural change. The 'crossroads' theme was a metaphorical representation, intentionally selected for our conference title for several reasons. One aim of our conference was to assemble a group of scholars who had never been brought together before, despite the fact that their research centers on, or directly bears upon, the biological and cultural record of South Asia. Researchers were brought together who had specialized experience in fields as diverse as archaeology, ethnography, biological anthropology, population genetics, linguistics, geology, and mammalian paleontology. Hence, by encouraging an inter-disciplinary examination of bio-cultural processes on the subcontinent, the editors felt that we could achieve a new 'crossroads' of understanding. Research both in and outside of the subcontinent have demonstrated that the region served as a dispersal route during multiple periods in the Plio-Pleistocene as well as a central region for culture contacts amongst agricultural communities to the west and east. Therefore, in this sense, the subcontinent certainly served as a 'crossroads' for hominin dispersals, population exchanges, and cultural interactions. And, finally, the 'crossroads' theme was selected to represent our feeling that while past and current investigations have shown that the prehistory of the subcontinent is remarkable in many respects, the record remains under-studied in comparison to other regions, and, unfortunately, the results continue to be downplayed in global syntheses. Thus, the editors feel that research on the biological and cultural history of the region is at a critical threshold, or 'crossroads', which requires fresh outlooks and new inter-disciplinary investigations to extend, improve and widen the scope of further research.

The "South Asia at the Crossroads" conference would not have been implemented without the administrative and financial assistance of the Ancient India and Iran Trust and the Leverhulme Centre for Human Evolutionary Studies. For their organizational help in the planning and execution stages of the conference, we would particularly like to thank James Cormick, Jose John, Ursala Sims-Williams, Madeline Watt, and Richard Wielechowski. The financial support of the British Academy and the Society for South Asian Studies was invaluable for the development of the conference, providing us the opportunity to invite scholars from Europe, North America, and South Asia.

The production of *The Evolution and History of Human Populations in South Asia* benefited from the guidance of Eric Delson, one of the Senior Editors of the Vertebrate Paleontology and Paleoanthropology Series. Eric's enthusiasm from the start of this project paved the way for a speedy publication. Tom Plummer, an Advisory Editor of the Series, introduced us to Eric at the Montreal meetings of the Paleoanthropology Society, and is thus responsible for initiating contacts. While on sabbatical from Cambridge, Petraglia made use of the facilities of the Maison de l'Archéologie et de l'Ethnologie (Universite de Paris, Nanterre). Françoise Audouze and Hélène Roche are thanked for providing office space during the height of our editorial review.

A number of scholars helped in the execution of the conference and academic enhancement of this book. We would like to particularly thank Ken Kennedy for providing opening remarks at the conference, including his fascinating personal reflections about the growth of bioanthropological research in South Asia. François Jacquesson provided us with an excellent introduction and overview of languages of the subcontinent. Conference chairs (Ken Kennedy, Greg Possehl, Graeme Barker, Martin Jones) helped to keep the sessions organized and conference discussants (Rob Foley, Paul Mellars, Colin Renfrew, Chris Stringer) provided their views on the contributions of the South Asian record and how it fit with the global picture of human evolution and culture change.

Chapters in this book underwent formal peer-review, and we thank the following individuals for providing comments on draft papers: P. Ajithprasad, Stan Ambrose, Mumtaz Baig, Larry Barham, Chris Clarkson, Eric Delson, Nick Drake, Peter Forster, John Gowlett, Colin Groves, Sudeshna Guha, Brian Hemphill, Christopher Henshilwood, Joel Irish, Lynn Jorde, Kriti Kapila, Mark Kenoyer, K. Krishnan, Curtis Larsen, Bob Layton, Richard Meadow, V.N. Misra, Rabi Mohanty, Rajeev Patnaik, Martin Richards, Phillip Rightmire, Derek Roe, Jerry Rose, Garth Sampson, Franklin Southworth, Richard Steckel, Sarah Grey Thomason, Martin Williams, and Michael Witzel. We thank Greg Possehl for providing concluding remarks in the closing section of this book.

Our spouses, Raymond Allchin and Nicole Boivin, deserve special recognition for their unwavering support of our research and their intellectual contributions to our research in South Asia. We dedicate this volume to them.

Bridget Allchin
Michael D. Petraglia
Cambridge, England

List of Contributors

Bridget Allchin
Ancient India & Iran Trust
Brooklands House
23 Brooklands Avenue
Cambridge, CB2 2BG
England
indiran@aol.com

Sheela Athreya
Department of Anthropology
Texas A&M University College Station
Texas, 77843
USA
athreya@tamu.edu

Nicole Boivin
Leverhulme Centre for Human
Evolutionary Studies
The Henry Wellcome Building
Fitzwilliam Street
University of Cambridge
Cambridge, CB2 1QH
England
nlb20@cam.ac.uk

Daniel G. Bradley
Smurfit Institute
Department of Genetics
Trinity College
Dublin 2
Ireland
dbradley@tcd.ie

Robin Dennell
Department of Archaeology
University of Sheffield
Northgate House
West Street
Sheffield, S1 4ET
England
r.dennell@sheffield.ac.uk

Phillip Endicott
Henry Wellcome Ancient
Biomolecules Centre
Department of Zoology
University of Oxford
Oxford, OX1 4AU
England
phillip.endicott@zoology.oxford.ac.uk

Dorian Q Fuller
Institute of Archaeology
University College London
31–34 Gordon Square
London, WC1H 0PY
England
d.fuller@ucl.ac.uk

Hannah V.A. James
Leverhulme Centre for Human
Evolutionary Studies
The Henry Wellcome Building
Fitzwilliam Street
University of Cambridge
Cambridge, CB2 1QH
England
h.james@human-evol.cam.ac.uk

Sacha C. Jones
Leverhulme Centre for Human
Evolutionary Studies
The Henry Wellcome Building
Fitzwilliam Street
University of Cambridge
Cambridge, CB2 1QH
England
s.jones@human-evol.cam.ac.uk

Toomas Kivisild
Leverhulme Centre for Human
Evolutionary Studies
The Henry Wellcome Building
Fitzwilliam Street
University of Cambridge
Cambridge, CB2 1QH
England
t.kivisild@human-evol.cam.ac.uk

Ravi Korisettar
Department of History and Archaeology
Karnatak University
Dharwad, 580 003
India
korisettar@yahoo.com

Samanti Kulatilake
Department of Earth Sciences
Faculty of Science and Technology
Mount Royal College
4825 Mount Royal Gate
S.W., Calgary, Alberta, T3E 6K6
Canada
skulatilake@mtroyal.ca

Marta Mirazón Lahr
Leverhulme Centre for Human
Evolutionary Studies
The Henry Wellcome Building
Fitzwilliam Street
University of Cambridge
Cambridge, CB2 1QH
England
m.mirazon-lahr@human-evol.cam.ac.uk

John R. Lukacs
Department of Anthropology
University of Oregon
Eugene, Oregon 94703
USA
jrlukacs@oregon.uoregon.edu

David A. Magee
Smurfit Institute
Department of Genetics
Trinity College
Dublin 2
Ireland
mageeaa@tcd.ie

Hideyuki Mannen
Laboratory of Animal Breeding and
Genetics, Faculty of Agriculture
Kobe University
1-1 Rokkoudai Kobe 657
Japan
mannen@kobe-u.ac.jp

April McMahon
Linguistics and English Language
University of Edinburgh
14 Buccleuch Place
Edinburgh, EH8 9LN
Scotland
april.mcmahon@ed.ac.uk

Robert McMahon
Molecular Genetics Laboratory
Western General Hospital
Edinburgh, Crewe Road
Scotland
rmcmahon@staffmail.ed.ac.uk

Mait Metspalu
Department of Evolutionary Biology
University of Tartu, Estonian Biocentre
Riia 23b, 51010, Tartu
Estonia
mait@ebc.ee

Kathleen D. Morrison
The University of Chicago
Department of Anthropology
1126 East 59th Street Chicago
Illinois, 60637
USA
k-morrison@uchicago.edu

Hannah J. O'Regan
School of Biological and Earth Sciences
Liverpool John Moores University
Liverpool L3 3AF
England
h.j.o'regan@ljmu.ac.uk

K. Paddayya
Department of Archaeology
Deccan College
Pune, 411 006
India
deccancollege@vsnl.com

Shanti Pappu
Sharma Centre for Heritage Education
28 I Main Road
C.I.T. Colony, Mylapore
Chennai, 60004
India
spappu@vsnl.com

Michael D. Petraglia
Leverhulme Centre for Human
Evolutionary Studies
The Henry Wellcome Building
Fitzwilliam Street
University of Cambridge
Cambridge, CB2 1QH
England
m.petraglia@human-evol.cam.ac.uk

Gregory L. Possehl
Department of Anthropology
325 University Museum
University of Pennsylvania
Philadelphia, Pennsylvania 19104-6398
USA
gpossehl@sas.upenn.edu

Jay T. Stock
Leverhulme Centre for Human
Evolutionary Studies
The Henry Wellcome Building
Fitzwilliam Street
University of Cambridge
Cambridge, CB2 1QH
England
j.stock@human-evol.cam.ac.uk

Alan Turner
School of Biological and Earth Sciences
Liverpool John Moores University
Liverpool, L3 3AF
England
a.turner@ljmu.ac.uk

S.R. Walimbe
Department of Archaeology
Deccan College
Pune, 411 006
India
walimbes@vsnl.com

1. Human evolution and culture change in the Indian subcontinent

MICHAEL D. PETRAGLIA

Leverhulme Centre for Human Evolutionary Studies
The Henry Wellcome Building
Fitzwilliam Street
University of Cambridge
Cambridge, CB2 1QH
England
m.petraglia@human-evol.cam.ac.uk

BRIDGET ALLCHIN

Ancient India & Iran Trust
Brooklands House
23 Brooklands Avenue
Cambridge, CB2 2BG
England
indiran@aol.com

Given the extraordinary discoveries of human fossils in Africa, the fascinating finds of cave art in Western Europe, and the antiquity of agriculture in the Fertile Crescent, one may wonder, why study the South Asian record at all? The simple answer is that South Asia has its own remarkable finds, and an archaeological record that rivals in richness those in better known regions of the world. The more complicated and important reply, however, is that South Asia in addition has a distinctive archaeological record that challenges many of the models and theoretical frameworks that have emerged on the basis of findings made in these other regions. South Asia provides the opportunity to re-evaluate, refine and in some cases revise a number of major conclusions concerning our evolutionary history, including the evolution of human behavior, 'Out of Africa' models, the origins of sedentism and domestication, and the emergence of social complexity and urbanization.

South Asia is of course not just of interest to archaeologists. It is a land of incredible cultural, linguistic and ethnic diversity, and its contemporary populations have constituted the focus of a wide range of disciplines, including anthropology, linguistics, history, and genetics. In these disciplines too, South Asia has much to offer general theoretical

1

M.D. Petraglia and B. Allchin (eds.), The Evolution and History of Human Populations
in South Asia, 1–20.
© 2007 *Springer.*

models and frameworks, which again have often focused on better-studied parts of the world. Nonetheless, all of these research areas, like archaeology, suffer from a key problem, which is their isolation and lack of engagement with other disciplines investigating South Asia. This volume therefore constitutes a crucial and novel attempt to bring together a variety of disciplines, in particular archaeology, biological anthropology, linguistics and genetics, in the study of South Asia's past and current populations. *The Evolution and History of Human Populations in South Asia* is in this sense a historical undertaking, and presents the beginnings of what is hoped will be sustained and fruitful mutual engagement between those disciplines involved in the study of South Asia's rich cultural and biological diversity.

Introduction to South Asia

The term 'South Asia' is a political designation, meant to describe an area containing the modern nations of Bangladesh, Bhutan, India, Nepal, Pakistan, Sri Lanka, and the Maldives. South Asia is a large landmass, measuring about 4.4 million square kilometers in extent. India is the largest of the seven countries that make up the region, measuring 3.3 million square kilometers, or six times than the size of France. The size of the South Asian landmass in itself suggests that there is much to be gained from examining the history of human geography in the region, including population dispersals and cultural interactions of various ages.

South Asia contains nearly 1.5 billion inhabitants, or more than one in five people in the world today. Though four main language groups are differentiated by linguists (i.e., Indo-Aryan, Dravidian, Austroasiatic, Tibeto-Burmese), people across the region speak at least 657 languages and language dialects. The linguistic diversity of the subcontinent is matched by a wide and impressive cultural diversity. In India alone, for example, the 2001 census indicates that approximately 8% of the population, or 84 million individuals, belonged to scheduled tribes (though as Morrison notes in this volume, caution needs to be exercised in interpreting such administrative classifications). In recent years, geneticists have been particularly enthusiastic about tracing the history of agriculturalists, tribes, and castes, though as indicated by Boivin (this volume) and Endicott, Metspalu and Kivisild (this volume), some serious methodological and interpretive considerations remain to be taken into account.

The term, the 'Indian subcontinent' is a geographic unit, meant to differentiate the landmass from the rest of the Eurasian continent. One of the most striking features of the subcontinent is the Himalayan mountains, which rise to a height of 8,850 meters at Mount Everest, and provide a nearly impenetrable wall of mountains in the north (Figure 1). The Western and Eastern Ghats are other distinctive mountain ranges aligned along the western and eastern flanks of the peninsula. The Deccan plateau is a distinctive geographic province in the peninsula, characterized by extensive flows of volcanic rocks. Large river valleys, including the Indus, Narmada, Ganges, Bhramaputra, Godavari, and Krishna, cross the region, emptying their waters into the Arabian Sea and the Bay of Bengal. A number of geological basins are located across the subcontinent, with distinctive rock types and plentiful natural resources in the form of streams, springs and animal and plant communities (Korisettar, this volume). Certain geographic areas were therefore attractive on account of their natural resources and ease of travel, whereas other zones, such as large mountain chains, were barriers to human settlement and communication.

South Asia is a semi-arid, subtropical environment with characteristic summer and winter monsoons. The monsoon is of fundamental importance for sustaining life on the subcontinent. Rainfall patterns shape the distribution and abundance of flora and fauna, offer

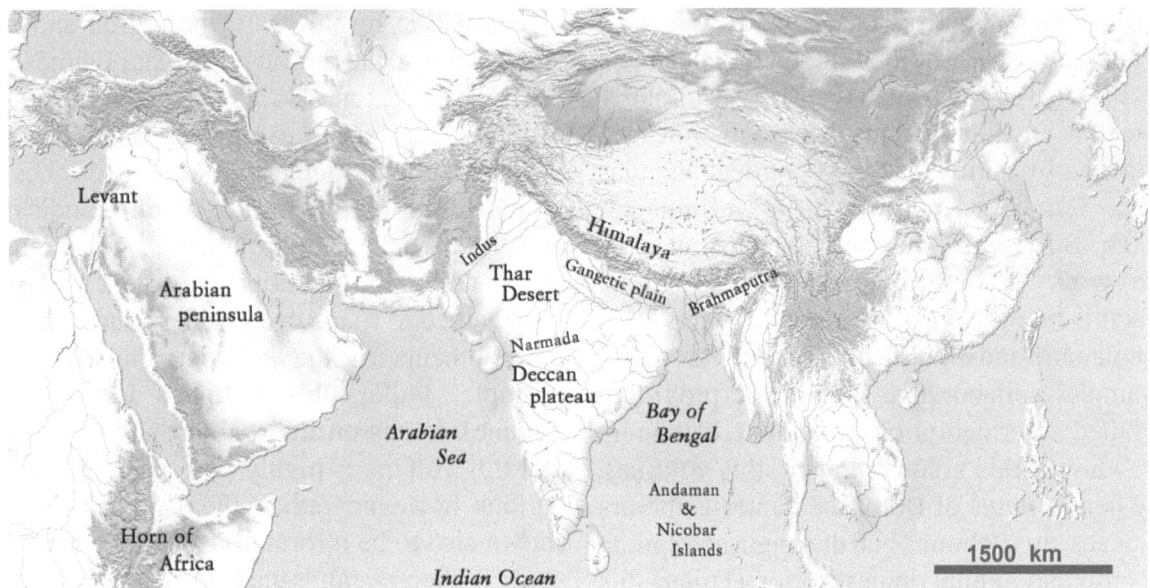

Figure 1. The Indian subcontinent showing key geographic features

essential nourishment to stressed ecologies during dry seasons, and present vital water supplies for sustaining domesticated plants and animals. The subcontinent has been a monsoonal environment since the Miocene, although fluctuations and shifts in its intensity are registered through time, in part due to the Himalayan-Tibetan uplift (e.g., Retallack, 1995; An et al., 2001). The stratigraphic record in the Thar Desert, or Great Indian Desert, demonstrates that phases of aridity (and dune formation) are sharply interspersed with periods of wetter, ameliorated climate (e.g., Andrews et al., 1998; Kar et al., 2001; Deotare et al., 2004). Monsoonal shifts during the Pleistocene and marked seasonal changes in wet and dry periods are thought to have structured hominin settlement behaviors and the survival of populations (Allchin et al., 1978; Paddayya, 1982; Korisettar and Rajaguru, 1998; James and Petraglia, 2005).

The prehistory and protohistory of the subcontinent is rather remarkable, as has been documented in many comprehensive books (e.g., Sankalia, 1974; Sharma, 1980; Allchin and Allchin, 1982; Misra and Bellwood, 1985; Dhavalikar, 1997; Lal, 1997; Kennedy, 2000;

Settar and Korisettar, 2002). These compendia clearly demonstrate the presence of a large number and wide range of archaeological sites, many providing significant cultural information. Among the better known of these sites are the Mode I occurrences in the Upper Siwaliks, the Acheulean quarry of Isampur, the Paleolithic and Mesolithic rockshelters of Bhimbetka, the Late Pleistocene caves of Kurnool, the Mesolithic cemeteries in the Ganga Plains, the early Neolithic settlement of Merhgarh, the Harappan cities in the Indus Valley, the Iron age centers in the Ganga Valley, and the architectural landscapes of the Vijayanagara Empire. Though some significant field work and research has been conducted in South Asia, serious problems hamper a fuller understanding of human evolution and societal change, including poor representation of hominin fossils, too few well-excavated sites revealing behavioral and social information, a poverty of detailed inter-disciplinary studies to recover paleoecological data, and poor chronological controls. One example illustrates the general problem: though there are more than 5,000 radiocarbon dates for Oxygen Isotope Stage 3 sites in western

Europe, only about 150 chronometric dates are available for the whole of the Late Pleistocene of India! To make matters worse, scholarly research and interactions may sometimes be impeded by nationalism and political agendas (e.g., Lamberg-Karlovsky, 1997; Ratnagar, 2004; Boivin, 2005). Yet, as illustrated by the work of contributors in this volume, much is being gained by current collaborative scholarship and inter-disciplinary research. For example, collaborative research is providing detailed information on Paleolithic behaviors (Paddayya, this volume; Pappu, this volume), the health status of Holocene hunter-gatherers (Lukacs, this volume), and the origin and spread of crop and animal domestication (Fuller, this volume).

The Evolution and History of Human Populations in South Asia examines the evolution of hominins and changes in human populations by employing information collected from different fields of inquiry, including archaeology, biological anthropology, linguistics and genetics. Contributors have provided information from their own areas of expertise and have drawn upon information from other research areas, thereby enriching interpretations about how the modern South Asian world of foragers, farmers, pastoralists, and urban industrialists was shaped by the past. An aim of this volume is to provide information from diverse disciplines to inform us about past processes, and in many cases, diverse information sources have converged to indicate research connections. However, as demonstrated in the following pages of this volume, sources of information sometimes show little correspondence and even provide conflicting interpretations. Naturally, such interpretive conflicts can not be rectified in this volume – instead, disagreements in conclusions suggest exciting directions for future research.

As is obvious, *The Evolution and History of Human Populations in South Asia* focuses on one region of the world, and as noted in our preface to this volume, we believe that such attention is justified in light of the wealth

of prehistoric resources in the region and its under-study in the field and in global syntheses. Focus on a region also has research merit as it allows analysts to examine how processes are shaped by particular circumstances. Thus, various contributions in this volume indicate that the particular environmental, geographic and demographic conditions prevalent in South Asia helped to shape evolutionary and cultural developments in the subcontinent (see for example, Fuller, this volume; James, this volume). Focus on the region does not mean that this volume is highly particularistic in its outlook however; rather, the regional data is shown also to be informative for investigating general theoretical issues and inter-regional relationships, including hominin dispersals and population interactions and exchanges, to name just two of the key issues addressed. In focusing on a large and culturally complex region, it will become apparent that we do not adequately report on all geographic areas, time periods and potential subjects. As exemplified in the three parts of this volume, however, a range of time periods and subjects is covered: Part I, 'Setting the Foundation', presents fossil and behavioral evidence for the earliest periods of hominin occupation; Part II, 'The Modern Scene', examines anatomically modern humans and the evolution of their behavior; and, Part III, 'New Worlds in the Holocene', concentrates on cultural and linguistic diversity and population interactions in the last 10,000 years.

Setting the Foundation

One of the most intriguing questions in paleoanthropology today concerns the timing of the earliest exit of hominins out of Africa. Exciting fossil discoveries in the Transcaucus date the earliest dispersal to ca. 1.8 Ma. The Dmanisi locality indicates that the earliest colonization of Eurasia was by relatively small-brained hominins that manufactured tools reminiscent of those found in the Oldowan of Africa (Gabunia et al., 2001). Archaeological evidence in the Levant indicates the presence of hominins by 1.4 Ma (Bar-Yosef, 1998). While fossil localities in Indonesia indicate that *H. erectus* reached Java by 1.8 Ma (Swisher et al., 1994), the earliest hominin occupation of the Indian subcontinent remains poorly known and controversial (Table 1).

One of the best contenders for an early occupation of the Indian subcontinent comes from the Upper Siwaliks of Pakistan

(e.g., Rendell et al., 1989; Dennell, 2004). As reviewed by Dennell in this volume, localities in the Soan Valley and Pabbi Hills consist of unstandardized cores and flakes, indicating typological and temporal affinity with the Oldowan. Although the findings of the Upper Siwaliks research has not been conducted without criticism (e.g., Hemingway and Stapert, 1989; Petraglia, 1998; Klein, 1999),

Table 1. Key events of Indian subcontinent, addressed in human evolution and cultural diversity of South Asia

Approximate Age	Issues and Questions	Nature of the Evidence
2–1.4 Ma	Colonization of South Asia by early hominins, presumably early *Homo erectus*	Dates of presence of *Homo erectus* in Indonesia by 1.8 Ma and in Eastern Asia by 1.36 Ma. In subcontinent, only potentially convincing evidence is at Riwat, Soan Valley, and in the Paabi Hills, Pakistan.
1.2 Ma	Colonization of South Asia by Acheulean hominins, presumably *Homo erectus*	Acheulean hominins present in West Asia by 1.4 Ma, suggesting the date of initial dispersal. Dating at Isampur Quarry suggests the possibility for a long chronology. If Acheulean hominins present, population sizes were not likely large.
500 ka	Colonization of South Asia by Acheulean hominins, possibly *Homo erectus* or *Homo heidelbergensis*	Acheulean evidence is geographically widespread, indicating adaptive success, although spatial and temporal lacunae may indicate small population sizes. Narmada hominin represents earliest hominin find to date.
250–200 ka	Transition from Late Acheulean to Middle Paleolithic	A transition from large cutting tool industries to prepared core and flake technologies appears gradually. This likely indicates adaptive changes and does not support an abrupt change as part of a dispersal event from Africa. Though the transition is present (e.g., Bhimbetka III-F-23, Kaladgi Basin), the age of the technological transition is not radiometricaly dated.
200–74 ka	Who were the makers of the Middle Paleolithic?	Middle Paleolithic localities are geographically widespread. Flake technologies dominate assemblages, though techniques of manufacture vary, including regular core and prepared core techniques. The early phases of the Middle Paleolithic of the subcontinent are coincident with the

(Continued)

Table 1. (Continued)

Approximate Age	Issues and Questions	Nature of the Evidence
		African MSA, the Mousterian of West Asia and Western Europe, and the pebble core industries of Eastern Asia.
74–55 ka	What was the effect of the Toba volcanic super-eruption on hominin populations? Did modern humans drive archaic hominins to extinction?	Evidence for the Toba volcanic super-eruption is present in marine and terrestrial settings of South Asia. This volcanic event would have directly impacted resident populations. Genetic studies indicate the arrival of modern humans into the region by 73–55 ka. Arrival of moderns is hypothesized to take place during the Middle Paleolithic.
55–10 ka	What is the evidence for "modern human behavior"?	Cultural patterns of South Asia show their own developmental trajectory in comparison to other regions. A wide array of core technologies is present in the Late Paleolithic (i.e., regular flake-, prepared-, blade-, bladelet-core). Modified space use occurs by 45 ka, and decorated objects appear by 28.5 ka. Microlithic tools and skeletons of *Homo sapiens* occur by 30–28.5 ka. Genetic data indicate a demographic expansion in the subcontinent between 30–20 ka. The Thar Desert expands to its maximum extent in the LGM. Microlithic site assemblages increase in frequency and abundance.
8 ka to present	What is the evidence for sedentism? Why do communities become sedentary?	A substantial population increase is hypothesized during the Mesolithic. The Ganga Plains Mesolithic sites are situated along extinct oxbow lakes. Some degree of permanence indicated by shelters with plastered floors, cemeteries, and reliance on lake resources.
	Are new crops, languages and populations introduced? Are developments indigenous or do they come from outside the region? What is the relationship to hunters and gatherers? Why do villages, trade networks and states develop?	Mehrgarh provides the earliest evidence for plant and animal domestication and village life. Other Neolithic transitions occur over the subsequent millennia in diverse regions of the subcontinent. While some appear to reflect the importation of crops, animals and/or people, other appear more autochthonous. Possible period of arrival of Indo-Aryan languages. Development of bronze and iron technology, emergence of state-level societies, first in the Indus Valley and then on the Ganges Plain. Iron Age megalithic cultures in south India.

the research is key to our understanding of the spread of hominins in the Late Pliocene and Early Pleistocene. Independent support for Dennell's prolonged research is provided by Turner and O'Regan (this volume), who argue that the most probable period for mammalian movements out of Africa are in the Late Pliocene, thus making a hominin dispersal distinctly possible.

Although Mode I tools are often described as 'simple' industries, their role in hominin adaptations was crucial to their survival. The development of such technologies in Africa by 2.5 Ma ensured access to a greater range of vital resources, such as meat and marrow (Plummer 2004). If hominins were tool-dependent creatures, as is usually assumed by archaeologists, then the availability of useable stone on the landscape is of critical importance. In contemplating the presence of tool-dependent hominins in the subcontinent, Dennell (this volume) has highlighted geographic differences in stone availability between the Siwalik Hills and the Indo-Gangetic drainage system. Dennell plausibly reasons that the wide floodplains of the Indo-Gangetic belt, with scarce workable stone, presented serious challenges to hominin adaptation, thereby accounting for the lack of archaeological sites. In this view, the conditions in the large river systems of the Indo-Gangetic region were not conducive to hominin settlement until deposition of boulder conglomerates in the early Middle Pleistocene.

Hominin movements from Africa to Eastern Asia are usually depicted by employing large, illustrative arrows through broad corridors (e.g., Bar-Yosef, 1998: Figure 8.1). Although the Indian subcontinent is in a central and southern margin position in Eurasia, it is unclear how, and even whether, the landmass was traversed by hominin populations. In thinking about the earliest hominin dispersals, three distinct possibilities about routes arise. One rather negative possibility is that hominins avoided the subcontinent altogether, employing

northern routes for expansion. A northern route would suggest that there were barriers to migration and it might account for the absence of Mode I sites in the Indian subcontinent. A second alternative is that Mode I hominins ranged across the subcontinent in the Late Pliocene or the Early Pleistocene. In this scenario, sites of this period are either not yet recognized or, alternatively, the Soan Valley and Pabbi Hill localities represent evidence for dispersals in the Siwalik Hills and the sub-Himalayan zone. This would suggest that hominins were able to either circumvent or range across potential barriers including the Himalayas and the Ganges-Brahmaputra river system. A third alternative is that early hominins avoided inland continental areas and spread via coastlines. Turner and O'Regan (this volume) provide evidence that indicates hominin use of the littoral zones in different continents, pointing out that hominins may have had the ability to cross water bodies to reach islands. Unfortunately, South Asia presents no supporting evidence for the use of littoral zones by early hominins, though sites relatively close to coastlines have been identified (Marathe, 1981). Korisettar (this volume) states that coastal routes were not likely employed by hominins, arguing instead that movements would have been transcontinental owing to the attractive environments offered by inland basins. Though Korisettar provides an intriguing hypothesis, concerted efforts should also be made to survey near- and off-shore settings as sites may be present along the continental shelf (Flemming, 2004).

Whatever routes were chosen, predictions are that Mode I sites in the Indian subcontinent will likely remain relatively rare even if additional and focused surveys are performed. Though early hominins have been viewed as successful colonizers once they dispersed out of Africa (e.g., Antón and Swisher, 2004), Dennell (2003) has argued instead that Mode I localities throughout Eurasia indicate sporadic and discontinuous occupations. In this view, the handful of Late Pliocene and Early Pleistocene localities indicate that hominin occupations were not permanent and successful, which would be in line with the nature of the South Asian evidence.

In contrast with the rather scant evidence for Mode I sites, Acheulean localities are relatively abundant and found in many different geographic areas of the subcontinent (e.g., Petraglia, 1998; Pappu, 2001; Paddayya, this volume; Pappu, this volume). South Asia represents the eastern-most location for the expansion of Acheulean hominins, yet little is known about the date of their initial entry into the subcontinent. The first application of the ESR method on Acheulean assemblages yielded preliminary dates of 1.2 Ma at the Isampur Quarry (Blackwell et al., 2001). Otherwise, chronometric dates in Pakistan place the Acheulean to shortly before 780 ka (Dennell, 1998). All other Acheulean sites on the subcontinent indicate a short chronology, samples dating to between ca. 350 and 200 ka (e.g., Petraglia, 1998), though several of these samples provide a minimum estimate.

The Acheulean record of the subcontinent has been considered to be extremely 'rich' based on the presence of hundreds of identified Acheulean sites and the large number of artifacts present in many of these localities (Petraglia, 1998). While such a characterization is warranted from a general perspective, the number of Paleolithic sites in areas does not accord well with an inference for continuous occupation. Evaluation of the Acheulean sites in the Hunsgi-Baichbal Valley illustrate this point (see Paddayya, this volume for background). At first blush, the discovery of 196 Acheulean localities appears to indicate significant occupation, but a simple evaluation by time indicates otherwise. In a long chronology model of hominin occupation (i.e., ca. 750,000 years in duration), an average of one site would be produced every 3,826 years, whereas in a short chronology (i.e., ca. 250,000 years in duration) one site is produced

every 1,275 years. If site density is an indication of occupation permanency, then it must be concluded that the 196 sites distributed over 500 km^2 is a sign of little use. This information, together with the presence of Acheulean sites in a number of basins of peninsular India, likely indicates that populations shifted their locations over the course of the Pleistocene. Such an interpretation would account for the presence of Acheulean sites in many places across the subcontinent, though not in the density that would be expected if occupation was penecontemporaneous and continuous in each place. While the Acheulean evidence supports the notion that hominin populations were present in the region over hundreds of thousands of years, the record does not indicate that populations were large. Rather, Acheulean populations may have been rather small, either going extinct in certain areas or shifting their locations between various basins over time.

Although archaeological surveys have been performed in many parts of the subcontinent, successfully identifying Acheulean sites, too few excavations have been performed which aim to recover environmental and behavioral information. A recent exception has been the high-quality excavations at Attirampakkam, which have documented stratified paleolithic deposits extending to a depth of 9 m (Pappu et al., 2003; Pappu, this volume). Attirampakkam is now among the best stratified sequences in the subcontinent next to the 10 m deep sequence at Patne (Sali, 1989) and the 20 m sequence at the 16R dune (Misra, 1995). In this volume, Pappu documents the presence of large cutting tools and changes in the presence of gravels through time. Study of stone tool reduction strategies indicated that hominins shifted their procurement and transport practices in response to changing on-site conditions. Another significant project in recent years has been the excavation of the Acheulean quarry at Isampur (Paddayya and Petraglia, 1997; Paddayya, this volume). Isampur is remarkable in that it represents a spot where Acheulean hominins manufactured large cutting tools from limestone slabs. The site is so well-preserved that the actions of hominins can sometimes be observed, which is of course, a rare phenomenon for any Lower Paleolithic site. Study of space use and stone tool reduction techniques have provided insights into hominin learning, cognition and transport behaviors (Petraglia et al., 1999, 2005). Based on the encouraging results of the Isampur excavations, Paddayya (this volume) makes a plea to local archaeologists, calling for a closer examination of Acheulean society and social habits.

The wealth of the Lower Paleolithic evidence of the Indian subcontinent is not matched by finds of the artifact manufacturers themselves. Despite the large number of archaeological surveys and excavations over the subcontinent, the Narmada calvarium represents the only hominin find to date (Sonakia, 1985; Athreya, this volume). The fossil has been variably classified as *H. erectus*, archaic *H. sapiens* and *H. heidelbergenis* (e.g., Sonakia, 1985; Kennedy et al., 1991; Cameron et al., 2004). In inter-regional comparisons, analysts view the Narmada fossil as part of wide-ranging *H. heidelbergenis* populations (Rightmire 2001) or as descendents of an early *H. sapiens* clade from Europe, with little relationship to Asian *H. erectus* or *H. pekinensis* (Cameron et al., 2004).

A wide geographic reading of the fossil evidence indicates that Mode I hominins of the subcontinent would have been part of the initial Out of Africa expansions, evidenced by early Asian *H. erectus* populations (assuming the legitimacy of a Mode I in the subcontinent and a dispersal link between South and Eastern Asia). Hypotheses have been put forward that suggest that *H. erectus* speciated in Eurasia, though the lack of Pliocene fossil evidence precludes further consideration (see Turner and O'Regan, this volume). Early dates of 1.2 Ma and 780 ka for Isampur and the Upper Siwaliks may indicate expansions of later populations of *H. erectus* carrying an Acheulean technology.

On the other hand, if a short chronology for the subcontinent is favored, i.e., less than 500 ka, *H. heidelbergensis* populations may be responsible for the spread and manufacture of Acheulean technologies. Lending evidence for this hypothesis, the Narmada fossil is associated with late Acheulean technology and mammalian fossils dating to more than 236 ka (Kennedy, 2000; Cameron et al., 2004).

The dearth of Pleistocene hominin fossils in the subcontinent has been sometimes viewed as a product of survey intensity, leading Paddayya (this volume) to remark that concerted efforts should be made to find hominin fossils in probable locations. Indeed, the subcontinent does contain Pleistocene mammalian faunas in river valleys (Badam, 1979) and occasional mammalian teeth in Acheulean contexts (Paddayya, this volume; Pappu, this volume), giving paleoanthropologists hope that hominin fossils may be discovered. Yet, as demonstrated by Dennell in this volume, the mammalian fauna in the Pabbi Hills is biased in favor of the recovery of animals that are larger than humans, indicating that identifiable hominin fossils would be rare finds.

Late Middle Pleistocene sites in the subcontinent indicate a technological change from the manufacture of large cutting tools to core and flake technologies of the Middle Paleolithic. This technological transition has been described as a profound adaptive change, as hominins supplemented hand-held implements with hafted technologies (McBrearty and Brooks, 2000). A number of sites in the subcontinent indicate that the transition from Acheulean technology to core and flake technologies occurred gradually (e.g., Misra, 1985; Petraglia et al., 2003). If the subcontinent contains a local developmental sequence from the Lower to Middle Paleolithic, it would imply technological convergence and independent transitions amongst different continents (e.g., Debono and Goren-Inbar, 2001; White and Ashton, 2003) rather than development of prepared core technology in Africa, and a subsequent technological spread to other areas, including South Asia. Since there are no human fossils to mark this transition, it is presumed that Narmada hominins or their descendents developed the core-flake technologies. It should be noted that in the same period, pebble core technologies in Eastern Asia were being used (Gao and Norton, 2002), providing a potentially significant biogeographical and cultural contrast with South Asia.

The Modern Scene

One of the most interesting questions in human evolution concerns the origins of modern humans and their interaction with archaic hominins as they spread around the world. Though Dennell (2005) makes the case that modern humans may have speciated in Eurasia, the most parsimonious reading of the genetic and fossil evidence indicates that anatomically modern humans first arose in Africa (e.g., Cann et al., 1987; Stringer, 2001; Endicott et al., this volume). Though Neanderthals and early modern humans occupied the Levant between 250–47 ka, the Upper Paleolithic revolution is thought to mark the expanding behavioral niche of modern humans (Bar-Yosef, 1998b; Shea, 2003). Modern humans appear to have colonized uninhabited regions such as Australia by 50 ka (e.g., Bowler et al., 2003), though the most conservative dates place their presence there by 45–42 ka (O'Connell and Allen, 2004). The extinction of geographically isolated populations of *H. neanderthalensis* in western Europe (Mellars, 2005) and *H. floresiensis* in Indonesia (Brown et al., 2004) was likely influenced by expansions of modern humans and increased levels of competition. To account for the early spread of modern humans to Australasia, a southern dispersal model has been hypothesized (Lahr and Foley, 1994; Stringer, 2000).

As reviewed by Endicott, Metspalu and Kivisild in this volume, Y-chromosome and MtDNA data support the colonization of South Asia by modern humans originating in Africa. South Asian lineages, like others in Eurasia, belong to haplogroups M and N, apparently descended from the L3 haplogroup that arose in Africa ca. 85,000 years ago (e.g., Metspalu et al., 2004; Forster and Matsumura, 2005). Coalescence dates for haplogroup M, which is shared by most non-European populations, average to between 73–55 ka (Kivisild et al., 2000). Based on a reading of the MtDNA and Y loci data, Endicott, Metspalu and Kivisild indicate that a single, early migration was responsible for the initial settlement of Eurasia and Australia. Recent study of the Andamanese mtDNA have support the interpretation for a rapid colonization of the region towards Australia (Endicott et al., 2003; Macaulay et al., 2005; Thangaraj et al., 2005). Hence, the genetic evidence of South Asia indicates a predominantly late Pleistocene heritage with no major population replacement events in later periods (Endicott et al., this volume). Inter-regional analysis of contemporary and recent crania provide support for this hypothesis, indicating that South Asian populations form a relatively distinct and homogeneous cluster, which suggests that demographic expansion within the subcontinent was in situ throughout much of prehistory (Stock et al., this volume).

The genetic evidence has been persuasive, but unfortunately there are no fossils to corroborate the presence of modern humans in South Asia, other than the skeletons of *H. sapiens*, dated to ca. 31,000 years ago at Fa Hien Cave and ca. 28,500 years before present at Batadomba-lena (Deraniyagala, 1992). Nevertheless, the genetic coalescence date for the arrival of modern humans in the subcontinent, together with the earlier range of the Australian dates, plausibly indicate the presence of modern humans in South Asia between 70–55 ka. If such dates for South Asia are valid, it would imply that modern humans expanded into the region using Middle Paleolithic technologies (James and Petraglia, 2005; James, this volume). And if this is the case, modern humans would have likely encountered archaic hominins also using Middle Paleolithic technologies. Such a hypothesis has more in common with the interchange of modern and archaic populations in West Asia, and contrasts with models which indicate that modern humans expanded out of Africa after 50 ka, employing an 'Upper Paleolithic' package (Klein, 1999, 2000).

With respect to the dispersal of modern humans, one major question concerns the role of the Toba super-eruption. This 74,000 year old volcanic super-eruption is hypothesized to have led to global climatic deterioration and the decimation of human populations (Ambrose, 1998), though as described by Jones in this volume, the severity of the impact has been contested by various scholars. South Asia is an ideal place to examine the direct impact of the volcanic event on hominins as geologists have found terrestrial and marine deposits of ash (e.g., Acharyya and Basu, 1993; Westgate et al., 1998). Though tephra deposits are associated with archaeological assemblages in India, the age of the ash and its precise relationship with Acheulean tools has been vociferously debated (e.g., Acharyya and Basu, 1994; Mishra and Rajaguru, 1994; Shane et al., 1995). Jones' chapter reviews the geological, environmental, paleontological and archaeological evidence from global and regional records and examines whether the regional evidence provides evidence for population continuity, extinction or survivals in refugia. Though it is clear that more precise field work needs to be conducted to understand the impact on local populations, the 74 ka date of the Toba event provides a lower limit of the hypothesized colonization of the subcontinent by modern humans (Kivisild et al., 2003; James and Petraglia, 2005; Endicott et al., this volume). As indicated by Jones, a number of questions therefore remain to be addressed, including precisely how the volcanic supereruption impacted local populations, whether the local environments and food chains were disrupted by the fall-out of ash and subsequent erosional cycles, and if the volcanic event in some way provided a competitive advantage of modern humans over their archaic ancestors.

The movement of modern humans from Africa towards Australia is thought to have involved coastlines (Lahr and Foley, 1994; Stringer, 2000). Coastal hypotheses assume that populations would have been able to move rapidly along the coastline towards Australia. Recent terrain analyses through GIS analysis supports the liklihood of coastal dispersals (Field et al., 2006). The GIS-based analyses indicate that entry into South Asia is more likely to have employed a coastal corridor that originated to the west. Once present in South Asia, populations may have followed a number of routes, which included both coastal and terrestrial contexts. Yet, Korisettar (this volume) places considerable doubt on coastal hypotheses for human movements along the continental shelf. Korisettar points out that archaeological evidence for exploitation of littoral contexts is so far absent in the subcontinent. Instead, Korisettar argues that dispersals would have been transcontinental, as supported by the presence of archaeological sites in many inland basins. Regardless of the precise routes undertaken by modern humans, geographic factors would have favored local adaptations over movements, as populations would have encountered barriers (e.g., Himalayan mountains, Ganges-Brahamaputra drainage system; see Stock et al., this volume) and zones with high resource diversity (Korisettar, this volume).

While fossil and genetic evidence indicate that our origins lie in Africa sometime around 200 ka, the evolution of modern human behavior has been debated. The archaeological evidence has been used to support either a gradual behavioral assembly of modern behavior from the Middle Pleistocene onwards (McBrearty and Brooks, 2000) or a rapid acceleration of behavioral traits afte 50 ka (Klein, 1999). Though anatomically modern humans appear to have expanded to western Europe by ca. 45–40 ka, introducing new adaptive features and innovations (Mellars, 2005), others counter that Neanderthals may have independently developed advanced behavioral characterisitics before the arrival of *H. sapiens* (D'Errico, 2003). As indicated by James in this volume, a significant problem in current studies of the evolution of modern behavior is that the Middle-Upper Paleolithic transition of Europe is typically contrasted against the MSA and LSA of Africa. Yet, such a comparison has severe limitations as the adaptations and material culture of modern human populations evolving in Africa will differ from the behaviors of populations spreading towards western Europe at a later date. Thus, to understand how cultural traits and behaviors evolved or were expressed in archaic and modern human populations, it is vital to know how the European and West Asian record compares to South Asia, Eastern Asia, and Australia where *H. sapiens* were immigrants.

The hypothesized spread of modern humans to the subcontinent before 50 ka has important demographic and cultural implications (James and Petraglia, 2005). If an early entry date in the subcontinent is valid, anatomically modern humans would have likely been spreading with Middle Paleolithic technology and meeting indigenous populations with similar technologies. The use of Middle Paleolithic technology by both modern and archaic populations in South Asia is analogous to the situation in West Asia, where Mousterian tools were used by *H. sapiens* and Neanderthals. Modern humans were, of course, the only populations to survive in South Asia, though much is to be learned about the degree to which the two populations interacted and whether archaic populations were eventually driven to extinction by direct or indirect competition. As James illustrates, the Middle Paleolithic of South Asia was followed by increasing technological diversity (i.e., flake, blade, bladelets) in the Late Paleolithic. The initial transition to the Late Paleolithic is gradual and marked

by increasing usage of blades alongside flake core industries, signalling a range of adaptations through time and space by local populations. At 45 ka, intentional site modifications are evidenced, indicating modification of living spaces in open-air and rockshelter contexts (Kenoyer et al., 1983; Misra, 1989; Dennell et al., 1992). Geometric microliths and beads are found in the Late Paleolithic at ca. 30 ka, indicating the introduction of sophisticated hafting technologies and explicit forms of symbolism (Deraniyagala, 1992).

With respect to the evolution of 'modern human behavior', it is clear that the trajectory and patterning of cultural traits in South Asia differs from the African and West European record. As James points out in this volume, the archaeological record of South Asia shows no rapid or sudden appearance of the modern behavioral package that can be considered equivalent to the Aurignacian in Europe, nor does it indicate the spread of an Upper Paleolithic package at around 50 ka. While the sporadic and gradual appearance of explicitly symbolic artifacts and microlithic technology are similar to cultural developments in the MSA of Africa, such traits occur in South Asia at a much later date. Hence, the trajectory of 'modern human behavior' in South Asia is distinct in comparison to other regions. For the earliest periods, it is currently difficult to differentiate between the behavior of modern humans and archaic populations. After 45 ka, the archaeological record shows a gradual shift in technology and rare introductions of traits considered to be part of the modern human behavioral package. The gradual and mosaic-like pattern of change in cultural traits in time and space indicates that material culture and symbolic objects developed in fundamentally different ways as modern humans spread around the globe.

The development of new cultural innovations in the Late Paleolithic of South Asia may be related, in part, to fluctuations in the environment and demographic changes.

The wetter and more stable conditions during Oxygen Isotope Stage 3 (60–25 ka) may correspond with a demographic expansion proposed on the basis of mitochondrial DNA analysis between 30–20 ka (Kivisild et al., 1999; Endicott et al., this volume). Occupation in the Thar Desert becomes increasingly sparse after ca. 25 ka, reflecting the heightened aridity and loss of available water sources (e.g., Deotare et al., 2004; Misra, 2001). Increasing aridity and expansion of the Thar Desert would have reduced and fragmented existing populations. Such demographic and enviromental conditions may be tied to some cultural innovations occurring between 30–20 ka, including the introduction of microlithic assemblages and the manufacture of symbolic artifacts (i.e., beads and 'art').

New Worlds in the Holocene

Mesolithic sites are well represented in the Indian subcontinent, leading archaeologists to infer that the large number of sites relates to marked growth in human populations (e.g., Misra, 2001:498). The dramatic increase in Mesolithic sites and the presence of settlements in previously unoccupied areas are thought to be related to the wetter climate at the beginning of the Holocene and the abundance and diversity of plants and animals. Mesolithic settlements have fascinated archaeologists as they often contain thousands of tiny microlithic bladelets (with retouched tools often fashioned into geometric forms), and they are often found across landscapes containing dozens of rockshelters adorned with artistic depictions of various animals and humans.

Though microliths have been found in Late Pleistocene contexts in Sri Lanka, the advent of the Mesolithic is usually placed in the Early Holocene. As Morrison notes (this volume), the end of the Mesolithic is not at all precise, as researchers use the term to refer variously to sites with microlithic technology, foraging lifeways, or as a particular archaeological period between the Paleolithic and Neolithic. Serious problems prevent an understanding of foraging lifeways in the Early and Middle Holocene as most microlithic sites are not dated and there have been very few studies which aim to retrieve archaeobotanical and faunal remains. As a result, interpretations about the activity and adaptations of these hunting and gathering societies are severely hampered. Further complicating the situation, it is clear that agriculturalists use microlithic technology and trade with foragers, thus making it difficult to clearly differentiate and discern foraging patterns from the task-oriented activities of farmers and pastoralists.

The increased food supply available in the Mesolithic is thought to have led to a reduction in mobility, as reflected in the large size of sites, the appearance of more substantial habitation deposits, and the presence of cemeteries, particularly in the Ganga Valley (Misra, 2001). Many of the 200 Mesolithic sites that have been mapped in the Ganga Valley occur adjacent to oxbow lakes (Sharma, 1973). The degree of mobility for these groups is debated, though they appear to have been semi-sedentary based on the presence of huts, communal hearths and burial areas (e.g., Pal, 1994; Chattopadhyaya, 1996; Lukacs, 2002). The Mesolithic cemeteries of the Ganga plains are among the most remarkable anywhere in Asia (Pal, 1992; Kennedy, 2000). The sites of Sarai-Nahar-Rai, Mahadaha, Damdama, thought to range from the period 10–4 ka, have provided some of the best samples of Mesolithic skeletons in the subcontinent (see Lukacs, this volume). The human remains from these cemeteries have been described as having massive skulls, a tall stature and a high degree of musculoskeletal robusticity. Though the Ganga Valley crania fall within South Asian diversity, the crania tend to be larger than those of contemporary and recent populations, sharing most morphological similarity with the Vedda and Sri Lankan populations (Stock et al., this volume). As the analysts describe, the craniometric data may therefore point to more diversity in prehistory. Stock et al. also note that the later homogenization of populations may be due to adaptation to similar environmental and cultural conditions, which is a point that Walimbe makes in his chapter. Lukacs (this volume) comes to slightly different conclusions concerning biological distance amongst Holocene populations in South Asia. Suggesting the possibility for a similar genetic heritage, Lukacs demonstrates a relatively close relationship between the populations of the Ganga Mesolithic sites, Neolithic Mehrgarh, and Chalcolithic Inamgaon.

The conventional outline of agricultural origins in South Asia is that plant harvesting and settled life appeared in the north-western part of the subcontinent at around 8 ka, as evidenced by · the settlement at Mehrgarh (Jarrige, 1984). The first urban centers appeared somewhat later, at around 5 ka in the Indus Valley (Kenoyer, 1998; Possehl, 1999). Contemporaneous with and post-dating Harappan developments, smaller agricultural and pastoral communities were established across the subcontinent, as indicated by abundant Neolithic and Chalcolithic settlements at this period (e.g., Korisettar et al., 2002; Panja, 2002; Singh, 2002). The significance of food production should not be underestimated, as intensification and surpluses fuelled the expansion of pastoralists and farmers, the development of trade and exchange networks, the diversification of society into occupations, and eventually, to the development of towns and cities. As is indicated by Lukacs and Walimbe in this volume, the transition from a hunting and gathering way of life to an agricultural subsistence pattern had serious biological costs. Lukacs demonstrates that the hunting and gathering societies of the Ganga Plain were generally healthy and well adapted to their environment, showing low dental caries, absence of evidence for nutritional and infectious diseases, and rare evidence for trauma and markers of occupational stress. In contrast, Walimbe notes that agriculture and sedentism is associated with nutritional deficiencies, infectious disease, trauma, degenerative pathologies and episodic stress.

With respect to the origins of domestication, archaeologists have long argued that plants and animals were introduced to South Asia from centres to the west and east, with the dispersal of domesticates being at times accompanied by the human colonizers themselves (e.g., MacNeish, 1992; Bellwood, 2005). Fuller (this volume) challenges such conventional thinking, arguing for a much more complex situation involving the development and exchange of domesticates by extra-regional and regional populations. Fuller contends that rice, water buffalo and chickens may have been separately domesticated in South Asia and East Asia, and suggests that additional varieties of these domesticates may have been introduced to these regions at a later date. In addition to agricultural imports from other regions, Fuller makes a persuasive argument for the domestication of certain animals (e.g., zebu, sheep) and varieties of plants (e.g., cotton, millets, pulses) in various geographic areas of the subcontinent, including the Indus basin, the middle Ganges, Gujarat, Orissa and south India. Based on genetic analysis of *Bos indicus*, Magee, Mannen and Bradley indicate the complexity of domestication, pointing out that cattle may have been domesticated in more than one place in the subcontinent.

Fuller's arguments in this volume also have a bearing on one of the most contentious and long-term research problems in South Asian prehistory, i.e., the origin and spread of languages, including hypotheses which envision that Indo-European languages were imported by farming or later pastoral communities from the west (e.g., Allchin and Allchin, 1982; Renfrew, 1987). Though genes and languages have been correlated with demographic expansions of farming populations and migrations towards South Asia (Cavalli-Sforza et al., 1994), recent studies of mitochondrial DNA argue against a strong differentiation of peoples speaking Indo-Aryan and Dravidian languages (Metspalu et al., 2004; Endicott et al., this volume) and no support for the entry of 'Aryan' populations is found in physical anthropological data (Kennedy, 1995; Walimbe, this volume). Genetic studies have, however, recently been used to support the idea of migrations of Tibeto-Burman and Austro-Asiatic speaking groups from East and Southeast Asia into India (see Endicott et al., this volume), which is consistent with archaeological hypotheses which infer that the Austro-Asiatic Munda languages were introduced by Neolithic populations from the Northeast (e.g., Bellwood, 2005) (though see Fuller, this volume, for a contrasting opinion).

While South Asia contains a great diversity of languages that may be used to discern population histories and relationships, the process of language change is a complex subject, as linguistic patterns may be combined and recombined by complex social and cultural processes. Languages, of course, may diverge as a product of social mechanisms and geographic distances, indicating how problems may arise for the linguist in recordation and cross-comparison. To overcome such interpretive difficulties, and to determine the relationship of the large array of languages that linguists are confronted with in South Asia, McMahon and McMahon (this volume) illustrate how quantitative methods and network programming can help to sort out this situation. An alternative way of examining the historical relationship of South Asian languages meanwhile is offered by Fuller (this volume), who examines the archaeological evidence for plant and animal domestication and compares this with linguistic loanwords and terms. Fuller argues for a number of separate transitions from hunting and gathering to domestication, with subsequent demographic expansions of farmer languages in different regions (i.e., Proto-Munda in Eastern India, Proto-Dravidian in south India, Para-Munda in the Greater Indus,

Language X in the Ganges basin, and proto-Nahali in Gujarat or south Rajasthan). As Fuller notes, this implies that there was cultural and linguistic diversity amongst hunter-gatherer populations in South Asia prior to the Neolithic. In his linguistic-domesticates model, Fuller argues that subsequent language changes (e.g., introduction of Indo-Aryan) were as much social and political in origin as they were demographic.

While linguistic analyses will continue to inform prehistorians about the complex relationships between farmers and foragers in South Asia, there can be little doubt that the spread of agriculture had profound effects on hunting and gathering societies. Certain foraging groups are likely to have adopted agricultural and pastoral lifeways, while in other cases, the mobility of hunter-gatherers was likely reduced, leading to relict populations in more marginal areas. Though tribals and mobile communities are often thought to be descendents of Pleistocene and Mesolithic populations, little firm skeletal evidence has been found to discriminate between forager and food producing populations in South Asia (Kennedy, 2000). In contrast, recent genetic data has suggested that tribal populations and agriculturalists may be differentiated, supporting the notion of different population sources and histories (e.g., Cordaux et al., 2004a, 2004b) (though such divisions are disputed on genetic and anthropological grounds; see Endicott et al., this volume; Boivin, this volume; Morrison, this volume). A recent claim for the lack of genetic interchange between foragers and agriculturalists comes from the isolated tribes of the Andaman Islands, who show deep time depths in their MtDNA, thus indicating that they are descendents of an ancient human substratum (Endicott et al., 2003; Thangaraj et al., 2005) (though other geneticists dispute this interpretation, see Cordaux and Stoneking, 2003). Craniometric comparisons do not support the genetic differentiation of the Andamanese, but some morphological divergence does exist for the Veddas, whose phenotypic characteristics have always been recognized (Kennedy, 2000; Stock et al., this volume). Archaeological support can be found for interactions between foraging populations and Harappan communities (Possehl, 1976; Morrison, this volume). In this case, the similarity of crania between forager and farming communities has been considered to be the result of genetic exchange (Kennedy et al., 1984; see also Stock et al., this volume). It is interesting that most Indian 'tribals', including hunter-gatherer groups (e.g., Chenchus, Juangs), speak agricultural languages (either Dravidian or Munda), implying either that these populations are descended from agriculturalists (and have specialized as foragers) or they have in the past adopted neighboring agricultural languages, thus implying a long history of close interaction. Though it may be assumed that many foraging groups were assimilated into agriculturally based societies, it has also been suggested that some agriculturalists may have shifted to foraging (Morrison, this volume). Genetic arguments have also been presented to suggest that foragers were integrated into caste-based society, on lower levels (Cordaux et al., 2004a) though this is a contentious claim. Contrary to views that portray foragers in a negative and marginalized light, Morrison (this volume) shows that foraging communities were flexible and diverse in their economic and social strategies. Morrison illustrates that trade-based foraging with settled communities was pivotal in developing historical economies, and in some cases, even responsible for underwriting the formation of coastal polities.

The modern world of South Asia consists of a rich cultural legacy with a diverse range of lifestyles, ranging from the nomadic Veddas of Sri Lanka, to the Raika camel herders of Rajasthan, and the city dwellers of Mumbai with its 18 million inhabitants. The rich, cultural tapestry of the modern era is matched by an incredibly rich and complicated prehistoric heritage that is clearly evident in archaeological, biological and linguistic sources. It is hoped that chapters in *The Evolution and History of Human Populations in South Asia* will not only demonstrate that the subcontinent can provide useful information about our evolutionary history, but also that an integrated, and multidisciplinary approach to history and prehistory has its advantages. While a comparison of the viewpoints of different disciplines does not lead to easy answers, it does suggest a wealth of new directions for future study.

References

Acharyya, S.K., Basu, P.K., 1993. Toba ash on the Indian subcontinent and its implications for correlation of Late Pleistocene alluvium. Quaternary Research 40, 10–19.

Acharyya, S.K., Basu, P.B., 1994. Reply to comments by S. Mishra and S.N. Rajaguru and by G.L. Badam and S.N. Rajaguru on 'Toba ash on the Indian subcontinent and its implication for the correlation of Late Pleistocene alluvium'. Quaternary Research 41, 400–402.

Allchin, B.A., Goudie, A., Hegde, K., 1978. The Prehistory and Palaeogeography of the Great Indian Desert. Academic Press, London.

Allchin, B., Allchin, R., 1982. The Rise of Civilization in India and Pakistan. Cambridge University Press, Cambridge.

Ambrose, S., 1998. Late Pleistocene human population bottlenecks, volcanic winter and differentiation in modern humans. Journal of Human Evolution 34, 623–651.

An, Z., Kutzbach, J.E., Prell, W.L., Porter, S.C., 2001. Evolution of Asian monsoons and phased uplift of the Himalaya-Tibetan plateau since Late Miocene times. Nature 411, 62–66.

Andrews, J., Singhvi, A., Kuhn, R., Dennis, P., Tandon, S.K., Dhir, R., 1998. Do stable isotope data from calcrete record late Pleistocene monsoonal climate variation in the Thar Desert of India? Quaternary Research 50, 240–251.

Antón, S.C., Swisher, C.C. III, 2004. Early dispersals of *Homo* from Africa. Annual Review of Anthropology 33, 271–296.

Badam, G.L., 1979. Pleistocene Fauna of India. Deccan College, Pune.

Bar-Yosef, O., 1998a. Early colonizations and cultural continuities in the Lower Palaeolithic of western Asia. In: Petraglia, M.D., Korisettar, R. (Eds.), Early Human Behaviour in Global Context: The Rise and Diversity of the Lower Palaeolithic Record. Routledge Press, London, pp. 221–279.

Bar-Yosef, O. 1998b. The chronology of the Middle Palaeolithic of the Levant. In: Akasawa, T., Aoki, K., Bar-Yosef, O. (Eds.), Neanderthals and Modern Humans in Western Asia. Plenum Press, New York, pp. 39–57.

Bellwood, P. 2005. First Farmers: The Origins of Agricultural Societies. Blackwell, Oxford.

Blackwell, B.A.B., Fevrier, S., Blickstein, J.I.B., Paddayya, K., Petraglia, M., Jhaldiyal, R., Skinner, A.R. 2001. ESR dating of an Acheulean quarry site at Isampur, India. Journal of Human Evolution 40, A3.

Boivin, N. 2005. Orientalism, ideology and agency: examining caste in South Asian archaeology. Journal of Social Archaeology 5(2), 225–252.

Bowler, J.M., Johnston, H., Olley, J., Prescott, J., Roberts, R., Shawcross, W., Spooner, N., 2003. New ages for human occupation and climate change at Lake Mungo, Australia. Nature 421, 837–840.

Brown, P., Sutikna, T., Morwood, M.J., Soejono, R.P., Jatmiko, Wayhu Saptama, E., Rokus Awe Due, 2004. A new small bodied hominin from the Late Pleistocene of Flores, Indonesia. Nature 431, 1055–1061.

Cann R., Stoneking, M., Wilson, A. 1987. Mitochondrial DNA and human evolution. Nature 325, 31–36.

Cameron, D., Patnaik, R., Sahni, A., 2004. The phylogenetic significance of the Middle Pleistocene Narmada hominin cranium from central India. International Journal of Osteoarchaeology 14, 419–447.

Cavalli-Sforza, L., Menozzi, P., Piazza, A., 1994. History and Geography of Human Genes. Princeton University Press, Princeton.

Chattopadhyaya, U.C., 1996. Settlement pattern and the spatial organization of subsistence and mortuary practices in the Mesolithic Ganges Valley, north-central India. World Archaeology 27(3), 461–476.

Cordaux, R., Aunger, R., Bentley, G., Nasidze, I., Sirajuddin, S.M., Stoneking, M., 2004a. Independent origins of Indian caste and tribal paternal lineages. Current Biology 14, 231–235.

Cordaux, R., Deepa, E., Vishwanathan, H., Stoneking, M. 2004b. Genetic evidence for the demic diffusion of agriculture to India. Science 304, 1125.

Cordaux, R., Stoneking, M., 2003. South Asia, the Andamanese and the genetic evidence for an 'early'

human dispersal out of Africa. American Journal of Human Genetics 72, 1586–1590.

Debono, H., Goren-Inbar, N., 2001. Note of a link between Acheulian handaxes and the Levallois method. Journal of the Israel Prehistoric Society 31, 9–23.

Dennell, R.W., 1998. Grasslands, tool making and the hominid colonization of southern Asia: a reconsideration. In: Petraglia, M.D., Korisettar, R. (Eds.), Early Human Behaviour in Global Context: The Rise and Diversity of the Lower Palaeolithic Record. Routledge Press, London, pp. 280–303.

Dennell, R.W., 2003. Dispersal and colonisation, long and short chronologies: how continuous is the Early Pleistocene record for hominids outside East Africa? Journal of Human Evolution 45, 421–440.

Dennell, R.W., 2004. Early Hominin Landscapes in Northern Pakistan: Investigations in the Pabbi Hills. British Archaeological Reports, International Series, 1265.

Dennell, R.W., 2005. Comments on 'modern human origins and the evolution of behavior in the later Pleistocene record of South Asia' by H.V.A. James and M.D. Petraglia. Current Anthropology, in press.

Dennell, R., Rendell, H., Halim, M., Moth, E., 1992. A 45,000-year-old open-air Paleolithic site at Riwat, northern Pakistan. Journal of Field Archaeology 19, 17–33.

Deotare, B., Kajale, M., Rajaguru, S., Basavaiah, N., 2004. Late Quaternary geomorphology, palynology and magnetic susceptibility of playas in western margin of the Indian Thar Desert. Indian Geophysical Union 8(1), 15–25.

Deraniyagala, S.U., 1992. The Prehistory of Sri Lanka: An Ecological Perspective. Department of the Archaeological Survey, Government of Sri Lanka, Colombo.

D'Errico, F., 2003. The invisible frontier: a multiple species model for the origin of behavioral modernity. Evolutionary Anthropology 12, 188–202.

Dhavalikar, M.K., 1997. Indian Protohistory. Books and Books, New Delhi.

Endicott, P., Gilbert, M., Stringer, C., Lalueza-Fox, C., Willerslev, E., Hansen, A., Cooper, A., 2003a. The genetic origins of the Andaman Islanders. American Journal of Human Genetics 72, 178–184.

Flemming, N.C., 2004. Submarine prehistoric archaeology of the Indian continental shelf: a potential resource. Current Science 86, 1225–1230.

Forster, P., Matsumura, S., 2005. Did early humans go north or south? Science 308, 965–966.

Gabunia, L., Antón, S.C., Lordkipanidze, D., Vekua, A., Swisher, C.C., Justus, A., 2001. Dmanisi and dispersal. Evolutionary Anthropology 10, 158–170.

Gao, X., Norton, C., 2002. A critique of the Chinese 'Middle Palaeolithic'. Antiquity 76, 397–412.

Hemingway, M.F., Stapert, D., 1989. Early artifacts from Pakistan? Some questions for the excavators. Current Anthropology 30, 317–322.

James, H.V.A., Petraglia, M.D., 2005. Modern human origins and the evolution of behavior in the later Pleistocene record of South Asia. Current Anthropology, in press.

Jarrige, J.-F., 1984. Chronology of the earlier periods of the Greater Indus as seen from Mehrgarh, Pakistan. In: Allchin, B. (Ed.), South Asian Archaeology 1981. Cambridge University Press, Cambridge, pp. 21–28.

Kar, A., Singhvi, A., Rajaguru, S., Juyal, N., Thomas, J., Banerjee, D., Dhir, R., 2001. Reconstruction of the late Quaternary environment of the lower Luni plains, Thar Desert, India. Journal of Quaternary Science 16(1), 61–68.

Kennedy, K.A.R., 1995. Have Aryans been identified in the prehistoric skeletal record from South Asia? Biological anthropology and concepts of ancient races. In: Erdosy, G. (Ed.), The Indo-Aryans of Ancient South Asia: Language, Material Culture and Ethnicity. Walter de Gruyter, Berlin, pp. 32–66.

Kennedy, K.A.R., 2000. God Apes and Fossil Men: Paleoanthropology in South Asia. University of Michigan Press, Ann Arbor.

Kennedy, K.A.R., Chiment, J., Disotell, T., Meyers, D., 1984. Principal-components analysis of prehisotric South Asian Crania. American Journal of Physical Anthropology 64, 105–118.

Kennedy, K.A.R., Sonakia, A., Climent, J., Verma, K.K., 1991. Is the Narmada hominid an Indian *Homo erectus*? American Journal of Physical Anthropology 86, 475–496.

Kenoyer, J.M. 1998. Ancient Cities of the Indus Valley Civilization. Oxford University Press, Karachi.

Kenoyer, J., Clark, J., Pal, J., Sharma, G., 1983. An Upper Palaeolithic shrine in India? Antiquity LVII, 88–94.

Kivisild, T., Rootsi, S., Metspalu, M., Mastana, S., Kaldma, K., Parik, J., Metspalu, E., Adojaan, M., Tolk, H-V., Stepanov, V., Gölge, M., Usanga, E., Papiha, S.S., Cinnioğlu, C., King, R., Cavalli-Sforza, L., Underhill, P.A., Villems, R., 2003. The genetic heritage of the earliest settlers persists both in Indian tribal and caste populations. American Journal of Human Genetics 72, 313–332.

Klein, R., 1999. The Human Career. Chicago University Press, Chicago.

Klein, R., 2000. Archeology and the evolution of human behavior. Evolutionary Anthropology 9, 17–36.

Korisettar, R., Rajaguru, S., 1998. Quaternary stratigraphy, palaeoclimate and the Lower Palaeolithic of India. In: Petraglia, M.D., Korisettar, R. (Eds.), Early Human Behaviour in Global Context: The Rise and Diversity of the Lower Palaeolithic Record. Routledge Press, London, pp. 304–342.

Korisettar, R., Venkatasubbaiah, P.C., Fuller, D.Q., 2002. Brahmagiri and beyond: the archaeology of the southern Neolithic. In: Settar, S., Korisettar, R. (Eds.), Indian Archaeology in Retrospect: Prehistory, Volume 1. Indian Council of Historical Research and Manohar Publishers, New Delhi, pp. 150–237.

Lahr, M.M., Foley, R., 1994. Multiple dispersals and modern human origins. Evolutionary Anthropology 3, 48–60.

Lal, B.B., 1997. The Earliest Civilization of South Asia. Aryan Books International, New Delhi.

Lamberg-Karlovsky, C.C., 1997. Politics and archaeology. Colonialism, nationalism, ethnicity, and archaeology, Part I. The Review of Archaeology 18(2), 1–14.

Lukacs, J.R., 2002. Hunter-gatherers, pastoralists, and agriculturalists in prehistoric India: a biocultural perspective on trade and subsistence. In: Morrison, K., Junker, L. (Eds.), Forager Traders in South and Southeast Asia. Cambridge University Press, Cambridge, pp. 41–61.

Macaulay, V., Hill, C., Achilli, A., Rengo, C., Clarke, D., Meehan, W., Blackburn, J., Semino, O., Scozzari, R., Cruciani, F., Taha, A., Shaari, N.K., Maripa Raja, J., Ismail, P., Zainuddin, Z., Goodwin, W., Bulbeck, D., Bandelt, H.-J., Oppenheimer, S., Torroni, A., Richards, M., 2005. Single, rapid coastal settlement of Asia revealed by analysis of complete mitochondrial genomes. Science 308, 1034–1036.

MacNeish, R.S., 1992. The Origins of Agriculture. University of Oklahoma, Norman.

Marathe, A.R., 1981. Geoarchaeology of the Hiran Valley, Saurashtra, India. Deccan College Postgraduate and Research Institute, Poona.

McBrearty, S., Brooks, A., 2000. The revolution that wasn't: a new interpretation of the origin of modern human behaviour. Journal of Human Evolution 39, 453–563.

Mellars, P., 2005. The impossible coincidence. A single-species model for the origins of modern human behavior in Europe. Evolutionary Anthropology 14, 12–27.

Metspalu, M., Kivisild, T., Metspalu, E., Parik, J.,
Hudjashov, G., Kaldma, K., Serk, P., Karmin, M.,
Behar, D.M., Gilbert, M.T.P., Endicott, P.,
Mastana, S., Papiha, S.S., Skorecki, K., Torroni, A.,
Villems, R., 2004. Most of the extant mtDNA
boundaries in South and Southwest Asia were likely
shaped during the initial settlement of Eurasia by
anatomically modern humans. BMC Genetics 5, 26.

Misra, V.N., 1985. The Acheulean succession at
Bhimbetka, central India. In: Misra, V.N.,
Bellwood, P. (Eds.), Recent Advances in Indo-
Pacific Prehistory. Oxford IBH, New Delhi,
pp. 35–47.

Misra, V.N., 1989. Stone Age India: an ecological
perspective. Man and Environment XIV(1), 17–64.

Misra, V.N., 1995. Geoarchaeology of the Thar Desert,
north west India. In: Wadia, S., Korisettar, R.,
Kale, V.S. (Eds.), Quaternary Environments and
Geoarchaeology of India. Geological Society of
India, Bangalore, pp. 210–230.

Misra, V.N., Bellwood, P. (Eds.), 1985. Recent
Advances in Indo-Pacific Prehistory. Oxford IBH,
New Delhi.

Mishra, S., Rajaguru, S.N., 1994. Comment on 'Toba
ash on the Indian subcontinent and its implication
for the correlation of Late Pleistocene alluvium'.
Quaternary Research 41, 396–397.

O'Connell, J.F., Allen, J., 2004. Dating the colonization
of Sahul (Pleistocene Australia-New Guinea): a
review of recent research. Journal of Archaeological Science 31, 835–853.

Paddayya, K., 1982. Achuelian Culture of Hunsgi
Valley. Deccan College, Pune.

Paddayya, K., Petraglia, M.D., 1997. Isampur: an
Acheulian workshop in the Hunsgi Valley,
Gulbarga District, Karnataka. Man and
Environment 22, 95–100.

Pal, J.N., 1992. Mesolithic human burials in the Ganga
Plain, north India. Man and Environment 17(2),
35–44.

Pal, J.N., 1994. Mesolithic settlements in the Ganga
Plain. Man and Environment 19, 91–101.

Panja, S., 2002. Research on the Deccan Chalcolithic. In:
Settar, S., Korisettar, R. (Eds.), Indian Archaeology
in Retrospect: Prehistory, Volume 1. Indian Council
of Historical Research and Manohar Publishers,
New Delhi, pp. 263–76.

Pappu, R.S., 2001. Acheulian Culture in Peninsular
India. D.K. Printworld Ltd., New Delhi.

Pappu, S., Gunnell, Y., Taieb, M., Brugal, J.-P.,
Touchard, Y., 2003. Excavations at the Palaeolithic
site of Attirampakkam, south India: preliminary
findings. Current Anthropology 44, 591–598.

Petraglia, M.D., 1998. The Lower Paleolithic of
India and its bearing on the Asian record.
In: Petraglia, M.D., Korisettar, R. (Eds.), Early
Human Behaviour in Global Context: The Rise
and Diversity of the Lower Palaeolithic Record.
Routledge, New York, pp. 343–390.

Petraglia, M., LaPorta, P., Paddayya, K., 1999. The first
Acheulean quarry in India: stone tool manufacture,
biface morphology, and behaviors. Journal of
Anthropological Research 55, 39–70.

Petraglia, M.D, J. Schuldenrein, Korisettar, R., 2003.
Landscapes, activity, and the Acheulean to Middle
Paleolithic transition in the Kaladgi Basin, India.
Journal of Eurasian Prehistory 1(2), 3–24.

Petraglia, M.D., Shipton, C., Paddayya, K., 2005.
Life and mind in the Acheulean: a case study
from India. In: Gamble, C., Porr, M. (Eds.),
The Hominid Individual in Context. Routledge,
London, pp. 197–219.

Plummer, T. 2004. Flaked stones and old bones:
biological and cultural evolution at the dawn of
technology. American Journal of Physical Anthropology 125 (S39), 118–164.

Possehl, G.L., 1976. Lothal: a gateway settlement of
the Harappan Civilization. In: Kennedy, K.A.R.,
Possehl, G.L. (Eds.), Ecological Backgrounds of
South Asian Prehistory. Occasional Papers and
Theses No. 4, Cornell South Asia Program, Ithaca,
pp. 118–31.

Possehl, G.L., 1999. Indus Age: The Beginnings.
University of Pennsylvania Press, Philadelphia.

Ratnagar, S., 2004. Archaeology at the heart of political
confrontation: the case of Ayodhya. Current
Anthropology 45, 239–59.

Rendell, H.M., Dennell, R.W., Halim, M., 1989. Pleis-
tocene and Palaeolithic Investigations in the Soan
Valley, Northern Pakistan. British Archaeological
Reports, International Series, 544.

Renfrew, C., 1987. Archaeology and Language: The
Puzzle of Indo-European Origins. Cambridge
University Press, Cambridge.

Rettalack, G.J., 1995. Palaeosols of the Siwalik Group
as a 15 Ma record of South Asia palaeoclimate.
In: Wadia, S., Korisettar, R., Kale, V.S. (Eds.),
Quaternary Environments and Geoarchaeology of
India. Geological Society of India, Bangalore,
pp. 123–135.

Rightmire, G.P., 2001. Comparison of Middle Pleis-
tocene hominids from Africa and Asia. In:
Barham, L., Robson-Brown, K. (Eds.), Human
Roots: Africa and Asia in the Middle Pleistocene.
Western Academic and Specialist Press, Bristol,
pp. 123–135.

Sali, S.A., 1989. The Upper Palaeolithic and Mesolithic Cultures of Maharashtra. Deccan College Post Graduate and Research Institute, Pune.

Sankalia, H.D., 1974. Prehistory and Protohistory of India and Pakistan. Poona: Deccan College.

Settar, S., Korisettar, R. (Eds.), 2002. Indian Archaeology in Retrospect (Volumes 1–4). Indian Council of Historical Research and Manohar Publishers, New Delhi.

Shane, P., Westgate, J., Williams, M., Korisettar, R., 1995. New geochemical evidence for the youngest Toba tuff in India. Quaternary Research 44, 200–204.

Sharma, G.R., 1980. History to Prehistory: Archaeology of the Ganga Valley and the Vindhyas. University of Allahabad, Allahabad.

Sharma, G. R. 1973. Mesolithic Lake Cultures in the Ganga Valley, India. Proceedings of the Prehistoric Society 39, 129–146.

Shea, J., 2003. The Middle Paleolithic of the East Mediterranean Levant. Journal of World Prehistory 17, 313–394.

Singh, P., 2002. The Neolithic cultures of northern and eastern India. In: Settar, S., Korisettar, R. (Eds.), Indian Archaeology in Retrospect: Prehistory, Volume 1. Indian Council of Historical Research and Manohar Publishers, New Delhi, pp. 127–150.

Sonakia, A., 1985. Skull cap of an early man from the Narmada valley alluvium (Pleistocene) of central India. American Anthropologist 87, 612–616.

Stringer, C., 2000. Coasting out of Africa. Nature 405, 24–27.

Stringer, C., 2001. The morphological and behavioural origins of modern humans. In: Crow, T. (Ed.), The Speciation of Modern *Homo sapiens*. Oxford University Press, Oxford, pp. 23–30.

Swisher, C.C. III, Curtis, G.H., Jacob, T., Getty, A.G., Suprijo, A., Widiasmoro, 1994. Age of the earliest known hominids in Java, Indonesia. Science 263, 1118–1121.

Thangaraj, K., Chaubey, G., Kivisild, T., Reddy, A.G., Singh, V.K., Rasalkar, A.A., Singh, L., 2005. Reconstructing the origin of Andaman Islanders. Science 308, 996.

Westgate, J.A., Shane, P.A.R., Pearce, N.J.G., Perkins, W.T., Korisettar, R., Chesner, C.A., Williams, M.A.J., Acharyya, S.K., 1998. All Toba Tephra Occurrences across Peninsular India Belong to the 75,00 yr B.P. Eruption. Quaternary Research 50, 107–112.

White, M., Ashton, N., 2003. Lower Palaeolithic core technology and the origins of the Levallois method in North Western Europe. Current Anthropology 44, 598–609.

PART I
SETTING FOUNDATIONS

2. Afro-Eurasian mammalian fauna and early hominin dispersals

ALAN TURNER AND HANNAH J. O'REGAN

School of Biological and Earth Sciences
Liverpool John Moores University
Liverpool, L3 3AF
England
a.turner@ljmu.ac.uk
h.j.o'regan@ljmu.ac.uk

Introduction

Although doubts about the African origins of the human lineage have been raised in recent years, particularly over the origins of *Homo erectus* (White, 1995; Dennell, 2004), the continent is conventionally seen as the home of the Hominini, or hominins, the tribe to which we and our closest fossil relatives belong, with most discussion centred on the timing of earliest movements out (Turner, 1999; Roebroeks, 2001; Dennell, 2003). The development of stone tool technology there from around 2.5 Ma (Semaw, 2000) undoubtedly made various aspects of life more efficient and perhaps easier for hominins, and must have given them the ability to occupy new and more challenging habitats, while the archaeological 'fact' that earliest hominin dispersals appear to have originated in Africa makes the establishment of a timescale for emergence an obvious and legitimate topic for investigation. It seems self-evident that the increasing freedom technology would have brought from the constraints imposed by circumstances could have laid the ground for dispersals into untried and perhaps previously unoccupiable areas, and may even suggest to us that tools might have provided a substantial stimulus to emigration. But while developing technology may have made large-scale movements possible, and the presence of artifacts (in the absence of fossils) outside Africa may point to places passed and points reached by a given date, such things by themselves tell us little about actual timescale, or reasons for movement. Nor can they really bear on the question of direction, at least without more precise dating. For insight into such matters we argue that a wider perspective is needed, one that takes account of the larger picture of the paleobiogeography of the terrestrial mammalian fauna of which the African hominins were a part. It will allow us to see that the order Primates certainly does not originate in Africa, and that the Hominidae, the family to which living and fossil great apes and humans

M.D. Petraglia and B. Allchin (eds.), The Evolution and History of Human Populations in South Asia, 23–39.

belong, may even have a rather complicated biogeographic past involving movement out of and back into Africa during the Miocene.

Our aim here is to set the scene for understanding the geographic origins of the hominins – and thus our emphasis throughout is on the larger patterns of Afro-Eurasian movement, rather than on events within South Asia *per se*. We are less concerned here with detailed changes in habitat pattern. We would argue that physical barriers offer a more useful basis on which to examine movements and possible constraints on movement, although of course we take account of conditions in regions that would have been traversed on any particular route. It should also go without saying that any such attempt to provide an overview is based on interpretations of the patterns we can currently discern in the fossil record.

We begin, therefore, with a review of the background pattern of mammalian movement into and out of Africa from Miocene times onwards, in an effort to discern the earliest most likely period for movement by a terrestrial primate equipped with what, to begin with, was a relatively unsophisticated toolkit. In the process we discuss the implications of an early timescale for the species of hominin making the earliest dispersals. We then go on to discuss the likely routes and direction of such dispersals, and from there assess the possible impetuses to movement that may have existed and which may go some way to tackling the deceptively simple question of why such dispersals took place. We then consider the likely attractions of coastal areas for early hominins, and examine the possibility that at least some may have been used as migration routes.

Africa – Geography and the Pattern of Mammalian Dispersals

Africa and southern Arabia joined the Eurasian plate in the early Miocene, around 25 to 18 Ma, making a land connection at the eastern end of

the Tethys Seaway in what is now the Mediterranean (Rögl, 1999). Primates, which appear to have originated in the northern hemisphere (Fleagle, 1999; Bloch and Silcox, 2006), are known in the Oligocene-age Fayum deposits of Egypt from as early as 33 to perhaps even 40 Ma, and an even earlier appearance around 45 Ma is suggested by material from Algeria (Godinot and Mahboubi, 1992), implying some form of island hopping across the shortening gap, although these may be minimum ages. Tabuce and Marivaux (2005) also summarize a number of lines of evidence for early movements by mammals, particularly of anthracotheriid artiodactyls from Asia to the Fayum and to Bir El Atar in Algeria, and clearly there is more to be known about events during these earlier periods of dispersal. Much of the later Miocene movement of faunas into and out of Africa probably took place across southern Arabia (Tchernov, 1992), although we know little of the Miocene mammalian fauna of that region (Whybrow and Clements, 1999). The movement of Africa, and of India to the east, into the Eurasian landmass also formed the mountain chains of the Taurus and Zagros, and across Afghanistan into the Himalayas, placing further control on movements into and out of Africa (Tchernov, 1992). Towards the end of the Miocene contact with the Atlantic was shut and the Mediterranean began to dry up during the Messinian Crisis (Kirjksman et al., 1999), but while some form of land bridge therefore existed between southern Spain and northern Africa at that time there is no compelling evidence for one across the Gibraltar Straits since then. Most, if not all, subsequent mammalian movements between Africa and Eurasia during Plio-Pleistocene times are therefore likely to have been via the Levant and perhaps Arabia (see below). During the early Pliocene the Mediterranean re-filled, while the Red Sea widened as the Arabian plate swung away and eventually broke the Bab-el-Mandeb land bridge in the Late Pliocene, part of the rifting process that changed the topography of eastern Africa massively, and which continues to the present day.

The Fayum primates were diverse (Kirk and Simons, 2001), with at least 17 genera, although their relationship to modern taxa are still unclear, but primitive apes of the family Proconsulidae have a reasonably good record from deposits of late Oligocene to Mid-Miocene age (Andrews and Humphrey, 1999). More advanced apes appear in a further radiation in the period 17 to 12 Ma, but their taxonomic placement is problematic, to say the least. In a summary treatment, Turner and Antón (2004) referred them to the subfamily Dryopithecinae as the tribes Afropithecini, Kenyapithecini and Dryopithecini within the family Hominidae, although Andrews (1996) formerly regarded them as subfamilies and now (Andrews, in litt) suggests recognising them as families within the Hominoidea. Although the Afropithecini occur in Africa and Arabia, the Kenyapithecini mostly occur in Turkey and southeastern Europe with some known from Kenya while the Dryopithecini are European in distribution, and we seem to have evidence for a major dispersal from Africa. This early dispersal from Africa may have implications for our understanding of the later geographic origins of the Homininae, the subfamily containing the African great apes and humans, a point to which we shall return briefly below.

The Late Eocene appearance of the Primates in Africa seems to have been accompanied only by a dispersal of the archaic predator order Creodonta, but a clearer pattern emerges in the early Miocene. Other immigrants from Eurasia at that latter time included the first rhinos and the bizarre-looking chalicotheres, giraffoid climactocerids and first antelopes as well as primitive pigs of the genus *Nguruwe*, which must have traversed the continent since they are known from Namibia and Kenya at around 17.5 Ma (Turner and Antón, 2004). True Carnivora entered the continent with the first appearance of cats and the amphi-cyonid bear-dogs as well as of mustelids (Morales et al., 1998) and, at least in North Africa, of members of the extinct cat-like family Nimravidae, although the creodonts continued to prosper as the dominant meat eaters. In the other direction, the probable dispersal of dryopithecine apes was perhaps preceded by the appearance of anthracotheres in Europe and possibly accompanied by a dispersal of primitive catarrhines of the genus *Pliopithecus*, the creodont *Hyainailouros* and the first movement of the proboscideans from Africa (Agustí and Antón, 2002).

The dryopithecine hominoid apes that dispersed from Africa are well-represented in Eurasia, including at times the Siwaliks, until the late-Miocene (Andrews and Humphrey, 1999; van der Made and Ribot, 1999; Begun et al., 2003). But in Africa, later Miocene fossil apes are relatively scarce (Andrews and Humphrey, 1999; Leakey and Harris, 2003). If that is a real pattern rather than an accident of discovery (a problem that always besets paleontology) then the ancestor for the later great ape and human lineage, the subfamily Homininae, may lie among those Eurasian apes as Solounias et al. (1999) have suggested, arguing that many of the savanna-dwelling mammals of Africa may well have originated in the Pikermi Biome and moved into Africa by around 8 Ma following a major shift in climate and a turnover in the western European fauna (Agustí et al., 1999). Movement of early hominin apes back into Africa is therefore entirely plausible as part of this larger pattern of dispersal, which would of course indicate a complex origin for the hominins rather than a simple evolution within Africa since earliest appearance there of the Primates, a point argued some years ago by Begun (1994) based on postcranial similarities between great apes and *Dryopithecus*.

The Miocene patterns of dispersal increased in complexity during the Messinian Crisis, with various African reduncine and hippo-tragine bovids and the more primitive hippo genus *Hexaprotodon* appearing in western Eurasia (Agustí and Antón, 2002), and continued into the Pliocene. Elephants of the genus *Mammuthus* exited Africa by 3.0 Ma, dogs of the genus *Canis* entered and the African bovids *Damalops, Oryx, Hippotragus* and *Kobus* appeared in Tadzhikistan and Siwalik deposits around 3–2.5 Ma. In the other direction, true, single-toed horses of the genus *Equus* entered Africa by 2.3 Ma, as did the bovids *Hemibos, Antelope, Rabaticerus, Pelea* and an ovibovine. The spotted hyaena, *Crocuta crocuta*, may have exited and *Parahyaena brunnea* may have entered *or* exited, around 2.5–2.0 Ma (Turner, 1992, 1995; Turner and Wood, 1993; Turner and Antón, 2004).

However, once we reach the Pleistocene, the rate of dispersals into and out of Africa slows down and the pattern becomes much less apparent. Lower Pleistocene movements out seem to consist only of the living large hippo genus *Hippopotamus*, the large baboon *Theropithecus*, perhaps the leopard, *Panthera pardus*, and possibly an African variety of the sabretoothed cat *Megantereon*. Middle Pleistocene emigrants consist of the lion, *Panthera leo*, and perhaps *Panthera pardus* if it did not move earlier (Turner, 1995), together with the African hunting dog *Lycaon*, although the latter only appears to have got as far as Israel (Stiner et al., 2001). Events of the Late Pleistocene involve perhaps the striped hyaena, *Hyaena hyaena* and of course *Homo sapiens*.

So far as the Indian subcontinent is concerned, we have referred above to the presence of African bovids of the genera *Damalops, Oryx, Hippotragus* and *Kobus* in Tadzhikistan and Siwalik deposits of late Pliocene age, and to earlier appearances of hominoid apes there. But despite a long history of work on the fossil remains from the area, and evidence that dispersals have taken place in both directions between India and the rest of Eurasia and even Africa since at least the mid Miocene (van der

Made, 1999; van der Made and Ribot, 1999, Begun et al., 2003), it remains difficult to give an overall and coherent account of Miocene-Pleistocene dispersal patterns that integrates the region fully, particularly in relation to Pliocene and Pleistocene events. Dennell (2004) has recently provided a brief overview, based on work by authors such as Barry (1995; Barry et al., 2002) and the now perhaps rather dated review by Lindsay et al. (1980). It seems clear that true horses of the genus *Equus* were also mid-late Pliocene immigrants, as were elephants and perhaps deer (Barry, 1995). The recent work in the Pabbi Hills of northern Pakistan reported by Dennell and co-authors substantiates the continued presence of *Damalops*, while the presence of the giant hyaena *Pachycrocuta*, the smaller *Crocuta* and the machairodont cat *Megantereon* (Turner, 2004) all almost certainly betoken dispersals into the Siwalik region in the late Pliocene and into the earliest Pleistocene. But the timescale of their movements, and indeed the routes of their dispersals, remain unclear.

What we can say with certainty, however, is that the overall feature of the Afro-Eurasian mammal fauna is therefore one of dispersals, such that even the Primates have not had a single pattern of movement into any given area and a subsequent indigenous evolution. From a purely paleontological perspective, the most probable period for hominin movements out of Africa might therefore logically be seen as late Pliocene, a time when a number of other species were managing to disperse into or out of the continent. What do we know of the earliest African hominins, the likely candidates for such movements, and of their distributions?

African Hominins

Currently, the earliest known putative hominins are the 7–6 million year old *Sahelanthropus tchadensis* from Toros Menalla in the Chad

Basin (Brunet et al., 2002) and the 6–5.5 million year old *Orrorin tugenensis* from Lukeino in Kenya (Senut et al., 2001). These are followed by the specimens of what is now named *Ardipithecus kadaba* from deposits of 5.8–5.2 Ma in the Middle Awash of Ethiopia, which Haile-Selassie et al. (2004) consider is very similar to *Orrorin* and *Sahelanthropus*, and the 4.5 Ma material of *Ardipithecus ramidus* from the Middle Awash (White et al., 1995) and to the west of the Awash (Semaw et al., 2005). Of course these represent only a small number of localities and specimens, so that drawing inferences about distributions is at best hazardous.

For later hominins, evidence suggests that between 3.5 and perhaps 2.0 Ma the genus *Australopithecus* was geographically split, and represented by *A. afarensis* in eastern Africa and *A. africanus* in the south (Turner and Wood, 1993). The genus *Paranthropus* is represented by *P. robustus* and perhaps *P. crassidens* in the south and by *P. aethiopicus* and *P. boisei* in the east and perhaps as far south as Malawi (Bromage and Schrenk, 1995). *Australopithecus anamensis*, a more primitive species, has been identified at East Turkana and Kanapoi in deposits dated to around 4.0 Ma (Leakey et al., 1995) while another, *A. bahrelghazali*, has been recognized at Koro Toro in Chad in deposits of slightly later age (Brunet et al., 1996). A third, *A. garhi*, has been identified in deposits of the Middle Awash Valley (Asfaw et al., 1999), and a fourth, placed in a new genus as *Kenyanthropus platyops* (Leakey et al., 2001), has been identified from deposits dated to 3.5 Ma at West Turkana.

The genus *Homo* has been conventionally identified in Africa back to about 2.5 Ma, first represented in eastern Africa by the species *H. habilis* and *H. rudolfensis*, although the latter has also been identified in Malawi by Bromage and Schrenk (1995). The fact that stone tools appear in the archaeological record at about the same time has led to speculations about the relationship between evolutionary

change and the development of tool-making abilities, and has of course been linked to ideas about earliest dispersals by our own genus. However, early species of *Homo* are a rather strange collection (Wood, 1991, 1992), and proposals to remove them from *Homo* altogether and leave *H. erectus* as the earliest clear member of the genus have recently re-emerged (Wood and Collard, 1999). This would place the first and seemingly abrupt appearance of our genus, in the form of *H. erectus*, after the earliest appearance of stone tools and with no obvious African antecedents, and has led to the suggestions of a possible origin outside Africa referred to above (White, 1995; Dennell, 2004). This would also make the earliest African tool-makers members of another hominin genus, but of course that is not inherently implausible.

The known distribution of Pliocene and perhaps Late Miocene hominins now stretches from Chad down through eastern Africa to South Africa, perhaps even from the Atlantic coast of the western Sahel down to the Cape (Brunet et al., 1995), although occupation of the northwestern region is unknown prior to ca 1.0 Ma (Raynal et al., 2001). But the true limits of distribution remain unclear, and Dennell (2004) has even suggested that if Pliocene hominins were in Chad some 2,500 Km west of the Rift Valley by 3.5 Ma then there is no obvious reason for them not to have reached as far to the north or east by the same period, which of course would place them in Arabia or even southwestern Asia. But if they did emerge so early, then what route(s) might they have taken?

Ways Out of (or Into) Africa

The only plausible routes out of Africa since the end Miocene have been by land across Sinai and the Levant, across the Bab-el-Mandeb Straits at the south of the Red Sea and then across the Arabian Peninsula proper or by the Gibraltar Straits. So far as the last is concerned, as we summarize elsewhere (O'Regan et al., in press), the vast bulk of the evidence argues firmly against this route, however superficially attractive it may appear. Of the extant and Holocene terrestrial mammals present in North Africa and Iberia, such as wild boar (*Sus scrofa*), red deer (*Cervus elephas*), otter (*Lutra lutra*) and the red fox (*Vulpes vulpes*), there are none that are not also present in the Levant, suggesting either a circum-Mediterranean route (Dobson and Wright, 2000) or a much greater degree of dispersal than assumed previously. We therefore believe that the Arabian Peninsula and the Levant remain the only established route of a two-way movement between Africa and Eurasia during the Pliocene and Pleistocene, as indicated by the mixed Afro-Eurasian nature of the fauna of the region since the later part of the Pliocene, and in particular by the Early Pleistocene deposits at 'Ubeidiya in Israel (Tchernov, 1992; Turner, 1999, Belmaker et al., 2002), and by the extensive – although undated – evidence of Lower Paleolithic industry there (Petraglia, 2003). Late Pliocene African bovids and giraffids are known to the north of the Taurus-Zagros mountains at Kuabebi in the Caucasus and Wolacks in Greece (Sickenberg, 1967), the Oltet Valley in Romania (Radulesco and Samson 1990) and Huélago in southern Spain (Alberdi et al., 2001), part of a larger faunal turnover in Eurasia (Azzaroli et al., 1988). However, there are few such elements at Dmanisi (Gabunia et al., 2000, 2001; Vekua et al., 2002).

Choosing between the Sinai-Levant route and the Bab-el-Mandeb Straits is difficult, as summarized by Petraglia (2003), although we have suggested elsewhere (Turner, 1999; Turner and O'Regan, in press) that the Straits may have offered the less attractive route for terrestrial mammals during the Lower and Middle Pleistocene. This is largely because the times when the route was most obviously available would have been when low sea level during glacial periods was probably offset by increased aridity in the region, making Arabia inhospitable (Glennie and Singhvi, 2002) although the coastal area of the eastern Red Sea might have remained attractive (see below). However, movement during the later Pliocene, before the Straits had fully formed and the crossing was dependent on sea level fall, may have been an entirely different matter, as previously pointed out (Turner, 1999). In combination, either synchronically or serially, the Levant and the Straits are therefore likely to have been the scene of significant movements of terrestrial mammals, and suggestions of possible hominin dispersals into Eurasia during the later Pliocene (Bonifay and Vandermeersch, 1991; Boitel et al., 1996), while unsupported by critical assessments of the evidence within Eurasia, cannot be dismissed *a priori* as impossible or even unlikely, paralleling some of the conclusions reached by Mithen and Reed (2002) and Dennell (1998, 2004). But whether such a view offers support for an extra-African origin for *H. erectus* remains less clear. There is nothing that we know of in the fossil record that would support massive early dispersals out of Africa by hominins and nothing that suggests intensive occupation, even if toeholds were gained. The known distribution of Lower Paleolithic material in Arabia remains frustratingly undated (Petraglia, 2003). Whatever the earliest appearance of hominins outside Africa, a consensus view would favour a shorter chronology for more extensive and intensive occupation of Europe in particular and of Eurasia in general – that is after around 0.5 Ma – with a tail of more sporadic appearances back to and perhaps even prior to the Plio-Pleistocene boundary (Turner, 1999; Roebroeks, 2001; Dennell, 2003, 2004). Whether these earliest dated appearances represent one or more discrete dispersal events from Africa is a matter for debate (Aguirre and Carbonell, 2001; Dennell, 2003; Petraglia, 2003), but the reasons for such a pattern are likely to include both the physical difficulty of getting out of Africa and the difficulty of maintaining a presence in Eurasia during the Early Pleistocene. Among the factors affecting the latter are likely to have been the presence of large predators and the resultant availability of resources for more primitive hominins coupled with the problems of maintaining a foothold in temperate regions subjected to periodic glaciations (and intense seasonal cold even during interglacials) and where food acquisition may have been a problem (see Turner, 1992, for a detailed discussion of these issues). However, the problems of getting out in the first place should not be overlooked. So what might have encouraged, or provoked, such movements? One possibility, touched on by Petraglia (2003), lies in the apparent attraction of coastal regions and the resources on offer there in particular as a means of getting into Arabia if not beyond. It is to that point we now turn.

Coastal Activities of Early Hominins

Modern humans have made extensive and intensive use of coastal resources, often causing changes to local ecosystems, both in the past (Mannino and Thomas, 2002) and the present (Jackson et al., 2001). This exploitation of coastal resources has hitherto largely been regarded as a facet of the behavior of *H. sapiens*, with the systematic usage of littoral environments coinciding with the movement out of Africa between 50–60,000 years ago (Klein, 1999). However, aquatic (and in some cases coastal) localities and resources were being used by modern humans for some time before this. At Katanda in Zaire a number of bone points or harpoons possibly related to fishing activities and dated to ∼90, 000 years BP have been recovered (Brooks et al., 1995; Yellen et al., 1995), while the coastal sites Klasies River Mouth in South Africa (Klein, 1999)

and the more recently reported site of Abdur on the Red Sea coast in Eritrea (Walter et al., 2000) date to the last interglacial, some 125 ka. The dating of Australasian colonisation to some 50,000 years ago (Bowler et al., 2003) suggests that people were confident enough to be using boats of some description by this time as no land crossings were possible, and the successful completion of this journey implies that they were familiar with water.

Walter et al. (2000:69) suggest, based on the evidence of Abdur and Klasies River Mouth, that 'an adaptation to coastal marine environments in Africa by 125 ka marks a significant, new behavior for early humans'. But is this assumption of a behavioral landmark at that time a reasonable one? *H. erectus* was a very successful animal, spreading through the Old World certainly by one million years ago, whatever the direction of travel. Whether it was a hunter or a scavenger, it must have been a creature with good powers of observation, able to identify and kill its own prey or to find and exploit the kills of other predators. Given that this animal travelled many thousands of kilometres and adapted to several different environments, is it likely that the advantages and opportunities offered by a coastal way of life would have been completely overlooked? Petraglia (2003:171) points out that many of the Acheulean sites of Arabia are associated with coastal zones, river terraces and lake shores, as they are in Africa, and while this patterning may have something to do with taphonomic processes the presence of water in arid landscapes can scarcely have been irrelevant to early hominins. Of course not all *H. erectus* populations need have exploited coasts, but surely some local groups are likely to have exploited them at different times. There is no reason for these hominins to have become physiologically adapted to coastal living – we are most emphatically not entering the scientifically absurd realm of the "aquatic ape" debate here – but small groups may have been behaviorally adapted to foraging along shorelines, whilst other populations lived inland.

One difficulty with any argument for coastal living and exploitation as a long-term feature of hominin behavior has been the paucity of 'smoking guns', of unambiguous evidence of early hominin activity on a beach, let alone exploitation of marine resources. One site that appears to show close association of stone tools and an aquatic mammal is that of Dungo V in Angola (Guitterez et al., 2001). Here, 57 Acheulean artifacts were found with a whale carcass on a paleoshoreline. The whale skeleton is encrusted with a concretion and cannot therefore be examined for cut marks, but experiments with modern large mammal butchery using elephants have shown that the meat and skin can be removed from the skeleton leaving no cut marks at all (Haynes, 1991). Given the amount of soft tissue on a whale, it is perfectly plausible that it could be scavenged without leaving marks.

Other early sites do show evidence of at least use of a littoral locality. One such is Boxgrove in Sussex in southern England, equated with Marine Oxygen Isotope Stage 13, where carcasses of large terrestrial mammals were butchered on a paleobeach (Roberts and Parfitt, 1998). The excavators suggested that such localities may have been ideal for catching and dealing with large ungulates. At Thomas Quarry 1 in Morocco, *H. erectus* fossils have been found in a coastal context (Raynal et al., 2001; Geraads, 2002), and at the lowest level of the *H. erectus* cave a mandible and an isolated upper premolar have been found in association with both land animals such as equids and marine species like the Mediterranean monk seal (*Monachus monachus*). The site is thought to represent a marine cave in association with a shoreline related to a very high sea level (Raynal and Texier, 1989) and has been dated to approximately 0.7–0.6 Ma (Raynal et al., 2001). Geraads et al. (2004) have subsequently suggested an earlier age of 1.0–1.5 Ma, but

Sahnouni et al. (2004) have criticized the basis for this dating. Whatever the precise age, the juxtaposition of *H. erectus* fossils with a marine species is very suggestive.

The site of Meta Menge in Flores, Indonesia, with stone tools associated with an early Pleistocene fauna (Sondaar et al., 1994) has been dated using fission track analysis to between 0.88–0.80 Ma (Morwood et al., 1998). Flores has been separated from other islands and mainland Asia by a deep channel throughout the Pleistocene, meaning that hominins would have to have swum, sailed or floated across a minimum of 19 km of water to reach it (Morwood et al., 1998). This has important implications for the behavior of *H. erectus* elsewhere in the world, since if South-East Asian populations were capable of crossing to islands then, unless it was a regional adaptation, other members of the same species should also have been capable of such a development. More recent reports of a small variety (or species, depending on one's view) of Late Pleistocene hominin, *H. floresiensis*, on the island (Brown et al., 2004; Morwood et al., 2004) also imply a sufficiently long presence for dwarfing to have taken place and thus perhaps a fairly early date for the dispersal.

Stone tools are often found at sites such as Berekhat Ram, Israel (Goren-Inbar, 1985) and Saffāqah, Syria (Whalen et al., 1984) where no fauna has been preserved to indicate the nature of exploitation. Bar-Yosef (1998) has noted that in the Levant different tool types are found at coastal sites in comparison with those found inland. The sites along the Lebanese coast (Fleisch and Sanlaville, 1974) have a higher proportion of amygdaloid and oval bifaces, while inland sites such as Latamne in Syria (Clark, 1967) have lanceolate and trihedral picks (Bar-Yosef, 1998). If the tool kits of coastal peoples were different from those used inland, which we might expect if they were exploiting alternative resources (although not if they were simply scavenging whale and seal carcasses or smashing shells with hammer stones), then this may provide a way of looking at resource use through a study of tool types and their possible uses. Thus far, however, the differences between coastal and inland sites have been interpreted as chronological (with inland sites being earlier than those on the coast) or as a result of diverse raw materials being used to make the tools (Bar-Yosef, 1998). However, it is also a plausible consequence of exploiting a different resource and it may be possible to test this by examining other assemblages from coastal and inland regions to see if this patterning is found elsewhere. These spatial contrasts in tool type have been reported from sites in Lebanon, Syria and Turkey (Bar-Yosef, 1998).

Parallels with Other Coastal Exploiters

In many non-hominin species some individuals exploit the coasts whilst their conspecifics do not. For example, the brown hyaena (*Parahyaena brunnea*) of southern Africa forages along the strand lines looking for carrion that may then be transported back to their dens (Klein et al., 1999). A recent comprehensive review of evidence for coastal exploitation by Moore (2002) lists species from 11 orders of mammals that have been observed eating maritime/littoral resources, from porcupines eating seaweed to sheep grazing on seabird chicks. Included amongst these are seven species of non-human primate that have been observed eating coastal animal items such as molluscs and crabs (Moore, 2002). Birds also commonly utilize aquatic resources, with a brief review by Erlandson and Moss (2001) listing 14 bird species that have been observed feeding on crabs, fish and molluscs.

If hominins were attuned to watching the skies to see where vultures were descending as an aid to finding carcasses the likelihood is that they would have been attracted to coastal debris (such as whale or seal carcasses) just as much as those inland. We are not suggesting a diet of purely animal resources, but that *H. erectus* and perhaps even earlier hominins may have found food on the beaches a useful addition to their diets. Indeed there is some evidence from another omnivore, the coyote (*Canis latrans*), that coastal animals have a wider dietary range and have a higher population density than those inland because of the input of aquatic resources (Rose and Polis, 1998). An earlier study of material along 4.2 km of highly productive beaches on islands in the Gulf of California found that approximately 2.5 large animals (fish, birds and sea mammals) per week were washed up in the study area over a two year period (Polis and Hurd, 1996). In addition, mass stranding sometimes occurred, in one case 3,981 squid were washed up over a 2 km length of beach on one night (Rose and Polis, 1998). This example indicates the amount of food that may be available for coastal scavengers in one relatively small area, which compares well with some estimates of *H. erectus* home range size (e.g. 73–452 ha [0.73–4.52 km^2] [Antón et al., 2002]). However, Baja California is an area of high productivity because of the upwelling of cold ocean currents, whereas the areas that hominins may have been exploiting would be somewhat impoverished in comparison. Mass strandings are an unpredictable resource, whilst other animals such as crabs, limpets, mussels, etc. that live in coastal areas can also be exploited and may be more reliable. Savannah baboons (*Papio cynocephalus*) have been seen exploiting freshwater oysters at times of low water levels in the Serengeti (Estes, 1991), whilst Hall (1962) reports that coastal populations of the chacma baboon (*Papio ursinus*) eat mussels, limpets, crabs and sand hoppers. They were not observed to use tools to open the shellfish, rather they used their teeth to crush them in their shells (Hall, 1962).

Topographic Changes and the Search for Sites of Hominin Coastal Activity

A major obstacle to finding early coastal sites of hominin activity is the large-scale changes of sea level during the Pleistocene, with extensive continental shelves being exposed during glaciations and submerged during interglacials. In many cases coastlines during glacials will have been many miles from where they are found now, especially in shallow, shelving waters. During the last glacial maximum, for example, the northern coastline of Libya was some 200km further out into the Mediterranean than it is today (Shackleton et al., 1984). All these sites are not necessarily lost as submerged archaeological landscapes have been identified in several areas around the world, although all but one date to the last 50,000 years (Flemming, 1998). In Table Bay, South Africa, three Acheulean handaxes have been recovered from the sea bed, with an estimated age (on typological grounds) of between 0.3–1.4 Ma, and this discovery raises the prospect of other submerged early sites being discovered in the future (Flemming, 1998). In other areas, tectonic activity has led to Pleistocene shorelines being preserved as raised beaches, in which artifacts and sites are sometimes found, although many of terraces only date back as far as the last Interglacial ($\sim 125,000$ BP), such as Abdur on the Red Sea (Walter et al., 2000) or are in areas where hominins did not disperse until relatively late, such as Peru (Zazo, 1999).

In a survey of early sites with evidence for aquatic exploitation, Erlandson (2001) found that the majority were associated with steep bathymetry. Clearly, a given fall in sea level along a steeply shelving coast results in the former waterline moving a shorter distance out from the current position than on a more gently shelving shore, where ancient sites may now be several miles out beneath the sea. This finding may allow us to select these areas for

surveying, as suggested by Stringer (2000) to find evidence of *H. sapiens* movement out of Africa, but traces of earlier movements may also be found. Such a project is currently underway to study the area around the Red Sea (see, for example, Flemming et al., 2003).

Thus while the evidence for early prehistoric coastal exploitation is limited there are some interesting pointers. First, and most importantly, there are several archaeological sites that have been found on islands, in a coastal context or containing artifacts associated with marine mammals. Secondly, different tool types have been found at coastal and inland sites in the Levant, usually explained in terms of site chronology or availability of raw materials, but perhaps also the result of different exploitation strategies. Thirdly, there is much circumstantial evidence that many other mammals (including Primates) exploit beaches and that *H. erectus* and even earlier hominins may also have done so.

Coastal Migration Routes

If coasts were indeed being used by early hominins, could they also have been used as migration routes? The idea of inland regions, such as Arabia, being difficult to traverse because of deserts, mountain ranges or large rivers has led to a view of occupation or migration only being an option when the routes were available during temperate or interglacial stages. However, migration along coastlines of the Red Sea may have allowed an animal adapted to using coastal resources to use a route that avoided many of the problems associated with the inland routes. Coastlines can provide a relatively flat, linear feature with an abundance of food and fairly good visibility, which may have made hominins and other species less vulnerable to predators. The beach would also allow movement relatively quickly, even if unintentionally, as hominins moving along the shore would really only be able to move in one of two directions unless they wished to go inland or into the sea. That is not to say that inland routes would have been ignored where suitable. Clearly, both hominins and other elements of the terrestrial fauna were present in the central parts of the Arabian Peninsula, at least from time to time, as shown by the number of archaeological sites (Petraglia, 2003; Petraglia and Alsharekh, 2004) and the Early Pleistocene faunal locality of An Nafud (Thomas et al., 1998).

Of course while the examples of non-hominin species exploiting coastal resources all suggest that coastal areas can be a significant source of abundant food, even at times a super-abundance, they also show that such areas are unreliable providers with unpredictable availability of meals. Such problems might be thought to argue against the significance of coastal routes. But we are not arguing here that such areas necessarily supported standing populations, whether of hominins or of any other seekers of a free lunch. Indeed the very fact of unreliability might even have ensured that exploiters took and then moved on, serving as it were as a further inducement to migrations along the shores. However, the one constraining factor, particularly for large bodied animals, would be the availability of fresh water, and increasing aridity inland may have made freshwater sources scarce along the coasts as well.

Although the Arabian peninsula may not have been an ideal place to enter and stay during glacial periods when lowered sea level made the crossing more obviously possible, it is conceivable that any species that did cross could have then moved northwards along the eastern shore of the Red Sea, eventually reaching the Levant without going through the internal part of the Peninsula. However, whether such a route would have taken or attracted more traffic than a more conventional one through Sinai during interglacial periods remains unclear. And of course getting into Arabia in the first place

says little about routes between Arabia and the Eurasian landmass. Whether coastal movements beyond Arabia and round into South Asia and the Indian subcontinent itself by a coastal route were possible is a much larger question, and one that lies beyond both the scope of this paper and our present knowledge. We have the impression, from general considerations of present-day topography, that while movement around the coast of India may have been reasonably attractive, the possibilities for movement along the Makran coast of southern Iran and Pakistan may have been rather more limited.

Conclusions

Early movements between Africa and Eurasia point to much interchange throughout the Miocene and into the Pliocene, and the fact that both predators and prey were among these dispersers suggests perhaps the wholesale expansion of ecosystems, possibly along the lines suggested by Solounias et al. (1999) at least so far as Miocene events are concerned. These movements appear to have included an initial dispersal of dryopithecine apes out of Africa during the mid Miocene, perhaps including movements into the Siwaliks, and a possible return in the later stages of that epoch to found the Homininae and the fossil and modern great apes and hominins of Africa.

However, in the later stages of hominin evolution the seemingly great expansion in our knowledge of Pliocene taxa that has developed in recent years has done little to clarify the origins of the genus *Homo*. Suggestions that late Pliocene taxa be removed from the genus leave *H. erectus* without an obvious ancestral grade, and have perhaps fuelled the debate about an origin outside Africa for this species. Paleontologically, a late Pliocene dispersal from Africa that included hominins and thus perhaps an ancestor for *H. erectus* is not impossible or implausible, although before we run away with that idea it is perhaps worth pointing out that simply because numerous Eurasian mammalian species got to America during the mid Pleistocene (Kurtén and Anderson, 1980) we cannot take that to argue for a human presence there at that time. The convenient fact that no Pliocene hominins are actually known from Eurasia is a two-edged sword in the argument for an extra-African origin for *H. erectus*.

It may be useful to see dispersal from Africa, if that is the direction we should investigate, as at least a two part problem both for the early hominins involved and for our understanding of the nature of movements. The first part involved getting into Arabia and the Levant and maintaining a foothold there, no easy matter in itself. The second involved getting beyond Arabia and into Eurasia, and of course maintaining a foothold there in the very different world of temperate and eventually glacial climates and a highly competitive carnivore guild in an environment where meat is likely to have been a major element in the diet. Perhaps dispersals to both were highly episodic, with earliest populations dying off and being later replaced by new groups many times, with Arabia in particular being colonized from intermittent movements along the Red Sea coasts as well as across Sinai and (perhaps) the Bab-el-Mandeb Straits. Such dispersals would have been constrained by circumstances that would have kept groups moving, and their intermittent nature would explain something of the patterning seen in the Early Pleistocene archaeological and hominin fossil records.

Acknowledgments

Part of the research that underpins this paper was funded by a NERC grant under the EFCHED Thematic Programme. Turner is grateful to Bridget Allchin and Michael Petraglia for the invitation to present the paper at the Cambridge meeting in 2004 and for their hospitality during the conference.

References

Aguirre, E., Carbonell, E., 2001. Early human expansion into Eurasia: the Atapuerca evidence. Quaternary International 75, 11–18.

Agustí, J., Antón, M., 2002. Mammoths, Sabertooths and Hominids: 65 Million Years of Mammalian Evolution in Europe. Columbia University Press, New York.

Agustí, J., Cabrera, L., Garcés, M., Llenas, M., 1999. Mammal turnover and global climatic change in the late Miocene terrestrial record of the Vallès-Penedès basin (NE Spain). In: Agustí, J., Rook, L., Andrews, P. (Eds.), Hominoid Evolution and Climatic Change in Europe, Volume 1. The Evolution of Neogene Terrestrial Ecosystems in Europe. Cambridge University Press, Cambridge, pp. 397–412

Alberdi, M.T., Alonso, M.A., Azanza, B., Hoyos, M., Morales, J., 2001. Vertebrate taphonomy in circum-lake environments; three cases in the Guadix-Baza Basin (Granada, Spain). Palaeogeography, Palaeoclimatology, Palaeoecology 165, 1–26.

Andrews, P., 1996. Palaeoeocology and hominoid palaeoenvironments. Biological Reviews 71, 257–300.

Andrews, P., Bernor, R., 1999. Vicariance biogeography and paleoecology of Eurasian Miocene hominoid Primates. In: Agustí, J., Rook, L., Andrews, P. (Eds.), Hominoid Evolution and Climatic Change in Europe, Volume 1. The Evolution of Neogene Terrestrial Ecosystems in Europe. Cambridge University Press, Cambridge, pp. 454–487.

Andrews, P., Humphrey, L., 1999. African Miocene environments and the transition to early hominines. In: Bromage, T., Schrenk, F. (Eds.), African Biogeography, Climatic Change and Early Hominid Evolution. Oxford University Press, New York, pp. 282–300.

Antón, S.C., Leonard, W.R., Robertson, M.L., 2002. An ecomorphological model of the initial hominid dispersal from Africa. Journal of Human Evolution 43, 773–785.

Asfaw, B., White, T., Lovejoy, O., Latimer, B., Simpson, S., Suwa, G., 1999. *Australopithecus garhi*: a new species of early hominid from Ethiopia. Science 284, 629–635.

Azzaroli, A,, De Giuli, C., Ficcarelli, G., Torre, D., 1988. Late Pliocene to Early Mid-Pleistocene mammals in Eurasia: faunal succession and dispersal events. Palaeogeography, Palaeoclimatology, Palaeoecology 66, 77–100.

Barry, J.C., 1995. Faunal turnover and diversity in the terrestrial Neogene of Pakistan. In: Vrba, E.S., Denton, G.H., Partridge, T.C., Burckle, L.H (Eds.), Paleoclimate and Evolution with Emphasis on Human Origins. Yale University Press, New Haven, pp. 115–134.

Barry, J.C., Morgan, M.E., Flynn, L.J., Pilbeam, D., Behrensmeyer, A.K., Raza, S.M., Khan, I.A., Badgley, C., Hicks, J., Kelly, J., 2002. Faunal and environmental change in the late Miocene Siwaliks of northern Pakistan. Paleobiology Memoirs 3, 1–55.

Bar-Yosef, O., 1998. Early colonizations and cultural continuities in the Lower Palaeolithic of western Asia. In: Petraglia, M.D., Korisettar, R. (Eds.), Early Human Behaviour in Global Context: The Rise and Diversity of the Lower Palaeolithic Record. Routledge, London, pp. 221–279.

Begun, D.R., 1994. Relations among the great apes and humans: new interpretations based on the fossil great ape Dryopithecus. Yearbook of Physical Anthropology, 37, 11–64.

Begun, D.R., Gülec, E. Geraads, D., 2003. Dispersal patterns of Eurasian hominoids: implications from Turkey. Deinsea 10, 23–39.

Belmaker, M., Tchernov, E., Condemi, S., Bar-Yosef, O., 2002. New evidence for hominid presence in the Lower Pleistocene of the southern Levant. Journal of Human Evolution 43, 43–56.

Bloch, J.I., Silcox, M.T., 2006. Cranial anatomy of the Paleocene plesiadapiform *Carpolestes simpsoni* (Mammalia, Primates) using ultra high-resolution X-ray computed tomography, and the relationships of plesiadapiforms to Euprimates. Journal of Human Evolution 50, 1–35.

Boitel, F., Dépont, J., Tourenq, J., Lorenz, J., Abelanet, J., Pomerol, J., 1996. Découverte d'une industrie lithique dans le Pliocène supérieur du sud du bassin parisien (Formation des sables et argiles du Bourbonnais). Comptes rendus de l'Academie des Sciences de Paris 322, 507–514.

Bonifay, E., Vandermeersch, B. (Eds.), 1991. Les Premiers Européens.: Editions du CTHS, Paris.

Bowler, J.M., Johnston, H., Olley, J.M., Prescott, J.R., Roberts, R.G., Shawcross, W. Spooner, N.A., 2003. New ages for human occupation and climate change at Lake Mungo, Australia. Nature 421, 837–840.

Bromage, T.G., Schrenk, F., 1995. Biogeographic and climatic basis for a narrative of early hominid evolution. Journal of Human Evolution 28, 109–114.

Brooks, A.S., Helgren, D.M., Cramer, J.S., Franklin, A., Hornyak, W., Keating, J.M., Klein, R.G., Rink, W.J., Schwarcz, H., Leith Smith, J.N., Stewart, K., Todd, N.E., Verniers, J., Yellen, J.E., 1995 Dating and context of three middle stone age sites with bone points in the Upper Semliki Valley, Zaire. Science 268, 548–552.

Brown, P., Sutikna, T., Morwood, M.J., Soejono, R.P., Jatmiko, R., Saptomo, E.W., Awe Due, R., 2004. A new small-bodied hominin from the Late Pleistocene of Flores, Indonesia. Nature 431, 1055–1061.

Brunet, M., Beauvillain, A., Coppens, Y., Heinz, E., Moutaye, A.H.E., Pilbeam, D., 1995. The first australopithecine 2500 kilometres west of the Rift Valley (Chad). Nature 378, 273–275.

Brunet, M., Beauvilain, A., Coppens, Y., Heintz, E., Moutaye, A.H.E., Pilbeam, D., 1996. *Australopithecus bahrelghazali*, a new species of early hominid from Koro Toro region, Chad. Comptes rendus de l'Academie des Sciences Paris 322, 907–913.

Brunet, M., Guy F., Pilbeam, D., Mackaye, H.T, Likius, A., Ahounta, D., Beauvillain, A., Blondel, C., Bocherens, H., Boisserie, J.R.,.de Bonis, L., Coppens, Y., Dejax, J., Denys, C., Duringer, P., Eisenmann, V., Fanone, G., Fronty, P., Geraads D, Lehmann T, Lihoreau F, Louchart A, Mahamat A, Merceron G, Mouchelin, G., Otero, O., Campomanes, P.P., Ponce de Leon, M., Rage, J.C., Sapanet, M., Schuster, M., Sudre, J., Tassy, P., Valentin, X., Vignaud, P., Viriot, L., Zazzo, A., Zollikofer, C., 2002. A new hominid from the Upper Miocene of Chad, Central Africa. Nature 418, 145–151.

Clark, J.D., 1967. The middle Acheulian occupation site at Latamne, Northern Syria. Quaternaria 9, 1–68.

Dennell, R.W., 1998. Grasslands, tool making and the hominid colonization of southern Asia: a reconsideration. In: Petraglia, M.D. Korisettar, R. (Eds.), Early Human Behaviour in Global Context: The Rise and Diversity of the Lower Palaeolithic Record. Routledge, London, pp. 280–303.

Dennell, R.W., 2003. Dispersal and colonisation, long and short chronologies: how continuous is the Early Pleistocene record for hominids outside East Africa? Journal of Human Evolution 45, 421–440.

Dennell, R.W., 2004. Early Hominin Landscapes in Northern Pakistan. BAR International Series 1265, Oxford.

Dobson, M., Wright, A., 2000. Faunal relationships and zoogeographical affinities of mammals from north-west Africa. Journal of Biogeography 27, 417–424.

Erlandson, J.M., 2001. The archaeology of aquatic adaptations: paradigms for a new millennium. Journal of Archaeological Research 9, 287–350.

Erlandson, J.M., Moss, M.L., 2001. Shellfish feeders, carrion eaters, and the archaeology of aquatic adaptations. American Antiquity 66, 413–432.

Estes, R.D., 1991. The Behavior Guide to African Mammals. University of California Press, Berkeley.

Fleagle, J.C., 1999. Primate Adaptation and Evolution, 2nd. ed. Academic Press, San Diego.

Fleisch, H., Sanlaville, P., 1974. La plage de +52 m et son Acheuléen à Ras Beyrouth et à l'Ouadi Aabet (Liban). Paléorient 2, 45–85.

Flemming, N.C., 1998. Archaeological evidence for vertical movement on the continental shelf during the Palaeolithic, Neolithic and Bronze Age periods. In: Vita-Finzi, C. (Ed.), Coastal Tectonics Geological Society of London Special Publication 146, London, pp. 129–146.

Flemming, N.V., Bailey, G.N., Courtillot, V., King, G., Lambeck, K., Ryerson, F., Vita-Finzi, C., 2003. Coastal and marine palaeo-environments and human dispersal across the Africa-Eurasia boundary. In: Brebbia, C.A. Gambin, T. (Eds.), Maritime Heritage. WIT Press, Southampton, pp. 61–74.

Gabunia, L., Vekua, A., Lorkipanidze, D., 2000. The environmental contexts of early human occupation of Georgia (Transcaucasia). Journal of Human Evolution 38, 785–802.

Gabunia, L., Antón, S.C., Lordkipanidze, D., Vekua, A., Justus, A., Swisher, C.C., 2001. Dmanisi and dispersal. Evolutionary Anthropology 10, 158–170.

Geraads, D., 2002. Plio-Pleistocene mammalian biostratigraphy of Atlantic Morocco. Quaternaire 13, 43–53.

Geraads, D., Raynal, J.-P., Eisenmann, V., 2004. The earliest occupation of North Africa: a reply to Sahnouni et al. Journal of Human Evolution. 46, 751–761.

Glennie, K.W., Singhvi, A.K., 2002. Event stratigraphy, paleoenvironment and chronology of SE Arabian deserts. Quaternary Science Reviews 21, 853–869.

Godinot, M., Mahboubi, M., 1992. Earliest known simian primate found in Algeria. Nature 357, 324–326.

Goren-Inbar, N., 1985. The lithic assemblage of the Berekhat Ram Acheulean site, Golan Heights. Paleorient 11, 7–28.

Gutierrez, M., Guérin, C., Lena., M., Piedade da Jesus, M., 2001. Human exploitation of a large stranded whale in the Lower Palaeolithic site of Dungo V at Baia Farta (Benguela, Angola). Comptes rendus de l'Academie des Sciences Paris 332, 357–362.

Haile-Selassie, Y., Suwa, G., White, T.D., 2004. Late Miocene teeth from Middle Awash, Ethiopia, and early hominid dental evolution. Science 303, 1503–1505.

Hall, K.R.L., 1962. Numerical data, maintenance activities and locomotion of the wild chacma baboon, Papio ursinus. Proceedings of the Zoological Society of London 139, 181–220.

Haynes, G., 1991. Mammoths, Mastodonts and Elephants. Biology, Behaviour and the Fossil Record. Cambridge University Press, Cambridge.

Jackson, J.B.C., Kirby, M.X., Berger, W.H., Bjorndal, K.A., Botsford, L.W., Bourque, B.J., Bradbury, R.H., Cooke, R., Erlandson, J.M., Estes, J.A., Hughes, T.P., Kidwell, S., Lange, C.B., Lenihan, H.S., Pandolfi, J.M., Peterson, C.H., Steneck, R.S., Tegner, M.J., Warner, R.R., 2001. Historical overfishing and the recent collapse of coastal ecosystems. Science 293, 629–638.

Kirjksman, W.F., Hilgen, J., Raffi, I., Sierros, F.J., Wilson, D.S., 1999. Chronology, causes and progression of the Messinian salinity crisis. Nature 400, 652–655.

Kirk, E.C., Simons, E.L., 2001. Diets of fossil Primates from the Fayum depression of Egypt: a quantitative analysis of molar shearing. Journal of Human Evolution 40, 203–229.

Klein, R.G., 1999. The Human Career, 2nd. ed. University of Chicago Press, Chicago.

Klein, R.G., Cruz-Uribe, K., Halkett, D., Hart, T., Parkington, J.E., 1999. Paleoenvironmental and human behavioral implications of the Boegoeberg 1 Late Pleistocene hyena den, Northern Cape Province, South Africa. Quaternary Research 52, 393–403.

Kurtén, B., Anderson, E., 1980. Pleistocene Mammals of North America. Columbia University Press, New York.

Leakey, M.G., Feibel, C.S., McDougall, I., Walker, A., 1995. New four-million-year-old hominid species from Kanapoi and Allia Bay, Kenya. Nature 376, 565–571.

Leakey, M.G., Harris, J.M., 2003. Lothagam: its significance and contributions In: Leakey, M.G., Harris, J.M (Eds.), Lothagam: The Dawn of Humanity in Eastern Africa. Columbia University Press, New York, pp. 625–659.

Leakey, M.G., Spoor, F., Brown, F.H., Gathogo, P.N., Kiarie, C., Leakey, L.N., McDougall, I., 2001. New hominin genus from eastern Africa shows diverse middle Pliocene lineages. Nature 410, 433–440.

Lindsay, E.H., Opdyke, N.D., Johnson, N.M., 1980. Pliocene dispersal of the horse *Equus* and late Cenozoic mammalian dispersal events. Nature 287, 135–138.

Made, J. van der, 1999. Intercontinental relationships Europe-Africa and the subcontinent. In: Rössner, G.E., Heissig, K. (Eds.), The Miocene Land Mammals of Europe. F. Pfeil, Munich, pp. 457–472.

Made, J. van der, Ribot, F., 1999. Additional hominoid material from the Miocene of Spain and remarks on hominoid dispersals into Europe. Contributions in Tertiary and Quaternary Geology 36, 25–39.

Mannino, M.A., Thomas, K.D., 2002. Depletion of a resource? The impact of prehistoric human foraging on intertidal mollusc communities and its significance for human settlement, mobility and dispersal. World Archaeology 33, 452–474.

Mithen, S., Reed, M., 2002. Stepping out: a computer simulation of hominid dispersal from Africa. Journal of Human Evolution 43, 433–462.

Moore, P.G., 2002. Mammals in intertidal and maritime ecosystems: interaction, impacts and implications. Oceanography and Marine Biology: an Annual Review 40, 491–608.

Morales, J., Pickford, M., Soria, D., Fraile, S., 1998. New carnivores from the basal Middle Miocene of Arrisdrift, Namibia. Eclogae Geologicae Helvetiae 91, 27–40.

Morwood, M.J., O'Sullivan, P.B., Aziz, F., Raza, A., 1998. Fission-track ages of stone tools and fossils on the east Indonesian island of Flores. Nature 392, 173–176.

Morwood, M.J., Soejono, R.P., Roberts, R.G., Sutikna, T., Turney, C.S.M., Westaway, K.E., Rink, W.J., Zhao, J., van den Bergh, G.D., Awe Due, R., Hobbs, D.R., Moore, M.W., Bird, M.I., Fifield, L.K., 2004. Archaeology and age of a new hominin from Flores in eastern Indonesia. Nature 431, 1087–1091.

O'Regan, H., Bishop, L., Lamb, A., Elton, S. Turner, A., (in press). Afro-Eurasian mammalian dispersions of the late Pliocene and early Pleistocene, and their bearing on earliest hominin movements. Courier Forschungsinstitut Senckenberg.

Petraglia, M.D., 2003. The Lower Paleolithic of the Arabian peninsula: occupations, adaptations

and dispersals. Journal of World Prehistory 17, 141–179.

Petraglia, M.D., Alsharekh, A., 2004. The Middle Palaeolithic of Arabia: implications for modern human origins, behaviour and dispersals. Antiquity 78, 671–684.

Polis, G.A., Hurd, S.D., 1996. Linking marine and terrestrial food webs: allochthonous input from the ocean supports high secondary productivity on small islands and coastal land communities. American Naturalist 147, 396–423.

Radulesco, C., Samson, P., 1990. The Plio-Pleistocene mammalian succession of the Oltet Valley, Dacic Basin, Romania. Quartärpaläontologie 8, 225–232.

Raynal, J.P., Sbihi Alaoui, F.Z., Geraads, D., Magoga, L., Mohi, A., 2001. The earliest occupation of North-Africa: the Moroccan perspective. Quaternary International 75, 65–75.

Raynal, J.P., Texier, P-J., 1989. Découverte d'Acheuléen ancien dans la carrière Thomas 1 à Casablanca et problème de l'ancienneté de la présence humaine au Maroc. Comptes rendus de l'Academie des Sciences Paris 308, 1743–1749.

Roberts, M., Parfitt, S., 1998. Boxgrove: A Middle Pleistocene Hominid Site at Eartham Quarry, Boxgrove, West Sussex. Oxbow Books, Oxford.

Roebroeks, W., 2001. Hominid behaviour and the earliest occupation of Europe: an exploration. Journal of Human Evolution. 41, 437–461.

Rögl, F., 1999. Mediterranean and Paratethys palaeogeography during the Oligocene and Miocene. In: Agustí, J., Rook, L., Andrews, P. (Eds.), Hominoid Evolution and Climatic Change in Europe, Volume 1. The Evolution of Neogene Terrestrial Ecosystems in Europe. Cambridge University Press, Cambridge, pp. 8–22.

Rose, M., Polis, G.A., 1998. The distribution and abundance of coyotes: the effects of allochthonous food subsidies from the sea. Ecology 79, 998–1007.

Sahnouni, M., Hadjouis, D., van der Made, J., Derradji, A., Canals, A., Medig, M., Belahrech, H., Harichane, Z., Rahbi, M., 2004. On the earliest human occupation in North Africa: a response to Geraads et al. Journal of Human Evolution 46, 763–775.

Semaw, S., 2000. The world's oldest stone artefacts from Gona, Ethiopia: their implications for understanding stone technology and patterns of human evolution between 2.6–1.5 million years ago. Journal of Archaeological Science 27, 1197–1214.

Semaw, S., Simpson, S.W., Quade, J., Renne, P.R., Butler, R.F., McIntosh, W.C., Levin, N., Dominguez-Rodrigo, M., Rogers, M., 2005. Early Pliocene hominids from Gona, Ethiopia. Nature 433, 301–305.

Senut, B., Pickford, M., Gommery, D., Mein, P., Cheboi, K., Coppens, Y., 2001. First hominid from the Miocene (Lukeino Formation, Kenya). Comptes rendus de l'Academie des Sciences Paris 332, 137–144.

Shackleton, J.C., Van Andel, T.H., Runnels, C.N., 1984. Coastal paleogeography of the Central and Western Mediterranean during the last 125,000 years and its archaeological implications. Journal of Field Archaeology 11, 307–314.

Sickenberg, O., 1967. Die unterpleistozäne Fauna von Wolaks (Griechland-Mazedonien). I. Eine neue Giraffe (*Macedonitherium martinii* nov. gen., nov. spec.) aus dem unteren Pleistozäne von Griecheland. Annales Géologiques des Pays Helléniques 18, 34–54.

Solounias, N., Plavcan, J.M., Quade, J., Witmer, L., 1999. The paleocecology of the Pikermian biome and the savanna myth. In: Agustí, J., Rook, L., Andrews, P. (Eds.), Hominoid Evolution and Climatic Change in Europe, Volume 1. The Evolution of Neogene Terrestrial Ecosystems in Europe. Cambridge University Press, Cambridge, pp. 436–453.

Sondaar, P.Y., van den Bergh, G.D., Mubroto, B., Aziz, F., de Vos, J., Batu U.L., 1994. Middle Pleistocene faunal turnover and colonization of Flores (Indonesia) by *Homo erectus*. Comptes rendus de l'Academie des Sciences Paris 319, 1255–1262.

Stiner, M.C., Howell, F.C., Martínez-Navarro, B., Tchernov, E., Bar-Yosef, O., 2001. Outside Africa: Middle Pleistocne *Lycaon* from Hayonim Cave, Israel. Bulletino della Societe Paleontologia Italica 40, 293–302.

Stringer, C.B., 2000. Coasting out of Africa. Nature 405, 4–27.

Tabuce, R., Marivaux, L., 2005. Mammalian interchanges between Africa and Eurasia: an analysis of temporal constraints on plausible anthropoid dispersals during the Paleogene. Anthropological Science 113, 27–32.

Tchernov, E., 1992. Eurasian-African biotic exchanges through the Levantine corridor during the Neogene and Quaternary. In: Von Koenigswald, W., Werdelin, L. (Eds.), Mammalian Migration and Dispersal Events in the European Quaternary. Courier Forschungsinstitut Senckenberg 153, 103–123.

Thomas, H., Gerrards, D., Janjou, D., Vaslet, D., Memesh, A., Billiou, D., Bocherens, H., Dobigny, G., Eisenmann, V., Gayet, M., de Lapparent de Broin, F., Petter, G., Halawani, M., 1998. First Pleistocene faunas from the Arabian peninsula: An Nafud desert, Saudi Arabia. Comptes rendus de l'Academie des Sciences Paris 326, 145–152.

Turner, A., 1992. Large carnivores and earliest European hominids: changing determinants of resource availability during the Lower and Middle Pleistocene. Journal of Human Evolution 22, 109–126.

Turner, A., 1995. Plio-Pleistocene correlations between climatic change and evolution in terrestrial mamals: the 2.5 Ma event in Africa and Europe. Acta Zoologia Cracoviensia 38, 45–58.

Turner, A., 1999. Assessing earliest human settlement of Eurasia: Late Pliocene dispersals from Africa. Antiquity 73, 563–570.

Turner, A., 2004. Vertebrate fossils: Carnivora. In: Dennell, R.W. (Ed.), Early Hominin Landscapes in Northern Pakistan. BAR International Series 1265, Oxford, pp. 404–411.

Turner, A., Antón, M., 2004. Evolving Eden. An Illustrated Guide to the Evolution of the African Large Mammal Fauna. Columbia University Press, New York.

Turner, A., O'Regan, H., (in press). Zoogeography – primate and early hominin distribution and migration patterns. In: Henke, W., Rothe, H., Tattersall, I. (Eds.), Handbook of Palaeoanthropology, Volume 1: Principles, Methods and Approaches.

Turner, A., Wood, B., 1993. Taxonomic and geographic diversity in robust australopithecines and other African Plio-Pleistocene mammals. Journal of Human Evolution 24, 147–168.

Vekua, A., Lordkipanidze, D., Rightmire, G.P., Agusti, J., Ferring, R., Maisuradze, G., Mouskhe-lishivili, A., Nioradze, M., Ponce de Leon, M., Tappen, M., Tvalchrelidze, M., Zollikofer, C., 2002. A new skull of early *Homo* from Dmanisi, Georgia. Science 297, 85–89.

Walter, R.C., Buffler, R.T., Bruggemann, J.H., Guillaume, M.M.M., Berhe, S.M., Negassi, B., Libsekal, Y., Cheng, H., Edwards, R.L., Cosel, R., Neraudeau, D., Gagnon, M., 2000. Early human occupation of the Red Sea coast of Eritrea during the last interglacial. Nature 405, 65–69.

Whalen. N., Siraj-Ali, J.S., Davis, W., 1984. Excavation of Acheulean sites near Saffâqah, Saudi Arabia, 1403 AH 1983. Atlal 8, 9–17.

White, T.D., 1995. African omnivores: global climatic change and Plio-Pleistocene hominids and suids. In: Vrba, E.S., Denton, G.H., Partridge, T.C., Burckle, L.H (Eds.), Paleoclimate and Evolution with Emphasis on Human Origins. Yale University Press, New Haven, pp. 369–385.

White, T.D., Suwa, G., Asfaw, B., 1995. *Australopithecus ramidus*, a new species of early hominid from Aramis, Ethiopia. Corrigendum. Nature 375, 88.

Whybrow, P.J., Clements, D., 1999. Arabian Tertiary fauna, flora, and localities. In: Whybrow, P.J., Hill, A. (Eds.), Fossil Vertebrates of Arabia. Yale University Press, New Haven, pp. 460–473.

Wood, B.A., 1991. Koobi Fora Research Project, Volume 4. Hominid Cranial Remains. Clarendon Press, Oxford.

Wood, B.A., 1992. Origin and evolution of the genus *Homo*. Nature 355, 783–790.

Wood, B.A., Collard, M., 1999. The human genus. Science 284, 65–71.

Yellen, J.E., Brooks, A.S., Cornelissen, E., Mehlman, M.J., Stewart, K., 1995. A Middle Stone Age worked bone industry from Katanda, Upper Semliki Valley, Zaire. Science 268, 552–556.

Zazo, C., 1999. Interglacial sea levels. Quaternary International 55, 101–113.

3. "Resource-rich, stone-poor": *Early hominin land use in large river systems of northern India and Pakistan*

ROBIN DENNELL

Department of Archaeology
Northgate House
West Street
University of Sheffield
Sheffield, S1 4ET
England
r.dennell@sheffield.ac.uk

Introduction

This chapter explores two related issues. The first is whether Early Pleistocene hominins were successful in colonizing the Indo-Gangetic floodplains of northern India and Pakistan; and the second is whether the paucity of evidence that they did so might help explain why the evidence for hominins in peninsula India dates to the Middle Pleistocene (Petraglia, 1998), with the exception of one recent, and unconfirmed, date of 1.27 Ma from Isampur, Karnataka (Paddayya et al., 2002). In an earlier paper (Dennell, 2003), I pointed out that current evidence indicates several major discontinuities in regional hominin records across Asia in the Early Pleistocene (see Figure 1). Peninsula India currently has one of the longest, as hominins were present at Dmanisi, Georgia, to the west at 1.75 Ma (Gabunia et al.,

2000a), and Java to the east by ca. 1.6 Ma (Larick et al., 2001) and possibly by ca.1.8 Ma (Swisher et al., 1994). However, apart from a small amount of material that remains controversial from Riwat (Dennell et al., 1988) and the Pabbi Hills, Pakistan (Dennell, 2004; Hurcombe, 2004), there is no incontrovertible evidence that hominins were living in the northern part of the Indian subcontinent in the Early Pleistocene, even though it is the obvious corridor route between Southwest and Southeast Asia.

In this chapter, I suggest that Early Pleistocene hominins would have found it very difficult to colonize successfully extensive floodplains such as those of northern India and Pakistan, and current evidence suggests that if they were there at all, it was probably on an intermittent basis and at very low densities of population. Important geological changes in this region towards the end of the Early

41

M.D. Petraglia and B. Allchin (eds.), The Evolution and History of Human Populations in South Asia, 41–68.

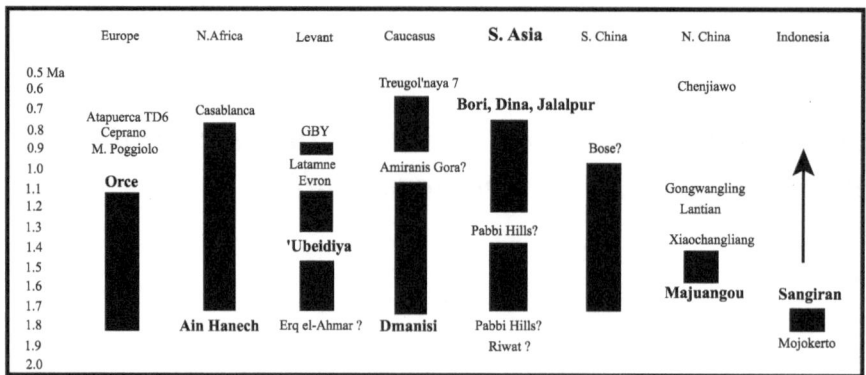

Figure 1. Discontinuities in the fossil and archaeological record for hominins in Eurasia during the Early and Middle Pleistocene prior to ca. 500 ka. Updated and adapted from Dennell, 2003: Figure 4. Names in **bold** indicate the earliest site or discovery in a region that is widely accepted as unambiguous

Pleistocene had potentially important consequences for hominins, notably in increasing the availability of stone for tool-making, and may have been a contributory factor in enabling them to colonize peninsula India in the Middle Pleistocene. We can begin by considering the characteristics of the modern Indo-Gangetic drainage system and its Early Pleistocene predecessors, and then the opportunities and problems that Early Pleistocene hominins might have encountered in these floodplain environments.

The Modern Indo-Gangetic Drainage System

The alluvial plains of the northern subcontinent cover some 770,000 sq km (roughly the same area as Spain and the U.K. combined), and include most of Sind, northern Rajasthan, most of the Punjab, Uttar Pradesh, Bihar, Bengal and half of Assam. The Ganges Plain is ca.1000 km from west to east, and the width varies from 500 km in the west, to < 150 km in the east. Figure 2 shows the modern drainage of the Indus, Ganges and Brahmaputra rivers. Its features were usefully summarized by Wadia (1974:364–365): "the whole of these plains, from one end to the other, is formed, with unvarying monotony, of Pleistocene and sub-

Recent alluvial deposits of the Indo-Gangetic system, which have completely shrouded the old land surface to a depth of several thousands of metres. ….. The deposition of this alluvium commenced after the final phase of the Siwaliks [see below] and has continued all through the Pleistocene up to the present". He continues: "the Indo-Gangetic depression is a true fore-deep, a downwarp of the Himalayan foreland, of variable depth, converted into flat plains by the simple process of alluviation. On this view, a long-continued vigorous sedimentation, loading a slowly sinking belt of the Peninsula shield from Rajasthan to Assam. ….. the deposition keeping pace with subsidence, has given rise to this great tectonic trough of India". A more recent, and slightly divergent, view over the relative effects of deposition and subsidence is taken by Srivastava et al. (2003:18): "The sediment input to the Ganga Plain occurs at a rate in excess of the down flexing, causing sedimentation rate to exceed the subsidence rate. Hence the basin surface remained above sea level". Both agree over the extent of sedimentation, the depth of which is estimated at < 1000–2000 m, with the greatest depth in the northern part of the syncline. The sediments are primarily "massive beds of clay, either sandy or calcareous, corresponding to the silts, mud and sand of the modern

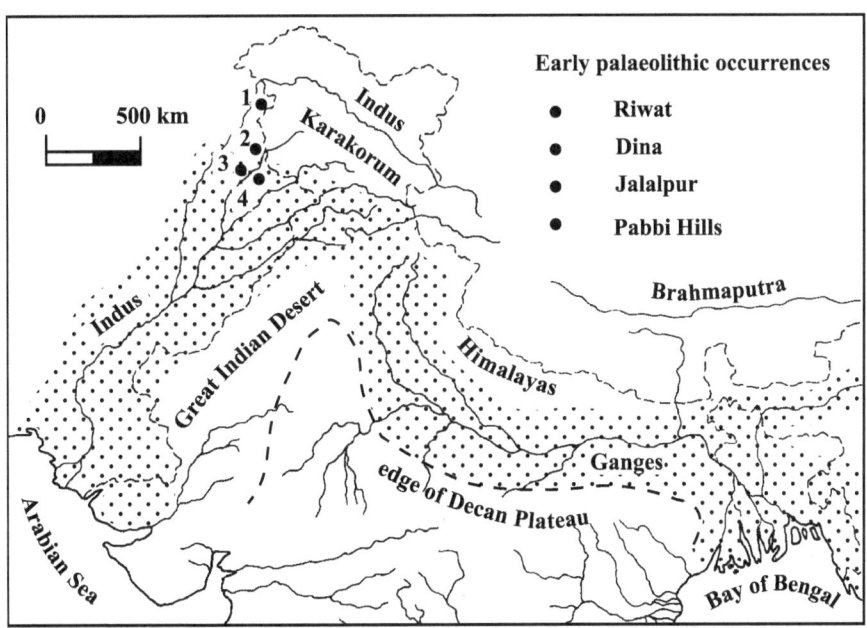

Figure 2. The Indo-Gangetic Plains, and location of archaeological occurrences mentioned in the text

rivers. Gravel and sand become scarcer as the distance from the hills increases. At some depth from the surface there occur a few beds of compact sands and even gravelly conglomerates" (Wadia, 1974:369). Figure 3 shows a schematic view of present-day topography and land-forms.

The highest part of the north Indian Plains, ca. 275 m a.s.l. between Saharanpur, Ambala and Ludhiana, separates the drainage systems of the Indus and Ganges. In addition to depositing an enormous thickness of sediment, both rivers and their major tributaries have frequently altered their courses. In the 16th century, the Chenab and Jhelum joined the Indus at Uch, instead of (as now) at Mithankot, 100 km downstream. Multan was then on the Ravi, whereas now it is 60 km from the confluence

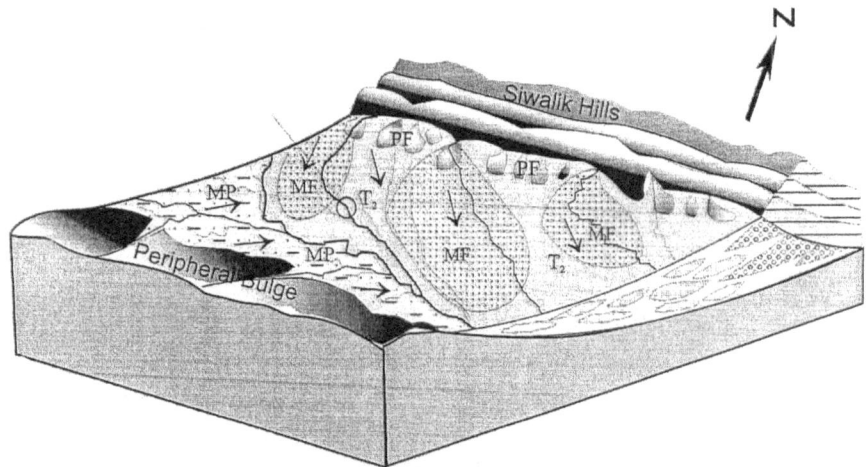

Figure 3. Schematic diagram showing geomorphic features of the modern Ganges (Ganga) Plain. Source: Srivastava et al., 2003: Figure 3. Key: PF – Piedmont Fan Surface; MF – Megafan Surface; T_1 River valley terrace surface; T_2 – Upland interfleuve surface; T_0 Active flood plain surface

of the Ravi and the Chenab. In the third century BC, the Indus was 130 km east of its present course, and its westward and often dramatic migration in subsequent periods is well documented (see e.g., Snelgrove, 1979). Similar changes have occurred in Bengal, notably the growth of the delta since 1750 to its current size of 130,000 sq km (roughly twice the size of Ireland), and a 60 km westward shift by the Brahmaputra (Wadia, 1974:369). Even more dramatically, the Kosi River has shifted 113 km westwards by at least 12 episodic changes of course in only the last 250 years (Wells, 1987).

The absence of evidence for occupation in the floodplains of the Indus and Ganges before the Middle Pleistocene can easily be explained as a consequence of massive sedimentation that has since occurred. However, the history of the Indo-Gangetic drainage system differs from that of many other large rivers in that its earlier history is well known as a result of tectonic uplift. This gives us one of the few windows we have on what these large river systems were like in the Late Pliocene and Early Pleistocene.

The Miocene to Middle Pleistocene Precursors of the Indo-Gangetic Drainage System

Our chief source of information about the Late Pliocene and Early Pleistocene history of the Indo-Gangetic drainage system are Upper Siwalik deposits (ca. 3.3–0.6 Ma). These, like the Lower and Middle Siwaliks, are predominantly fluvial in origin, and resulted from the deposition of several kilometers of sediments in and along rivers that drained southwards from the Karakorum and Himalayas. The reason why so much is known about the Siwaliks is that they have been tilted and uplifted along large sections of the Himalayan forefront, and thus form low and often deeply dissected hills. Uplift ceased between ca. 2 Ma and 400 ka, and thus subsequent, post-Siwalik fluvial deposits are horizontally bedded. For the most part, Siwalik deposits record the history of second- and third-order tributaries of the modern Indus, Ganges and Brahmaputra, and thus provide information about the upper parts of these drainage systems after these rivers left the Karakorum and Himalayas.

The Upper Siwaliks comprises three stages, the Tatrot, Pinjor and Boulder Conglomerate (see Figure 4). Of these, the Pinjor Formation is the longest and most important paleoanthropologically, as its maximum span is from 2.5 to 0.6 Ma. The sediments are primarily, as with most Siwalik formations, sands, silts and clays; the finer sands and silts are often overbank deposits, and paleosols are also common but rarely well-developed. Soil carbonate analyses indicate that vegetation was overwhelmingly open grassland (Quade et al., 1993). There is no influx from loessic or glacial deposits. The loess over northern Pakistan is post-Siwalik; most of that preserved dates from 75-18 ka and was probably derived from fans along the Indus River (Rendell et al., 1989:92). Most Upper Siwalik deposits are too far from the Himalayan forefront to receive glacial debris.

The Upper Siwaliks have been investigated for over a century, and large amounts of vertebrate fossils have been collected from them, although often with little detailed attention to their provenance. Compared with other areas of southern Asia, there is a large amount of data on the fossil vertebrate record of the Late Pliocene and Early Pleistocene: the first monograph on Siwalik paleontology was published as early as 1845 (Falconer and Cautley, 1845); the first anthropoid apes were found in the 1870's (Lydekker, 1879): a large amount of fossil material was collected and studied in the British period (e.g., Pilgrim, 1913, 1939; Matthew, 1929; Colbert, 1935); and Indian paleontologists have been very active in the last 30 years (e.g., Badam, 1979; Sahni and Khan, 1988; Nanda, 2002). The absence of hominin remains cannot therefore be attributed solely to a lack of fieldwork. Why then have no hominin remains been found in the Upper Siwaliks? To answer

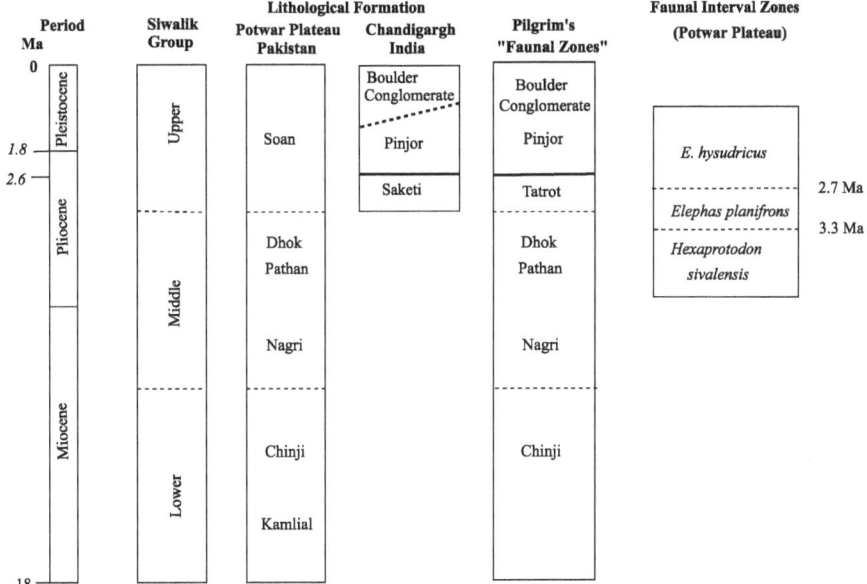

Figure 4. Zonation of the Upper Siwaliks. Source: Dennell, 2004: Figures 2.2 and 2.6, and Hussain et al., 1992: Figure 6

this question, we need first to consider the advantages and drawbacks of the types of floodplain environments presented by the Upper Siwaliks for early hominins.

Large-Scale Fluvial Systems, Early Pleistocene Hominins and the Availability of Stone

At first sight, the extensive floodplains of the Early Pleistocene ancestors of the Indus, Ganges, and their major tributaries should have been attractive areas for *H. erectus* to colonize, as water, a large range of mammalian, and probably also plant and other resources were widely available. There are several reasons, however, why these landscapes might have been beyond their capability to colonize successfully on a long-term basis.

Floodplains and Natural Hazards

The first drawback of large floodplains for early hominins is the summer monsoon, when most of the annual rainfall occurs. During this time, river levels can rise dramatically, and flood extensive parts of the floodplain. This is not in itself a hazard, unless one is unfortunate enough to be trapped on a channel bar or mid-stream island, or behind a levée when it breaks, but high flood levels are disruptive to movement, either through the formation of temporary bodies of water, or extensive water-logging. A second and related problem comes from water-borne infections and illnesses, particularly after the main monsoonal rains. Bar-Yosef and Belfer-Cohen (2001) rightly suggested that epidemiological factors may have played an important, if invisible, part in influencing early hominin settlement and dispersal, and may have included malaria near water logged and flooded areas. There is also in South Asia a wide range of water-borne parasites and diseases, and large floodplains may also have been unhealthy places to stay in late summer. A third problem is that Indian rivers experience episodic major flood events on an average of every 20–25 years or so (Gupta, 1995). Although the flood waters do not persist, these flood events can have long term consequences on the geometry of the river bed and the direction of flow; rivers may change course afterwards, and among the effects noted are a widening of the stream channel, erosion of bars, and scouring of floodplains and stream beds (Baker et al., 1995). As explained below, these changes may have had consequences on lithic procurement.

Floodplains and Predator Avoidance

Wide, flat floodplains that were predominantly open grassland would also have afforded hominins few vantage points from which to assess risks and opportunities, and little protection (such as trees as places of refuge) from large predators such as *Pachycrocuta brevirostris*. Evidence from Zhoukoudian, China, indicates that this giant hyaenid frequently ate *H. erectus* in the Middle Pleistocene (Boaz et al., 2000), and Turner (1992) suggested that the abundance of large carnivores in Europe during the Early Pleistocene was an important factor in delaying the entry of hominins into that continent. Data from the Pabbi Hills, Pakistan (Dennell et al., 2005a), and from other Upper Siwalik localities Nanda (2002) indicate several large Early Pleistocene predators: the giant hyaenid *Pachycrocuta brevirostris*, the sabre-toothed *Megantereon*, the pantherine *Panthera uncia*, a large canid *Canis cautleyi*, as well as the hyaenids *Crocuta crocuta* and probably *Hyaenictis or Lycyaena*. The giant felid *Homotherium* may also have been present, as it is evidenced in neighboring regions in the Early Pleistocene. Crocodiles and/or gavials were also significant riverine predators. Analysis of the three fossil accumulations (localities 73, 362 and 642) that were excavated in the Pabbi Hills did not provide any indication that hominins were able to compete with large carnivores for prey or carcass segments in the Early Pleistocene (Dennell et al., 2005b, 2005c); as example, at locality 642 (1.4–1.2 Ma), *Pachyrocuta* was able to select prime adult *Damalops palaeoindicus* as its main prey (Dennell et al., 2005b).

Floodplains and Hominin Home Range Sizes

Because Early Pleistocene hominins appear to have had small home ranges, large floodplains that might commonly have been > 50 km wide and uniform over large distances might have been too large for them to exploit efficiently. Estimates of their average home range are usually based on inferences of the estimated height, weight and brain size of fossil hominins, and linked to known home range sizes of extant primates and modern gatherer-hunters. These estimates can be only approximate, and much depends on assumptions of their diet (particularly over how much meat was eaten). Nevertheless, they are useful aids to modeling early hominin behavior.

The extraordinarily complete 1.75 Ma old specimens from Dmanisi, Georgia, currently provide our best insights into the earliest known inhabitants of Asia. As Table 1 shows, they were remarkably small-brained. The most recent, and very detailed, taxonomic assessment is that these individuals represent the most primitive form yet found of *H. erectus* sensu lato, and that they may form part of the source population of *H. erectus* (a.k.a. *H. ergaster*) in East Africa, and *H. erectus* sensu stricto in Java (Rightmire et al., 2005). The one published post-cranial specimen (D2021, a proximal right third metatarsal) also indicates that they were short, with an estimated stature of only 1.48 ± 0.65 m (Gabunia et al., 2000:31). As might be expected for such primitive forms of *Homo*, their behavior also appears to have been very primitive. The large (4,446 pieces) and associated lithic assemblage has recently been classified as "pre-Oldowan", in the sense that it lacks the small retouched tools that feature in East African Oldowan assemblages (de Lumley et al., 2005). Although no details are yet available on their subsistence behavior, it is probable that the cognitive abilities of these hominins were very limited relative to later populations of *Homo*, and that they were

Table 1. Cranial capacities of specimens of early H. erectus *from Dmansi and later* H. erectus *from Java and Zhoukoudian (China)*

Specimen	Description	Cranial Capacity (cc^3)
D2280	Adult braincase, possibly male	775
D2282/D211	Partial cranium of a young adult	650–660
D2700/D2735	Complete skull of small subadult	600
D3444/D3900	Edentulous old cranium	625
Sangiran	Average of Sangiran 2,4,10,12,17, IX and Trinil 2	918
Zhoukoudian	Average of crania II,III,V,VI,X–XII	1029

Sources: Rightmire et al., 2005 for Dmanisi and Antón, 2002:Table 1 for Java and Zhoukoudian

not particularly sophisticated at dealing with complex subsistence strategies.

Estimates based on body weight estimates for the Dmanisi hominins and early African *H. erectus* suggest home range sizes of up to 413 hectares (assuming a diet at the low end of the range of modern tropical foragers), but only 331 hectares if estimates are based on the third metatarsal (specimen 2021) from Dmanisi (Antón and Swisher, 2004:288). This figure implies an annual home range of ca. 82–100 sq km for a group of 25 early *H. erectus* (i.e., the smallest number assumed to be viable for mating and child-rearing), or an operating radius of only 5.1–5.6 km. As will be shown below, this estimate agrees very closely with estimates based on the distances over which hominins transported stone in the Early and Middle Pleistocene. On large floodplains that were uniform in relief and vegetation over several tens of kilometers, a foraging radius of this size could have been too small to include all the resources (such as stone, water, carcasses, plant foods, etc.) that were needed on a daily or weekly basis. While they might have offset those disadvantages by increased mobility and frequent relocation of their home range, the risks would have been high if the distances involved were considerable, especially as the weakest members of the group, such as the very young,

pregnant females, and the infirm would have been vulnerable to predators. It is probably because of their small foraging ranges and limited cognitive abilities that Early Pleistocene *H. erectus* appears to have preferred areas such as small lake basins where a wide variety of resources were available within a small area. Examples are the lake basins at Dmanisi (Gabunia et al., 2000b), Erq el-Ahmar (if the flaked stones are accepted as artifacts, and assumed to be Early Pleistocene in age), 'Ubeidiya and Gesher Benot Ya'aqov in the Jordan Valley (Feibel, 2004) and Nahal Zihor, Israel (Ginat et al., 2003); Dursunlu, Turkey (Güleç et al., 1999); Kashafrud, Iran (Arai and Thibault, 1975/77) (if its dating to the Early Pleistocene is accepted); and sites such as Majuangou and Xiaochangliang in the Nihewan basin, China (Zhu et al., 2003, 2004). Other favored locations (and for which the evidence for hominins is predominantly Middle Pleistocene) were small river valleys and stream channels, such as Evron, Israel (Ronen, 1991); Latamne, Syria (Clark, 1969); numerous Oldowan and Acheulean sites in Saudi Arabia (Petraglia, 2003); the Hunsgi-Baichbal valleys (Paddayya, 2001); in Spain, the upper parts of tributaries of the Duero (but not the main river itself) and along the Somme, Thames, Solent and Ouse in northern Europe (Gamble, 1999:143).

Floodplains and the Scarcity of Stone

The floodplains of northern India and Pakistan would also have been very deficient in one item that hominins appear to have depended upon after 2.5 Ma, namely flakeable stone. As Misra (1989a:18) commented, "In the case of the Ganga valley, the non-availability of stone, the basic raw material for making tools, may have been responsible for man avoiding this region". We noted above that the history of fluvial deposition in the Siwaliks, from 18 Ma to ca. 1 Ma, and thereafter, is primarily one of sand, silt and clay. Stone is virtually absent from most Upper Siwalik sequences before the Boulder Conglomerate Stage, when thick coarsely-sorted conglomerates were deposited on top of the predominantly fine-grained sediments of the Pinjor Stage (see below). When found in Pinjor Stage deposits, stone tends to occur as "stringers", or as thin layers a few meters long, in or by the active river channel. Stone also tends to be rare on the margins of floodplains, as often the nearest elevated ground (often several kilometers from the main river course) is usually formed of uplifted, earlier Siwalik exposures of silts and sands.

For stone-dependant hominins during the Late Pliocene and Early Pleistocene, access to what stone there may have been along the river channel is likely to have been highly seasonal. Stone, being the heaviest part of a river's bed load, would have been found in only the active, year-round channel, and most easily obtained when the river was at its lowest, i.e., immediately before the summer monsoonal rains, or in mid-winter. When river levels rose, after heavy spring rains and during the monsoon, it would have been inaccessible (see Figure 5). As noted above, the rivers in this region are prone to changes of course, so there need not have been any guarantee that stone would be available at a particular place from decade to decade, or even year to year. Major flood events could therefore have had important local consequences in re-arranging the location and amount of stone available along stream channels – for example, a previously-used source might have been buried, or scoured away; or other stone sources might have been exposed further downstream.

Two factors – transport distances, and coping strategies – need to be considered when assessing the competence (or otherwise) of early Pleistocene hominins to deal successfully with the problems of inhabiting an environment where stone was extremely scarce and not readily-available year-round.

Transport Distances Available data suggest that stone was carried in the Early and Middle Pleistocene over very short distances, typically less than 5 km. At Olduvai (Beds I and II), for example, most artifacts were made from stone obtainable within 2–4 km, and occasionally 8–10 km (Hay, 1976:183 quoted in Isaac, 1989:171). Petraglia et al. (2005) report that stone was rarely transported beyond 2 km in the Hunsgi Valley. The distances over which stone was obtained appear similar throughout the Lower Paleolithic in Europe (see Roebroeks et al., 1988; Gamble, 1999). At the Caune d'Arago (France), for example, 80% of lithics came from < 5 km; the maximum distance reported was 35 km (Gamble, 1999:126). The same pattern continues into the Middle Paleolithic: in Southwest France, 55–98% of stone came from < 5 km and 2–20% from 5–20 km. For all of Europe, only 15 of 94 raw material transfers from 33 sites/layers were > 30 km, and the average maximum distance was 28 km (see Gamble, 1999:126–127). Feblot-Augustins (1999) also observed that at 33 European lower and 19 Middle Paleolithic sites, > 90% of utilized stone came from < 3 km, and the rest usually from 5–12 km.

The same pattern appears to hold for Early and early Middle Pleistocene Asia: at Dmanisi, stone came from two nearby rivers (de Lumley et al., 2005:3), at 'Ubeidiya (Bar-Yosef and Goren-Inbar, 1993:121), and Gesher Benot Ya'aqov (Goren-Inbar et al., 2000), it was

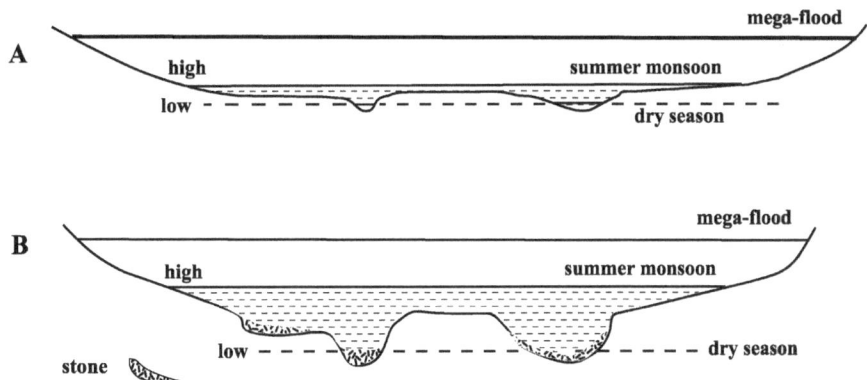

Figure 5. River levels and the probable seasonal availability of stone. Source: Dennell, 2004: Figure 11: 8. In A, the vertical and horizontal scales are the same. In B, the vertical scale has been exaggerated to show more clearly the effect of changes in river level on the accessibility of stone. As shown, stone might have been inaccessible for a substantial part of the year during and immediately after the summer monsoon. River levels can also rise dramatically in spring after heavy rains, and when snow melts in mountainous areas further upstream. Megaflood events could have major consequences on channel profiles, including burying or removing sources of stone, or exposing new ones

immediately available; at Dongutuo, Nihewan basin, North China, there was a nearby chert outcrop (Schick et al., 1991; Pope and Keates, 1994); at Xaiochangliang, also in the Nihewan basin, stone was available nearby (Zhu et al., 2001); at Dursunlu (Turkey), stone probably came from nearby hills (Güleç et al., 1999); at Dawādmi in Saudi Arabia, at least 24 Acheulean sites are known from along the northern side of an andesite dike near which there may have been a low-lying lake (Petraglia, 2003). At Evron (Israel), most of the lithics were made from stones in the adjacent stream channel, but a few from calcite geodes 5 km away (Ronen, 1991).

As might be expected, there are a few exceptions. Gamble (1999:126–127) notes that at the Caune d'Arago, a small amount of stone came from 35 km away, and as far as 80 km from Labastide d'Anjou; and Mishra (1994:61) notes that at Yedurwadi, one quartzite spheroid probably from a distance of > 50 km. Nevertheless, the pattern for Early Paleolithic sites is always that most stone was obtained, used and discarded within 5 km of where it was found, and often less. This figure

strengthens the estimates cited above that these hominins operated within a 5 km radius. As Gamble (1999:144) notes, they probably also lacked the social and exchange networks that enabled them to learn about and obtain resources associated with other groups. It is not until the late Pleistocene that stone (and other exotic items such as shell) was routinely transported > 80 km and even > 200 km from its source (Gamble 1999:315).

Coping Strategies How then did hominins cope with landscapes that were "resource rich, stone poor"? Options available were using very small stones; curation; caching; substitution of stone by other materials; and avoidance of stone-poor areas. One example of using very small and poor quality stone is the 2.3 Ma old assemblages from the Omo River, Ethiopia, where the tools are little more than smashed quartz pebbles (Merrick, 1976). Another example is the loess landscape of southern Tajikistan, where stone was extremely scarce and found mostly in stream beds. Stone tools were made from small pebbles of quartzite, limestone, schists, cornelian, porphyry and poor quality flint and chert. At Kuldara, ca. 955-880 ka and the oldest site in the region, 25% of flaked pieces were < 2 cm in length, and 50% were 2–4 cm long. At Kuratau I (620–570 ka), 18% were 2–3 cm long, and ca. 70% of all flaked stones showed flake removals of < 5 cm. Artifacts are also usually found in very low densities, or as isolated finds. At Kuldara, for example, only 96 items (of which only 40 are tools) were found in an area of 62 sq m, and at an estimated density of only one find per 4 m³ of deposit. At Karatau, the density of flaked pieces was only 1.9 per sq m, although at Lakhuti 1 (530–475 ka), it was as high as 10.5 per sq m (Ranov, 2001; Ranov and Dodonov, 2003).

Evidence for curation in the Early Paleolithic is very rare, and most artifacts seem to have been made and used expediently. One example is from Mudnur VIII at Hunsgi-Baichbal valley, where ca. 24 massive handaxes of limestone were found without any associated blocks of raw material or/and debitage, and which may have been a cache meant for future use (Paddayya and Petraglia, 1995:349). A third is the piece of antler of *Megaceros* from Boxgrove (524–476 ka) that was probably used as a soft hammer (Roberts and Parfitt, 1999:395). Another example may be a type of hemispherical or polyhedral core that was found in the Pabbi Hills and represented by 14 examples, and which showed evidence of wear on one face (Hurcombe, 2004:231). Otherwise, curation does not appear to have provided Early Pleistocene hominins with a solution to the problem of dealing with extreme scarcities and/or seasonal shortages of stone.

Evidence that in some situations, hominins used stone sparingly is provided by recent work at Hata in Ethiopia (Heinzelin et al., 1999). Here, the explanation offered is that late Pliocene hominins were operating in areas where stone was very scarce, in contrast with areas such as Gona, where stone was naturally very abundant (as was evidence of tool-making). It is worth quoting the discussion of this evidence at some length:

Nearly contemporary deposits at Gona, only 96 km to the north, produced abundant surface and in situ 2.6-Ma Oldowan artifacts. In contrast, surveys and excavations of the Hata beds have so far failed to reveal concentrations of stone artifacts. Rare, isolated, widely scattered cores and flakes of Mode I technology appearing to have eroded from the Hata beds have been encountered during our surveys. Most of these surface occurrences are single pieces. Where excavations have been undertaken, no further artifacts have been found ... At the nearby Gona site, abundant Oldowan tools were made and discarded immediately adjacent to cobble conglomerates that offered excellent, easily accessible raw materials for stone-tool manufacture. It has been suggested that the surprisingly advanced character of this earliest Oldowan technology was conditioned by the ease of access to appropriate fine-grained raw materials at Gona. Along the Karari escarpment at Koobi Fora, the basin margin at Fejej, and the lake margin at Olduvai Gorge, hominids also had easy access to nearby outcrops of raw material. In contrast, the diminutive nature of the Oldowan assemblages in the lower Omo [made on tiny quartz pebbles] was apparently conditioned by a lack of available large clasts. The situation on the Hata lake margin was even more difficult for early toolmakers. Here, raw materials were not readily available because of the absence of streams capable of carrying even pebbles. There were no nearby basalt outcrops. The absence of locally available raw material on the flat featureless Hata lake margin may explain the absence of lithic artifact concentrations ... The paucity of evidence for lithic artifact abandonment at these sites suggests that these early hominids may have been curating their tools (cores and flakes) with foresight for subsequent use ... The Bouri discoveries show that the earliest Pliocene archaeological assemblages and their landscape patterning are strongly conditioned by the availability of raw material (Heinzelin et al., 1999:628–629).

As we shall see below, these conclusions could have been written for the evidence from the Pabbi Hills, notably the absence of nearby large sources of flakeable stone, the rarity of flaked stone across the landscape, the preponderance of isolated finds, and the failure of excavations to find artifacts in situ.

There is very little evidence that Early or Middle Pleistocene hominins used materials other than stone when the latter was scarce or not available. Fossilized wood is effectively a type of stone, so its use in the (undated)

Anyathian Early Paleolithic of Myanmar is not strictly an example of substitution, especially as it is widely available throughout Myanmar (Oakley, 1964:232). Even if bone was very occasionally flaked (as with, for example, the bone handaxe from Castel di Guido, Italy [Gamble, 1999:137]), stone was still used to flake it. Arguments that bamboo was used instead of stone in Indonesia and mainland southeast Asia (Pope, 1989) are unconvincing, as in all other areas used by hominins, stone was always used if available along with organic materials such as wood. Nevertheless, one interesting example that may show the substitution of bone for stone comes from Kalpi, on the Ganges Plain in the Yamuna Valley: here, Middle Paleolithic lithic artifacts, dated to ca. 45 ka, were made from small (2–4 cm) quartzite pebbles, and were greatly outnumbered by a variety of bone artifacts, including scrapers, points and burins (Tewari et al., 2002). This is the first Middle Paleolithic evidence from the Ganges Plain, and may indicate how hominins were later able to overcome the scarcity of workable stone in these large floodplain environments. At this point, we can consider the fossil and archaeological record of the Upper Siwaliks.

The Mammalian Fossil Record of the Upper Siwaliks

The Pinjor Stage is one of the most detailed for the Early Pleistocene, and includes at least 49 vertebrate taxa (Nanda, 2002). Although fossil collecting has at times been haphazard and unsystematic, we can assume that after a century or so, the main animals have probably been recorded. Nevertheless, consideration of the adult body size and types of animals recorded indicate that the full range has yet to be established. The genera listed in the Pinjor Stage are overwhelmingly medium (> 50 kg) to large (> 250 kg) herbivores. Figure 6 shows the mammals represented in the Pabbi Hills (Dennell, 2004; Dennell

et al., 2005a), which has a long and often fossil-rich sequence spanning 2.2–0.9 Ma that is broadly similar to other Upper Siwalik sequences of the same age-range. As can be seen, very few taxa were found that are within the size range of hominins, and the few that were (e.g., gazelle, small pigs and small carnivores) were also very rare. A second feature of the Upper Siwalik vertebrate record is that rare taxa are sampled very poorly. For example, *Pachycrocta, Megantereon, Canis cautleyi, Ursus* and anthracotheres are recorded (but by only a few specimens each) in the Pabbi Hills sequence in Pakistan, but not in India; conversely, *Camelus* (Opdyke et al., 1979), *Theropithecus* (Delson, 1993) and small primates (Barry, 1987) are recorded in India but not Pakistan; and in neither country is *Homotherium* recorded, even though it was present at Dmanisi to the west (Gabunia et al., 2000b), Kuruksay, Tajikistan, (Sotnikova et al., 1997) to the north, and Longuppo, China, to the east (Wanpo et al., 1995). The absence of hominins is not therefore as well established as might at first sight appear.

Hominin Remains in Fluvial Contexts

Hominin remains are very rare from fluvial deposits, and most of those known are from rivers that were probably smaller than many of those indicated by the Upper Siwaliks. The principal finds are shown in Table 2. As shown, most finds have been of crania or mandibles; post-cranial remains are rare, although six femora were found at Trinil, two tibiae at Solo (Ngandong) (Day, 1986), a tibia fragment at Sambungmachan, and a clavicle at Narmada in India (Sankhyan, 1997). Most discoveries of hominin remains in fluvial deposits have tended to be "one-offs", in the sense that further searching in the same deposits has rarely led to the discovery of other hominin remains. There are a few exceptions: a partial one is Swanscombe, where a third piece of cranium was found in 1955 that miraculously joined

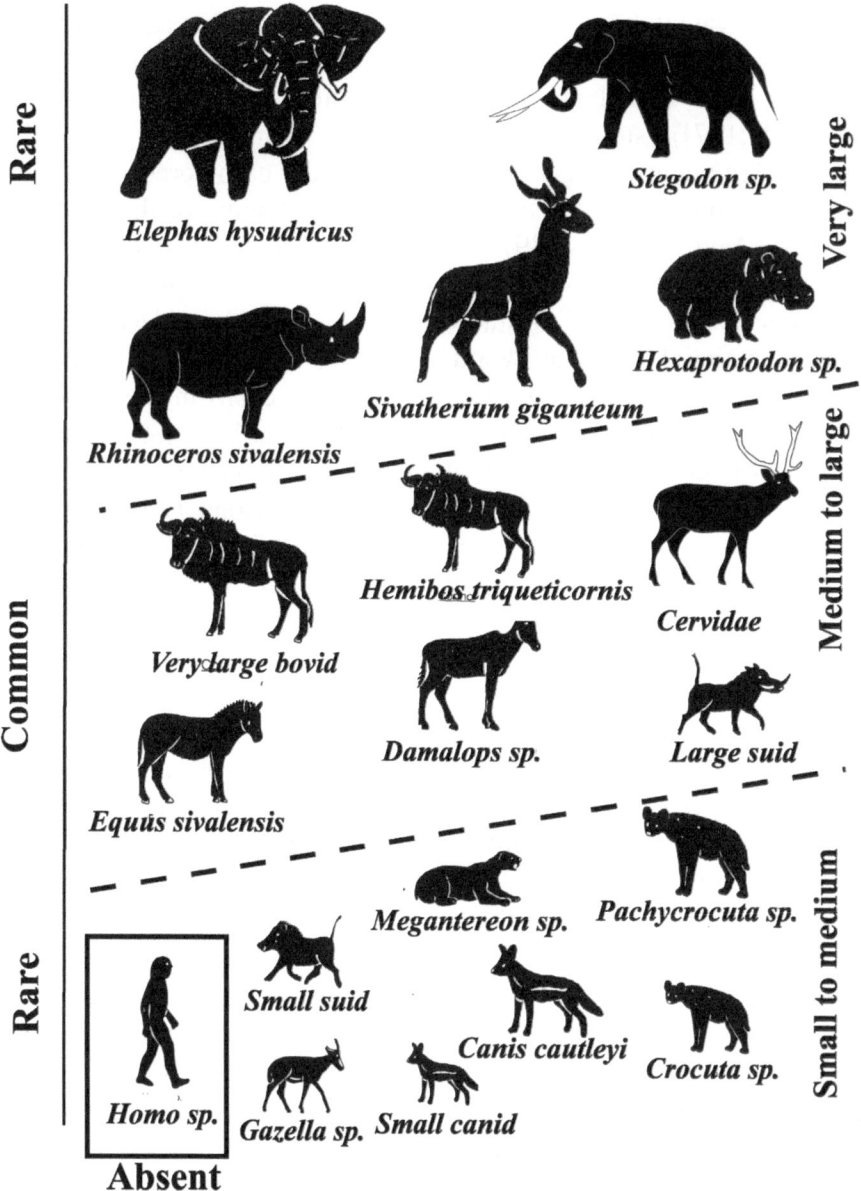

Figure 6. Type and body size of mammals represented in the Pabbi Hills, Pakistan. The taxa shown are broadly similar to those from comparable Indian exposures. As shown, most taxa are considerably larger than humans. Those nearest in body size – the mammalian carnivores, gazelle and small pigs – were very rare. The commonest taxa were medium to large ungulates, particularly bovids: note that there were at least two types of cervid, and probably two other types of medium-sized bovids in the Pabbi Hills that have not been shown here. Very large mammals were rare, especially if counts of fragmented tooth and tusk are ignored. Current absence of hominins from the Upper Siwaliks can be explained in large part by the bias towards the preservation of taxa larger than humans. Data derived from Dennell (2004)

Table 2. Pliocene and Pleistocene hominin skeletal remains from fluvial deposits

Site	Date	Element(s)	Age
Binshof, Germany	1974	Skull	$21,300 \pm 320$ BP; now 3090 ± 45 BP
Ceprano, Italy	1994	Skull fragment	Middle Pleistocene
Dali, China	1978	Cranium	Middle Pleistocene
Hahnöfersand, Germany	1973	Skull fragment	$36,300 \pm 600$ BP; now 7500 ± 55 BP
Lantian, China	1963	Mandible	Lower/Middle Pleistocene
Mauer, Germany	1907	Mandible	Middle Pleistocene
Narmada, India	1982	Skull fragment, clavicle	Middle Pleistocene
Olduvai OH9, Tanzania	1960	Calvarium	Lower Pleistocene
Omo SL7A, Ethiopia	1967	Mandible	Late Pliocene
Paderborn, Germany	1976	Skull fragment	$27,400 \pm 600$ BP; now 238 ± 39 BP
Saccopastore I, Italy	1929	Cranium	Upper Pleistocene, level 5
Saccopastore II, Italy	1935	Cranium	Upper Pleistocene, level 7
Sambungmachan, Indonesia	1973 onwards	3 calvaria, 1 tibia fragment	Middle or Upper Pleistocene
Steinheim, Germany	1933	Calvaria	Middle Pleistocene
Swanscombe, U.K.	1935–6, 1955	3 conjoining skull fragments	Middle Pleistocene
Hadar, AL-333, Ethiopia	1975–1977	13 individuals; > 200 pieces	Pliocene
Ngandong (Solo), Indonesia	1931–3, 1976–80	14 crania, 2 tibiae, 1 innominate fragment	Late Pleistocene
Trinil, Indonesia	1891–1900	Calotte, 2 teeth and perhaps 5 femora	Early Pleistocene

Sources: Day, 1986, except for: Sankhyan, 1997 and Sonkalia, 1985 for Narmada; Johanson et al., 1982 for AL-333; Bronk Ramsey et al., 2002 for Hahnöfersand, Paderborn and Binshof; these are included because they were thought to be Pleistocene in age when discovered. Ascenzi et al. (1996) for Ceprano; Oakley K et al. (1971: 254) for Saccopastore; and Oakley K. et al. (1975: 79) for Lantian.

Notes: the context is problematic at Sambunmachan. The *Homo ergaster* skeleton WT15000 (Kenya) is excluded as that was derived from the edge of a swamp, not a river. Two notorious fakes from fluvial contexts have been excluded but are otherwise consistent with the above: the Moulin Quignon mandible of 1863, and the Piltdown skull cap and mandible of 1913–15.

the other two, conjoining pieces found in 1935 and 1936; even so, all three can be counted as part of just one skeletal element. Two of the most unusual sets of hominin remains from fluvial contexts are the calotte and six femora from Trinil, and the 14 crania and two tibiae from Solo (Ngandong). The integrity of these assemblages is questionable: the femora at Trinil were probably from an overlying layer (Bartsiokas and Day, 1993), and the Ngandong remains may have been reworked from an earlier mass-drowning event (Dennell, 2005). The most startling exception is the "first family" from Hadar, with at least 13 individuals

represented by virtually all skeletal elements; this find is unique, and wholly unlike other discoveries of hominins in fluvial deposits.

There are two other taphonomic issues that may also be relevant to why hominin remains have yet to be found in the Upper Siwaliks. Both are indicated by data from the Pabbi Hills, Pakistan, where over 18 months of fieldwork was dedicated to looking for hominin remains from both surface exposures and by the excavation of fossil concentrations, and so the absence of hominins cannot be attributed to lack of fieldwork or searching. These investigations also paid more attention to taphonomic issues than previous studies of the Upper Siwaliks, and provide some further insights into the type of fossil record from this type of fluvial landscape (Dennell, 2004:341–362).

Fragmentation Fossil specimens (as from other Siwalik exposures) were often very fragmented, and this clearly affected the preservation and identification of rare and small taxa such as hominins. Carnivores, for example, were also very rare, as were remains of gazelle. Overall, only ca. 30% of the fossil material ($> 40,000$ specimens) collected on modern erosional surfaces could be identified to taxon and anatomical part, but this proportion varied (exceptionally) from 50% at a few localities to (commonly) $< 10\%$. Fragmentation rates tended to be highest on flat surfaces (where fossils are fully exposed to heavy rain and trampling), and lowest on steep, actively eroding slopes. The material collected was heavily biased to larger taxa, and the most robust parts of the skeleton.

Where animals are preserved A very striking feature of the Pabbi Hills data is that large areas contained very little fossil material. Although $> 40,000$ fossils (including non-diagnostics) were collected from over 600 places, over half came from 20 concentrations. Two types were recognized. The first were channel bar deposits, where fossils had accumulated through down-stream transport, and from predation (probably by crocodile) near the stream margins. The second were found away from the active river margins, in abandoned channels and on the flood-plain, and were accumulated by carnivores. These concentrations were not only the largest source of data, but also the richest in that they account for almost the full range of taxa found. The only exception was an anthra-cothere, represented by three specimens. Some taxa were better represented outside these concentrations, such as *Elephas*, *Stegodon* and *Sivatherium*, and large felids. On the whole, however, the Pabbi Hills is a record that

is biased towards those animals that died near or at an active stream margin, or those eaten by a predator. What are missing are those in woodlands, such as primates; small mammals such as hare; ones perhaps smart enough not to be caught by a hyaena or crocodile; and those that died away from stream margins. Although some hominins died in streams or were eaten by hyaena (as at Zhoukoudian, see Boaz et al., 2000) and possibly by crocodile (as with the 1953 ["*Meganthropus*"] mandible from Sangiran [von Koenigswald, 1968] or Olduvai OH7 [Davidson and Solomon, 1990]), there are sadly (for paleoanthropologists) no indication of similar, but probably rare, fates in the Pabbi Hills.

In summary, the absence of hominin remains from the Upper Siwaliks is not simply due to insufficient fieldwork, although more fieldwork is obviously desirable. While the absence of fossil evidence for hominins before 1.7 Ma in the Pabbi Hills sequence could be seen as genuine evidence of absence, the same argument is harder to apply to the material collected from Sandstone 12 (1–4–1.2 Ma), which comprised half the total amount found, and was often extremely well preserved. Hominins should have been in South Asia by this time, and the recent date (if accepted) of 1.27 Ma from Isampur in south India (Paddayya et al., 2002) indicates that they may have been in peninsula India by this date. The main reasons for the absence of hominin fossils appear to be the bias towards the preservation of animals larger than hominins; the bias against carnivores and other rare taxa; the extent of fragmentation which lessens even further the preservation of smaller animals; and the limited circumstances under which most fossils were preserved. Additionally, hominins may also have been very scarce in these large floodplains. Though we might now better understand the nature of the haystack, the hominin needle is still elusive.

The Archaeological Evidence for Hominins in the Upper Siwaliks

There are only three sets of archaeological evidence for hominins from Upper Siwalik deposits. The first and least controversial were the discovery of three handaxes in conglomerates at Dina and Jalalpur (Rendell and Dennell, 1985), in contexts just above the Brunhes-Matuyama boundary and thus ca. 0.6–0.78 Ma. These are still the earliest definite indications of the Acheulean in South Asia, apart from the recent ESR date of 1.27 Ma from Isampur (Paddayya et al., 2002). The other sets may be Early Pleistocene in age, and are derived from Riwat, in the Soan Valley, and the Pabbi Hills. Each can be briefly summarized.

The Soan Valley (Riwat)
(33⁰ 40′N; 73⁰ 20′E)

The artifact assemblage from Riwat, in the Soan Valley, has been described elsewhere (Rendell et al., 1987, 1989; Dennell et al., 1988; Dennell and Hurcombe, 1989:105–127), and only the key issues of identification, context and dating will be considered here.

Identification The assemblage is extremely small, and discussion normally focuses on three pieces (see Figure 7). The main one (R001) is a large core ($168 \times 118 \times 74$ mm) that was struck eight or nine times in three directions; there are clear impact points and ripple scars on at least three of the flake removals (see for example, Dennell and Hurcombe, 1989:113, Figures 7.6, 7.8). The size of flakes removed (average 6.6×6.2 mm) is within the range of those seen at Olduvai Gorge, Bed I. Several Paleolithic archaeologists (including several far more authoritative on early Paleolithic lithic technology than the author) who have seen a resin cast of this object (unfortunately the only one so replicated) have accepted it as demonstrating unambiguously intentional flaking.

A second piece (R014) is a large flake ($132 \times 79 \times 58$ mm) that had been struck from a cobble; there is a clear bulb of percussion and associated ripple marks on the dorsal face, and at least three flakes were struck along the side (see Dennell and Hurcombe, 1989:Figure 7.10), creating an edge straight in side view. There were eight scar surfaces resulting from flaking in three directions (Dennell and Hurcombe, 1989:115). A third piece (R88/1) is a flake ($59 \times 45 \times 20$ mm) with a clear positive flake scar on one face, a negative one on the other (see Dennell and Hurcombe, 1989:116, Figure 7.15), and evidence of flaking from three directions.

Context Core R001 was found firmly embedded in 1983 in an outcropping gritstone/-conglomerate horizon near the base of a gully 70 m. When found, it was obvious that some flake scars extended into the outcrop, and thus the piece could not have been flaked after exposure. After the piece had been removed, the socket showed the flake scars that were on the core (see Dennell and Hurcombe, 1989: Figure 7.22). A resin replica was made of the socket and surrounding gritstone, and the replica of the core can be re-inserted into it. Piece R014 was chiselled out of a gritstone block that had been detached from the same gritstone/conglomerate section nearby (Hurcombe and Dennell, 1989:102, 105). Flake R88/1 was found in 1988 in a freshly-eroding vertical section 50 m from the core R001. The intervening section was drawn at a scale of 1:20, and all stones > 2 cm were drawn and recorded (see Dennell and Hurcombe, 1989:116; Figure 7.16). Of the 1,264 stones in that section, not a single one showed any signs of flaking (Hurcombe and Dennell, 1989:121). Contra Klein (1999:329) it is therefore wholly inaccurate to say that the claimed artifacts "represent simply one extreme along a continuum of naturally flaked pieces".

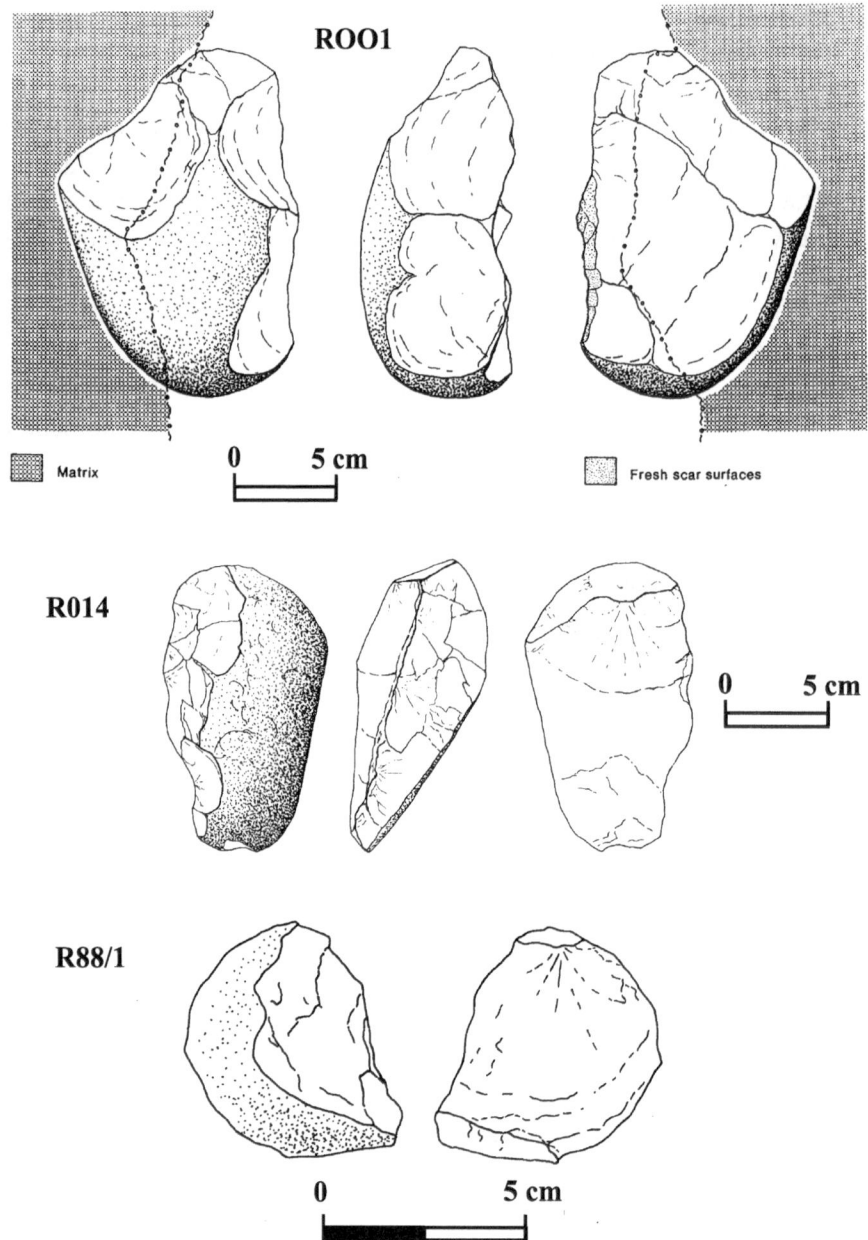

Figure 7. Flaked pieces R001, R014 and R88/1 from the lower conglomerate horizon at Riwat, Soan Valley, Pakistan. Source: Dennell and Hurcombe, 1989: Figures 7.8, 7.10 and 7.15

Dating The Soan Valley consists of a syncline that dips gently at ca. 10–15° on its southern side, but rears up almost vertically on its northern limb (see Figure 8). Its stratigraphic sequence and age were investigated very thoroughly by American geologists whose primary interest was in the evolution of the Himalayan forelands. They concluded that the Soan Syncline formed in the late Pliocene (Burbank and Johnson, 1982, 1984; Johnson N.M. et al., 1982; Burbank and Raynolds, 1984; Raynolds and Johnson, 1985; Johnson, G.D. et al., 1986). This age estimate was based on paleomagnetic evidence that showed that the basal deposits of the syncline belonged to the early Matuyama Chron; and by the observation that the vertical layers of the northern limb of the syncline were

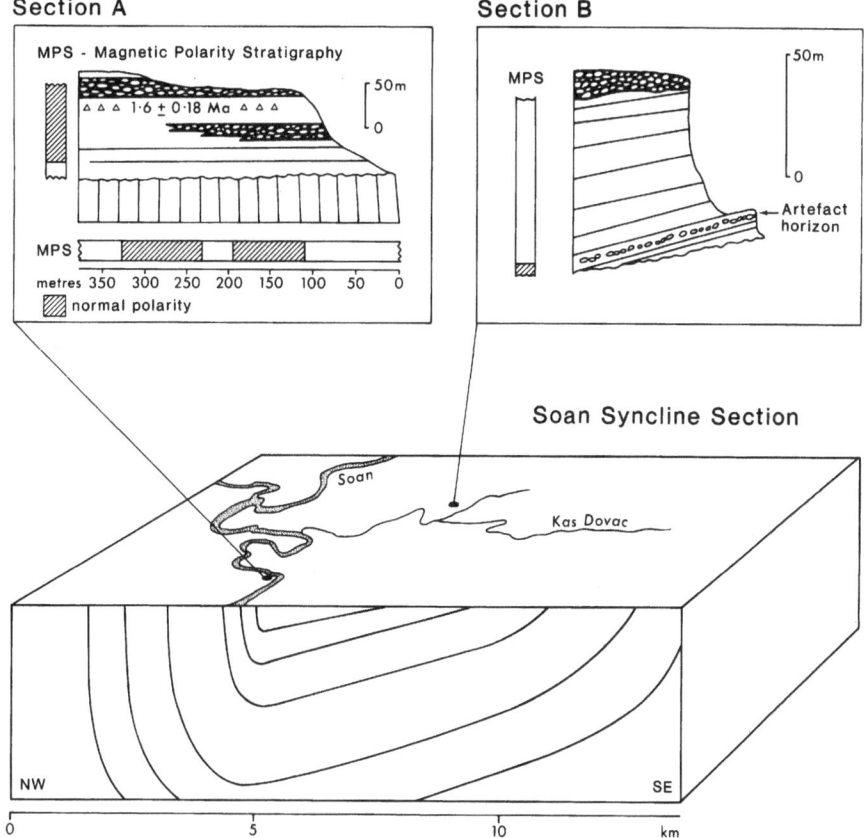

Figure 8. The dating of the artifact-bearing lower conglomerate horizon at Riwat, Soan syncline, Pakistan. The sediments containing the artifact-bearing horizon (section B) slope at 10–15°, but are folded vertically in the part containing section A. Here, they are overlain unconformably by horizontal deposits containing a volcanic ash dated at ca. 1.6 Ma. Source: Dennell et al., 1988: Figure 4

truncated, and unconformably overlain by horizontally-bedded fluvial deposits. These had a normal polarity, as well as a volcanic tuff that was dated by K/Ar to 1.6 ± 0.2 Ma (see Figure 8). This age estimate was consistent with assigning the surrounding horizontal deposits with normal magnetic polarity to the Olduvai Event. If one allowed for the time needed for the folding of the northern limb of the syncline through almost 90°, the truncation of the exposed deposits, and the subsequent deposition of overlying horizontally bedded fluvial deposits ca. 1.6 Ma, a late Pliocene age was entirely convincing for the deposition of the fluvial deposits of the Soan Syncline. No one has ever questioned the dating of the Soan Syncline sequence.

Rendell et al. (1989:71–75; Rendell et al., 1987) demonstrated that the artifact-bearing horizon was integral to the Soan Syncline, and not part of a later channel fill. She also showed through very close sampling (280 samples from 71 sampling points with a mean spacing of 1.7 meters) that all the deposits above the artifact-bearing horizon had a reversed polarity, as would be expected if they were deposited in the Matuyama Chron. An additional important, but rarely noticed, point was that these deposits (including the artifact bearing horizon) had all been rotated by 30° during the tilting and folding that had taken place. In contrast, no rotation was observed in the overlying, 1.6 Ma-old horizontal strata that capped the Soan Formation. This clearly indicates that the rotated deposits were older than 1.6 Ma.

Whether or not this evidence is accepted as indicating that hominins were present in South Asia by or before 1.9 Ma depends upon the criteria by which evidence of stone tool making is considered convincing. The obvious limitations of the Riwat assemblage is that it is very small; was found in a secondary context even if there is little indication of abrasion and rolling; and is not associated with any other evidence of hominins, such as cut-marked bone or hominin remains. If the evidential threshold is set at, for example, a minimum of 100 unambiguous artifacts, in a primary context, and preferably with associated cut-marked bones, the Riwat assemblage is clearly unacceptable as indicating the presence of hominins. For this author, the main reason for not rejecting the small assemblage from Riwat is the opinion of all those who have seen the cast of core R001 and are more knowledgeable than himself about early lithic technology that it could not have been flaked naturally. Additionally, no sceptic has ever demonstrated that stones found in stratified contexts that clearly pre-date the first appearance of hominins or humans have similar flaking characteristics to the pieces found in context at Riwat: such contexts might be, for example, an Oligocene conglomerate in the Old World, a Middle Pleistocene conglomerate in the Americas, or one a few millennia old in New Zealand or Malagasy.

The Pabbi Hills (32⁰ 50′N, 73⁰ 50′E)

The flaked stone assemblage that was found during the surveys for fossil material has been described in detail by Hurcombe (2004:222–292). Overall, 607 pieces of flaked stone were considered as artifacts (i.e., intentionally flaked). The density of flaked stone was extremely low. Although flaked stones were found in 211 places, they were found as isolated pieces in 45% of all cases, and in 78% of cases, not more than three were

found. Approximately half (n = 307) were found on exposures of Sandstone 12; 102 on exposures younger than Sandstone 12 and probably 1.2–0.9 Ma; and 198 on exposures of deposits that belonged to, or were earlier than, the Olduvai Subchron, and thus ca. 2.2–1.7 Ma. Most of the artifacts were simple cores (41%) and flakes (58%); a selection is shown in Figure 9. Almost all (96%) of the lithic assemblage was made of quartzite, and only 2.8% showed any signs of deliberate retouch. The non-quartzite component consisted of 12 small pieces of flint, including six micro-cores, four hammerstones and six fragments of polished stone axes: these are all probably neolithic or later. The quartzite assemblage is typologically consistent with the very simple, unstandardized type of assemblages that are elsewhere classed as Oldowan, and is also broadly similar to the much large assemblage from Dmanisi (de Lumley et al., 2005). Significantly, there were no examples of the type of Acheulean bifaces, prepared cores, or blades that are common on Middle and Upper Pleistocene exposures that we have examined elsewhere in northern Pakistan.

Because these artifacts are all surface finds, there is of course no direct indication of their age. Various possibilities can be considered as a series of probabilities. The least likely is that the flaked stones were recent in origin, and the result of, for example, shepherds flaking stones (if available) out of boredom, or sharpening their axes by pounding the blade on cobbles, as suggested (but not observed) by Stiles (1978:139). We saw no evidence of any recent tradition of flaking quartzite, and the behavior of bored shepherds that we observed is most unlikely to have resulted in the type of flaked stones that we found. They would in any case have had to carry stone with them in anticipation of allaying their tedium with some knapping as naturally occurring stone is virtually absent in the Pabbi Hills. Contra Mishra (2005), the flaked assemblage cannot

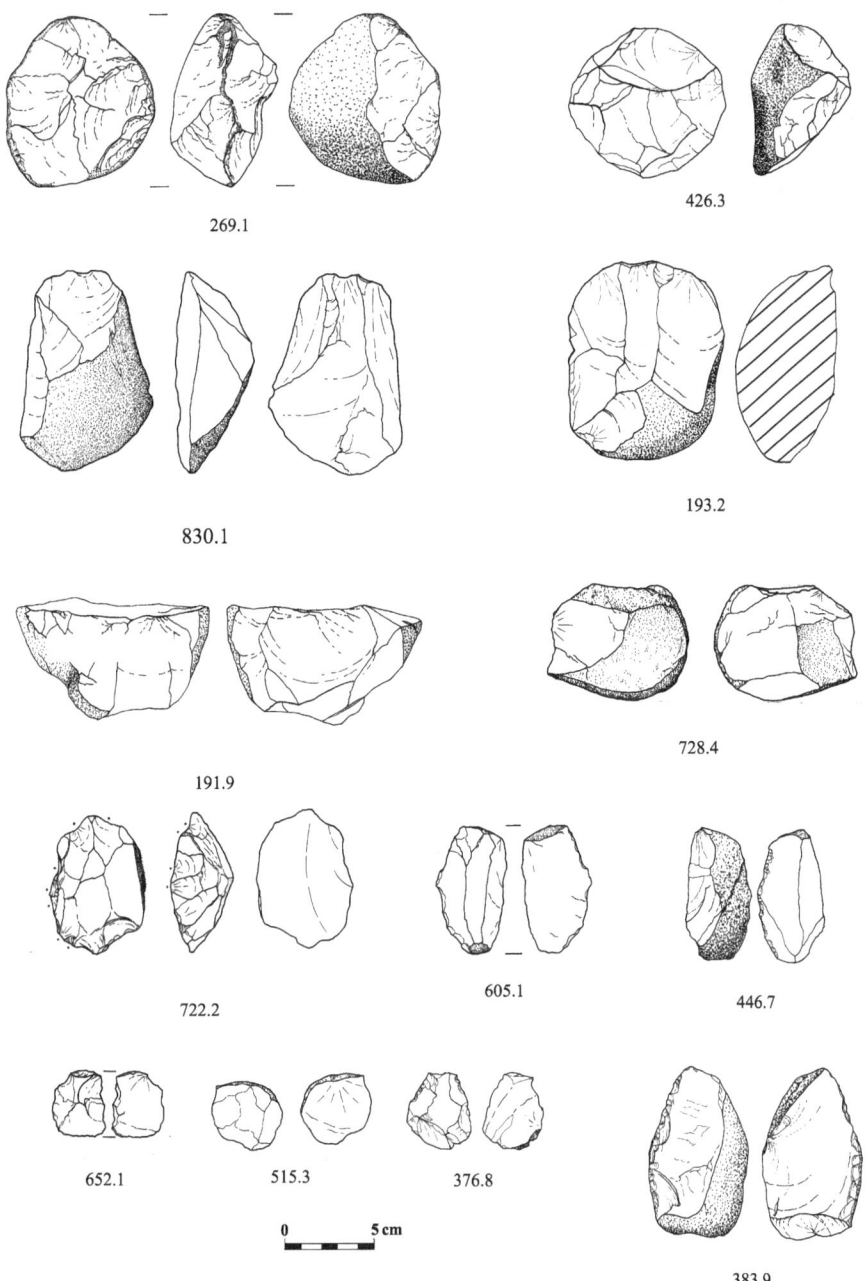

269.1

426.3

830.1

193.2

191.9

728.4

722.2

605.1

446.7

652.1   515.3   376.8

0   5 cm

383.9

Figure 9. A selection of stone tools from the Pabbi Hills. Source: Hurcombe, 2004

be confused with or derived from rail and road ballast from the Peshawar-Lahore railway line and the Grand Trunk (or G.T.) road that run alongside each other and cut through the Pabbi Hills. Hurcombe's field observations showed that ballast is smashed, not flaked, and typically comprises small angular fragments with none of the flaking characteristics seen in the artifact assemblage. The artifacts were also found at higher elevations than the road and railway, and usually several kilometers away, as indicated in the numerous maps showing where material was collected (see Dennell, 2004). Additionally, areas near the road and railway were not surveyed because of dense vegetation.

A second possibility is that the quartzite artifacts were derived from reworked residues of deposits that formed after the anticline was formed 400 ka. This too is thought very unlikely. First, there is no evidence that the fossil material found on the surveys is a mixture that includes fossils from the last 400 ka. Secondly, there is no evidence of the type of flaked stone (for example, Acheulean bifaces or Levallois cores) that we found elsewhere in northern Pakistan on Middle and Upper Pleistocene exposures. Given the size of the areas surveyed for stone tools and vertebrate fossils (often several times), and the thoroughness of collecting (with the smallest item weighing only 1 gm [Hurcombe, 2004:224])), it is inconceivable that objects as distinctive as a handaxe would have been missed. Known later types (such as the polished stone axes and flint micro-cores that are probably neolithic) account for only a dozen or so pieces. Thirdly, it is most unlikely that stone (or fossil) would have remained on erosional surfaces in the Pabbi Hills throughout the Middle and Upper Pleistocene. All of the evidence accumulated during the surveys indicated that fossils were eroded very rapidly, and either destroyed shortly after exposure, or washed into gullies and thence out of the Pabbi Hills during the summer monsoon. Our own experiments of placing marked stones on such surfaces and monitoring their movement over a 10-year period implied that stones on the flatter areas at the base of slopes might have remained there for perhaps tens but not hundreds of years; on slightly steeper slopes, for a few to many tens of years; and on the steeper slopes, they would have moved quickly once exposed but then been reburied, or incorporated into part of a steep but stepped slope which acted as a series of small terraces. It might therefore have taken a stone a few hundred, or even a few thousand years to work its way down a 20 meter slope of this kind into a gully. However, it does not seem likely that even in these stable areas the erosion processes would have taken scores of millennia. These findings strongly

imply that most of the stone artifacts on slope surfaces in the Pabbi Hills are highly unlikely to be derived from residues of post-Siwalik material from the last 400 ka (see Hurcombe, 2004:245–249).

The final possibility (and in our view, the least implausible) is that the quartzite flaked stone assemblage eroded from the underlying Upper Siwalik strata and are thus (depending on the age of the exposures on which they were found) between 2.2 and 0.9 Ma. As none of this material was found in situ, the case for dating it to the Early Pleistocene remains circumstantial. Nevertheless this type of field survey data forms an important part of the archaeological literature, and those readers who might reject this evidence on the grounds that it was found on the surface might reflect how much other data collected by field surveys elsewhere should also be rejected.

Implications of the Evidence from Riwat and the Pabbi Hills

The evidence from Riwat and the Pabbi Hills can be interpreted in two ways. If it is rejected on the grounds that the Riwat assemblage is too small, and in a secondary context, or that none of the Pabbi Hills material was found in a datable context, the obvious conclusion to be drawn is that there is no evidence that hominins occupied the Indo-Gangetic drainage basin during the Early Pleistocene. If so, the absence of hominins from the subcontinent during the Early Pleistocene might indeed be genuine. (This conclusion also implies that hominins may have entered Java ca. 1.6–1.8 Ma without crossing South Asia). An alternative possibility is that the evidence from Riwat and the Pabbi Hills is consistent with the observations made earlier on the limited availability of stone in these Upper Siwalik landscapes. In the Riwat area of the Soan Valley, stone was available on only one occasion in a 70-meter sequence of sands and silts, and some were used for making tools. In the Pabbi Hills, stone was probably scarce at all times, especially when rivers rose in the monsoon and covered the few sources of stone available. The type and patterning of stone tools across that landscape is consistent with what has been found in other stone-poor landscapes such as Hata, Ethiopia. Another relevant example, this time from the Ganges Plain, comes from Anangpur near Delhi. Here, Sharma (1993) reported Acheulean handaxes in a gravel deposit at the base of sands and silts from a former paleochannel of the Yamuna: in other words, in the one part of this sequence when stone was available, it was used by hominins.

The Boulder Conglomerate, Tectonic Uplift and the Availability of Stone

The end of the Pinjor Stage is marked by the Boulder Conglomerate Stage, which is composed of large, poorly sorted clasts that often include the type of quartzites suitable for flaking stone. As researchers working in Pakistan (e.g., Opdyke et al., 1979:32; Rendell et al., 1989:41) have pointed out, this is not a "stage" as it is not synchronous across northern India; rather, it marks the inception of coarse conglomeratic deposition following a steepening of river gradients in local river basins, and thus its age varies considerably. As shown in Figure 10, the timing of this change in bed-load varies in India from 1.72 Ma at Nagrota-Jammu to 0.6 Ma at Parmandal-Utterbeni (see Nanda, 2002). In Pakistan, its age ranges from 1.9 Ma in the Soan Valley and Rohtas anticline, to shortly after the Olduvai Subchron (i.e., post 1.77 Ma) at Mangla-Samwal, to Late Matuyama times in the Chambal area, ca. 0.7 Ma at Dina, and only < 0.5 Ma in the Pabbi Hills (see Opdyke et al., 1979:31; Rendell et al., 1989:41).

The deposition of these conglomerates has a two-fold significance for understanding the hominin colonization of India. The first was that it introduced large amounts of flakeable stone into landscapes that had previously been stone-poor or even stone-free, thus providing hominins with readily available stone. As these conglomerates – or "coarse pebbly phases" (Opdyke et al., 1979) during the early Pleistocene were probably relatively short events, and usually followed by the more familiar deposition of sands and silts, the opportunities for hominins were probably local and short-lived; i.e., while stone was available along a particular river. Obvious examples are the Acheulean handaxes that were found in early Middle Pleistocene conglomerates at Dina and Jalalpur (Rendell and Dennell, 1985). The second and much more significant consequence was the uplift that occurred in the late Early and early Middle Pleistocene across much of northern India and Pakistan (Amano and Taira, 1992): a very dramatic example is the 1300–3000 m of uplift that resulted in the formation of the Pir Panjal, or "lesser Himalaya" in Kashmir (Burbank and Raynolds, 1984:118; Valdiya, 1991). This uplift marked the end of the Siwalik Series and the on-set of post-Siwalik deposition, and exposed previously buried conglomerates. Thereafter, the year-round availability of stone along many river systems – particularly their middle and upper parts – was no longer problematic. In many areas (such as the Potwar Plateau, Pakistan), these conglomerates are either exposed in river sections, or as sheets, where their overlying finer sediments have been eroded. As evidenced by numerous finds of handaxes, prepared cores and blades, these exposures were commonly used as sources of raw material during the Middle and Upper Pleistocene. A good example is from the Thar Desert of India. Here, the earliest formation is the Jayal, which is composed of coarse gravels and cobbles, and was probably deposited in the Late Tertiary to Early Pleistocene, before the arrival of hominins. In the Middle Pleistocene, it was uplifted to form an extensive ridge up to 50 m above the surrounding plain, and was used as a source of raw material

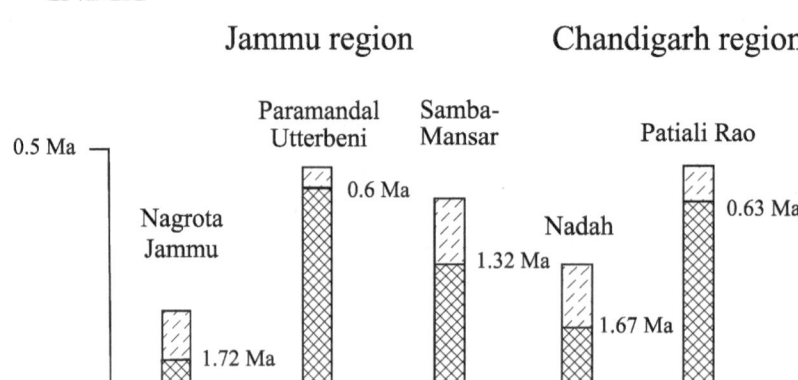

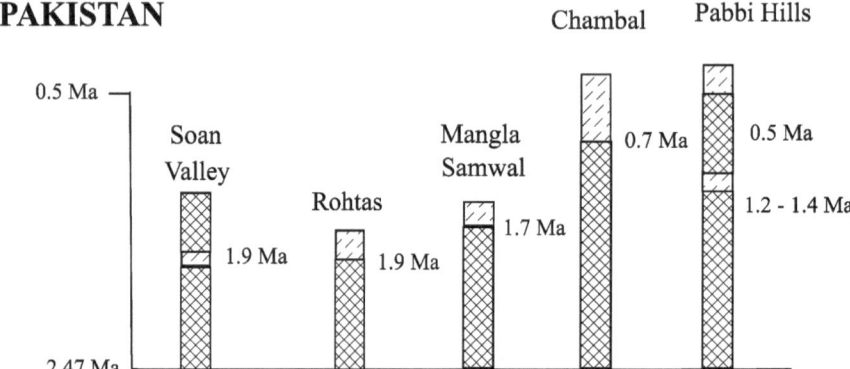

Figure 10. The timing of the deposition of conglomerates across northern India and Pakistan.
The Indian data are from Nanda (2002: Figure 3); the Pakistani data are from Opdyke et al. (1979)
and Rendell et al. (1989:41)

(Misra, 1989b; Misra and Rajaguru, 1989), as evidenced by early Paleolithic artifacts found on and in its surface.

I thus suggest that the increased availability of stone resulting from the deposition of conglomerates and (more importantly) their subsequent uplift and re-exposure may have been a key factor that facilitated the sustained occupation of northern India during the Middle Pleistocene, and possibly also the colonization of the Indian peninsula at that time. We should also bear in mind that Middle Pleistocene hominins had larger brains (averaging 918–1029 cc^3 in the case of *H. erectus* at Sangiran and Zhoukoudian (Antón, 2002; see Table 1), and were undoubtedly better at dealing with complex environmental situations than their Early Pleistocene counterparts. Estimates of home range size based on body and brain sizes indicate a marked increase, to ca. 452 hectares for later *H. erectus*, and 471 hectares for early *H. sapiens* (Antón and Swisher, 2004:288), or operational radii of 11.9 and 12.2 km respectively. Nevertheless, as noted above, the distances over which stone was routinely transported did not increase significantly until the Late Pleistocene.

Discussion

The paucity of archaeological evidence from the fluvial landscapes of northern India is not atypical of other large river systems. The Nile is an excellent example, particularly as it is often cited as the obvious corridor by which hominins left Africa. It is especially telling that Bar-Yosef (1994, 1998), one of the staunchest advocates of this idea, was able to cite only a report by Bovier-Lapierre (1926) as evidence for this early dispersal. This report, on the plain of Abbassieh outside Cairo, mentions but does not describe the presence of "eoliths" at the base of a sand-dominated section that includes at higher levels uncontroversial examples of developed Acheulean bifaces. The eoliths are given short shrift in the text, and of course none of this material is dated. The use of the Nile as a corridor during the late Pliocene and Lower Pleistocene is otherwise uncorroborated (see e.g., Wendorf and Schild, 1975:162, 1976), despite its appearance on numerous maps showing the alleged migration of hominins out of Africa in the Early Pleistocene. There is no fossil or archaeological evidence for Early Pleistocene hominins from the Tigris-Euphrates in Iraq, or the great rivers of South East Asia, such as the Irrawaddy, Chao Phraya, Mekong and Yangtse. Previous explanations of the absence of evidence have tended to cite the accumulation of silt as our chief restriction on finding evidence (e.g., Robson-Brown, 2001:198). While not denying the importance of this factor, I suggest that the prevalence of silt and sand would also have been problematic for the early hominins that we are trying to investigate, particularly because of the scarcity of workable stone (particularly when stone sources were submerged), and the inability of hominins to transport stone more than a few kilometers from its source. Early small-brained but tool-dependent hominins might have encountered severe problems in scheduling on the one hand their access to static, patchy and seasonally-available resources of the stone they needed to deflesh carcasses before other carnivorous competitors intervened, and on the other hand, to the mobile resources of meat upon which they depended. This suggestion carries two implications. One is that hominin dispersal across the river systems of northern India and Pakistan was not simply a matter of foraging uninterruptedly through the floodplains of these major river systems, but involved a complex set of local adaptations that may well have restricted mobility because of the need to stay near localized sources of stone. These large rivers might not have been corridors so much as widely spaced and frequently changing stepping-stones, depending upon where and when stone was easily available. Secondly, when stone did become available prior to the uplift of the Middle Pleistocene – as when conglomerates were deposited, for example, in the Soan Valley, and later, at Dina and Jalapur – hominins made use of it, but probably then sought out new areas when it became unavailable. Accounts of hominin dispersals across southern Asia need therefore to recognize that large-scale sedimentary systems may have proved more challenging to early stone-dependent hominins than to other mammals. After all, they were not in the same situation as hyaenas or large cats that can rely upon their teeth, claws, power and speed to kill their prey; instead, it was the ability to flake stone that enabled them to access meat and bone-marrow, make a wide range of implements from wood, and thus increase the quantities and types of plant and other types of foods that they could eat. As Petraglia (1998:381) remarked:

If preservation conditions or archaeological sampling are shown not to be the cause for the absence of early sites, it is also possible that the lack of identified occurrences in Early Pleistocene or Early Middle Pleistocene contexts, and the wealth of sites in the mid-Middle Pleistocene, may be due to ... the relative success of hominids in adapting to environments.

The floodplains of the Indus and Ganges may provide one example where early hominins were generally unsuccessful.

Acknowledgments

This paper owes much to discussions and arguments over the years with numerous friends and colleagues interested in human origins in Asia, and particularly its southern part. I would like also to thank Bridget Allchin for her sustained support over many years, and Mike Petraglia for his informed stimulation in many informal discussions of the Asian paleolithic. The British Academy is very gratefully thanked for granting the author a three-year research professorship so that he can research the early paleolithic of Asia.

References

Amano, K., Taira, A., 1992. Two-phase uplift of Higher Himalayas since 17 Ma. Geology 20, 391–394.

Antón, S., 2002. Evolutionary significance of cranial variation in Asian *Homo erectus*. American Journal of Physical Anthropology 118, 301–323.

Antón, S., Swisher, C.C. III, 2004. Early dispersals of *Homo* from Africa. Annual Review of Anthropology 33, 271–296.

Arai, A. and Thibault, C., 1975/1977. Nouvelles précisions à propos de l'outillage paléolithique ancien sur galets du Khorassan (Iran). Paléorient 3, 101–108.

Ascenzi, A., Biddittu, I., Cassoli, P.F., Segre, A.G., Segre-Naldini, E., 1996. A calvarium of late *Homo erectus* from Ceprano, Italy. Journal of Human Evolution 31, 409–423.

Badam, G.L., 1979. Pleistocene Fauna of India. Deccan College Post-Graduate and Research Institute, Pune.

Baker, V.R., Ely, L.L., Enzel, Y., Kale, V.S., 1995. Understanding India's rivers: late Quaternary palaeofloods, hazard assessment and global change. In: Wadia, S., Korisettar, R., Kale, V.S. (Eds.), Quaternary Environments and Geoarchaeology of India. Geological Society of India Memoir 32, Bangalore, pp. 61–77.

Barry, J.C., 1987. The history and chronology of Siwalik Cercopithecids. Human Evolution 2 (1), 47–58.

Bartsiokas, A., Day, M.D., 1993. Electron probe energy dispersive X-ray microanalysis (EDXA) in the investigation of fossil bone: the case of Java man. Proceedings of the Royal Society of London, Series B 252, 115–123.

Bar-Yosef, O., 1994. The Lower Palaeolithic of the Near East. Journal of World Prehistory 8 (3), 211–265.

Bar-Yosef, O., 1998. Early colonizations and cultural continuities in the Lower Palaeolithic of western Eurasia. In: Petraglia, M. D., Korisettar, R. (Eds.), Early Human Behavior in Global Context: The Rise and Diversity of the Lower Palaeolithic Record. Routledge, London, pp. 221–27.

Bar-Yosef, O., Belfer-Cohen, A., 2001. From Africa to Eurasia - early dispersals. Quaternary International 75, 19–28.

Bar-Yosef, O., Goren-Inbar, N., 1993. The Lithic Assemblages of 'Ubeidiya, a Lower Palaeolithic Site in the Jordan Valley. Hebrew University Monographs of the Institute of Archaeology (Qedem) 34, Jerusalem, pp. 1–266.

Boaz, N.T., Ciochon, R.L., Xu Qinqi, Liu Jinyi, 2000. Large mammalian carnivores as a taphonomic factor in the bone accumulation at Zhoukoudian. Acta Anthropologica Sinica 19 (Supplement), 224–234.

Bovier-Lapierre, R.P.P., 1926. Les gisements paléolithiques de la Plaine de l'Abbassieh. Bulletin de l'Institut d'Égypte 8, 257–275.

Bronk Ramsey, C., Higham, T.F.G., Owen, D.C., Pike, A.W.G., Hedges, R.E.M., 2002. Radiocarbon dates from the Oxford AMS system: Archaeometry datelist 31. Archaeometry 44 (3) Supplement 1, 1–149.

Burbank, D.W., Johnson, G.D., 1982 Intermontane-basin development in the past 4 Myr in the north-west Himalaya. Nature 298, 432–436.

Burbank, D.W., Johnson, G.D., 1983. The late Cenozoic chronologic and stratigraphic devrelopment of the Kashmir intermontane basin, northwestern Himalaya. Palaeogeography, Palaeoclimatology, Palaeoecology 43, 205–235.

Burbank, D.W., Raynolds, R.G.H., 1984. Sequential late Cenozoic structural disruption of the northern Himalayan foredeep. Nature 311, 114–118.

Clark, J.D., 1969. The Middle Acheulean occupation site at Latamne, Northern Syria (second paper). Quaternaria 10, 1–60.

Colbert, E.H., 1935. Siwalik mammals in the American Museum of Natural History. Transactions of the American Philosophical Society 26, 1–401.

Davidson, I., Solomon, S., 1990. Was OH7 the victim of a crocodile attack? In: Solomon, S., Davidson, I., Watson, D. (Eds.), Problem Solving in Taphonomy. *Tempus* 2, 229–239.

Day, M.H., 1986. Guide to Fossil Man (4th edition). Cassell, London.

Delson, E., 1993. *Theropithecus* fossils from Africa and India and the taxonomy of the genus. In: Jablonski, N.G. (Ed.), *Theropithecus*: The Rise and Fall of a Primate Genus. Cambridge University Press, Cambridge, pp. 157–189.

Dennell, R.W., 2003. Dispersal and colonisation, long and short chronologies: how continuous is the Early Pleistocene record for hominids outside East Africa? Journal of Human Evolution 45, 421–440.

Dennell, R.W., 2004. Early Hominin Landscapes in Northern Pakistan: Investigations in the Pabbi Hills. British Archaeological Reports International Series 1265, Oxford.

Dennell, R.W., 2005. The Solo (Ngandong) *Homo erectus* assemblage: a taphonomic assessment. Archaeology in Oceania 40, 81–90.

Dennell, R.W., Hurcombe, L.M., 1989. The Riwat assemblage. In: Rendell, H.M., Dennell, R.W., Halim, M. (Eds.), Pleistocene and Palaeolithic Investigations in the Soan Valley, Northern Pakistan. British Archaeological Reports International Series 544, pp. 105–127.

Dennell, R.W., Coard, R., Turner, A., 2005a. The biostratigraphy and magnetic polarity zonation of the Pabbi Hills, northern Pakistan: an Upper Siwalik (Pinjor Stage) Upper Pliocene-Lower Pleistocene fluvial sequence. Palaeogeography, Palaeoclimatology and Palaeoecology 234, 168–185.

Dennell, R.W., Coard, R., Beech, M., Anwar, M., Turner, A., 2005b. Locality 642, an Upper Siwalik (Pinjor Stage) fossil accumulation in the Pabbi Hills, Pakistan. Journal of the Palaeontological Society of India 50 (1), 83–92.

Dennell, R.W., Rendell, H.R., Hailwood, E., 1988. Early tool-making in Asia: two-million-year-old artefacts in Pakistan. Antiquity 62, 98–106.

Dennell, R.W., Turner, A, Coard, R., Beech, M., Anwar, M., 2005c. Two Upper Siwalik (Pinjor Stage) fossil accumulations from localities 73 and 362 in the Pabbi Hills, Pakistan. Journal of the Palaeontological Society of India 50 (2), 101–111.

Falconer, H., Cautley, P.T., 1845. Fauna Antiqua Sivalensis, Being the Fossil Zoology of the Sewalik Hills in the North of India. Smith, Elder and Co., London.

Feblot-Augustins, J., 1999. Raw material transport patterns and settlement systems in the European Lower and Middle Palaeolithic: continuity, change and variability. In: Roebroeks, W., Gamble, C. (Eds.), The Middle Palaeolithic Occupation of Europe. University of Leiden Press, Leiden, pp. 193–214.

Feibel, C.S., 2004. Quaternary lake margins of the Levant Rift Valley. In Goren-Inbar, N., Speth, J.D. (Eds.), Human Paleoecology in the Levantine Corridor. Oxbow, Oxford, pp. 21–36.

Gabunia, L., Lumley, M.-A. de, Berillon, G., 2000. Morphologie et fonction du troisième métatarsien de Dmanissi, Géorgie Orientale. In: Lordkipanidze, D., Bar-Yosef, O., Otte, M. (Eds.), Early Humans at the Gates of Europe: Les Premiers Hommes aux Portes de l'Europe. ERAUL, Liège, pp. 29–42.

Gabunia, L., Vekua, A., Lordkipanidze, D., Swisher, C.C., Ferring, R., Justus, A., Nioradze, M., Tvalcherlidze, M., Anton, S.C., Bosinski, G., Jöris, O., Lumley, M.-A. de, Majsuradze, G., Mouskhelishvili, A., 2000a. Earliest Pleistocene hominid cranial remains from Dmanisi, Republic of Georgia: taxonomy, geological setting, and age. Science 288, 1019–1025.

Gabunia, L., Vekua, A., Lordkipanidze, D., 2000b. The environmental contexts of early human occupation of Georgia (Transcaucasia). Journal of Human Evolution 38, 785–802.

Gamble, C., 1999. Palaeolithic Societies of Europe. Cambridge University Press, Cambridge.

Ginat, H., Zilberman, E., Saragusti, I., 2003. Early Pleistocene lake deposits and Lower Palaeolithic finds in Nahal (wadi) Zihor, Southern Negev desert, Israel. Quaternary Research 59, 445–458.

Goren-Inbar, N., Feibel, C.S., Versoub, K.L., Melamed, Y., Kislev, M.E., Tchernov, E., Saragusti, I., 2000 Pleistocene milestones on the Out-of-Africa corridor at Gesher Ya'aqov, Israel. Science 289, 944–947.

Güleç, E., Howell, F.C., White, T.D., 1999. Dursunlu – a new Lower Pleistocene faunal and artifact-bearing locality in southern Anatolia. In: Ulrich, H. (Ed.), Hominid Evolution: Lifestyles and Survival Strategies. Edition Archaea, Gelsenkirchen, pp. 349–364.

Gupta, A., 1995. Fluvial geomorphology and Indian rivers. In: Wadia, S., Korisettar, R., Kale, V.S. (Eds.), Quaternary Environments and Geoarchaeology of India. Geological Society of India Memoir, Bangalore, 32, 52–60.

Heinzelin, J.de, Clark, J.D., White, T., Hart, W., Renne, P., Wolde-Gabriel, Beyene, Y., Vrba, E., 1999. Environment and behaviour of 2.5-million-year-old Bouri hominids. Science 284, 625–629.

Hurcombe, L.M., 2004. The stone artefacts from the Pabbi Hills. In: Dennell, R.W. (Ed.), Early Hominin Landscapes in Northern Pakistan: Investigations in the Pabbi Hills. British Archaeological Reports International Series 1265, pp. 222–292.

Hussain, S.T., Bergh, G.D. van den, Steensma, K.J., Visser, J.A. de, Vos, J. de, Arif, M., Dam, J. van, Sondaar, P.Y., Malik, S.B., 1992. Biostratigraphy of the Plio-Pleistocene continental sediments (Upper Siwaliks) of the Mangla-Samwal Anticline, Azad Kashmir, Pakistan. Proceedings of the Koninkijke Nederlandse Akademie van Wetenschappen 95 (1), 65–80.

Isaac, G.L., 1989. The archaeology of human origins: studies of the Lower Pleistocene in East Africa 1971–1981. In: Isaac, B. (Ed.), The Archaeology of Human Origins: Papers of Glynn Isaac. Cambridge University Press, Cambridge, pp. 120–187.

Johanson, D.C., Taieb, M., Coppens, Y., 1982. Pliocene hominids from the Hadar Formation, Ethiopia (1973–1977): stratigraphic, chronologic and palaeoenvironmental contexts, with notes on hominid morphology and systematics. American Journal of Physical Anthropology 57, 373–402.

Johnson, G.D., Raynolds, R.G.H., Burbank, D.W., 1986. Late Cenozoic tectonics and sedimentation in the north-western Himalayan foredeep: I. Thrust ramping and associated deformation in the Potwar region. In: Allen, P.A., Homewood, P. (Eds.), Foreland Basins. Special Publication International Association of Sedimentologists 8, 273–291.

Johnson, N.M., Opdyke, N.D., Johnson, G.D., Lindsay, E.H., Tahirkheli, R.A.K., 1982. Magnetic polarity stratigraphy and ages of Siwalik group rocks of the Potwar Plateau, Pakistan. Palaeogeography, Palaeoclimatology, Palaeoecology 37, 17–42.

Klein, R.G., 1999. The Human Career: Human Biological and Cultural Origins. University of Chicago Pres, Chicago.

Larick R., Ciochon, R.L., Zaim, Y., Suminto, S., Izal, Y., Aziz, F., Reagan, M., Heizler, M. 2001., Early Pleistocene Ar-40/Ar-39 ages for Bapang Formation hominins, Central Java, Indonesia. Proceedings of the National Academy of Sciences USA 98 (9), 4866–4871.

Lumley, H. de, Nioradzé, M., Barsky, D., Cauche, D., Celiberti, V., Nioradzé, G., Notter, O., Zvania, D., Lordkipanidze, D., 2005. Les industries lithiques préoldowayennes du début du Pléistocène inférieur du site de Dmanissi en Géorgie. L'Anthropologie 109, 1–182.

Lydekker, R., 1879. Notices of Siwalik mammals. Records of the Geological Survey of India 12, 33–52.

Matthew, W.D., 1929. Critical observations upon Siwalik mammals (exclusive of Proboscidea). Bulletin of the American Museum of Natural History 56, 437–560.

Merrick, H.V., 1976. Recent archaeological research in the Plio-Pleistocene deposits of the Lower Omo, southwestern Ethiopia. In: Isaac, G.L., E.R. McCown (Eds.), Human Origins: Louis Leakey and the East African Evidence. Staples Press, Menlo Park, California, pp. 461–482.

Mishra, S., 1994. The South Asian Lower Palaeolithic. Man and Environment 10, 57–71.

Mishra, S., 2005 Review of Dennell, R.W. 2004. Early Hominin Landscapes in Northern Pakistan: Investigations in the Pabbi Hills. British Archaeological Reports International Series 1265. Quaternary International (in press).

Misra, V.N., 1989a. Stone age India: an ecological perspective. Man and Environment 14 (1), 17–64.

Misra, V.N., 1989b. Human adaptations to the changing landscape of the Indian arid zone during the Quaternary period. In: Kenoyer, J.M. (Ed.), Old Problems and New Perspectives. Wisconsin Archaeological Reports, Madison, pp. 3–20.

Misra, V.N., Rajaguru, S.N., 1989. Palaeoenvironment and prehistory of the Thar Desert, Rajasthan, India. In: Frifelt, K., Sørensen, P. (Eds.), South Asian Archaeology 1985. Curzon Press, London, pp. 296–320.

Nanda, A.C., 2002. Upper Siwalik mammalian faunas of India and associated events. Journal of Asian Earth Sciences 21, 47–58.

Oakley, K.P., 1964. Frameworks for Dating Fossil Man. Weidenfeld and Nicolson, London.

Oakley, K., Campbell, B.G., Molleson, T.V., 1971. Catalogue of Fossil Hominids Part II: Europe. London: British Museum (Natural History).

Oakley, K., Campbell, B.G., Molleson, T.V., 1975. Catalogue of Fossil Hominids Part III: Americas, Asia, Australasia. London: British Museum (Natural History).

Opdyke, N.M., Lindsay, E., Johnson, G.D., Johnson, N., Tahirkheli, R.A.K., Mirza, M.A.I., 1979. Magnetic polarity stratigraphy and vertebrate palaeontology of the Upper Siwalik subgroup of northern Pakistan. Palaeogeography, Palaeoclimatology, Palaeoecology 27, 1–34.

Paddayya, K., 2001. The Acheulean Culture Project of the Hunsgi and Baichbal Valleys, peninsular India. In: Barham, L., Robson-Brown, K. (Eds.), Human Roots: Africa and Asia in the Middle Pleistocene. Western Academic and Specialist Press Ltd., Bristol, pp. 235–258.

Paddayya, K., Petraglia, M.D., 1995. Natural and cultural formation processes of the Acheulean sites of the Hunsgi and Baichbal valleys, Karnataka. In: Wadia, S., Korisettar, R., Kale, V.S. (Eds.), Quaternary Environments and Geoarchaeology of India. Geological Society of India Memoir, Bangalore, 32, 333–351.

Paddayya, K., Blackwell, B.A.B., Jhaldiyal, R., Petraglia, M.D., Fevrier, S., Chaderton, D.A., Blickstein, J.I.B., Skinner, A.R., 2002. Recent findings on the Acheulean of the Hunsgi and Baichbal valleys, Karnataka, with special reference to the Isampur excavation and its dating. Current Science 83 (5), 641–647.

Petraglia, M.D., 1998. The Lower Palaeolithic of India and its bearing on the Asian record. In: Petraglia, M. D., Korisettar, R. (Eds.), Early Human Behavior in Global Context: The Rise and Diversity of the Lower Palaeolithic Record. Routledge, London, pp. 343–390.

Petraglia, M.D., 2003. The Lower Palaeolithic of the Arabian peninsula: occupations, adaptations, and dispersals. Journal of World Prehistory 17 (2), 141–179.

Petraglia, M.D., Shipton, C., Paddayya, K., 2005. Life and mind in the Acheulean: a case study from India. In: Gamble, C., Porr, M. (Eds.), The Hominid Individual in Context: Archaeological Investigations of Lower and Middle Palaeolithic Landscapes, Locales and Artefacts. London and New York: Routledge. pp. 197–219.

Pilgrim, G.E., 1913. The correlation of the Siwaliks with mammal horizons of Europe. Geological Survey Records of India 43, 264–326.

Pilgrim, G.E., 1939. The fossil Bovidae of India. Memoirs of the Geological Survey of India (Palaeontologica Indica) 26, 1–356.

Pope, G.C., 1989. Bamboo and human evolution. Natural History 98 (10), 49–57.

Pope, G.C., Keates, S.G., 1994. The evolution of human cognition and cultural capacity. In: Corruchini, R., Ciochon, S. (Eds.), Integrative Paths to the Past: Paleoanthropological Advances in Honor of F. Clark Howell. Prentice Hall, New Jersey, pp. 531–567.

Quade, J., Cerling, T.E., Bowman, J.R., Jah, A., 1993. Paleoecologic reconstruction of floodplain environments using palaeosols from Upper Siwalik Group sediments, northern Pakistan. In: Schroder, J.F. (Ed.), Himalaya to the Sea: Geology, Geomorphology and the Quaternary. Routledge, London and New York, pp. 213–226.

Ranov, V.A., 2001. Loess-paleosoil formation of southern Tajikistan and the loess palaeolithic. Praehistoria 2, 7–27.

Ranov, V.A., Dodonov, A.E., 2003. Small instruments of the Lower Palaeolithic site Kuldara and their geoarchaeological meaning. In: Burdukiewicz, J.M., Ronen, A. (Eds.), Lower Palaeolithic Small Tools in Europe and Asia. British Archaeological Reports International Series 1115, pp. 133–14.

Raynolds, R.G.H., Johnson, G.D., 1985 Rates of Neogene depositional and deformational processes, north-west Himalayan foredeep margin, Pakistan. In: Snelling, N.J. (Ed.), The Chronology of the Geological Record. Blackwells Scientific Publications Memoir 10, Oxford, pp. 297–311.

Rendell, H., Dennell, R.W., 1985. Dated lower Palaeolithic artefacts from northern Pakistan. Current Anthropology 26 (3), 393.

Rendell, H., Hailwood, E., Dennell, R.W., 1987. Magnetic polarity stratigraphy of Upper Siwalik Sub-Group, Soan Valley, Pakistan: implications for early human occupance of Asia. Earth and Planetary Science Letters 85, 488–496.

Rendell, H.M., Dennell, R.W., Halim, M., 1989. Pleistocene and Palaeolithic Investigations in the Soan Valley, Northern Pakistan. British Archaeological Reports International Series 544, Oxford.

Rightmire, G.P., Lordkipanidze, D.,Vekua, A., 2005. Anatomical descriptions, comparative studies and evolutionary significance of the hominin skulls from Dmanisi, Republic of Georgia. Journal of Human Evolution 50, 1–27.

Roberts, M.B., Parfitt, S.A., 1999. Boxgrove: A Middle Pleistocene Hominid Site at Eartham Quarry, Boxgrove, West Sussex. English Heritage Archaeological Report 17, London.

Robson-Brown, K., 2001. The Middle Pleistocene record of mainland southeast Asia. In: Barham, L., Robson-Brown, K. (Eds.), Human Roots: Africa and Asia in the Middle Pleistocene. Western Academic and Specialist Press, Bristol, pp. 187–202.

Roebroeks, W., Kolen, J., Rensink, E., 1988. Planning depth, anticipation and the organisation of Middle Palaeolithic technology: the "archaic natives" meet Eve's descendants. Helinium 28 (1), 17–34.

Ronen, A., 1991. The lower palaeolithic site Evron-Quarry in western Galilee, Israel. Sonderveröffentlichungen Geologisches Institut der Universität zu Köln 82, 187–212.

Sahni, M.R., Khan, E., 1988. Pleistocene Vertebrate Fossils and Prehistory of India. Books and Books, New Delhi.

Sankhyan, A.R., 1997. Fossil clavicle of a Middle Pleistocene hominid from the Central Narmada Valley, India. Journal of Human Evolution 32, 3–16.

Schick, K., Toth, N., Wei, Q., Clark, J.D., Etler, D., 1991. Archaeological perspectives in the Nihewan Basin, China. Journal of Human Evolution 21, 13–26.

Sharma, A.K., 1993. Prehistoric Delhi and its Neighbourhood. Aryan Books International, New Delhi.

Snelgrove, A.K., 1979. Migration of the Indus River, Pakistan, in response to plate tectonic motions. Journal of the Geological Society 20, 392–403.

Sonkalia, A., 1985. Skull cap of an early man from the Narmada Valley alluvium (Pleistocene) of Central India. American Anthropologist 87, 612–616.

Sotnikova, M.V., Dodonov, A.E., Pen'kov, A.V., 1997. Upper Cenozoic bio-magnetic stratigraphy of Central Asian mammalian localities. Palaeogeography, Palaeoclimatology, Palaeoecology 133 (3–4), 243–258.

Srivastava, P., Singh, I.B., Sharma, M., Singhvi, A.K., 2003 Luminescence chronometry and Late Quaternary geomorphic history of the Ganga Plain, India. Palaeogeography, Palaeoclimatology, Palaeoecology 197, 15–41.

Stiles, D., 1978. Palaeolithic artefacts in Siwalik and post-Siwalik deposits of northern Pakistan. Kroeber Anthropological Society Papers 53–54, 129–148.

Swisher III C.C. Curtis G.H. Jacob T. Getty A.G. Suprijo A., Widiasmoro, 1994 Age of the earliest known hominids in Java Indonesia. Science 263, 1118–1121.

Tewari, R., Pant, P.C., Singh, I.B., Sharma, S., Sharma, M., Srivastava, P., Singhvi, A.K., Mishra, S., Tobschall, H.J., 2002. Middle palaeolithic human activity and palaeoclimate at Kalpi in Yamuna valley, Ganga Plain. Man and Environment 27 (2), 1–13.

Turner, A., 1992. Large carnivores and earliest European hominids: changing determinants of resource availability during the Lower and Middle Pleistocene. Journal of Human Evolution 22, 109–126.

Valdiya, K.S., 1991. Quaternary tectonic history of northwest Himalaya. Current Science 61, 664–668.

von Koenigswald, G.H.R., 1968. Observations upon two *Pithecanthropus* mandibles from Sangiran, Central Java. Proceedings of the Academy of Sciences, Amsterdam B71, 99–107.

Wadia, D.N., 1974. Geology of India. Tata McGraw Hill, New Delhi.

Wanpo H., Ciochon, R., Yumin, G., Larick, R., Qiren, F., Schwarcz, H., Yonge, C., Vos J. de, Rink, W., 1995 Early *Homo* and associated artefacts from Asia. Nature 378, 275–278.

Wells, N.A., 1987. Shifting of the Kosi River, northern India. Geology 15, 204–207.

Wendorf, F., Schild, R., 1975. The paleolithic of the Lower Nile Valley. In: Wendorf, F., Marks, A.E. (Eds.), Problems in Prehistory: North Africa and the Levant. SMU Press, Dallas, pp. 127–169.

Wendorf, F., Schild, R., 1976. The Prehistory of the Nile Valley. Academic Press, London.

Zhu, R., Zhinsheng An, Potts, R., Hoffman, K.A., 2003. Magnetostratigraphy of early humans in China. Earth-Science Reviews 6, 341–359.

Zhu, R.X., Hoffman, K.A., Potts, R., Deng, C.L., Pan, Y.X., Guo, B., Shi, C.D., Guo, Z.T., Yuan, B.Y., Hou, Y.M., Huang, W.W., 2001. Earliest presence of humans in northeast Asia. Nature 413, 413–417.

Zhu, R.X., Potts, R., Xie, F., Hoffman, K.A., Deng, C.L., Shi, C.D., Pan, Y.X., Wang, H.Q., Shi, R.P., Wang, Y.C., Shi, G.H., Wu, N.Q., 2004. New evidence on the earliest human presence at high northern latitudes in northeast Asia. Nature 431, 559–562.

4. Toward developing a basin model for Paleolithic settlement of the Indian subcontinent: *Geodynamics, monsoon dynamics, habitat diversity and dispersal routes*

RAVI KORISETTAR

Department of History and Archaeology
Karnatak University
Dharwad, 580 003
India
korisettar@yahoo.com

Introduction

The Indian subcontinent is a continental scale landmass that contains a wide range of physiographic zones and geomorphic features. The Himalayan mountains on the north, the Baluchistan Mountains in the West, the Burma Mountains in the east, and the Indian Ocean on the south provide a distinct geographic boundary to the subcontinent (Figure 1). The mountain ranges and the Indian Ocean effectively isolate the landmass from the rest of Asia. An understanding of hominin dispersal patterns and adaptations requires a proper understanding of the geologic, geomorphic, and environmental features of the region. This chapter attempts to provide a comprehensive geological and ecological framework for delineating Paleolithic colonization and settlement patterns in the Indian subcontinent. The documented evidence of Paleolithic sites reveals a distinctive pattern that helps to explain the existence of zones that may be classified as 'core', 'peripheral' and 'isolated areas'. This brings into focus the debate on dispersal routes, i.e., whether movements were primarily along the coast or whether they were transcontinental. In a wider view, it is important to consider whether hominin populations of the subcontinent were at the crossroads of movements or whether populations remained isolated for long periods of time.

Subbarao (1958) constructed the first geographic model for Indian archeology. Subbarao's synthesis, *The Personality of India* (1958), was set in a geographic and environmental framework that adopted the revised three-tier system of Early, Middle and Late Stone Ages. This work identified areas of 'attraction', 'relative isolation', and 'isolation' based on the evidence and distribution of prehistoric sites in diverse geographic areas of the subcontinent. Geographical deterrents governing the patterns of habitation were

M.D. Petraglia and B. Allchin (eds.), The Evolution and History of Human Populations
in South Asia, 69–96.

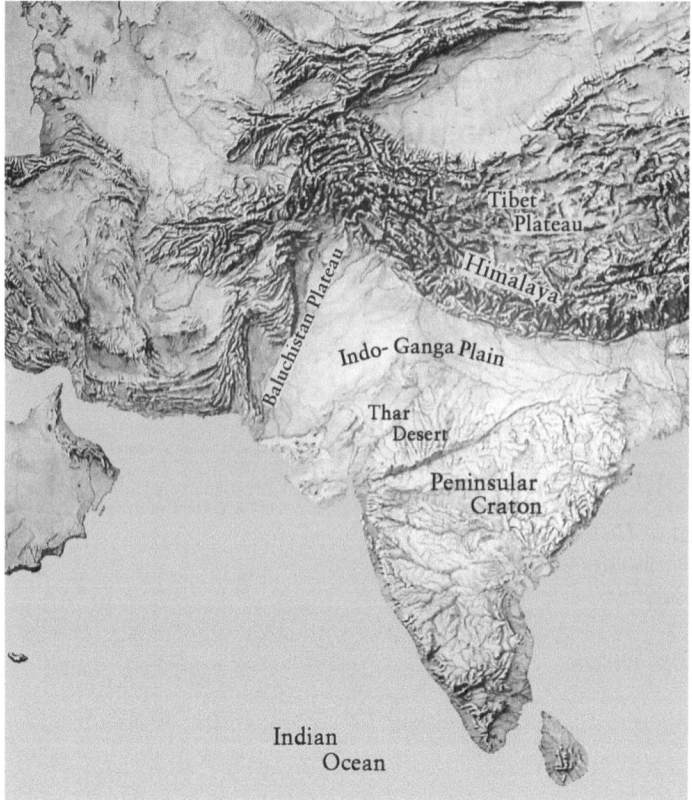

Figure 1. Physiography of the Indian subcontinent showing major orographic features, plateaus, the peninsular craton and the Indian Ocean

emphasized. This model, of course, did not make provision for inclusion of significant evidence brought to light after 1958. Much archaeological research has been carried out during the last five decades that warrants a fresh understanding of the patterns of Paleolithic colonization and to determine the subcontinent's place along Pleistocene dispersal routes of the Old World.

Corridors and the Paleolithic Record

The physical distribution of mountain ranges and arid zones in the subcontinent's have influenced hominin colonization and dispersal patterns through time. The northwestern part of the subcontinent constitutes a critical area in the study of hominin dispersal patterns. The region represents a potential entry corridor of early populations and connects the Paleolithic habitats of Pakistan such as the Rohri Hills, the Salt Range and the Subhimalayan plateaus with those in Rajasthan (Figure 2). This zone is potentially connected with the transcontinental routes in North Africa, Arabia and central Asia (Petraglia, 2003, 2005). The Sulaiman and Kirthar Ranges of Baluchistan and the valleys that traverse them (e.g., Khyber, Bolan), were potential passes for hominin dispersals from the Central Asian plateau. If early historical movements into the Indus Basin are a sign, the role of such passes in prehistory cannot be overlooked.

The evidence for early human colonization of the Indian subcontinent is mainly archeological. The archeological record clearly reveals that the Acheulean was the earliest stage of hominin occupation of the subcontinent (R.S. Pappu, 2001). Although early hominin presence in the Potwar region of Pakistan has been dated to 1.9 Ma (Dennell

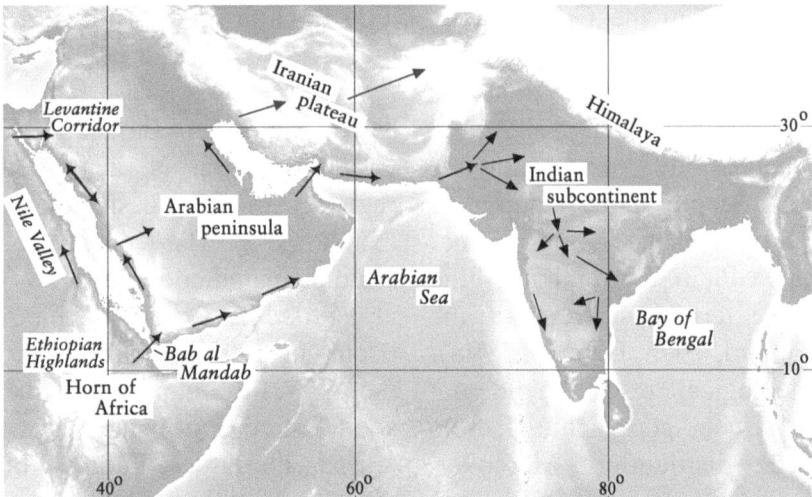

Figure 2. Out of Africa dispersal routes of early hominins. The current evidence in peninsular India does not support the suggested coastal route across the Indian subcontinent to Southeast Asia (base map from Petraglia 2005). The Sindh Plain in Pakistan and the Thar desert in India occupy significant geographic positions along potential dispersal routes

et al., 1988; Rendell et al., 1989; Dennell, this volume), additional archaeological evidence from this region has not been brought forth. Well-dated Acheulean occurrences in Pakistan range between 800 and 700 ka. Although an ESR date of 1.2 Ma has been reported for the Acheulean at Isampur (Paddayya et al., 2002), the vast majority of radiometric determinations place the Acheulean in the age range of 400 to 300 ka (see Korisettar, 2002). To date, the only hominin fossil evidence is from the Narmada River in the Vindhyan Basin, though indicating that archaic hominins were certainly present in the subcontinent (Sonakia, 1985; see Athreya, this volume).

It has been generally believed that the differential distribution of Paleolithic sites across the subcontinent was due to a lack of intensive surveys in those areas where archeo-logical localities were not well documented. Yet, survey intensity alone is not likely to be an adequate explanation. For instance, sector-wise intensive survey in the region of 'Dharwar Batholith', comprising granite-gneiss and greenstone formations in southern India, did not reveal any Paleolithic occurrences. Similarly the granulite/charnokite terrain of south Tamil Nadu plain, the region south

of the Kaveri River (including the Nilgiris and sub-Kaveri region) is not known for Lower Paleolithic occupations. The known Acheulean sites in this region are associated with the coastal Gondwana Basins, north of the Kaveri River (S. Pappu, 1996). This has led to identifying the 'Kaveri Line' as the southern boundary of the Acheulean occupation area. Although Acheulean sites are being reported from the Dang-Deokhuri valleys of Nepal (Corvinus, 1995, 1998) and occasionally from the Ladakh province in the Central Himalaya (Sharma, 1995), there are generally very few sites and artifact counts are generally low. Although such evidence indicates the need to slightly reposition the 'Movius Line', the evidence for Acheulean occupation is not abundant, indicating differential preferences in habitation on the part of hominins.

Another example of differential site distribution occurs in northern India. In the transitional region between the northern slopes of the Vindhyas, in the Son and Belan valleys, and the alluvial plains of the Ganga River, Acheulean sites are well preserved (Sharma and Clark, 1983; Williams and Clarke, 1995), suggesting that hominins adapted to basin margin geomorphic environments. Further north of the Vindhyas, in the Ganga Plain, there is no record of Acheulean occupation. Some archeologists tend to argue that the sites lay deeply buried in the Ganga alluvium. While this is a possibility, this inference requires a more in-depth understanding of the geomorphic history of the region. Further north of the Ganga Plain, Lower Paleolithic sites occur on the terrace surfaces of the Siwaliks (see Gaillard, 1995; Dennell, 1998).

Previous archeological research did not take account of such geographic factors in Paleolithic site distribution. Moreover, researchers often did not consider Quaternary processes that accounted for landform development and change. Thus, Paleolithic sites and hominin ranging behaviors have been studied independent of an understanding of basin geometry, geology, and geomorphic history. The current approach places emphasis on the Purana-Gondwana Basins and the Quaternary history of the areas in order to account for the location and concentration of archeological sites. An argument presented here is that the observed spatial variation in the patterns of Paleolithic settlement across the subcontinent is likely to be a product of hominin presence in particular zones and not simply a product of survey effort. It is predicted that these geographic patterns will persist, even with additional surveys in the future. An attempt is made here to examine the processes responsible for conditioning hominin occupation, placing particular emphasis on the Purana-Gondwana Basins in order to arrive at a basin model. The model emphasizes on the reconstruction of Quaternary paleogeography and paleoecology for understanding hominin distributions and adaptations to preferred settings. An understanding of the Purana and Gondwana Basins as favorable habitats is critical to an understanding of dispersal processes and hominin adaptations as they provided critical resources, including a reliable water supply, a high animal and plant biomass, and raw materials for stone tool making.

The Purana and Gondwana Basins of Peninsular India

There are seven Purana Basins and seven Gondwana Basins across the peninsular region with diverse shapes and areal extent (Tables 1, 2). The Coastal Gondwana Basins are much smaller and occur as discrete basins along the East Coast of India. All these basins occur in a geographic continuity between the Aravalli Mountains in the west and the southeast coast of peninsular India (Figure 3). Among the largest Purana-Gondwana Basins are the Vindhyachal Basin (60,000 sq km) and the Cuddapah Basin (45,000 sq km). The Bhima Basin is the smallest (5,000 sq km) and, the Indravati, Chhattisgarh and Abujhmar are even smaller (Radhakrishna, 1987).

The rocks of the Purana Basins fall within the time bracket of the Middle to Late Proterozoic period. The lithosequences in these basins are comprised of conglomerates, arkoses, quartzites, sandstones, shales, limestones and dolomites. Sediments underwent diagenetic lithification and metamorphic changes through time and rocks are generally highly indurated. The rocks of the Purana Basins are classified into: a) igneous rocks, including volcanics and sills of all compositions; b) siliciclastic sediments such as conglomerates, breccias, arkoses and quartzitic sandstones; c) argillites such as shale, siltstone and mudstone; and, d) carbonates including limestone, dolomite, and cherty equivalents. Each of the seven basins has an abrupt geographical boundary, often corresponding with significant geophysical anomalies. Their boundaries are commonly marked by faults or shear zones. Each is a distinct basin with well-defined structural boundaries and each is unique in character and content.

The Phanerozoic geological record of the Gondwana Basins is mainly preserved in the form of a thick pile of sediments comprising conglomerates, sandstones, shales and coal-bearing strata deposited in narrow, linear

Table 1. Proterozoic Purana Basins of Peninsular India

Basin	Area Exposed (sq km)	Estimated Maximum Area Including Loss Due to Erosion & Concealment (sq km)	Maximum Aggregate Thickness (m) of Stratigraphy	Maximum Estimated Thickness (m)(Geo-physically Determined)	Shape of Exposed Outline	Lithosequence	Tectonic Structure
Vindhyachal	60,000	1,60,000	4,500	2,400	Arcuate	Semri, Kaimur, Rewa, Bhander Groups	
Chattisgarh	33,000	40,000	2,000	4,000	Semicircular (crescent shaped), irregular southern margin	Lower Chanderpur and upper Raipur Groups	
Indravati	9,000				Roughly circular	Tiratgarh, Kanger, and Jagdalpur formations	
Sukma	700				Large triangular	Similar to that of Indravati Basins formations	
Abujhmar Bastar	3000 15,000	50,000	800		Nearly rectangular Disjointed patchy outcrops along arcuate belt		
Pranhita-Godavari (Pakahal Basin)	15,000	30,000	6,000	4,000	Linear belt	Mallampalli (825 m), Mulug (2830 m), and Albaka (1,000 m) Groups	Western sector- Folded into a major anticline. Eastern sector- Folded into anticlines and synclines
Cuddapah	45,000	50,000	12,000	10,000	Crescent	Cuddapah Supergroup (Papaghni, Chitravati, and Nallamalai Groups) and Kurnool Group (Banganpalle, Narji, Owk, Koilakuntla, and Nandyal Formations	The eastern margin of the Basin is faulted and affected by thrusting
Bhima	5,000	10,000	200	1,500	En echelon array of narrow strips	Sahabad, Rabanpalli, Hulikal, and Harwal Formations	
Kaladgi	8,000	20,000	3,800	7,000	Elliptical to arcuate	Mesoproterozoic Bagalkot and yonger Badami Groups	

Table 2. Phanerozoic Gondwana Basins of Peninsular India

Basin	Area Exposed (sq km)	Maximum Aggregate Thickness of Stratigraphy (m)	Shape of Exposed Outline	Lithosequence	Tectonic structure
Damodar				Facies A- Talchir Formation. Facies B- Barakar, Barren Measures, and Ranigunj. Facies C- Panchet Formation. Facies D- Mahadeva Formation.	
South Rewa				Talchir, Damuda, Pali-Tike, Parsora and Jabalpur Formations	
Talchir		280	Triangular	Talchir	
Satpura	12,000	50,000	Spindle shaped	Lower Grops: Talchir, Barakar, and Motur Formation. Middle Group: Panchmarhi and Matkuli Formation. Upper Group: Jabalpur and Pre-trappean sediments	Pre-rift, Syn-rift, and Post-rift (presently in interior rift)
Pranhita-Godavari (Pakhal Basin)	7,000	2,700		Talchir, Barakar, Barren Measures, and Kamthe Formations	
Krishna Basin				Vemavaram and Vejendla Formations	
Gundlakamma		150		Kundukur, Satyavedu and Sriperambadur (Chittoor District)	
Chintalapudi		3000		Talchir, Barakar, Kamthi, Kota and Gangapur	
Vagai			Graben structure		NW-SE trending plunging syncline. longitudinal and transverse faults.
Palar	2500			Satyavedu and Sriperambadur Formations	

subsiding basins (Table 2). The sequence of rocks has been designated the Gondwana Supergroup and divided into the Lower and Upper Gondwanas. They are mainly exposed in three river basins: a) the Koyal-Damodar Valley Basin; b) the Son-Mahanadi Basin; and, c) the Pranhita-Godavari Basin. The Godavari and Mahanadi Basins extend from the Eastern Ghats Mobile Belt to the Narmada-Son lineament in central India. The Koyal-Damodar Basin extends toward the Bengal Basin. The majority of these basins are located in the eastern half of the peninsula covering the Chhotanagpur plateau, Jharkhand, Madhya Pradesh, Orissa and Andhra Pradesh. The Satpura and the Pranhita-Godavari Basins are the larger basins. Scattered Upper Gondwana Basins are located along the east coast in Orissa, Andhra Pradesh and Tamil Nadu. They are represented by a series of outcrops: the Athgarh sandstone in the Cuttack area of Orissa; the Gollapalli sandstone, the Raghava-puram shale and the Tirupati sandstone in the Godavari-Krishna sub-basin (Krishna and West Godavari districts); the Budavada

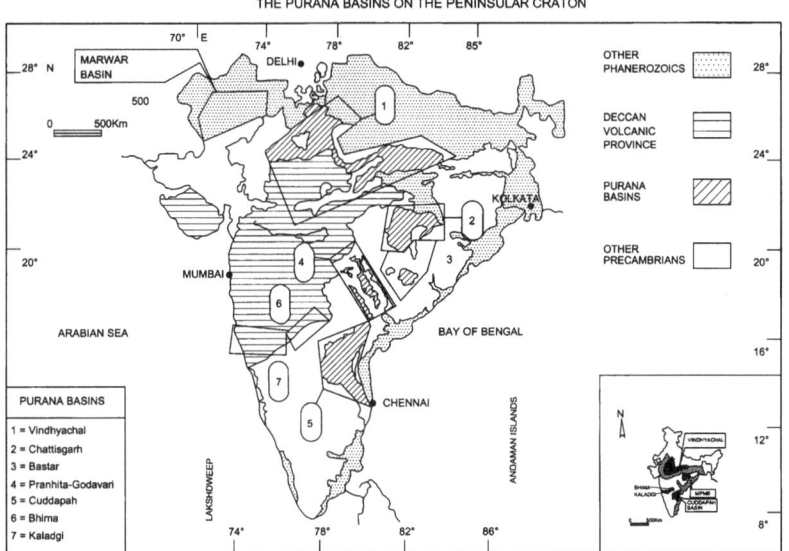

Figure 3. Map showing the seven major Purana Basins of peninsular India. These basins are scattered across the peninsular craton. Some of them are partially covered by the Deccan Volcanic Province. The smaller Gondwana Basins occur in contiguity with these basins. Together they constitute the core area of Paleolithic occupation in the Indian subcontinent

sandstone (Prakasam and Nellore districts) in Andhra Pradesh; the Satyavedu and Sriperambadur Formation in northern Tamil Nadu and the Uttatur Beds in the Tiruchirappalli district of Tamil Nadu. Archeological investigations in these areas indicate a significant relationship between these basins and early hominin occupations. Interestingly, the southern boundary of Acheulean site distribution is co-terminous with the southernmost Upper Gondwana Basin in Tamil Nadu. Small areas of the Upper Gondwana outcrops are also known from the Garo Hills of northeast India, though there is no well defined Paleolithic evidence from this region.

Evolution of Monsoon Environment and Habitats across the Indian Subcontinent

The differential distribution of Paleolithic sites is suggestive of hominin preference to specific habitats that are co-terminous with geological provinces. The following sections trace the historical evolution of the habitats during the Cenozoic that set the stage for hominin occupation. The timing of the Paleolithic colonization of the subcontinent has been controversial. Understanding the dynamics of landscape and habitat evolution in the northern and northwestern regions of the subcontinent is critical to the development of the basin model as well as retracing of the dispersal routes of both grazing mammals and hominins into the subcontinent.

Habitat evolution is intimately linked to Cenozoic tectonic activity and climatic history of the subcontinent. These developments can be seen through three distinctive phases of the Cenozoic, i.e., during the Paleogene, the Neogene and the Quaternary. Though the onset of the monsoon climate dates back to the Pliocene, the timing of its stability is crucial to understanding of ecosystems and colonization processes. As discussed below, the Middle Pleistocene ushered in the formation of stable ecosystems coinciding with the expansion of hominins across the subcontinent. The monsoon over the subcontinent fluctuated in consonance with the global climatic cycles during the Pleistocene (Korisettar and Ramesh, 2002). Therefore, it is important to discuss the evolution of the subcontinent's environment and the concomitant formation of ecosystems in order to assess the suitability of geographic areas for hominin settlement.

Himalayan Geodynamics and Montane Habitats

The Neogene was the period of intense crustal deformation along the leading edge of the Indian plate that set in motion a series of tectonic upheavals giving rise to the formidable Himalayan orogen. This initiated the evolution of India's geography, physiography and the monsoon environment, including the creation of a network of diverse habitats. The episodic rise of the Himalaya and the attainment of an altitude of 8,000 m AMSL were key factors in the initiation of a monsoon circulation over the subcontinent. The Himalaya occupies a 2,400 km long and 300–250 km wide area along the northern frontier of the subcontinent and constitutes a formidable mountain barrier. This snow-covered barrier prevents the flow of cold winter winds from Siberia and causes relatively warm winters over north India. Though the Himalaya had attained an altitude of orographic prominence by the Mid-Miocene, it could not prevent the movement of hippopotamus, rhinoceros and elephant across the foreland lake basins of Potwar, Kashmir and Kathmandu (Valdiya, 1999). Apparently it maintained a moderate topographic relief with wide-open savanna valleys, and a thick soil mantle that was generated under a hot-humid or a hot-dry climate under a prolonged tectonic quiescence.

Between 11 and 7.5 Ma ago, abrupt tectonic movements triggered further uplift of the plateau and by 8 Ma the Himalaya had attained its present altitude, defining the physiographic configuration of the subcontinent as seen today. This spurt of tectonic activity affected the Indian Ocean crust that was deformed simultaneously. This tectonic resurgence was of great consequence inasmuch as it was accompanied by abrupt climatic changes over the subcontinent. New landscapes emerged. The senile topography was rejuvenated. The foreland basins were filled with detritus (the Siwalik molasses) eroded from uplifting higher terraces. According to Valdiya (1999), climate change was characterized by the onset of the seasonal monsoon including alternating strong wet and dry spells, causing intense weathering, erosion and bed load. Such processes were effective in reshaping of landscapes and modification of ecosystems in the mountains, plains and the oceans. Ecosystems were characterized by a shift in vegetation from C3 to C4. Consequently grasslands came into existence in the foreland basins, creating a habitat for grazing mammals from outside the subcontinent. This is said to have caused a faunal turnover, resulting in the marginalization or extinction of indigenous animals. The faunal population increased considerably and was three times richer than at present. There were 30 species of elephant compared to just one at present, and 15 genera of anthropoid apes, including *Ramapithecus* (Valdiya, 1999). Around 9.5 Ma, the European three-toed horse, *Hipparion*, and pigs appeared in the Potwar. Around 7.5 Ma, *Stegodon*, hippo, *Hexaprotodon*, and *Elephas planifrons* colonized the foreland Siwalik terrain. This was also the period when the Himalaya witnessed accelerated erosion as a result of accelerated uplift and the accompanying heavy rainfall. After a short period of quiescence, the Plio-Pleistocene transition was accompanied by tectonic uplift (between 4 and 2 Ma). The uplifted mountain ranges created large cool areas and induced precipitation of snow. This coincided with the global cooling and the onset of the Pleistocene glacial at 2.5 Ma (see Vrba, 1996).

Between 1.7 and 1.5 Ma, the Himalaya experienced increased glaciation and a revival of tectonic uplift. This spurt of tectonic movements caused crustal disturbances in the Lesser Himalaya and the Siwalik foreland basins, coinciding with the Olduvai Magnetic Polarity Event and the onset of a glacial phase. During the transition from the Early to Middle Pleistocene, around 800–700 ka (coinciding with the reversal of geomagnetic polarity from reversed Matuyama to normal Bruhnes Polarity Epochs), another major tectonic episode occurred that led to increased sediment load into the Bengal Basin. The Himalayan habitats were greatly affected by the vast expanse of snow that spread far and wide over Potwar, Kashmir, Kangra and Tibet including the Central Himalaya. Consequently the southwest monsoon weakened causing spells of dry and cold conditions over the pre-existing savanna ecosystems in the Siwaliks and in the Lesser Himalaya in north India. Powerful bed load streams emerged (i.e., the Boulder Conglomerate phase), marking the end of the Siwalik lithosequence (Valdiya, 1999). The large grazing mammals were forced to immigrate southward into the peninsular cratonic basins, i.e., in the relatively stable peninsular Purana-Gondwana Basins. Though these basins were unaffected by Himalayan tectonism, they were affected by changed climatic conditions. Habitats across the central Indian Purana-Gondwana Basins remained relatively stable, attracting Subhimalayan fauna. These forces rendered the Subhimalayan terrain inhospitable during the Lower and early Middle Pleistocene, not only because of strong fluvial dynamism but also because of severe cold conditions and low biomass of food resources. The immigrant mammals and hominins from the warmer latitudes could not have adapted to this extreme environment. This could explain the lack of evidence for a long chronology of hominin occupation.

The present network of drainage basins of the Indus, Ganga and Brahmaputra is a southern extension of the foreland basins that have emerged in the post-Siwalik phase, i.e., since 500 ka. These drainage systems have carved terraces on the Siwaliks through down cutting and lateral shifting. The drainages also created vast alluvial floodplains in their middle and lower reaches and fan systems along the Siwalik Frontal Range and in the continental shelf, under the influence of the Indian Ocean monsoon. These fan systems preserve an excellent record of evolving landforms, monsoonal variations and terrestrial biodiversity (see Valdiya, 1999; Korisettar and Ramesh, 2002). The record indicates the emergence of a stable monsoon ecosystem that began to attract Siwalik grazing mammals, and eventually hominins.

The Evolution of the Foreland Plains As noted earlier, Acheulean sites are not known from the foreland plains, and any identified sites are restricted to the rocky elevated landforms in the Sindh Plain (the Rohri Hills) and the Siwaliks. The Middle Ganga alluvial plain does not contain Acheulean sites and the identified sites are toward the circum-Vindhyan and Gondwana Basins. An understanding of the geological evolution of the foreland plains is relevant to answer why the heartland of the alluvial plains was inhospitable for hominin occupation.

The Sindh Plain on the west and the Bengal Plain on the east complete the Indo-Ganga-Brahmaputra plains along the Siwalik front. As an active Himalayan foreland basin, the Ganga Plain exhibits all its major components, namely an orogen (the Himalaya), a deformed and uplifted foreland basin (the Siwalik molasses), an active depositional basin (the Ganga Plain) and a cratonic peripheral bulge (Bundelkhand-Vindhyan plateau or the northern edge of the peninsular craton). It has also been argued that different sectors of the Siwalik Frontal Thrust experienced uplift between 800–500 ka. This led to the expansion of the southern part of the basin over the peninsular bulge. However, the last major tectonic activity in the Himalaya occurred around 500 ka, causing uplift of the Upper Siwalik sediments (see Singh and Bajpai, 1989; Singh, 2004).

The Ganga Plain exhibits an asymmetrical sediment wedge, only a few tens of meters thick toward the peninsular craton (toward the Yamuna and northern Vindhyan-Bundelkhand ridge) and up to 5 km thick near the Himalayan orogen. The down-flexing lithosphere below the Ganga Plain shows many inhomogenities in the form of ridges and basement faults. The basement highs and faults have controlled the thickness of the alluvial fill and also have affected the behavior of the river channels on the surface. The sediment input to the Ganga occurs at a rate in excess of down-flexing, causing the sediment rate to exceed the rate of subsidence (Singh, 1996, 2004). The northern slopes of the Vindhyan region were apparently in the distal area of Ganga fluvial dynamism during the Middle Pleistocene.

The basement highs of the Ganga Plain have affected the geomorphic evolution of this region, during the post-500 ka period. Positioning of the Ganga and Yamuna rivers at a higher altitude, and the location of the Ghaghara River at a low altitude, is a strong indication of uplift of the southern margin of the Ganga Plain. The axial rivers have not shifted orogenward but have responded by a vertical incision in alluvial deposits. There is accumulation of sediments in the peripheral bulge region by the drainage originating in the Vindhyas. The peripheral bulge region is an area of extensional tectonics, causing down-flexing and creation of space leading to preservation of sedimentary deposits. Consequently, the thickness of the Ganga alluvium is much greater toward the Subhimalayan Thrust. This could be an explanation for the absence of hominins in this region during the Middle Pleistocene. In the western margin of the Ganga Plain, the Aravalli-Delhi Ridge

extends below the alluvium (Singh, 1996, 2004). At present, the Ganga Plain is in a relaxation phase and there is evidence of prominent tectonic activity in the northern margin of the peninsular craton (i.e., the Vindhya-Bundelkhand bulge).

The Ganga and Punjab Plains are presently experiencing subsidence. A network of horst and graben structures characterizes the Siwalik landforms that are drained by the Himalayan rivers. The autochthonous rivers drain the intermontane graben or *dun* valleys (e.g., the Dang and Deokhuri Valleys of Nepal; Corvinus, 1995). Two sets of Lower Paleolithic industries have been documented in the *dun* valleys, i.e., Acheulean and Soanian assemblages. The latter is generally known from the Siwalik terraces carved by rivers emerging out of the Siwaliks into the Plains during the last 500 ka, i.e., since the last major uplift of the Siwaliks.

The Western Ghats and the Rise of Monsoon Habitats on the Peninsular Craton

The present structural configuration of peninsular India took place in conjunction with developments along the leading edge of the Indian plate during the consequent collision of the Indian and Asian plates. As noted above, this was a period of intense tectonic activity even in the peninsular region, though differing in magnitude and scale in comparison with the extra-peninsular region.

The uplift of the Western Ghats in the Early Neogene effectively changed the peninsular ecosystem from tropical woodland in the Early Tertiary to a network of savanna ecosystems between the West Coast and the plateau. The east-west oriented secondary divides have given rise to the formation of broad shallow valleys drained by the rivers Godavari, Krishna, Kaveri and their tributaries. The drainage basins of these rivers cut through a network of geological provinces, including the Purana-Gondwana Basins and Eastern Ghats, terminating in the Bay of Bengal.

Savanna ecosystems became well-defined in the rocky triangle between the Western and Eastern Ghats and within the Purana-Gondwana Basins. This orographic configuration resulted in a progressive decrease in the intensity of rainfall along the path of the southwest monsoon, the basic element controlling the habitability of the area. Obviously, the structure of food resources between the coastal, upland and plateau ecosystems differed substantially. The key controlling factor would have been the secular variation in the intensity of the southwest monsoon across the peninsular region. The occurrence of massive lithic calcretes on Late Tertiary laterites and ferricretes in the upland Deccan Volcanic Province is good evidence for climatic fluctuations and the onset of a semi-arid monsoon ecosystem during the Late Neogene to Early Pleistocene (Rajaguru, 1997). This is consistent with the onset of global cooling at the beginning of the Pleistocene (Vrba, 1996).

Faunal remains have been recovered from fluvial gravels in the peninsular valleys (Badam, 2002). The Narmada fauna has been broadly divided into Lower and Upper Groups of the Mid-Late Pleistocene periods. The Lower Group of the Narmada has yielded Acheulean artifacts along with Middle Pleistocene fossils of *Bos namadicus, Sus namadicus, Hexaprotodon namadicus, Elephas hysudricus, Equus namadicus* and *Stegodon-insignis-ganesa*. Similar fauna is also reported from the Deccan basins. Existence of this fauna attests to the prevalence of ponds and swampy areas within the larger savanna-woodland ecosystems of the Purana-Gondwana Basins. This faunal assemblage continued into the Late Pleistocene (Badam, 2002).

The widespread occurrence of similar animals in central and southern India was possible because of comparable ecological and climatic conditions across the network of basins on the peninsular shield. These animals seem to have had a zonal distribution in these areas without any definite ecological barriers between them. A majority of these forms are late survivals from the Siwaliks, having migrated from the Himalayan foreland basins into the southern basins. Many of the species became extinct during the course of such migrations, while some survived and evolved into advanced forms adapting to the Terminal Pleistocene/Holocene climatic amelioration.

It should be noted that this faunal composition included both indigenous and exotic elements. The migratory routes lay east and west of the Himalaya (Pilgrim, 1925; Turner and O'Regan, this volume). Most of the larger mammals migrated from North Africa, Arabia, central Asia and North America through routes across Alaska, Siberia and Mongolia. Hippopotamus and elephant, which had their early origin in central Africa, emigrated outward and entered India during the Tertiary period through Arabia and the Iranian plateau. Rhinoceros, horse and camel, all originating in North America, evolved in some countries of central and western Asia before migrating to India.

In view of the evidence, it can be argued that stable savanna-woodland environments prevailed in diverse geomorphic systems over peninsular south India during the Pleistocene, i.e., between the Vindhyas in the north and the Nilgiris in the south. These ecosystems were the home of many native and exotic grazing mammals. The latter emigrated from the foreland basins at times of severe climatic conditions and by the Middle Pleistocene had colonized the entire region. However, this fossil record is not older than the Middle Pleistocene. Early hominins would have subsequently dispersed into the peninsular ecosystems ranging across in search of ideal habitats within the larger geological provinces.

Coastal Habitats

The emergence of coastal lowlands along the western seaboard was concomitant with the rise of the Western Ghats during the Neogene. This was followed by an extensive phase of lateritization of pediment surfaces when climatic conditions were conducive for deep weathering. The lowlands along the coast formed under relative tectonic stability and relative semi-aridity (Kale and Rajaguru, 1987).

On the East Coast, recent study of Holocene sediments and the associated paleoclimate in the Koratallaiyar-Cooum Basin (Tamil Nadu) is relevant to an understanding of habitat development during the Quaternary in the circum-Gondwana Basin area (i.e., an area known as the Palar Basin; S. Pappu, 2001; Nagalakshmi and Achyuthan, 2004). Landform drainage evolution in the Palar Basin largely relates to the presence of faults and lineaments as well as changes in sediments as a product of fluvial regime changes. The presence of slack water ponded facies and fining upward sand-mud units, in Holocene sequences is a clue to the prevalence of perennial ponded environments in Pleistocene landscapes, as is attested by the occurrence of prehistoric sites in fine-grained deposits and ferricrete surfaces. Paleolithic sites also occur as discontinuous patches and can be related to the old Palar courses. Hominin activity was related to fluvial activity originating in the upland Gondwana formations (Allikulli Hills), from where denudation processes gave rise to circum-basin landforms to which the hominins successfully adapted. Lineaments, faults and the lithosequence favored spring activity. Paleolithic localities are not only related to the paleochannel but also to other landform units that represent extinct water bodies. This relation is not accidental; apparently, the Koratallaiyar has incised into older landform units. Regional and local tectonic disturbance during the Neogene was also responsible for abrupt changes in spring activity, with implications for landform changes and drainage pattern modification (Figure 4). Additionally, long- and short-term changes in climate and associated base level changes also triggered water table fluctuations.

Isolated Northeast Humid Landforms

While the Subhimalaya and the Bengal Basin experienced severe tectonic disturbance and dynamic fluvial environments during the Early and Middle Pleistocene, the northeastern mountain tracts and the Burma Mountains attained their altitudes and received higher monsoon precipitation in comparison to the northern and northwestern Himalaya. Owing to the present physiographic configuration of the subcontinent, this region receives high precipitation from the Indian Ocean monsoon system, thereby harboring a forest ecosystem comparable to the equatorial rainforests. This, in combination with mountainous terrain, acted as barriers across the southern dispersal route of modern humans into Southeast Asia.

The modern states of Assam, Meghalaya, Tripura, Nagaland, Mizoram and Manipur constitute the northeastern region of the subcontinent. The Manipur Valley connects northeast India with northern Burma through the Hukuang Valley. Archeological research in this region has not yielded Paleolithic artifacts, though some have been reported from the Garo Hills in Meghalaya (T.C. Sharma, 1991). The current evidence from this region does not support the idea of a significant dispersal route for hominins. It should, however, be noted that dense evergreen forest ecosystems in the subcontinent and elsewhere in the Old World have not yielded substantial evidence for Paleolithic occupation.

Fresh Water Resources in the Peninsular Basins

The Southwest summer monsoon is the major source of fresh water over the Indian landmass and has a dominant role in the terrestrial

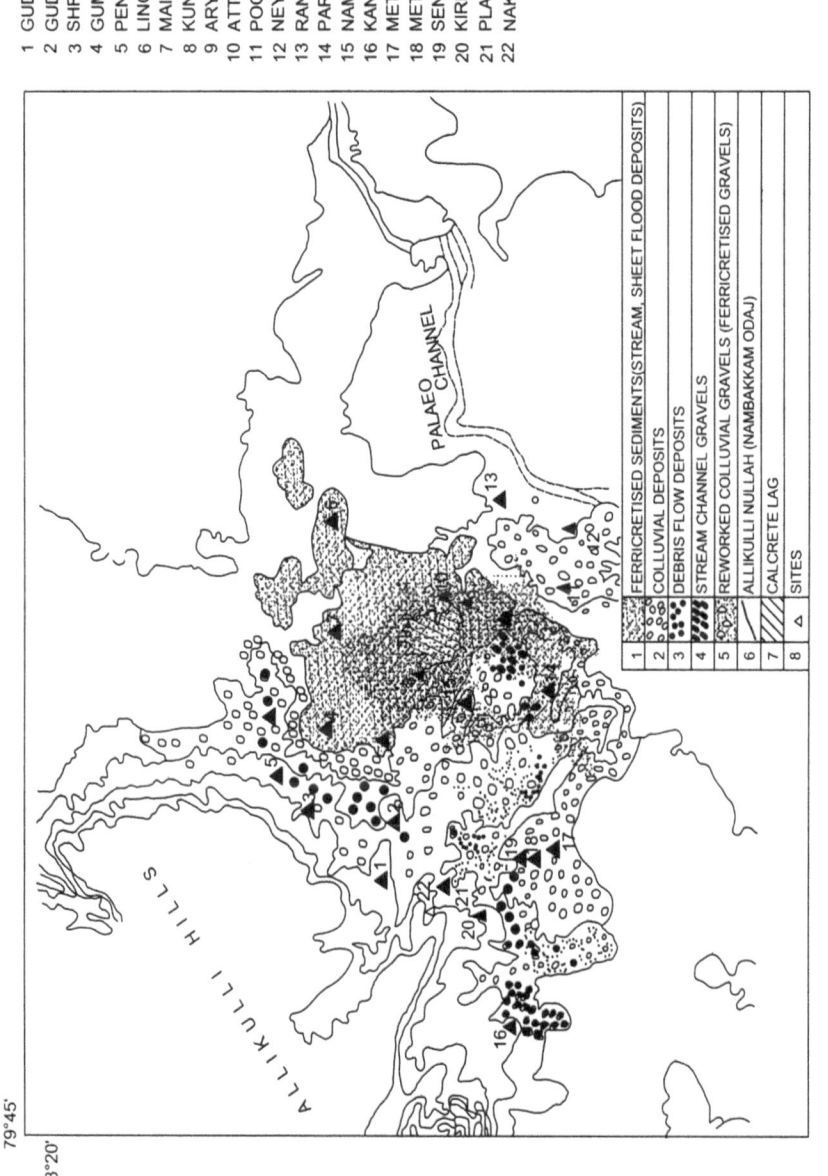

1 GUDIYAM
2 GUDIYAM VILLAGE
3 SHRIKRISHNAPURAM
4 GUMIPALAYAM
5 PENNALURPET
6 LINGAMMANAIDUPALLI
7 MAILAPUR
8 KUNJARAM
9 ARYATHUR
10 ATTRAMBAKKAM- 1,2,3
11 POONDI
12 NEYVELLI
13 RANGAVARAM
14 PARIKULUM
15 NAMBAKKAM
16 KANJIPADI
17 METTUPALAYAM
18 METTUR
19 SENRAYANPALAYAM
20 KIRINAYATTAM
21 PLACEPALAYAM
22 NAKALKONA

SEDIMENTARY CONTEXTS OF ARCHAEOLOGICAL SITES IN THE KORATALLAIYAR BASIN

1	FERRICRETISED SEDIMENTS(STREAM, SHEET FLOOD DEPOSITS)
2	COLLUVIAL DEPOSITS
3	DEBRIS FLOW DEPOSITS
4	STREAM CHANNEL GRAVELS
5	REWORKED COLLUVIAL GRAVELS (FERRICRETISED GRAVELS)
6	ALLIKULLI NULLAH (NAMBAKKAM ODAJ)
7	CALCRETE LAG
8	SITES

Figure 4. The Koratallaiyar Valley Paleolithic sites and their sedimentary contexts in the circum-Gondwana Basin, Allikulli Hills, northern Tamil Nadu. The basin represents the southern-most Acheulean habitat in south India (after S. Pappu 1996)

hydrological cycle. The Himalayan Mountains and the Indian Ocean define the continental geography of the region and strongly influence the onset of monsoon regimes. However, internal orographic features, including the Western and Eastern Ghats (Hills), the Deccan plateau, the Subhimalaya and the Aravallis have played a significant role in the latitudinal and longitudinal variation of the intensity of monsoon precipitation across the landmass. This has given rise to a mosaic of landforms and ecosystems, which were governed by the distribution and relative variation in the intensity of rainfall. These ecosystems range from the arid Thar, the semi-arid Deccan plateau, the sub-humid coastal landforms, and the north-eastern humid landforms (see Meher-Homji, 2002; Fuller and Korisettar, 2004). Glacial melt-waters, orographic monsoon rainfall, and post-monsoon ground-water movement fed the peninsular river systems.

Stronger southwest monsoon regimes during warm and humid interglacials favored high lake levels, increasing flow in rivers and causing flooding. A decreased monsoon during cold and dry glacials altered these conditions, resulting in lowered sea levels and increased continentality (see Kale et al., 2004, for a summary). Consequently, the monsoon reach over the continental area was affected. It is increasingly clear that rainfall patterns governed the habitability of a given area, thus it is necessary to identify perennial habitats under these conditions. Obviously, climates, physical and hydrological environments and the distribution of resources exerted great influence on the habitability of areas during the Pleistocene.

Large sectors of the Purana-Gondwana Basins exhibit sub-horizontal sequences. Folding is observed in discrete zones within them. These deformed zones occur along the margins of the individual basins. Synsedimentary deformational features have also been documented (Kale, 1991:246). The Purana-Gondwana physiography presents a sub-montane environment, where geomorphic relief in the intra-basin area is much higher than the circum-basin areas. These basins constitute watersheds and catchments while being part of the larger allochthonous drainage network of monsoonal streams. Post-monsoon fresh water resources were governed by spring discharge into the antiformal valleys, where surface water was retained either in topographic lows or as gentle stream discharge. Similar processes operated simultaneously in the circum-basin areas.

The widespread occurrence of ferricretes, spring tufas and travertine indicate that water tables were much higher during the Pleistocene. Such features constitute conspicuous landscape units, representing distinctive phases in the way in which landforms have evolved both in intra-basin and circum-basin areas. The best examples of such features are clearly seen in the circum-basin area of the Gondwana Basin in northern Tamil Nadu (the Koratallaiyar-Palar Basin) (Figure 4).

The majority of the basins constituting the core areas of Paleolithic settlements is covered by the monsoon cycle (both the southwest and northeast monsoon) and possesses higher rates of natural recharge. While the Western Ghats and the western Deccan are not covered by the northeast monsoon, occasionally winter cyclonic storms in the Bay area cause rainfall in this area. During glacial periods, the northeast monsoon was stronger than the southwest (see Korisettar and Ramesh, 2002; Kale et al., 2004 and references therein). Under increased continentality of glacial periods, the western parts of peninsular India experienced arid conditions affecting the habitability of the western Deccan ecosystems. Eastern and central parts of the peninsula remained unaffected by glacial aridity, which favored stronger northeast monsoon regimes. They remained the most attractive habitats and hence the core habitats of Paleolithic settlement. Moreover, the two streams of southwest monsoon one from the Arabian Sea and another from the Bay intersect over central India, westwards of the Eastern Ghats, causing higher rainfall. This is the region of Vindhyachal, one of the Paleolithic core habitats. The Vindhyan and Gondwana (Gond forest) deciduous vegetation has been sustained under this regime.

The Deccan Volcanic Province, occupying the major part of western peninsular India, is a key geological and geographical region for our understanding of hominin ranging habits and their dispersal patterns. The Deccan basalts exhibit a multilayered aquifer system comprising aquicludes of massive units and aquifers of vesicular units and fractured zones. Water resources can vary owing to differences in lithology. Compact basalt being well-jointed can store ground water, whereas the amygdaloidal basalts do not usually make good aquifers due to their massive nature and tendencies for lesser jointing. The degree and development of openings govern the movement of ground water within the lithology. In the upland Deccan Volcanic Province, characterized by adequate relief, provided for water resources in the post monsoon season. Though the province covers a vast area, the density and continuity of Paleolithic sites is very low in comparison with the Purana-Gondwana Basins.

In the region west of the Aravalli Mountains lacustral deposits representing saline and fresh water conditions in lakes are good examples of the fluctuating monsoon and ground water emerging from the Aravalli Fold Belt. The saline lakes received fresh water from ground water seepage during the Quaternary. Monsoon precipitation along the Aravallis and their deformation has largely facilitated ground water movement on either side of the mountain range (Sinha et al., 2004). The Thar was well watered during the major part of the Pleistocene, as demonstrated by the vast expanse of alluvial sands (later subject to desert aeolian processes), boulder beds and the large number of Paleolithic sites (Misra, 1995).

The Basin Model: Core, Peripheral and Isolated Habitats

The following subsections examine the relationship of sites to geographic zones. The discussion is divided into habitats inferred to be 'core', 'peripheral' and 'isolated' zones. Examination of these areas helps to trace the dispersal routes and to establish zones of maximum habitability.

Core Areas

The Purana-Gondwana basins constitute core areas. These areas contain evidence for long-term occupation. Hominin occupations are evident in caves, rock-shelters and along perennial springs and low order streams. The geomorphic environment and the lithological unconformities facilitated high permeability of ground water in the rocks, including ground water discharge through springs. Perennial water resources were present and such environments would have maintained a high biomass index of plants and animals, likely adapted to a deciduous woodland-savanna ecosystem. The stone outcrops yield hard rocks as well as cryptocrystalline varieties that were used for stone tool manufacture.

Archaeological surveys have been carried out in the Purana and Gondwana basins, revealing a wealth of sites and a continuity of occupation (e.g., Foote, 1880, 1914, 1916; Korisettar, 2002; Pal, 2002; Raju and Venkatasubbaiah, 2002). The Vindhyan Basin of central India has received the greatest attention and includes integrated archaeological and Quaternary studies (e.g., R. Singh, 1956; Khatri, 1962; Joshi, 1964, 1978; Misra et al., 1983; Sharma and Clark, 1983; Jacobson, 1985; Mishra, 1985; Sonakia, 1985; Supekar, 1985; Acharyya and Basu, 1993;

Williams and Clarke, 1995; V.D. Misra, 1997). Archaeologists have examined other basins, including the Bhima Basin (e.g., Foote, 1876; Paddayya, 1982, 1989), the Kaladgi Basin (e.g., Joshi, 1955; R.S. Pappu, 1974; Korisettar and Petraglia, 1993; Korisettar et al., 1993; Pappu and Deo, 1994; Petraglia et al., 2003), and the Cuddapah and Kurnool Basins (e.g., Soundara Rajan, 1952, 1958; Thimma Reddy, 1968, 1977; Murty, 1974, 1979; Raju, 1981, 1988; Varaprasada Rao, 1992).

The major rivers that have been surveyed through the Purana-Gondwana Basins are allochthonous. Archeologists have usually focused their attention on the river valleys that cut across the basins. Though surveys in river channels were productive, previous investigations were not extended into inland areas. That such an approach would prove more productive has been amply demonstrated by the investigations carried out in the Bhima and Kaladgi Basins of northern Karnataka (Korisettar, 2004a, 2004b). Scattered but intensive surveys have been carried out in the Cuddapah Basin of Andhra Pradesh, yet the focus was on the river valleys. Future investigations in the inland areas of Purana-Gondwana Basins are likely to be productive.

Peripheral Areas

These are intermediate between core areas, and are in fact geologically contiguous with them. Peripheral areas represent zones which experienced hominin range expansion during wetter climatic conditions. The geological and geotectonic framework, and associated landform units, are markedly different from those of core areas. The areas maintained high water tables in some pockets of high relief, facilitating perennial spring activity that might have supported a fairly high biomass of food resources. Paleolithic occupation is not widespread as revealed by discontinuous Paleolithic succession. The Deccan Volcanic Province and the greenstone belts in the Western Ghats, which receive only the southwest monsoon, constitute peripheral areas. The basaltic province of western India is physiographically characterized by a north to south disposition of Western Ghats, with secondary divides running west to east that have given rise to broad shallow valleys dominantly sculpted by fluvial processes. Interglacial phases provided adequate perennial water and other food resources for small hominin populations. Middle Paleolithic occupation coincides with the Last Interglacial, and was relatively widespread. The Upper Paleolithic is not well represented in peripheral areas. Terminal Pleistocene Epipaleolithic evidence coincides with the onset of Early Holocene southwest monsoon climatic amelioration. The density of sites is higher in the upland regions where the landscape has higher relief that facilitates ground water discharge.

Isolated Areas

Isolated areas consist of peninsular gneissic and granitic terrain with low relief. The areas are characterized by a tor-inselberg landscape, with low ground water potential, low suitable raw material resources, poor and shallow surface water resources, and low plant and animal food biomass. Though characterized by seasonal grasses, it is largely a scrubland ecosystem. Pediments around inselbergs are monotonous plains, between the piedmont and the topographic low or denudation troughs, where ephemeral streams give rise to swamps. Dolerite dyke swarms cutting through the inselbergs are potential springs. Paleolithic communities occasionally ranged into these areas. Hominins may have ranged across these landscapes to enter adjacent perennial habitats or to pursue seasonal food resources. The greenstone belts, occurring within the larger gneissic landscapes, are the prominent orographic belts that break the monotony of the rolling plains. These belts are as yet not well surveyed. While the Shimoga-Goa

Greenstone Belt is part of the Western Ghats, the Nellore Schist Belt is part of the Nalla-malai Fold Belt of the Cuddapah Basin. Though these belts comprise suitable raw materials and other basic resources they are not prominently exposed. Paleolithic populations do not seem to have dispersed into the heart of these basins. The marginal areas in Goa, along the Western Ghats, have yielded evidence for Acheulean occupation. Similarly along the leeward in the Shimoga-Goa Green-stone Belt (comprising metavolcanics, iron formations and ferruginous quartzite) in south-western Karnataka, Acheulean localities are associated with ground water seepage areas. Vein quartz occurring in the granite, though an intractable material, was utilized for stone tool manufacture. These sites appear to be in isolated clusters that warrant an organized program to assess their significance. The northeastern humid forest ecosystems were also isolated during much of the Pleistocene.

Further Insights into Basins and their Archaeology

Presently little is known about how hominins adapted to the range of geomorphic environments in the subcontinent and the role of the Pleistocene monsoon in structuring food resources across basins. Hominin populations appear to have colonized most geographical areas of the subcontinent with preference for the Purana-Gondwana Basins. Under prevailing monsoon circulation patterns, these basins were characterized by savanna vegetation, with perennial springs and ponds. The density and continuity of Paleolithic occupations suggest that even under reduced monsoon regimes, these basins were covered by the stronger northeastern monsoons and that fresh water and subsistence resources were not depleted. The distribution of Acheulean sites suggests that the hominin populations ranged into numerous basins during wetter phases and returned to more dependable Purana-Gondwana

Basins during drier phases. A further investigation of this environmental-behavioral hypothesis is needed.

Surveys have established the archeological richness of many regions, demonstrating the occurrence of Paleolithic sites away from the channeled watercourses, including Jacobson (1985) in the Vindhyas, Paddayya (1982) in the Bhima Basin and Pappu and Deo (1994) in the Kaladgi Basin. Region-specific survey clearly indicated that a wide range of landform units across the basin provided opportunities to hominins. Modeling Quaternary alluvial stratigraphies in the northern Vindhyan Son and Belan valleys and the associated cultural succession (Clark and Williams, 1990; Williams and Clarke 1995) is significant in the context of the suggested basin model (Figure 5).

In the Kaladgi Basin, Pappu and Deo (1994) documented as many as 452 Paleolithic sites over an area of approximately 8,000 sq km. The inadequacy of the geomorphometric methods for interpreting site location was brought out by subsequent investigations (Korisettar and Petraglia, 1993; Korisettar et al., 1993; Korisettar, 1994; Petraglia et al., 2003). The southern parts of the Kaladgi Basin have revealed that the Lower Paleolithic yielding gravel bodies along the middle course of the Malaprabha and Ghataprabha are a series of coalescent alluvial fans. Previous studies incorrectly identified these gravel deposits as former bed load regimes of rivers. Recent research has shown that an extensive calcrete-tufa surface developed over the older laterite surface followed by the Late Quaternary alluvial sequence in the Malaprabha channel. The artifact bearing conglomerates exposed in the bed of the Malaprabha can be traced away from the channels up into the piedmont region away from the river. These represent former alluvial fan systems originating from the Kaladgi ridges upon which the Late Quaternary alluvial deposits are superimposed, in part

THE PROTEROZOIC VINDHYAN (VINDHYACHAL) BASIN

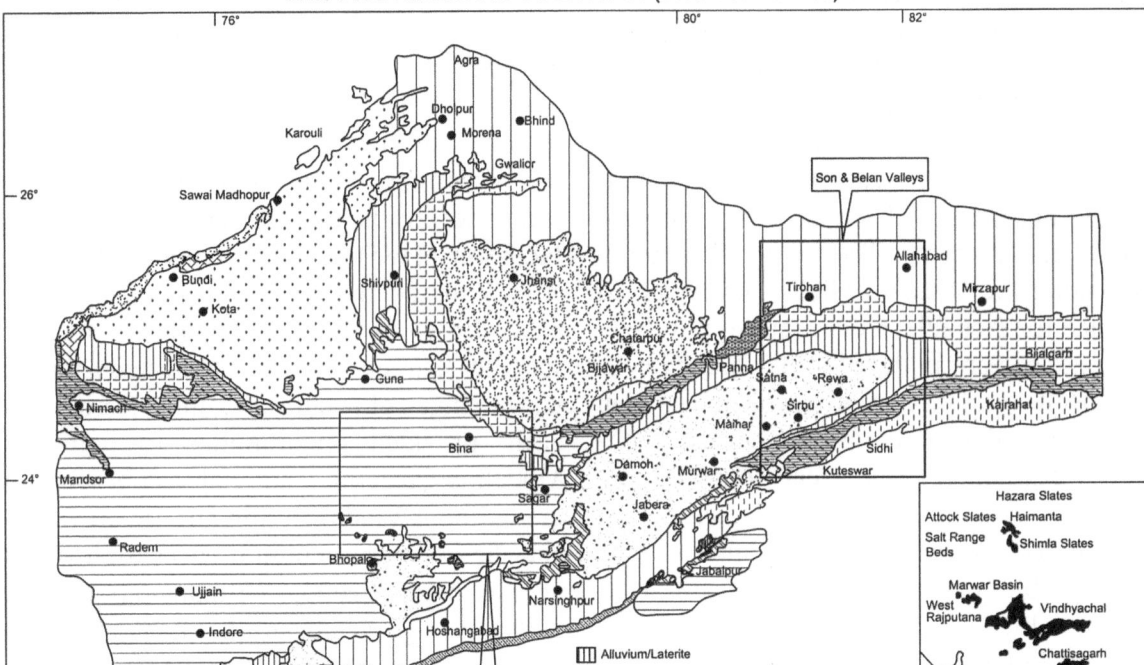

Figure 5. Map of the Vindhyan Basin (Vindhyachal), the largest among the Purana Basins. The boxes demarcate the areas investigated by Jacobson (1985) and Misra (1985) in Raisen District of Madhya Pradesh and Sharma and Clark (1983) in the Son and Belan valleys of the northern Vindhyan Basin including the Gondwanas. This is a major Paleolithic complex zone among the Proterozoic-Phanerozoic Basins

burying cultural material (Korisettar et al., 1993). Acheulean occupation post-dates the fan formation in the foothill regions and is associated with perennial spring systems. Such springs are now extinct as attested by travertine and spring tufa in the proximity of Paleolithic sites (Figure 6). The tectonic framework of local faults and folds under high water table regimes of the Pleistocene governed the hominin activity in these basins. Our investigations have revealed that the Acheulean sites do not correspond to the present drainage network. The Lower and Middle Paleolithic activity areas on the pediments were associated with ponds and

springs, clays and fine-grained silts. These settlements were located by the now extinct watercourses and suggest that hominin activity loci were not related to channel networks as seen at present (see also Korisettar, 1995).

Based on such observations, the Lakhmapur area was selected for a detailed study. The Lakhmapur Lower and Middle Paleolithic complex occurs on the northern slopes of the quartzite ridge, on the southern margin of the Kaladgi Basin. The surface scatters and buried stone line of Lower and Middle Paleolithic artifacts have been mapped in relation to extinct springs in the area (tufaceous beds) (Figure 7). Further examination of the thin

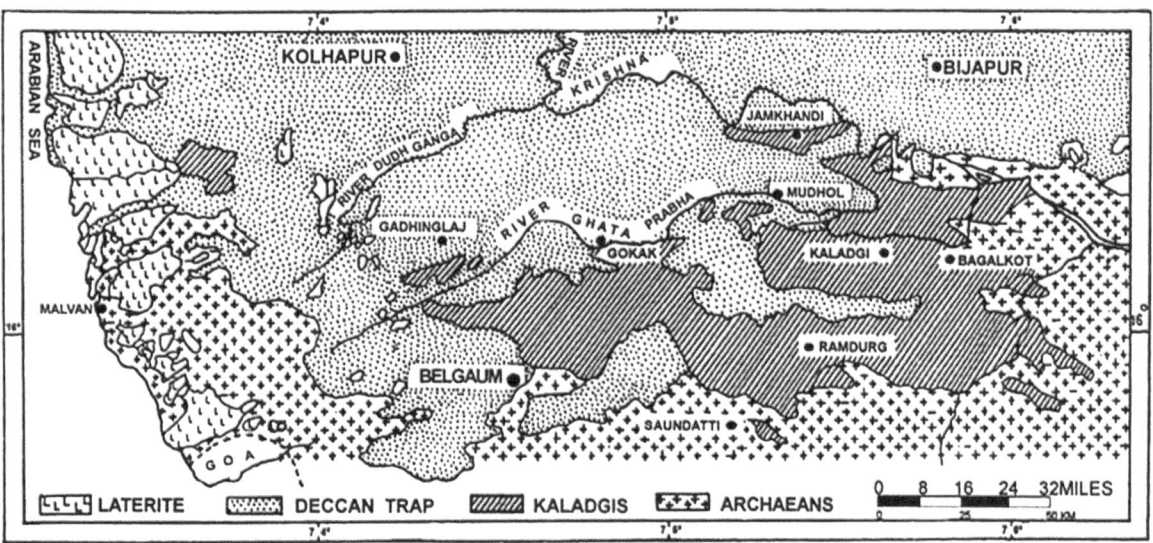

Figure 6. Map of the Kaladgi Basin showing its geological formations. Nearly 400 Paleolithic sites have been documented from different parts of the basin

sediment mantle in the region revealed that the transitional Lower to Middle Paleolithic facies can also be documented which is further separated from the typical Middle Paleolithic occurrence on the pediment surface. These two archeological horizons are on two different surfaces with an intervening depositional facies (Petraglia et al., 2003). In addition to spring tufas around the scatter of artifacts, the existence of chert breccias has also been documented. These fault-related breccias are known to be zones of spring activity resulting in ponded water resources.

The Hunsgi-Baichbal Valley excavations and other related inter-disciplinary investigations of the Acheulean localities are of profound importance in our understanding of the earliest phase of occupation in peninsular India (Paddayya, 1977a, 1977b, 1982, 1989; Paddayya and Petraglia, 1997; Paddayya, this volume). Just over 200 Acheulean sites have been documented from this part of the Bhima Basin (Figure 8). The basin is currently drained by numerous small rivers, which combine at various points on the valley floor to give rise to the Hunsgi Nala, a feeder on the northern bank of the Krishna River. The majority of Lower Paleolithic sites are situated between the channel banks, pediments and foothill regions. Although the current distribution of Paleolithic sites has been shown with respect to a modern drainage network (Figure 9), the tectonic framework of the basin and the prevalent monsoon climate favored the formation of springs (i.e., tufa, travertine deposits). The valley floor has preserved extensive vertisol deposits, which generally form under low energy fluvial and ponded environments, suggesting a fresh look into geomorphological and environmental processes operating in the region. The geological junctions (between basalts, Puranas and the Archaeans) and the associated unconformities and faults explain the factors and the processes governing the habitability of the Bhima Basin.

The observed distribution of the Paleolithic sites in Karnataka indicates that among the Coastal, Upland and Maidan ecosystems, the latter was most preferred. Within the Maidan there exist different geological basins, the greenstone belts, the granite batholiths, the Purana Basins and the Deccan Volcanic Province. Though these sub-regions possess a uniform thorn and a scrubland and dry deciduous ecosystem, the Bhima and Kaladgi Basins were the most favored habitats

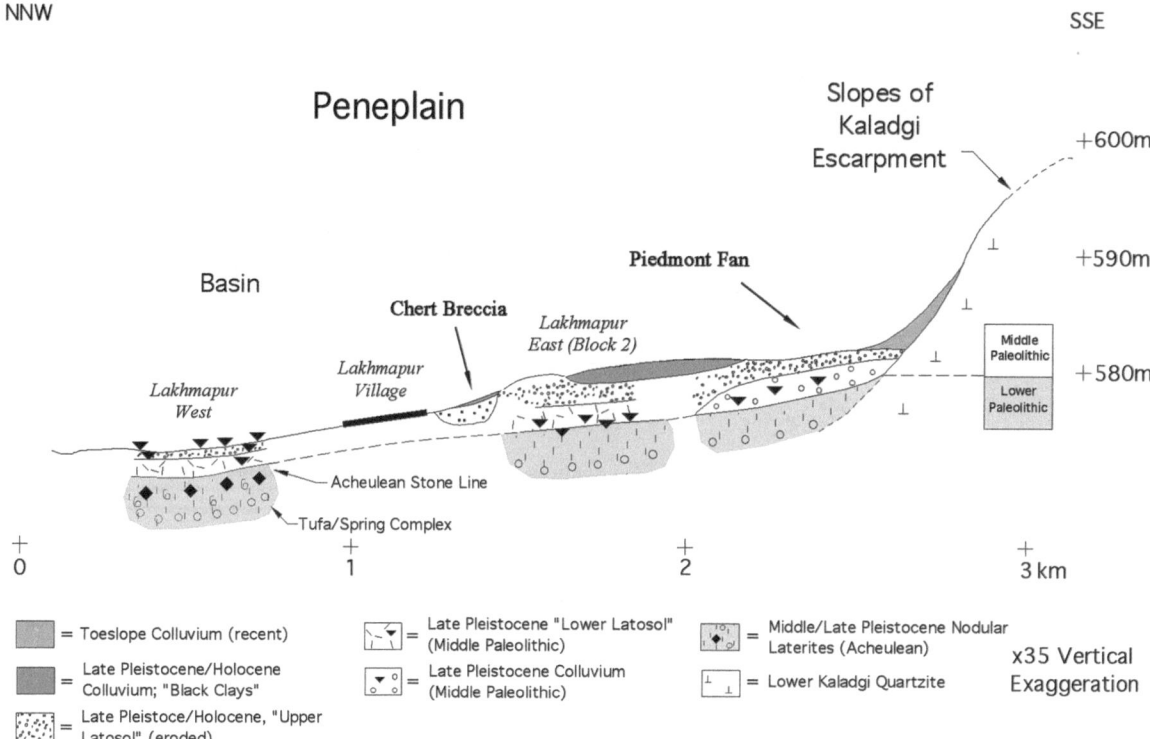

Figure 7. Landscape relations and stratigraphy at Lakhmapur. Box to right of section illustrates separate stratigraphic contexts of Acheulean and Middle Paleolithic deposits. The Acheulean assemblages are associated with the nodular laterite, which is covered with calcrete tufa (extinct spring) and chert breccia (modified from Petraglia et al., 2003)

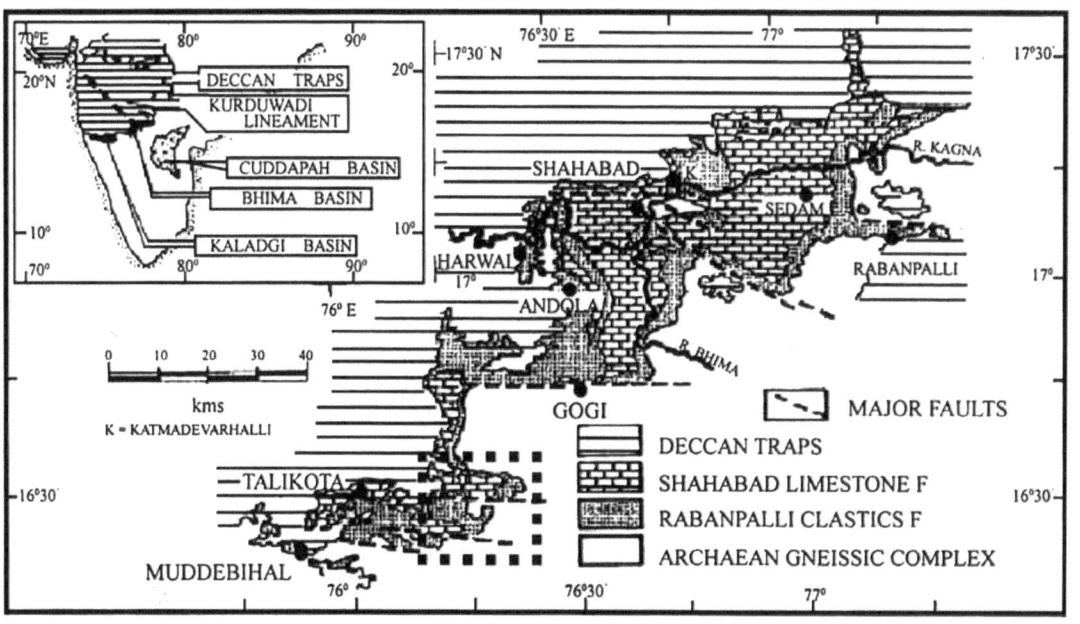

Figure 8. Map of the Bhima Basin with its geological formations. This is the smallest among the seven major Purana Basins (after Kale and Peshwa, 1995). The box demarcates the investigated area that has yielded nearly 300 Paleolithic sites

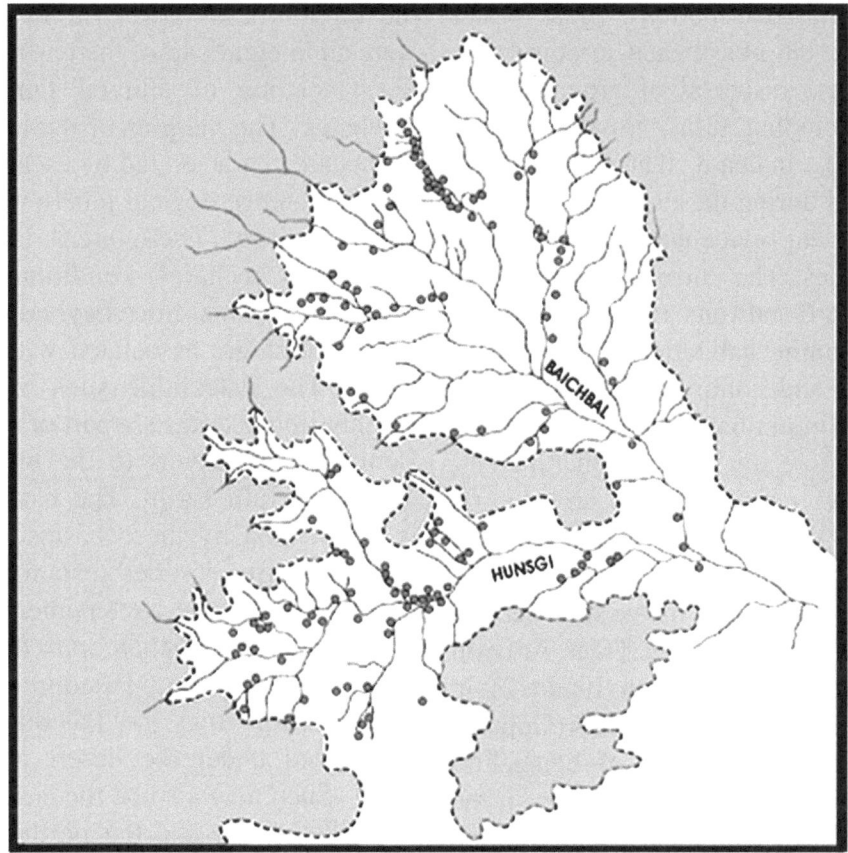

Figure 9. Distribution of Acheulean sites in relation to the modern drainage in the Hunsgi-Baichbal valleys. The dense network of low order ephemeral streams appears to date from the Holocene. Such streams are undergoing periodic reorganization under the prevailing seasonal monsoon regime. The basin morphology and its geological structure favored spring activity under Pleistocene high water tables. The Hunsgi-Baichbal valleys would have been covered with a high biomass of plant and animal resources with spring-fed pools and ponds in topographic lows across the valley floor (after Petraglia, 1998)

(Figures 6, 8). As many as 300 Lower Paleolithic sites, 150 Middle Paleolithic sites and about 50 Upper Paleolithic sites have been documented from the three regions. Among the known Paleolithic sites, more than 80% are in the Bhima and Kaladgi Basins of the northern Maidan, suggestive of a preference for these geomorphic environments to the surrounding Archaean-greenstone terrain. This distribution pattern helps us identify the Purana Basins as the core area of Paleolithic occupation. The availability of a variety of suitable rocks and minerals and suitable geomorphic environments attain significance in this context. The Vindhyan

Paleolithic record is the most impressive and continuous and the sites have been documented from a variety of geomorphic contexts. More than 90 Lower Paleolithic occurrences have been identified in the Raisen district (Jacobson, 1985). Though not identified during archaeological survey, it is probable that there were seasonal springs in the basin during the Pleistocene.

The Case for the Basin Model

The basin model has been developed on the basis of the geotectonic framework and considers associated ground water resources and raw material distributions. Slack water deposits in the intra-basin and circum-basin areas, and dense dispersal of ground water carbonates (including tufas, travertines and artesian wells), indicate that fresh water systems existed during the Pleistocene. These features are clear evidence of Pleistocene high-water tables. The current model argues that the Purana-Gondwana Basins were core areas for hominin habitation, as reflected in the density and continuity of Paleolithic occupations. Circum-basin areas are transitional between core areas and peripheral areas, and are to be considered as an integral part of zones that facilitated movement of hominins under certain ecological conditions. This situation is present in the western sectors of the Aravallis and in the Trans-Aravalli Vindhyas (Marwar) in Rajasthan (Figure 2). In the region southwest of the Aravallis (Gujarat) is the central basaltic massif of Saurashtra and the known littoral Paleolithic sites in the Hiran, Bhadar, and Kalubhar valleys. Such sites are in the marginal areas of the massif and few sites are present on the massif itself. The Thar is not a core area despite its significant geographic position. The region experienced extreme dry and cold conditions during the Late Pleistocene, thereby limiting Upper Paleolithic occupation.

Acheulean sites in the Son, Belan and Narmada valleys are within the contiguous Purana-Gondwana Basins of the Vindhyachal. While the circum-Vindhyan escarpments served as watersheds for streams that gave rise to alluvial/colluvial fans, the tectonic structure of the basin gave rise to antiformal valleys in the intra-basin areas. Thus, depositional basins occurred both within and on the circum-basin area of the Vindhyas. These basins have experienced relative tectonic stability during the post-Proterozoic. While the Deccan Volcanic Province later covered some of them, some remained outside the influence of Deccan volcanism. The Basin Model does not exclude the circum-basin areas; in fact these areas were an integral part of the basins that favored the formation of alluvial fans and spring discharge. The margins of the majority of the basins are characterized by escarpments/scarp surfaces and geological junctions which facilitated aquifers. These areas favored higher monsoonal recharge conditions relative to inland sub-basins. Sites beyond the Vindhyas in Rajasthan are associated with the Marwar Basin. The Paleolithic sites in Nagaur and neighboring districts are part of this basin. The Luni system belongs to the western margin of the Aravalli Basin. The basaltic province is surrounded by these basins and provided adequate surface water resources during the Middle Pleistocene wetter interglacial conditions. Beyond Rajasthan up to the Rohri Hills there is no record of Paleolithic occupation. It is possible that the Paleolithic evidence lay buried under the desert and the Indus sands. Such may also be the case between the Siwalik terraces and the northern margin of the Vindhyas. However, the increasing depth of the alluvium toward the orogen renders it difficult to determine whether or not this is the case.

The Paleolithic sites on the Subhimalayan terraces are post-Boulder Conglomerate in age. The older terraces of the Siwaliks, either in the Potwar, Kangra or Nepal, are younger than 500 ka. Between the Siwaliks and the northern edge of the peninsular bulge (the northern margins of the Vindhyas) is the Indo-Ganga Plain that is sterile of Paleolithic evidence. On the margins and within these Proterozoic Basins smaller Gondwana Basins are situated. Some of the sites in Bihar are part of the circum-basin ecosystems of the Purana-Gondwana Basins.

Conclusions

Between the western frontiers in Pakistan and the Aravallis in India, it appears as though Acheulean occupation took place in scattered geological basins, owing perhaps to burial of Middle Pleistocene landscapes. Disruption of drainage networks during the later Pleistocene aeolian geomorphic environments caused inhospitable conditions for hominin occupation. The early occupations in the Upper and Lower Sindh represent milestones along hominin dispersal routes from where they ranged northwards into the Potwar Subhimalaya, eventually moving along into the *dun* valleys of Punjab, Himachal and Nepal during late Middle Pleistocene. In Rajasthan, Paleolithic sites along the 27° latitude belt indicate a dispersal route into the Aravallis and beyond, including the Paleolithic sequence around Budha Pushkar (Allchin and Allchin, 1997). The Luni Valley could not have been a corridor owing to the extension of an arm of the sea into the Great Rann, the region of the Luni's mouth. This was a major barrier along the coastal route from the Makran Coast of Pakistan. Unlike the shallow burial of Pleistocene landscapes in Rajasthan, the Ganga and Punjab Plains are made of a deep wedge of alluvium, going to a depth of 5 km toward the Siwalik Frontal Range to tens of meters toward the Vindhya-Aravalli ridge. The Aravallis separate the Punjab and Ganga Plains. The Sindh Plain lies in the distal area of the Himalayan orogen, whereas the Punjab and Ganga Plains lie in the proximal region that experienced intense geodynamism. Though there is a good record of neotectonic activity in the Sindh-Kachchh province, this did not affect the habitability of the region.

The Siwalik *duns* constitute inter-montane valleys. During the course of the uplift and formation of Siwalik Frontal Range, broad intermontane valleys (grabens) were formed, such as Dehradun (Uttaranchal) and the Dang and Deokhuri Duns (Nepal). While the *duns* in the Punjab have been less productive in terms of Acheulean evidence, the terrace surfaces of the Siwaliks have yielded numerous Acheulean sites. In the *duns* proper, Soanian assemblages are common (Mohapatra, 1997).

Technologically both these assemblages have been shown to be similar (Gaillard, 1995). Though more evidence is needed the *duns*, the Nepal Siwaliks have been found to be Acheulean habitats (Corvinus, 1995). The Paleolithic occupation on the Siwalik terraces is post-Boulder Conglomerate in age. The older terraces of the Siwaliks, either in the Potwar or Kangra or Nepal, are younger than 500 ka.

Evidently the plains of the Indus, Punjab, Ganga and Brahmaputra remained unoccupied, owing largely to ecological deterrents. While the Indus Plain is interrupted by rocky uplands, for instance the Rohri Hills, the plains of Punjab and Ganga have buried the low-lying Aravalli and Vindhyan outcrops. The Brahmaputra Valley of northeast India constitutes a humid tropical environment and appears to have been densely vegetated through much of the Pleistocene. But for the Paleolithic evidence from Sindh (Rohri Hills) and the Potwar, the remainder of the foreland plains remained unoccupied. Pleistocene occupation towards the Siwalik foothills is as yet not evident, whereas towards the southern margin (along the Yamuna Plain) the Middle Paleolithic site of Kalpi-on-Yamuna dates back to 40–30 ka (Tewari et al., 2002). Most parts of the Punjab and Ganga Plains, whether or not densely vegetated, were unoccupied by hominins who preferred upland Siwalik terraces and the *duns* to the dynamic fluvial plains that were devoid of raw material resources (see Dennell, this volume). In the Ganga-Yamuna Plain, the evidence of Paleolithic occupation is associated with the Aravalli-Delhi ridges. Although the ethnographic present is not an exact parallel of Paleolithic adaptations, it is possible to suggest that hunter-gatherers preferred rocky valleys with a monsoonal, deciduous-savanna ecosystem. In view of Early and Middle Pleistocene geodynamic conditions in the Himalaya the foreland plains experienced intense fluvial activity leading to deep burial of mineral resources along their southern margins and restricted the hominin range of activity to the Aravalli-Vindhya-Siwalik tract, i.e., in the circum-Ganga Plain.

Entry of hominins into and across the subcontinent need to be assessed against the basin model. The Purana-Gondwana Basins were areas of habitation for Acheulean hominins as well as later populations. The basins assume importance in understanding population expansions and extinctions as the basins would have been potential zones of interaction between archaic and modern humans (James and Petraglia, 2005; Korisettar 2005). Future research in the Gondwana-Purana Basins is likely to reveal important information about hominin dispersal patterns, adaptations and settlement history.

Acknowledgments

The basin model developed over three seminar presentations, two in the UK (Cambridge and London) and one in India. It has been strengthened by the encouraging response from senior colleagues and friends in India and abroad. I appreciate stimulating discussions with Michael Petraglia, Nicole Boivin and Dorian Fuller over the years. I thank Bridget and Raymond Allchin for their long-term support. I wish to acknowledge my graduate students at Karnatak University for their assistance during the preparation of the manuscript. Tejasvi Lakundi and Balu Kattimani produced several figures included here. Gratitude is extended to Vice Chancellor Professor M. Khajapeer and the authorities of Karnatak University for academic support.

References

Acharyya, S.K., Basu, P.K., 1993. Toba ash on the Indian subcontinent and its implications for correlation of late Pleistocene alluvium. Quaternary Research 40, 10–19.

Athreya, S., 2007. Was *Homo heidelbergensis* in South Asia? A test using the Narmada fossil from Central India. In: Petraglia, M.D., Allchin, B. (Eds.), The Evolution and History of Human Populations in South Asia: Inter-disciplinary Studies in Archaeology, Biological Anthropology, Linguistics and Genetics. Springer, Netherlands, pp. 137–170.

Allchin, B., Allchin, F.R., 1997. Origins of a Civilization: the Prehistory and Early Archaeology of India. Viking Penguin Books India, New Delhi.

Badam, G.L., 2002. Quaternary vertebrate palaeontology in India: fifty years of research. In: Settar S., Korisettar, R. (Eds.), Indian Archaeology in Retrospect, Vol. III. Archaeology and Interactive Disciplines. Manohar-Indian Council of Historical Research, New Delhi, pp. 209–45.

Clark, J.D., Williams, M.A.J., 1990. Prehistoric ecology, resource strategies and culture change in the Son valley, northern Madhya Pradesh, central India. Man and Environment XV (1), 13–24.

Corvinus, G., 1995. Quaternary stratigraphy of the intermontane dune valleys of Dang-Deokhuri and associated prehistoric sites in western Nepal. In: Wadia, S., Korisettar, R., Kale, V. (Eds.), Quaternary Environments and Geoarchaeology of India. Geological Society of India, Bangalore, pp. 124–49.

Corvinus, G., 1998. The Lower Palaeolithic occupations in Nepal in relation to South Asia. In: Petraglia, M.D, Korisettar, R. (Eds.), Early Human Behaviour in Global Context: The Rise and Diversity of the Lower Palaeolithic Record. Routledge, London-New York, pp. 391–417.

Dennell, R.W., Rendell, H., Hailwood, E., 1988. Early tool making in Asia: two-million- year artefacts in Pakistan. Antiquity 62 (234), 98–106.

Dennell, R.W., 1998. Grasslands, tool making and the hominid colonization of southern Asia: a reconsideration. In: Petraglia, M.D., Korisettar, R. (Eds.), Early Human Behaviour in Global Context: The Rise and Diversity of the Lower Palaeolithic Record. Routledge, London-New York, pp. 280–303.

Dennell, R., 2007. "Resource-rich, stone poor": early hominin land use in large river systems in northern India and Pakistan. In: Petraglia, M.D., Allchin, B. (Eds.), The Evolution and History of Human Populations in South Asia: Inter-disciplinary Studies in Archaeology, Biological Anthropology, Linguistics and Genetics. Springer, Netherlands, pp. 41–68.

Dutta, P., 2002. Gondwana lithostratigraphy of peninsular India. Gondwana Research 5(2), 540–53.

Foote, R.B., 1876. The geological features of the south Mahratta country and adjacent districts. Memoirs of Geological Survey of India 12, 240–243.

Foote, R.B., 1880. Notes on the occurrence of stone implements in the coastal laterite, south of Madras, and in the high level gravels and other formations in the south Mahratta country. Geological Magazine 7, 542–546.

Foote, R.B., 1914. The Foote Collection of Indian Prehistoric Antiquities: Catalogue Raisonne. Government Museum, Madras.

Foote, R.B., 1916. The Foote Collection of Prehistoric and Protohistoric Antiquities Notes on their Ages and Distribution. Government Museum, Madras.

Fuller, D.Q., Korisettar, R., 2004. The vegetational context of early agriculture in south India. Man and Environment XXIX (1), 7–27.

Gaillard, C., 1995. An early Soan assemblage from the Siwaliks: a comparison of processing sequences between this assemblage and as Acheulian assemblage from Rajasthan. In: Wadia, S., Korisettar, R., Kale, V. (Eds.) Quaternary Environments and Geoarchaeology of India. Geological Society of India, Bangalore, pp. 231–245.

Jacobson, J., 1985. Acheulian surface sites in central India. In: Misra, V.N., Bellwood, P. (Eds.), Recent Advances in Indo-Pacific Prehistory. Oxford-IBH, New Delhi, pp. 49–57.

James, H., Petraglia, M.D., 2005. Modern human origins and the evolution of behavior in the later Pleistocene record of South Asia. Current Anthropology 46, S3–S27.

Joshi, R.V., 1955. Pleistocene studies in the Malaprabha basin. Deccan College-Karnatak University, Poona-Dharwad.

Joshi, R.V., 1964. Acheulian succession in Central India. Asian Perspectives VIII (1), 150–163.

Joshi, R.V., 1978. Stone Age Cultures of Central India: Report of the Excavations of Rock-shelters at Adamgarh, Madhya Pradesh. Deccan College, Poona.

Kale, V.S., 1991. Constraints on the evolution of the Purana basins of peninsular India. Journal of Geological Society of India 38 (September), 231–252.

Kale, V., Peshwa, V.V., 1995. The Bhima Basin. The Geological Society of India, Bangalore.

Kale, V.S., Rajaguru, S.N., 1987. Late Quaternary alluvial history of northwest Deccan upland region. Nature 325, 612–614.

Kale, V.S., Gregory, K, Joshi, J., Veena, U., 2004. Progress in palaeohydrology: focus on monsoonal areas- an introduction. Journal of Geological Society of India 64 (4), 381–82.

Khatri, A. P., 1962. 'Mahadevian': an Oldowan pebble culture of India. Asian Perspectives 6, 186–197.

Korisettar, R., 1994. Quaternary alluvial stratigraphy and sedimentation in the upland Deccan region, Western India. Man and Environment XIX (1–2), 29–41.

Korisettar, R., 1995. Review of 'Man-land relationships during Palaeolithic times in the Kaladgi basin, Karnataka', by R.S. Pappu and S.G. Deo. Man and Environment XX (1), 123–125.

Korisettar, R., 2002. The archaeology of the South Asian Lower Palaeolithic: history and current status. In: Settar, S., Korisettar, R. (Eds.), Indian Archaeology in Retrospect, Vol. I. Prehistoric Archaeology of South Asia. Manohar-Indian Council of Historical Research, New Delhi, pp. 1–65.

Korisettar, R., 2004a. Purana basins of South India as Stone Age habitats: with special reference to the Kaladgi-Bhima basins, Karnatak. In: Srivastava, V.K., Singh, M.K. (Eds.), Issues and Themes in Anthropology. Palaka Prakashan, New Delhi, pp, 83–100.

Korisettar, R., 2004b. Geoarchaeology of Purana and Gondwana Basins of peninsular India: Peripheral or paramount. Presidential Address, Section of Archaeology. Sixty- Fifth Session of the Indian History Congress. Mahatma Jyotiba Phule Rohilkhand University, Bareilly, December 28–30.

Korisettar, R., 2005. Comments on modern human origins and the evolution of behavior in the later Pleistocene record of South Asia by James, H., Petraglia, M.D. Current Anthropology 46, S19–S20.

Korisettar, R., Petraglia, M.D., 1993. Explorations in the Malaprabha Valley, Karnataka. Man and Environment XVIII, 43–48.

Korisettar, R., Petraglia, M.D., 1998. The archeology of the Lower Palaeolithic: background and overview. In: Petraglia, M.D., Korisettar, R. (Eds.), Early Human Behaviour in Global Context: The Rise and Diversity of the Lower Palaeolithic Record. Routledge, London-New York, pp. 1–22.

Korisettar, R., Ramesh, R., 2002. The Indian monsoon: roots, relation and relevance. In: Settar, S., Korisettar, R. (Eds.), Indian Archaeology in Retrospect, Indian Archaeology in Retrospect, Vol. III. Archaeology and Interactive Disciplines. Manohar-Indian Council of Historical Research, New Delhi, pp. 23–60.

Korisettar, R., Gogte, V.D., Petraglia, M.D., 1993. 'Calcareous Tufa' at the site of Banasankari in the Malaprabha valley, Karnataka: revisited. Man and Environment XVIII (2), 13–21.

Meher-Homji, V.M., 2002. Bioclimatology and Quaternary palynology in India. In: Settar, S., Korisettar, R. (Eds.), Indian Archaeology in Retrospect, Vol. III. Archaeology and Interactive Disciplines. Manohar-Indian Council of Historical Research, New Delhi, pp. 61–68.

Mishra, S., 1985. Early man and environments in Western Madhya Pradesh. Ph.D. Dissertation, University of Poona.

Misra, V.D., 1997. Lower and middle palaeolithic cultures of Northern Vindhyas. In: Misra, V.D., Pal, J.N. (Eds.), Indian Prehistory 1980. Department of Ancient History, Culture and Archaeology, University of Allahabad, Allahabad, pp. 61–74.

Misra, V.D., Rana, R.S., Clark, J.D., Blumenshine, R.J., 1983. Preliminary excavations of Son river sections at Nakjhar Kurd. In: Sharma, G.R., Clark, J.D., (Eds.), Palaeoenvironments and Prehistory in the Middle Son Valley. Abinash Prakashan, Allahabad, pp. 101–15.

Misra, V.N., 1985. The Acheulian succession at Bhimbetka,Central India. In: Misra, V.N., Bellwood, P. (Eds.), Recent Advances in Indo-Pacific Prehistory. Oxford-IBH, New Delhi, pp. 35–47.

Misra, V.N., 1995. Geoarcheology of the Thar desert. In: Wadia, S., Korisettar, R., Kale, V.S. (Eds.), Quarternary Environments and Geoarchaeology of India. Geological Society of India, Bangalore, pp. 210–30.

Mohapatra, G.C., 1997. Reidentification of Acheulian elements in the Western sub-Himalayan lithic complex in the light of new discoveries. In: Misra, V.D., Pal, J.N. (Eds.), Indian Prehistory 1980. Allahabad University, Allahabad, pp. 43–50.

Murty, M.L.K., 1974. A Late Pleistocene cave site in southern India. Proceedings of American Philosophical Society 118 (2), 196–230.

Murty, M.L.K., 1979. Recent research on the Upper Palaeolithic phase in India. Journal of Field Archaeology 6, 301–320.

Nagalakshmi, T., Achyuthan, H., 2004. Radiocarbon dating and Holocene episodes of alluvial sedimentation in the Koratallaiyar-Cooum river basin, South India. Journal of Geological Society of India 64, 461–469.

Paddayya, K., 1977a. The Acheulian culture of the Hunsgi valley (Shorapur Doab), peninsular India. Proceedings of American Philosophical Society 121, 383–406.

Paddayya, K., 1977b. An Acheulian occupation site at Hunsgi, penisnular India: a summary of results of two seasons of excavation (1975–6). World Archaeology 8, 94–110.

Paddayya, K., 1982. The Acheulian Culture of the Hunsgi Valley (Peninsular India): A Settlement System Perspective. Deccan College, Pune.

Paddayya, K., 1989. The Acheulian culture localities along the Fatehpur Nullah, Baichbal valley, Karnataka (Peninsular India). In: Kenoyer, J.M. (Ed.), Old Problems and New Perspectives in the Archaeology of South Asia. Prehistory Press, Madison, pp. 21–28.

Paddayya. K., 1994. Investigation of man-environment relationship in Indian archeology: some theoretical considerations. Man and Environment XIX (1–2), 1–28.

Paddayya, K., 2007. The Acheulean of peninsular India with special reference to the Hunsgi and Baichbal Valleys of the Lower Deccan. In: Petraglia, M.D., Allchin, B. (Eds.), The Evolution and History of Human Populations in South Asia: Inter-disciplinary Studies in Archaeology, Biological Anthropology, Linguistics and Genetics. Springer, Netherlands, pp. 97–119.

Paddayya, K., Petraglia, M.D., 1997. Isampur: an Acheulian workshop site in the Hunsgi valley, Gulbarga district, Karnataka. Man and Environment XXII (2), 95–100.

Paddayya, K., Blackwell, B.A.B., Jhaldiyal, R., Petraglia, M.D., Fevrier, S., Chaderton II, D.A., Blickstein, J.I.B., Skinner, A.R., 2002. Recent findings on the Acheulian of the Hunsgi and Baichbal valleys, Karnataka, with special reference to the Isampur excavation and its dating. Current Science 83 (5), 101–107.

Pal, J.N., 2002. The Middle Palaeolithic culture of South Asia. In: Settar, S., Korisettar, R. (Eds.), Indian Archeology in Retrospect, Vol. I. Prehistory Archeology of South Asia. Manohar-Indian Council of Historical Research, New Delhi, pp. 67–83.

Pappu, R.S., 1974. Pleistocene Studies in the Upper Krishna Basin. Deccan College, Poona.

Pappu, R.S., 2001. Acheulian Culture in Peninsular India. D. K. Printworld (Pvt.) Ltd., New Delhi.

Pappu, R.S., Deo, S.G., 1994. Man-land relationships during Palaeolithic times in the Kaladgi basin, Karnataka. Deccan College, Pune.

Pappu, S., 1996. Reinvestigations of the prehistoric archaeological records in Kortallayar basin, Tamil Nadu. Man and Environment XXI (1), 1–23.

Petraglia, M.D., 2003. The Lower Palaeolithic of the Arabian peninsula: occupations, adaptations and dispersal. Journal of World Prehistory 17 (2), 141–80.

Petraglia, M.D., 2005. Hominin response to pleistocene environmental change in Arabia and South India. In: Head, M.J., Gibbard, P.L. (Eds.), Early-Middle Pleistocene Transitions: the Land-Ocean Evidence. Geological Society, London special Publication 247, London, pp. 305–19.

Petraglia, M.D., Schuldenrein, J., Korisettar, R., 2003. Landscapes, activity, and the Acheulean to Middle Palaeolithic transition in the Kaladgi basin, India. Eurasian Prehistory 1(2), 3–24.

Pilgrim, G.E., 1925. The migration of Indian mammals. Presidential Address, Section of Geology 12th Indian Science Congress.

Radhakrishna, B.P., (Ed.), 1987. Purana Basins of Peninsular India. Geological Society of India, Bangalore.

Rajaguru, S.N., 1997. Climatic changes in western India during the Quaternary: a palaeopedological approach. DST- Project Completion Report (Unpublished).

Raju, D.R., 1981. Early settlement patterns in Cuddapah district, Andhra Pradesh: a palaeoanthropological study. Ph.D. Dissertation, University of Poona.

Raju, D.R., 1988. Stone age Hunter-gatherers: an Ethnoarchaeology of Cuddapah region, South-East India. Ravish, Pune.

Raju, D.R., Venkatasubbaiah, P.C., 2002. The archeology of the Upper Palaeolithic phase in India. In: Settar, S., Korisettar, R. (Eds.), Indian Archaeology in Retrospect, Vol. I. Prehistory Archaeology of South Asia. Manohar-Indian Council of Historical Research, New Delhi, pp. 85–109.

Reddy, K.T., 1977. Billasurgam: an Upper Palaeolithic cave site in south India. Asian Perspectives 20(2), 206–227.

Rendell, H. M., Dennell, R. W., Halim, M. A., 1989. Pleistocene and Palaeolithic Investigation in the Soan valley, Northern Pakistan. In: Allchin, F.R., Allchin, B., Ashfaque, S.M. (Eds.), British Archaeological Mission to Pakistan. British Archaeological Reports International Series 544, Oxford.

Sharma, G.R., Clark, J.D., (Eds.), 1983. Palaeoenvironment and Prehistory in the Middle Son valley (Madhya Pradesh, North Central India). Abinash Prakashan, Allahabad.

Sharma, K.K., 1995. Quaternary stratigraphy and prehistory of the upper Indus valley, Ladakh. In: Wadia, S., Korisettar, R., Kale, V.S. (Eds.), Quaternary Environments and Geoarchaeology of India. Geological Society of India, Bangalore, pp. 98–108.

Sharma, T.C., 1991. Prehistoric situation in north-east India. In: Singh, J.P., Sengupta, G. (Eds.), Archaeology of North-Eastern India. Vikas Publishing House, New Delhi, pp. 41–58.

Sharma, S., 2005. Geomorphological contexts of the Stone Age record of the Ganol and Rongram valleys in Garo hills, Meghalaya. Man and Environment XXVII (2), 15–29.

Singh, I.B., 1996. Geological evolution of Ganga plain – an overview. Journal of Palaeontological Society of India 41, 99–137.

Singh, I.B., 2004. Late Quaternary history of the Ganga plain. Journal of Geological Society of India 64 (4), 431–454.

Singh, I.B., Bajpai, V. N., 1989. Significance of syndepositional tectonics in facies development, Gangetic alluvium near Kanpur, Uttar Pradesh. Journal of Geological Society of India 34, 61–66.

Singh, R., 1956. The Palaeolithic industry of Budelkhand. Ph.D. Dissertation, University of Poona.

Sinha, R., Stueben, D., Berner, Z., 2004. Palaeohydrology of the Sambhar Playa, Thar Desert, India, using geomorphological and sedimentological evidences. Journal of Geological Society of India 64 (4), 419–430.

Sonakia, A., 1985. Early homo from Narmada valley, India. In: Delson, E. (Ed.), Ancestors: The Hard Evidence. Alan Liss, Inc., New York, pp. 334–38.

Soundara Rajan, K.V., 1952. Stone Age industries near Giddalur, District Kurnool. Ancient India 8, 64–92.

Soundara Rajan, K.V., 1958. Studies in the Stone Age of Nagarjunakonda and its neighbourhood. Ancient India 14, 49–113.

Subbarao, B., 1958. The Personality of India. M.S. University of Baroda, Baroda.

Supekar, S.G., 1985. Some observations on the Quaternary stratigraphy of the central Narmada valley. In: Misra, V.N., Bellwood, P. (Eds.), Recent Advances in Indo-Pacific Prehistory. Oxford-IBH, New Delhi, pp. 19–27.

Tewari, R., Pant, P. C., Singh, I.B., Sharma, S., Sharma, M., Srivastava, P., Singhvi, A.K., mISHRA, P.K., Tobschall, H.J., 2002. Middle Palaeolithic human activity and palaeoclimate at Kalpi in Yamuna valley, Ganga plain. Man and Environment XXVII (2), 1–13.

Thimma Reddy, K., 1968. Prehistory of Cuddapah District. Ph.D. Dissertation, University of Sagar.

Turner, A., O'Regan, H.J., 2007. Afro-Eurasian mammalian fauna and early hominin dispersals. In: Petraglia, M.D., Allchin, B. (Eds.), The

Evolution and History of Human Populations in South Asia: Inter-disciplinary Studies in Archaeology, Biological Anthropology, Linguistics and Genetics. Springer, Netherlands, pp. 23–39.

Valdiya, K.S., 1999. Rising Himalaya: advent and intensification of monsoon. Current Science 76 (4), 514–524.

Varaprasada Rao, J., 1992. Prehistoric environments and archeology of the Krishna Tungabhadra Doab. Ph.D. Dissertation, University of Poona.

Vrba, E.S., 1996. Climate, heterchrony and human evolution. Journal of Anthropology and Archaeology 52 (1), 1–28.

Williams, M.A.J., Clarke, M.F., 1995. Quaternary geology and prehistoric environments in the Son and Belan valleys, north central India. In: Wadia, S., Korisettar, R., Kale, V.S. (Eds.), Quaternary Environments and Geoarchaeology of India. Geological Society of India, Bangalore, pp. 282–308.

5. The Acheulean of peninsular India with special reference to the Hungsi and Baichbal valleys of the lower Deccan

K. PADDAYYA

Department of Archaeology
Deccan College
Pune, 411 006
India
deccancollege@vsnl.com

Introduction

The Indian subcontinent occupies an important place in the study of the Lower Paleolithic of the Old World. Recent discoveries at Riwat near Peshawar in Pakistan (Dennell et al., 1988a, 1988b; Dennell, this volume) and Isampur in lower Deccan suggest that the age of the subcontinent's Lower Paleolithic stretches beyond 1.0 Ma. Intensive regional surveys and systematic excavations over the last half-century have brought to light an incredibly large number and variety of Lower Paleolithic localities from diverse geographical settings ranging from the sub-Himalayan zone to coastal tracts of the Bay of Bengal and the Arabian Sea (Sankalia, 1974; Paddayya, 1984; Misra, 1987, 1989; Petraglia, 1998, 2001; Pappu, 2001; Dennell, 2002) (Figure 1). The Indian subcontinent and its Lower Paleolithic record have found a place in recent reviews devoted to hominin origins and dispersals in the Old World (Bar-Yosef, 1998; Clark, 1998; Dennell, 1998, 2003; Petraglia, 2005).

The Indian Lower Paleolithic has two different lithic tool traditions. The Soanian, the lesser known among the two, is characterized by unifacial and bifacial pebble tools (i.e., choppers, chopping tools). Lithic assemblages representing this tradition are known from the sub-Himalayan zone – the well-known Soan culture sequence worked out by De Terra and Paterson (1939; Paterson and Drummond, 1962). Pebble-tool assemblages have also been reported from the Siwaliks (Sharma, 1977; Verma, 1991) and from a few localities in peninsular India (Khatri, 1962; Ansari, 1970; Armand, 1983; Rajendran, 1989). But one must admit that all these lithic assemblages are derived from secondary contexts and are devoid of chronometric dates. Serious doubts have also been expressed about the Soan culture-sequence put forward by De Terra and Paterson (Dennell and Rendell, 1991).

M.D. Petraglia and B. Allchin (eds.), The Evolution and History of Human Populations in South Asia, 97–119.
© 2007 *Springer.*

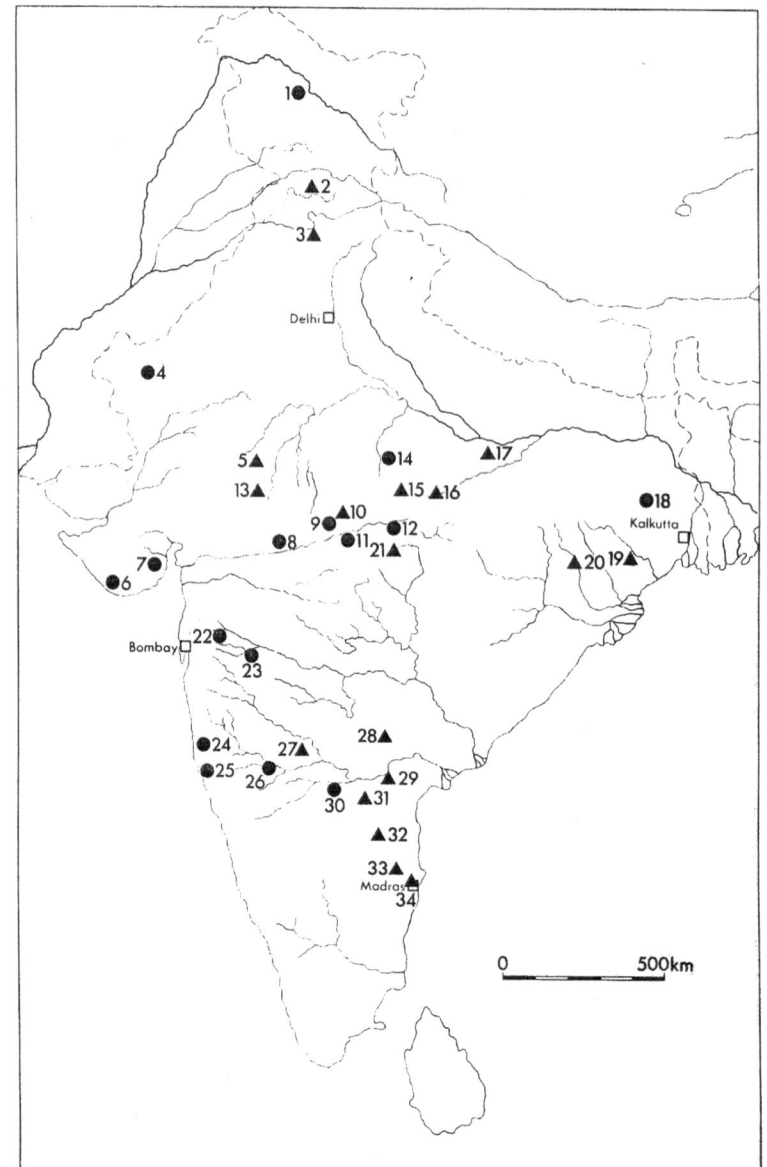

Figure 1. Important Lower Paleolithic sites (●) and site complexes (▲) of India. 1 Pahlgam;
2 Beas-Banganga complex; 3 Sirsa-Ghaggar complex; 4 Jaisalmer-Pokaran Road; 5 Berach complex;
6 Umrethi; 7 Samadhiala; 8 Durkadi; 9 Bhimbetka; 10 Raisen complex; 11 Adamgarh; 12 Mahadeo
Piparia; 13 Chambal complex; 14 Lalitpur; 15 Damoh complex; 16 Son complex; 17 Belan complex;
18 Sisunia; 19 Singhbhum complex; 20 Brahmani complex; 21 Wainganga complex; 22 Gangapur;
23 Nevasa-Chirki; 24 Mahad; 25 Malwan; 26 Anagwadi; 27 Hunsgi-Baichbal basin complex;
28 Nalgonda complex; 29 Nagarjunakonda complex; 30 Nittur; 31 Kurnool complex; 32 Cuddapah
complex; 33 Rallakalava complex; 34 Kortallayar complex. After Paddayya, 1984

This chapter is concerned with the more widely known Acheulean biface tool-tradition of peninsular India. After an overview of the current knowledge of the Acheulean in this part of the subcontinent, I provide an account of my quarter-century-long research on the Acheulean of the Hunsgi and Baichbal valleys in the lower Deccan region, pointing out how theoretical and methodological perspectives were revised during the course of this project. I will place particular emphasis on more recent excavations at the quarry and occupation site of Isampur. I point out the need to make the Acheulean studies in peninsular India more comprehensive, emphasizing the need to examine regional variability in Lower Paleolithic adaptations and their cognitive and social dimensions.

Prehistoric research commenced in peninsular India almost simultaneously with the epoch-making discoveries of Boucher de Perthes in the Somme valley of northern France. Robert Bruce Foote of the Geological Survey of India found Paleolithic implements at Pallavaram near Madras (now Chennai) in May 1863 and, more importantly, assiduously followed up this finding in the next quarter-century with the discovery of a large number of Paleolithic sites scattered over many parts of India (Foote, 1916). The next major advance was the four-tier cultural sequence (Series I to IV) put forward by Cammiade and Burkitt (1930) for the southeast coast on the basis of the stratigraphical record preserved in river sections. The investigators postulated a sequence of wet and dry climatic episodes and inferred that Early Man preferred dry climatic phases when the landscape was covered with open vegetation. Series I to IV are broadly equivalent to what we now call the Lower-, Middle- and Upper-Paleolithic and Mesolithic periods. Thus, the basic framework of Indian prehistory was established in the 1930s.

F.E. Zeuner's (1950) investigations in 1949 in the Sabarmati, Narmada, Godavari and Krishna valleys, and H.D. Sankalia's (1956) work in the Deccan further confirmed the stratigraphic and climatic cultural sequence proposed by Cammiade and Burkitt. This framework was adopted in the next 20 years by Sankalia himself and many other workers in their respective regional studies in peninsular India. These studies greatly expanded the regional distribution of sites, and clarified the typo-technological features of the three Paleolithic cultural stages and their stratigraphical positions in alluvial sequences. However, with the advent and adoption of the New Archaeology in the 1970s, flaws became apparent in this scheme, as awareness increased that artifacts recovered from secondary contexts like alluvial deposits could not serve as data on which to base behavioral interpretations. It was also realized that climatic interpretations based on river deposits were untenable (see Korisettar and Rajaguru, 1998).

Against this background, a new approach was felt to be necessary in Indian Stone Age research (Paddayya, 1978). The chief components of such an approach were: a) region-based intensive field surveys; b) emphasis on in situ sites, carefully excavated with piece plotting; c) use of both biological and earth sciences in paleoenvironmental reconstruction; and, d) use of experimental, ethnographic and other comparative approaches for interpreting the Paleolithic record. In an accompanying paradigm shift, culture was redefined as a systemic whole and as a means of human adaptation, rather than a mere vehicle for classifactory and descriptive definitions, as used in earlier schemes.

In the ensuing quarter-century, projects employing some or all of these strategies were carried out in open air and rockshelter sites in the Siwaliks, central India covering the Narmada, the Belan and Son valleys, and in the arid tract of Rajasthan, Saurashtra, the Deccan uplands, South India and Chota Nagpur (e.g., Singh, 1965; Murty, 1966; Pappu, 1974; Chakrabarti, 1977; Jacobson, 1975, 1985; Corvinus, 1981, 1983; Misra, 1978, 1995; Sharma and Clark, 1983; Rendell et al., 1989; Pant and Jayaswal, 1991; Williams and Clarke, 1995). In recent years, some important sites in the Kaladgi, Cuddapah and Kortallayar basins of South India have been investigated by the same standard (Pappu and Deo, 1994; Raju 1988; Petraglia et al., 2003; Pappu et al., 2004). A number of absolute dates obtained by the use of uranium-series, potassium-argon and thermoluminescence methods are now available for the Acheulean of peninsular India (Mishra, 1982, 1995), bracketing sites to between ca. 600 to 200 ka.

My own project on the Acheulean of the Hunsgi and Baichbal valleys of the lower Deccan was a response to two kinds of doubt prevailing in the 1970s among prehistorians in India. Back then we were weighed down by what I call the 'Olduvai complex', the notion that good prehistory could be done in India only when sites of the age and order of preservation of Olduvai Gorge were found. Secondly, some of us were haunted by the acerbic comments of archaeologists like Sir Mortimer Wheeler who, having become impatient with the endless efforts to catalogue and classify stone artifacts, had dubbed our whole endeavor as nothing more than a 'concern with a multitude of stones'. Thus in *Early India and Pakistan* (1960:34, 63), Wheeler headed two chapters dealing with the Lower Paleolithic and later cultures as simply 'Stones' and 'More Stones'! The challenge of these mocking titles was, could one erase this impression of Stone Age studies in India by realigning theory and method, as to draw from stone artifacts some meanings in terms of settlement patterns, food economy, social organization and other aspects of hominin behavior? An attempt was deemed necessary.

Hunsgi and Baichbal Valleys

Area and Cultural Sequence

The Hunsgi and Baichbal valleys are part of the Shorapur Doab (lying between the parallels of 76^0 18' and 77^0 18' east longitude and between the parallels of 16^0 15' and 17^0 12' north latitude) and constitute a single erosional basin of Tertiary age. The Doab has a long tradition of archeological investigations. Meadows Taylor was the first to publish three well-illustrated articles on the Iron Age burial sites in the nineteenth

century, and these were reprinted in 1941 by G. Yazdani as *Megalithic Tombs and Other Ancient Remains in the Deccan*. Robert Bruce Foote found a few Paleolithic implements at a place called Yedihalli in the Baichbal valley (Foote, 1876:247, 1916:122). In the 1930s, Mahadevan (1941) and Mukherjee (1941) found many other archeological sites comprising Neolithic ashmounds and Iron Age burials during their geological surveys in the Doab. After initial studies dealing with Neolithic sites (Paddayya, 1973), my own Lower Paleolithic research started in 1970 with the chance finding of bifacial tools of limestone at Gulbal in the Hunsgi valley.

The Hunsgi and Baichbal valleys are separated from each other by a narrow, remnant plateau strip of shale and limestone. This twin basin covers an area of some $500 \, km^2$ and is surrounded by shale and limestone tablelands and low hills of Dharwar schist and granite which extend for another $500 \, km^2$ and beyond. The basin floor slopes gently from west to east (i.e., from 480 to 420 m AMSL), while the surrounding uplands and hills are 20 to 60 m higher.

The Hunsgi valley forms the headweater of a minor left bank tributary of the Krishna named the Hunsgi stream. The drainage of the valley is palmate and consists of many minor nullahs which originate in the surrounding hills and plateaus and flow for distances up to 15 km. Four of these nullahs, each with a network of first, second and third order nullahs, join together at the village of Hunsgi. Likewise, the four nullahs draining the Baichbal valley unite at the village of Baichbal to form the Baichbal stream which in turn joins the Hunsgi stream after completing a course of 12 km. The Hunsgi stream itself flows into the Krishna after a course of 15 km.

The average annual rainfall of 50 to 60 cm is received wholly from the southwest monsoon and is thus markedly seasonal, with the wet season spanning between July and November-December. Seasonality notwithstanding, perennial water pools exist at bends and meanders of nullah beds. It is equally important to note that both basins have seep springs at several places. These emanate from the junction of sedimentary rocks and underlying Archaean formations. The springs at Isampur and Wajal feed the Hunsgi nullah with a shallow but perennial flow. Likewise, the springs at Mudnur in the Baichbal basin contribute a steady flow to the local nullah. Extensive travertine deposit occurs at Devapur, Kaldevanahalli and Mudnur, clearly showing that local spring activity must antedate hominin occupation.

Today the area is dominated by thorn scrub of the *Hardwickia-Anogeissus* series. Surveys of natural plant and animal life of the area yielded about 55 types of wild plant foods (fruits, seeds, berries, gums, leafy greens) and some 30 species of wild fauna comprising mammals, birds, fishes, and insects (Paddayya, 1982:63–81). Until 1985 the two valleys were part of a large dry farming belt in the Deccan with *jowar, bajra* and groundnut as the major crops. Irrigation then opened the area to cash crops such as sunflower, chillies, and cotton. In more recent years paddy cultivation has been introduced on an extensive scale, leading to large scale destruction of archeological sites.

The basin's enclosed, amphitheatre-like form, its gently undulating floor, the low surrounding hills and tablelands, the availability of perennial seep springs, the occurrence of various suitable rocks for tool-making, and the availability of various plant and animal foods all combined to promote long term, continuous human occupation. My surveys and excavations have brought to light over 300 archeological sites ranging from the Lower Paleolithic to the Iron Age. Table 1 shows the sequence of archeological cultures of the basin in relation to sedimentary stratigraphy and their dating.

Acheulean

Surveys undertaken since 1974 led to the discovery of over 200 Acheulean localities in the area, making it one of the densest concentrations of Lower Paleolithic sites in the Old World (Figure 2) (Paddayya 2001). In age, these sites range from 0.2 to 0.35 Ma (Szabo et al., 1990). One of their unique features is the use of limestone as the major raw material for tool making, although other rocks such as granite, dolerite, basalt

Table 1. Archaeological culture sequence in relation to sedimentary stratigraphy and age estimates

Culture	Stratigraphic Context	Age Estimate
Iron Age Megalithic	Post-black soil	Early half of first millennium B.C.
Neolithic	Post-black soil	2500–1000 B.C.
Mesolithic	Top surface or upper portion of black silt	8000–2500 B.C.
Upper Paleolithic	Lower portion of black silt	30–10 ka
Middle Paleolithic	Brown silt (upper part)	70–30 ka
Developed Acheulean	Brown silt (lower part)	200–70 ka
Early Acheulean	As a discrete level on weathered bedrock/travertine/kankar conglomerate/fluviatile deposit	1.2 Ma – 200 ka

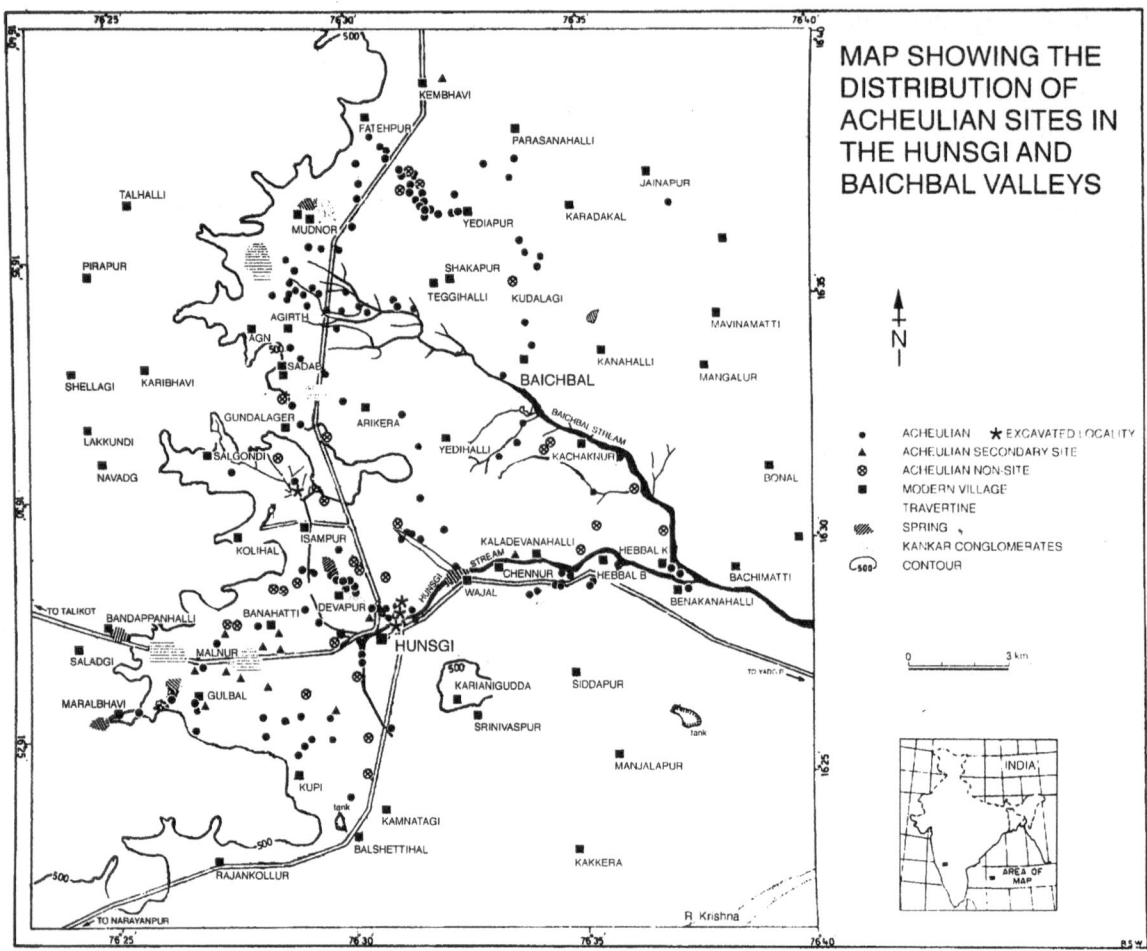

Figure 2. Map of Hungsi and Baichbal valleys showing the distribution pattern of Acheulean
localities. After Paddayya, 2001

and chert were also occasionally used (see Paddayya and Petraglia, 1993). The major artifact types include handaxes and cleavers of various forms, knives, scrapers, polyhedrons, and cores and flakes of different types (Figure 3). Some of the localities also yielded small amounts of fossil fauna comprising wild cattle, horse, elephant and deer (Paddayya, 1985). Sites yielding discrete clusters of artifacts were called 'primary occurrences' because they were thought to contain cultural material in relatively undisturbed condition, useful for the reconstruction of Stone Age behavioral patterns.

At many sites, stone artifacts were found on or close to present surfaces, raising doubts about their stratigraphical context. To test for contextual integrity, excavations were conducted at Hunsgi localities V and VI in the Hunsgi valley, and at Yediyapur locality VI in the Baichbal valley. Here well-defined levels yielding raw material blocks, finished tools and debitage were found in brown/black silts (15–50 cm below current ground surface). In more recent excavations at Isampur, the Acheulean level was found below brown/black silts, at a depth of 1.5 m below surface. This suggests that the shallow depths of Hunsgi V, VI and Yediyapur VI, and the surface condition of many other sites in the region resulted from recent loss of overlying sediments due to sheet wash and rill erosion, promoted by wood-cutting and ploughing (Paddayya and Jhaldiyal, 2001).

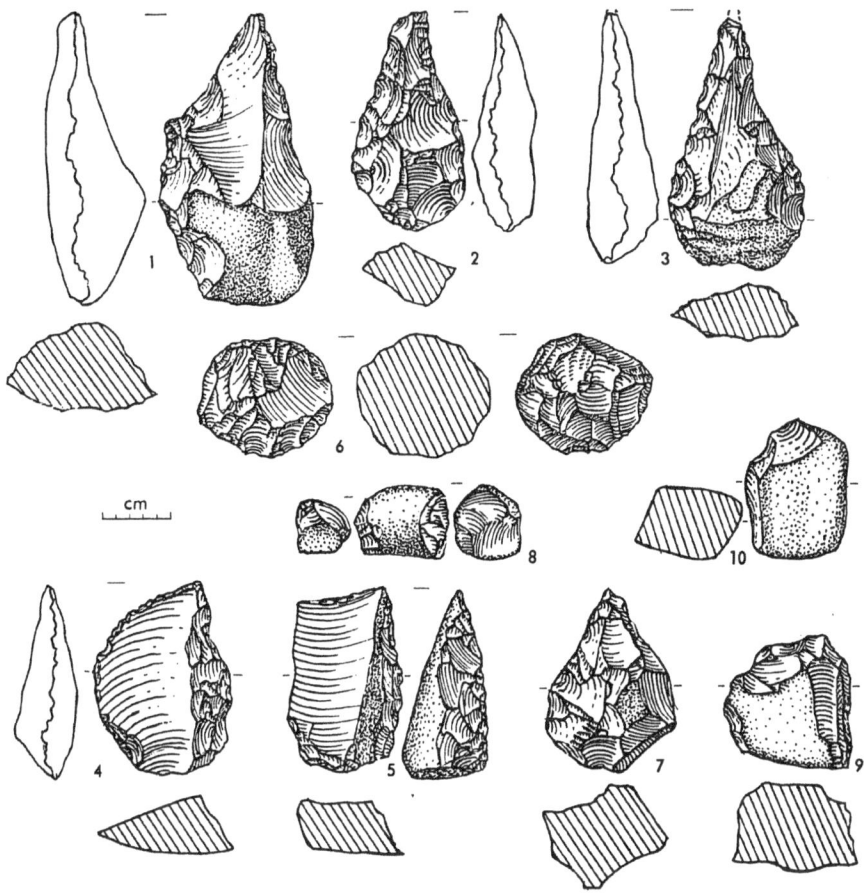

Figure 3. Important tool types found in the Acheulean of the Hunsgi and Baichbal valleys. 1–3
Handaxes; 4 Knife; 5 Cleaver; 6 Polyhedron; 7 Pick; 8 Hammestone; 9 Anvil stone; 10 Chopper.
After Paddayya, 1984

Based on artifact numbers, three arbitrary categories were recognized among the Acheulean sites: small, medium and large. In addition, spots with a few artifacts were called non-sites. Two major site concentrations (each made up of 15 to 20 sites and spread over distances of one to two kilometers) were recognized – one along the Hunsgi stream (Figure 4) and the second along the Fatehpur nullah in Baichbal valley. The remaining sites were found in a dispersed condition all over the basin floor. The late Glynn Isaac (1981) observed a similar differential distribution of Lower Paleolithic sites in East Africa and called it the 'scatter between the patches'.

Using a settlement system perspective, I then pooled the data on site size and relation of sites to topographic features, water and rock sources, and other local resource availability within a site catchment, as inferred from modern ethnic wild plant and animal exploitation. From these I derived a potentially testable model of an Acheulean settlement system for this region that hinges on two resource management strategies: a) dry season aggregation for large game hunting that resulted in the concentrations of archaeological sites near perennial water pools in the Hunsgi stream and Fatehpur nullah; and b) wet season dispersal to acquire small fauna and plant foods at random points across the basin resulting in the widely scattered small occurrences and non-sites (Paddayya, 1982). This is an admittedly coarse-grained

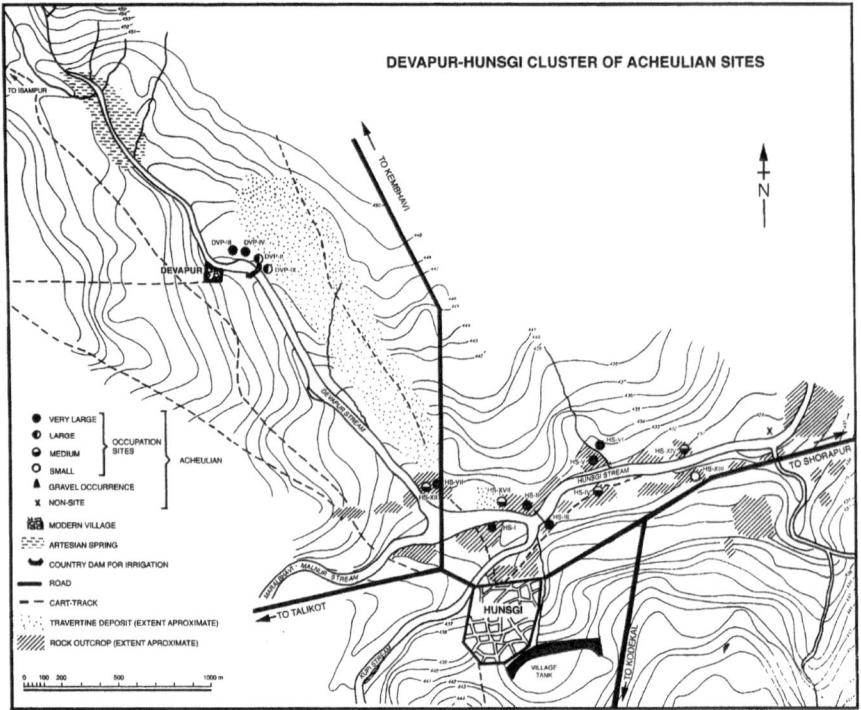

Figure 4. Cluster of Acheulean localities along the Hunsgi stream. After Paddayya, 2001

model, and its testing will require good faunal preservation at both large and small sites.

Since 1987, jointly with Michael Petraglia and Richa Jhaldiyal, I undertook a more detailed examination of the sedimentary and topographic contexts of the sites, employing the then novel perspective of site formation processes (Paddayya and Petraglia, 1993, 1995; Jhaldiyal, 1997). Sites were classified according to topographic settings, sedimentary matrix, local lithology, and drainage features. The goal was to reconstruct contemporary land-use patterns. Among the potential site settings were foothill/pediment zones of shales with limestone uplands, and channel banks and slopes of the interfluves of the basin floor. In both settings sites occur on hard substrates, either weathered or unweathered. These may be calcretes, travertine or clayey silt deposits. Sites also occur on the surface of colluvial and fluvial gravels and occasionally within them. Various categories emerged from the analysis: a) fluvial occurrences;

b) colluvial occurrences; c) sheetwash occurrences; and, d) deflationary occurrences.

We also undertook studies of the stone tool technology and typology, the results of which were combined with the above analysis. From this exercise we derived several hypotheses about specific site functions. Lithic criteria were proposed for the recognition of manufacturing sites, occupation sites, food processing sites, single event activity sites and stone tool caches. The lithic analysis also led to the conclusion that there are two very distinct kinds of Acheulean assemblages, tentatively labeled "early" and "evolved". This division is based on differences in stratigraphical contexts, and tool technology and typology. It is intra-regional in scope, without reference to Acheulean sites recorded from other parts of India and elsewhere in the Old World. At one end we have bedrock assemblages from sites like Hunsgi V and VI, and Yediyapur VI, dominated by large and roughly made bifacial implements. At the other end there are sites like Mudnur X in the Baichbal valley, where the cultural assemblage

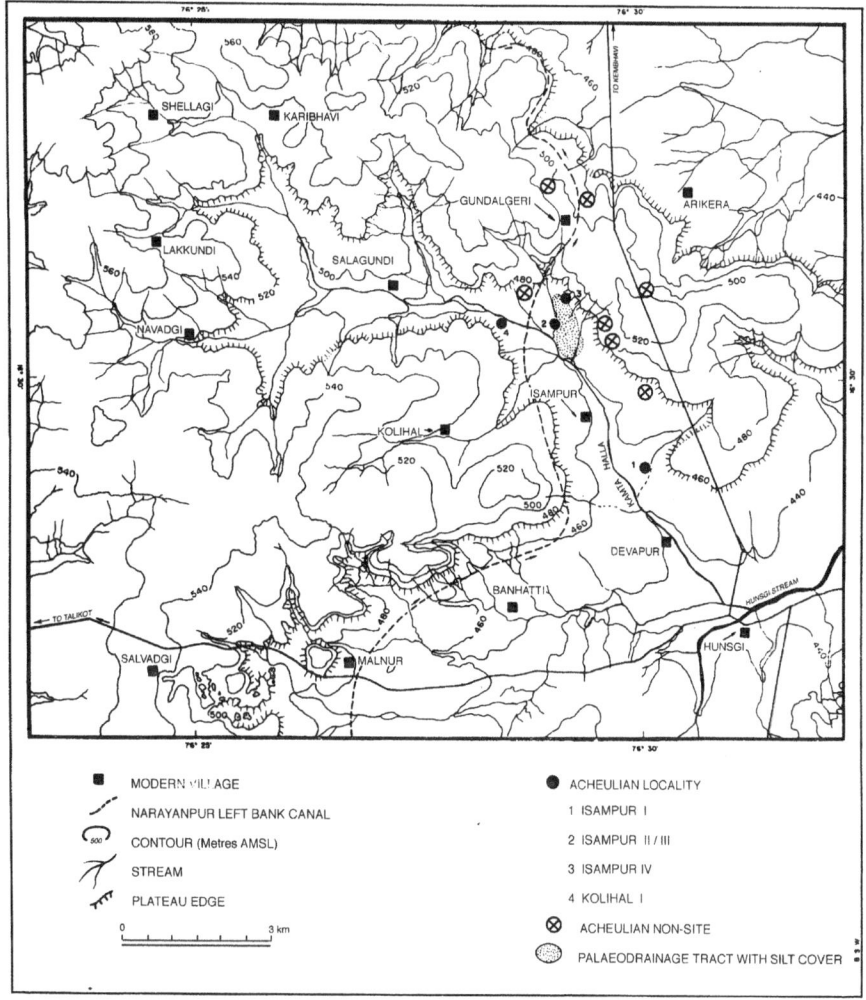

Figure 5. Map of Isampur valley showing the location of Isampur Acheulean quarry in relation to other localities and non-sites. After Paddayya, 2001

is found in soft brown silt overlying bedrock and the stone artifacts show unmistakable refinements in manufacturing techniques and also noteworthy reduction in size and thickness. Sites like Kolihal in the Hunsgi valley, which show evidence of prepared core technology, may represent an intermediate stage (Paddayya and Jhaldiyal, 1999).

Isampur Excavation As part of renewed study of the Acheulean sites of the basin, Isampur (16° 30′ N and 76° 29′ E) was selected for detailed field investigation. Isampur is located in the northwestern corner of the Hunsgi valley (Figure 5). A few artifacts were originally discovered in 1983 when the Irrigation

Department had quarried away much of the 1.5 to 2 m thick brown/black silt overlying the cultural level. Subsequent land modification activities and soil erosion led to the discovery of the quarry in 1994. The site was found to contain dense concentrations of finished tools and waste material, which consisted of limestone raw material blocks, cores, hammerstones, and flakes and chips (Figures 6, 7). Five seasons of systematic surface studies and excavation (1997–2001) and an examination of geoarchaeological features of the surrounding area revealed that this is among the very few Lower Paleolithic sites with excellent spatial integrity (Paddayya

Figure 6. Close view of artifacts exposed to surface at Isampur Acheulean quarry

Figure 7. View of Acheulean level exposed below 1.5 m thick brown/black silt in a stratigraphical
cutting at Isampur Acheulean quarry

and Petraglia, 1997, 1998; Paddayya et al., 1999, 2000, 2001).

Nine stratigraphical cuttings (6 to 8 m² in extent) and 30 trial pits (1 m²) revealed that the site was associated with a weathered outcrop of silicified limestone, typically made up of triangular, rectangular and square slabs typically measuring ca. 30 to 40 cm in plan view and 10 to 12 cm in thickness. These blocks were an attraction to Acheulean knappers, as was the closeness of the spot to the edge of a 2 to 3 m deep paleochannel which probably held perennial water. From here they could have had an excellent view of the surrounding uplands and the valley floor, and the movement of game and hominin groups on these surfaces.

Five large additional trenches (covering a total area of 159 m²) were excavated, exposing a well-preserved, 20 to 30 cm thick Acheulean level in a well-cemented matrix of brownish calcareous silt, overlain by 10 to 50 cm thick recently reworked soil overburden. All these surface studies and the excavation at the Isampur site warrant the following observations:

1) The site covers a total area of three-quarters of a hectare, divisible into four dense patches (called sub-localities I to IV) of cultural material comprising cores, debitage, finished tools, and hammerstones of local chert, basalt and quartzite. The patches are separated from each other by diffuse scatters of lithics (Figure 8).

2) The area was a major quarry-workshop geared to the exploitation of a weathered bed of silicified limestone comprising blocks of shapes and sizes suited to the needs of Acheulean knappers.

3) Small quantities of dental and bone remains of bovids and cervids, and turtle-shell fragments were found along with lithic material, suggesting that the site also may have been used for food-processing and food-consumption.

4) In Trench 1 (70 m² in extent) excavated in sub-locality I, a number of chipping clusters were present, made up of cores, large flake blanks, finished tools, hammerstones, and debitage were identified (Figures 9–12) (see Petraglia et al., 2005 for a preliminary spatial analysis of these clusters). The clusters had potential arrangements of large limestone slabs that may have served as seats for the knappers.

The Isampur excavation yielded over 15,000 artifacts. This large, relatively undisturbed sample has provided new insights into the quarrying and lithic reduction strategies adopted by Acheulean groups (see Petraglia et al., 1999, 2005 for details):

1) Suitable limestone slabs were selected and, in some cases, even pried from the outcrop. Slabs were prepared into cores by flaking off irregular projections from corners and sides.

2) Large flakes were removed from cores and transformed with a minimum of secondary chipping into knives and chopping tools.

3) Some flakes were shaped into bifacial implements (oftentimes cleavers), through more elaborate secondary flaking and chipping.

4) Some thinner limestone slabs were used for manufacturing handaxes.

In Trench 1, the Acheulean level measured about 20 cm in thickness. The bottom 10 cm thick portion of this level (40–50 cm below present ground surface) yielded 13,043 stone artifacts, comprising cores (n = 198, 1.51%), flakes (n = 301, 2.31%), shaped tools (n = 169, 1.30%), utilized/modified pieces (n = 279, 2.14%), and debitage (n = 12,096, 92.74%). Table 2 provides a break-down of the shaped tools.

The occurrence of knives as a regular type, the high proportions of scrapers and modified/utilized pieces, the overwhelming majority of debitage authenticating on-the-spot chipping, and the presence of perforators (large *zinken*-like artifacts) are some of the noteworthy features of the assemblage. The occurrence of large hollow scrapers

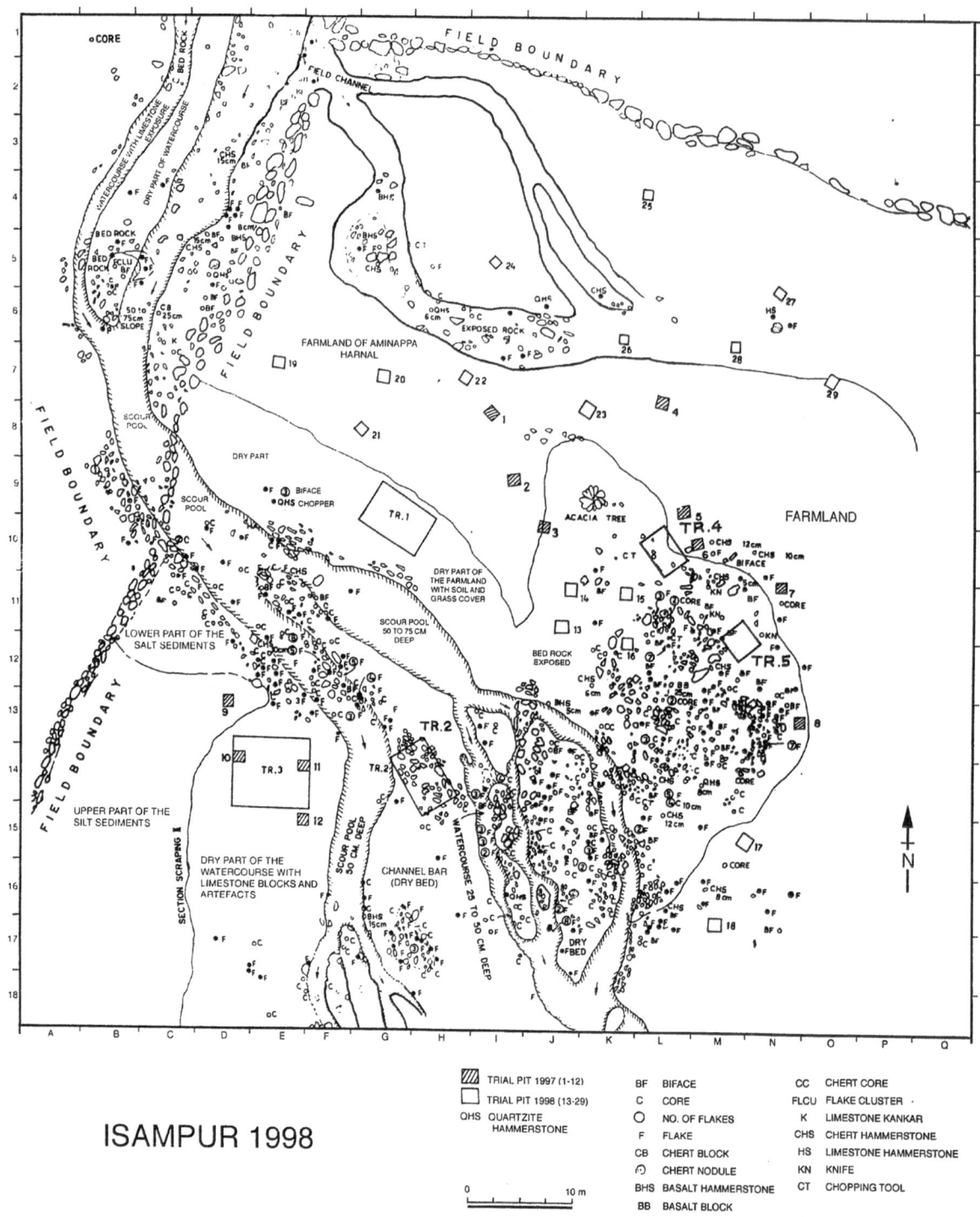

ISAMPUR 1998

Figure 8. Map of Acheulean quarry at Isampur showing surface scatter of limestone slabs and artifacts exposed due to modern quarrying of overlying silt deposit. The entire site is gridded at 5 m intervals. Note the positions of Trenches 1 to 5 and trial pits

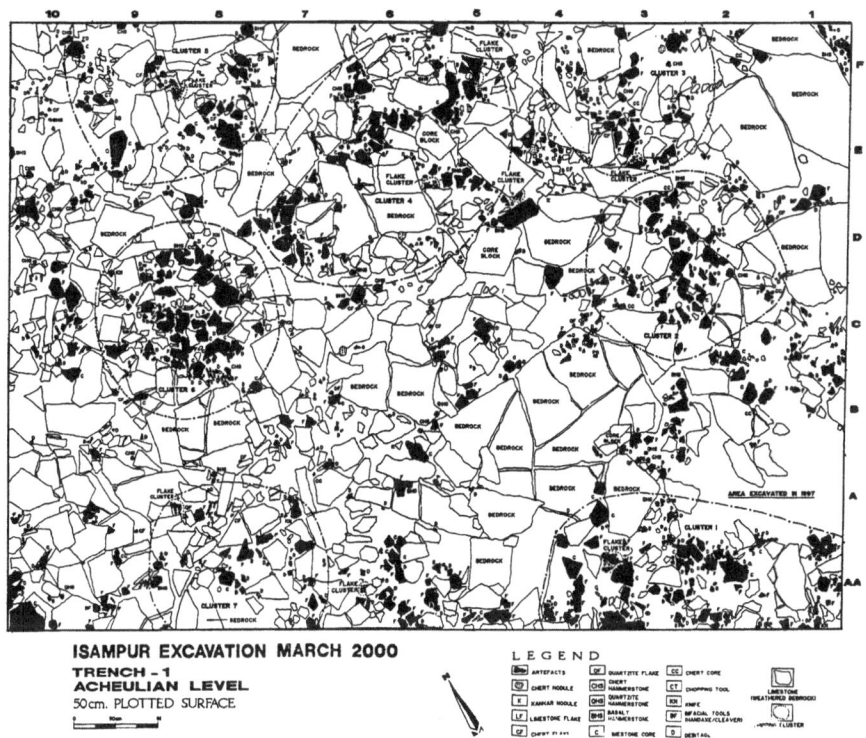

Figure 9. Plot of Acheulean surface in Trench 1 of Isampur Quarry

Figure 10. View of limestone blocks, cores, flakes, and a handaxe exposed in grid AA-6 of Trench 1 (45 cm depth) at Isampur Acheulean quarry

Figure 11. View of a large, flaked limestone core exposed in grid D-5 of Trench 1 (50 cm depth) at Isampur Acheulean quarry. The scale shows intervals of 5 cm

Figure 12. Large hammerstone of basalt in Trench 1 (50 cm depth) at Isampur Acheulean quarry

Table 2. Shaped tools from trench 1, Isampur (40–50 cm level)

Tool type	Number (n)	Per cent (%)
Handaxes	48	28.4
Cleavers	15	8.9
Knives	18	10.6
Scrapers	65	38.5
Chopping tools	14	8.2
Discoids	3	1.8
Perforators	5	3.0
Indeterminate	1	0.6

and perforators implies the use of organic materials like wood for making artifacts (Figures 13, 14).

The Isampur research permitted insights into Acheulean settlement systems. Ten sites (small sites and non-sites), yielding artifacts of limestone similar to that of Isampur, were found within a radius of 5 to 6 km from Isampur. One may infer that Isampur served as a localized hub for manufacturing and occupation activities, from where hominins radiated onto the uplands and across the valley floor as part of their daily foraging activities.

Isampur has established the long-suspected high antiquity of the Acheulean in India. Three samples of enamel extracted from bovine teeth found in excavation were measured by electron spin resonance in the Chemistry Laboratory of Williams College, Boston, U.S.A., resulting in an average age of 1.2 Ma (Blackwell et al., 2001; Paddayya et al., 2002). Isampur is thus the oldest archaeological site known thus far in India.

The prolonged research on the Acheulean of the Hunsgi and Baichbal valleys has been accompanied by a readiness to introduce changes in theory and method. In this way we elevated the Lower Paleolithic studies in the subcontinent far beyond Wheeler's much deplored "concern with a multitude of stones" and well beyond mere tool typology and description. It is now clear that the

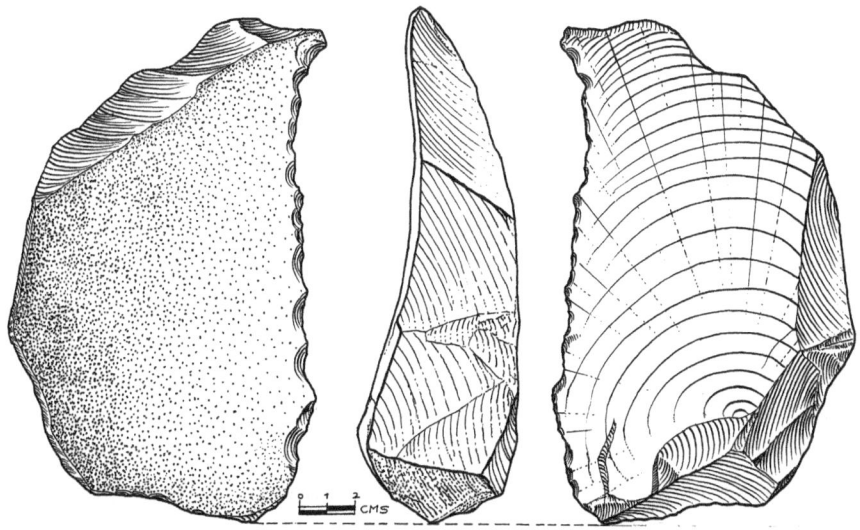

Figure 13. Concave edged scraper on a large limestone flake from grid A-1 of Trench 1 (50 cm depth) at Isampur Acheulean quarry

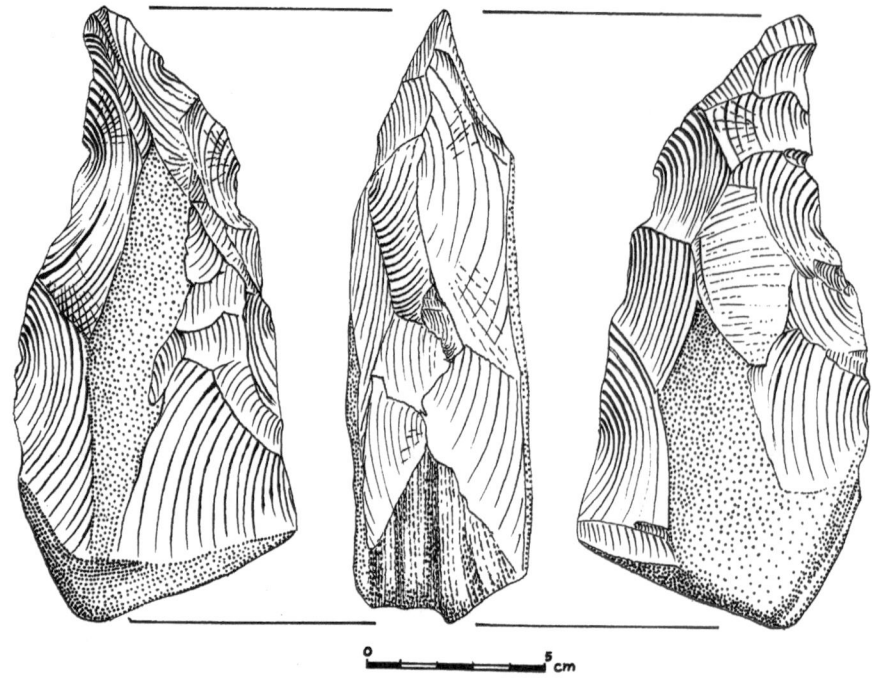

Figure 14. Large perforator of limestone with a zinken-like working end from grid AA-2 of Trench 1 (45 cm depth) at Isampur Acheulean quarry

surface context of some of the Acheulean sites in peninsular India is of recent origin, both natural and man-made. Hence surface sites, hitherto relegated to the background in Indian Stone Age research, should be examined more carefully for their histories of exposure before being dismissed as out of context. It is also clear that stone artifacts and bone fragments can provide important insights into hominin behavior (Paddayya, 2003). The Hunsgi and Baichbal valleys project has further demonstrated that Acheulean knappers already possessed evolved cognitive skills and abilities, and used them to exploit their surroundings and its resources.

Conclusion

In closing, I have a few general comments to make on the state of Paleolithic research in peninsular India. Sir Mortimer Wheeler was right – the Stone Age record still comprises stone artifacts, with only a precious few organic associations. The actors themselves

are still largely missing from the stage except for the chance finds at Hathnora on the Narmada (Sonakia, 1985; Sankhyan, 1997). We cannot wait indefinitely for more such chance discoveries. New discoveries will only come through concerted efforts to recover hominin remains in the vicinity of major excavated Paleolithic sites.

Archaeologists have gone some way towards building up a spatio-temporal framework for the Indian Stone Age, and we have some preliminary grasp of settlement patterns and adaptations. We now routinely apply techniques of high standard for recovery and absolute dating, and also employ regional surveys and associated paleoenvironmental studies. But we have only just begun to think about how to reconstruct hominin behavioral patterns because we still treat the material record as something to be described and classified, leaving the functional aspect almost untouched. We tend to think too often on a pan-Indian level and also treat both the Paleolithic cultures and their respective environmental contexts in isolation from each other. In short, the study of human cultures as adaptive systems is yet to find a secure place in Paleolithic research. Recently, Korisettar (2004; see also this volume) has drawn attention to the concentration of Paleolithic sites in the various Purana basins of peninsular India. These site-concentrations were influenced by the availability of rocks and minerals, prevalence of savannah vegetation, and presence of perennial springs and ponds. Yet, it is important to realize that the geography of peninsular India has immense variety and that this variety has a close bearing on our understanding of Stone Age adaptations. The way was already shown by Robert Bruce Foote (1916:36–48) who sought to understand the distribution of Paleolithic sites in terms of landforms, rainfall, vegetation, and raw material sources. Thus he opined that the Western Ghats, a mountainous zone of very high rainfall and thick vegetation, were probably avoided by Early Man. This hypothesis has yet to be challenged.

My concern here is whether we can go still further. I think we can and should. We should give a serious thought to examining Paleolithic "cultures" as regional adaptive systems. One would like to consider seriously the implications of the distribution of Acheulean sites in different geographical settings of the subcontinent – the Pabbi hills of the Siwalik zone of Pakistan; the dun valleys of Nepal (Corvinus, 1998); the arid zone of Rajasthan; the hilly and forested zones of central India and Chota Nagpur; the open landscape of upland Deccan; the Bhima, Kaladgi, Cuddapah and Kortallayar basins of southern India; and the coastal tracts of the Arabian Sea and the Bay of Bengal (Chakrabarti, 1977; Joshi et al., 1979; Marathe, 1995). We need to understand the finer aspects of these diverse environmental settings – their terrain features, the relative abundance of rockshelters and caves, and the nature and seasonal availability of water sources, plant foods and prey animals. Important too are the possible implications of the use of different rocks for tool making – quartzites in southern India, limestone in the Hunsgi-Baichbal basin, dolerites in the Deccan uplands, granite in Bundelkhand, and quartz in western Rajasthan. Only then can the regional peculiarities of various typo-technological systems of the Paleolithic be understood in a meaningful way. Once we start looking again at our evidence along these lines I have no doubt that some headway will be made in identifying regional adaptations.

I would further argue that we may try to rise above the level of reconstructing region-based Stone Age adaptations and even aim for the cognitive, social and aesthetic levels of hominins which have hitherto been regarded as intractable. Sharply deviating from the prevailing concerns with typological and technological aspects of lithic artifacts, some years ago Wynn (1993, 1995) sought to employ some of the tenets of Jean Piaget's well-known genetic epistemology for extracting information about the hominins' cognitive abilities, as reflected in the spatial and morphological attributes noticed on the stone implements. Wynn inferred that our hominin ancestors (particularly the authors of the Acheulean) possessed cognitive abilities closer to those of modern humans such as reversibility, whole-part relations, etc. The work on the Acheulean of the Hunsgi and Baichbal valleys offers further clues. At the general level, one must first refer to the selection by the hominins of an amphitheatre-like topographical form as the habitat and their recognition of the spatial and seasonal attributes of food, water and raw material resources. At the site specific level, Isampur offers a good example of the hominins' choice of the location on the basis of favourable factors like the presence of a raw material outcrop, closeness to a water source and a clear view of the surroundings up to a distance of two or three kilometers. Aspects like procurement of hammerstones from distances of one or two kilometers, use of hammerstones of different sizes (less than a kilogram to 2 or 3 kg in weight) for different purposes of flaking, and preference for limestone slabs of medium thickness (4 to 8 cm) for making handaxes clearly suggest that the Isampur hominins possessed the attributes of prior planning and decision-making (Petraglia et al., 1999).

Coming to the social level, I must in particular refer to the work of Clive Gamble, who in 1986, produced his synthesis of the European Paleolithic, entitled *The Palaeolithic Settlement of Europe*. It is rather startling that Gamble, after a gap of little more than a decade, has published a revised synthesis entitled, *The Palaeolithic Societies of Europe* (1999). This shift of emphasis from settlement to society is to be noted. In the more recent publication, Gamble desires to rise above what he calls the 'stomach-led and brain-dead' interpretations of the European Paleolithic and bemoans the widely held view that Paleolithic Europe was peopled by ecological creatures rather than social actors. What Gamble proposes is that, rather than restricting our studies to food economy, spatial analysis of settlements and typology, we must aim at social prehistory. He proposes that: "social life in the Palaeolithic involved hominids in the continuous and different construction of their surrounding environment" (Gamble, 1999:96).

For this purpose, Gamble proposes a conceptual scheme employing the spatial scales of locales and regions. Locales are encounters, gatherings and social occasions or places and substitute the more familiar terms like camp sites, home bases, satellite camps, etc. The local and regional scales are linked by rhythms, illustrated by the paths and tracks trodden by the hominins, the operational sequences by which they made their tools and the task-scapes where they attended to one another. Following the ideas of Leroi-Gourhan, Gamble stresses the social nature of technical acts (be it stone tool making at a particular spot, or procurement and transport of raw materials from one spot to another, or killing a wild animal). These operational sequences are seen as social productions which evolved their rhythms and forms as the material action progresses. This social approach covers all aspects of mobility, production, consumption and discard. The bottom-up approach advocated by Gamble treats cultural entities as the products of individuals repeating the technical gestures they learnt rather than the collective mind of a group producing a pattern of culture. This last aspect has been further developed in a more recent publication entitled, *The Hominid Individual in Context*, edited by Gamble and Porr (2005).

This is another facet of the new horizon towards which Paleolithic archeology in peninsular India must journey. The Acheulean record of the Hungsi and Baichbal valleys, coarse-grained though it is at present, once again offers some helpful clues about the working of the individual mind and social interactions of the hominins (Shipton, 2003; Petraglia and Paddayya, 2005; Petraglia et al., 2005). In addition to the abilities of spatial mapping of the landscape and anticipatory planning already referred to above, the Isampur quarry excavation offers useful clues about structuring of tool-making strategies. In some cases, large limestones blocks (over a metre long) were pried out from bedrock and were reduced into smaller blocks with the help of large hammerstones. This, and the accompanying actions of preparing the actual spots for tool-making (chipping clusters identified in Trench 1) and selection of suitable hammerstones, obviously required the participation of two or more individuals. Likewise, from the evidence of large limestone blocks set securely on the ground (seats for the knappers?) and occurrence of cores and hammerstones of various dimensions on the chipping clusters, it would seem that joint efforts of two or

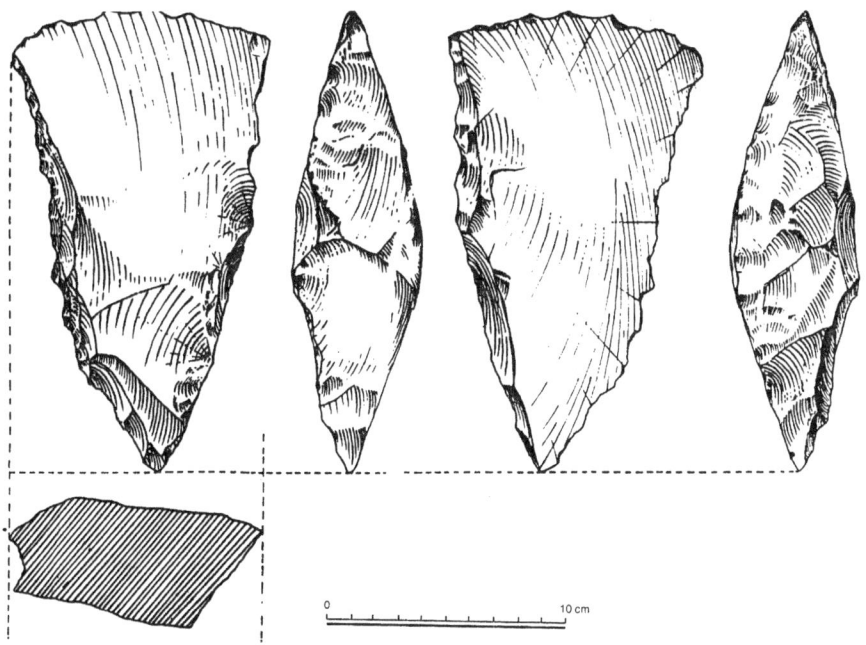

Figure 15. Limestone cleaver with a symmetric form from Trench 3 at Isampur Acheulean quarry

more individuals are required to knock off large flakes from limestone blocks and to prepare roughouts of large bifacial tools, particularly handaxes from slab pieces or large flakes. The commonness of certain tool shapes like handaxes, cleavers and knives among a large number of sites argues for the existence of a tradition of learning by mimicking and use of gestures for transmission of ideas.

Finally, one should not lose sight of the possible aesthetic dimension involved in the manufacture and use of artifacts by the hominins. There are some examples of bifaces (handaxes and cleavers) in the Acheulean record of the Hunsgi and Baichbal valleys which by virtue of being perfectly symmetrical are also artistic in character, immediately evoking one's sense of aesthetic appreciation (Figure 15). Indeed, aesthetics represents the latest theoretical turn in archeology. I would like to think that it is the interplay of these adaptive, cognitive, social and aesthetic faculties of the hominins which not only allowed them to devise a stable way of life spanning long stretches of time, but also permitted them to evolve into higher biological and cultural forms.

Acknowledgments

I gratefully acknowledge the financial support extended by the Deccan College, Pune, and University Grants Commission, New Delhi, to the Hunsgi and Baichbal valleys project. I also thank Richa Jhaldiyal and Michael Petraglia for their valuable collaboration in the second phase of the project. Mr. Devadatta D. Phule and Mr. Sunil Jadhav prepared the drawings and photographs. Mr. Sayyed Amin patiently typed the manuscript. I have benefited enormously from the critical comments received from three anonymous reviewers. Shortcomings, if any, are entirely my responsibility.

References

Ansari, Z.D., 1970. Pebble tools from Nittur (Mysore state). Indian Antiquary 4, 1–7.

Armand, J., 1983. Archaeological Excavations in Durkadi Nala: An Early Palaeolithic Pebble-Tool

Workshop in Central India. Munshiram Manoharlal, New Delhi.

Bar-Yosef, O., 1998. Early colonizations and cultural continuities in the Lower Paleolithic of western Asia. In: Petraglia, M.D., Korisettar, R. (Eds.), Early Human Behaviour in Global Context: The Rise and Diversity of the Lower Paleolithic Record. Routledge, London and New York, pp. 221–279.

Blackwell, B.A.B., Fevrier, S., Blickstein, H.B., Paddayya, K., Petraglia, M.D., Jhaldiyal, R., Skinner, A.R., 2001. ESR dating of an Acheulean quarry site at Isampur, India. Journal of Human Evolution 40 (A3) (Abstract).

Cammiade, L.A., Burkitt, M.C., 1930. Fresh light on the Stone Age of southeast India. Antiquity 4, 327–339.

Chakrabarti, S., 1977. The prehistory of Bhavnagar district. Saurashtra, Gujarat state. Ph.D. dissertation, University of Pune.

Clark, J. D., 1998. The Early Palaeolithic of the eastern region of the Old World in comparison to the West. In: Petraglia, M.D., Korisettar, R. (Eds.), Early Human Behaviour in Global Context: The Rise and Diversity of the Lower Palaeolithic Record. Routledge, London and New York, pp. 437–449.

Corvinus, G., 1981. A Survey of Pravara River System in Western Maharashtra, India, Volume 1: The Stratigraphy and Geomorphology of the Pravara River System. Institut fur Urgeschichte, Tubingen.

Corvinus, G., 1983. A Survey of the Pravara River System in Western Maharashtra, India, Volume 2: The Excavations of the Acheulian Site of Chirki-on-Pravara. Institut fur Urgeschichte, Tubingen.

Corvinus, G., 1998. Lower Palaeolithic occupations in Nepal in relation to South Asia. In: Petraglia, M.D., Korisettar, R. (Eds.), Early Human Behaviour in Global Context: The Rise and Diversity of the Lower Palaeolithic Record. Routledge, London and New York, pp. 391–417.

Dennell, R.W., 1998. Graslands, tool making and the hominid colonization of southern Asia: a reconsideration. In: Petraglia, M.D., Korisettar, R. (Eds.), Early Human Behaviour in Global Context: The Rise and Diversity of the Lower Paleolithic Record. Routledge, London and New York, pp. 280–303.

Dennell, R.W., 2002. The Indian Lower Paleolithic since independence: an outsider's assessment. In: Paddayya, K. (Ed.), Recent Studies in Indian Archaeology. Munshiram Manoharlal, New Delhi, pp. 1–16.

Dennell, R.W., 2003. Dispersal and colonization, long and short chronologies: how continuous is the Early Pleistocene record for hominids outside Africa? Journal of Human Evolution 43, 421–440.

Dennell, R., 2007. "Resource-rich, stone poor": early hominin land use in large river systems of northern India and Pakistan. In: Petraglia, M.D., Allchin, B. (Eds.), The Evolution and History of Human Populations in South Asia: Interdisciplinary Studies in Archaeology, Biological Anthropology, Linguistics and Genetics. Springer, Netherlands, pp. 41–68.

Dennell, R.W., Rendell, H.M., 1991. De Terra and Paterson and the Soan flake industry: a new perspective from the Soan valley, northern Pakistan. Man and Environment 16, 91–99.

Dennell, R.W., Rendell, H.M., Hailwood, E., 1988a. Late Pliocene artifacts from northern Pakistan. Current Anthropology 29, 495–498.

Dennell, R.W., Rendell, H.M., Hailwood, E., 1988b. Early tool-making in Asia: two-million-year old artifacts in Pakistan. Antiquity 62, 98–106.

De Terra, H., Paterson, T.T., 1939. Studies on the Ice Age in India and Associated Human Cultures. Carnegie Institute of Washington, Washington,, D.C.

Foote, R.B., 1876. The geological features of the south Mahratha country and adjacent districts. Memoirs of the Geological Survey of India 12, 1–268.

Foote, R.B., 1916. The Foote Collection of Indian Prehistoric and Protohistoric Antiquities: Notes on their Ages and Distribution. Government Museum, Madras.

Gamble, C., 1986. The Palaeolithic Settlement of Europe. Cambridge University Press, Cambridge.

Gambe, C., 1991. The Palaeolithic Societies of Europe. Cambridge University Press, Cambridge.

Gamble, C., Porr, M. (Eds.), 2005. The Hominid Individual in Context: Archaeological Investigations of Lower and Middle Palaeolithic Landscape, Locales and Artefacts. Routledge, London and New York.

Isaac, G., 1981. Stone Age visiting cards: approaches to the study of early land use patterns. In: Hodder, I., Isaac, G., Hammond, N., (Eds.), Pattern of the Past: Studies in Honour of David Clarke. Cambridge University Press, Cambridge, pp. 131–155.

Jacobson, J., 1975. Early Stone Age habitation sites in eastern Malwa. Proceedings of the American Philosophical Society 119, 280–297.

Jacobson, J., 1985. Acheulian surface sites in central India. In: Misra, V.N., Bellwood, Peter (Eds.), Recent Advances in Indo-Pacific Prehistory. Oxford-IBH, New Delhi, pp. 49–57.

Jhaldiyal, R., 1997. Formation processes of the prehistoric sites in the Hunsgi and Baichbal basins, Gulbarga district, Karnataka. Ph.D. Dissertation, University of Pune.

Joshi, R.V., Subrahmanyam, R., Rao, V.V.M., 1979. Late Acheulian culture of the Paleru river valley, coastal Andhra Pradesh. Bulletin of the Deccan College Research Institute 38, 28–36.

Khatri, A.P., 1962. Mahadevian: an Oldowan pebble culture in India. Asian Perspectives 6, 186–196.

Korisettar, R., 2004. Geoarchaeology of the Purana and Gondwana basins of peninsular India: peripheral or paramount – a suggested core-periphery model. Presidential address of Archaeology Section of 65th Session of Indian History Congress, Bareilly, pp. 1–40.

Korisettar, R., 2007. Toward developing a basin model for Paleolithic settlement of the Indian subcontinent. In: Petraglia, M.D., Allchin, B. (Eds.), The Evolution and History of Human Populations in South Asia: Inter-disciplinary Studies in Archaeology, Biological Anthropology, Linguistics and Genetics. Springer, Netherlands, pp. 69–96.

Korisettar, R., Rajaguru, S.N. (1998). Quaternary Stratigraphy, Palaeoclimate and the Lower Palaeolithic Evidence of India. In: Petraglia, M.D., Korisettar, R. (Eds.), Early Human Behaviour in Global Context: The Rise and Diversity of the Lower Palaeolithic Record. Routledge, London and New York, pp. 304–342.

Mahadevan, C., 1941. Geology of the south and southwestern parts of Surapur taluk of Gulbarga district. Journal of the Hyderabad Geological Survey 4, 102–161.

Marathe, A.R., 1995. Prehistory and Quaternary sea level changes along the west coast of India: a summary. In: Wadia, S., Korisettar, R., Kale, V.S. (Eds.), Quaternary Environments and Geoarchaeology of India. Geological Society of India, Bangalore, pp. 405–413.

Mishra, S., 1992. The age of the Acheulian in India: new evidence. Current Anthropology 33, 325–328.

Mishra, S., 1995. Chronology of the Indian Stone Age: the impact of recent absolute and relative dating attempts. Man and Environment 20, 11–16.

Misra, V.N., 1978. The Acheulian industry of rock shelter III F-23 at Bhimbetka, central India: a preliminary study. Australian Archaeology 8, 63–106.

Misra, V.N., 1987. Middle Pleistocene adaptations in India. In: Soffer, O. (Ed.), The Pleistocene Old World. Plenum Press. New York, pp. 99–119.

Misra, V.N., 1989. Stone Age India: an ecological perspective. Man and Environment 14, 17–64.

Misra, V.N., 1995. Geoarchaeology of the Thar desert, northwest India. In: Wadia, S., Korisettar, S., Kale, V.S., (Eds.), Quaternary Environments and Geoarchaeology of India. Geological Society of India, Bangalore, pp. 210–230.

Mukherjee, S.K., 1941. Geology of parts of Surapur and Shahpur taluks of Gulbarga district. Journal of the Hyderabad Geological Survey 4, 9–54.

Murty, M.L.K., 1966. Stone Age cultures of Chittoor district, Andhra Pradesh. Ph.D. Dissertation, University of Pune.

Paddayya, K. 1973. Investigations into the Neolithic Culture of the Shorapur Doab, South India. E.J. Brill, Leiden.

Paddayya, K., 1978. New research designs and field techniques in the Palaeolithic archaeology of India. World Archaeology 10, 94–110.

Paddayya, K. 1982. The Acheulian Culture of the Hunsgi Valley (Peninsular India): A Settlement System Perspective. Deccan College, Pune.

Paddayya, K., 1984. Stone Age India. In: Mueller-Karpe, H. (Ed.), Neue Forschungen zur Altsteinzeit. C.H. Beck Verlag, Munich, pp. 345–403.

Paddayya, K., 1985. Acheulian occupation sites and associated fossil fauna from the Hunsgi-Baichbal valleys. Anthropos 80, 653–658.

Paddayya, K., 2001. The Acheulian culture project of the Hunsgi and Baichbal valleys, peninsular India. In: Barham, L., Robson-Brown, K. (Eds.), Human Roots: Africa and Asia in the Middle Pleistocene. Western Academic and Specialist Press, Bristol, pp. 235–258.

Paddayya, K., 2003. Artifacts as texts. Presidential Address of Archaeology Section 64th Session of Indian History Congress, Mysore, pp. 1–19.

Paddayya, K., Jhaldiyal, R., 1999. A new Acheulian site at Koihal in the Hunsgi valley, Karnataka. Puratattva 29, 1–7.

Paddayya, K., Jhaldiyal, R., 2001. The role of surface sites in the Indian Palaeolithic research: a case study from the Hunsgi and Baichbal valleys, Karnataka. Man and Environment 26, 29–42.

Paddayya, K., Petraglia, M.D., 1993. Formation processes of Acheulian localities in the Hunsgi and Baichbal valleys, peninsular India.

In: Goldberg, P., Nash, D.T., Petraglia, M.D. (Eds.), Formation Processes in Archaeological Context. Prehistory Press, Madison, pp. 61–82.

Paddayya, K., Petraglia, M.D., 1995. Natural and cultural formation processes of the Acheulian sites of the Hunsgi and Baichbal valleys, Karnataka. In: Wadia, S., Korisettar, Ravi, Kale, V.S. (Eds.), Quaternary Environments and Geoarchaeology of India. Geological Society of India, Bangalore, pp. 333–351.

Paddayya, K., Petraglia, M.D., 1997. Isampur: an Acheulean workshop site in Hunsgi valley, Gulbarga district, Karnataka. Man and Environment 22 (2), 95–110.

Paddayya, K., Petraglia, M.D., 1998. Acheulian workshop at Isampur, Hunsgi valley, Karnataka: a preliminary report. Bulletin of the Deccan College Research Institute 56–57, 3–26.

Paddayya, K., Jhaldiyal, R., Petraglia, M.D., 1999. Geoarchaeology of the Acheulian workshop at Isampur, Hunsgi valley, Karnataka. Man and Environment 24, 167–184.

Paddayya, K., Jhaldiyal, R., Petraglia, M.D., 2000. The significance of the Acheulian site of Isampur, Karnataka, in the Lower Palaeolithic of India. Puratattva 30, 1–24.

Paddayya, K., Jhaldiyal, R., Petraglia, M.D., 2001. Further field studies at the Lower Palaeolithic site of Isampur, Karnataka. Puratattva 31, 9–13.

Paddayya, K., Blackwell, B.A.B., Jhaldiyal, R., Petraglia, M.D., Fevrier, S., Chanderton II, D.A., Blickstein, I.I.B., Skinner, A.R., 2002. Recent findings on the Acheulian of the Hunsgi and Baichbal valleys, Karnataka, with special reference to the Isampur excavation and its dating. Current Science 83 (5), 641–657.

Pant, P.C., Jayaswal, V., 1991. Paisra: The Stone Age Settlement of Bihar. Agam Kala Prakashan, Delhi.

Pappu, R.S., 1994. Pleistocene Studies in the Upper Krishna Basin. Deccan College, Pune.

Pappu, R.S., 2001. Acheulian Culture in Peninsular India: An Ecological Perspective. D.K. Printworld, New Delhi.

Pappu, R.S., Deo, Sushma G., 1994. Man-Land Relationships during Palaeolithic Times in the Kaladgi Basin, Karnataka. Deccan College, Pune.

Pappu, S., Gunnell, Y., Taieb, M., Kumar, A., 2004. Preliminary report on excavation at the Palaeolithic site of Attirampakkam, Tamil Nadu (1999–2004). Man and Environment 29 (2), 1–17.

Paterson, T.T., Drummond, H.J.H., 1962. Soan, the Palaeolilthic of Pakistan. Department of Archaeology, Karachi.

Petraglia, M.D., 1998. The Lower Palaeolithic of India and its bearing on the Asian record. In: Petraglia, M.D., Korisettar, R. (Eds.), Early Human Behaviour in Global Context: The Rise and Diversity of the Lower Palaeolithic Record. Routledge, London and New York, pp. 343–390.

Petraglia, M.D., 2001. The Lower Palaeolithic of India and its behavioural significance. In: Barham, L., Robson-Brown, K. (Eds.), Human Roots: Africa and Asia in the Middle Pleistocene. Western Academic and Specialist Press, Bristol, pp. 217–233.

Petraglia, M.D., 2005. Hominin responses to Pleistocene change in Arabia and South Asia. In: Head, M.J., Gibbard, P.L., (Eds.), Early-Middle Pleistocene Transitions: The Land-Ocean Evidence. Geological Society of London, London, pp. 305– 319.

Petraglia, M.D., Paddayya, K., 2005. New insights on the Acheulian: a case study from south India. Humankind 1, 61–75.

Petraglia, M.D., LaPorta, P., Paddayya, K., 1999. The first Acheulian quarry in India: stone tool manufacture, biface morphology, and behaviors. Journal of Anthropological Research 55, 39–70.

Petraglia, M.D., Schuldenrein, J., Korisettar, R., 2003. Landscapes, activity, and the Acheulian to Middle Palaeolithic transition in the Kaladgi basin, India. Eurasian Prehistory 1 (2), 3–24.

Petraglia, MD., Shipton, C., Paddayya, K., 2005. Life and mind in the Acheulian: a case study from India. In: Gamble, C., Porr, M. (Eds.), The Hominid Individual in Context: Archaeological Investigations of Lower and Middle Palaeolithic Landscapes, Locales and Artifacts. Routledge, London and New York, pp. 197–219.

Rajendran, P., 1989. The Prehistoric Cultures and Environments of Kerala. Classical Publishing Company, New Delhi.

Raju, D.R. 1988. Stone Age Hunter-gatherers: An Ethnoarchaeology of Cuddapah Region, Southeast India. Ravish Publishers, Pune.

Rendell, H., Dennell, R.W., Halim, M.A., 1989. Pleistocene and Palaeolithic Investigations in the Soan Valley, Northern Pakistan. BAR International Series, Oxford.

Sankalia, H.D., 1956. Animal fossils and Palaeolithic industries from the Pravara basin at Nevasa, district Ahmednagar. Ancient India 12, 35–52.

Sankalia, H.D., 1974. Prehistory and Protohistory of India and Pakistan. Deccan College, Pune.

Sankhyan, A.R., 1977. A new human fossil find from the central Narmada basin and its chronology. Current Science 73: 1110–1111.

Sharma, G.R., Clark, J.D. (Eds.), 1983. Palaeoenvironment and Prehistory in the Middle Son Valley (Madhya Pradesh, north central India). Abinash Prakashan, Allahabad.

Sharma, J.C., 1977. Palaeolithic tools from Pleistocene deposits in Panjab. Current Anthropology 18, 94–95.

Shipton, C., 2003. Sociality and cognition in the Acheulian: a case study on the Hunsgi-Baichbal basin, Karnataka, India. M.Phil Dissertation, University of Cambridge.

Singh, R., 1965. The Palaeolithic industry of northern Bundelkhand. Ph.D. Dissertation, University of Pune.

Sonakia, A., 1985. Skull cap of an early man from the Narmada valley alluvium (Pleistocene) of central India. American Anthropologist 87, 612–616.

Szabo, S.J., Mckinney, C., Dalbey, T.S., Paddayya, K., 1990. On the age of Acheulian culture of the Hunsgi-Baichbal valleys, peninsular India. Bulletin of the Deccan College Research Institute 50, 317–332.

Taylor, M., 1941. Megalithic Tombs and Other Ancient Remains in the Deccan (edited by G.Y. Yazdani). The Nazam's State, Hyderabad.

Verma, B.C., 1991. Siwalik Stone Age sequence. Current Science 61, 496.

Wheeler, R.E.M., 1960. Early India and Pakistan. Thames and Hudson, London.

Williams, M.A.J., Clarke, M.F., 1995. Quaternary geology and prehistoric environments in the Son and Belan valleys, north central India. In: Wadia, S., Korisettar, R., Kale, V.S. (Eds.), Quaternary Environments and Geoarchaeology of India. Geological Society of India, Bangalore, pp. 282–308.

Wynn, T., 1993. Two developments in the mind of early Homo. Journal of Anthropological Archaeology 12, 299–322.

Wynn, T., 1995. Handaxe enigmas. World Archaeology 27, 10–24.

Zeuner, F.E., 1950. Stone Age and Pleistocene Chronology in Gujarat. Deccan College, Pune.

6. Changing trends in the study of a Paleolithic site in India: *A century of research at Attirampakkam*

SHANTI PAPPU

Sharma Centre for Heritage Education
28, I Main Road
C.I.T. Colony
Mylapore, Chennai, 60004
India
spappu@vsnl.com

To account for this immensely numerous collection of implements in a small space is a question more easily proposed than solved (Foote, 1869:234)

Introduction

The above quote was made by the British geologist Robert Bruce Foote, while discussing his discovery of stone tools at the prehistoric site of Attirampakkam, Tamil Nadu, South India. Prior to this, Foote had identified the first stone tool in the subcontinent (Foote, 1866); a find which not only pushed back the known antiquity of human occupation in India, but which also generated awareness of what stone tools looked like. Thus, on 28th September 1863, when Foote and his colleague William King were investigating the 'Stri Permatoor' (Sriperumbudur) shale beds in the gully around 'Atrampakkam nullah' (Attirampakkam), King was able to recognize and pick up two "well-shaped oval implements" on a terrace of 'quartzite shingle' in the gully bed (Foote, 1866:3). Little did they realize that this discovery would initiate a century of research into the Paleolithic archaeology of Tamil Nadu; which was later, often referred to as constituting the 'Madras Paleolithic industries'. Subsequent sporadic research at Attirampakkam and neighboring sites in the basin of the river Kortallaiyar (a part of the Palar river basin), led to the emergence of concepts that came to influence much of later Indian prehistory. Reviews of the Indian Lower Paleolithic, and studies in various parts of the country, invariably included references to these industries, including the site of Attirampakkam; despite the fact that only two research papers and short notes on the early excavations constituted almost all what was known of this region. This paper briefly discusses these early studies, against which ongoing excavations at Attirampakkam may be situated.

The site of Attirampakkam is located around 47 km inland from the current shoreline

121

M.D. Petraglia and B. Allchin (eds.), The Evolution and History of Human Populations in South Asia, 121–135.

(13° 13′ 50″ N and 79° 53′ 20″ E; 37.5 AMSL), around 1 km to the north of the river Kortallaiyar, Northern Tamil Nadu (Figure 1). To the west lie the NNE-SSW-trending Allikulli hills (200–380 m AMSL), which are cobble-to-boulder size fanglomerates or paleodeltas of early Cretaceous age (Muralidharan et al., 1993; Kumaraguru and Trivikrama Rao, 1994; see Pappu, 2001b). The lower-lying areas of the eastern Cuddapah piedmont in the vicinity of the Allikulli Hills, are underlain by a shaly marine formation (Avadi formation) which is coeval and inter-tonguing with the conglomerate beds (Kumaraguru and Trivikrama Rao, 1994). The region falls in an area of seasonally dry tropical conditions, receiving 105 to 125 cm of annual rainfall with a major peak occurring from September to November (National Commission on Agriculture, Rainfall and Cropping Patterns, 1976; Pappu, 2001b). In the Kortallaiyar basin as a whole, Acheulean to Middle Paleolithic sites occur in ferricretes (1.5–2 m thick) resting on shales. Younger ferricretes, which appear to represent eroded gravels sourced by the older outcrops during the Late Pleistocene, contain Middle Paleolithic artifacts; while microliths occur on the surface.

The Discovery of Attirampakkam

The discovery of tools at Attirampakkam was not a matter of chance but of careful observation by King and Foote, in the course of their geological surveys. The area that they studied ranges from the point where the Attirampakkam gully discharges into the river Kortallaiyar, to around 2 km upstream to the village of Nambakkam. The fresh condition of artifacts lying in the gully, prompted Foote to trace their origin to nearby lateritic gravels seen in the gully sections; within which he was able to locate in-situ tools (Table 1). This was confirmed by a small test pit excavated in the lateritic gravels, which yielded further tools. Variability in the nature of these lateritic gravel beds in terms of the presence of ferruginous matter and pebbles, was considered to have affected the condition of the implements in terms of weathering and staining (Foote, 1869:233). The laterites rested on rolling shale beds, which he believed, were subjected to extreme erosion prior to the deposition of the laterite. Foote did however note the presence of a large implement on the surface of these 'plant shales' underlying the lateritic conglomerate, a fact that assumes importance in the light of our recent excavations.

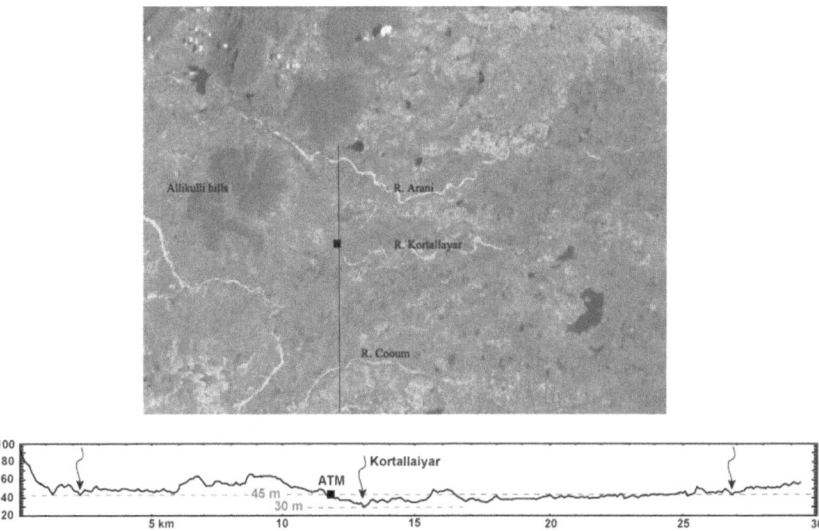

Figure 1. A part of the Kortallaiyar river basin showing location of Attirampakkam (ATM)

Table 1. History of archaeological investigations at Attirampakkam

Investigator	Stratigraphic Context	Lithics
R.B. Foote (1866, 1869, 1870, 1916)	• Soil and made ground 6′ • Lateritic conglomerate with layers of quartzite pebbles 11/2–3′ • Sandy-clay with quartzite pebbles and kunkar in strings 51/2–6 ½′ • Lateritic conglomerate full of quartzite pebbles 3′ • Grey shales (plant shales) with ferruginous stains 1′8″ • Nearby in bed 'c' he found a hatchet shaped implement 10–11′ below the surface. In another section he notes a tool lying on the face of 'e' i.e. plant shale.	Class I: Implements with one blunt or truncated edge a. Pointed weapons (spear heads); b. wedge shaped weapons (axes-hatchets; etc.) Class II: a) Implements with a cutting edge all around. Implements pointed at one or both ends; b. Oval or almond shapes implements; c. discoidal implements. Class III: Flakes.
De Terra and Paterson (1939)	Tools occur in basal lateritic gravels of terrace T2. A few rolled specimens of earlier date were similar typologically to those from Vadamadurai tank. After deposition of the detrital laterite over a white boulder conglomerate it was dissected producing a set of three terraces, of which Terrace 2 at 20′ is aggradational and preserved at Attirampakkam.	Late Acheulean handaxes and cleavers some in situ in the basal lateritic gravel, and a few rolled specimens. **Cleavers**: numerous cleavers, made on flakes with the flake surface untouched or partially flaked. Existence of Vaal River variant, with a parallelogramatic cross section. Cleaver shapes are rectangular, with straight or convex butt; or with sides converging slightly, and some are triangular in outline with pointed butts. The working end is usually straight and at right angles to the axis of the tool; but in some cases it is oblique and in a few cases convex or concave. **Handaxes**: mostly on flakes though the flake surface is partly or wholly trimmed. The flaking is by the step technique with small, flat and neat steps with small step retouch at the edge. Mostly pear to tongue shaped, the latter having fairly straight and slightly convergent rather than convex sides. They range from 8–6 inches down to 2 × 2 inches and small

(Continued)

Table 1. (Continued)

Investigator	Stratigraphic Context	Lithics
		and large forms are found in fairly equal numbers. Some S-twist examples are seen. **Cores**: discoidal type, with more or less alternate flaking some being retouched and used as tools. Some are more oval in shape with alternate flaking resemlbling unfinished crude handaxes. Some were retouched to form notched, steep or ordinary side scrapers.
V.D.Krishnaswami (1938a)	• Sandy loam • Pellety loam • White sandy-clay with kunkar • Hard lateritic conglomerate (Lower to Middle Acheul and Late Acheul) • Sriperumbudur shale	Industry V: It shows a stratigraphic evolution of the Acheulean culture from the lateritic basal gravels of this terrace to the loam on the top in exposed sections. He notes a derived series (both rolled Abbevillian coup-de-poings and the lateritised tools being early Acheul) and a contemporary series (fresh from the lateritic conglomerate upwards of T2), and fresh of lateritic patination. The coup de poings and cleavers predominate and compare well with Late Acheul forms of Europe and Africa. Victoria West handaxes, and those made on the double Vaal principle were also seen. Some handaxes simulate Micoquian types of the Somme valley. Towards the end of T2, levallois flakes appear. Handaxes and ovates: 61%, cleavers: 27%, cores: 12%
Zeuner (1949: 45) Attirampakkam gully section	• Orange soil • Pink grey weathering • Implimentiferous • Shale • Note: the stratigraphy diagram attached clearly indicates the presence of an implimentiferous gravel.	
K.D. Banerjee (1969:20–22)	• Top soil (microliths) • Compact brownish soil with lateritic pellets (sterile)	He found that the detrital lateritic gravel had an industry which he termed post-Acheulean (comprising

(Continued)

Table 1. (Continued)

Investigator	Stratigraphic Context	Lithics
	• Redeposited detrital laterite (post-Acheulean) • Redeposited Shaly-clay (sterile) • Detrital laterite (post-Acheulean) • Weathered shale (Acheulean) • Shale	points, scrapers and longish flake-blades). He contradicted the work of previous scholars that the lateritic gravel was the Acheulean horizon. No further analysis was done.

These observations not only established the context of artifacts, but also inspired a debate between Foote and King, which is possibly amongst the earliest discussions on 'site formation processes' at prehistoric sites (Foote, 1866:27; Pappu, 1991–1992). A significant feature of the site, observed through the years, is the high density of artifacts per unit area as compared to any other site in the region or elsewhere in India. This struck Foote and King who suggested various explanations for the same. King attempted to explain the presence of the vast number of artifacts by suggesting that these were localities of manufacture of tools for purposes of barter, as inferred from the presence of unfinished artifacts and flakes along with complete tools. He further argued that there was no visible system of distribution of tools across the ground, in lines or groups, which could imply long-distance fluvial transport, although some local drifting might have been possible (King in Foote, 1866: Appendix:ix). Thus the site was thought to be primary in nature. Writing in 1866, Foote believed that this matter, though not improbable, required further study. He however differed from King and other geologists who believed that the lateritic conglomerate at this site was derived from higher lateritic beds located elsewhere (Foote, 1866:9). In Foote's opinion, the laterites of the valley and upland, belonged to the same formation and were structured by the slope of the underlying plant-bearing shales (Foote,

1866:9). This could also be judged by the perfect state of preservation of tools, which did not correspond with the theory of their long-distance transportation. By 1869, Foote however disagreed with King, and argued that the site could not have been a manufacturing centre owing to the paucity of chips and flakes, and also wondered why so many perfect implements were left behind (Foote, 1869:234). He argued that for 'savage people', such foresight in preparing tools long before their use was improbable and that improvidence was rather to be argued for. He further argued against in situ manufacture, arguing that the size ranges of the gravel clasts available were not sufficient for tool manufacture for even the middle-sized implements. Rather, he believed that the area when submerged under the sea, was used for fishing by people in rafts and catamarans and that their weapons were washed off during sudden squalls, and lost along the surf line, to be finally embedded by tidal action. Foote was also the first to observe the fact that tools were vertically embedded; possibly stuck in the mud on sinking under water (Foote, 1869:234–235). Subsequently, many of his views changed; and he declared himself wrong as regards the marine origin of laterites. Instead, he invoked powerful pluvials that corresponded with the Pleistocene Ice ages. Tools were now believed to be transported to greater or lesser degrees, along with lateritic deposits, within which they were finally embedded (Foote, 1916:181).

Classification of implements followed that of John Evans (Foote, 1866:17), with the addition of forms that he believed were not found in Europe as per the books available to him in India (see Table 1). In particular, the class termed as the 'Madras Type', was thought to be unique to this region. Following classification, Foote also put forward hypotheses on the use of some of these tools, as also the effect that varied uses would have on the edge. He was perhaps one of the earliest scholars to suggest that flakes were not only waste products, but could also be utilized as effective tools.

Foote's monumental study of the prehistoric archaeology of southern India (Pappu, 2001b), enabled him to place Attirampakkam within the context of other sites in the region. His insightful observations on stratigraphy and tool contexts; the presence of tools on the surface of the shales, vertically embedded artifacts; and thoughts on the causes for the high density of tools at this site; are as relevant today as they were in the 19th century. Following his death, interest in prehistory in this region lapsed, save for a sole publication by P.T. Srinivasa Ayyanger (1988 reprint) who wrote on the Stone Age of India, summarizing and interpreting data available at the time.

Linking 'Cultures': Attirampakkam as part of the 'Grand Sequence'

This phase characterized principally by the work of T.T. Paterson and V.D. Krishnaswami, is one where the foundations of modern Indian prehistory were laid, and where discoveries made during the 19th century were critically re-examined. The age of the 'Amateur' archaeologist yielded to that of the 'Professional'; (Levine, 1986) and while much was gained in terms of new information, the breadth of vision in interpreting data so characteristic of the work of Foote, was often sacrificed in the search for hard facts.

L.A. Cammiade and M.C. Burkitt's (1930) classic study linking sequences of pluvials and interpluvials with evolving archaeological 'cultures' along the southeast coast of India; closely followed by the search for hominins in the subcontinent, culminating in the Yale-Cambridge expedition under De Terra and Paterson (1939), stimulated research in the Kortallaiyar basin. This expedition aimed at understanding "the Ice Age cycle in the Himalaya and to unravel the Pleistocene history of Stone Age man in other parts of India" (De Terra and Paterson, 1939:1). Influenced by concepts of 'Evolution progressive' as exemplified in the work of Brueil and popularized by de Terra and Paterson (Dennell, 1990:553), the Kortallaiyar basin was chosen as an ideal example of a coastal river system where prevalent ideas on stratigraphy, chronology, climate and industries could be tested; and form another link in the attempt to build up a 'Grand Sequence' (Dennell, 1990). A geological approach involving construction of river terraces and cultural sequences across the country, linking of these sequences; and correlating them with Pleistocene pluvials and interpluvials, formed the foundation of this work. The work of V.D. Krishnaswami (1938a, 1938b, 1947), who participated in this research may also be viewed within the context of the growing nationalist movement in the subcontinent, and reflected his desire to establish the cultural heritage of India on a firm basis, to place it in the perspective of changes occurring all over the world, to establish the great antiquity of man in India, and to draw together the Soan and the Acheulean into a pan-Indian sequence. The major issues characterizing this phase of research are discussed elsewhere (see Pappu, 2001b).

The basis of these studies lay in the identification of four terraces of the river Kortallaiyar; TD to T3 at elevations of 100′ (30 m), 60′ (18 m), 20′ (6 m) and 8′ (2 m); which were a logical extension of the terrace sequences built up in the Soan and

Narmada regions (De Terra and Paterson, 1939). Terraces were either aggradational or erosional and were associated with a definite stratigraphic sequence, climatic phase and with an 'evolving culture sequence' ranging from the 'Abbevillio-Acheulean' to the 'Upper Paleolithic' (Krishnaswami, 1938a; De Terra and Paterson, 1939). Of the four terraces identified, Attirampakkam fell on Terrace T2 (20′). This was described as an aggradational terrace with thicker gravels and covered by silts and sands. The interval between T1 and T2 was interpreted as a period of erosion and aridity followed by deposition of coarse gravel on T2 pointing to a definite resumption of increased pluvial conditions. Terrace formation was further linked to eustatic sea level changes and phases in the migration of the river Palar along the present course of the river Kortallaiyar.

Krishnaswami (1938a) noted the presence of a lateritic conglomerate overlying the Sriperumbudur shales and capped by loams separated by lenses of pellety laterites. The industry within this terrace was classified as Industry V, defined as having Late Acheulean, Micoquian and Levallois elements. No difference in assemblage composition was noted throughout the terrace, either in gully sections or in excavations. This fact was confirmed by a study of collections from different levels made by his colleague Drummond, also a member of the Yale-Cambridge expedition. Despite these preliminary observations, Krishnaswami was careful to note that further sub-division on "grounds of stratigraphy though it has not been yet made and this may bring forth minor differences in the industry as to how the handaxes and cleavers have developed here" (Krishnaswami, 1938:71). In describing this industry he noted two series of tools: a) a derived series of tools, comprising rolled Abbevillian 'coups-de-poing' and early Acheul lateritized types; and, b) a contemporary series, above the lateritic conglomerate and devoid of patination. 'Coups-de-poing' and cleavers

were predominant and had a variety of forms comparable with that of Africa and Europe. He stated that some handaxes were flat with such a thinly elongated point that they simulated Micoque types from the Somme valley. He noted the presence of Victoria West type handaxes, in addition to handaxes and cleavers made on the single and double Vaal principle. Towards the top of the terrace, Levallois flakes were noted (Krishnaswami, 1938a:75).

Paterson chose to term industries in this region as the 'Madras Paleolithic industries', a term that seems to have been applied in a general way to collections from this area distributed in various parts of the world (Petraglia and Noll, 2001). In his sequence of four-fold terraces, Paterson joined Krishnaswami in placing 'Attrampakkam', on terrace T2 (20′), of the river Kortallaiyar. Tools occurred within the basal laterites and were largely late Acheulean handaxes and cleavers. He also noted the presence of a few rolled tools, similar to those of the first two groups at the site of Vadamadurai (i.e., 'Abbevillian, earliest Acheulean, and Middle Acheulean). Like Foote, he remarked on the extreme freshness of most tools. The presence of flake cleavers, and occurrence of the Vaal river variant of South Africa were documented by Paterson (Table 1). The industry also contained pear to tongue shaped handaxes, and discoidal cores (some retouched into tools) (Paterson, 1939:329). He also noted the presence of a small group of cores and flakes, which were entirely fresh and unpatinated, and some of which were found in the overlying silt, while others were on the surface.

These studies established a standard against which other industries were compared. Analysis of other lithic assemblages were invariably within a framework of how they could be situated within the Madras biface or pebble tool family, and until a re-examination of this region in the 1990's (Pappu, 1997, 2001), cultures were compared with the Kortallaiyar basin terrace sequences.

The Archaeological Survey of India and Excavations at Attirampakkam

From 1957 to 1979, excavations were conducted by K.D. Banerjee of the Archaeological Survey of India at Gudiyam, Attirampakkam, Vadamadurai, Poondi and Neyvelli. Barring brief notes (I.A.R., 1962–63, 1963–64, 1964–65, 1966–67), this work was never published. Banerjee denied the existence of fluvial terraces and put forward a hypothesis of marine terraces at elevations of 73 m, 45 m, 30 m, and 17 m AMSL (Agrawal, 1982:53) (Table 1). Like Foote, Banerjee also observed Acheulean artifacts on the surface of, and slightly within the weathered shale. He was unclear as to whether the horizon of this industry was on the surface of the shale or in an overlying deposit, which was subsequently washed away (Banerjee, 1969:20–22). Unlike previous scholars, he associated the laterite with a post-Acheulean industry (comprising points, scrapers and longish flake-blades). He also confirmed the presence of a redeposited laterite higher in sequence with a post-Acheulean industry capped by a sterile brownish soil.

Excavations at Attirampakkam (1999–2004)

A re-examination of the archaeology of the Kortallaiyar river basin was initiated in 1991 (Pappu, 1996, 1999, 2001a, 2001b). In this study, the ferricretes or ferruginous gravels (corresponding to the laterites of previous scholars), bearing Acheulean tools were noted resting disconformably on Sriperumbudur shales (also classified as the Avadi formation), and were capped by silty-clays. A study of the gully sections did not reveal any artifacts in the shales or clays. Preliminary observations indicated that the principle geomorphic processes include weathering of the bedrock clasts, winnowing of the siliceous and ferruginous matrix of the Sriperumbudur and Satyavedu formations, and erosion of Tertiary ferricretes, which contributed source material in the form of gravels, silts, sands and clays constituting Pleistocene deposits in this region. Subsequently, transport and redeposition took place due to colluvial processes, stream and sheet floods and stream channel processes; this was followed by weathering of the profiles and ferricritization. Further, deposits in the low-lying areas on either side of the Kortallaiyar, formed part of the flood plain of the Old Palar/Kortallaiyar, while the northern areas were under the influence of the river Arani. While terraces were noted, they did not correspond with the 'four-fold' terrace sequence put forward by previous scholars. The surface assemblages were predominantly Middle Paleolithic. Only three possible Acheulean tools were collected from the surface. These were stained white owing to contact with the bedrock (shale) and were rolled and abraded, which hinted at the possibility of a pre-lateritic archaeological horizon. Studies of site formation processes and lithic technology were conducted, and the site was situated with the broader context of the regional archaeological record (Pappu, 2001b).

These preliminary studies of the site were based only on surface collections and the study of gully sections. It was subsequently decided to expand this preliminary study into a project for examining the Pleistocene archaeology and paleoenvironments of the region. Excavations (1999–2004) were initiated at Attirampakkam by the author, with geochronological and geomorphological studies under the direction of M. Taieb and Y. Gunnell (Pappu et al., 2003a, 2003b, 2004; Pappu and Kumar, 2005). The ongoing research project aims at investigating questions related to hominin behavior in the context of changing Pleistocene environments, to situate the site within the broader geomorphic context; to obtain a series of dates; to study lithic technology and the nature of cultural transitions through time; and to situate these studies within the regional archeological landscape, and within the context of South Asian prehistory.

Artifacts were mapped eroding over an area of around $50,000\,\mathrm{m}^2$. A contour map of the site and surface deposits was prepared at 1 m intervals. Following this, an area of $220\,\mathrm{m}^2$ was excavated in the form of test-pits, geological step trenches and horizontal trenches, the methodology of which is discussed elsewhere (Pappu et al., 2004). Emphasis was laid on meticulous recording of all artifacts and features (e.g., three-dimensional measurements, orientation, inclination, nature in which the tool was embedded in the sediment); sieving all excavated sediments, digital photography, drawing and videography; and collection of samples for sedimentological, paleobotanical and micropaleontological and rock magnetic studies; and for obtaining paleomagnetic measurements, ESR, cosmogenic Be and OSL dates (Figure 2).

The site has yielded a stratified cultural sequence (maximum depth of 9 m) comprising Lower, Middle and possible Upper Paleolithic deposits with a microlithic component in the upper layers and on the surface. Six sedimentary units were recognized, comprising laminated clays (Layer 6), disconformably overlain by a thick sequence of ferruginous gravels (Layer 5) capped by clayey-silts (Layers 3, 4), which were in turn overlain by fine ferruginous gravels (Layer 2) and clayey-silts (Layer 1). Acheulean industries were noted in Layers 5 and 6 with industries possibly transitional to the Middle Paleolithic in Layers 3 and 4. Middle Paleolithic assemblages were noted in Layer 2, with a possible component of an early Upper Paleolithic. Microliths were noted eroding out of an overlying ferruginous gravel, which is capped by sands. Three significant aspects of this stratigraphic sequence included: a) discovery of Acheulean artifacts within clays (previously assigned to Lower Cretaceous shales); b) occurrence of Acheulean levels within the lateritic gravels (contradicting Banerjee, who assigned these to the post-Acheulean industries); and, c) occurrence of assemblages within the clayey-silts of Layers 3, 4 (previously thought to be archaeologically sterile) (Figure 3).

The most significant discovery was that of the occurrence of Acheulean assemblages within a sequence of laminated clay deposits previously classified as a Lower Cretaceous shale of the Avadi or Sriperumbudur series (Foote, 1870; Krishnaswami, 1938a; Banerjee, 1969). An important question that arose was whether Acheulean tools were or have been sinking into the 'Cretaceous shales of the Avadi series'

Figure 2. General view of the excavations (Trenches T7A, B,C)

Figure 3. Stratigraphic sequence (Trench T7A)

Figure 4. A handaxe within the clays of Layer 6 (Trench T8)

(Figure 4). This issue is under investigation and is being examined using geochemical, geochronological and micropaleontological studies. Geomorphological studies under the direction of Y. Gunnell (Pappu et al., 2003a, 2004) indicate that the laminated clay of Layer 6 is a Pleistocene floodplain deposit of fluvial origin, sourced by an Avadi shale outcrop, and aggraded during site occupation. Sedimentological studies (Pappu et al., 2003a, 2004), indicate a fluvial context, with the site located < 1 km from a large meander in the Kortallaiyar river cutting into its former floodplain. The negligible content in organic matter (< 0.2%) of Layer 6 suggests episodic flooding rather than a perennial swamp with high biological productivity. Sedimentation was never interrupted for sufficiently long periods of time for paleosols to develop in the profile. This is also supported by the geochemical homogeneity of the sediment, which suggests stable paleoenvironmental conditions throughout the history corresponding to Layer 6. These studies are supported by those comparing artifacts and natural clast sizes between Layers 5 (ferricrete gravel) and Layer 6 (clay) in terms of their percentages, size ranges, and patterns of abrasion. The absence of pebbles and ferricrete pisoliths predominating in Layer 5 and in Layer 6, and the absence of evidence of ferruginous staining or patination on tools in Layer 6 (Pappu et al., 2003a), tend to eliminate the possibility that tools are sinking into the clays of Layer 6 from Layer 5. This does not imply that vertical movement of tools did not occur, as a conjoinable pieces were noted in Trench T3, over 60 cm, which represents around ∼ 10-12% of the total thickness of Layer 6. This does not seem to imply a large scale downward movement of artifacts. A total of 16 oriented samples of sediments from Layer 6, was collected in the trench T3 at 50 cm intervals from the surface to −7 m. Although no Pleistocene magnetic reversals could be identified due to insufficiently clear patterns in magnetic declination, the consistently low inclination values exclude a Cretaceous age for Layer 6. No major difference between magnetic directions from Layers 1–5 and Layer 6 further suggests that these formations are all of Pleistocene age (Pappu et al., 2003a).

It is inferred that Acheulean tools were periodically used at the site and left lying until buried by overwash. The overwash was generated by laminar flow overtopping the paleo-Kortallaiyar alluvial levees at a time when the river bed was 10–15 m higher than today, and the critical shear stress of such flow depths was insufficient to entrain or disturb the discarded artifacts. As episodic sedimentation proceeded, new tools continued to be discarded onto the fresh depositional surfaces. The laminations are typical of sediment settled by low-energy sheet flow in crevasse splays, floodplain ponds or abandoned channels. As per studies (Pappu et al., 2003a), the ferricrete colluvium of Layer 5 may coincide with a time of meander incision and erosional response of the hillslopes in the left bank of the meander bend. Ongoing studies will help to address this issue.

In the earliest phases of occupation (Layer 6, laminated clays), Acheulean hominins occupying the site utilized the area for specific tasks principally associated with the use of Large Cutting Tools (Figure 5). The deposit is devoid of cobble-sized clasts, nor are there any cores in these layers suitable for the detachment of large flakes. Preliminary manufacture of large flake blanks or trimming of cobbles was carried out off-site, possibly within a radius of 3–4 km of the site, where outcrops of quartzite cobble-boulder sized fanglomerates and gravel beds of the Allikulli hills and their outliers occur. Secondary trimming and retouch was carried out at the site, as is seen from the high percentage of debitage. In addition to finished tools, large flakes were also transported here, and were either shaped into other tools or utilized with minimum retouch or without retouching (Pappu and Kumar, 2005).

Acheulean assemblages noted in Layer 5 (ferruginous gravels) have evidence of extensive manufacturing activities, with cores, hammerstones, debitage and tools in all stages of manufacture, as also refitting artifacts (Figure 6). Raw material in the form of cobbles and pebbles in channel gravels were utilized. Assemblages in Layers 3 and 4 once again do not point to any intensive manufacture but represent largely a finished tool component, indicating use of the site for specific tasks. A complete sequence of manufacture including a few cores, debitage, tools and refitting artifacts are seen in Layer 2 (ferruginous gravels). A significant fact is the presence of more than 50% vertically embedded tools in certain levels of Layers 2, 3, 4 and 6, reasons for which are as yet unclear. The high resolution data recovered from the trenches is being analyzed to this end using Geomatics and GIS packages to investigate questions related to the spatial clustering of artifacts, and natural and cultural factors structuring their horizontal and vertical distribution. In addition to artifacts, three fossil faunal teeth, a set of impressions, possibly representing animal hoof-impressions, and fragmentary shells were also recovered (Pappu et al., 2004).

Figure 5. Handaxe from Layer 6

Figure 6. Cores, debitage and tools from Layer 5 (ferruginous gravels)

Discussion

In the last 142 years since Foote's studies, research into the Indian Paleolithic has led to the discovery of sites, shed new light on Pleistocene environments, led to developments in chronology, and provided new perspectives in the study of lithics and in concepts regarding prehistoric behavior. Periodic reviews of these studies (Sankalia, 1974; Jayaswal, 1978; Jacobsen, 1979; Paddayya, 1984; Misra, 1989; Mishra, 1994, 1995; Korisettar and Rajaguru, 1998; Petraglia, 1998; Pappu R.S., 2001; Corvinus, 2004), and efforts to situate the Indian Paleolithic in a global perspective are however hampered by several factors. Much of what is written on the Indian Paleolithic is based on research conducted on surface occurrences in a few regions. Few large scale excavations of sites have been undertaken, and these are reported often briefly only in the official journal of the Archaeological Survey of India (*Indian Archaeology: A Review*), or published in the form of brief articles in journals (see Mishra, 1994, Pappu R.S., 2001; Petraglia, 1998). Doctoral dissertations often deal with selected aspects of the excavation (Alam, 1990). Comprehensive excavation reports are even fewer (Joshi, 1978; Armand, 1983; Corvinus, 1983; Sharma and Clark, 1983; Pant and Jayaswal, 1991). This is compounded by problems associated with the study of surface scatters (questions related to site contexts, occurrence of palimpsets and lack of stratigraphic or chronological control); while the study of older collections in institutes or museums, is often marred by a lack of contextual information. There has also been an excess emphasis on the study of gully sections which requires to be treated with caution. In this context, valuable lessons can be learnt from our research at Attirampakkam, where the study of gully sections do not reveal artifacts within the clays of Layer 6 or the silty-clays of Layers 3 and 4, as seen in the excavations; thus leading to a stratigraphy which prevailed untested for almost a century. Further, excavations across the site also indicate that the surface assemblages are a result of the differential erosion of artifacts from varying layers, and thus need to be studied with caution. In this context, systematic excavations at Attirampakkam, have yielded high-resolution data which is being used to answer a wide range of questions on hominin behavior and paleoenvironmental change over the Pleistocene. With the exception of the 16R dune profile at Didwana, Rajasthan (Misra and Rajaguru, 1989), few open-air excavated Paleolithic sites in India have revealed similar thick sequences of Lower and Middle Paleolithic industries. In particular, the thick sequence of stratified deposits at Attirampakkam provides a unique opportunity for examining long-term changes during the Lower, Middle and Upper Paleolithic. Although dates are still awaited, this stratified

sequence of Acheulean industries at the site assumes significance when examining early dates for the Acheulean elsewhere in India (Mishra, 1995). Further, few studies in the Indian Paleolithic have specifically addressed questions related to hominin behavior. In South India, the classic study of the Hungsi-Baichbal complex (Paddayya, 1982), followed by recent excavations at Isampur (Petraglia et al., 1999), revealed patterns of hominin movement across the landscape, variability and choices exercised in raw material usage, caching behavior, and well-planned quarrying activities. At Attirampakkam, analysis is still in process and thus at present brief observations can only be made at present. Changing patterns of raw material exploitation are noted with import of finished tools to the site in the lowest levels of the Acheulean (Layer 6), followed by on-site manufacture in the upper levels (Layer 5) with import of boulders for use as giant cores; in accordance with changing depositional environments and raw material accessibility over the Pleistocene. The site was occupied for over much of the Middle to Late Pleistocene, with evidence of abandonment only seen in Layer 1. Situating this site in a global perspective, may as yet be premature as results of various studies are still awaited; but our studies will certainly add important information to existing knowledge on change and continuity in the South Asian Paleolithic.

Acknowledgments

Excavations at Attirampakkam were funded at various points in time by the Homi Bhabha Fellowships Council, the Leakey Foundation and the Earthwatch Institute, for which we are very grateful. Permits were granted by the Archaeological Survey of India, and the Department of Archaeology, State Government of Tamil Nadu, for which I am grateful. Institutional and financial aid was provided by the Sharma Centre for

Heritage Education. I am deeply grateful
to my colleagues Kumar Akhilesh, Yanni
Gunnell and Maurice Taieb for archaeological and geomorphological inputs in this
project.

References

Agrawal, D.P., 1982. The Archaeology of India.
Scandinavian Institute of Asian Studies,
Monograph Series No. 46.

Alam, Md. S., 1990. A morphometric study of
the Palaeolithic industries of Bhimbetka, Central
India. Ph.D. Dissertation, University of Poona.

Armand, J., 1983. Archaeological Excavation in
Durkhadi Nala: An Early Palaeolithic Pebble
Tool Workshop in Central India. Munshiram
Manoharlal, New Delhi.

Banerjee, K.D., 1969. Excavation at Attirampakkam, Distirct Chingleput. Indian Archaeology:
A Review 1964–65, 20–22.

Cammiade, L.A., Burkitt, M.C. 1930. Fresh light on the
Stone Age of southeast India. Antiquity 4, 327–39.

Corvinus, G., 1983. A Survey of the Pravara River
System in Western Maharashtra, India. Volume 2.
The Excavations of the Acheulian Site of Chirki-
on-Pravara, India. H. Muller-Beck, Tubingen.

Corvinus, G., 2004. *Homo erectus* in East and Southeast
Asia, and the questions of the age of the species
and its association with stone artifacts, with special
attention to handaxe-like tools. Quaternary International 117, 141–151.

De Terra H., Paterson T.T., 1939. Studies on the
Ice Age in India and Associated Human Culture.
Carnegie Institution of Washington Publication
No. 493, Washington:.

Dennell, R.W., 1990. Progressive gradualism, imperi-
alism and academic fashion: Lower Palaeolithic
archaeology in the 20th century. Antiquity 64,
549–558.

Foote, R.B., 1916. The Foote Collection of Prehis-
toric and Protohistoric Antiquities: Notes on Their
Ages and Distribution. Government Museum,
Madras.

Foote, R.B., 1866. On the occurrence of stone implements in lateritic formations in various parts of the
Madras and North Arcot Districts. Madras Journal
of Literature and Science 3rd, series II, 1–35, with
an appendix by William King, 36–42.

Foote, R.B., 1869. On quartzite implements of Palaeolithic types from the laterite formations of the east
coast of southern India. International Congress of

Prehistoric Archaeology: Transactions of the Third
Session, Norwich 1868, 224–239.

Foote, R.B., 1870. Notes on the geology of neighbourhood of Madras. Records of the Geological
Survey of India III (1), 11–17.

Gaillard, C., Misra, V.N., Rajaguru, S.N., 1983.
Acheulian occupation at Singi Talav in the
Thar Desert: a preliminary report on the 1981
excavation. Man and Environment 7, 112–130.

Indian Archaeology: A Review (I.A.R.), 1962–
63, 1963–64, 1964–65, 1966–67. Archaeological
Survey of India, New Delhi.

Jacobsen, J., 1979. Recent developments in South Asian
prehistory and protohistory. Annual Reviews of
Anthropology 8, 467–502.

Jayaswal, V., 1978. Palaeohistory of India (A study
of the Prepared Core Technique). Agam Kala
Prakashan, Delhi.

Joshi, R.V., 1978. Stone Age Cultures of Central
India: Report on Excavations of Rock Shelters
at Adamgarh, Madhya Pradesh. Deccan
College, Pune.

Korisettar, R., Rajaguru, S.N., 1998. Quaternary stratig-
raphy, palaeoclimate and Lower Palaeolithic of
India. In: Petraglia, M.D. Korisettar R. (Eds.),
Early Human Behaviour in Global Context.
Routledge, London and New York, pp. 304–342.

Krishnaswami, V.D., 1938a. Environmental and
cultural changes of prehistoric man near Madras.
Journal of the Madras Geographic Association
13, 58–90.

Krishnaswami, V.D., 1938b. Prehistoric man around
Madras. Indian Academy of Sciences, Madras
Meeting 3, 32–35.

Krishnaswami, V.D., Stone Age India. Ancient India
3, 11–57.

Kumaraguru, P., Trivikrama Rao A., 1994.
A reappraisal of the geology and tectonics of
the Palar basin sediments, Tamil Nadu. In Ninth
International Gondwana Syposium, Hyderabad,
Jan. 1994, 2, Rotterdam, Geological Survey of
India and Balkema, pp.821–831.

Levine, P., 1986. The Amateur and the Professional. Antiquarians, Historians and Archaeologists in Victorian England 1838–1886. Cambridge
University Press, Cambridge.

Mishra, S., 1994. The South Asian Lower Palaeolithic.
Man and Environment XIX (1–2), 57–72.

Mishra, S., 1995. Chronology of the Indian Stone Age:
the impact of recent absolute and relative dating
attempts. Man and Environment 20(2), 11–16.

Misra V.N., Rajaguru, S.N., 1989. Palaeoenvironments
and prehistory of the Thar Desert, Rajasthan,

India In: Frifelt, K., Sorensen, R. (Eds.), South Asian Archaeology 1985, Scandinavian Institute of Asian Studies, Occasional Papers No. 4. Copenhagen, pp. 296–320.

Misra, V.N., 1989. Stone Age India: an ecological perspective. Man and Environment XIV (1), 17–64.

Muralidharan, P.K., Prabhakar, A., Kumarguru, P., 1993. Geomorphology and evolution of the Palar Basin, in Workshop on Evolution of East Coast of India, April 18–20, Abstract of Papers. Tamil University, Tanjore.

National Commission on Agriculture, Rainfall and Cropping Patterns. Tamil Nadu Volume XIV. 1976. Government of India, Ministry of Agriculture and Irrigation, New Delhi.

Paddayya K., 1984. India: Old Stone Age. In: Bar-Yosef, O. (Ed.), Neue forschungen zur Altsteinzeit. Sonderdruck. Verlag C.H, Munich, pp. 345–403.

Paddayya, K., 1982. The Acheulian Culture of the Hunsgi Valley (Peninsular India): A Settlement System Perspective. Poona: Deccan College Postgraduate and Research Institute.

Pant, P.C., Jayaswal, V., 1991. Paisra: The Stone Age Settlement of Bihar. Agam Kala Prakashan, Delhi.

Pappu, R.S., 2001. Acheulian Culture in Peninsular India. D.K. Printworld, New Delhi.

Pappu, S., 1991. Robert Bruce Foote and the formation processes of the archaeological record. Bulletin of the Deccan College Post-Graduate and Research Institute 51–52, 647–654.

Pappu, S., 1996. Reinvestigation of the prehistoric archaeological record in the Kortallaiyar Basin, Tamil Nadu. Man and Environment XXI (1), 1–23.

Pappu, S., 2001a. Middle Palaeolithic stone tool technology in the Kortallaiyar Basin, South India. Antiquity 75, 107–117.

Pappu, S., 2001b. A Re-examination of the Palaeolithic Archaeological Record of Northern Tamil Nadu, South India. BAR International Series 1003, Oxford.

Pappu, S., Gunnell, Y., Taieb, Brugal, J-P., Anupama, K., Sukumar, R., Kumar, A. Excavations at the Palaeolithic site of Attirampakkam, South India. Antiquity 77(297), online project gallery.

Pappu, S., Gunnell, Y., Taieb, M., Brugal, J-P., Touchard, Y., 2003. Ongoing excavations at the Palaeolithic site of Attirampakkam, South India: preliminary findings. Current Anthropology 44, 591–597.

Pappu, S., Kumar, A., 2005. Preliminary observations on the Acheulian assemblages from Attirampakkam, Tamil Nadu. Paper Presented at the Workshop On Acheulian Large Cutting Tools, The Hebrew University Of Jerusalem, The Institute For Advanced Studies.

Pappu, S., Gunnell, Y, Taieb, M., Kumar, A., 2004. Preliminary report on excavations at the Palaeolithic site of Attirampakkam, Tamil Nadu (1999–2004). Man and Environment XXIX (2), 1–17.

Pappu, S., 1999. A study of natural site formation processes in the Kortallaiyar Basin, Tamil Nadu, South India. Geoarchaeology 14 (2), 127–150.

Paterson, T.T., 1939. Stratigraphic and typologic sequence of the Madras Palaeolithic Industries. In: De Terra, H., Paterson, T.T. (Eds.), Studies on the Ice Age in India and Associated Human Cultures. Carnegie Institution of Washington Publication No. 493, Washington, pp. 327–330.

Petraglia, M.D., 1998. The Lower Palaeolithic of India and its bearing on the Asian record. In: Petraglia, M.D., Korisettar, R. (Eds.), Early Human Behaviour in Global Context: The Rise and Diversity of the Lower Palaeolithic Record. Routledge, London, pp. 343–390.

Petraglia, M.D., LaPorta, P., Paddayya, K., 1999. The first Acheulian quarry in India: stone tool manufacture, biface morphology and behavior. Journal of Anthropological Research 55, 39–70.

Petraglia, M.D., Noll, M., 2001. Indian Palaeolithic collections in the Smithsonian Institution: international exchanges and their archaeological potential. Man and Environment XXVI (1), 57–68.

Sankalia, H.D., 1974. Prehistory and Protohistory of India and Pakistan. Deccan College, Pune.

Soundara Rajan, K.V., 1966. Prehistoric and Protohistoric Madras (unpublished document). Deccan College Postgraduate and Research Institute, Pune.

Srinivasa Ayyangar, P.T., 1988 (reprint). The Stone Age in India. Asian Educational Services, New Delhi.

Swami, A., 1976. Archaeology of the Chingleput District (Prehistoric Period). Ph.D Dissertation, University of Madras.

Zeuner, F.E., 1949. Extracts from unpublished diary of F.E. Zeuner (Box #6, Diary 13). Library, University College, London.

7. Was *Homo heidelbergensis* in South Asia? *A test using the Narmada fossil from central India*

SHEELA ATHREYA

Department of Anthropology
Texas A&M University
College Station, Texas, 77843
USA
athreya@tamu.edu

Introduction

Substantial archaeological evidence from South Asia confirms the fact that hominin occupation of the Indian subcontinent was widespread by the Middle Pleistocene (Mishra et al., 1995; Petraglia, 2001; Paddayya et al., 2002). Only one South Asian specimen, a partially complete cranium, is considered a definitive hominin from this time period: the Narmada fossil. This specimen was discovered in 1982 (Sonakia, 1984) and subsequently described by de Lumley and Sonakia (1985) and Kennedy et al. (1991). The taxonomic classification of Narmada has been a matter of contention since its discovery. Initially assigned to *H. erectus*, Narmada has also been classified as a member of the "archaic" grade of *H. sapiens* (Kennedy and Chiment, 1991; Cameron et al., 2004).

Debate over the taxonomic affinity of Narmada is due in part to its insecure dating as well as its mosaic morphology. However, shifting theoretical perspectives on the taxonomy of Middle Pleistocene *Homo* have also contributed to the debate. While previous studies have sought to determine whether Narmada was *H. erectus* or *H. sapiens*, the resurrection of the taxon *H. heidelbergensis* to encompass many intermediary fossils between these two groups reconfigures the question of Narmada's taxonomic status.

In this study I examine the morphological features of the Narmada fossil using a comparative sample of Middle Pleistocene specimens designated by many as *H. heidelbergensis*. Early and Late Pleistocene specimens are included, as well as a sample of early Holocene specimens from archaeological sites throughout the subcontinent. Finally, I include a sample of Andaman Islanders for comparison. Narmada's classification is evaluated using a statistical technique, multinomial logistic regression. This method is comparable to discriminant function analysis which is often used to evaluate the taxonomic affinities of a new specimen. However, unlike discriminant function analysis, multinomial

M.D. Petraglia and B. Allchin (eds.), The Evolution and History of Human Populations
in South Asia, 137–170.

logistic regression is robust against viola-tions of the assumptions of large group sizes and equal covariance matrices among groups (Lesaffre and Albert, 1989), and is thus useful for fossil hominin studies that almost always deal with small data sets and uneven group sizes. The goal is to identify the morpho-logical affinities of the Narmada fossil relative to comparable temporal and regional groups.

Background

Previous studies of the Narmada hominin have focused on the taxonomic affinity of this fossil in light of a question that has been of interest since the first reports of fossil finds of early humans by Black (1926) in Asia, Weiden-reich (1943) in China, and Dubois (1937) in Indonesia. The issue pertains to the nature of evolutionary trajectories in Asia during the Pleistocene. Though these authors considered their finds to be ancestral to *H. sapiens*, a paradigm shift occurred in the late 1940s. It was arguably facilitated by Hallam Movius, then at Harvard, who reviewed the archae-ological record and, based largely on the absence of handaxes suggested that:

It seems very unlikely that this vast area [Asia] could ever have played a vital and dynamic role in early human evolution, although very primitive forms of Early Man apparently persisted there long after types at a comparable stage of physical evolution had become extinct elsewhere. (Movius, 1948:411)

Thus, the notion that Asia was occupied by an evolutionary dead-end lineage of *H. erectus* has a long history in paleoan-thropology. According to this model, the Narmada hominin, as an Asian specimen, would automatically belong to *H. erectus* and therefore would be unrelated to subsequent modern *H. sapiens* populations.

Today, Asian *H. erectus* populations are seen by some as having a role in the ancestry of modern Asian populations (Pope, 1988; Etler, 1990) and by others as part of an evolutionary dead end lineage (Wang 1979 in

Wu, 1981; Stringer, 1984; Rightmire, 1990). The fate of *H. erectus* has relevance for Narmada, since it was ultimately assigned to this taxonomic category three years after its discovery by de Lumley and Sonakia (1985) in their detailed description of the fossil.

However, both prior and subsequent to its designation as an Asian *H. erectus*, the Narmada hominin's status was considered indeterminate. This was due in part to its "transitional" morphology. In his initial publi-cation of the specimen in Western scien-tific literature, Sonakia (1985a) described it simply as *Homo* sp. Following that, Kennedy et al. (1991) revisited the assignment of the Narmada hominin to *H. erectus* and demon-strated that it possessed a suite of *H. erectus* and *H. sapiens* traits, much like other Middle Pleistocene finds such as Petralona (Murrill, 1981), Ndutu (Rightmire, 1983) and Dali (Wu, 1981). These authors suggested that Narmada belonged to the transitional group of early "archaic" *H. sapiens* though they never explicitly supported the use of this term as describing a proper taxonomic unit, and left out a formal proposal for nomenclature from their conclusions.

Narmada's insecure taxonomic designation reflected a larger discussion taking place in paleoanthropology in the 1980s that perceived the terms "*H. erectus*," "Neandertal" and "*H. sapiens*" as inadequate to deal with the range of morphology observed in the fossils of the Middle and early Late Pleistocene, and particularly among Middle Pleistocene *Homo* (Santa Luca, 1978; Tattersall, 1986). In the early 1980's the term "archaic" *H. sapiens* was applied by Howells (1980) to a growing group of specimens that possessed a mosaic of *H. erectus* and *H. sapiens* traits. This was actually a revival of a term that had first been used by Mayr (1950) several decades earlier in a paper that ironically was a criticism of the excess of species names present in human phylogeny.

The term "archaic" *Homo sapiens* is not formal scientific nomenclature, lacking the requisite suite of defining features to classify specimens grouped in this category. In general, those that have resisted more traditional classification are included. They share a temporal niche, a larger-than-*erectus*-like cranial capacity, but nothing more (Bilsborough, 1976; Howell, 1994). Though not a proper taxonomic designation, this term was adopted and applied as formal scientific nomenclature in studies for the next decade (Santa Luca, 1978; Cronin et al., 1981; Bräuer, 1984; Clarke, 1990). Tattersall (1986:169) noted the weakness of this system, saying:

[T]he urge to consider forms as diverse as Petralona, Steinheim, Neandertal and you and me in the single species *H. sapiens* must be sociological in origin...[T]he only rational explanation for the taxonomic corralling together of these widely differing fossils...is the setting of an unconscious "cerebral Rubicon" perhaps at somewhere around 1200 ml. The fact that such a Rubicon, unconscious or otherwise, would exclude a good many members of the living species from *H. sapiens* has gone unremarked.

That is, the emphasis on a single trait (brain size) allowed fossils from the Middle Pleistocene to be included in the species *H. sapiens* with the "archaic" qualification denoting that they were morphologically unlike modern humans in many other ways. The widespread use of this terminology virtually elevated it to a formal taxonomic designation. Although flawed, it served as a heuristic device and reflected the dominant impression of the time regarding the complex mosaic features of these specimens and the importance of cranial capacity in defining *H. sapiens*.

Within this paradigm, Kennedy (1992) reviewed the debate over Narmada's status as *H. erectus* (and qualifications therein such as "evolved" *erectus*) vs. *H. sapiens*. They argued that it belonged with specimens such as Arago and Petralona, which were then being assigned to the species *H. sapiens*. The Narmada hominin's morphological combination of *erectus*- and *sapiens*-like traits plus its modern-like estimated cranial capacity of 1208 ml placed it in the company of these

similar fossils that were designated "archaic" *Homo sapiens* because of their estimated brain size.

More recently, Cameron et al. (2004) applied a cladistic analysis to the Narmada specimen. Their paper was significant in that it used cladistic methods and the identification of uniquely derived traits relative to sister taxa and outgroups to evaluate this specimen's classification. Based on this, they supported the results of Kennedy and others who argued that Narmada possessed phylogenetically significant traits other than just brain size to distinguish it from *H. erectus*. They concluded that the Narmada hominin should be classified as an "archaic" *H. sapiens*, not *H. erectus*.

Recently, the term "*H. heidelbergensis*" has become a common way to classify many of these specimens (Rightmire, 1988; Carbonell et al., 1995). The resurrection of this taxon to refer to many of the transitional Middle Pleistocene fossils known from Eurasia and Africa was initially proposed around the time of the discovery of the Narmada hominin (Rightmire, 1985) and was gradually adopted over the subsequent two decades. The almost universal abandonment of the term "archaic *Homo sapiens*" today reflects the fact that this was inadequate and unsatisfactory nomenclature in discussions of Pleistocene *Homo* phylogeny and systematics.

These previous studies illustrate the extremely complex mosaic nature of craniofacial variability during the Middle Pleistocene and the sometimes inadequate taxonomic terminology to deal with it. However, they also raise further questions about Middle Pleistocene hominin phylogeny, particularly with reference to the Narmada specimen. The last publication to address the taxonomic position of Narmada is Cameron et al.'s (2004) paper. But given the fact that the term "archaic" *Homo sapiens* is not actually a proper taxonomic designation, the question remains: to what taxon does the Narmada hominin belong? Likewise, the resurrection of the term *H. heidelbergensis* demands an inquiry as to whether Narmada belongs to this taxon. Yet, the adoption of a species name for these fossils indicates that some researchers envision a speciation event in the Middle Pleistocene, and if this is the case, what is the role of India in such an evolutionary scenario?

These questions are indicative of two conflicting paradigms that are relevant for understanding the taxonomic status of Narmada and India's role in Middle Pleistocene evolutionary scenarios. The first, as discussed earlier, is the question of whether *all* transitional fossils with relatively large cranial capacities should be considered *H. heidelbergensis*. At present, Rightmire, who has studied the African, European, and East Asian Middle Pleistocene material most extensively, argues for an Asian presence of this species (Rightmire, 2004). The alternative paradigm is based on the legacy of Movius, and considers the Asian Middle Pleistocene fossils to belong to an evolutionary dead-end lineage—either *H. erectus*, or a derived (but nonetheless extinct population of) *H. erectus*. Given the position of the Indian subcontinent between the geographic regions of Asia and Africa, it is unclear whether the East/Southeast Asian model is necessarily the most appropriate one by which to interpret the Indian fossil material. Therefore, a broad geographic comparative review is warranted.

Predictions

The various models that have been developed to evaluate the evolutionary history of *H. heidelbergensis* share the fact that they are based on an Asian sample derived almost exclusively from Northeast and Southeast Asia. It remains unclear where the Indian subcontinent, and more specifically the Narmada hominin, is situated within these scenarios of Middle Pleistocene hominin evolution. In an attempt to resolve this ambiguity, I examine the Narmada specimen

with respect to four current configurations of the taxonomic groups of the Middle Pleistocene.

The four phylogenetic scenarios used here represent various views on the geographic distribution of *H. heidelbergensis* and *H. erectus*. The main question being addressed is whether Narmada shares its strongest morphological affinities with *H. heidelbergensis* vs. *H. erectus*, and evaluates this question by testing a combination of distribution patterns of these species. My premise in this study rests upon the acceptance of *heidelbergensis* as a valid taxonomic and evolutionary species. Since an evaluation of the validity of *H. heidelbergensis* is beyond the scope of this chapter, I assume here is that it is a valid species.

The answer to the question of Narmada's taxonomic status is likely contingent upon which definition of *H. heidelbergensis* is used. Some believe that *H. heidelbergensis* is a pan-Old World phenomenon (Rightmire, 2004) and classify "transitional" specimens as belonging to this species. In contrast, others (Carbonell et al., 1995; Hublin, 1996; Dean et al., 1998) argue that it is restricted to Europe and is exclusively ancestral to the Neandertals, who are conventionally believed to be absent in East Asia.

My predictions are based on analyses that are designed to evaluate morphological affinity, probability of classification within each group, and typicality of a given group. The results are expected to show that Narmada will show the strongest affinities with *H. heidelbergensis* when that group is defined to include other Asian "transitional" Middle Pleistocene specimens such as Dali and Jinniushan. When *H. heidelbergensis* is defined only to include the African and European, or only the European "transitional" fossils, Narmada is not expected to show a strong affinity with this group. This prediction is based on the mosaic *erectus-sapiens* morphology that I believe is in part a reflection of a mosaic of African and Asian traits, given this fossil's provenience at the geographic crossroads between these two regions. Though Narmada is not typical of either region in its overall morphology, its stronger affinities are predicted to lie with the Asian fossils.

Materials

Sample

The Narmada specimen (Figures 1, 2) consists of the complete right half of a skullcap with part of the left parietal attached. Most of the facial skeleton is missing with the exception of part of the right orbit. No dental material was recovered nor was a mandible or any postcranial material found at the time of the initial excavation. The gracility of the cranial features strongly suggests that it is a female. The specimen was discovered in a boulder conglomerate along the Narmada River in the village of Hathnora, in the Hoshangabad district of the central Indian state of Madhya Pradesh in 1982 (Sonakia, 1984, 1985b). Recently a minimum age of 236 ka was obtained for a bovid scapular fragment found with the

original cranial specimen based on gamma spectrometric U-series dating (Cameron et al., 2004). Although unverified by further chrono-metric dating, this age is consistent with the faunal material found in association with the specimen as well as studies of the age of the Surajkund geological formation within which the specimen was found (Agrawal et al., 1988; Tiwari and Bhai, 1997). It places the fossil at the latter part of the Middle Pleistocene. Further details of the specimen's morphology and preservation and the context of its discovery can be found in the literature (Sonakia, 1984; de Lumley and Sonakia, 1985; Kennedy and Chiment, 1991; Kennedy et al., 1991).

Tables 1a and 1b list the fossil and modern human specimens, respectively, that are included in this study. The relevant comparative fossil material dates to the Middle Pleistocene and has a large Old World distribution, from Western Europe to North and Southeast Asia. Several of the Asian fossils date to both the Middle and Late Pleistocene and are considered *H. erectus*. A set of temporally neighboring Early and Late Pleistocene specimens from all regions of the Old World, from Western Europe to Australia, is

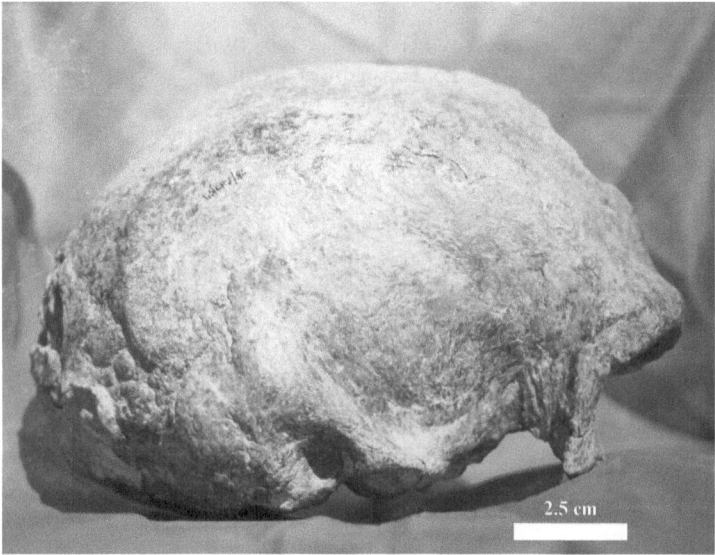

Figure 1. Lateral view of Narmada cranium. Photograph courtesy of A. Sonakia

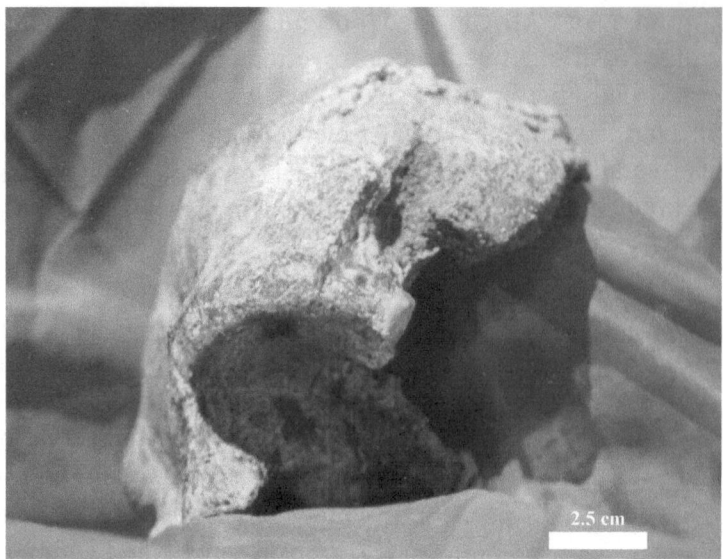

Figure 2. Frontal view of Narmada cranium. Photograph courtesy of A. Sonakia

Table 1a. Fossil sample

Site	References
Narmada	Kennedy et al., 1991; Cameron et al., 2004
Amud	Suzuki and Takai, 1970; Kennedy et al., 1991
Arago 21	Kennedy et al., 1991
Bodo	Murrill, 1981; Adefris, 1992; Conroy et al., 2000
Cohuna	Brown, 2005
Combe Capelle	Vandermeersch, 1981
Coobool Creek (N = 28)	Brown, 1989, 2005
Cro Magnon 1, 2, 3	Vandermeersch, 1981; Kennedy et al., 1991
Daka VP-2/66	Asfaw, 1983
Dali	Wu and Poirier, 1995; Wu, 1981
Dmanisi 2280, 3444	Lordkipanidze et al., 2005
Elandsfontein	Singer, 1954, 1958
ER3733, 3883	Kennedy et al., 1991; Adefris, 1992
Florisbad	Cast
Forbes Quarry	McCown and Keith, 1939; Suzuki and Takai, 1970; Smith, 1976; Vandermeersch, 1981
Herto VP-16/1	White et al., 2003
Hexian	Wu, 1982; Wu and Poirier, 1995
Irhoud 1	Hublin, 1992
Jinniushan	Wu, 1988; Wu and Poirier, 1995
Kabwe	Singer, 1954, 1958; Santa Luca, 1980; Rightmire, 1990
Kanalda	Storm, 1995; Brown, 2005
Keilor 12	Wunderly, 1943; Weidenreich, 1945; Storm, 1995
Kow Swamp 3, 5	Brown, 2005
Krapina C, E	Radovci'c et al., 1988
La Chapelle aux Saints	Santa Luca, 1980; Murrill, 1981; Kennedy et al., 1991; Piperno and Scichilone, 1991

(Continued)

Table 1a. (Continued)

Site	References
La Ferrassie	Santa Luca, 1980; Murrill, 1981; Kennedy et al., 1991; Piperno and Scichilone, 1991; Lahr and Wright, 1996
La Quina 5	Suzuki and Takai, 1970; Vandermeersch, 1981
Laetoli 18	Cohen, 1996; Lahr and Wright, 1996
Lake Nitchie 15	Brown, 1989
Liujiang	Woo, 1959; Wu and Braüer, 1993; Wu and Poirier, 1995; Shen et al., 2002
Minatogawa 1, 2, 4	Suzuki and Hanihara, 1982
Mladec 1, 2, 5, 6	Matiegka, 1934; Frayer, 1986
Monte Circeo	Piperno and Scichilone, 1991
Nacurrie 2	Brown, 2005
Nahal Ein Gev	Hershkovitz et al., 1995; Lahr and Wright, 1996; Stock et al., 2005
Nanjing 1	Zhao et al., 2001; Wu et al., 2002
Ndutu	Clarke, 1990; Rightmire, 1990
Neandertal	Piperno and Scichilone, 1991
Ngandong 1, 6, 7, 10, 11, 12	Weidenreich, 1951; Santa Luca, 1980; Delson et al., 2001
Ngawi 1	Widianto and Zeitoun, 2003
OH9	Kennedy et al., 1991; Delson et al., 2001
Ohalo2	Hershkovitz et al., 1995; Lahr and Wright, 1996; Stock et al., 2005
Omo 1, 2	Haile-Selassie et al., 2004
Petralona	Stringer et al., 1979; Murrill, 1981; Rightmire, 1990; Kennedy et al., 1991
Predmost 1, 3, 4, 9, 10	Matiegka, 1934
Qafzeh 6, 9	Vandermeersch, 1981; Lahr and Wright, 1996
Saccopastore 1	Howell, 1960; Kennedy et al., 1991
Sambungmacan 1, 3	Delson et al., 2001
Sambungmacan 4	Baba et al., 2003
Sangiran 2, 17	Weidenreich, 1951; Santa Luca, 1980; Rightmire, 1990; Aziz et al., 1996; Delson et al., 2001
Shanidar 1	Murrill, 1981; Trinkaus, 1983
Sima de los Huesos 4, 5	Arsuaga et al., 1997
Skhul 4, 5, 9	Snow, 1953; Suzuki and Takai, 1970; Vandermeersch, 1981; Lahr and Wright, 1996
Spy 1, 2	Suzuki and Takai, 1970; Vandermeersch, 1981
Steinheim	Howell, 1960; Kennedy et al., 1991
Swanscombe	Kennedy et al., 1991
Tabun 1	McCown and Keith, 1939; Vandermeersch, 1981
Trinil	Rightmire, 1990; Delson et al., 2001
Wadjak 1	Santa Luca, 1980; Storm, 1995; Brown, 2005
Zhoukoudian 2, 3, 5, 10, 11, 12	Wu and Poirier, 1995
Ziyang	Woo, 1958; Li and Zhang, 1984; Wu and Braüer, 1993; Wu and Poirier, 1995
ZKD Upper Cave 101, 102, 103	Kamminga and Wright, 1988; Kennedy et al., 1991; Wu and Poirier, 1995

Table 1b. South Asian archaeological and recent modern human sample

Site	References
Zuttiyeh	Cast
Bagor IV	Kennedy, 1982
BellanPalassa (N = 4)	Kennedy, 1965
Brahmagiri (N = 4)	Sarkar, 1960
Burzahom BZH 1, 6	Basu and Pal, 1980
Chanhu-Daro	Krogman and Sassaman, 1943
Harappa R37, H1	Bernhard, 1998
Harappa (N = 26)	Reddy and Reddy, 1987
Kodumanal (N = 4)	Reddy and Reddy, 1987, 2004
KumharTekri (N = 16)	Gupta and Basu, 1970
Langhnaj (N = 4)	Ehrhardt and Kennedy, 1965
Mahadaha (N = 5)	Kennedy et al., 1992
Mohenjo-Daro 1	Sewall and Guha, 1973
Nagarjunakonda 1a	Gupta et al., 1970
Ruamgarh (N = 2)	Kennedy, 1972
SaraiNaharRai1970–4	Dutta et al., 1972; Kennedy et al., 1986
Tekkalakota 15	Reddy and Reddy, 1987
Timargarha (N = 18)	Bernhard, 1967
Yeleswaram 1/61	Kennedy, 2000
Andaman Islanders (N = 54)	Howells, 1996
Onge L1, L5	Gupta et al., 1960

included to evaluate the strength of taxonomic categories that are considered ancestral or descendant to these populations.

In addition, two groups of post-Pleistocene modern human groups are included, both from South Asia. Several authors (Wolpoff et al., 1984; 2004; Etler, 1988, 2004) argue that Asia demonstrates regional morphological continuity between local Middle and Late Pleistocene populations and modern human inhabitants. The morphology of the Narmada hominin is thus compared not just to other Middle Pleistocene fossils, but also to subsequent early Holocene and recent South Asian populations to investigate the presence of regional continuity and evaluate trends through time. The early Holocene data are derived from Kennedy's work on archaeological populations in the subcontinent, from sites dating from the Neolithic through the Bronze Age and Roman times.

The recent modern human sample consists of a mixed-sex series of Andaman Islanders from Howells' data set (1973) (see also Stock et al., this volume, for an analysis). I included this population based on the findings of recent genetic studies (Thangaraj et al., 2003, 2005) suggesting their close affinities to modern Africans and hence their possible status as descendants of the earliest modern African immigrants into South Asia (see Endicott et al., this volume). Studies proposing a link between the postcranial morphology of Narmada and that of Andaman Islanders presume morphological continuity between these two populations (Sankhyan, 1997), a possibility that requires further testing. Therefore, this sample was included to allow for a test of this hypothesis.

The variables consist of published linear craniometric caliper measurements (Table 2) taken from original reports and in some cases from secondary studies that used published measurements. All measurements are taken on the original specimens with the exception of Zuttiyeh and Florisbad for which high-quality first-generation casts were used for some of the measurements. Since the Narmada fossil is the specimen of interest in this study, only measurements available on this specimen are included in the analysis.

The majority of studies of South Asian archaeological material follow the measurement protocols of Martin and Saller (1957) while the studies of the Pleistocene fossil material follow these as well as similar standard morphometric protocols of Howells (1973), though occasionally the technique was not reported. All individuals included in this study are non-pathological adults, defined as having an erupted third molar and/or basilar suture closure. In some cases when the face and cranial base were not preserved, age estimation was accomplished by evaluating cranial suture closure. The sample is mixed sex and size correction, reviewed below, accounted for size differences due to sex. Shape differences due to sex account for a portion of the variation observed in the results and will be considered in the discussion of results.

Methods

Data Analysis

Since my main purpose in this study is to investigate the morphological affinities of the Narmada hominin, I use two techniques designed to evaluate similarity and distance. The first analysis of the data is a set of multinomial logistic regressions (MLR or multinomial logit) used to identify the affinities of the Narmada hominin within predefined regional and temporal categories of Pleistocene *Homo*. Like discriminant function analysis (DFA), MLR is a method of allocating unknown cases to one of the *a priori* defined groups in an analysis and generating probabilities that an unclassified case belongs to a predefined group. Compared to DFA, multinomial logit is considerably more robust against violations of the assumptions of large and equal sample group sizes (Long and Freese, 2001), so it is preferable in paleoanthropological studies where small and unequal group sizes are common.

Multinomial Logistic Regression vs. Discriminant Function Analysis The original objective of discriminant function analysis developed by Fisher (1936) was to produce linear combinations of variables that maximized the ratio of the among-group to within-group variance and thus provide maximum separation between two groups. The method was then extended to allocate unknown specimens to one of multiple groups in biological classification (Rao, 1948). Discriminant function analysis in paleoanthropology is a commonly used method, and is usually applied to the discovery of a new specimen of unknown taxonomic affinity (e.g., Trinkaus et al., 2003). It is based on the assumption that within-group covariance matrices are the same for all populations, and that the data are normally distributed (Reyment et al., 1984). In addition, the method is sensitive to small and uneven group sizes (Manly, 1994).

In contrast, multinomial logit uses a maximum-likelihood estimation procedure to maximize the probability of getting the observed results. This method is similar to linear regression in that it measures the contribution of an independent variable to variations in the dependent variable. However, it is different in two key ways. First, rather than an ordinary least-squares model, MLR is based on a nonlinear (logistic) model of the relationship between independent and dependent variables. Second, the probability of a given outcome is always estimated using a maximum likelihood function. Multinomial logit is an extension of a basic logistic regression, which is assumes a binary outcome (Long, 1997), with the multinomial model allowing for more than two outcomes. In both the basic and multinomial models, the probability values of a given outcome fall between 0.0 and 1.0 (Bull and Donner, 1987) in the same way that posterior probabilities in a discriminant function analysis do (Campbell, 1984).

The use of MLR is desirable in paleoanthropological studies such as this for a number of reasons. First, the maximum-likelihood estimator is superior to an ordinary least squares regression model in studies where the sample sizes are smaller (Long, 1997), provided they are not too small. The dataset here is sufficiently large (i.e., total $n >$ 100; each group n is not less than the number of parameters being estimated) to achieve statistically unbiased results (Long and Freese, 2001). The posterior probability values of DFA and MLR are virtually identical when the assumptions of DFA are met (Efron, 1975). However, when covariance matrices are unequal and/or data are nonnormally distributed, DFA has a tendency to produce biased estimates of posterior probabilities; thus, MLR and the use of the maximum likelihood method are preferable (Press and Wilson, 1978).

Imputation of Missing Data In this study, as in many paleoanthropological analyses, it is necessary to deal with the fact that the data set was incomplete due to poor preservation of fossil specimens. The number of variables with missing data in this analysis is shown in Table 2. The vast majority of multivariate statistical techniques require a complete data matrix and various options exist for dealing with missing data. The most commonly known methods for dealing with missing data are listwise deletion (eliminating cases with missing data) and pairwise deletion (eliminating particular combinations of specimens for different statistical analyses). Both of these are undesirable options in paleoanthropology because most specimens are missing at least some data points, and the resulting data matrix would be so small as to be untenable for statistical analysis. Another approach is to fill in (impute) missing data using mean substitution, but this has a tendency to create artificial similarities among fossils even when applied under the strictest of conditions (Holt and Benfer, 2000). Perhaps more importantly, from a statistical standpoint this method tends to create underestimated variances and covariances, as does the multiple regression method of imputing missing data (Marc, 2002).

A preferable approach is the expectation-maximizing (EM) procedure for predicting missing data values. This method uses the maximum likelihood estimation to predict missing values based on the covariance matrix of complete characters. The principal components method also uses the covariance

Table 2. Information on missing data

Variable	Howells' Abbreviation	Martin and Saller Number	Number Missing	% Missing
BASION-BREGMA HEIGHT	BBH	17	64	23.19
BIASTERION BREADTH	ASB	12	86	31.16
FRONTAL SAGITTAL ARC	FAA	26	47	17.03
FRONTAL SAGITTAL CHORD	FAC	29	15	5.43
MAXIMUM CRANIAL BREADTH	XCB	8	37	13.41
MAXIMUM CRANIAL LENGTH	GOL	1	4	1.45
MINIMUM FRONTAL BREADTH	WFB	9	99	35.87
OCCIPITAL SAGITTAL ARC	OAA	28	75	27.17
OCCIPITAL SAGITTAL CHORD	OCC	31	80	28.99
ORBIT BREADTH	OBB	51	61	22.10
ORBIT HEIGHT	OBH	52	57	20.65
PARIETAL SAGITTAL ARC	PAA	27	45	16.30
PARIETAL SAGITTAL CHORD	PAC	30	18	6.52

Number of Cases = 276

matrix of known characters to predict missing values by regressing those with missing values onto the principal components of the complete-character covariance matrix. The EM approach, though similar, has been shown to be superior in sets with a larger percentage of missing data (Strauss et al., 2003) because it continually re-estimates of the parameters of the data set and uses those new parameters to re-estimate missing values until the two converge.

I used the EM algorithm to obtain maximum-likelihood estimates of the parameters of the incomplete data set with the software package NORM (Schafer and Olsen, 1998; Schafer, 1999). These parameters were then used to create imputations for the missing data. Because the data in this study span the entire Pleistocene and Holocene, the data set was initially broken down into subsets based on geological period before imputation. However, for most subsets, the model did not converge due to the combined problems of too few cases and too many missing data points. It was therefore necessary to use the entire data set to estimate the parameters for the missing data. I used dummy variables of geological period (corresponding to Early, Middle, Late Pleistocene, Historic Modern and Recent Modern), geographic region (Western Europe, Eastern Europe, West Asia, East Asia, Southeast Asia, South Asia, Australia and Africa), and sex as predictor variables in parameter estimation and the imputation of missing values. This was done to avoid having the larger Historic and Modern South Asian population values unduly influence the imputation of missing values for the smaller Pleistocene and non-South Asian populations.

I then created four data sets of random imputations from the parameters estimated with the EM algorithm. Each data set was created from 5,000 iterations, and the results of the four imputed data sets were subject to the same data analysis routines used in the

final study to determine if the outcomes were similar with respect to the posterior probability assignments of Narmada and other fossil specimens. In each case, the results were consistent. While the probability values varied somewhat, each specimen's highest posterior probability consistently assigned it to the same group with each data set, demonstrating the stability of this method in producing repeatable results. I averaged the four imputed data sets to form the data set used in the final analyses.

It is important to reiterate that the fossil in question here, the Narmada specimen, did not contain any missing values. Thus, these analyses are based solely on comparisons of measurements taken on the neurocranium and upper facial/orbital regions. More importantly, none of the data points for Narmada were imputed from other specimens in the study.

Principal Components Analysis Like DFA, MLR assumes that variables are uncorrelated. To accommodate this assumption, the data were first size-standardized following Darroch and Mossiman's (1985) procedure and then reduced using principal components analysis to create orthogonal variables that redescribed the data set. Of the thirteen principal components obtained the first eight, representing 92% of the total variance in the data set, were scored.

Looking at the eigenvector loadings for principal components (Table 3), the first principal component (PC) exhibits high loadings for measures of maximum breadth and length, reflecting variance in overall shape. The second PC loads highly for parietal and occipital arc measurements, and the third for frontal and occipital measurements. The fourth and fifth PC likewise seem to be capturing variance in the frontal and facial measurements used here. An evaluation of these loadings enables an interpretation of the morphological traits that are being captured by each PC and will assist in interpreting the final results.

Table 3. Eigenvalues of principal components

Component	Eigenvalue	Difference	Proportion	Cumulative
1	4.37973	2.39531	0.3369	0.3369
2	1.98442	0.41648	0.1526	0.4895
3	1.56794	0.14176	0.1206	0.6102
4	1.42617	0.51923	0.1097	0.7199
5	0.90694	0.08815	0.0698	0.7896
6	0.81879	0.32210	0.0630	0.8526
7	0.49669	0.03621	0.0382	0.8908
8	0.46048	0.07784	0.0354	0.9262
9	0.38264	0.13075	0.0294	0.9557
10	0.25189	0.04319	0.0194	0.9751
11	0.20871	0.09911	0.0161	0.9911
12	0.10959	0.10358	0.0084	0.9995
13	0.00602		0.0005	1.0000

Relying on a small number of primary principal components to evaluate geographical patterns and relationships has been shown to be a useful and powerful method (Cavalli-Sforza et al., 1993) since the first PC's are those that maximize variance among the variables. Following Wagner (1984), the number of independent traits discarded by failing to include all of the principal components as variables can be calculated by summing the eigenvalues of the unused PC's. The retention of eight principal components discarded less than one independent trait (.96).

Multinomial Logistic Regression Analysis Multinomial logistic regression analyses were performed using Intercooled Stata 8.0 (StataCorp, 2004). One limitation of multinomial logistic regression is that it requires the data set to consist of fewer variables than *a priori* groups (N). These six analyses thus used different numbers of principal components as variables. Models I and III had six predefined groups, Model II and the Regional analysis each had eight, Model IV had seven, and the analysis by Time (geological period) used five *a priori* groups. The number of PC's used was N-1 for each analysis.

Typicality Probabilities and Discriminant Function Analysis Unlike discriminant function analysis, MLR does not allow for the evaluation of typicality probabilities. Typicality probabilities assess the likelihood that a given case is typical of any of the groups regardless of its *a priori* designation, unlike posterior probabilities where each case is assumed to belong to one of the *a priori* defined groups (Albrecht, 1992). Typicality probability is calculated from the cumulative χ^2 of the absolute value of the Mahalanobis's distance with v (number of discriminant functions) degrees of freedom. The typicality probability is also known as conditional probability, and can be understood as the probability of a given discriminant function score given membership in the most likely group. Posterior probabilities evaluate the relative affinities of an individual for each of the predefined groups, while typicality probabilities evaluate the affinity of an individual to a single group without reference to the other populations. Therefore, the sum total of all posterior probabilities for a single individual across groups must add up to 1.0 while the typicality probabilities do not.

The difference between these two values reflects the distinction that exists between "forced" vs. "unforced" allocation procedures (Campbell, 1984). In the former, a case will be given a high posterior probability of belonging to a particular group if it is closer to that group's centroid than any other centroid—even if it is actually atypical of that group. In the latter, such a case will be given a low typicality probability reflecting the fact that, though a case is closer to one group centroid than any other, it is still not typical of that group and may in fact belong to another, undefined group.

In order to evaluate the typicality probability of the cases in this study, a formal discriminant function analysis was performed using SPSS. Despite the fact that the properties of this data set are undesirable for a DFA, the biases are mainly in overestimations of the posterior probability values. For this stage of the analysis only the DFA was used to compute typicality probabilities. These were evaluated in order to get a sense of how typical Narmada is of the groups to which it is assigned. While posterior probability values are computed from the discriminant function, which are affected by unequal covariance matrix size, typicality probability values are computed directly from the Mahalanobis' distance and as such unequal covariance matrices are less influential on their value (Reyment et al., 1984).

As indicated earlier, a Box's M test demonstrated that the groups defined in these analyses were, in all cases, significantly different with respect to their means and covariances. Therefore, the DFA was performed with separate-groups covariance matrices using SPSS. Since a separate-groups analysis was performed, cross-validation was not possible. Manly (1994) has suggested that failing to use cross-validation does not substantially increase the rate of misclassification, and Campbell (1980) suggests that it increases misclassification by at most 10%. Since the focus of this stage of the analysis was mainly to identify the position of the Narmada hominin relative to various group centroids, as opposed to evaluating the

posterior probability values for all specimens, the imperfections associated with using a separate-groups analysis was not considered a significantly confounding factor in the typicality probability values. Nevertheless, a consideration of the fact that they may be slightly flawed as a result of being computed from a DFA with unpooled matrices that was not cross-validated is taken into account in the discussion.

Evolutionary Models Tested

As stated earlier, the answer to the question of Narmada's taxonomic affinity is contingent upon the way each species is defined. Because of the methods used, the *a priori* assignment of the other fossils to which Narmada is being compared will determine how Narmada is classified. Therefore, I test four separate definitions of *H. heidelbergensis* and *H. erectus*. However, only one model can be correct and it is possible that none of these models capture the true phylogeny of Middle Pleistocene *Homo*, and I will consider this possibility when discussing the results. The *a priori* assignments of the comparative sample for each model are listed in Table 4. Narmada remained unclassified in all analyses.

The first model (**Model I**), following Rightmire (2004), considers *H. heidelbergensis* to have been in Asia. This is based on his proposal to classify certain Asian specimens such as Dali and Jinniushan as *H. heidelbergensis*. Therefore, this *a priori* designation is given to "transitional" fossils from *all* Old World regions occupied in the Middle Pleistocene. The African and Asian members of *H. erectus* are considered one species and are therefore given the same *a priori* classification. In this model, one possible but not necessary implication is that neither the Asian *H. erectus* nor *H. heidelbergensis* populations are dead-end lineages. This model is the only one that allows such a possibility.

Table 4. A priori classifications for each analysis

Fossil	Geol Period	Region	MODEL 1	MODEL 2	MODEL 3	MODEL 4
Narmada						
Amud	LATE PLEIST	W ASIA	NEANDERTAL	NEANDERTAL	NEANDERTAL	NEANDERTAL
Arago 21	MID PLEIST	W EUROPE	HEIDELBERG	HEIDELBERG	HEIDELBERG	HEIDELBERG
Bodo	MID PLEIST	AFRICA	HEIDELBERG	ARCHAIC	HEIDELBERG	HEIDELBERG
Cohuna	LATE PLEIST	AUSTRALIA	AMHS	AMHS	AMHS	AMHS
Combe Capelle	LATE PLEIST	W EUROPE	AMHS	AMHS	AMHS	AMHS
Coobool Creek (N = 28)	LATE PLEIST	AUSTRALIA	AMHS	AMHS	AMHS	AMHS
Cro Magnon 1, 2, 3	LATE PLEIST	W EUROPE	AMHS	AMHS	AMHS	AMHS
Daka VP-2/66	EARLY PLEIST	AFRICA	ERECTUS	ERGASTER	ERECTUS	ERGASTER
Dali	MID PLEIST	E ASIA	HEIDELBERG	ERECTUS	ERECTUS	ERECTUS
Dmanisi 2280, 3444	EARLY PLEIST	W ASIA	ERECTUS	ERGASTER	ERECTUS	ERGASTER
Elandsfontein	MID PLEIST	AFRICA	HEIDELBERG	ARCHAIC	HEIDELBERG	HEIDELBERG
ER3733, 3883	EARLY PLEIST	AFRICA	ERECTUS	ERGASTER	ERECTUS	ERGASTER
Florisbad	MID PLEIST	AFRICA	HEIDELBERG	ARCHAIC	HEIDELBERG	HEIDELBERG
Forbes' Quarry	LATE PLEIST	W EUROPE	NEANDERTAL	NEANDERTAL	NEANDERTAL	NEANDERTAL
Herto BOU VP-16/1	MID PLEIST	AFRICA	AMHS	AMHS	AMHS	AMHS
Hexian	MID PLEIST	E ASIA	ERECTUS	ERECTUS	ERECTUS	ERECTUS
Irhoud 1	MID PLEIST	AFRICA	HEIDELBERG	ARCHAIC	HEIDELBERG	HEIDELBERG
Jinniushan	MID PLEIST	E ASIA	HEIDELBERG	ERECTUS	ERECTUS	ERECTUS
Kabwe	MID PLEIST	AFRICA	HEIDELBERG	ARCHAIC	HEIDELBERG	HEIDELBERG
Kanalda	LATE PLEIST	AUSTRALIA	AMHS	AMHS	AMHS	AMHS
Keilor 12	LATE PLEIST	AUSTRALIA	AMHS	AMHS	AMHS	AMHS
Kow Swamp 3, 5	LATE PLEIST	AUSTRALIA	AMHS	AMHS	AMHS	AMHS
Krapina C, E	MID PLEIST	E EUROPE	NEANDERTAL	NEANDERTAL	NEANDERTAL	NEANDERTAL
La Chapelle aux Saints	LATE PLEIST	W EUROPE	NEANDERTAL	NEANDERTAL	NEANDERTAL	NEANDERTAL
La Ferrassie	LATE PLEIST	W EUROPE	NEANDERTAL	NEANDERTAL	NEANDERTAL	NEANDERTAL
La Quina 5	LATE PLEIST	W EUROPE	NEANDERTAL	NEANDERTAL	NEANDERTAL	NEANDERTAL
Laetoli 18	MID PLEIST	AFRICA	HEIDELBERG	ARCHAIC	HEIDELBERG	HEIDELBERG
Lake Nitchie 15	LATE PLEIST	AUSTRALIA	AMHS	AMHS	AMHS	AMHS
Liujiang	LATE PLEIST	E ASIA	AMHS	AMHS	AMHS	AMHS
Minatogawa 1, 2, 4	LATE PLEIST	E ASIA	AMHS	AMHS	AMHS	AMHS
Mladec 1, 2, 5, 6	LATE PLEIST	E EUROPE	AMHS	AMHS	AMHS	AMHS
Monte Circeo	LATE PLEIST	W EUROPE	NEANDERTAL	NEANDERTAL	NEANDERTAL	NEANDERTAL
Nacurrie 2	LATE PLEIST	AUSTRALIA	AMHS	AMHS	AMHS	AMHS
Nahal Ein Gev	LATE PLEIST	W ASIA	AMHS	AMHS	AMHS	AMHS

(Continued)

Table 4. (Continued)

Fossil	Geol Period	Region	MODEL 1	MODEL 2	MODEL 3	MODEL 4
Nanjing 1	MID PLEIST	E ASIA	ERECTUS	ERECTUS	ERECTUS	ERECTUS
Ndutu	MID PLEIST	AFRICA	HEIDELBERG	ARCHAIC	HEIDELBERG	HEIDELBERG
Neandertal	LATE PLEIST	W EUROPE	NEANDERTAL	NEANDERTAL	NEANDERTAL	NEANDERTAL
Ngandong 1, 6, 7, 10, 11, 12	LATE PLEIST	SE ASIA	ERECTUS	ERECTUS	ERECTUS	ERECTUS
Ngawi 1	LATE PLEIST	SE ASIA	ERECTUS	ERECTUS	ERECTUS	ERECTUS
OH9	EARLY PLEIST	AFRICA	ERECTUS	ERGASTER	ERECTUS	ERGASTER
Ohalo2	LATE PLEIST	W ASIA	AMHS	AMHS	AMHS	AMHS
Omo 1, 2	MID PLEIST	AFRICA	AMHS	AMHS	AMHS	AMHS
Petralona	MID PLEIST	E EUROPE	HEIDELBERG	HEIDELBERG	HEIDELBERG	HEIDELBERG
Predmost 1, 3, 4, 9, 10	LATE PLEIST	E EUROPE	AMHS	AMHS	AMHS	AMHS
Qafzeh 6, 9	LATE PLEIST	W ASIA	AMHS	AMHS	AMHS	AMHS
Saccopastore 1	LATE PLEIST	W EUROPE	NEANDERTAL	NEANDERTAL	NEANDERTAL	NEANDERTAL
Sambungmacan 1	MID PLEIST	SE ASIA	ERECTUS	ERECTUS	ERECTUS	ERECTUS
Sambungmacan 3, 4	LATE PLEIST	SE ASIA	ERECTUS	ERECTUS	ERECTUS	ERECTUS
Sangiran 2, 17	EARLY PLEIST	SE ASIA	ERECTUS	ERECTUS	ERECTUS	ERECTUS
Shanidar 1	LATE PLEIST	W ASIA	NEANDERTAL	NEANDERTAL	NEANDERTAL	NEANDERTAL
Skhul 4, 5, 9	LATE PLEIST	W ASIA	AMHS	AMHS	AMHS	AMHS
Sima de los Huesos 4, 5	MID PLEIST	W EUROPE	HEIDELBERG	HEIDELBERG	HEIDELBERG	HEIDELBERG
Spy 1, 2	LATE PLEIST	W EUROPE	NEANDERTAL	NEANDERTAL	NEANDERTAL	NEANDERTAL
Steinheim	MID PLEIST	W EUROPE	HEIDELBERG	HEIDELBERG	HEIDELBERG	HEIDELBERG
Swanscombe	MID PLEIST	W EUROPE	HEIDELBERG	HEIDELBERG	HEIDELBERG	HEIDELBERG
Tabun 1	LATE PLEIST	W ASIA	NEANDERTAL	NEANDERTAL	NEANDERTAL	NEANDERTAL
Trinil	EARLY PLEIST	SE ASIA	ERECTUS	ERECTUS	ERECTUS	ERECTUS
Wadjak 1	LATE PLEIST	AUSTRALIA	AMHS	AMHS	AMHS	AMHS
Ziyang	LATE PLEIST	E ASIA	AMHS	AMHS	AMHS	AMHS
Zhoukoudian 2 3, 5, 10, 11, 12	MID PLEIST	E ASIA	ERECTUS	ERECTUS	ERECTUS	ERECTUS
ZKD Upper Cave 101, 102, 103	LATE PLEIST	E ASIA	AMHS	AMHS	AMHS	AMHS
Zuttiyeh	MID PLEIST	W ASIA	HEIDELBERG	ARCHAIC	HEIDELBERG	HEIDELBERG
Bagor 4	HISTORIC MOD	S ASIA	HISTORIC MOD	HISTORIC MOD	HISTORIC MOD	HISTORIC MOD
Harappa	HISTORIC MOD	S ASIA	HISTORIC MOD	HISTORIC MOD	HISTORIC MOD	HISTORIC MOD
Kodumanal A1, 2; B2, 3	HISTORIC MOD	S ASIA	HISTORIC MOD	HISTORIC MOD	HISTORIC MOD	HISTORIC MOD
Kumhar Tekri 5, 11, 12, 13, 14, 15. 16, 17, 18, 19, 26, 27, 31, 32, 40	HISTORIC MOD	S ASIA	HISTORIC MOD	HISTORIC MOD	HISTORIC MOD	HISTORIC MOD
Langhnaj 1, 2, 3, 5	HISTORIC MOD	S ASIA	HISTORIC MOD	HISTORIC MOD	HISTORIC MOD	HISTORIC MOD

(Continued)

Table 4. (Continued)

Fossil	Geol Period	Region	MODEL 1	MODEL 2	MODEL 3	MODEL 4
Mahadaha 12, 19, 23, 24, 26	HISTORIC MOD	S ASIA	HISTORIC MOD	HISTORIC MOD	HISTORIC MOD	HISTORIC MOD
Mohenjo-Daro 1	HISTORIC MOD	S ASIA	HISTORIC MOD	HISTORIC MOD	HISTORIC MOD	HISTORIC MOD
Nagarjunakonda 1a	HISTORIC MOD	S ASIA	HISTORIC MOD	HISTORIC MOD	HISTORIC MOD	HISTORIC MOD
Ruamgarh 1, 2	HISTORIC MOD	S ASIA	HISTORIC MOD	HISTORIC MOD	HISTORIC MOD	HISTORIC MOD
Sarai Nahar Rai 1970–4	HISTORIC MOD	S ASIA	HISTORIC MOD	HISTORIC MOD	HISTORIC MOD	HISTORIC MOD
Tekkalakota 2, 15/5	HISTORIC MOD	S ASIA	HISTORIC MOD	HISTORIC MOD	HISTORIC MOD	HISTORIC MOD
Timargarha 1,2, 3, 4, 5, 6, 101a,b,c; 134, 139, 142a, 142b, 144, 165a, 173a,b, 197	HISTORIC MOD	S ASIA	HISTORIC MOD	HISTORIC MOD	HISTORIC MOD	HISTORIC MOD
Yeleswaram 1/61, 4/61, 1/62	HISTORIC MOD	S ASIA	HISTORIC MOD	HISTORIC MOD	HISTORIC MOD	HISTORIC MOD
Onge L1, L5	RECENT MOD	S ASIA	RECENT MOD	RECENT MOD	RECENT MOD	RECENT MOD
ANDAMAN 1509–1545; 3176–3190	RECENT MOD	S ASIA	RECENT MOD	RECENT MOD	RECENT MOD	RECENT MOD

In contrast, Hublin and others (Carbonell et al., 1995; Hublin, 1996; Dean et al., 1998) feel that *H. heidelbergensis* was restricted to Europe and was an ancestral species to Neandertals, who are conventionally believed to be absent in East Asia. The best-known configuration of this proposal is the Accretion Hypothesis, which suggests that the origin of *H. sapiens* is to be found in an ancestral stock in Africa, and the origin of Neandertals is in the *H. heidelbergensis* lineage in Europe (see Hawks and Wolpoff, 2001 for detailed discussion and critique). This implicitly places all Asian Middle Pleistocene specimens on an evolutionary dead end lineage of *H. erectus*, though the Accretion Model was

not developed for and does not purport to deal directly with the Asian evidence.

In order to apply this hypothesis to the second model, **Model II**, only the European Middle Pleistocene specimens are classified *a priori* as *H. heidelbergensis*. This reflects the supposition that this group evolved in Europe and gave rise only to Neandertals. The African members of this morphological grade are considered ancestral to *Homo sapiens* and are given an *a priori* classification of "archaic *H. sapiens*" following the informal use of this term in other studies (Dean et al., 1998; Antón, 2003). The Asian specimens are placed in *H. erectus*, an inference based on the fact that while this scenario does not explicitly name them as such, it nevertheless argues that these populations do not belong to either *H. heidelbergensis* or the ancestral stock of *H. sapiens* in Africa. To accommodate the assumption that the Asian fossils have been a distinct lineage from the Early Pleistocene, the term "*H. erectus*" is applied only to specimens from this region. Early Pleistocene Africans are classified as *H. ergaster* as are the Dmanisi materials from Georgia, based on interpretations that they resemble the African more strongly than the Asian material (Rosas and Bermúdez de Castro, 1998; Gabunia et al., 2000). Model II therefore assumes that Asian *H. erectus* populations are a regionally unique, deep-rooted dead-end evolutionary lineage with respecxt to modern human origins.

No other formal proposals specific to the mid- and late- Middle Pleistocene have been constructed and tested by researchers. However, several publications of Pleistocene evolutionary scenarios encompass worthy discussions of the fossils of this time period. Therefore, the final two models attempt to extract predictions for Middle Pleistocene populations from theories devised to explain the phylogenetic history of other populations or single regions.

The third model I test (**Model III**) reflects the possibility that *H. heidelbergensis* was present in both Europe and Africa, but not Asia (Aguirre and Carbonell, 2001; see also discussion in Stringer, 2002). In this configuration of the data, the African and Asian *H. erectus* specimens are grouped together as one population, but the African "transitional" Middle Pleistocene specimens are designated *H. heidelbergensis* while the Asian fossils are not. The appropriate classification of comparable Asian fossils such as Jinniushan and Dali in such a model is unclear from the published literature. Presumably they would belong to a separate species, either *H. erectus* or an "archaic" grade of *H. sapiens*, although the latter is doubtful given that, in this model, similar African and European fossils are not being assigned to *H. sapiens*.

Therefore these two late-Middle Pleistocene Chinese specimens are placed in *H. erectus* following the implicit assumption that all Asian non-modern fossils represent an evolutionary dead end trajectory of *H. erectus*. The Early Pleistocene African fossils are considered *H. erectus*, not *H. ergaster*, reflecting the supposition that *H. erectus* was not exclusive to Asia.

The fourth model (**Model IV**) similarly recognizes *H. heidelbergensis* in Europe and Africa but not Asia. However, this model assumes that *H. erectus* is exclusively Asian. Therefore, *H. ergaster* is again used to classify the African erectine fossils. Like Model II, this scenario tests the notion that *H. erectus* populations from Asia and Africa are separate evolutionary lineages. But here the model allows for *H. heidelbergensis* to be present in Africa as opposed to an "archaic" *H. sapiens* lineage.

A final configuration of the data set classifies the fossils only by region. Although this stage of the analysis telescopes fossils artificially by time, my purpose is to investigate underlying patterns that would not be apparent when dividing the sample by various taxonomic schemes. In particular, I explore the similarity between Narmada and post-Pleistocene South Asian populations to determine if Narmada shows any affinity with the archaeological and indigenous South Asians as compared to other groups in the study. If so, the question of regional continuity would again be brought into discussion. In addition, even the taxonomic models being tested here artificially compress groups by time, though not as broadly as this stage of the analysis. While extreme in its configuration, this final analyses by region serves to bracket the taxonomic analyses by providing a comparative basis by which to evaluate the strength of Narmada's affinities with respect to the evolutionary models.

In the first four models, the Zuttiyeh Middle Pleistocene fossil from Israel is given an *a priori* classification consistent with the

African specimens. Though this specimen possesses Asian affinities as well (Sohn and Wolpoff, 1993), the African classification is based on the assumption that for much of the Pleistocene Israel was part of the Northern African biogeographic zone. None of the European fossils in this study date to the earliest phase of the Middle Pleistocene, thus the role of the proposed species *H. antecessor* is not considered in the models as they are defined here. In all analyses, the modern South Asian sample is divided into Historic and Recent groups, reflecting the fact that they are separate populations not just temporally but also demographically, with the archaeological sample representing a broad swath of historic Indian populations and the Andaman Islanders representing a relic surviving island group.

While membership in the various Pleistocene taxonomic groups was variable for many fossils—particularly the Middle Pleistocene ones—the sample sizes for the other temporal/taxonomic groups was relatively stable. The number of specimens given an *a priori* classification of anatomically modern *H. sapiens* (AMHS) was, in all analyses, sixty-seven: thirty-one non-Australian specimens and thirty-six crania from Australia, including twenty-eight terminal Pleistocene specimens from Coobool Creek. For the early Holocene South Asians (Historic Moderns), the sample size was ninety-eight. The number of Andaman Islanders, referred to here as Recent Moderns, was fifty-four.

Results

Model I

In the MLR analysis, the first five principal components representing 79% of the total variance were used. The *a priori* groups were configured such that *H. heidelbergensis* was recognized as being present in Asia as a distinct population from the *H. erectus* specimens from Ngandong and Zhoukoudian. The results

give Narmada a 67.8% posterior probability of belonging to *H. heidelbergensis*. Of the 276 specimens total, 26.8% were misclassified or ambiguously classified. In contrast, in the DFA Narmada had a 32.1% typicality probability relative to *H. erectus* and only a 1% typicality probability relative to *H. heidelbergensis*. This was one of two models in which the two analyses gave conflicting results. The results of all analyses are shown in Table 5.

Model II

The first seven principal components, representing 89% of the variance of the original data, were analyzed. In this analysis, none of the African specimens were placed in *H. erectus*, and none of the Asian fossils—with the exception of the AMHS ones—were placed in any other taxon but *H. erectus*. Only the European fossils were defined as *H. heidelbergensis*, with comparable African ones being designated "archaic" *H. sapiens*. By this configuration of the data, the Narmada hominin emerged in the MLR as having a 98.5% posterior probability of being *H. erectus*. A total of 18.8% of the sample was misclassified in this analysis. In the DFA, Narmada again had the highest typicality probability of being classified as *H. erectus*, defined here as an exclusively Asian lineage, but that value was extremely low at less than 1.0% in the separate-groups analysis. Thus, while Narmada is closer to the group centroid of an exclusively Asian *H. erectus* population than any other, as configured here, in absolute terms it is not typical of that group.

Model III

In analysis of the third model, five principal components were again used to determine if Narmada was more similar to Asian *H. erectus* than *H. heidelbergensis* when the African specimens were included in the latter category and the Asians were excluded. The results of the MLR give Narmada a 63.7% posterior probability of belonging to the *H. erectus*

Table 5. Results: Probability values of classification for Narmada by model

| | MODEL I | | MODEL II | | MODEL III | | MODEL IV | | Regional Analysis | |
	MLR	DFA	MLR	DFA	MLR	DFA	MLR	DFA	MLR	DFA
Highest probability of group membership	*H. heidelbergensis* 67.8%	*H. erectus* 32.1%	*H. erectus* 98.5%	*H. erectus* < 1.0%	*H. erectus* 63.7%	*H. erectus* 48.5%	*H. erectus* 93.71%	*H. ergaster* 6.0%	East Asia 36.27%	Southeast Asia 3.0%

Multinomial logistic regression (MLR) results are posterior probabilities
Discriminant function analysis (DFA) results are typicality probabilities based on a separate-groups analysis

group and a 30.25% posterior probability of belonging to *H. heidelbergensis*. Twenty-five percent of the total cases were misclassified or ambiguously classified. In the DFA, Narmada had a 48.5% typicality probability relative to the group centroid of *H. erectus* in the separate-groups analysis, and a < 1.0% typicality probability relative to the centroid of *H. heidelbergensis*.

Model IV

This configuration of taxonomic units again recognized *H. heidelbergensis* in Europe and Africa but not Asia. Asia was treated as a distinct evolutionary lineage, with the *a priori* classification "*H. erectus*" used only for these fossils. *H. ergaster* was used to designate the African erectine specimens. Following this model, the MLR was conducted using 6 PC's which represented 85% of the total variance in the data set. The results of the MLR give Narmada a 93.7% posterior probability of belonging to *H. erectus*. A total of 22% of the sample was misclassified in this analysis. In the DFA, Narmada is given a 6% typicality probability relative *H. ergaster*, and this is its highest value relative to any group.

Regional Groups

The final analysis conducted here divided the sample up by region, regardless of the temporal period or proposed taxonomic group to which it belonged. In the MLR, Narmada had a higher posterior probability of being classified as East Asian (44%) vs. South Asian (< 1.0%). There was an overall misclassification rate of 22% for this configuration of the data. In the DFA, the Narmada specimen's highest typicality probability was relative to the Southeast Asian group, though it was extremely low (3%). It had a 1% typicality probability of grouping with the East Asians.

Plots of the distribution of groups used in this study were constructed from the first and second discriminant functions of the separate-groups covariance matrix DFA. These illustrate the range of differentiation and overlap encompassed by these specimens and their respective group centroids. They are presented in Figures 3 to 7.

Methodological Considerations in Evaluating Results

Although the analyses yielded fairly consistent results, there are nonetheless sources of error that could also be affecting the results. The data were compiled from published reports, and there could be inter-observer error in identifying craniometric landmarks or applying Martin and Saller's vs. Howells' protocols. Sex differences that are not related to size are also a source of variation in the sample that have not been taken into account, and are an important source of variation.

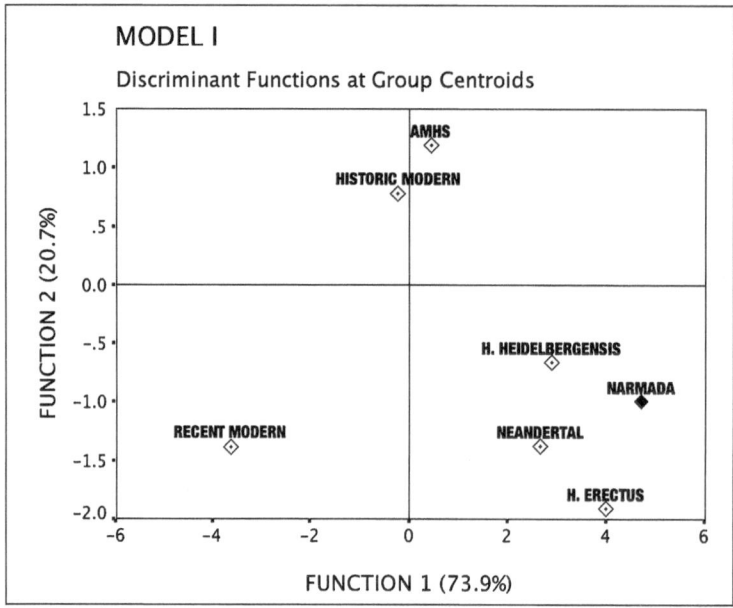

Figure 3. Scatterplot of discriminant functions 1 and 2 for Model I

The imputation procedure for dealing with missing data is also a source of potential error that could be influencing the results (Norell and Wheeler, 2003). The expectation-maximizing algorithm uses the parameters of the entire data set to determine missing values. A few characteristics of this data set may have affected those values. First, the specimens span over a million years of evolution as well as a geographical range consisting of the entire Old World thus corralling together of widely temporally and geographically disparate specimens. The use of dummy variables to represent a specimen's time period and geographic region as well as sex, and incorporated in the parameter

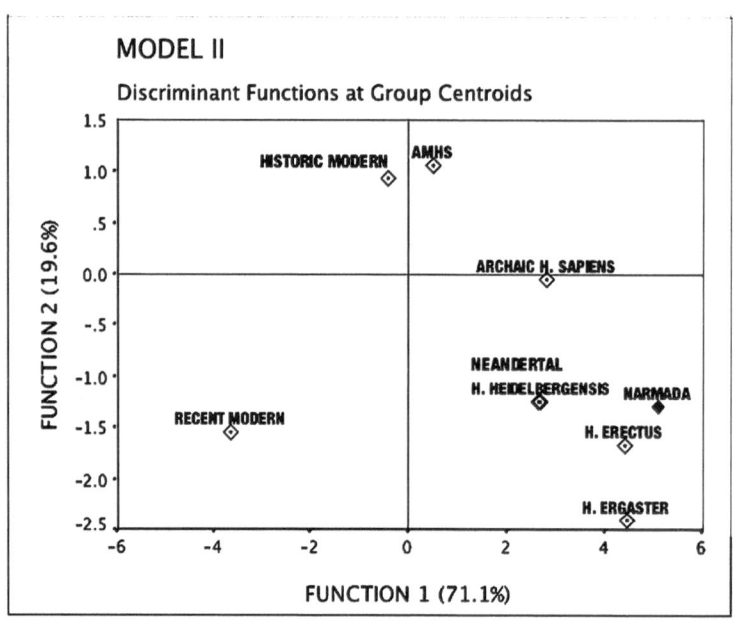

Figure 4. Scatterplot of discriminant functions 1 and 2 for Model II

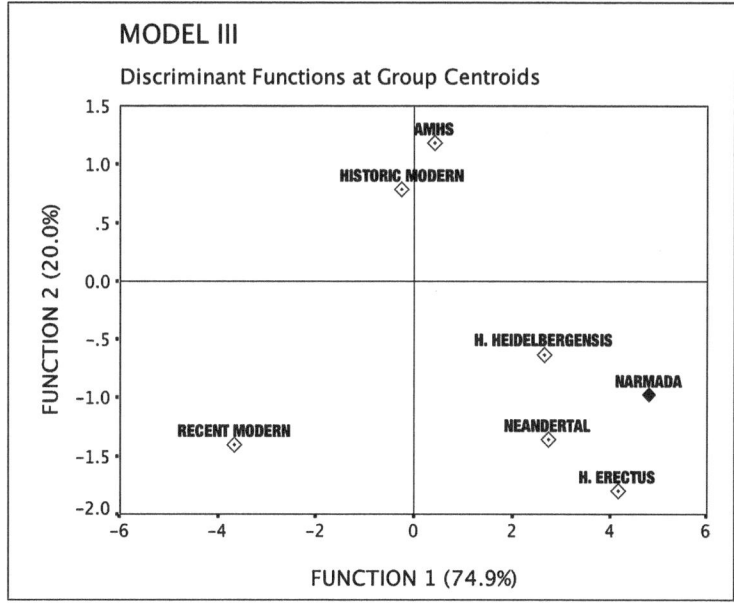

Figure 5. Scatterplot of discriminant functions 1 and 2 for Model III

estimation to determine missing data values, would have minimized the influence of other regions and time periods on the imputation of missing values for a given subsample. That is, fossils from the Early Pleistocene of Africa had missing values imputed with a weighted consideration of other specimens from that region and time period. However, the param-

eters were ultimately determined by the entire data set, invariably influencing the overall final imputed values.

Second, the differences in subsample size may have led to the estimation of values that were biased towards the largest subset: South Asian historic and recent modern humans. Looking at the results of the MLR and

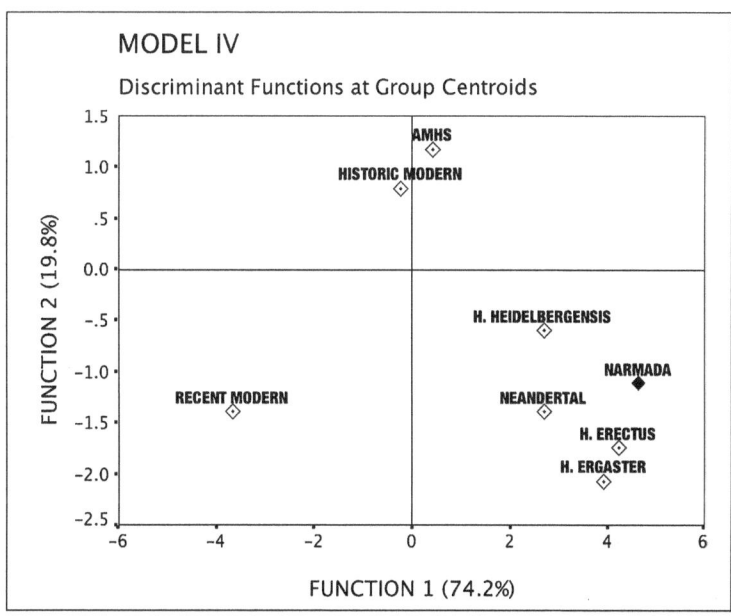

Figure 6. Scatterplot of discriminant functions 1 and 2 for Model IV

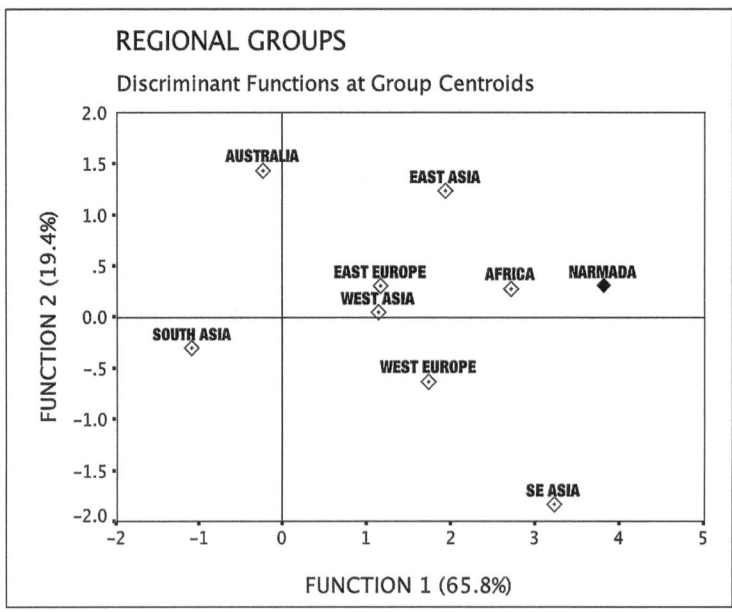

Figure 7. Scatterplot of discriminant functions 1 and 2 for Regional Analysis

DFA, this was likely the case particularly for the Regional analyses, where the Historic and Recent Modern samples were not subdivided in this analysis and therefore South Asian specimens comprised an disproportionately large group. Several specimens were incorrectly classified as South Asian in this analysis, due in part to the influence of the size of this regional sample on overall parameter estimation.

A final consideration in examining these results is the choice of statistical analysis. As mentioned earlier, the use of MLR and the examination of posterior probabilities falls under the "forced allocation" class of procedures (Campbell, 1984). It assumes that the unallocated individual must belong to one of the *a priori* defined groups. In one of the analyses here, that is a valid assumption: the geographical regions used are not arbitrary, but are comprised of real categories to which a specimen must belong. While different studies with different levels of refinement in category definition may yield different results, the notion that the Narmada hominin must belong to a particular region is clearly appropriate. However, the analyses of evolutionary models

are assuming that the specimen must belong to a particular predefined group. To address this, the examination of the typicality (in addition to posterior) probability of this specimen is essential because it directly addresses that assumption. However, the shortcomings of discriminant function analysis for this sample means that the scatterplots of groups by discriminant function, while reasonably illustrative of the overall distribution and overlap of groups as well as the position of Narmada relative to these points, do not present an entirely unbiased depiction of the relationship of these groups to each other.

Discussion

The results of these analyses suggest that the Narmada hominin falls between *H. heidelbergensis* and *H. erectus* in the morphology analyzed here, depending on how each of these species is defined. Model I was the only evolutionary scenario wherein certain Asian specimens were given an *a priori* designation as *H. heidelbergensis*. In that sense, it was a direct test of the question of whether this species was present in Asia at all, and the Indian subcontinent in particular. In this model, the MLR and DFA tests each classified Narmada differently and this was the only analysis in which their results conflicted. The MLR gave Narmada a higher probability of being *H. heidelbergensis* while the typicality probability from the DFA placed it closer to the *H. erectus* group centroid. The mixed results suggest that if Narmada is *H. heidelbergensis*, it is not typical of that population.

A plot of the first and second discriminant functions for this model (Figure 3), which coincide with the first and second principal components, illustrates that the statistical ambiguity may actually be a reflection of real morphological similarities between Narmada and both the *H. heidelbergensis* and *H. erectus* samples. In both cases, Narmada was given a relatively high probability value of belonging to the alternative group as well. That is, in the MLR its second highest posterior probability of group membership was a 25.4% value of classifying with the *H. erectus* group, while in the DFA it had a 28.4% typicality probability relative to its second highest group, *H. heidelbergensis*.

The overlap in these two groups' distributions is reflected in other misclassifications that occurred in this analysis. Four specimens that have been at the heart of the definition of *H. heidelbergensis* (Rightmire, 1998, 2001a, 2001b)—Bodo, Kabwe, Dali and Ndutu—all misclassified as *H. erectus*. Interestingly, the only two Asian specimens classified as *H. heidelbergensis*, Dali and Jinniushan, actually had higher posterior probabilities of being *H. erectus*. Jinniushan was marginally misclassified—it had a nearly equal value of belonging to *H. erectus* and *H. heidelbergensis* (49.7% and 48.4% respectively)—but the point still stands that its position was ambiguous in this model. The implication is that if *H. heidelbergensis* was in Asia, none of the East Asian specimens included in this analysis strongly support this. Thus, the imposition of an Asian model on

understanding the Narmada hominin's species allocation between *H. erectus* and *H. heidelbergensis* may not be entirely appropriate.

Though the rate of misclassification was similar for all models, the affinity and strength of Narmada's classification in each model offers clues as to which evolutionary scenario has the most support here. Of all models tested, Model III is the most definitive in terms of: a) the consistency of results between the MLR and DFA, and, b) the strength of both the posterior and typicality probabilities. In this scenario, Narmada has both a high posterior and typicality probability of grouping with *H. erectus*.

In contrast, in Models II and IV, Narmada classified in the MLR as *H. erectus* but had a very low typicality probability relative this group and in one case—the DFA of Model IV—Narmada actually had a higher typicality probability of classifying with *H. ergaster* though its value was extremely low, at 6%. These models share the fact that the *H. erectus* sample was divided into an African and an Asian species. If this division is valid, Narmada's typicality relative to *H. erectus*, defined as exclusively Asian, is extremely low. In addition, in Model II the group 'archaic' *H. sapiens* was defined to describe the African specimens only. If such a taxonomic unit exists, Narmada does not belong to it based on this analysis.

If Model III is not correct, and *H. ergaster* is a separate species, Narmada's low typicality probability relative to this and all other taxa, when this taxon is used in the model, could suggest that this specimen represents an entirely new species. However, my own analysis of the features of Narmada does not support such an interpretation, nor does that of any other scholar who has studied this specimen directly. It is not morphologically distinct from existing contemporaneous specimens to warrant a separate species name. Moreover, larger considerations of macroevolutionary theory

(Conroy, 2002, 2003) do not support the more speciose models of Pleistocene hominin evolution. I think more parsimonious explanations for the pattern of results can be found.

Narmada's classification and strength of association with other groups seems to be most strongly affected by whether a comparative sample with mixed Asian and African morphology is present. When the *H. heidelbergensis* category is comprised of a sample that mixed African and Asian Middle Pleistocene morphology, as in Model I, Narmada exhibits strong posterior probability values with respect to that group, though not strong typicality probabilities. When *H. erectus* is defined as both an African and Asian phenomenon, as in Model III, Narmada has both high posterior probability *and* high typicality probabilities relative to that taxon. However, when the model does not configure any populations as such, as in Models II and IV wherein the Africans are placed in *H. ergaster*, Narmada's typicality probability relative to *H. erectus* changes considerably, and drops to nearly zero. In addition, in Models II-IV in which the Asian "transitional" specimens were given an *a priori* assignment of *H. erectus*, none of them were then misclassified as *H. heidelbergensis*, nor did Narmada align with the latter group, regardless of how that species was defined. This illustrates the cohesiveness of the Asian sample, shaped largely by the Ngandong and Zhoukoudian *H. erectus* samples which influence the group centroid of this species.

The extreme decline in typicality values and, by association, Narmada's strength of classification when a mixed-African/Asian sample is not present supports the notion that Narmada's morphology is a reflection of mixed Asian and African features—an interpretation consistent with its geographic location between these two regions. The Asian "transitional" specimens such as Dali and Jinniushan are likewise not typical of the taxon *H. heidelbergensis* with respect to the morphology studied here, but interestingly enough neither are a few key African specimens such as Bodo and Kabwe. So while Narmada is aligned with these other Asian specimens in terms of showing ambiguous associations with *H. heidelbergensis* and stronger affinities with *H. erectus*, it likely does so because it possesses a mosaic of African and Asian morphology, which is consistent with its provenience.

The results of the Regional Analysis further support this interpretation. Although the *a priori* allocations given in this test artificially combine certain specimens that are temporally quite distant, the sample is restricted enough to allow certain interpretations to be made, particularly in light of the results that are obtained. As in Model I, the MLR and DFA results in this analysis are conflicting. Here, the MLR gives Narmada a higher posterior probability of being East Asian, while the DFA gives it a relatively equal—and low—typicality probability relative to the two Asian samples from the Southeast (3%) and East (1%). Again, the ambiguity is actually itself informative if for no other reason than it is consistent. In this case, the Southeast Asian sample is predominantly archaic in morphology with several fossils classified as *H. erectus*, almost all of which are considered Late Pleistocene (Barstra and Basoeki, 1989). The East Asian sample is likewise largely *H. erectus* with a few transitional specimens including Liujiang, a possible modern *H. sapiens* with "archaic" retentions (Hanihara, 1994). That Narmada's classification is weakly aligned with these Asian samples is consistent with the notion that it possesses transitional *H. heidelbergensis-H. erectus*-like traits, many of which are archaic/African pleisomorphic retentions. The prediction that a South Asian specimen, derived from a geographic crossroads between Asia and Africa, would possess a mosaic of traits but be slightly more aligned with the Asian sample is therefore supported here. I interpret this as a reflection of clinal variation during the Pleistocene, and these initial results from an important intermediary region indicate that such a model may be more appropriate than placing species-level limitations on regional groups. The latter may be elevating normal geographic variation to higher taxonomic significance than is warranted.

The results of the regional affinity analysis are also interesting in light of another question: that of continuity through time in one geographic region. Given the large sample size of South Asians in the Regional analysis, it is particularly notable that the Narmada hominin did not show affinities with other members of its region but rather with the smaller group of Africans and East Asians in the analysis. The lack of affinity between Narmada and modern South Asians could indicate that for the craniofacial traits examined here, there is not strong morphological continuity between the Middle Pleistocene and Holocene populations. The morphology of Narmada provides support for the notion of a dispersal with gene flow between African and East Asian populations within the subcontinent during the Middle Pleistocene.

However, the lack of a substantial record of the craniofacial morphology of Late Pleistocene modern humans in the Indian subcontinent makes it impossible to infer what the population dynamics were between these groups and subsequent populations that ultimately gave rise to the Holocene archaeological populations used in this study. Here, the grouping of the Historic Modern South Asians with the AMHS sample (Figures 3–7), which was derived from all regions of the Old World, suggests that there may have been some continuity between these Late Pleistocene and Holocene groups, although at this point such an inference is strictly conjecture. The only evidence we currently have between Late Pleistocene and recent South Asians is dental, based on a study by Hawkey (1998)

of the 32–35 ka material from Sri Lanka. Her results suggest that there is continuity between Late Pleistocene and modern South Asians, but the poor preservation of the Sri Lankan cranial material makes it difficult to assess its morphological affinities in other aspects of the skeleton between these and earlier populations such as the one represented by Narmada.

Of further interest is the marked distinctiveness of the Andaman Islanders in all analyses. The pattern and overall low rates of misclassification as well as the largely non-overlapping distribution of this population in the discriminant function scatterplots relative to other groups indicates the morphological distinctiveness of this population. In particular, there is very little overlap between this and the Historic Modern sample. The scatterplots suggest that the Andaman Islanders share certain similarities with the archaic Pleistocene sample, based on their relative position on Function 2. Likewise, in the various analyses the historical sample shares clear overlap with the Late Pleistocene sample on both Functions 1 and 2. However, the historic and Andaman samples are always distinct from each other on both Functions, encompassing on average 93% of the variance. This suggests that the Andamanese, a fairly endogamous group, possess a morphology shaped largely by genetic drift and the selective forces of island biogeography. As such, though they are genetically most likely descended from one of the initial dispersed African populations (Thangaraj et al., 2003, 2005), they are not necessarily a reasonable reflection of the average morphology of indigenous South Asians due to the several tens of thousands of years of evolution that have subsequently shaped their phenotypic makeup. Their distinctiveness, as well as the probable reasons for it, should be taken into consideration when using this population to represent modern South Asians in other comparative analyses.

H. heidelbergensis vs. H. erectus
in India?

So, was *H. heidelbergensis* in South Asia? The answer, based on the linear morphology the Narmada specimen, is that this species can not be said to be present here. Narmada appears to have a mosaic of African and Asian *H. erectus* traits but these results suggest that it is more similar to *H. erectus*. However, this limited set of measurements captures only continuous traits that may or may not be diagnostic at the species level. In addition, any taxonomic-level inferences are contingent upon a deeper understanding of the relationship between morphology and biology as well as a consideration of non-metric traits that may be diagnostic for these groups. The use of discriminant function analysis, though shown to be highly useful in biological classification, is nonetheless an analysis only of morphological similarity and difference and not a conclusive statement about phylogenetic relationships (Albrecht, 1992).

Perhaps the most important variables absent in these analyses are overall size and brain size. The fairly large cranial capacity of Narmada (estimates range from 1135–1421 cc) may entitle it to membership in a more "derived" species on that basis alone. As discussed earlier, there has been a long-standing notion that brain size is a diagnostic trait for assigning a specimen to our own species or likewise keeping them out. Tattersall's indictment of this tendency as quoted earlier captures the problems of such a practice, but its continued implementation, conscious or otherwise, has influenced our notions of where various specimens belong on the human phylogenetic tree.

However, if we reject a heavy weighting of brain size and accept that, despite its larger cranial capacity, the Narmada hominin is morphologically *H. erectus*, there are several implications. Namely, the results of studies by Kennedy et al., and Cameron et al., placing

Narmada in *H. sapiens*, would be wrong as would Rightmire's assessment of *H. heidelbergensis* being present in Asia. There are problems with rejecting these findings, not the least of which is that the detailed morphological studies conducted by these researchers are based on more variables than were covered here, and they are sound.

The main problem with naming Narmada *H. erectus* really lies in the historical underpinnings of how Asian *H. erectus* was first defined, and how it conflicts with beliefs as to how *H. sapiens* should be defined with respect to brain size. When the former model is adopted, Movius's interpretation of Asia as a marginal "region of cultural retardation" (Movius, 1948) based on the archaeological record and the absence of handaxes is brought to bear on our understanding of the Indian Middle Pleistocene, and much of the subcontinent is likewise seen as "marginal." I believe that Movius's interpretations of lithic assemblage variability have shaped studies of fossil populations in Asia, particularly in Indonesia where the emphasis has been on the relatively low cranial capacities and archaic features of *H. erectus* populations there. The interpretation of the region as home to an extinct evolutionary side branch continues, despite studies that have demonstrated the presence of many of the Ngandong traits in extant Australasian populations (Hawks et al., 2000; Wolpoff et al., 2001). For example, recent dating placing the Javanese specimens in the Late Pleistocene (Swisher et al., 1996) has been interpreted as evidence for contemporaneity between *H. erectus* and *H. sapiens* in the region, and the finding of a diminutive Late Pleistocene population in Flores has been taken as evidence for a late-surviving separate Australasian species (Brown et al., 1979).

These studies have served to perpetuate Movius's impression of Asia as being an evolutionary backwater and, by implication, of Asian *H. erectus* being an evolutionary dead-end. Our notions of what should be included in it today have thus been influenced by how this species has been understood historically. So to then call Narmada—a large-brained hominin with several derived features—"*H. erectus*" is untenable within this paradigm.

If the latter trait of brain size is the main one taken into account, Narmada shows affinities with anatomically modern Africans and Europeans as well as most Middle Pleistocene *H. heidelbergensis* specimens, and could be classified as such. However, as Tattersall (1986) points out, in emphasizing brain size, a number of extant human populations are left out. It then appears inappropriate to rest taxonomic definitions on this trait, most significantly of our own species.

A more realistic model would take into consideration the mosaic African/Asian morphology of Narmada and other Middle Pleistocene specimens. As the results of this study demonstrate, there is actually morphological overlap between the Asian and African Early and Middle Pleistocene *H. erectus* and *H. heidelbergensis* samples. The Narmada fossil, located in an important intermediate region between Africa and East/Southeast Asia, reflects the presence of continuity between these regions as well. If we are willing to accept a clinal model of variation for Middle Pleistocene *Homo*, it makes sense that we find mixed regional traits in a specimen that is from a geographically intermediate location.

Conclusion

At present, there are a few options for classifying Narmada. Morphologically, when all metric traits are taken into account, it should be considered *H. erectus*, provided that the definition of this species in South Asia is configured so as to include a mosaic of African and Asian features, without imposing the East/Southeast Asian paradigm that this species is an evolutionary dead-end onto the

material from Indian subcontinent. If only the subjective criteria of brain size and "transitional" morphology are used, it could be classified as *H. heidelbergensis*. Finally, it can simply be referred to as "Middle Pleistocene *Homo*" a term that is sufficiently descriptive without the historical baggage of nomenclature that comes from ascribing this specimen to Asian *H. erectus*. The temporal-based term is preferable because it allows for the possibility of a continuous morphological trajectory from the Early to Middle Pleistocene in certain regions of the Old World. While *H. erectus* is indeed a morphologically cohesive group when the African specimens are included, and there is no support for considering the Asian specimens to represent a distinctive and, by implication, separate biological lineage. Rather, these results suggest it is a population shaped by its geographic location at an 'edge' but it is nonetheless part of a clinal continuum of variation across the Old World.

The proposal to classify the Narmada specimen strictly according to its temporal group is reflective of a broader philosophy regarding Middle Pleistocene hominins. The designation *H. heidelbergensis* is extremely useful for dealing with a growing number of specimens dated to the Middle Pleistocene and exhibiting transitional features. However, based on these results it is unclear whether or not this species was present in Asia. More importantly, it is not clear whether the transitional specimens assigned to it deserve to be considered a separate evolutionary trajectory of the Pleistocene, distinct from that of *H. erectus* in Asia. The name may simply be a replacement for the less desirable term "archaic *Homo sapiens*." Some fossils from this time period such as the Sima de los Huesos specimens appear to foreshadow Neandertal morphology in the West; others such as Dali foreshadow modern human morphology in the East. In reality, then, this group of fossils may be filling in what were once convenient gaps in the temporal and regional post-*Homo erectus*/pre-*Homo sapiens* records of the Old World.

This study brings up the question of the cohesiveness of *H. heidelbergensis* as a species within the larger population of Pleistocene *Homo*. The inadequacy of our terminology may reflect a larger problem with how we define species in the fossil record and, more importantly, how we deal with variation in space and time. While giving a separate designation to fossils that share a suite of mosaic traits and a temporal niche, as do the specimens considered *H. heidelbergensis*, is a sensible and meaningful heuristic device, it is less clear that these fossils are also distinctive in terms of their position on the phylogenetic tree of human ancestry. They do not necessarily possess evidence of a speciation event and unique evolutionary trajectory in the Middle Pleistocene. More detailed studies of specific fossils, as I have tried to do here for the Narmada specimen, may serve to further clarify the role of this population in human evolutionary history.

Acknowledgments

A special thanks to Mica Glantz for her extensive feedback and help with various stages of this study. I am also grateful to Laura Shackelford for measuring Florisbad and Zuttiyeh, and also for talking through the theory with me; to Yang Jianping and Shang Hong for their help in translating the Chinese fossil reports and extracting measurements; and to Rajeev Patnaik for providing references on the Narmada fossil. Thanks to Adel Varghese for his comments on the manuscript, and to three anonymous reviewers whose feedback was enormously helpful. Thanks also to David Carlson, Andrew Scherer and Gene Albrecht for help with the statistical methods. Finally, I would like to acknowledge G. Philip Rightmire for discussing the possible contribution of such a study with me long ago, when it was barely an idea; Kenneth A.R. Kennedy for his consistent generosity with his knowledge and resources, particularly with the measurements of the Holocene skeletal material; and Michael Petraglia for inviting me to contribute to this volume, for his infinite patience, and his constant willingness to assist me in my work.

References

Adefris, T., 1992. A description of the Bodo cranium: an archaic *Homo sapiens* from Ethiopia. Ph.D. dissertation, New York University.

Agrawal, D.P., Kotlia, B.S., Kusumgar, S., 1988. Chronology and significance of the Narmada formations. Proceedings of the Indian National Science Academy 54(3), 418–424.

Aguirre, E., Carbonell, E., 2001. Early human expansions into Eurasia: the Atapuerca evidence. Quaternary International 75(1), 11–18.

Albrecht, G.H., 1992. Assessing the affinities of fossils using canonical variates and generalized distances. Human Evolution (Firenze) 7(4), 49–69.

Antón, S.C., 2003. Natural history of *Homo erectus*. Yearbook of Physical Anthropology 46, 126–170.

Arsuaga, J.-L., Martínez, I., Gracia, A., Lorenzo, C., 1997. The Sima de los Huesos crania (Sierra de Atapuerca, Spain): a comparative study. Journal of Human Evolution 33, 219–281.

Asfaw, B., 1983. New hominid parietal from Bodo, Middle Awash Valley, Ethiopia. American Journal of Physical Anthropology 61(3), 367–371.

Aziz, F., Baba, H., Watanabe, N., 1996. Morphological study on the Javanese *Homo erectus* Sangiran 17 skull based upon the new reconstruction. Geological Research and Development Center Paleontology Series No. 8, 11–24.

Baba, H., Aziz, F., Kaifu, Y., Suwa, G., Kono, R.T., Jacob, T., 2003. *Homo erectus* calvarium from the Pleistocene of Java. Science 299(5611), 1384–1388.

Barstra, G., Basoeki, 1989. Recent work on the Pleistocene and the Palaeolithic of Java. Current Anthropology 30(2), 241–244.

Basu, A., Pal, A., 1980. Human remains from Burzahom. Anthropological Survey of India, Calcutta, India.

Bernhard, W., 1967. Human skeletal remains from the cemetery of Timargarha. Ancient Pakistan 3, 291–407.

Bernhard, W., 1998. Two previously undescribed human skeletons from prehistoric Harappa, Pakistan. Homo 49(1), 21–54.

Bilsborough, A., 1976. Patterns of evolution in Middle Pleistocene hominids. American Journal of Physical Anthropology 5, 423–439.

Black, D., 1926. Tertiary man in Asia: the Chou Kou Tien deposit. Palaeontologica Sinica, Geological Survey of China, Series D 7(1), 1–26.

Bräuer, G., 1984. The "Afro-European *sapiens* hypothesis," and hominid evolution in East Asia during the Late Middle and Upper Pleistocene. Courier Forshungsinstitut Senckenberg 69, 145–165.

Brown, P., 1989. Coobool Creek: A Morphological and Metrical Analysis of the Crania, Mandibles and Dentitions of a Prehistoric Australian Human Population. Department of Prehistory, Australian National University, Canberra.

Brown, P., 2005. Australian Terminal Pleistocene and Early Holocene Cranial, Mandibular and Dental Metrics. University of New England, Australia.

Brown, T., Pinkerton, S.K., Lambert, W., 1979. Thickness of the cranial vault in Australian aboriginals. Archaeology and Physical Anthropology in Oceania 14(1), 54–71.

Bull, S.B., Donner, A., 1987. The efficiency of multinomial logistic regression compared with multiple group discriminant analysis. Journal of the American Statistical Association 82(400), 1118–1122.

Cameron, D., Patnaik, R., Sahni, A., 2004. The phylogenetic significance of the Middle Pleistocene Narmada hominin cranium from central India. International Journal of Osteoarchaeology 14(6), 419–447.

Campbell, N.A., 1980. On the study of the Border Cave remains: statistical comments. Current Anthropology 21(4), 532–535.

Campbell, N.A., 1984. Some aspects of allocation and discrimination. In: Van Vark, G.N., Howells, W.W. (Eds.), Multivariate Statistical Methods in Physical Anthropology. Reidel, Dordrecht, Netherlands, pp. 177–192.

Carbonell, E., Bermudez de Castro, J.M., Arsuaga, J.L., Diez, J.C., Rosas, A., Cuenca-Bescos, G., Sala, R., Mosquera, M., Rodriguez, X.P., 1995. Lower Pleistocene hominids and artifacts from Atapuerca-TD6 (Spain). Science 269, 826–829.

Cavalli-Sforza, L.L., Menozzi, P., Piazza, A., 1993. Demic expansions and human evolution. Science 259(5095), 639–646.

Clarke, R.J., 1990. The Ndutu cranium and the origin of *Homo sapiens*. Journal of Human Evolution 19, 699–736.

Cohen, P., 1996. Fitting a face to Ngaloba. Journal of Human Evolution 30(4), 373–379.

Conroy, G.C., 2002. Speciosity in the early *Homo* lineage: too many, too few, or just about right? Journal of Human Evolution 43(5), 759–766.

Conroy, G.C., 2003. The inverse relationship between species diversity and body mass: do primates play by the "rules"? Journal of Human Evolution 45(1), 783–795.

Conroy, G.C., Weber, G.W., Seidler, H., Recheis, W., Zur Nedden, D., Haile Mariam, J., 2000. Endocranial capacity of the Bodo cranium determined from three-dimensional computed tomography. American Journal of Physical Anthropology 113(1), 111–118.

Cronin, J.E., Boaz, N.T., Stringer, C.B., Rak, Y., 1981. Tempo and mode in hominid evolution. Nature 292, 113–122.

Darroch, J., Mossiman, J., 1985. Canonical and principal components of shape. Biometrika 72, 241–252.

de Lumley, M.-A., Sonakia, A., 1985. First discovery of a *Homo erectus* on the Indian sub-continent, at Hathnora, in the middle valley of the Narmada river. Anthropologie–Paris 89(1), 3–12.

Dean, D., Hublin, J.-J., Holloway, R.L., Ziegler, R., 1998. On the phylogenetic position of the pre-Neandertal specimen from Reilingen, Germany. Journal of Human Evolution 34, 485–508.

Delson, E., Harvati, K., Reddy, D., Marcus, L.F., Mowbray, K., Sawyer, G.J., Jacob, T., Marquez, S., 2001. The Sambungmacan 3 *Homo erectus* calvaria: A comparative morphometric and morphological analysis. Anatomical Record 262(4), 380–397.

Dubois, E., 1937. On the fossil human skulls recently discovered in Java, and Pithecanthropus erectus. Man (N.S.) 37(1), 1–7.

Dutta, P.C., Pal, A., Biswas, J.N., 1972. Late Stone Age human remains from Sarai Nahar Rai: the earliest skeletal evidence of man in india. Bulletin of the Anthropological Survey of India 21(1–2), 114–138.

Efron, B., 1975. The efficiency of logistic regression compared to normal discriminant analysis. Journal of the American Statistical Association 70(352), 892–898.

Ehrhardt, S., Kennedy, K.A.R., 1965. Excavations at Langhnaj: 1944–63 Part III, the Human Remains. Deccan College Postgraduate and Research Institute, Poona, India.

Endicott, P., Metspalu, M., Kivisild, T., 2007. Genetic evidence on modern human dispersals in South Asia: Y chromosome and mitochondrial DNA perspectives. In: Petraglia, M.D., Allchin, B. (Eds.), The Evolution and History of Human Populations in South Asia: Inter-disciplinary Studies in Archaeology, Biological Anthropology, Linguistics and Genetics. Springer, Netherlands, pp. 229–244.

Etler, D., 2004. *Homo erectus* in East Asia: human ancestor or evolutionary dead-end? Athena Review 4(1), 37–50.

Etler, D.A., 1990. Case study of the "erectus-sapiens" transition: Asian hominid remains from Hexian and Chaoxian counties, Anhui province, China. Kroeber Anthropological Society Papers 71–71, 1–19.

Fisher, R.A., 1936. The use of multiple measurements in taxonomic problems. Annals of Eugenics London 10, 422–429.

Frayer, D., 1986. Cranial variation at Mladec and the relationship between Mousterian and Upper Paleolithic hominids. In: Novotny, V.V., Mizerova, A. (Eds.), Fossil Man: New Facts-New Ideas: Papers in Honor of Jan Jelinek's Life Anniversary. Anthropos Institute-Moravian Museum, Brno, pp. 243–256.

Gabunia, L., Vekua, A., Lordkipanidze, D., Swisher, C.C., III, Ferring, R., Justus, A., Nioradze, M., Tvalchrelidze, M., Antón, S.C., Bosinski, G., Jöris, O., Lumley, M.-A.-d., Majsuradze, G., Mouskhelishvili, A., 2000. Earliest Pleistocene hominid cranial remains from Dmanisi, Republic of Georgia: taxonomy, geological setting, and age. Science 288(5468), 1019–1025.

Gupta, P., Basu, A., 1970. Early historic crania from Kumhar Tekri, Ujjain. In: Gupta, P., Basu, A., Dutta, P.C. (Eds.), Ancient Human Remains. Anthropological Survey of India, Calcutta, pp. 37–74.

Gupta, P., Basu, A., Dutta, P.C., 1970. A study of the Nagarjunakonda skeletons. In: Gupta, P., Basu, A., Dutta, P.C. (Eds.), Ancient Human Remains. Anthropological Survey of India, Calcutta, pp. 1–33.

Gupta, P., Basu, A., Gupta, A., 1960. A study on Onge skeletons from Little Andaman. Bulletin of the Anthropological Survey of India 9(1), 27–40.

Haile-Selassie, Y., Asfaw, B., White, T.D., 2004. Hominid cranial remains from Upper Pleistocene deposits at Aduma, Middle Awash, Ethiopia. American Journal of Physical Anthropology 123(1), 1–10.

Hanihara, T., 1994. Craniofacial continuity and discontinuity of Far Easterners in the Late Pleistocene and Holocene. Journal of Human Evolution 27(5), 417–441.

Hawkey, D.E., 1998. Out of Asia: Dental evidence for affinities and microevolution of early populations from India/Sri Lanka. Ph.D. Dissertation, Arizona State University.

Hawks, J., Oh, S., Hunley, K., Dobson, S., Cabana, G., Dayalu, P., Wolpoff, M.H., 2000. An Australasian test of the recent African origin theory using the WLH-50 calvarium. Journal of Human Evolution 39(1), 1–22.

Hawks, J.D., Wolpoff, M.H., 2001. The Accretion Model of Neandertal evolution. Evolution 55(7), 1474–1485.

Hershkovitz, I., Speirs, M.S., Frayer, D., Nadel, D., Wish-Baratz, S., Arensburg, B., 1995. Ohalo II H2: A 19,000-year-old skeleton from a water-logged site at the sea of Galilee, Israel. American Journal of Physical Anthropology 96(3), 215–234.

Holt, B., Benfer, J.R.A., 2000. Estimating missing data: An iterative regression approach. Journal of Human Evolution 39(3), 289–296.

Howell, F.C., 1960. European and Northwest African Middle Pleistocene hominids. Current Anthropology 1(3), 195–232.

Howell, F.C., 1994. A chronostratigraphic and taxonomic framework of the origins of modern humans. In: Nitecki, M.H., Nitecki, D.V. (Eds.), Origins of Anatomically Modern Humans. Plenum Press, New York, pp. 253–319.

Howells, W.W., 1973. Cranial Variation in Man: A Study by Multivariate Analysis of Patterns of Difference Among Recent Human Populations. Papers of the Peabody Museum, Harvard University 67.

Howells, W.W., 1980. *Homo erectus*–who, when, and where: a survey. Yearbook of Physical Anthropology 23, 1–23.

Howells, W.W., 1996. Howells' craniometric data on the internet. American Journal of Physical Anthropology 101(3), 441–442.

Hublin, J.-J., 1992. Recent human evolution in Northwestern Africa. In: Aitken, M.J., Stringer, C.B., Mellars, P.A. (Eds.), The Origin of Modern Humans and the Impact of Chronometric Dating. Princeton University Press, Princeton, pp. 118–131.

Hublin, J.-J., 1996. The first Europeans: evolution of *Homo sapiens* in Europe. Archaeology 49, 36–44.

Kamminga, J., Wright, R.V.S., 1988. The upper cave at Zhoukoudian and the origins of the Mongoloids. Journal of Human Evolution 17, 739–767.

Kennedy, K.A.R., 1965. Human skeletal material from Ceylon, with an analysis of the island's prehistoric and contemporary populations. Bulletin of the British Museum of Natual History, Geology 2(4), 14–213.

Kennedy, K.A.R., 1972. Anatomical description of two crania from Ruamgarh: an ancient site in Dhalbhum, Bihar. Journal of the Indian Anthropological Society 7, 129–141.

Kennedy, K.A.R., 1982. Part II: Biological anthropology of human skeletal remains from Bagor: Osteology. In: Lukacs, J.R., Misra, V.N., Kennedy, K.A.R. (Eds.), Bagor and Tilwara: Late Mesolithic Cultures of Northwest India. Deccan College Postgraduate and Research Institute, Poona, pp. 27–60.

Kennedy, K.A.R., 1992. The fossil hominid skull from the Narmada valley: *Homo erectus* or *Homo sapiens*? In: Jarriage, C. (Ed.), South Asian Archaeology 1989. Prehistory Press, Madison, WI.

Kennedy, K.A.R., 2000. Yeleswaram revisited: the skeletal record. Man and Environment 35(1), 35–57.

Kennedy, K.A.R., Chiment, J., 1991. The fossil hominid from the Narmada Valley, India: *Homo erectus* or *Homo sapiens*? Indo-Pacific Prehistory Association Bulletin 10(1), 42–58.

Kennedy, K.A.R., Lovell, N.C., Burrow, C.B., 1986. Mesolithic human remains from the Gangetic plain: Sarai Nahar Rai. South Asia Occasional Papers and Theses 10.

Kennedy, K.A.R., Lukacks, J.R., Pastor, R.F., Johnston, T.L., Lovell, N.C., Pal, N., Burrow, C.B., 1992. Human skeletal remains from Mahadaha: a Gangetic Mesolithic site. Cornell University, Ithaca, NY.

Kennedy, K.A.R., Sonakia, A., Chiment, J., Verma, K.K., 1991. Is the Narmada hominid an Indian *Homo erectus*? American Journal of Physical Anthropology 86, 475–496.

Krogman, W.M., Sassaman, W.H., 1943. Chapter XVI: Skull found at Chanhu-Daro. In: Mackay, E.J.H. (Ed.), Chanhu-daro Excavations 1935–1936. American Oriental Society, New Haven, pp. 252–265.

Lahr, M.M., Wright, R.V.S., 1996. The question of robusticity and the relationship between cranial size and shape in *Homo sapiens*. Journal of Human Evolution 31, 157–191.

Lesaffre, E., Albert, A., 1989. Multiple-group logistic regression diagnostics. Applied Statistics 38(3), 425–440.

Li, X.N., Zhang, S.S., 1984. Paleoliths discovered in Ziyang man Locality B. Acta Anthropologica Sinica 3, 215–224 (in Chinese with English abstract).

Long, J.S., 1997. Regression Models for Categorical and Limited Dependent Variables. Sage Publications, Thousand Oaks, CA.

Long, J.S., Freese, J., 2001. Regression Models for Categorical Dependent Variables Using Stata. Stata Press, College Station, TX.

Lordkipanidze, D., Vekua, A., Ferring, R., Rightmire, G.P., Agusti, J., Kiladze, G., Mouskhelishvili, A., Nioradze, M., de Leon, M.S.P., Tappen, M., Zollikofer, C.P.E., 2005. Anthropology: the earliest toothless hominin skull. Nature 434(7034), 717.

Manly, B.F.J., 1994. Multivariate Statistical Methods: A Primer. Chapman and Hall, London.

Marc, R.F., 2002. Classification trees as an alternative to linear discriminant analysis. American Journal of Physical Anthropology 119(3), 257–275.

Martin, R., Saller, K., 1957. Lehrbuch der Anthropologie. G. Fischer, Stuttgart.

Matiegka, J., 1934. *Homo predmostensis*, Fosilni Clovek z Predmosti na Morave (Fossil Man from Predmosti in Moravia, Czechoslovakia). Nakladem Ceska Akademie Ved a Umeni (Czech Academy of Sciences and Arts), Prague.

Mayr, E., 1950. Taxonomic categories in fossil hominids. Cold Spring Harbor Symposia on Quantitative Biology. The Biological Laboratory, Cold Spring Harbor, New York, pp. 109–118.

McCown, T.D., Keith, A., 1939. The Stone Age of Mt. Carmel: The Fossil Human Remains from the Levalloiso-Mousterian. Clarendon Press, Oxford.

Mishra, S., Venkatesan, T.R., Rajaguru, S.N., Somayajulu, B.L.K., 1995. Earliest Acheulian industry from peninsular India. Current Anthropology 36, 847–851.

Movius, H.L., 1948. Lower Paleolithic cultures of southern and eastern Asia. Proceedings of the American Philosophical Society, pp. 330–420.

Murrill, R.I., 1981. Petralona Man: A Descriptive and Comparative Study, With New and Important Information on Rhodesian Man. Charles C. Thomas, Springfield, IL.

Norell, M.A., Wheeler, W.C., 2003. Missing entry replacement data analysis: a replacement approach to dealing with missing data in paleontological and total evidence data sets. Journal of Vertebrate Paleontology 23(2), 275–283.

Paddayya, K., Blackwell, B.A.B., Jhaldiyal, R., Petraglia, M., Fevrier, S., Chaderton, D.A.I., 2002. Recent findings on the Acheulian of the Hunsgi and Baichbal Valleys, Karnataka, with special reference to the Isampur excavation and its dating. Current Science 83, 641–647.

Petraglia, M., 2001. The Lower Palaeolithic of India and its behavioural significance. In: Barham, L., Robson-Brown, K. (Eds.), Human Roots: Africa and Asia in the Middle Pleistocene. Western Academic and Specialist Press, Bristol, pp. 217–233.

Piperno, M., Scichilone, G. (Eds.), 1991. The Circeo 1 Neandertal skull: Studies and Documentation. Istituto Poligrafico e Zecca Dello Stato, Rome.

Pope, G.G., 1988. Recent advances in Far Eastern paleoanthropology. Annual Review of Anthropology 17, 43–77.

Press, S.J., Wilson, S., 1978. Choosing between logistic regression and discriminant analysis. Journal of the American Statistical Association 73(364), 699–705.

Radovci'c, J., Smith, F.H., Trinkaus, E., Wolpoff, M.H., 1988. The Krapina Hominids: An Illustrated Catalog of Skeletal Collection. Croatian Natural History Museum, Zagreb.

Rao, C.R., 1948. The utilization of multiple measurements in problems of biological classification. Journal of the Royal Statistical Society. Series B (Methodological) 10(2), 159–203.

Reddy, V.R., Reddy, B.K.C., 1987. Human skeletal remains from Kodumanal, Tamil Nadu: A craniometric study. Tamil Civilization 5(1–2), 100–117.

Reddy, V.R., Reddy, B.K.C., 2004. Morphometric status of human skeletal remains from Kodumanal, Periyar District, Tamil Nadu. Anthropologist 6(2), 105–112.

Reyment, R.A., Blackith, R.E., Campbell, N.A., 1984. Multivariate Morphometrics. Academic Press, London.

Rightmire, G.P., 1983. The Lake Ndutu cranium and early *Homo sapiens* in Africa. American Journal of Physical Anthropology 61, 245–254.

Rightmire, G.P., 1985. The tempo of change in the evolution of Mid-Pleistocene *Homo*. Ancestors: The Hard Evidence. Alan R. Liss, Inc., New York, pp. 255–264.

Rightmire, G.P., 1988. *Homo erectus* and later Middle Pleistocene humans. Annual Review of Anthropology 17, 239–259.

Rightmire, G.P., 1990. The Evolution of *Homo erectus*: Comparative Anatomical Studies of an Extinct Human Species. Cambridge University Press, Cambridge.

Rightmire, G.P., 1998. Human evolution in the Middle Pleistocene: the role of *Homo heidelbergensis*. Evolutionary Anthropology 6(6), 218–227.

Rightmire, G.P., 2001a. Morphological diversity in Middle Pleistocene *Homo*. In: Wood, B., Tobias, P.V. (Eds.), Humanity From African Naissance to Coming Millenia: Colloquia in Human Biology and Palaeoanthropology. Firenze University Press, Firenze, Italy, pp. 135–140.

Rightmire, G.P., 2001b. Patterns of hominid evolution and dispersal in the Middle Pleistocene. Quaternary International 75, 77–84.

Rightmire, G.P., 2004. Affinities of the Middle Pleistocene crania from Dali and Jinniushan. American Journal of Physical Anthropology Supplement 38, 167.

Rosas, A., Bermúdez de Castro, J.M., 1998. On the taxonomic affinities of the Dmanisi mandible (Georgia). American Journal of Physical Anthropology 107(2), 145–162.

Sankhyan, A.R., 1997. Fossil clavicle of a Middle Pleistocene hominid from the Central Narmada Valley, India. Journal of Human Evolution 32(1), 3–16.

Santa Luca, A.P., 1978. A re-examination of presumed Neandertal-like fossils. Journal of Human Evolution 7, 619–636.

Santa Luca, A.P., 1980. The Ngandong Fossil Humans: A Comparative Study of a Far Eastern *Homo erectus* Group. Yale University Press, New Haven.

Sarkar, S.S., 1960. Human skeletal remains from Brahmagiri. Bulletin of the Anthropological Survey of India 9(1), 5–24.

Schafer, J.L., 1999. Norm: Multiple Imputation of Incomplete Multivariate Data under a Normal Model. Pennsylvania State University, State College, PA.

Schafer, J.L., Olsen, M.K., 1998. Multiple imputation for multivariate missing-data problems: A data analysts's perspective. Multivariate Behavioral Research 33, 545–571.

Sewall, R.B.S., Guha, B.S., 1973. Chapter XXX: Human remains. In: Marshall, S.J. (Ed.), Mohenjo-Daro and the Indus Civilization. Indological Book House, New Delhi, pp. 599–648.

Shen, G., Wang, W., Wang, Q., Zhao, J., Collerson, K., Zhou, C., Tobias, P.V., 2002. U-series dating of the Liujiang hominid site in Guangxi, Southern China. Journal of Human Evolution 43, 817–829.

Singer, R., 1954. The Saldanha skull from Hopefield, South Africa. American Journal of Physical Anthropology 12(3), 345–362.

Singer, R., 1958. The Rhodesian, Florisbad and Saldanha skulls. In: von Koenigswald, G.H.R. (Ed.), Hundert Jahre Neanderthaler. Kemink en Zoon N. V., Utrecht, pp. 52–62.

Smith, F., 1976. The Neandertal Remains from Krapina: A Descriptive and Comparative Study. Reports of Investigation No. 15. University of Tennessee Department of Anthropology, Knoxville, TN, p. 359.

Snow, C.E., 1953. The ancient Palestinian: Skhul V reconstruction. In: Hencken, H. (Ed.), American School of Prehistoric Research Bulletin no. 17. American School of Prehistoric Research, Cambridge, MA, pp. 5–10.

Sohn, S., Wolpoff, M.H., 1993. Zuttiyeh face: A view from the east. American Journal of Physical Anthropology 91(3), 325–347.

Sonakia, A., 1984. The skull-cap of early man and associated mammalian fauna from Narmada valley alluvium, Hoshangabad area, Madhya Pradesh (India). Geological Survey of India Records 113(6), 159–172.

Sonakia, A., 1985a. Early *Homo* from Narmada valley, India. In: Delson, E. (Ed.), Ancestors: The Hard Evidence. Alan R. Liss, Inc., New York, pp. 334–338.

Sonakia, A., 1985b. Skull cap of an early man from the Narmada valley alluvium (Pleistocene) of central India. American Anthropologist 87, 612–615.

StataCorp, 2004. Stata. College Station, TX.

Stock, J., Pfeiffer, S.K., Chazan, M., Janetski, J., 2005. F-81 skeleton from Wadi Mataha, Jordan, and its bearing on human variability in the Epipaleolithic of the Levant. American Journal of Physical Anthropology 128(2), 453–465.

Stock, J., Lahr, M.M., 2007. Human dispersals and cranial diversity in South Asia relative to global patterns of human variation. In: Petraglia, M.D., Allchin, B. (Eds.), The Evolution and History of Human Populations in South Asia: Interdisciplinary Studies in Archaeology, Biological Anthropology, Linguistics and Genetics. Springer, Netherlands, pp. 245–268.

Storm, P., 1995. The Evolutionary Significance of the Wajak Skulls. Nationaal Natuurhistorisches Museum, Leiden.

Strauss, R.E., Atanassov, M.N., De Oliveira, J.A., 2003. Evaluation of the principal-component and expectation-maximization methods for estimating missing data in morphometric studies. Journal of Vertebrate Paleontology 23(2), 284–296.

Stringer, C., 1984. The definition of *Homo erectus* and the existence of the species in Africa and Europe. Courier Forschungsintitut Senckenberg 69, 131–143.

Stringer, C.B., 2002. Modern human origins: progress and prospects. Philosophical Transactions of the Royal Society of London B 357, 563–579.

Stringer, C.B., Howell, F.C., Melentis, J.K., 1979. The significance of the fossil hominid skull from Petralona, Greece. Journal of Archaeological Science 6, 235–253.

Suzuki, H., Hanihara, K. (Eds.), 1982. The Minatogawa Man: The Upper Pleistocene Man from the Island of Okinawa. University of Tokyo Press, Tokyo.

Suzuki, H., Takai, F. (Eds.), 1970. The Amud Man and his Cave Site. Academic Press of Japan, Tokyo.

Swisher, C.C., Rink, W.J., Antón, S.C., Schwarcz, H.P., Curtis, G.H., Suprijo, A., Widiasmoro, 1996. Latest *Homo erectus* of Java: potential contemporaneity with *Homo sapiens* in Southeast Asia. Science 274, 1870–1874.

Tattersall, I., 1986. Species recognition in human paleontology. Journal of Human Evolution 15, 165–175.

Thangaraj, K., Chaubey, G., Kivisild, T., Reddy, A.G., Singh, V.K., Rasalkar, A.A., Singh, L., 2005. Reconstructing the origin of Andaman Islanders. Science 308(5724), 996.

Thangaraj, K., Singh, L., Reddy, A.G., Rao, V.R., Sehgal, S.C., Underhill, P.A., Pierson, M., Frame, I.G., Hagelberg, E., 2003. Genetic affinities of the Andaman Islanders, a vanishing human population. Current Biology 13(2), 86.

Tiwari, M.P., Bhai, H.Y., 1997. Quaternary Stratigraphy of the Narmada Valley. Geological Survey of India Special Publication 46, 33–63.

Trinkaus, E., 1983. The Shanidar Neandertals. Academic Press, New York.

Trinkaus, E., Milota, S., Rodrigo, R., Mircea, G., Moldovan, O., 2003. Early modern human cranial remains from the Pestera cu Oase, Romania. Journal of Human Evolution 45(3), 245–253.

Vandermeersch, B., 1981. Les Hommes Fossiles de Qafzeh. Centre National de la Recherche Scientifique, Paris.

Wagner, G.P., 1984. On the eigenvalue distribution of genetic and phenotypic dispersion matrices: Evidence for a nonrandom organization of quantitative character variation. Journal of Mathematical Biology 21, 77–95.

Weidenreich, F., 1943. The Skull of *Sinanthropus pekinensis*: A Comparative Study on a Primitive Hominid Skull. Geological Survey of China, Pehpei, Chungking.

Weidenreich, F., 1945. The Keilor skull: a Wadjak type from Southeast Australia. American Journal of Physical Anthropology 3, 225–236.

Weidenreich, F., 1951. Morphology of Solo Man. American Museum of Natural History, New York.

White, T.D., Asfaw, B., DeGusta, D., Gilbert, H., Richards, G.D., Suwa, G., Clark Howell, F., 2003. Pleistocene *Homo sapiens* from Middle Awash, Ethiopia. Nature 423(6941), 742–747.

Widianto, H., Zeitoun, V., 2003. Morphological description, biometry and phylogenetic position of the skull of Ngawi 1 (East Java, Indonesia). International Journal of Osteoarchaeology 13, 339–351.

Wolpoff, M.H., Hawks, J., Frayer, D.W., Hunley, K., 2001. Modern human ancestry at the peripheries: a test of the replacement theory. Science 291(5502), 293–297.

Wolpoff, M.H., Wu, X., Thorne, A.G., 1984. Modern *Homo sapiens* origins: a general theory of hominid evolution involving the fossil evidence from East Asia. In: Smith, F.H., Spencer, F. (Eds.), The Origins of Modern Humans: A World Survey of the Fossil Evi-dence. Alan R. Liss, New York, pp. 411–483.

Woo, J.-K., 1958. Tzeyang Paleolithic man–earliest representative of modern man in China. American Journal of Physical Anthropology 16, 459–471.

Woo, J.-K., 1959. Human fossils found in Liukiang, Kwangsi, China. Vertebrata Palasiatica 3(3), 109–118.

Wu, R., 1982. Preliminary study of *Homo erectus* remains from Hexian, Anhui. Acta Anthropologica Sinica (Chinese with English Summary) 1(1), 2–13.

Wu, R., 1988. Reconstruction of the fossil human skull from Jinniushan, Yingkou, Liaoning province and its main features. Acta Anthropologica Sinica (Chinese with English Summary) 8(2), 97–101.

Wu, R.K., Xingxue, L., Xinzhi, W., Xinan, M. (Eds.), 2002. *Homo erectus* from Nanjing. Jiangsu Science and Technology Publishing House, Nanjing.

Wu, X., 2004. Fossil humankind and other anthropoid primates of China. International Journal of Primatology 25(5), 1093–1103.

Wu, X., Braüer, G., 1993. Morphological comparison of archaic *Homo sapiens* crania from China and Africa. Zeitschrift fur Morphologie und Anthropologie 79(3), 241–259.

Wu, X., Poirier, F.E., 1995. Human Evolution in China: A Metric Description of the Fossils and a Review of the Sites. Oxford University Press, New York.

Wu, X.-Z., 1981. The well preserved cranium of an early *Homo sapiens* from Dali, Shaanxi. Scientia Sinica 24(4), 531–541.

Wunderly, J., 1943. The Keilor Fossil Skull: Anatomical Description. Memoirs of the National Museum of Victoria 13, 57–69.

Zhao, J.-X., Hu, K., Collerson, K.D., Xu, H.-K., 2001. Thermal ionization mass spectrometry U-series dating of a hominid site near Nanjing, China. Geology 29(1), 27–30.

PART II
THE MODERN SCENE

8. The Toba supervolcanic eruption: *Tephra-fall deposits in India and paleoanthropological implications*

SACHA C. JONES

Leverhulme Centre for Human Evolutionary Studies
The Henry Wellcome Building
Fitzwilliam Street
University of Cambridge
Cambridge, CB2 1QH
England
s.jones@human-evol.cam.ac.uk

Introduction

The 74,000 year-old supereruption of the Toba volcano, located in northern Sumatra, is recognized as one of Earth's largest known eruptions and was certainly the largest of the Quaternary period (Smith and Bailey, 1968). It is hypothesized to have led to both global climatic deterioration (Rampino et al., 1988; Rampino and Self, 1992, 1993a) and the decimation of modern human populations (Rampino and Self, 1993b; Ambrose, 1998, 2003a, 2003b; Rampino and Ambrose, 2000). However, the severity of Toba's impact on climate and hominins has been contested (Oppenheimer, 2002; Gathorne-Hardy and Harcourt-Smith, 2003). Geological and archaeological evidence from the Indian subcontinent provides an excellent opportunity to address these issues. The scale of the Toba supereruption was so vast that

it led to the deposition of a blanket of volcanic ash over India, Malaysia, the Indian Ocean, and the Arabian and South China Seas. Resulting terrestrial tephra deposits have been documented in a number of river valleys throughout India (Acharyya and Basu, 1993; Shane et al., 1995; Westgate et al., 1998). A handful of these localities preserve archaeological, paleontological and paleoenvironmental evidence which is in direct association with the tephra. This chapter is the first study to pull together this evidence, looking at the relationship between Toba ash and archaeology from sites across the Indian subcontinent. It is proposed here that the specific local impacts of the Toba eruption can be analyzed by applying a multidisciplinary approach to the archaeological and geological evidence from certain localities in the Indian subcontinent. This data can be combined with additional evidence from a wider context

M.D. Petraglia and B. Allchin (eds.), The Evolution and History of Human Populations
in South Asia, 173–200.

in order to give a broader understanding of the regional impacts of the eruption on the Indian subcontinent as a whole. Current investigations of tephra-archaeology associations in the Middle Son valley (Madhya Pradesh) and at Jwalapuram (Kurnool District, Andhra Pradesh) are assessing the possible local impacts of the ~74 ka Toba eruption on hominin populations and their environments.

The Supervolcanic Eruption of Toba 74,000 Years Ago

Toba erupted four times during the Quaternary period: ~1.2 Ma (Haranggoal Dacite Tuff), ~840 ka (Oldest Toba Tuff), ~501 ka (Middle Toba Tuff) and ~75 ka (Youngest Toba Tuff) (Chesner et al., 1991). These eruptions culminated in the formation of a massive 100 km by 30 km caldera, now known as Lake Toba (Rose and Chesner, 1987; Chesner, 1998). The eruption of the Youngest Toba Tuff (YTT) at ~74 ka was significantly larger than the others, possessing a Volcanic Explosivity Index (VEI) of 8 (Newhall and Self, 1982), or magnitude $\geq$ M8 (Mason et al., 2004). The sheer scale of the YTT eruption renders it a "Supereruption", and Toba a "Supervolcano" (the concept of the "Supervolcano" and the frequency and magnitude of supereruptions have been discussed by Rampino [2002] and Mason et al. [2004]).

Dense Rock Equivalent (DRE) estimates of eruptive volume for the ~74 ka eruption have varied between 2,000 km^3 (van Bemmelen, 1939) and 3,000 km^3 (Aldiss and Ghazali, 1984). The most frequently quoted DRE estimate, calculated by Rose and Chesner (1987, 1990), is ~2,800 km^3 (7×10^{18} g) of erupted magma. Of this, ~2,000 km^3 consists of thick ignimbrite deposits, formed by devastating pyroclastic flows that covered 20,000 to 30,000 km^2 of northern Sumatra (van Bemmelen, 1939; Ninkovich et al., 1978a; Rose and Chesner, 1990). The remaining ~800 km^3 (2×10^{18} g) comprises ash-fall deposits. Rose and Chesner (1987, 1990) have specified that this total volume estimate of ~2,800 km^3 is a minimum value. The accumulation of new evidence over the past few years now suggests that a DRE estimate of 2,800 km^3 is most probably an underestimate. Recent discoveries of Toba tephra in marine cores from the Indian Ocean, Arabian Sea and South China Sea have greatly extended the previously known limits of this ash-fall (Schulz et al., 1998, 2002; Pattan et al., 1999, 2001, 2002; Bühring et al., 2000; Song et al., 2000).

The Role of the Toba Eruption in Late Pleistocene Climate Change and Hominin Evolution

The Relationship Between the Toba Eruption and Global Climate Change

Rampino et al. (1988) proposed that the after-effects of a volcanic eruption the size of Toba at 74 ka could have induced a volcanic winter, similar to predicted nuclear winter scenarios, as modeled by Turco et al. (1983, 1990). The injection of vast amounts of gaseous aerosols and dust into the atmosphere, which follow large volcanic eruptions, is predicted to have detrimental consequences for global climate. Past historical eruptions, such as Tambora in 1815 (Stothers, 1984) and Pinatubo in 1991 (McCormick et al., 1995), have provided evidence of post-eruption climatic deterioration. With the eruption of Toba having been far larger than both of these historical eruptions, its consequences are therefore assumed to have been far more devastating.

The presence of volcanic ash in the atmosphere would have led to temporarily darkened skies. Ash, however, has a relatively short residence time in the atmosphere when compared to the longevity of sulphuric acid aerosols. It is the latter which are predicted to have had the most damaging repercussions for global climate. Dense concentrations of H_2SO_4 aerosols in the stratosphere can prevent radiation reaching the Earth from the Sun, absorbing as well as backscattering incoming solar radiation (Rampino et al., 1988). This process results in cooling of both the atmosphere and the surface of the Earth.

Rampino and Self (1992) estimated that Toba's volcanic dust and aerosol clouds would have caused a brief but dramatic cooling event (i.e., volcanic winter) that lasted for a few years immediately following the eruption. During this time, it is predicted that land temperature fell to ~5–15 °C less than normal between latitudes of 30° to 70°N. This is thought to have been coupled with widespread hard freezes at mid-latitudes, a decrease in rainfall, very low summer temperatures, and immediate coolings of ~10 °C at lower latitudes. It has been predicted that average hemispheric surface temperatures continued to exhibit a decrease of 3–5 °C for several years after this abrupt cooling period (Rampino and Self, 1992, 1993a, 1993b; Rampino and Ambrose, 2000). The environmental implications of this volcanic winter involve possible obliteration of cold-sensitive tropical vegetation as well as severe droughts in monsoonal and tropical rainforest zones. At higher latitudes, reduced temperatures during the early growing season could have destroyed forests, coupled with greatly delayed recovery times (Rampino and Ambrose, 2000).

Evidence from Greenland ice-core data lends support to the theory of post-Toba climatic perturbation. Zielinksi et al. (1996a, 1996b) identified the existence of a six-year-long peak in volcanic sulphate concentration in the GISP2 core, dated to 71 ± 5 ka. This peak has been attributed to Toba because of the magnitude of this sulphate signal and the absence of associated tephra shards, the latter implying a distant source for the signal (Zielinksi et al., 1996a, 1996b). In addition, the GISP2 data shows that this sulphate peak is followed by a 200-year-long peak in calcium (Zielinksi et al., 1996b), indicative of large quantities of wind-blown dust, possibly caused by decreased vegetation cover and/or

exposure of sediments after a drop in sea-level (Ambrose, 2003).

Earlier investigations of tephra deposits in Indian Ocean marine cores placed the YTT event at the transition from Oxygen Isotope Stage (OIS) 5a to OIS 4 (Ninkovich et al., 1978a). This evidence was used to support the theory that the after-effects of the Toba eruption contributed to the onset of the glacial period of OIS 4 (Rampino and Self, 1992, 1993a, 1993b). However, the GISP2 data indicates that although the volcanic sulphate peak is coincident with a ~1,000 year-long hypercold stadial, this was separated from the onset of OIS 4 by interstadial 19, a warm period, lasting ~2,000 years (Zielinksi et al., 1996b). This evidence led to criticism of the volcano-climate forcing models of Rampino and Self (1992, 1993a, 1993b), suggesting that the Toba eruption may not have directly contributed to the onset of OIS 4 (Kerr, 1996; Rampino and Ambrose, 2000; Oppenheimer, 2002). This conclusion is supported by evidence from further afield. Marine cores, preserving YTT, indicate a 1 °C cooling period in the South China Sea for ~1–3 ka after the eruption. This cool interval is again followed by a warm period, presumed to be Interstadial 19, making it further increasingly unlikely that YTT forced the onset of OIS 4 (Huang et al., 2001). Nor, indeed, is it necessarily the case that the eruption initiated the 1,000-year-long stadial at 71 ka (Huang et al., 2001; Oppenheimer, 2002). Evidence from the GISP2 core indicates that the cooling process in between interstadials 19 and 20 was already underway by the time of the eruption. However, it is argued that the effects of Toba magnified this process, leading to "enhanced cooling" for several centuries after the eruption (Zielinski, 2000).

In a detailed criticism of the evidence that has been used to link the Toba event to climate change, Oppenheimer (2002) questions the notion that the eruption had such a detrimental impact on global climate. Estimates of eruption duration, intensity and plume height are described as "poorly constrained", as are those for the total amount of sulphur aerosols released. Further, correlation of the volcanic sulphate peak in the GISP2 core with the YTT eruption is criticized. Oppenheimer (2002) concludes that it is difficult to reliably establish Toba's climatic consequences. Overall global cooling resulting from the eruption may be significantly less than the 3–5 °C drop estimated by Rampino and Self (1993a). Instead, he suggests a more conservative global average cooling of 1 °C.

Evidence of a Late Pleistocene Human Population Bottleneck in Mitochondrial DNA

For the last two decades, population geneticists have been studying the pattern of modern human variation, captured in both mitochondrial and nuclear DNA. A large body of literature on the subject has grown over the years. Of particular note is a line of research which conducts pairwise comparisons of modern human mitochondrial DNA in order to obtain information on past demographic processes, recorded in mtDNA mismatch distributions (Rogers and Harpending, 1992). These studies resulted in the formulation of the "weak Garden of Eden" hypothesis of Harpending et al. (1993). This states that modern humans spread into separate regions from a restricted source ~100 ka; these populations consequently became geographically and hence genetically isolated. During this time, the number of humans is estimated to have fallen to only a few thousand breeding females (Harpending et al., 1993).

Populations subsequently expanded out of this bottleneck some time after separation and isolation. Numerous estimates have been provided for the timing of this demographic growth, including ~50 ka (Harpending et al., 1993), 30 to 65 ka (Sherry et al., 1994), ~60 ka (Rogers and Jorde, 1995), 100 to 50 ka (Harpending et al., 1998), 30 to 130 ka

(Harpending et al., 2000), or ~60 to 80 ka (Watson et al., 1997).

It has been pointed out that "most" of these separate populations underwent a bottleneck. These regional bottlenecks could have been either brief and severe, or extended and mild (Rogers and Jorde, 1995). However, the evidence does suggest that these bottlenecks (or expansions out of the bottlenecks) were simultaneous. Three explanations are offered for this. First, population growth and expansion was driven by the appearance of an advantageous cultural innovation, in the form of Late Stone Age and Upper Paleolithic technologies that swept across the species (Harpending et al., 1993, 1998). Second, climate change could have been a causal factor (Rogers and Jorde, 1995). Third, the YTT event and its supposed devastating consequences has been used as an explanation for simultaneous population bottlenecks, whereby subsequent population growth occurred only after climatic amelioration (Rampino and Self, 1993b; Rogers and Jorde, 1995; Ambrose, 1998). The second and third explanations are supported by the identification of a more or less contemporary bottleneck in *Pan troglodytes schweinfurthii* and *H. sapiens* (Rogers and Jorde, 1995); the former undergoing an expansion ~ 67 ka (Ambrose, 1998). However, Goldberg and Ruvolo (1997) argue for a wide date range of between 20 and 61 ka for this expansion, stating that the evidence does not strongly support contemporaneous expansions in humans and eastern chimpanzees. Furthermore, it should be noted that the same pattern is not seen in any other *Pan* or *Gorilla* taxa (Gagneux et al., 1999).

Toba, Volcanic Winter, and Human Population Decline

Ambrose (1998) hypothesized that the Toba eruption was an underlying cause behind this Late Pleistocene demographic crash (Ambrose, 1998; later revised and elaborated by Rampino and Ambrose, 2000). He proposed that initial population expansions occurred with climatic improvement after the eruption and not with the arrival of Late Stone Age or Upper Paleolithic technologies, as suggested by Harpending et al. (1993) in their "weak Garden of Eden" model. Ambrose argues that technological change occurred later than the population expansions, suggesting an earlier climatic release from the bottleneck either ~70 ka or ~60 ka, and later population growth ~ 50 ka (or possibly earlier [Ambrose, 2002]) due to improved technology as indicated by the appearance of innovations in the Late Stone Age of Africa (Ambrose, 1998, 2002, 2003a, 2003b).

Ambrose (1998) and Rampino and Ambrose (2000) propose that the supposed volcanic winter and harsh glacial climate that followed the eruption caused a human population bottleneck. This bottleneck would have greatly reduced modern human diversity as well as population size. With climatic amelioration, population explosion out of this bottleneck would have occurred, either ~70 ka at the end of a hypercold millennium, located between interstadials 19 and 20, or ~10 ka later with the transition from OIS 4 to warmer OIS 3. Post-Toba populations would have reduced in size such that founder effects, genetic drift and local adaptations occurred, resulting in rapid population differentiation (Ambrose, 1998). In this way, the Toba eruption of ~74 ka would have shaped the diversity that is seen in modern human populations today.

Oppenheimer (2002) has criticized this theory by stating that there is no firm evidence linking the YTT event to a human demographic crash. He questions the validity of linking the relatively precisely dated Toba eruption to the "putative" human population bottleneck given the broad time-frame within which it is supposed to have occurred. Additionally, he argues that the climate ~74 ka was not uniquely cold for the Quaternary period. However, he does suggest that the eruption could have had negative consequences for human populations if it occurred at a sensitive time in human evolution, or that it was the suddenness of the eruption's after-effects that had a significant impact. Oppenheimer does not deny that the direct effects of the eruption would have been destructive, but concludes that the impacts would have been regional, not global. Consequently, he stresses that further research is necessary to produce well-constrained estimates of eruption intensity, duration, height, and sulphur emissions, in order to facilitate reliable predictions of how the atmosphere, and hence global climate, respond to large eruptions.

A second criticism is made by Gathorne-Hardy and Harcourt-Smith (2003) following studies of fauna on the Mentawai islands, ~350 km south of Lake Toba. The Mentawai islands have been isolated from the mainland for ~3 Ma and the presence of nine endemic rainforest-obligate animals (four primates, four squirrels and one mouse species), together with termite data, indicate that the islands have been continually covered by rainforest throughout this period. Due to the occurrence of these species, they conclude that the maximum radius of direct destruction by heavy tephra must therefore have been less than 350 km. They state the following:

It is possible that heavy tephra traveled further in other directions than south east, but the Mentawai data indicate that it is unlikely that its destructive effects reached as far as India or Indochina to the north, or Java to the south (Gathorne-Hardy and Harcourt-Smith, 2003:228).

In fact, YTT deposits are found throughout India and off the coast of Indochina, but they have not been recorded in Java, suggesting that YTT traveled in all directions apart from to the south-east. Therefore, YTT may not have reached the Mentawai islands which might explain the faunal composition described by Gathorne-Hardy and Harcourt-Smith (2003). However, it is worth pointing out that the presence of nine endemic species and termite data alone do not necessarily indicate the absence of the destructive effects of tephra.

According to the theories outlined above, tropical rainforests should have been adversely affected by the proposed volcanic winter which may, in turn, have had a detrimental impact upon rainforest fauna. However, it is not necessarily the case that we should expect to see the extinction of endemic Mentawai Island species following the Toba eruption. Instead, rainforest may have contracted due to climate change, creating a refugium, or pockets of refugia (evidence suggests that areas of ancient rainforest refugia do exist in the Mentawai islands [Brandon-Jones, 1998; Gathorne-Hardy and Harcourt-Smith, 2003]). The Mentawai fauna may have experienced local population extinctions, but not species extinction, and are thus still present on the islands today.

In addition, it is possible that the species studied by Gathorne-Hardy and Harcourt-Smith (2003) reoccupied or were introduced into the islands via rafting at some point after the YTT eruption. In fact, Brandon-Jones (1998) does refer to dispersals from the Mentawai Islands to other areas of Southeast Asia during the Late Pleistocene. However, such dispersals are difficult to prove and would have to have involved all nine species. It is perhaps only through genetic studies of these fauna – specifically, whether the Mentawai islands faunal species in question carry the genetic signature of a Late Pleistocene bottleneck – that this issue can be clarified.

It is commonly argued that if Toba was truly devastating for *H. sapiens* then comparable bottlenecks should also be seen in many other species. This point has been contested for two reasons. First, it may simply be the case that the mitochondrial DNA of other large mammal species has not been investigated in as much detail as that of modern humans and chimpanzees. Ambrose highlights

the fact that such studies are a labor-intensive process of creating sequence data that takes a considerable amount of time (Ambrose, 1998). Second, Harpending et al. (2000) point out that bottlenecks should not necessarily be evident in every species because those species that survive the aftermath of the eruption in two or more refugia will retain greater genetic variation than a species that survives in only one refugium. They therefore argue that both *H. sapiens* and *Pan troglodytes schweinfurthii* must have survived in a single refugium. However, this contradicts previous interpretation of the mitochondrial evidence where it is said that human populations became geographically separated and relatively isolated from each other from ~100 ka until ~50 ka (Harpending et al., 1993). This would surely seem to suggest that *H. sapiens* survived the bottleneck in at least two or more refugia.

Widespread Geographic Dispersal of the Youngest Toba Tuff: Evidence from the Indian Subcontinent

Having reviewed the theories surrounding the effects of the Toba eruption on both global climate and human populations in general, it is now possible to turn our attention to evidence provided in the specific context of the Indian subcontinent. By way of background, it is important to recognize that ca. 800 km^3 of fine-grained volcanic ash was injected into the stratosphere during the 74 ka Toba eruption. The ash was carried by prevailing winds in all directions, apart from to the south-east. As a result, large areas of southern and southeastern Asia were covered in tephra, from the Arabian Sea in the west to the South China Sea in the East (Figure 1). YTT has been documented in marine cores from the northeast Indian Ocean and Bay of Bengal (Ninkovich et al., 1978a, 1978b; Dehn et al., 1991) and more recently from the Central Indian Ocean Basin (Pattan et al., 1999, 2002), Arabian Sea (Schulz et al.,

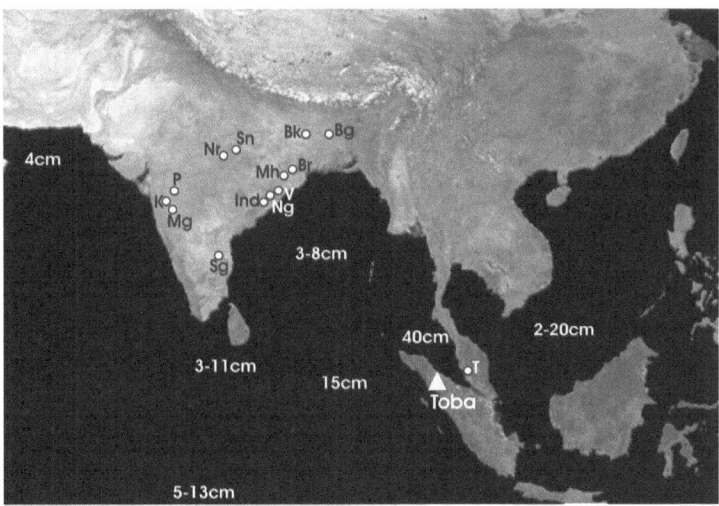

Figure 1. Marine and terrestrial distribution of the Youngest Toba Tuff in southern and south-east Asia. The thickness of marine ash layers at various localities is indicated (in white, in centimetres). The white circles in India, Bangladesh and Malaysia indicate the location of terrestrial tephra sites. The letters correspond to the following localities; K=Kukdi, P=Purna, Mg=Morgaon, Sg=Sagileru, Ind=Indravati, Ng=Nagavali, V=Vansadhara, Mh=Mahanadi, Br=Brahmani, Bg=Bogra, Bk=Barakar, Sn=Son, Nr=Narmada, T=Tampan

1998, 2002; Pattan et al., 2001) and South China Sea (Bühring et al., 2000; Song et al., 2000). The thickness of the YTT bed in these deep-sea cores ranges from ~2 to ~40 cm and does not appear to reduce significantly with increasing distance from the Toba caldera. However, marine ash layers, as with terrestrial tephra beds, can experience reworking and bioturbation and thus some may not preserve a precise record of air-fall thickness.

Terrestrial YTT deposits have been identified in India (Rose and Chesner, 1987; Acharyya and Basu, 1993; Shane et al., 1995; Westgate et al., 1998), Malaysia (Ninkovich et al., 1978a; Rose and Chesner, 1987; Shane et al., 1995) and possibly in Bangladesh (Acharyya and Basu, 1993; Acharyya et al., 2000). Of these, Indian YTT occurrences are far more numerous and extensive, having been documented in river valleys throughout the subcontinent (Table 1). Well-described Toba ash localities include those from the Middle Son and Central Narmada valleys in Madhya Pradesh, the Kukdi river in Maharashtra, and the Indravati, Nagavali, Vansadhara,

Mahanadi and Brahmani river basins in Orissa. Less attention has been paid to YTT deposits from Barakar in Bihar, Sagileru in Andhra Pradesh, and Purna and Morgaon in Maharashtra. There are references to other tephra localities, including the "Karnool area" (Westgate et al., 1998), and "a few localities in Madhya Pradesh, Andhra Pradesh and Uttar Pradesh" (Acharyya and Basu, 1993). Yet very few details are provided, such as precise location information, stratigraphical data or tephra characteristics.

The documentation of occurrences of YTT deposits in India has been relatively recent. Williams and Royce (1982) reported the first discovery of volcanic ash in India, encountered during explorations in the Middle Son valley in 1980. This was followed by further discoveries of tephra deposits in the Middle Son valley, as well as in the central Narmada valley (Basu et al., 1987), the Kukdi river valley (Korisettar et al., 1988), and in the Vansadhara, Nagavali, Mahanadi, Brahmani and Indravati river basins of Orissa (Devdas and Meshram, 1991). Around the same time, a tephra layer was recorded

Table 1. A summary of Indian tephra localities, indicating the nature of the tephra deposits, those that have been identified as YTT and dated by absolute means, and those that have been found in association with Paleolithic artifacts

River Valley	Tephra Locality	Tephra Thickness (m)	Extension of Tephra (km)	Geochemically Characterized as YTT?	Absolute Dates	Archaeology-tephra Associations?
Middle Son	Ghogara	1.5	30	Yes[a,b,c,d]	No	Yes (Middle and Upper Paleolithic)
	Ramnagar	0.5				
	Nakjhar	0.7				
	Khutiali	1.5				
Central Narmada	Pawlaghat	0.24–0.4	20	Yes[b,c,d]	121 ± 22 ka[d]	Yes (Middle and Upper Paleolithic)
	Guruwara	0.8				
	Devakachar	<1				
	Hirapur	<1				
Kukdi	10 sections near Bori	0.2–1	3.5	Yes[b,c,d]	1.4 Ma[e]; 0.54 Ma, 0.64 Ma, 0.023 Ma[f]; 0.67 Ma[g].	Yes (Lower Paleolithic)
Vansadhara	Goguparu	1.25	2	Yes[b]	No	Yes (Lower Paleolithic, but stratigraphic association between ash and artifacts is unclear)
	Kareni	3–5				
Nagavali	Rayagada	1.5–6	17	Not analysed	No	Not reported
Mahanadi	Kumbia	2.1	45	Yes[d]	No	Not reported
	Krushnamohonpur	2				
	Pitamohul	3				
	Sonepur	1.5–2.5				
Brahmani	Samal	2.5	2.5	Not analysed	No	Not reported
Indravati	Damkelar	0.2–0.3	0.5	Not analysed	No	Not reported
Sagileru	(not specified)	0.3–2	90	Not analysed	No	Not reported
Barakar	Opposite Barakar town	1.83	0.1	Yes[b]	No	Not reported
Purna	e.g. "Gandhigram Place"	Not specified	Not specified	Yes[c,d]	84 ± 16 ka[d]	Yes Paleolithic (Acheulean?)[c]
Karha	Morgaon	0.17	0.54	Not analysed	No	Yes Lower Paleolithic (Acheulean)
(Andhra Pradesh?)	"Karnool area"[d]	Not specified	Not specified	Yes	No	Not reported

[a] Rose and Chesner (1987);
[b] Acharyya and Basu (1993);
[c] Shane et al. (1995);
[d] Westgate et al. (1998);
[e] Korisettar et al. (1989);
[f] Horn et al. (1993);
[g] Mishra et al. (1995)

from the Barakar river basin in West Bengal-Bihar (Basu and Biswas, 1990), and along the Sagileru river valley in Andhra Pradesh where an extensive tephra bed is traceable for 90 km (Anonymous, 1991; Acharyya and Basu, 1993).

None of these tephra deposits were identified as YTT until 1987 when Rose and Chesner compared the Middle Son tephra with YTT samples from Lake Toba and marine cores. They concluded that the Son tephra was evidence of an extensive co-ignimbrite ash-fall (Rose and Chesner, 1987). Further, the true extent of YTT distribution in India was not realized until the work of Acharyya and Basu (1993). They geochemically characterized tephras from five Indian river valleys (Kukdi, Narmada, Son, Vansadhara, Barakar), identifying all of them as YTT. Their work was later confirmed and built upon by two separate studies. First, by Shane et al. (1995) who analyzed tephras from the Kukdi, Narmada, Purna and Son valleys. Second, by Westgate et al. (1998) who confirmed the presence of YTT in the Vansadhara, Mahanadi, Middle Son, Central Narmada, Purna, and Kukdi river valleys, as well as in the Karnool area. They also obtained fission-track dates for ash samples from the Purna and Narmada basins, giving dates of 84 ± 16 ka and 121 ± 22 ka respectively (Westgate et al., 1998). It appears that samples of volcanic ash from the Indravati, Brahmani, Nagavali and Sagileru river valleys have not been geochemically characterized, although it is assumed by Acharyya and Basu (1993) that these are all deposits of YTT.

There is significant variation in the thickness of Indian tephra deposits between as well as within particular localities. In addition, the terrestrial deposits are considerably thicker than those from marine cores, probably due to reworking and redeposition of terrestrial ash. These deposits range in thickness from 0.2 to 6 m (Figure 2). Most are less than 3m thick, while those from the Vansadhara and Nagavali river valleys in Orissa are the exception, exhibiting a maximum thickness of

5m and 6m respectively. Acharyya and Basu (1993) believe that redeposition of the tephra into thicker deposits is more or less contemporary with the initial ash-fall.

Implications of the Toba Ash-Fall for Late Pleistocene Hominins in the Indian Subcontinent

The Toba eruption of 74 ka was clearly an event of great magnitude, far greater than any known historical eruption (Table 2; Figure 3), suggesting it had devastating repercussions. Questions exist, however, as to the scale of these repercussions. The effects of Toba may or may not have been far-reaching such that the eruption can be said to have had an impact on a global scale. Perhaps impacts were on a more regional scale, for example, at the level of Southeast Asia or the Indian subcontinent as a whole. Alternatively, the hazardous effects of Toba may have been localized, whereby individual habitats or ecosystems were affected, yet other areas in a region remained unscathed.

These questions necessitate consideration of two separate aspects of the eruption; first, the consequences of possible rapid global climatic deterioration and second, the direct effects of the ash-fall on hominins and their environments in India. An additional issue that needs to be addressed is the identification of those hominins that inhabited India $\sim$74 ka and thus those who may have been affected by Toba. In this regard, numerous questions need to be asked. How many hominin species occupied India at the time of the eruption, one species or more? Was *H. sapiens* present in the subcontinent $\sim$74 ka, or did initial colonization occur at a later date?

Unfortunately, hominin fossil evidence from the Indian subcontinent is limited and cannot currently provide answers to these questions. Only a few specimens are known, none of which belong to the time-frame of the Toba

Table 2. Eruptive volumes (DRE) of a number of well-known volcanic eruptions

Volcanic Eruption	Date of Eruption	Location	Dense Rock Equivalent (DRE) in km^3	Reference
Fish Canyon Tuff	27.8 Ma	La Garita caldera, Colorado (USA)	4500	Mason et al., 2004
YTT	74 ka	Toba caldera, Sumatra	2800	Rose and Chesner, 1987
Huckleberry Ridge Tuff	2 Ma	Yellowstone, Wyoming (USA)	2200	Mason et al., 2004
OTT[a]	788 ka	Toba caldera, Sumatra	800–1000	Lee et al., 2004
Oruanui	26.5 ka	Taupo, New Zealand	530	Mason et al., 2004
OTT[b]	840 ka	Toba caldera, Sumatra	500	Diehl et al., 1987; Chesner, 1998
Bishop Tuff	700 ka	Long Valley, California (USA)	450	Mason et al., 2004
Campanian Ignimbrite	33.5 ka	Campi Flegrei, Bay of Naples, Italy	80	Rosi et al., 1996
MTT	501 ka	Toba caldera, Sumatra	60	Chesner, 1998
Tambora	1815 AD	Lesser Sunda Islands, Indonesia	50	Self et al., 1996
Krakatau	1883 AD	Sunda Strait, Indonesia	10	Self et al., 1996
Pinatubo	1991 AD	Luzon, Philippines	5	Self et al., 1996
Vesuvius	79 AD	Bay of Naples, Italy	1.5	Gurioli et al., 2005
Mount St Helens	1980 AD	Washington (USA)	0.55	Carey and Sigurdsson, 1982

[a]The background behind the estimates obtained by Lee et al. (2004) is discussed in the text. Their DRE estimate combines the estimate of $500 \, km^3$ for terrestrial OTT, with an estimate of $310–590 \, km^3$ for distal air-fall ash. The latter estimate is calculated using evidence from layer D in the Indian Ocean deep-sea core, ODP Site 758, and 3 marine cores in the South China Sea. Lee et al. (2004) believe that layer D and the South China Sea ash are OTT.

[b]OTT date and volume estimates prior to the work of Lee et al. (2004). The date of 840 ka was obtained by Diehl et al. (1987), using $^{40}Ar/^{39}Ar$.

eruption. The oldest fossil from India is represented by the Narmada hominin (see Athreya, this volume), dated to not less than 236 ka (Cameron et al., 2004), or to some time in between 150 and 250 ka (Kennedy, 2001:167). Modern human remains have been discovered in an undated Late Paleolithic context at Bhimbetka rockshelter III-A-28 (Wakankar, 2002:5) and from three cave sites in Sri Lanka, dating from 27.7 ka (Kennedy 1999, 2001).

Investigations of modern human genetic variation have been providing some important insights into the timing of modern human dispersals into the Indian subcontinent (see Endicott et al., this volume), yet it does not provide information on other hominin species that may have inhabited India at the time of the Toba event. A synthesis of modern genetic data illustrates that the initial colonization of India by *H. sapiens* took place after the Toba eruption via a southern dispersal route from Africa, through southern Arabia and into India. Various dates have been obtained for the colonization of India; ~50 to 60 ka

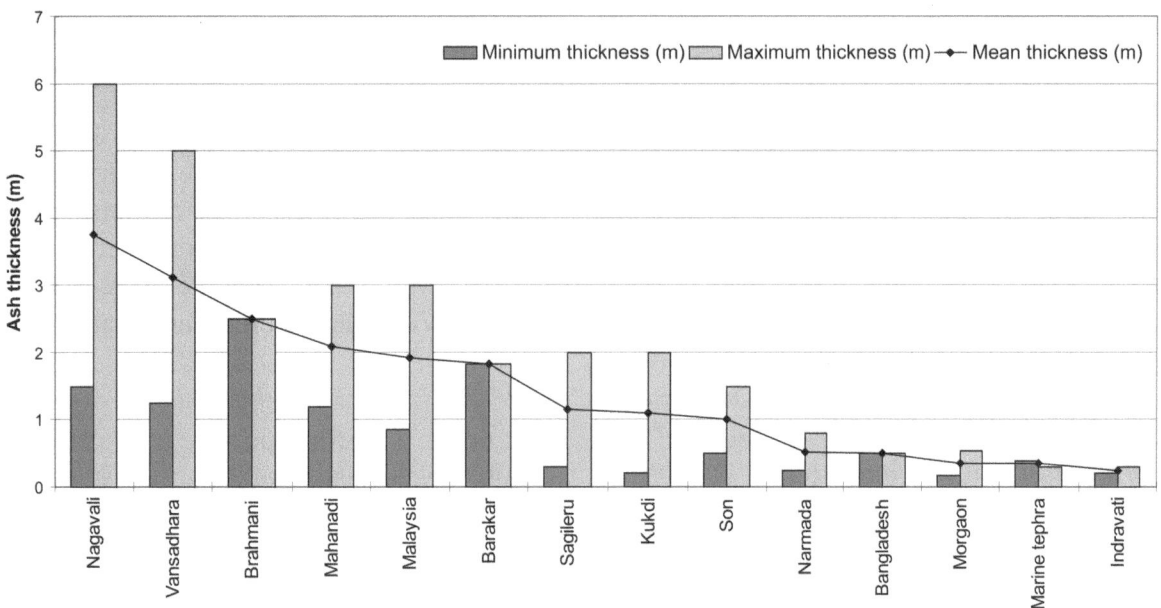

Figure 2. Thickness (in metres) of Toba ash deposits in the Indian Subcontinent

(Quintana-Murci et al., 1999), ~45 to 50 ka (Underhill et al., 2001), ~60 ka (Kivisild et al., 2003), 60 to 70 ka (Palanichamy et al., 2004), ~65 ka (Macauley et al., 2005), and ~50 to 70 ka (Thangaraj et al., 2005). It should be noted how close some of these quoted dates are to the age of the Toba eruption. Evidence for pre-Toba migrations to India is provided by Oppenheimer (2003) and possibly by Forster and Matsumura (2005), the latter giving a timeframe of between 85 and 55 ka for dispersals to India. Using phylogeographic data, Oppenheimer (2003) argues that *H. sapiens* occupied India before ~74 ka and may have undergone "mass extinction" as a result of the Toba eruption (Oppenheimer, 2003).

A comprehensive review of the Late Pleistocene archaeological record of the Indian

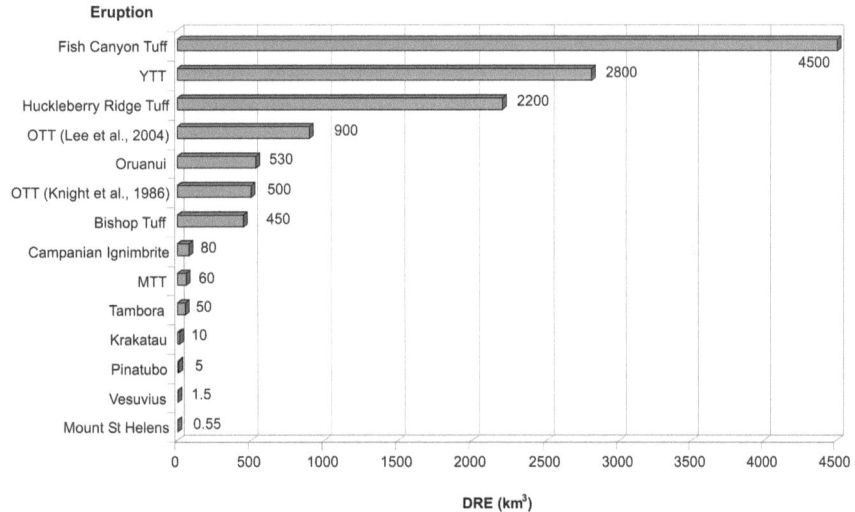

Figure 3. Eruptive volumes (Dense Rock Equivalent (DRE) in km^3) of a number of well-known volcanic eruptions

subcontinent has recently been provided by James and Petraglia (2005), with particular focus on the colonization of the subcontinent by *H. sapiens*. They argue that two distinct lithic industries characterize the Late Pleistocene in South Asia; an earlier Middle Paleolithic, beginning ~150 ka (Misra, 1995), and a later technological complex ("Late Paleolithic"), beginning ~45 ka. They find no evidence of a symbolic (or Upper Paleolithic) revolution in the Indian record that can be said to be associated with the appearance of the first modern humans. Instead, they tentatively date the colonization of India by *H. sapiens* to ~70-50 ka and indicate that these early colonizers may have used Middle Paleolithic technologies. In this scenario, distinguishing assemblages created by *H. sapiens* from those made by other hominin species, who also produced Middle Paleolithic artifacts, is considered problematic (James and Petraglia, 2005). Therefore, in the broadest sense it is possible to predict that hominins in India were practicing Middle Paleolithic technologies at the time of the Toba eruption, however, it is not possible to discern which hominin species manufactured them.

Archaeological evidence from the Indian subcontinent can only provide evidence of the impacts of the Toba eruption on hominins if Palaeolithic remains are preserved in either extremely well-dated contexts or in direct association with deposits of the Youngest Toba Tuff. Unfortunately, the former are currently a rarity in India; however, a number of sites exist in the Indian subcontinent where Toba tephra is preserved together with Paleolithic assemblages.

Associations between the Youngest Toba Tuff and the Indian Paleolithic

Several localities in India preserve tephra deposits that are directly associated with stone artifacts that pre-date and post-date the eruption. While Indian YTT deposits are

relatively well-described in the geological literature, the archaeological significance of any associations between ash and Paleolithic remains has yet to be explored. There is currently very little published information which describes these associations. Such information is only available for the following river valleys; Kukdi, Middle Son, central Narmada, Vansadhara, Purna and Karha.

Kukdi River Valley, Maharashtra

Kale et al. (1986) reported an Acheulean assemblage, consisting of 34 artifacts, located near the village of Bori in the Kukdi river valley, in Maharashtra. The presence of tephra was later described by Korisettar et al. (1988), who identified four localities along the Kukdi that preserved volcanic ash. Another Lower Paleolithic assemblage of 152 artifacts, that appeared (somewhat controversially) to post-date the tephra, has also been recorded (Korisettar et al., 1988). Further work in the area has revealed ten ash-bearing sections (Mishra et al., 1995) (Figure 4). Five of these preserve Lower Paleolithic artifacts derived from gravels, some of which contain remains of *Bos* sp. and *Elephas* sp. (Kale et al., 1986; Korisettar, 1994). In four of the ten sections, artifacts are found in gravel beds which are said to cut into the tephra. Only two artifacts (flakes) have been discovered beneath the tephra (Mishra et al., 1995). It is noteworthy that no Middle or Upper Paleolithic assemblages have been found at any of these localities. Instead, only Lower Paleolithic and "Epi-Paleolithic" artifacts have been recovered, existing as stratigraphically separated assemblages (Korisettar et al., 1989; Korisettar, 1994). Middle Paleolithic tools are documented nearby however, at the site of Ranjani only ~4 km away (Mishra and Ghate, 1990).

Several attempts have been made to date the Kukdi ash. These have produced a series of highly discrepant ages, probably because bulk tephra samples were dated. It was initially

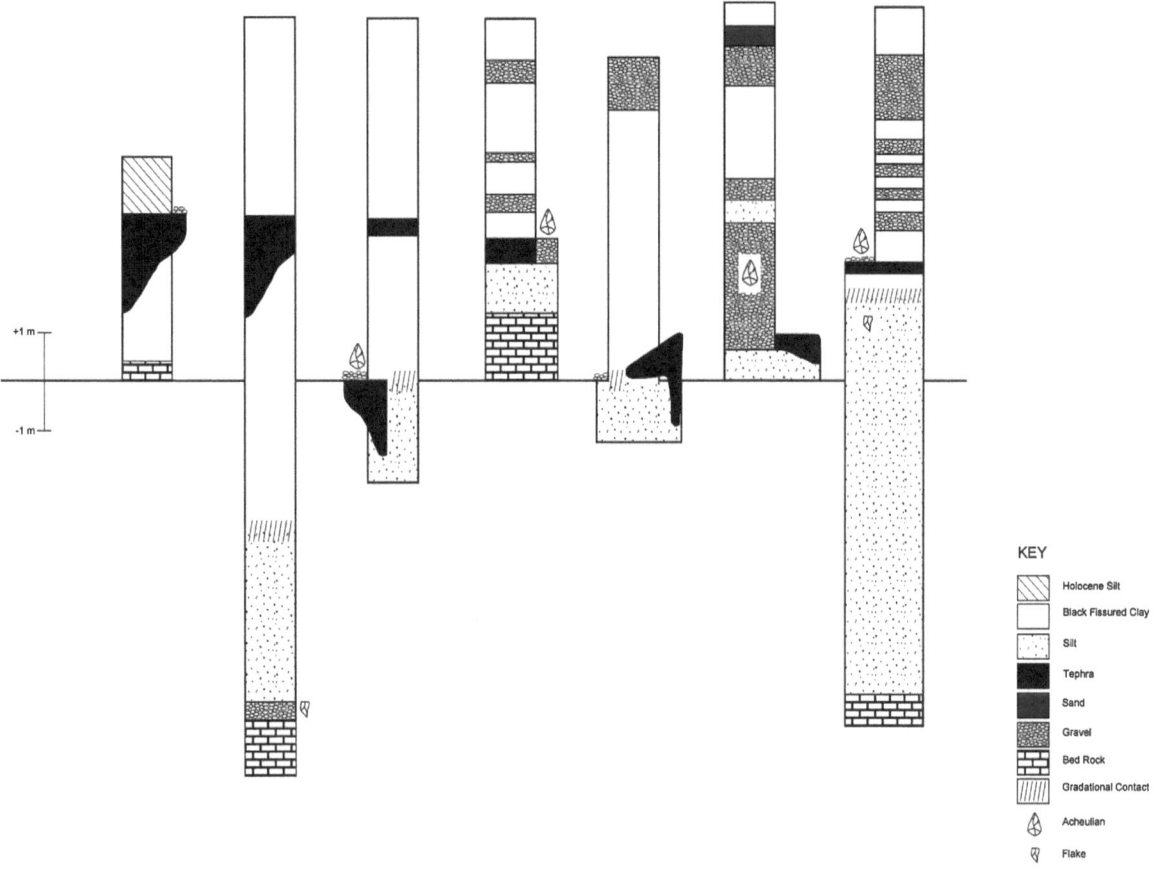

Figure 4. Sections exposed along the river Kukdi, in Maharashtra, indicating the associations between Toba tephra and Acheulean artifacts (after Mishra et al., 1995; Petraglia, 1998)

dated to ~1.4 Ma (Korisettar et al., 1989) and later to ~0.54 Ma, ~0.64 Ma, ~0.023 Ma (Horn et al., 1993) and ~0.67 Ma (Mishra et al., 1995). However, the ash has also been geochemically characterized as ~74 ka Toba tephra on the basis of comparisons with YTT from Sumatra and marine cores (Acharyya and Basu, 1993; Shane et al., 1995; Westgate et al., 1998). Following these contradictory findings, a debate persists as to whether all Toba tephra occurrences in peninsular India belong to YTT (Acharyya and Basu, 1993, 1994; Shane et al., 1995, 1996; Westgate et al., 1998), or if Toba tephras older than YTT exist at some Indian sites (Mishra and Rajaguru, 1994, 1996).

In spite of the similarity between the chemical composition of Kukdi tephra and YTT, Mishra and Rajaguru (1994) insist that both the early Acheulean assemblage associated with the

tephra and the dates listed above are sufficient evidence to date the tephra. In addition, they observe that no Middle Paleolithic tools were found associated with the tephra, indicating that this would be expected if the tephra is 74,000 years old. Noting similarities between the Early Acheulean of Africa and India, they state that the Early Acheulean from East Africa is dated to 1.6 Ma and therefore, "the dating of the Kukdi ash to 1.4 Ma is consistent with the associated archaeological evidence" (Mishra and Rajaguru, 1994). In response, however, Acharyya and Basu (1994) make the point that East African dates should not be extrapolated to provide a chronology for the Indian subcontinent.

It has been argued that the Early Acheulean artifacts at Kukdi may have been reworked into younger fluvial deposits, thus refuting an early date for the tephra (Acharyya and Basu, 1994; Shane et al., 1995; Westgate et al., 1998; Acharyya, 2003). The process of reworking and reintroduction of artifacts in alluvium is not uncommon (Acharyya and Basu, 1994). Unfortunately, the possible geomorphological mechanisms behind this kind of depositional scenario at Kukdi have not been explained, nor has any solid evidence been given to support this suggestion.

Others purport that the Kukdi artifacts are not reworked on the basis that the majority show no abrasion; this would be expected following fluvial transportation and long-term exposure of lithic artifacts (Mishra et al., 1995; Mishra and Rajaguru, 1996). Where "slight abrasion" is visible on the tools, it is indicated that this is only on the exposed surface, occurring after "recent exposure" and not during incorporation into the gravel. Further, the number of flakes in the assemblage matches that expected from counting flake scars on the cores and bifaces (Mishra and Rajaguru, 1996). Reworking would decrease the number of small artifacts. Finally, the angularity and local source of the lithocomponents are argued to indicate little fluvial transport (Mishra et al., 1995).

It has been suggested by Horn et al. (1993) that the tephra bed itself may have been disturbed, creating spatially separated ash layers and lenses. They conclude that this explains their thermoluminescence date of ∼23 ka for the tephra. Post-depositional re-exposure of sediments to sunlight would give a date younger than the actual age of the ash (Horn et al., 1993). Re-emplacement of ash deposits is not impossible, particularly if the tephra was initially deposited on top of weak and unconsolidated alluvial sediments. Such processes were recently observed in the Middle Son valley where local erosional events have caused large blocks of tephra to fracture from the tephra bed and fall downslope (Jones and Pal, 2005). It is conceivable that the Kukdi tephra underwent

a similar process at certain locations, where it became interred in a secondary context, thus raising the possibility that wrongful stratigraphical associations may have been made between artifacts and ash.

The debate concerning the identification and age of the tephra at Kukdi has recently resurfaced (Lee et al., 2004). Lee et al. (2004) use evidence from tephra layers in South China Sea marine cores to support claims that the Kukdi tephra is the Oldest Toba Tuff. They argue that the OTT event was wrongly correlated by Dehn et al. (1991) with tephra layer E in the Indian Ocean core, ODP Site 758. It is tephra from layer E which Shane et al. (1995) and Westgate et al. (1998) believed to be OTT and hence used in their analysis and identification of Indian tephras. Instead, Lee et al. (2004) chemically correlate layer D from ODP site 758 with a tephra from the South China Sea and conclude that OTT is the source, redating the OTT event from ∼840 ka (Diehl et al., 1987) to 788 ± 2.2 ka. Layer D is very similar in chemical composition to YTT, thus raising the possibility that the Kukdi tephra may have been mistakenly assigned to YTT.

While Lee et al. (2004) highlight some problems with the identification of deep-sea and terrestrial tephras, their findings alone are not enough to attest to the presence of OTT at Kukdi. Additionally, neither the highly discrepant ages for the Kukdi tephra (ranging from 0.023 Ma to 1.4 Ma), nor the erroneous argument that the associated Acheulean assemblage can be used to date the tephra, are valid reasons to support identification of the Kukdi ash as OTT. It is evident that the exact stratigraphic relationship between the Lower Paleolithic artifacts and tephra deposits at Kukdi remains unresolved (Petraglia, 1998). There is no doubt about the significance of the archaeological remains and tephra deposits at Kukdi; the Acheulean associations are certainly intriguing. However, it is unfortunate that none of these tephra localities have been excavated as this may help to resolve some of these issues. Furthermore, fission-track dating could also be applied to the Kukdi ash with the hope to finally clear up this controversy. This method has already provided Late Pleistocene dates for YTT shards from Purna and Narmada (Westgate et al., 1998).

The Middle Son Valley, Madhya Pradesh

The Middle Son valley, in Madhya Pradesh, preserves extensive alluvial sediments from the Quaternary period which contain a record that is both rich in archaeology and faunal remains. During surveys of the valley in the 1970s and 1980s, a total of 334 sites were discovered, ranging from the Lower Paleolithic through to the Neolithic periods. A number of these sites were subsequently excavated (Sharma, 1980; Sharma and Clark, 1983). In addition, valuable geological research in the valley culminated in the development of geomorphological models that described the depositional sequence of alluvial strata during the Pleistocene and Holocene periods (Williams and Royce, 1982, 1983; Williams and Clarke, 1984, 1995). These geological investigations lead to the discovery of volcanic ash within the alluvial deposits of the Middle Son valley (Williams and Royce, 1982). Later surveys highlighted the extent of the tephra, which was found to cover a distance of ~30 km. Exposures have been described at four localities; Ghogara, Ramnagar, Nakjhar and Khutiali, where the ash is 1.5 m, 0.5 m, 0.7 m and 1.5 m thick respectively (Basu et al., 1987; Acharyya and Basu, 1993).

Four Quaternary formations characterize the geological deposits of the Middle Son valley. In chronological order, these are the Sihawal, Patpara, Baghor and Khetaunhi formations, the latter being the most recent. The Baghor formation is divided into a lower coarse member and an upper fine member (Williams and Royce, 1982). More than 6,000 animal fossils have been collected from these deposits (Badam et al., 1989); however, the Baghor coarse member in particular preserves the richest fossil record (Dassarma and Biswas, 1977; Badam et al., 1989; Blumenschine and Chattopadhyaya, 1983). Stone artifacts are found throughout. Lower Paleolithic assemblages (bifaces and cleavers) are found in the Sihawal formation. Middle Paleolithic industries are represented in the Patpara formation; interestingly, some of these assemblages contain a notable diminutive biface and cleaver component. The overlying Baghor coarse member preserves abraded Middle Paleolithic artifacts towards the base as well as relatively unabraded Upper Paleolithic artifacts. Both Upper Paleolithic and microlithic assemblages are present in the Baghor fine member and the Khetaunhi formation preserves microliths as well as Neolithic pottery (Clark and Williams, 1987:25). The observation that the Middle Paleolithic artifacts from the Baghor coarse member appear abraded and rolled has led to the interpretation that they had been re-introduced into younger geological deposits (Williams and Royce, 1982, 1983; Sharma and Clark, 1983; Clark and Williams, 1987). On the other hand, others claim these artifacts are "possibly" contemporaneous, suggesting that the Middle Paleolithic artifacts from the Baghor coarse member have not been significantly reworked (Acharyya and Basu, 1993).

Descriptions of the stratigraphical position of the Youngest Toba Tuff in relation to these Quaternary formations have been somewhat inconsistent. Some studies have placed YTT within the coarse member of the Baghor formation (Williams and Royce, 1982; Basu et al., 1987; Acharyya and Basu, 1993). However, Williams and Clarke (1995) describe the ash bed as located beneath the Baghor coarse member. Identifying the precise geological context of the ash deposit in relation to the surrounding archaeological remains is crucial. Recent field research in the Middle Son valley suggests that the Toba tephra layer is situated at the junction of the Patpara formation and Baghor coarse member and not within the latter (Jones and Pal, 2005) (Figure 5; Figure 6).

To date, there has been minimal discussion of the relationship between YTT and Paleolithic assemblages in the valley. It should be clear, however, that the association between YTT and archaeological and faunal remains in the Middle Son valley provides an excellent opportunity to assess the extent of the impact of the Toba eruption on both local hominin populations and their environments. This problem is currently being addressed through detailed analysis of both lithic assemblages and sediments that pre-date and post-date the Toba eruption (Jones and Pal, 2005). Others are also assessing the environmental repercussions of the Toba event in the

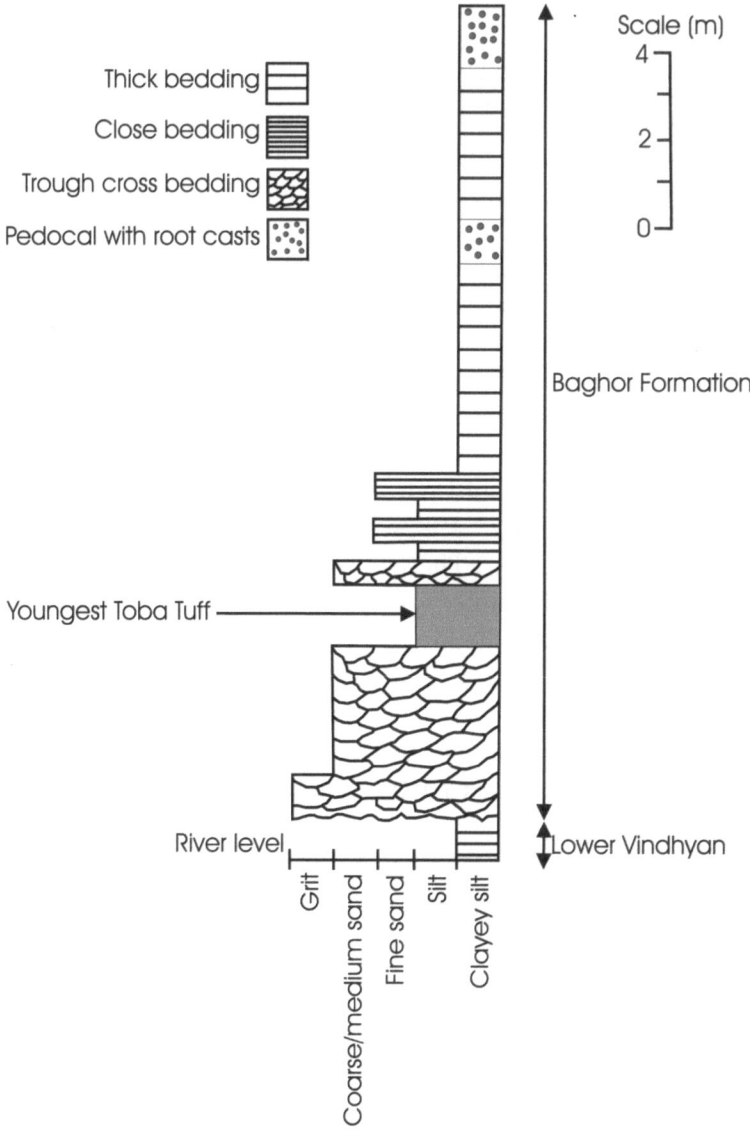

Figure 5. Ghogara section in the Middle Son valley. The stratigraphical position of the Youngest Toba as shown here by Basu et al. (1987) is not agreed upon. Recent field research in the Middle Son valley indicates that the YTT deposit at Ghogara is located at the junction between the underlying Patpara formation and the overlying coarse member of the Baghor Formation (Jones and Pal, 2005), as depicted in figure 6 (after Basu et al., 1987)

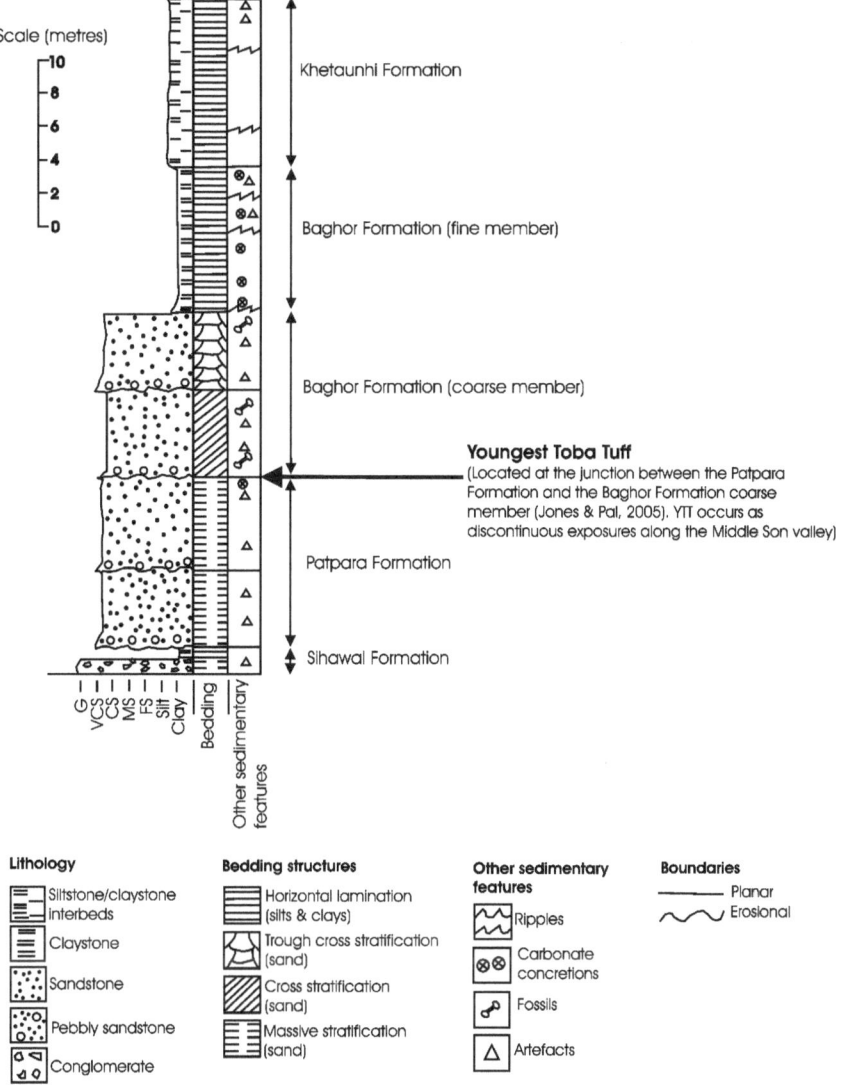

Figure 6. Composite geological column for the Quaternary alluvial sediments of the Middle Son valley. The position of the YTT layer as indicated here is a revised assessment of the exact stratigraphical placement of the Toba ash (Jones and Pal, 2005) (after Williams and Royce, 1983)

Middle Son valley (Williams, 2004). There is no doubt that further investigations in the valley promise to be revealing, particularly those which focus on excavating archaeological deposits associated with YTT. In addition, the importance of the faunal record of the valley should not be underestimated for two reasons. First, analysis of these faunal remains provides an opportunity for assessing the distribution of species before and after the Toba eruption and hence, the likelihood of faunal contractions or even

extinctions. Second, the wealth of paleontological evidence already recovered from the Middle Son valley surely suggests the future possibility of hominin fossil discoveries.

The Central Narmada Valley, Madhya Pradesh

The Quaternary alluvial sediments of the central Narmada valley stretch over ~300 km, from Handia in the west to Jabalpur in the east. The width of the valley varies from 20 to 55 km, bordered by the Vindhyan range to the north and the Satpura range to the south (Biswas and Dassarma, 1986). The Quaternary deposits are ~250m thick, however, only the upper 50m has been exposed by the flow of the Narmada. As with the Middle Son valley, these deposits preserve a wealth of both archaeological (e.g., Supekar, 1968; Kennedy, 2000) and faunal remains (Biswas and Dassarma, 1986; Poddar et al., 1991; Sonakia and Biswas, 1998) and have notably produced India's only undisputed Middle Pleistocene hominin remains (e.g., Sonakia, 1984).

Four alluvial formations have been recognized in the central Narmada deposits; from oldest to youngest these are the Sobhapur, Narsingpur, Devakachar, Jhalon (coarse and fine members) and Narmada formations (Basu et al., 1987; Acharyya and Basu, 1993). Based on the presence of both Toba tephra and the same assemblage of faunal species, the Quaternary formations of the central Narmada and Middle Son valleys have been correlated with one another. The Devakachar and Jhalon formations of the central Narmada have been associated with the Patpara and Baghor formations of the Middle Son respectively (Basu et al., 1987; Acharyya and Basu, 1993). It should be noted that others have applied a different terminology, assigning the names Nimsarya and Dhansi (Poddar et al., 1991) to formations that correspond with the Jhalon coarse member and Devakachar units respectively (Acharyya and Basu, 1993).

Toba tephra has been discovered at "several localities" (Poddar et al., 1991) in the central Narmada valley, for example, at Pawlaghat (Figure 7), Guruwara, Devakachar and Hirapur. In places, the tephra reaches a maximum thickness of almost 1m and is traceable as a discontinuous deposit for ~20 km (Basu et al., 1987; Poddar et al., 1991; Acharyya and Basu, 1993). It is argued that YTT from the Narmada has a more limited extent and reduced thickness when compared to the tephra deposits from the Middle Son (Basu et al., 1987).

In spite of the wealth of artifacts preserved in the central Narmada deposits, cultural remains found in association with YTT are poorly described. Underlying the volcanic ash, the Devakachar (Dhansi) formation preserves Lower Paleolithic (Acheulean) artifacts, apparently with more advanced artifacts in the upper levels (Acharyya and Basu, 1993); however, no further description of the latter is given. Artifacts are present in large numbers in this formation (Poddar et al., 1991). The tephra is said to occur in the lower levels of the Jhalon coarse member; Middle to Upper Paleolithic artifacts are represented in this formation (Acharyya and Basu, 1993).

Although descriptions of the associations between Paleolithic technologies and Toba tephra have been rather uninformative to date, the potential of the central Narmada for discerning the impacts of the eruption on hominins and their environments should not be overlooked. The extensive faunal and archaeological records of the valley permit the same approach as that described above for the Middle Son material. A reassessment of the precise position of the tephra deposit in relation to the surrounding alluvial formations is required. Excavation of sites preserving Toba ash should be planned with the aim to collect stratified artifacts. Such assemblages can then be analyzed to reveal the characteristics of pre- and post-eruption hominin behavior in the valley.

Other Archaeology-Tephra Associations

There are three other Indian river valleys where tephra deposits are discussed specifically in terms of their archaeological associations. First, the tephra deposits in the Vansadhara river valley in Orissa are described as being associated with "Acheulean-type tools and cores" (Anonymous, 1990; Acharyya and Basu, 1993); however, no precise or reliable details regarding this stratigraphical relationship are provided. Additionally, YTT deposits have also been identified in the Purna valley in

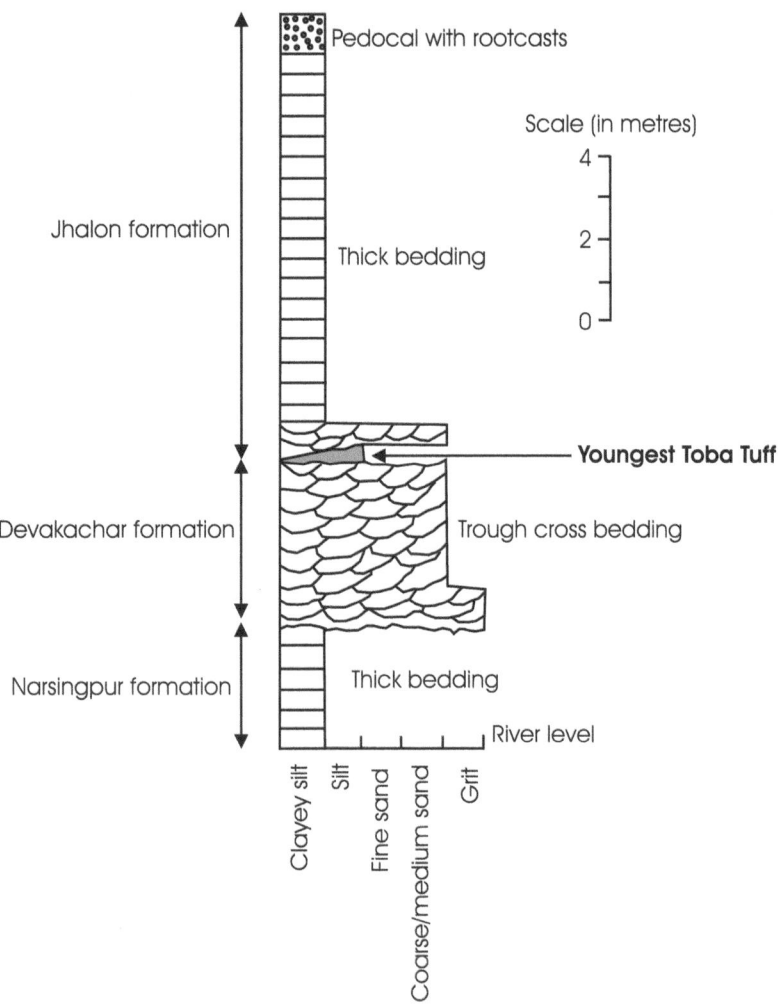

Figure 7. Pawlaghat section in the central Narmada valley. The position of the Youngest Toba Tuff lies at the base of the Jhalon Formation (after Basu et al., 1987)

Maharashtra (Shane et al., 1995; Westgate et al., 1998). A sample of this ash was dated to 84 ± 16 ka by the fission-track method (Westgate et al., 1998). The ash deposits are said to be associated with Paleolithic artifacts (Shane et al., 1995), but unfortunately, no further information has been published on the Purna ash and associated archaeology.

A tephra bed, 0.17 to 0.54 m thick, was discovered at Morgaon in the Karha valley, Maharashtra (Kale et al., 1993; Rajaguru et al., 1993). Acheulean artifacts apparently overlie the ash and on the basis of this and a presumed similarity to the Kukdi tephra, an early Pleistocene age has been assigned to

the tephra (Rajaguru et al., 2004). However, this early age is based only on the association between the tephra and the Acheulean and on the authors' acceptance of the controversial Middle Pleistocene dates for the Kukdi tephra. Others accept that the tephra at both Morgaon and Kukdi is YTT (Kale et al., 2004), yet the Morgaon ash has not been geochemically characterized, nor have radiometric dates been obtained for either the ash or the artifacts. As with the archaeological localities at Kukdi, no Middle or Upper Paleolithic artifacts are denoted as present at Morgaon. Instead, the deposits containing the Acheulean and those preserving overlying microlithic technologies seem to be separated by at least 3m of culturally sterile sediment.

Finally, an uncharacterized tephra deposit, again associated with Acheulean artifacts, is reported from 1 km north of Rajbag in the Karha valley, only a few kilometers from Morgaon (Deo et al., 2004). Although tephras from Rajbag and Morgaon have not been characterized, it is noteworthy that those from Kukdi and Vansadhara, also associated with Acheulean assemblages, have all been geochemically identified as YTT.

Discussion

In order to fully assess the impacts of the Toba supervolcanic eruption of ~74 ka, it is proposed here that its effects should be broken down into three components; global, regional and local. The potential global climatic effects of the Toba eruption have already been discussed and a significant amount of research has focused on this problem. The assertion that Toba resulted in climatic deterioration is not really questioned; the historical eruptions of Mount Pinatubo and Tambora attest to the fact that even eruptions that are significantly smaller than supervolcanic events can have an impact on global climate. However, it is the hypothesized severity of this impact that has been criticized (e.g., Oppenheimer, 2002).

The regional impacts of the Toba eruption can be defined as those that affect broad geographical zones, such as the Indian subcontinent, south-east Asia and sub-Saharan Africa. These effects have been discussed with a focus on Africa, where a harsh volcanic winter and further climatic perturbations are argued to have caused a population bottleneck in *H. sapiens*, particularly affecting those inhabiting areas outside tropical refugia (Ambrose, 1998). Others have argued against a population bottleneck based on evidence from the Mentawai islands in south-east Asia (Gathorne-Hardy and Harcourt-Smith, 2003). The regional impacts of the Toba

eruption elsewhere, namely in India, have not previously been addressed in any detail. This is surprising given the coverage of vast areas of southern and southeastern Asia in volcanic ash. In this case, when assessing the human and ecological impacts of Toba, it is necessary to consider the impact of volcanic ash on the Indian landscape as well as the proposed climatic deterioration.

Examination of the local effects of the eruption at the level of the habitat or ecosystem, and thus at the level of the hominin groups that occupy these areas, is also necessary. In this regard, a number of factors can be considered. These include an assessment of the detrimental impacts of the ash-fall on vegetation cover; whether vegetation and sources of drinking water were poisoned by fluoride in the volcanic ash; and consideration of how hazardous were the consequences of inhaling the fine-grained volcanic ash. In addition, the hypothesized volcanic winter may have resulted in the interruption of the monsoonal system, causing increased aridity in certain areas. All these factors could have had an impact on food resources, fauna and hominins, eventually disrupting entire ecosystems. Furthermore, consideration also needs to be given to whether or not certain geographical areas escaped the direct effects of the ash-fall. The evidence indicates that a blanket of tephra did cover a vast area of India, yet precisely how long it took for the ash to be redeposited by aeolian, pluvial and/or fluvial processes into the river valley deposits that can be seen today has not been previously assessed. It is possible that certain areas were left unscathed if the ash was rapidly flushed away. That said, it is the river valleys, with their grazing animal populations and sources of drinking water, that would probably have been the most attractive and resource-rich areas for hominin occupation. However, the possibility that regions of refugia survived should not be ignored. Refugia may have existed in the

Western and Eastern Ghats, in the southern tip of India, and in Sri Lanka (Brandon-Jones, 1998; Field et al., 2006).

Three broad outcomes can be postulated for hominins following the Toba eruption; population continuity, population extinction, or the extinction of some populations but survival of others in areas of refugia. Each of these possible outcomes would have involved a number of processes that may have been triggered by the Toba event and directly affected hominins. These include processes such as habitat contraction and fragmentation, intra- and inter-specific competition, genetic drift and founder effects, population and possibly species replacement, and forced population migrations. Consideration of these scenarios should address two key questions that pertain to the hominin inhabitants of India ~74 ka.

First, had *H. sapiens* colonized India before the eruption? The majority of genetic evidence seems to suggest that the initial colonization of India took place soon after the Toba event. It should be noted, however, that on the basis of this evidence, the hypothesis that modern human populations inhabited India before ~74 ka and underwent extinction as a result of Toba cannot be ruled out. If population extinction occurred, there would be no trace of their DNA in present-day humans. Although there is no fossil evidence of *H. sapiens* in the subcontinent until ~28 ka, early modern human remains that date to the start of the Late Pleistocene have been recovered from western Asia at the sites of Skhūl and Qafzeh. It is plausible that this early exodus of *H. sapiens* out of Africa and into western Eurasia also reached the Indian subcontinent.

A second question needs to be addressed; what other hominin species may have inhabited India before the Toba eruption? For this, even less evidence is available. Future investigations in India, in particular those that result in fossil hominin discoveries, will hopefully elucidate this problem. However,

until such time as this evidence is uncovered it is recognized that this question cannot be answered with any certainty. An intriguing pattern to recall here concerns the association between Acheulean artifacts and the Youngest Toba Tuff from four river valleys in India (Kukdi, Morgaon, Vansadhara and Karha). If all these tephra deposits are YTT and if these Acheulean artifacts have not experienced significant reworking, then the establishment of a direct association between Acheulean assemblages and ~74 ka Toba ash is certainly unexpected because it automatically implies a late survival of the Acheulean in the Indian subcontinent. If this scenario is correct, then this would have significant implications concerning the identity of hominins in parts of the Indian subcontinent ~74 ka (e.g., *H. erectus*, *H. heidelbergensis*). However, for the following three reasons, it must be strongly stated here that this is an issue that requires considerable future investigation. First, previous research which notes an association between Toba ash and the Acheulean presents either minimal information or, in the case of the evidence from Kukdi, is shrouded in controversy. Second, none of the tephra deposits in these river valleys have been excavated which is essential in order to correctly interpret stratigraphic relationships. Third, all Acheulean artifacts are found in fluvial contexts and this raises the possibility that they have been reworked into younger deposits.

Given the current problems associated with the identification of those hominin species that occupied the Indian subcontinent ~74 ka, it is more useful to address the three possible outcomes of the Toba eruption (extinction, continuity, survival in refugia) at the population level. This population-level approach involves an examination of the specific local impacts of the eruption. Evidence from India provides an excellent opportunity for addressing such impacts. Throughout the subcontinent, YTT forms a valuable isochronic marker bed to which any associated, and stratigraphically integrous, fossils or artifacts can be given a useful pre- or post-74 ka age; a rare chronological certainty in the Indian Paleolithic record. This evidence permits regional comparisons within India of more or less contemporary archaeological sites.

Analysis of excavated lithic assemblages and sediments that underlie and overlie YTT deposits result in a description of pre- and post-Toba hominin behavior and paleoenvironments. If behavioral change following the ash-fall exists, then, depending on the nature of the change, past population dynamics can be inferred. If change occurred as a result of the Toba eruption, it is not necessary for it to be manifested in a gross shift in cultural practices. Subtle but significant variation within the Paleolithic industries of the Late Pleistocene may, among a number of scenarios, suggest the appearance of new populations or even species. Alternatively, it may indicate population continuity but with adaptation to a changed post-tephra environment. To attempt to explain why such behavioral change arose is certainly a challenge, yet it is absolutely necessary. Within such attempts, it is further essential that a multidisciplinary approach is taken whereby the archaeological data is combined with that from sedimentology and other types of analysis that address paleoenvironmental and paleoclimatic change. A combination of such evidence will strengthen any conclusions regarding past populations dynamics in response to Toba.

This approach is currently being pursued in two areas of the Indian subcontinent. First, in the Middle Son valley in Madhya Pradesh (in collaboration with scholars from Allahabad University [Jones and Pal, 2005]), and second, at Jwalapuram, a site recently discovered in Kurnool District, Andhra Pradesh (a joint investigation with researchers from the University of Cambridge and Karnatak University, as well as other U.K. institutions).

Conclusion

The potential offered by the Indian fossil, geological and archaeological records towards answering questions related to the ~74 ka Toba supereruption and its human impact is substantial. The Son and Narmada valleys both preserve YTT and a record that is rich in both archaeology and Pleistocene fauna. Further explorations of these sites promise to be rewarding. The ash deposits of Kukdi and Morgaon remain controversial but would greatly benefit from further investigation, namely excavation. A number of river basins in Orissa preserve YTT deposits (some exceptionally thick at ~6 m) as well as Paleolithic assemblages. However, as with many other river valleys that have documented YTT and Paleolithic remains, these two forms of evidence are not discussed in concert. An approach that marries the geological and archaeological evidence should be employed in future studies. Furthermore, excavation of many Toba ash localities throughout India would also prove invaluable, facilitating India-wide comparisons of hominin and ecological responses to the ash-fall in varying habitats. Finally, in India and beyond, archaeologists and geologists should be aware of the possible presence of cryptotephra horizons of Toba ash in Late Pleistocene sites. Given that YTT layers, at least 4 cm thick, are present in the Arabian Sea, there should be no doubt that Toba ash dispersed beyond India. Accurate knowledge of the distribution of the Toba tephra blanket facilitates precise estimates of eruption magnitude and ultimately has important implications concerning the human and environmental impacts of this widespread ash-fall.

Acknowledgments

This research would not have been possible without the financial support of studentships (2002–2005) from Newnham

College, Cambridge, an Allen Meek and Read Scholarship (2004–2005) from the University of Cambridge, and grants from the Sir Richard Stapley Educational Trust (2003–2005). My sincere thanks go to my Ph.D. supervisor, Dr Michael Petraglia, for his generous help and guidance over the past few years. I am also grateful to Dr Marta Lahr and Professor Robert Foley for their continuing support and to Dr Clive Oppenheimer and Dr David Pyle for their advice concerning all things volcanological. I wish to thank Prof. J.N. Pal and Prof. R. Korisettar for their advice and assistance.

References

Acharyya, S.K., Basu, P.K., 1993. Toba ash on the Indian subcontinent and its implications for correlation of Late Pleistocene alluvium. Quaternary Research 40, 10–19.

Acharyya, S.K., Basu, P.K., 1994. Reply to comments by S. Mishra and S.N. Rajaguru and by G.L. Badam and S.N. Rajaguru on "Toba ash on the Indian subcontinent and its implication for the correlation of Late Pleistocene alluvium". Quaternary Research 41, 400–402.

Acharyya, S.K., Lahiri, S., Raymahashay, B.C., Bhowmik, A., 2000. Arsenic toxicity of groundwater in parts of the Bengal basin in India and Bangladesh: the role of Quaternary stratigraphy and Holocene sea-level fluctuation. Environmental Geology 39 (10), 1127–1137.

Acharyya, S.K., 2003. Recent findings of the Acheulian of Isampur excavations and its dating. Current Science 84 (2), 127–128.

Aldiss, D.T., Ghazali, S.A., 1984. The regional geology and evolution of the Toba volcano-tectonic depression, Indonesia. Journal of the Geological Society 141 (3), 487–500.

Ambrose, S.H., 1998. Late Pleistocene human populations bottlenecks, volcanic winter, and differentiation of modern humans. Journal of Human Evolution 34, 623–651.

Ambrose, S.H., 2002. Small things remembered: origins of early microlithic industries in Subsaharan Africa. In: Elston, R., Kuhn, S. (Eds.), Thinking Small: Global Perspectives on Microlithic Technologies. Archaeological Papers of the American Anthropological Association 12, pp. 9–29.

Ambrose, S.H., 2003a. Population bottleneck. In: Robinson, R. (Ed.), Genetics, volume 3. Macmillan, New York, pp. 167–171.

Ambrose, S.H., 2003b. Did the super-eruption of Toba cause a human population bottleneck? Reply to Gathorne-Hardy and Harcourt-Smith. Journal of Human Evolution 45 (3), 231–237.

Anonymous, 1990. Late Quaternary ash bed and Acheulian implements from Orissa – the implication. News Geological Survey of India, Eastern Region 10 (1), 7.

Anonymous, 1991. Volcanic ash in the Quaternary formations Sagileru valley. News Geological Survey of India, Southern Region 9 (2), 9.

Athreya, S., 2007. Was Homo heidelbergensis in South Asia? A test using the Narmada fossil from Central India. In: Petraglia, M.D., Allchin, B. (Eds.), The Evolution and History of Human Populations in South Asia: Inter-disciplinary Studies in Archaeology, Biological Anthropology, Linguistics and Genetics. Springer, Netherlands, pp. 137–170.

Badam, G.L., Misra, V.D., Pal, J.N., Pandey, J.N., 1989. A preliminary study of Pleistocene fossils from the Middle Son Valley, Madhya Pradesh. Man and Environment 13, 41–47.

Basu, P.K., Biswas, S., Acharyya, S.K., 1987. Late Quaternary ash beds from Son and Narmada Basins, Madhya Pradesh. Indian Minerals 41 (2), 66–72.

Basu, P.K., Biswas, S., 1990. Quaternary ash beds from Eastern India. Records of the Geological Survey of India 123 (2), 12.

Biswas, S., Dassarma, D.C., 1986. Stratigraphy of the Quaternary alluvial deposits of the Central Narmada Valley, Madhya Pradesh. Proceedings of the XI Indian Colloquium on Micropalaeontology and Stratigraphy, Bulletin Geological Mining Metallurgical Society of India 54, 18–27.

Blumenschine, R.J., Chattopadhyaya, U.C., 1983. A preliminary report on the terminal Pleistocene fauna of the Middle Son valley. In: Sharma, G.R., Clark, J.D. (Eds.), Palaeoenvironments and Prehistory in the Middle Son Valley. Abinash Prakashan, Allahabad, pp. 281–284.

Brandon-Jones, D., 1998. Pre-glacial Bornean primate impoverishment and Wallace's line. In: Hall, R., Holloway, J.D. (Eds.), Biogeography and Geological Evolution of SE Asia. Backbuys Publishers, Leiden, pp. 393–404.

Bühring, C., Sarnthein, M., Leg 184 Shipboard Scientific Party, 2000. Toba ash layers in the South China Sea: evidence of contrasting wind directions during eruption ca. 74 ka. Geology 20 (3), 275–278.

Cameron, D., Patnaik, R., Sahni, A., 2004. The phylogenetic significance of the Middle Pleistocene Narmada hominin cranium from central India. International Journal of Osteoarchaeology 14 (6), 419–447.

Carey, S.N., Sigurdsson, H., 1982. Influence of particle aggregation on deposition of distal tephra from the May 18, 1980, eruption of Mount St. Helens volcano. Journal of Geophysical Research 87 (B8), 7061–7072.

Chesner, C.A., Rose, W.I., Deino, A., Drake, R., Westgate, J.A., 1991. Eruptive history of Earth's largest Quaternary caldera (Toba, Indonesia) clarified. Geology 19, 200–203.

Chesner, C.A., 1998. Petrogenesis of the Toba Tuffs, Sumatra, Indonesia. Journal of Petrology 29 (3), 397–438.

Clark, J.D., Williams, M.A.J., 1987. Paleoenvironments and prehistory in North Central India: a preliminary report. In: Jacobsen, J. (Ed.), Studies in the Archaeology of India and Pakistan. Aris and Phillips Ltd, Warminster, pp. 19–41.

Dassarma, D.C., Biswas, S., 1977. Newly discovered late Quaternary vertebrates from the terraced alluvial fills of the Son valley. Indian Journal of Earth Sciences 4 (2), 122–136.

Dehn, J., Farrell, J.W., Schmincke, H.-U., 1991. Neogene tephrochronology from Site 758 on Northern Ninetyeast Ridge: Indonesian arc volcanism of the past 5 Ma. Proceedings of the Ocean Drilling Program, Scientific Results 21, 273–295.

Deo, S.G., Ghate, S, Rajaguru, S.N., Karmalkar, N., Kale, M., 2004. Discovery of an Acheulian site in association with tephra at Rajbag, Taluka Supa, District Pune, Maharashtra. Man and Environment 29 (1), 108.

Devdas. V., Meshram, S.N., 1991. Search for Quaternary ash bed in the Quaternary Basins of Orissa. Records of the Geological Survey of India 124 (3), 40–42.

Diehl, J.F., Onstott, T.C., Chesner, C.A., Knight, M.D., 1987. No short reversals of Brunhes age recorded in the Toba tuffs, north Sumatra, Indonesia. Geophysical Research Letters 14, 753–756.

Endicott, P., Metspalu, M., Kivisild, T., 2007. Genetic evidence on modern human dispersals in South Asia: Y chromosome and mitochondrial DNA perspectives. In: Petraglia, M.D., Allchin, B. (Eds.), The Evolution and History of Human Populations in South Asia: Inter-disciplinary Studies in Archaeology, Biological Anthropology, Linguistics and Genetics. Springer, Netherlands, pp. 229–244.

Field, J., Petraglia, M., Lahr, M., 2006. The southern dispersal hypothesis and the South Asian archaeological record: examination of dispersal routes through GIS analysis. Journal of Anthropological Archaeology, in press.

Forster, P., Matsumura, S., 2005. Did early humans go North or South? Science 308 (5724), 965–966.

Gagneux, P., Wills, C., Gerloff, U., Tautz, D., Morin, P.A., Boesch, C., Fruth, B., Hohmann, G., Ryder, O.A., Woodruff, D.S., 1999. Mitochondrial sequences show diverse evolutionary histories of African hominoids. Proceedings of the National Academy of Sciences 96, 5077–5082.

Gathorne-Hardy, F., Harcourt-Smith, W., 2003. The super-eruption of Toba, did it cause a human bottleneck? Journal of Human Evolution 45 (3), 227–230.

Goldberg, T.L., Ruvolo, M., 1997. The geographic apportionment of mitochondrial genetic diversity in East African chimpanzees, *Pan troglodytes schweinfurthii*. Molecular Biology and Evolution 14 (9), 976–984.

Gurioli, L., Houghton, B.F., Cashman, K.V., Cioni, R., 2005. Complex changes in eruption dynamics during the 79 AD eruption of Vesuvius. Bulletin of Volcanology 67, 144–159.

Harpending, H.C., Sherry, S.T., Rogers, A.R., Stoneking, M., 1993. The genetic structure of ancient human populations. Current Anthropology 34 (4), 483–496.

Harpending, H.C., Batzer, M.A., Gurven, M., Jorden, L.B., Rogers, A.R., Sherry, S.T., 1998. Genetic traces of ancient demography. Proceedings of the National Academy of Sciences 95 (4), 1961–1967.

Harpending, H., Rogers, A., 2000. Genetic perspectives on human origins and differentiation. Annual Review of Genomics and Human Genetics 1, 361–385.

Horn, P., Müller-Sohnius, D., Storzer, D., Zöller, L., 1993. K-Ar-, fission-track-, and thermoluminescence ages of Quaternary volcanic tuffs and their bearing on Acheulian artifacts from Bori, Kukdi Valley, Pune District, India. Zeitschrift der Deutschen geologischen Gesellschaft 144, 326–329.

Huang, C-Y., Zhao, M., Wang, C-C., Wei, G., 2001. Cooling of the South China Sea by the Toba eruption and correlation with other climate proxies ~71,000 years ago. Geophysical Research Letters 28 (20), 3915–3918.

James, H.V.A., Petraglia, M.D., 2005. Modern human origins and the evolution of behavior in the later

Pleistocene record of South Asia. Current Anthropology 46, S3–S27.

Jones, S.C., Pal, J.N., 2005. The Middle Son valley and the Toba supervolcanic eruption of 74 kyr BP: Youngest Toba Tuff deposits and Palaeolithic associations. Journal of inter-disciplinary Studies in History and Archaeology 2 (1), in press.

Kale, V.S., Ganjoo, R.K., Rajaguru, S.N., Ota, S.B., 1986. A discovery of an Acheulian site at Bori, Dist. Pune (Maharashtra). Bulletin of the Deccan College Postgraduate and Research Institute 45, 47–55.

Kale, V.S., Patil, D.N., Powar, N.J., Rajaguru, S.N., 1993. Discovery of a volcanic ash bed in the alluvial sediments at Morgaon, Maharashtra. Man and Environment 18 (2), 141–143.

Kale, V.S., Joshi, V.U., Hire, P.S., 2004. Palaeohydrological reconstructions based on analysis of a palaeochannel and Toba-ash associated alluvial sediments in the Deccan Trap region, India. Journal Geological Society of India 64, 481–489.

Kennedy, K.A.R., 1999. Paleoanthropology of South Asia. Evolutionary Anthropology 8 (5), 165–185.

Kennedy, K.A.R., 2000. God-apes and fossil men. Paleoanthropology in South Asia. The University of Michigan Press, Ann Arbor.

Kennedy, K.A.R., 2001. Middle and Late Pleistocene hominids of South Asia. In: Tobias, P.V., Raath, M.A., Moggi-Cecchi, J., Doyle, G.A. (Eds.), Humanity from African Naissance to Coming Millennia – Colloquia in Human Biology and Palaeoanthropology. Florence University Press, Florence, pp. 167–174.

Kerr, R.A., 1996. Volcano-Ice Age link discounted. Science 272, 817.

Kivisild, T., Rootsi, S., Metspalu, M., Mastana, S., Kaldma, K., Parik, J., Metspalu, E., Adojaan, M., Tolk, H-V., Stepanov, V., Gölge, M., Usanga, E., Papiha, S.S., Cinnioğlu, C., King, R., Cavalli-Sforza, L., Underhill, P.A., Villems, R., 2003. The genetic heritage of the earliest settlers persists both in Indian tribal and caste populations. American Journal of Human Genetics 72, 313–332.

Korisettar, R., Mishra, S., Rajaguru, S.N., Gogte, V.D., Ganjoo, R.K., Venkatesan, T.R., Tandon, S.K., Somayajulu, B.L.K., Kale, V.S., 1988. Age of the Bori volcanic ash and Lower Palaeolithic culture of the Kukdi Valley, Maharashtra. Bulletin of the Deccan College Postgraduate and Research Institute 48, 135–138.

Korisettar, R., Venkatesan, T.R., Mishra, S., Rajaguru, S.N., Somayajulu, B.L.K., Tandon, S.K., Gogte, V.D., Ganjoo, R.K., Kale, V.S., 1989.

Discovery of a tephra bed in the Quaternary alluvial sediments of Pune District (Maharashtra), Peninsular India. Current Science 58 (10), 564–567.

Korisettar, R., 1994. Quaternary alluvial stratigraphy and sedimentation in the Upland Deccan Region, Western India. Man and Environment 19 (1–2), 29–41.

Lahr, M.M., Foley, R., 1994. Multiple dispersals and modern human origins. Evolutionary Anthropology 3, 48–60.

Lee, M-Y., Chen, C-H., Wei, K-Y., Iizuka, Y, Carey, S., 2004. First Toba supereruption revival. Geology 32, 61–64.

Macaulay, V., Hill, C., Achilli, A., Rengo, C., Clarke, D., Meehan, W., Blackburn, J., Semino, O., Scozzari, R., Cruciani, F., Taha, A., Shaari, N.K., Raja, J.M., Ismail, P., Zainuddin, Z., Goodwin, W., Bulbeck, D., Bandelt, H-J., Oppenheimer, S., Torroni, A., Richards, M., 2005. Single, rapid coastal settlement of Asia revealed by analysis of complete mitochondrial genomes. Science 308 (5724), 1034–1036.

Mason, B.G., Pyle, D.M., Oppenheimer, C., 2004. The size and frequency of the largest explosive eruptions on Earth. Bulletin of Volcanology 66, 735–748.

McCormick, M. P., Thomason, L. W., Trepte, C. R., 1995. Atmospheric effects of the Mt Pinatubo eruption. Nature 373, 399–404.

Mishra, S., Ghate, S., 1990. Ranjani: A Middle Palaeolithic site from Maharashtra. Man and Environment 15 (2), 23–27.

Mishra, S., Rajaguru, S.N., 1994. Comment on "Toba ash on the Indian subcontinent and its implication for the correlation of Late Pleistocene alluvium". Quaternary Research 41, 396–397.

Mishra, S., Venkatesan, T.R., Rajaguru, S.N., Somayajulu, B.L.K., 1995. Earliest Acheulian industry from Peninsular India. Current Anthropology 36 (5), 847–851.

Mishra, S., Rajaguru, S.N., 1996. Comment on "New geochemical evidence for the Youngest Toba Tuff in India". Quaternary Research 46, 340–341.

Misra, V.N., 1995. Geoarchaeology of the Thar Desert, North West India. In: Wadia, S., Korisettar, R., Kale, V.S. (Eds.), Quaternary Environments and Geoarchaeology of India. Geological Society of India: Bangalore, pp. 210–230.

Newhall, C.G., Self, S, 1982. The Volcanic Explosivity Index (VEI): an estimate of the explosive magnitude for historical volcanism. Journal of Geophysical Research 87 (C2), 1231–1238.

Ninkovich, D. Shackleton, N.J., Abdel-Monem, A.A., Obradovich, J.D., Izett, G., 1978a. K-Ar age of the late Pleistocene eruption of Toba, north Sumatra. Nature 276, 574–577.

Ninkovich, D., Sparks, R.S.J., Ledbetter, M.T., 1978b. The exceptional magnitude and intensity of the Toba eruption, Sumatra: an example of the use of deep-sea tephra layers as a geological tool. Bulletin Volcanologique 41, 286–298.

Oppenheimer, C., 2002. Limited global change due to the largest known Quaternary eruption, Toba ≈74 kyr BP? Quaternary Science Reviews 21, 1593–1609.

Oppenheimer, S., 2003. Out of Eden: The Peopling of the World. Constable, London.

Palanichamy, M.G., Sun, C., Agrawal, S., Bandelt, H-J., Kong, Q-P., Khan, F., Wang, C-Y., Chaudhuri, T.K., Palla, V., Zhang, Y-P., 2004. Phylogeny of Mitochondrial DNA macrohaplogroup N in India, based on complete sequencing: implications for the peopling of South Asia. American Journal of Human Genetics 75, 966–978.

Pattan, J.N., Pearce, N.J.G., Banakar, V.K., Parthiban, G., 2002. Origin of ash in the Central Indian Ocean Basin and its implication for the volume estimate of the 74,000 year BP Youngest Toba eruption. Current Science 83 (7), 889–893.

Pattan, J.N., Shane, P., Pearce, N.J.G., Banakar, V.K., Parthiban, G., 2001. An occurrence of ~74 ka Youngest Toba Tephra from the Western Continental Margin of India. Current Science 80 (10), 1322–1326.

Pattan, J.N., Shane, P., Banakar, V.K., 1999. New occurrence of Youngest Toba Tuff in abyssal sediments of the Central Indian Basin. Marine Geology 155, 243–248.

Petraglia, M.D., 1998. The Lower Palaeolithic of India and its bearing on the Asian record. In: Petraglia, M.D., Korisettar, R. (Eds.), Early Human Behaviour in Global Context: The Rise and Diversity of the Lower Palaeolithic Record. Routledge, London and New York, pp. 343–390.

Poddar, B.C., Verma, K.K., Tewari, M.P., Rahate, D.N., Khan, A.A., Sonakia, A., Biswas, S., Nandi, A., Krishna, S.G., Dubey, U.S., Bhai, H.Y., Fahim, M., Sitaramaiah, Y., 1991. Narmada Valley Project. Records of Geological Survey of India 124 (6), 245–247.

Quintana-Murci, L., Semino, O., Bandelt, H-J., Passarino, G., McElreavey, K., Santachiara-Benerecetti, A.S., 1999. Genetic evidence of an early exit of Homo sapiens sapiens from Africa through eastern Africa. Nature Genetics 23, 437–441.

Rajaguru, S.N., Kale, V.S., Badam, G.L., 1993. Quaternary fluvial systems in Upland Maharashtra. Current Science 64, 817–822.

Rajaguru, S.N., Deo, S.G., Mishra, S., Ghate, S., Naik, S., Shirvalkar, P., 2004. Geoarchaeological significance of the detrital laterite discovery in the Karha Basin, Pune District, Maharashtra. Man and Environment 29 (1), 1–6.

Rampino, M. R., Stothers, R.B., Self, S., 1988. Volcanic winters. Annual Review of Earth and Planetary Sciences 16, 73–99.

Rampino, M. R., Self, S., 1992. Volcanic winter and accelerated glaciation following the Toba super-eruption. Nature 359, 50–52.

Rampino, M.R., Self, S., 1993a. Climate-volcanism feedback and the Toba eruption of ~74,000 years ago. Quaternary Research 40, 269–280.

Rampino, M.R., Self, S., 1993b. Bottleneck in human evolution and the Toba eruption. Science 262 (5142), 1955.

Rampino, M.R., Ambrose, S.H., 2000. Volcanic winter in the Garden of Eden: The Toba supereruption and the late Pleistocene human population crash. In: McCoy, F.W., Heiken, G. (Eds.), Volcanic Hazards and Disasters in Human Antiquity. Geological Society of America Special Paper 345, pp. 71–82.

Rampino, M.R., 2002. Supereruptions as a threat to civilisations on Earth-like planets. Icarus 156, 562–569.

Rogers, A.R., Harpending, H., 1992. Population growth makes waves in the distribution of pairwise genetic differences. Molecular Biology and Evolution 9, 552–569.

Rogers, A.R., Jorde, L.B., 1995. Genetic evidence on modern human origins. Human Biology 67 (1), 1–36.

Rose, W.I., Chesner, C.A., 1987. Dispersal of ash in the great Toba eruption, 75 ka. Geology 15, 913–917.

Rose, W.I., Chesner, C.A., 1990. Worldwide dispersal of ash and gases from earth's largest known eruption: Toba, Sumatra, 75 ka. Palaeogeography, Paleoclimatology, Palaeoecology 89, 269–275.

Rosi, M., Vezzoli, L., Aleotti, P., Censi, M. De., 1996. Interaction between caldera collapse and eruptive dynamics during the Campanian Ignimbrite eruption, Phlegraean Fields, Italy. Bulletin of Volcanology 57 (7), 541–554.

Schulz, H., von Rad, U., Erlenkeuser, H., 1998. Correlation between Arabian Sea and Greenland climate oscillations of the past 110,000 years. Nature 393, 54–57.

Schulz, H., Emeis, K-C., Erlenkeuser, H., von Rad, U., Rolf, C., 2002. The Toba volcanic event and interstadial/stadial climates at the Marine Isotopic Stage 5 to 4 transition in the Northern Indian Ocean. Quaternary Research 57, 22–31.

Self, S., Zhao, J-X., Holasek, R.E., Torres, R.C., King A.J., 1996. The atmospheric impact of the 1991 Mount Pinatubo eruption. In: Newhall, C.G., Punongbayan, R.S. (Eds.), Fire and Mud: Eruptions and Lahars of Mount Pinatubo, Philippines. University of Washington Press, Seattle, pp. 1089–1115.

Shane, P., Westgate, J., Williams, M., Korisettar, R., 1995. New geochemical evidence for the youngest Toba tuff in India. Quaternary Research 44, 200–204.

Shane, P., Westgate, J., Williams, M., Korisettar, R., 1996. Reply to Comments by S. Mishra and S.N. Rajaguru on "New Geochemical Evidence for the Youngest Toba Tuff in India". Quaternary Research 46, 342–343.

Sharma, G.R., 1980. History to prehistory. Archaeology of the Ganga Valley and the Vindhyas. Department of Ancient History, Culture and Archaeology, University of Allahabad, Allahabad.

Sharma, G.R., Clark, J.D., 1983. Introduction. In: Sharma, G.R., Clark, J.D. (Eds.), Palaeoenvironments and Prehistory in the Middle Son Valley. Abinash Prakashan, Allahabad, pp. 1–8.

Sherry, S.T., Rogers, A.R., Harpending, H., Soodyall, H., Jenkins, T., Stoneking, M., 1994. Mismatch distributions of mtDNA reveal recent human population expansions. Human Biology 66 (5), 761–775.

Smith, R.L., Bailey, R.A., 1968. Resurgent cauldrons. Geological Society of America Memoir 116, 613–662.

Sonakia. A., 1984. The skull-cap of early man and associated mammalian fauna from Narmada valley alluvium, Hoshangabad area, Madhya Pradesh (India). Records of the Geological Survey of India 113 (6), 159–172.

Sonakia, A., Biswas, S., 1998. Antiquity of the Narmada *Homo erectus*, the early man of India. Current Science 75 (4), 391–392.

Song, S.-R., Chen, C.-H., Lee, M.-Y., Yang, T.F. Iizuka, Y., Wei. K.-Y., 2000. Newly discovered eastern dispersal of the youngest Toba Tuff. Marine Geology 167, 303–312.

Stothers, R.B., 1984. The great Tambora eruption in 1815 and its aftermath. Science 224, 1191–1198.

Supekar, A.G., 1968. Pleistocene stratigraphy and prehistoric archaeology of the central Narmada basin. Ph.D. Dissertation, Deccan College Postgraduate and Research Institute, Poona.

Thangaraj, K., Chaubey, G., Kivisild, T., Reddy, A.G., Singh, V.K., Rasalkar, A.A., Singh, L., 2005. Reconstructing the origin of Andaman islanders. Science 308 (5724), 996.

Turco R.P., Toon O.B., Ackerman T.P., Pollack J.B., Sagan C., 1983. Nuclear winter: global consequences of multiple nuclear explosions. Science 222, 1283–1292.

Turco R.P., Toon O.B., Ackerman T.P., Pollack J.B., Sagan C., 1990. Climate and smoke: an appraisal of nuclear winter. Science 247,166–176.

Underhill, P.A., Passarino, G., Lin, A.A., Shen, P., Lahr, M.M., Foley, R.A., Oefner, P.J., Cavalli-Sforza, L.L., 2001. The phylogeography of Y chromosome binary haplotypes and the origins of modern human populations. Annals of Human Genetics 65, 43–62.

van Bemmelen, R.W., 1939. The volcano-tectonic origin of Lake Toba (North Sumatra). De Ingenieur Nederlandsch-Indie 6, 126–140.

Wakankar, V.S., 2002. Burial systems in Bhimbetka. In: Kennedy, K.A.R., Lukacs, J.R., Misra, V.N. (Eds.), The Biological Anthropology of Human Skeletal Remains from Bhimbetka, Central India. Indian Society for Prehistoric and Quaternary Studies, Pune, pp. 1–5.

Watson, E., Forster, P., Richards, M., Bandelt, H-J., 1997. Mitochondrial footprints of human expansions in Africa. American Journal of Human Genetics 61, 691–704.

Westgate, J.A., Shane P.A.R., Pearce, N.J.G., Perkins, W.T., Korisettar, R., Chesner, C.A., Williams, M.A.J., Acharyya, S.K., 1998. All Toba tephra occurrences across peninsular India belong to the 75,000 yr B.P. eruption. Quaternary Research 50, 107–112.

Williams, M.A.J., 2004. Environmental impacts of extreme events: the Toba mega-eruption, volcanic winter and the near demise of humans. Allahabad Journal of inter-disciplinary Studies in History and Archaeology 1 (1), 118–119.

Williams, M.A.J., Clarke, M.F., 1984. Late Quaternary Environments in north-central India. Nature 308, 633–635.

Williams, M.A.J., Clarke, M.F., 1995. Quaternary geology and prehistoric environments in the Son and Belan valleys, north central India. In: Wadia, S., Korisettar, R., Kale, V.S. (Eds.), Quaternary Environments and Geoarchaeology of India. Geological Society of India, Bangalore, pp. 282–308.

Williams, M.A.J., Royce, K., 1982. Quaternary geology of the Middle Son Valley, north central India: implications for prehistoric archaeology. Palaeogeography, Palaeoclimatology, Palaeoecology 38, 139–162.

Williams, M.A.J., Royce, K., 1983. Alluvial history of the Middle Son Valley, north central India. In: Sharma, G.R., Clark, J.D. (Eds.), Palaeoenvironments and Prehistory in the Middle Son Valley. Abinash Prakashan, Allahabad, pp. 9–21.

Zielinski, G.A., Mayewski, P.A., Meeker, L.D., Whitlow, S., Twickler, M., 1996a. A 110,000-year record of explosive volcanism from the GISP2 (Greenland) ice core. Quaternary Research 45, 109–118.

Zielinski, G.A., Mayewski, P.A., Meeker, L.D., Whitlow, S., Twickler, M., Taylor, K., 1996b. Potential atmospheric impact of the Toba mega-eruption ~71,000 years ago. Geophysical Research Letters 23, 837–840.

Zielinski, G.A., 2000. Use of paleo-records in determining variability within the volcanism-climate system. Quaternary Science Reviews 19, 417–438.

9. The emergence of modern human behavior in South Asia: *A review of the current evidence and discussion of its possible implications*

HANNAH V.A. JAMES

Leverhulme Centre for Human Evolutionary Studies
The Henry Wellcome Building
Fitzwilliam Street
University of Cambridge
Cambridge, CB2 1QH
England
h.james@human-evol.cam.ac.uk

Introduction

The human species combines a distinct anatomical form with a unique behavioral repertoire, and it is this combination that sets *H. sapiens* apart from the rest of the animal kingdom. One of the most dramatic behavioral differences is the accumulation over the last 2.5 million years of a complex and symbolic material culture. Aspects of this cultural package, such as the use of lithic technology, are characteristics shared by a number of hominin species, including the entire genus *Homo* and perhaps the australopithecines (e.g., Wood and Strait, 2004). Indeed, archaeological assemblages from the Middle and Late Pleistocene suggest a number of behavioral similarities between *H. sapiens* and our closest relatives (e.g., Shea, 2003; Brown et al., 2004; Morwood et al., 2004). Yet the exclusive survival of *H. sapiens* at the expense

of species such as *H. neanderthalensis*, *H. erectus* and *H. floresiensis* indicates that there were considerable differences between the behaviors and cognitive capabilities of these closely related species. The precise nature and the evolutionary processes behind the emergence of this uniquely modern behavioral package remain to be elucidated.

The comparative analysis of the archaeological record from a number of different regions represents the most appropriate methodological approach to investigating the evolution of modern human behavior. South Asia, a crucial geographical region between Africa and the rest of the Old World, with a vast and highly informative archaeological record, has potentially vital information for the study of behavioral evolution. By presenting the results of an initial analysis of South Asian archaeological assemblages from the last 250,000 years, this chapter

M.D. Petraglia and B. Allchin (eds.), The Evolution and History of Human Populations
in South Asia, 201–227.
© 2007 *Springer.*

discusses the ways in which the Paleolithic archaeology from the Indian subcontinent can help create a better understanding of how and why humans are behaviorally unique.

Current Perspectives on the Evolution of Modern Human Behavior

H. sapiens shares anatomical and behavioral characteristics with other members of the genus *Homo*. Although such "ancestral" traits provide valuable information on evolutionary relationships, the anatomical features used to characterize the human species are those that set it apart from its closest sister species, traits considered to be "derived" for *H. sapiens*. Such derived traits not only allow the recognition of *H. sapiens* within the fossil record, but because they provide information on the different way in which humans interacted with their environment (compared to other members of the genus *Homo*) may also provide evidence regarding how and why the speciation of *H. sapiens* occurred. Elucidating the behavioral package that is derived for *H. sapiens* would serve a similar purpose, and enhance the evidence available from the fossil record. Just as it is derived traits that characterize *H. sapiens* anatomically, derived traits should characterize *H. sapiens* behaviorally.

There is a great deal of debate in the archaeological literature concerning the traits that might represent the different cognitive and behavioral capabilities of anatomically modern humans and the extent to which they are present in other species, including– but not restricted to–the Neanderthals. Such debates confuse two issues, the evolution of the behaviors we see present in humans today and the evolution of the behaviors that, from an evolutionary perspective, define the human species. It is the latter, derived, behaviors that are investigated within this chapter, and that will be referred to as "modern human behaviors". Therefore modern human behavior can be defined as the actions and

abilities that set us apart from not only the extant primates but our closet sister species within the genus *Homo*. In many cases this will be a matter of degree rather than a clear presence/absence distinction. But to be classed as modern human behavior the precise degree of these archaeologically manifested actions must not occur in any other hominin species. Thus, if the Neanderthals, for example, are seen to have produced a behavior independently, then this behavior, while complex and interesting in its own right, can not be classed as a modern human behavior.

Modern Human Behavior and the Archaeological Record

In the last few decades a number of archaeological traits have been proposed that set anatomically modern humans apart from the rest of the genus *Homo* (e.g., Mellars, 1989, 1991, 2002, 2005; Klein, 1999; McBrearty and Brooks, 2000) (Table 1). Each proposed trait fulfills two criteria that enable it to be suggested as exclusive to modern humans. Firstly the traits used are those that characterize the appearance of the Upper Paleolithic within Europe and the Late Stone Age (LSA) within Africa. Secondly they reflect behaviors that have been shown through anthropological research to be represented in some form or other by all human cultures.

The Upper Paleolithic and the LSA represent very different packages of material culture to those that characterize the preceding Middle Paleolithic and the Middle Stone Age (MSA) of Western Europe and Africa respectively (e.g., Bar-Yosef, 2002; Mellars, 2002). There is an increase in the production of specialized tools with standardized forms, and such tools are made on a wide range of materials including stone blades, bone and antler. Blade production is intensified, with the majority punch-struck from prismatic cores. Microlithic stone artifacts are combined in composite tools. The combination of such technological changes allowed more

Table 1. The archaeological signatures of modern human behavior (compiled from Mellars, 1991; Klein, 1999; McBrearty and Brooks 2000)

Functional Attributes	Increased diet breadth and better environmental exploitation:
	• Specialized technology (standardized blank production, new lithic technologies such as blades, micro blades and hafted tools) in a variety of materials. An increase in formal tool categories. Geographic variation within formal tool categories.
	• Long distance procurement of raw materials. Exchange?
	• Scheduling and seasonality in resource exploitation
	• Structured use of domestic space
	• Colonization of new habitats
Symbolic Attributes	Identity:
	• Group and individual self identification through artifact style
	• Regional artifact styles
	• Self adornment, e.g. use of beads and ornaments
	• Use of pigment
	• Notched and incised objects (bone, egg shell, ochre, stone)
	• Image and representation
	• Burials with grave goods, ochre, ritual objects

specialized and difficult to attain foodstuffs to be incorporated into the diet. This increased technological ability to manipulate the environment was combined with the first expressions of a world outside the mundane. Both marine and ostrich egg shell were used in the production beads, presumably used for the decoration of clothes and as jewelry. Representational and idealized art appears, both in the form of carved figurines and painted on the walls of caves and rock shelters. Dead individuals begin to be buried with grave goods. The Upper Paleolithic of Europe seems to represent an explosion of symbolism and creativity (Mellars, 2005). Such a dramatic change does not seem to be represented in Africa, however, with a number of these archaeological traits being found within assemblages dating to the MSA to LSA transition (e.g., McBrearty and Brooks, 2000).

While not all Upper Paleolithic or LSA industries contain evidence for all of these specific behaviors, most provide evidence for a material culture that combined technological flexibility and specialization with explicitly symbolic artifacts and actions. Even functional objects may have varied stylistically (e.g., Sackett, 1977). Crucially, both of these intertwined elements are universal within human culture, even if the specific ways in which they are manifested vary. Specialized technologies clearly vary between different ecological zones and social groups, but the ability to utilize technology within culturally imposed limits, to extract the maximum gain from the environment, is shared by the majority of human cultures. Anthropological research indicates that all human use of material culture is symbolic, with even everyday objects being used as the means to define and manipulate individual and group identity (e.g., Malinowski, 1922; Mauss, 1970; Hodder, 1982; Okley, 1983). The extent to which such behaviors are exclusive to *H. sapiens* does, however, remain to be elucidated. Given the inherent flexibility of human behavioral capabilities, and the difficulties in interpreting a variably preserved archaeological record, it is the behaviors behind the specific actions that are concentrated on in current research (e.g., McBrearty and Brooks, 2000; D'Errico, 2003; Henshilwood and Marean, 2003). In particular, the capacity for symbolic thought (and its correlation with language) is increasingly seen as one of the most important differences between anatomically modern humans and their closest relatives (e.g., Henshilwood and Marean, 2003).

If there is now some consensus regarding the kind of material culture that could plausibly represent a human set of behaviors, the timing of the evolution of this behavioral package remains in question. Clearly, if these behaviors are unique to anatomically modern humans they should not be evident within the technological industries that were produced by other members of the genus *Homo*. Thus, if any of these behaviors are found in the Middle Paleolithic of Europe, solely inhabited by the genetically isolated Neanderthals (e.g., Ovchinnikov, 2000; Carameli et al., 2003), then they should not be classed as modern human behaviors. The same statement is not true for the archaeological assemblages belonging to the African MSA. Given that anatomically modern humans evolved within Africa at ca. 195–150 ka (e.g., Stringer, 2001; White et al., 2003; McDougall et al., 2005), at least some MSA assemblages were produced by anatomically modern humans. In theory, the origins of the modern behavioral package could date back to the origin of the human species. Problematically, few of the assemblages currently recovered from the African MSA combine all the aspects discussed above

(e.g., Klein, 1995, 2000). This has lead to a variety of views on the timing of the emergence of modern human behavior. Depending on the way in which various aspects of the MSA archaeological record are interpreted, modern human behavior either appeared relatively suddenly at ca. 50 ka (e.g., Klein, 2000) or gradually between ca. 200 and 50 ka (e.g., McBrearty and Brooks, 2000; Barham, 2001; Deacon and Wurz, 2001). In order to discuss the emergence of modern behavioral traits within South Asia, it is necessary to examine some of these disputed traits in closer detail.

The Functional aspects of Modern Human Behavior
While the functional behaviors listed within Table 1 are clearly representative of complex resource procurement strategies and a high degree of cognitive capability, they may not all be exclusive to anatomically modern humans. Here the case for including each trait as part of the exclusively modern human behavioral package is reviewed.

Blade Production. A growing body of research in the last decade has suggested that one of the most dramatic changes associated with the "Upper Paleolithic Revolution", i.e., the deliberate, standardized production of blades, should no longer be exclusively associated with anatomically modern humans (e.g., Bar-Yosef and Kuhn, 1999). Deliberately produced blades characterize the Acheulio-Yabrudian from the Levant, and are dated to between 330 and 270 ka at Tabun (Mercier et al., 1995; Bar-Yosef and Kuhn, 1999). Within Africa evidence for deliberate blade manufacture has also been recovered from Acheulean contexts, including the late Acheulean site of GnJh-03, in the 280 ka Kapthurin Formation, Kenya (McBrearty and Brooks, 2000). Numerous MSA assemblages provide evidence for significant amounts of blade production, including those recovered from Gademotta and Aduma in Ethiopia and Haua Fteah, Libya (McBrearty and Brooks, 2000). Blade production is also represented within Western Europe, within sites such as Tönchesberg, Seclin and Rheindalen (Conard, 1990). The presence of blade technology within the Levant, Western Europe and Africa indicates that a number of hominin species, including *H. sapiens* and *H. neanderthalensis*, were capable of blade production.

The blade production that characterizes the European Upper Paleolithic and the LSA may represent a different technological phenomenon to that seen during earlier industries. While the early industries of Africa, the Near East and Western Europe indicate the deliberate production of blades in significant proportions, blade-based assemblages are either restricted to defined technological industries (such as the Still Bay within Southern Africa), or are relatively rare. The Upper Paleolithic and the LSA represent an intensification of blade production, while the use of prismatic cores and indirect percussion allowed increasingly standardized blade manufacture (e.g., Mellars, 2005). It is this industrial and technological change that is unique to modern humans, not the production of blades per se.

Higher quality diet. The evidence for intensive food procurement and the exploitation of higher quality plant and animal resources prior to the LSA or Upper Paleolithic is equivocal. The Middle Paleolithic of Europe, the Near East and the MSA of Africa provide evidence of complex food procurement strategies (e.g., Stiner and Kuhn, 1992; Mellars, 1996; McBrearty and Brooks 2000; Shea, 2003), though the timing of resource acquisition shifts within the MSA has been debated (e.g., Klein, 1995, 2000; Milo, 1998; McBrearty and Brooks, 2000). Questions remain over the differences in diet and resource acquisition behavior between early African anatomically modern humans and their archaic relatives within the rest of the world.

Innovative food procurement technology. It is certainly the case that the MSA does provide evidence for a number of technological innovations that are currently missing from other Middle Paleolithic industries. The use of bone and the production of barbed points is attested as early as ca. 90 ka within assemblages from Katanda (Yellen et al., 1995; McBrearty and Brooks, 2000), with shaped bone tools being recovered from a number of other sites including Blombos (Henshilwood and Sealey, 1997). Backed artifacts also make an early appearance within the MSA (Barham, 2002). The earliest evidence for the production of complex composite tools is also African, and takes the form of the geometric microlithic Howiesons Poort Industry, dated to ca. 70–60 ka at Klasies River Mouth (Wurz, 2002) and other South African MSA sites, including Bloomplaas and Border Cave (McBrearty and Brooks, 2000). Klasies River Mouth has notably also produced anatomically modern human remains (Rightmire, 1989; McBrearty and Brooks, 2000), though they have been dated to earlier than the Howiesons Poort assemblages. The current lack of geometric microlithic technology within the Middle Paleolithic of Western Europe and the Near East supports the hypothesis that the production of complex composite tools is a behavior exclusive to anatomically modern humans. Though composite tools (e.g., Howiesons Poort, Still Bay) and bone tools are present over large areas, such industries and artifacts are restricted occurrences within the MSA (e.g., McBrearty and Brooks, 2000). It is not until the LSA that they occur regularly within archaeological deposits. Early anatomically modern humans may have invented such technology, but they did not use it consistently until much later on in the Upper Pleistocene.

A tool kit both standardized and specialized. An increase in both geographical and chronological variation in standardized artifact types has been argued to signal the appearance of modern human behavior (e.g., McBrearty and Brooks, 2000). Within the Upper Paleolithic and the LSA such spatial and temporal variation is present within particular tool forms. In theory, such variation in assemblage composition represents both the incorporation of more specialized resources within the diet of *H. sapiens*, but also the appearance of distinct regional traditions. As a marker for modern human behavior it has functional and symbolic significance.

The lithic industries of Middle Paleolithic Western Europe, the Near East and MSA Africa all exhibit spatial and temporal variation in assemblage composition (e.g., Mellars, 1996; Bar-Yosef, 1998; McBrearty and Brooks, 2000; Shea, 2003). Both Africa and Western Europe also show regional variations within techniques of artifact manufacture (e.g., Mellars, 1996; Van-Peer, 1991). A wide range of environments was exploited within Africa, including probable rainforests within the Congo (Barham, 2001) and coastal environments within Eritrea (Walter et al., 2000). Temporal variation is evident within sites such as Klasies River Mouth (Wurz, 2002) and Florisbad (Kuman et al., 1999). At Twin Rivers, such temporal variation in assemblage composition has been interpreted as a response to fluctuating climatic conditions (Clark and Brown, 2001). Similarly variable environments were inhabited by the Western European Neanderthals (e.g., Gamble, 1986) and there is evidence for both inter-and intra-site variations within European Middle Paleolithic assemblage composition (Mellars, 1996). While the variability, at least in environments occupied, appears greater within the MSA, it is difficult to determine how much this discrepancy is due to the differing size and environments of the two regions. From a purely functional perspective, it is clear that a number of hominin species were able to produce lithic technology that varied with need and resource availability. It has been claimed, however, that the MSA is unique amongst these

other prepared core industries as it is characterized by a number of regional traditions that show variations in artifact style (McBrearty and Brooks, 2000). For this to be a marker of modern human behavior such stylistic variations must be deliberate and symbolic, and they must be markers of group identity.

Group identification through artifact style. Style within lithic artifacts is difficult to determine within the archaeological record, due in part to the effect of functional considerations on artifact design (e.g., Sackett, 1977, 1982, 1986; Close 1978). Because of the duality of the functional and the symbolic within stone tool technology it is difficult to identify which part of the variation, if any, is due to social norms and which is the result of more practical factors. Raw material availability, frequencies of tool re-use (e.g., Dibble and Rowland, 1992) as well as the function of the artifact both affect and constrain the options available to the toolmaker (e.g., Schiffer and Skibo, 1997). A number of choices will remain during the manufacturing processes, and these will be made on the basis of social norms (e.g., Le Monnier, 1991; Dobres, 1995; Sinclair, 1995; White, 1997). In addition, variation will creep in due to cultural drift, the accumulation of random variation in the reproduction of a particular behavior as it gets farther (in time or distance) from originator of the technique or artifact type. Cultural drift occurs when random copying errors accumulate in an aspect of material culture that is not being acted upon by natural selection (e.g., Neiman, 1995; Shennan, 2002). Thus, even though the European Middle Paleolithic exhibits distinct regional traditions of tool manufacture (such as the MTA A and B) (Mellars, 1996) characterized by specific tools or different tool morphologies, it would be difficult to interpret them as deliberate variations in style or deliberately symbolic. Even the variation that could not be assigned to resource availability or function could be the result of cultural drift. While the means exist to test whether a particular stylistic variation is the result of cultural drift or deliberate imposition (e.g., Shennan, 2002), they require the variation to be non-functional, a distinction not easily made with lithic artifacts.

Items of material culture are, however, used as symbolic markers of cultural and individual identity (e.g., Hodder, 1982; Wiessner, 1991) and lithic artifacts are no exception. Archaeologists consider projectile point style, in particular, as a good indicator of group identity or the interaction of certain groups at the expense of others. The use of points in identity construction is attested within the ethnographic record (e.g., Wiessner, 1983). But even within this artifact category, the use of stylistic variation alone in determining interaction is problematic (e.g., Odess, 1998). While the laws of aerodynamics may well constrain any variation, whether random or deliberate in projectile point form, such variations can occur if they are selectively neutral (i.e., do not effect performance), or act to enhance the artifacts performance. In the former case such variation could either be due to cultural drift or to deliberately imposed style.

The MSA can be seen to be divided into a number of regionally and temporally distinct industries that are characterized by particular "styles" of projectile points (McBrearty and Brooks, 2000). But this variation does not necessarily have to be symbolic. Given the lack of chronological data for a number of the industries involved the effects of functional variation and cultural drift may still explain part of the variation. Given the larger landmass and more variable environments within Africa relative to Europe, it is possible that the effects of functional variation and cultural drift on archaeological assemblages would lead to a greater diversity of technological industries within the former.

Such arguments should be discounted before variation between lithic artefacts can be argued to be purely stylistic.

Ritual treatment of the dead.   Equally problematic is the issue of burial (as a symbolic act) versus the simple internment of hominin remains. Claims for the deliberate internment of both Neanderthal and anatomically modern human remains have been made from Western Europe, the Near East and Africa. Most of these internments contain little in the way of grave goods or other evidence for a symbolic marking of the grave, leading to suggestions that the majority of such finds may well be the result of natural processes (Gargett, 1999). The best early evidence for symbolic burials has been recovered from the Near East. At Qafzeh, anatomically modern human remains have been buried with antelope remains placed over the head, while at Shanidar, a Neanderthal skeleton has been recovered from a context associated with the remains of flowers (Solecki, 1975; 1989; Vandermeersch, 1981; Bar-Yosef and Vandermeersch 1993). While the ritualistic treatment of modern human remains may be represented by the charred and cutmarked bones from Klasies River Mouth, such claims need further substantiation (Mitchell, 2002). It is not until later that modern human remains associated with explicitly symbolic material culture (i.e., beads) are found in numbers. Indeed, it is interesting to note that burials from the early Aurignacian are rare. The nature of the evidence makes it difficult to argue that there is a significant difference between early modern humans and their contemporaries, although there does seem to be a change in the evidence as the Upper Paleolithic progresses.

The non-utilitarian use of ochre.   Ochre has been recovered from a large number of archaeological sites dating from the Late Acheulean upwards within regions as disparate as Africa, South Asia, Europe and the Near

East (e.g., Paddayya, 1982; Bednarik, 1990; Barham, 1998; Hovers et al., 2003). Ochre may have had a partly functional use, perhaps in hafting (e.g., Lombard, 2005), or it may have been used for non-utilitarian, symbolic purposes (Hovers et al., 2003). For whatever reason, however, it seems to have been utilized by a number of hominin species, and the use of ochre cannot, at present, be said to definitely distinguish early modern humans from their contemporaries.

Explicitly symbolic artifacts (beads and art). There is, however, one probable symbolic use of ochre within the MSA, the piece of incised ochre recovered from Blombos cave, South Africa in levels dated to ca. 75 ka (Henshilwood et al., 2002). This find represents the earliest evidence of art any where in the world. From the same levels shell beads have also been recovered, suggesting that the inhabitants of Blombos ca. 75 ka were capable of symbolic thought (Henshilwood et al., 2004). The evidence for symbolism within Middle Paleolithic Europe is much less clear (though see D'Errico, 2003 for the opposing view point). While the Chatelperronian does represent the production of beads by *H. neanderthalensis* (D'Errico et al., 1998), such finds seem chronologically later than the earliest Aurignacian (Mellars, 2005), suggesting that this was not an independent innovation.

The capacity for symbolic thought may represent one of the best candidates for an ability that distinguishes anatomically modern humans from our closest sister species. The Upper Paleolithic represents an explosion of symbolic expression, evidenced by carved figurines and cave art that are incomparable to anything recovered from Middle Stone Age and Middle Paleolithic contexts. Explicitly symbolic artifacts have been recovered from African Middle Stone Age assemblages indicating that the ability to produced such artifacts dates to at least 75 ka. But though the African Middle Stone Age does show

evidence for the ability, artifacts such as beads and art are relatively rare, indicating that the initial appearance of symbolic artifacts is not the same phenomenon as that seen with the arrival of the Upper Paleolithic.

The Story of the Evolution of Modern Human Behavior So Far

A number of traits clearly reveal themselves to be exclusively associated with anatomically modern humans, including the production of explicitly symbolic artifacts (i.e., beads and art) and the production of composite, hafted technology. Such artifacts appear sporadically in Africa from ca. 75 ka and become considerably more common with the arrival of the LSA and the Upper Paleolithic. Interestingly this still leaves a considerable gap in time between the emergence of anatomically modern humans (ca. 195–150 ka) and the appearance of behavioral modernity. As the above discussion also suggests, however, there are a number of traits that may stretch this behavior back further within the MSA, including the chronological, geographical and possible stylistic variation between assemblages. What remains unclear is whether the variation exhibited here, which does seem more significant than that in the Western European Middle Paleolithic, reflects a uniquely modern human behavioral and cognitive capacity or is a reflection of the differing ecology and size of Africa and Europe.

There is a startling disparity between the African and European record regarding the tempo of the appearance of the modern human behavioral package. A gradual, sporadic, appearance of modern behavioral traits, within a region in which the modern human anatomical form evolved is contrasted with the sudden appearance of a complete package as anatomically modern humans enter Western Europe. On the basis of such evidence it is tempting to conclude that the arrival of the modern behavioral package can be directly correlated with the arrival of anatomically modern humans within a new region. The gradual record within Africa can be argued to represent the slow process of the origin of the human species. In contrast, the sudden appearance of modern human behavior within Europe represents the entrance of the complete modern human package, i.e., anatomically and behaviorally modern humans. But this ignores the other factors that affect the expression of mental capabilities within the archaeological record. The very nature of human material culture is that it is manipulated through societal pressures as well reflecting the functions that it is used for. The material culture packages of different societies are never exactly the same, even though their manufacturers are all modern humans. The Upper Paleolithic of Western Europe must be looked at as a particular societies response to a particular set of conditions. To be able to tease apart the effects of these environmental and demographic conditions from the evolutionary process it is vital to know how similar the European record is to the other regions in which *H. sapiens* is an immigrant species. As the effects of environment, demography and social norms, may be affecting the archaeological record within Africa as well, then it is necessary to examine the uniqueness of the African record. In order to understand how behavioral modernity evolved, the emergence of modern human behavior within regions of the world such as South Asia needs to be investigated.

The Later Pleistocene Archaeological Record of South Asia

Archaeological assemblages from the South Asian Later Pleistocene (i.e., from ca. 250 to 10 ka) have significant implications for understanding the processes behind the evolution of modern human behavior, as well as providing an illuminating picture of Paleolithic life in their own right. Artifacts have been recovered from throughout the subcontinent, and provide evidence for both chronological and spatial variation within lithic technology.

During most of the Later Pleistocene, the lithic technology of hominin populations consisted of the prepared core based industries of the Middle Paleolithic (e.g., Paddayya 1984). The exact date of its earliest occurrence is currently unknown. Evidence from the Kaladgi Basin and elsewhere suggests, however, that the Middle Paleolithic developed from the Late Acheulean of the subcontinent (Petraglia et al., 2003). By ca. 150 ka typologically Middle Paleolithic industries were being produced at the 16R Dune, Rajasthan (Misra, 1995) (Figures 1, 2). The Middle Paleolithic industry at Patpara, in the Middle Son Valley, has been dated to less than ca. 103 ka (Blumenschine et al., 1983; Williams and Clarke, 1995). Dates of ca. 75 and > 60 ka are associated with flake based industries recovered from Samnapur (Narmada Valley) and Balotra (Luni River Valley) (Misra et al,. 1990; Mishra et al., 1999), while Middle Paleolithic assemblages from the Hiran Valley are dated to ca. 69–56 ka (Baskaran et al., 1986).

The earliest blade-based assemblage is that recovered from Site 55, Pakistan where the loess overlying the occupational horizon has been dated to ca. 45 ka before present (Dennell et al., 1992). Within India numerous blade-based industries, including those from Mehtakheri, Inamgoan, Chandrasal, Dharamouri and Nandipalle, have been dated to between ca. 40 and 20 ka (Mishra, 1995). At the 16R Dune, a blade-based assemblage is dated to ca. 26 ka (Misra, 1995). Within Sri Lanka, microlithic industries date from ca. 28.5 ka at Batadomba-lena, and from ca. 28 ka at Site 49 and Site 50 (Deraniyagala, 1992). Microlithic elements are combined with blades and burins at Patne within an assemblage dated to ca. 24.5 ka. Given the typological and chronological overlap of such blade-based, "Upper Paleolithic", assemblages and the early Microlithic industries of India

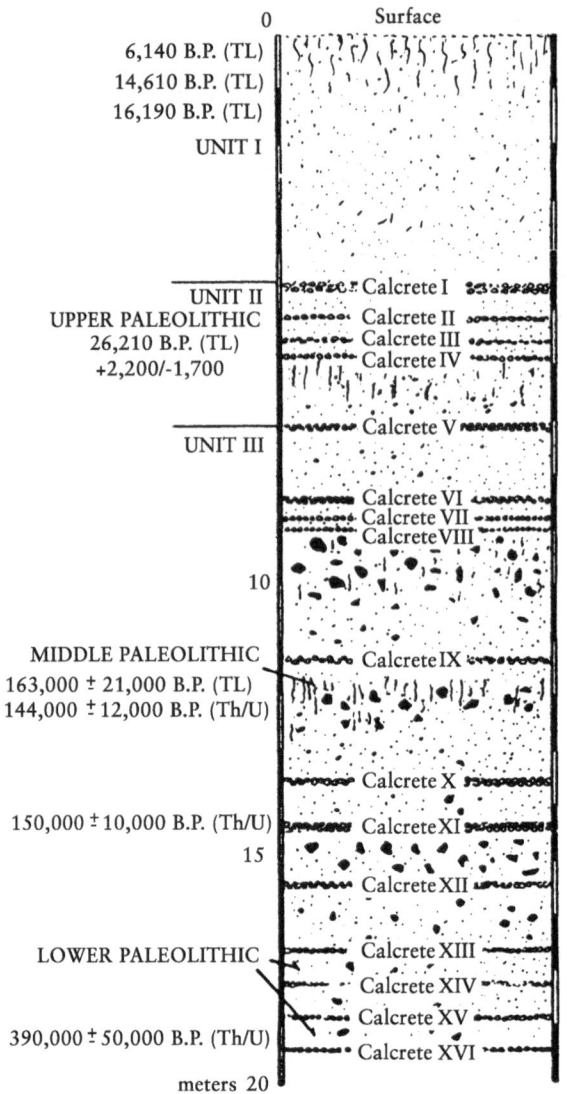

Figure 1. Sequence of the 16 R Dune (after Misra 1995b)

and Sri Lanka, the use of either term is confusing. Instead, it has been proposed that such assemblages may be better termed Late Paleolithic (James and Petraglia, 2005), and this terminology is followed here.

Throughout the Later Pleistocene, hominin populations occupied a wide range of environments throughout the subcontinent (Figure 3). Lithic assemblages have been recovered from coastal and esturine localities, including Ramayogi Agraharam (Rath et al., 1997) and the Hiran Valley (Marathe, 1981) as well as arid regions such as the Thar (e.g., Allchin et al. 1978; Misra, 1995). Occupation was clustered

within river basins that contained freshwater in the post-monsoon season (see Korisettar, 2004; Korisettar, this volume) making use of caves (Vijaya Prakash et al., 1995), rockshelters (e.g., Misra, 1985) and open air localities (e.g., Pappu, 2001, Pappu et al. 2003).

Climatic oscillations throughout the Pleistocene meant that hominin populations were subjected to environmental change. Cycles of arid and humid conditions were coupled with a trend towards increasing aridity as the Upper Pleistocene progressed. Increasingly open environments supplanted the mixed woodland and grassland ecosystems that

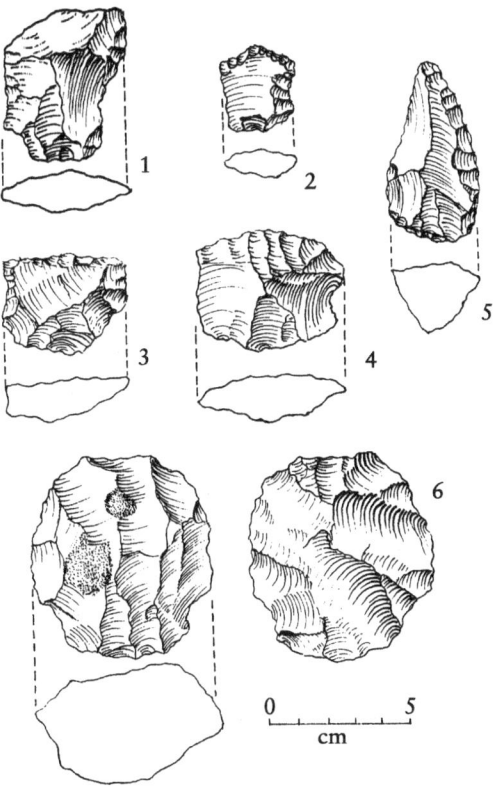

Figure 2. Middle Paleolithic artifacts from the 16R Dune, Thar Desert (after Misra 1995b: figures 16 and 17). 1–4, scrapers; 5, point; 6, core

characterized the Middle Pleistocene, with brackish swamps replacing plentiful fresh-water sources (Misra, 2001). Within marginal environments, such as the Thar, the effects of the fluctuating climate were especially dramatic, with periods of dune formation sharply interspersed with periods of wetter, ameliorated conditions (e.g., Andrews et al., 1998; Kar et al., 2001; Deotare et al., 2004). The changing environments affected South Asian population demographics, with popula-tions expanding during periods of plenty but decreasing in size when resources became sparse (James and Petraglia, 2005). Archaeo-logical evidence from the Thar suggests that the occupation of the region became increas-ingly sparse and isolated during the Late Paleolithic (Allchin et al., 1978), probably reflecting the heightened aridity and loss of available water sources with the region post-25 ka (e.g., Misra, 2001; Deotare et al., 2004).

It is against this backdrop of a changing, fluctuating environment that the industries of the South Asian Middle and Late Paleolithic were produced.

The Middle Paleolithic of the Subcontinent

The Middle Paleolithic of the Indian subcon-tinent can be characterized as a flake-based industry in which artifacts classified as scrapers, points, diminutive handaxes and borers dominate (Figures 4–6). Flakes are struck from a mixture of prepared and non-prepared cores, with blade production forming an important but not dominant method of tool manufacture. With the exception of bifacial tools, such as diminutive handaxes, the extent of retouch is relatively limited, especially in comparison to the European Middle Paleolithic.

Both the production of lithic artifacts and the lithic artifacts themselves show

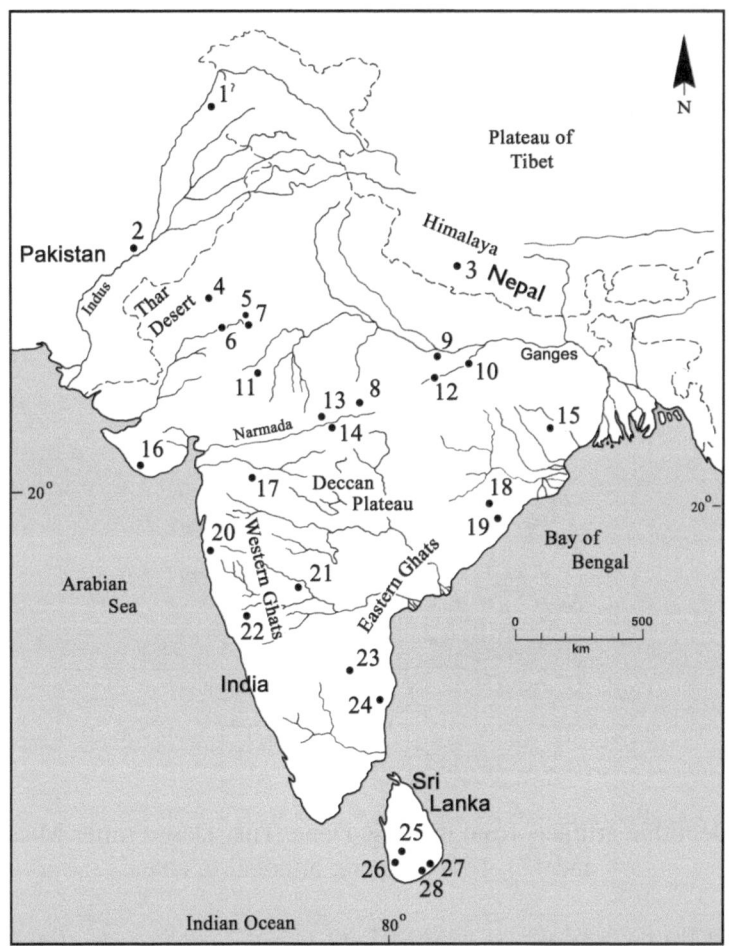

Figure 3. Principal Later Pleistocene Localities in South Asia (James and Petraglia, 2005). Locations represent site or site clusters. 1, Site 55; 2, Chancha Baluch; 3, Dang-Deokhuri complex; 4, Didwana complex; 5, Budha Pushkar; 6, Luni Valley complex; 7, Hokra; 8, Samnapur; 9, Belan Valley complex; 10, Middle Son Valley complex; 11, Beas-Berach complex; 12, Upper Son Valley complex; 13, Bhimbetka; 14, Adamgarh; 15, Singhbhum; 16, Hiran Valley complex; 17, Patne; 18, Bora; 19, Ramayogi Agraharam; 20, Konkan complex; 21, Shorapur Doab complex; 22, Kaladgi Basin complex; 23, Kurnool Caves; 24, Attirampakkam; 25, Badatomba-lena; 26, Fa Hien Cave; 27, Site 50; 28, Site 49

considerable variation across the subcontinent. The most common methods of core preparation are the "Levallois" and "Discoidal" techniques, with both methods showing a subcontinent wide distribution. Neither is ubiquitous however, and there appears to be distinct spatial variation in the relative popularity of each technique (James, 2003). The Levallois technique dominates within the river valleys of Uttar Pradesh (Pant, 1982) and the Kortallayar Basin, Tamil Nadu (Pappu,

2001). In contrast within the Wagan and Kadmali river basins (Misra, 1967, 1968) the use of the Discoidal technique is far more common. Such variation is apparent even within a single river valley. For example, the site of Siddhpur lacks evidence for either Levallois cores or flakes despite other sites within the Belan Valley producing evidence of the Levallois flake manufacture (Pant, 1982). Other techniques of core preparation have been described (i.e., cylinder cores,

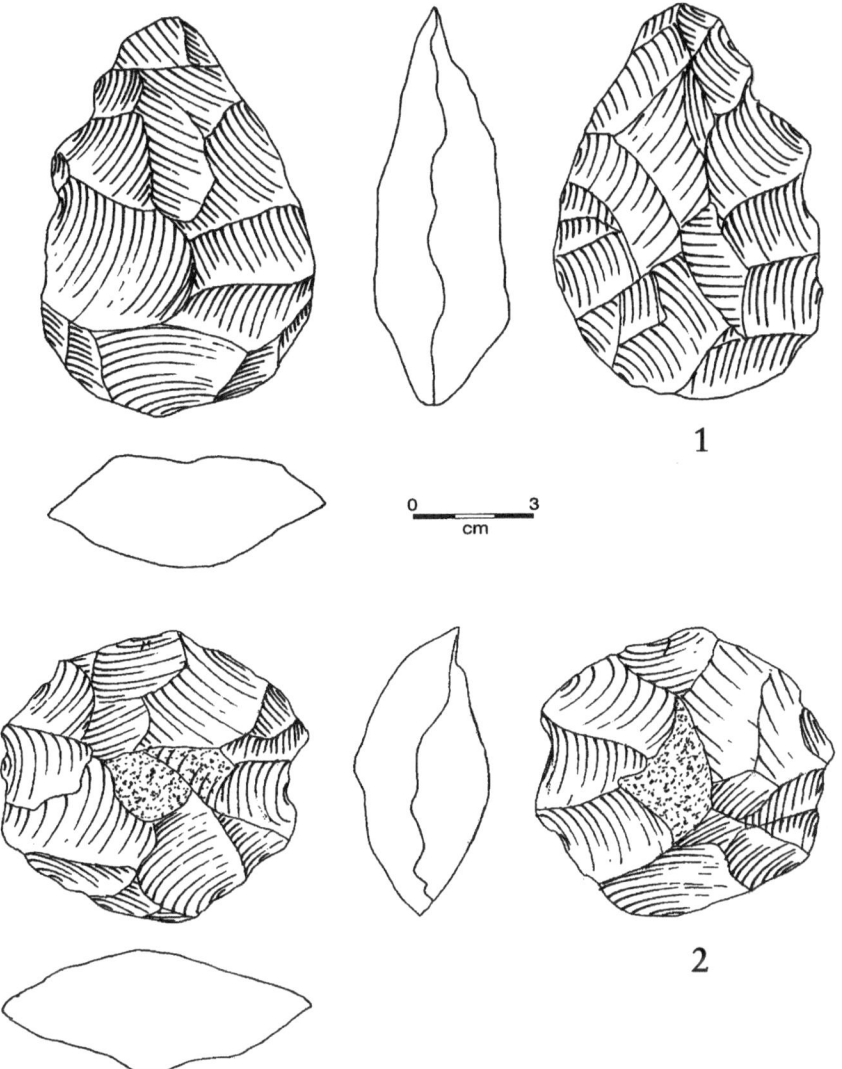

Figure 4. Early Middle Paleolithic artifacts from the Kaladgi Basin, Karnataka, India. 1, diminuitive biface; 2, prepared flake core (after Petraglia et al., 2003: figures 8, 9)

Misra, 1967) and numerous sites indicate the use of a range of differing techniques for prepared flake manufacture, including Hajiakheri (Misra, 1968), Lahchura 2 (Pant, 1982) and Attirampakkam (Pappu, 2001). Unprepared and regularly flaked cores are common in most assemblages. Within the South Asian Middle Paleolithic there does not seem to be one consistent method of flake manufacture, but instead a variety of techniques that were combined in different ways in different contexts. Such variation may well suggest that flake manufacture within the South Asian Middle Paleolithic was flexible to different ecological situations.

Blade and flake-blade cores are documented from Middle Paleolithic assemblages throughout the subcontinent, including at Bhimbetka (Misra, 1985) and Patpara (Blumenschine et al., 1983). Unidirectional cores, bearing the scars of a series of blade and flake-blade removals, such as those recovered from the Thar Desert at sites such as Hokra 1-a and Gurha (Allchin et al., 1978) indicate the deliberate production of blades as blanks (Figure 6). In the majority

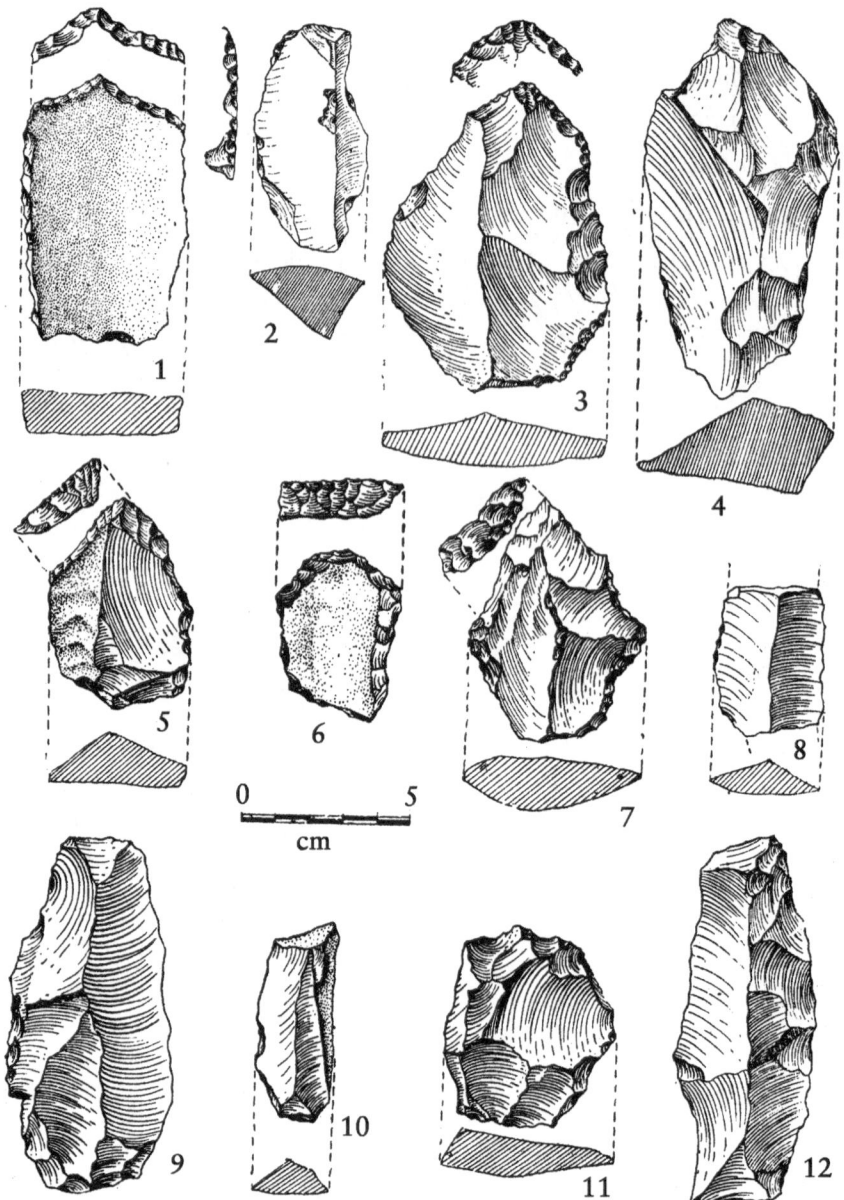

Figure 5. Middle Paleolithic artifacts from Bhimbetka III F-23, Madhya Pradesh, India (after Misra 1985: figure 4). 1–3, 5–7, scrapers; 8–10, 12, blades; 4, 11, flakes

of cases, a single flake is removed from a nodule or tablet of stone, creating a platform from which a number of blades and flakes can be struck. A small number of blades and flakes are then removed from the core before the core is discarded, indicating a lack of systematic reduction. Significantly, however, blades and flake blades represent the majority of removals from the core, suggesting that the production of narrow, elongated flakes and blades was the intention of the knapper. Interestingly, initial research suggests that the technique may have become more developed as the Middle Paleolithic progressed. Cores from the "advanced" Middle Paleolithic at Patne reflect a similar process of blade production. They are, however, more intensively reduced than cores

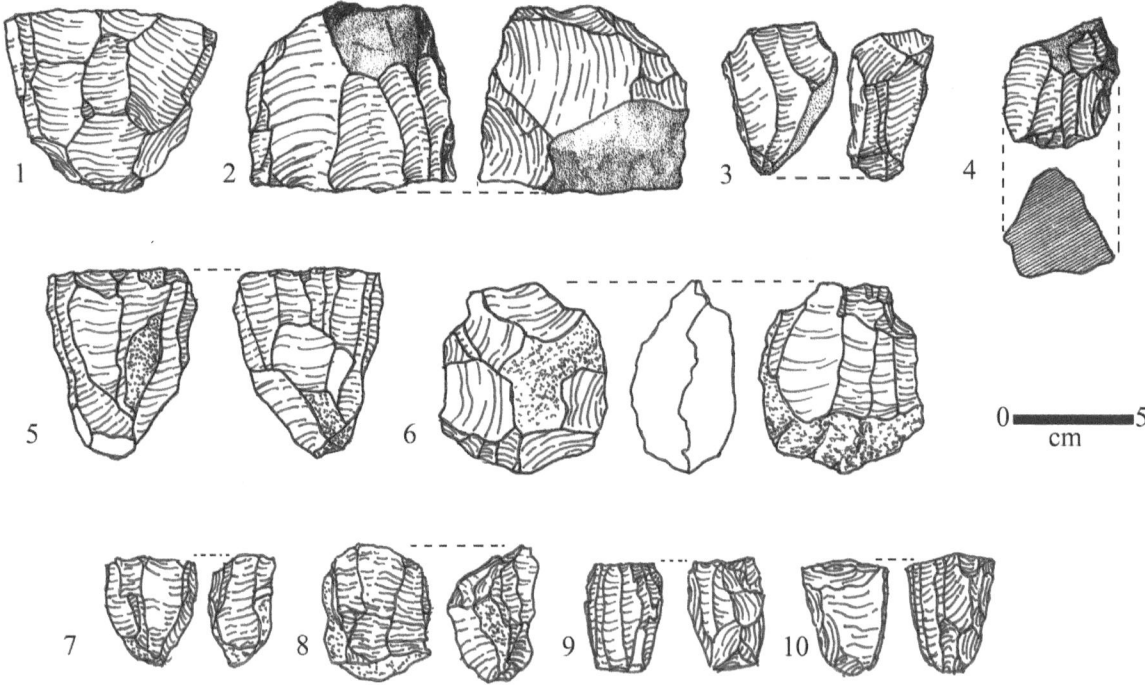

Figure 6. Middle and Late Paleolithic blade cores from the Indian subcontinent. Top: Middle Paleolithic cores. 1, core from Hokra (after Allchin et al., 1978: figure 4.6 no.1); 2, core from the Kadmali River basin (after Misra, 1967: figure 26 no.1); 3, core from Patpara (after Blumenschine et al., 1983: figure 8 no. 15); 4, core from the Wagan River basin (after Misra, 1967: figure 21 no. 5). Middle: cores from the Middle-Late Paleolithic transitional assemblage of Chancha Baluch. 5–6, cores from Chancha Baluch (after Allchin et al., 1978: figure 8.7 no. 1, 4). Bottom: Late Paleolithic cores. 7,8, cores from Visadi (after Allchin et al., 1978, figure 7.3 no. 1, 2); 9–10 cores from Budha Pushkar (after Allchin et al., 1978, figure 4.9 no. 4, 3)

from earlier Middle Paleolithic sites, with a higher number of blades (and indeed flakes) being removed from each core (James and Petraglia, 2005). The blades and flake blades struck from these cores typically form a small proportion of the blank production, but they are retouched into a number of finished tool forms, including various types of scraper. Such blade-based tools have been recovered from the Middle Paleolithic industries of Chancha Baluch (Allchin et al., 1978), the Panchmahals (Sonawane, 1984), the Godavari Valley (Joshi et al., 1979–80), and Bhagi Mohari (Paddayya, 1982–83) among others.

Though South Asian Middle Paleolithic artifacts do not tend to be heavily retouched, there is evidence for spatial variability in assemblage composition throughout the subcontinent (James, 2003). Scrapers dominate the majority of assemblages, but the production of points seems especially variable. Points are rare in some of the assemblages from the north of the subcontinent, and are particularly poorly represented in the assemblages recovered from Bhimbetka III F23 (Misra, 1985), Patpara (Blumenschine et al., 1983) and the Upper Son (Ahmed, 1984). Within the southeast of the subcontinent, such as in the Gunjana Valley (Raju, 1988), points appear to occur in higher proportions (Figure 7). Though diminutive handaxes have been identified within sites distributed throughout the subcontinent, they are not present within every Middle Paleolithic assemblage. Unfortunately, the chronological resolution is currently lacking

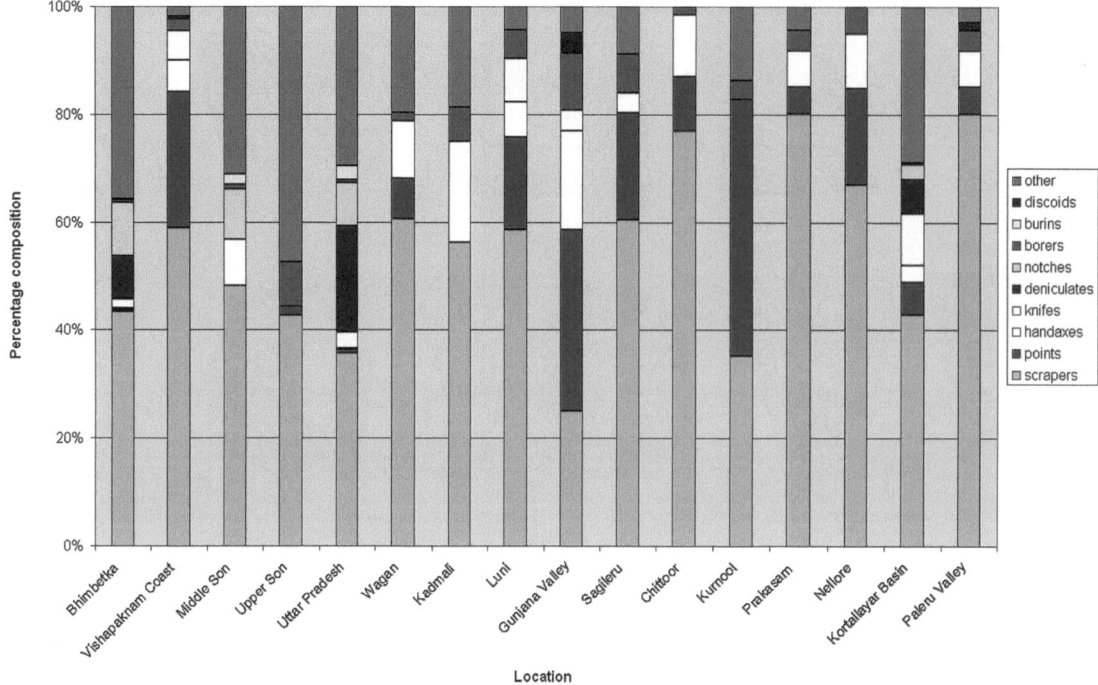

Figure 7. Variability in Middle Paleolithic assemblage composition

to test the hypothesis that this represents temporal variation (Paddayya, 1984). Borers and "knives" have been reported from a number of Middle Paleolithic sites, including Bhimbetka (Misra, 1985). In addition, denticulates, notches and, rarely, burins have been recovered from a small number of sites, including Parsidhia (Pant, 1982), Mangalpura (Misra et al., 1982) and Bhimbetka III F23 (Misra, 1985). A degree of temporal variation may be indicated by the distribution of some of the rarer Middle Paleolithic tool types. Polyhedrons seem to be exclusively found within earlier Middle Paleolithic assemblages such as Indola-ki-Dhani and Singi Talav, Didwana (Misra et al., 1982). A number of technological innovations also occur during the South Asian Middle Paleolithic, including the introduction of tanged points at Ramayogi Agraharam (Rath et al., 1997) and the Upper Son Valley (Ahmed, 1984) and the appearance of backed artifacts at sites such as Aryathur (Pappu, 2001), Chainpura (Pant, 1982) and Mahuli (Malik, 1959).

The Late Paleolithic of the Subcontinent

The Late Paleolithic is distinguished from the Middle Paleolithic by an increase in the production of blades and an increase in assemblage variability. This assemblage variability is evident both in the degree of reliance on blade technology, and in the particular forms of artifacts that are produced. With the exception of the geometric microliths, however, the standardization of retouched artifact forms is not comparable to those seen within the Aurignacian and later industries of Upper Paleolithic Europe.

As with the Middle Paleolithic, a variety of methods of lithic artifact production are found in the Late Paleolithic, with some assemblages maintaining an important flake component. Flakes are used in the manufacture of the scrapers, points and borers at Singhbum and Watru Abri (Murty, 1979; Paddayya, 1984), for example. The blades within these industries are relatively large in size, but at the site of Mehtakheri (Mishra, n.d.) scrapers on flakes are associated with small backed blades. Both micro-and macro-blades and cores are reported from Visadi (Allchin, 1973) while at Site 55, Pakistan, flake-blades are associated with blades small enough to be classed as microliths (Dennell et al., 1992) indicating that a variety of sizes of blades were produced at a number of Late Paleolithic sites. The microlithic element within the Late Paleolithic varies between sites. Sites such as Inamgoan, dated to ca. 25 ka to ca. 21 ka, combine geometric microliths with scrapers, blades and points (Murty, 1979). In contrast, the earlier Sri Lankan industries of Batadomba-Lena, Site 50 and Site 49 are completely based on the production of geometric microliths from small blades struck from small prismatic cores (Deraniyagala, 1992). At Patne, geometric microlithic technology appears to develop from an industry characterized by a few backed blades and burins (Sali, 1989). Intriguingly, Patne's microlithic industries date to ca. 24.5 ka, slightly post-dating their appearance within Sri Lanka. It is crucial to note that based on available evidence, there is no clear sequence of industrial sub-divisions within the Late Paleolithic of South Asia.

Patne's sequence, which contains assemblages that range from the Middle Paleolithic to the Mesolithic (Sali, 1989), is key in understanding the emergence of modern human behavior within South Asia. An initial comparison between cores within the Patne sequence suggests that there is no sudden shift to "classic" prismatic cores at the onset of the Late Paleolithic (James and Petraglia, 2005). Instead the most notable difference between cores from the early Late Paleolithic and those from the advanced Middle Paleolithic is in reduction intensity. Late Paleolithic cores exhibit evidence for a higher number of flake and blade removals. Prismatic blade cores are present within the Later Paleolithic assemblages from Patne, and increasingly dominate the assemblages as the period progresses. These prismatic cores are small in size and

seem to have been used in the production of blades, microlithic blades and bladelets. It is significant that these levels also include geometric microliths, such as lunates and triangles (Figure 8).

By the terminal Pleistocene, the makers of these Late Paleolithic industries were using a range of different materials to manufacture artifacts and were modifying their living spaces by constructing structures. Bone tools have been recovered from a number of Sri-Lankan microlithic contexts, of which the earliest is the assemblage at Batadomba-lena dated to ca. 28.5 ka (Deraniyagala, 1992) (Figure 9). At Site 55, Pakistan, a stone lined pit and low wall have been dated to 45 ka (Dennell et al., 1992) and structures have also been reported from Bhimbetka (Misra, 1989).

While claims have been made for evidence for symbolic thought within the South Asian Acheulean (e.g., Kumar, 1996; Bednarik, 2003), the most convincing evidence for explicitly symbolic artifacts dates to the Late Paleolithic. Ostrich egg shell beads have been recovered from the ca. 28.5 ka horizon at Batadomba-lena (Deraniyagala, 1992) and the ca. 24.5 ka strata at Patne (Sali, 1989). The Late Paleolithic bead-manufacturing site recently discovered in Madhya Pradesh (Mishra et al., 2004) indicates that such artifacts were produced within the subcontinent. The earliest evidence for art also comes from Patne, where the Late Paleolithic beads were associated with a geometrically incised fragment of ostrich eggshell (Sali, 1989). Another possible piece of mobile Late Paleolithic art is a carved and polished "goddess" figurine that was recovered from the Belan Valley of Uttar Pradesh (Misra, 1977), though this has recently been re-interpreted as a segment of a bone harpoon point (Bednarik, 2003). There may even be suggestions of a ritualistic element of Late Paleolithic life within South Asia. A sandstone platform and curiously patterned rock at Baghor I has been interpreted as the earliest

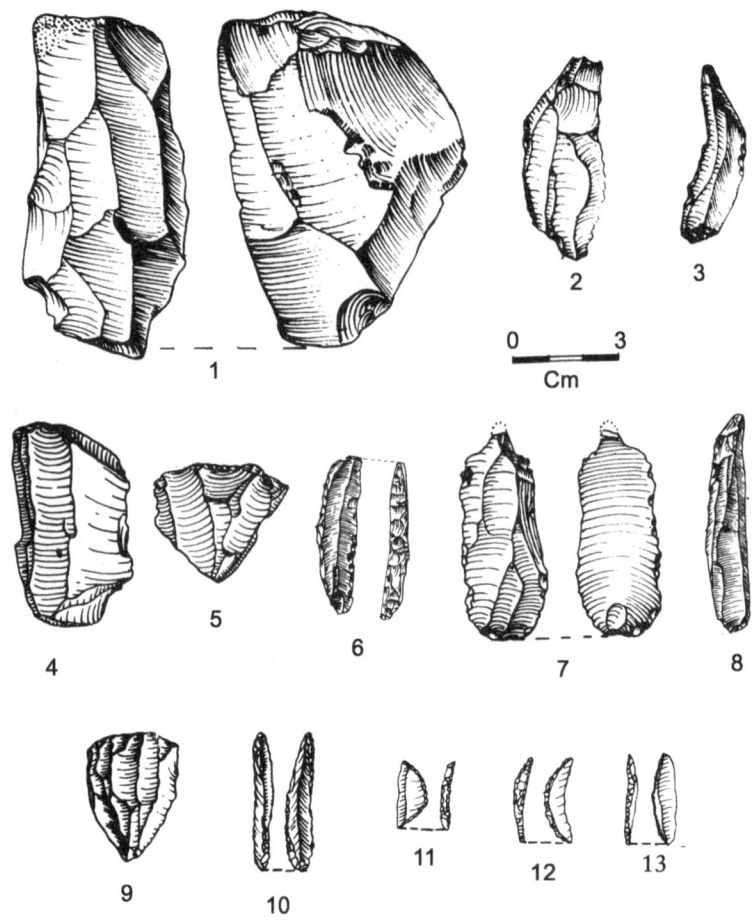

Figure 8. Technological transition of stone tool assemblages at Patne. Top: advanced Middle Paleolithic (phase 1); Middle: Early Upper Paleolithic (phase IIB); Bottom: Late Upper Palaeolithic (phase IID) (after Sali, 1989; figures 19, 21, 23). 1, 4, 5, 9, blade cores; 2, 3, 7, 10, retouched blades; 6, backed blades; 8, blade; 11–13, lunates

"mother goddess" shrine within the subcontinent (Kenoyer et al., 1983). More convincing evidence has been recovered from Bhimbetka, in the form a buried anatomically modern human wearing two ostrich egg shell beads near the neck (Bednarik, 2003).

The Emergence of Modern Human Behavior within South Asia

The earliest fossils of anatomically modern humans to be recovered from the Indian subcontinent are the human remains recovered from Fa-Hien Cave and Batadomba-Lena, Sri Lanka, dated to ca. 34 ka and ca. 28.5 ka respectively (Deraniyagala, 1992; Kennedy,

2000). Genetic data from both mitochondrial DNA and the Y-Chromosome suggests that *H. sapiens* was an intrusive species to the region, whose origins lay in Africa some 200 ka (e.g., Kivisild et al., 1999a, 1999b, 2003; Quintana-Murci et al., 1999; Ingman et al., 2000; Cann, 2001). The timing of the first dispersal of anatomically modern humans into South Asia is currently unclear, but mtDNA coalescence dates support the colonization of the Indian subcontinent sometime during the last 73–53 ka (Kivisild et al., 2000). This raises the possibility that modern humans may have been present within South Asia prior to the first fossil evidence of their arrival. An early colonization of the Indian subcontinent

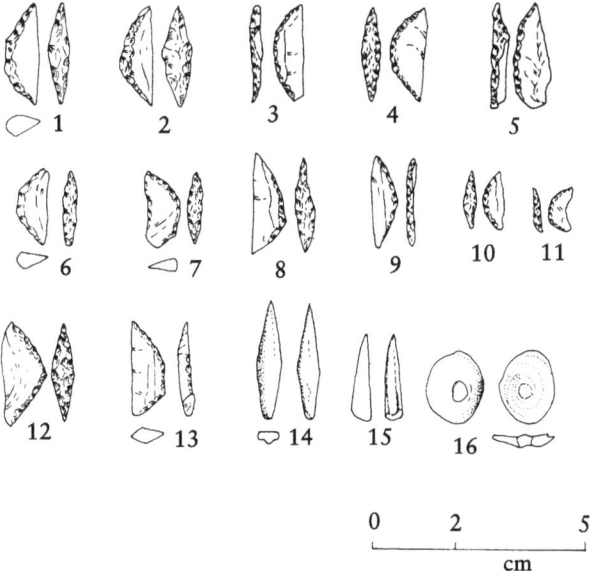

Figure 9. Microlithic artifacts from Sri Lanka (after Deraniyagala, 1992: figure 59). 1–13, geometric microliths; 14, 15 bone points; 16, bead

would fit with the current dates for the colonization of Australia (e.g., Bowler et al., 2003; O'Connell and Allen, 2004), and is thus highly plausible.

If anatomically modern humans did enter South Asia between or immediately after 73–55 ka, then it is probable that they carried with them a Middle Paleolithic toolkit (James and Petraglia, 2005). The archaeological record from the region supports such a conclusion, as the innovations of the Late Paleolithic all appear to have local origins. Blade technology, geographical and chronological variation in artifact types and form first appear within South Asia during the Middle Paleolithic. The Late Paleolithic, at least in its initial stages, merely represents an intensification of these behaviors, not the sudden introduction of them. Of course the intensification of a highly flexible tool kit does not necessarily indicate the appearance of fully modern behavior. But crucially, the first evidence for symbolic thought, and complex composite technology, is contained within a material culture package that developed from these initial Late Paleolithic assemblages.

There is a change in material culture after ca. 30 ka, and that change represents the full expression of the modern human behavior package. Complex lithic technology, including geometric microliths, and the creation of structures are combined with the first explicitly symbolic artifacts. Beads are manufactured and used, and the period produces the first evidence of artistic expression. Yet the change is not to the same degree as the Upper Paleolithic in Europe. There is no symbolic explosion; instead symbolic artifacts are spread sporadically in both time and space. Crucially, such symbolic artifacts are associated with a lithic technology that has its eventual roots somewhere within the South Asian Middle Paleolithic. At least some of the beads and microliths that signal the emergence of behavioral modernity within South Asia are local innovations (Figure 10).

The patterning of the emergence of modern human behavior within South Asia is different to that of Africa and Western Europe, despite the similarities the subcontinent shares with both regions. Like Europe, anatomically modern humans are an intrusive species. In terms of ecology and climate, the Indian subcontinent shows close similarities to Africa,

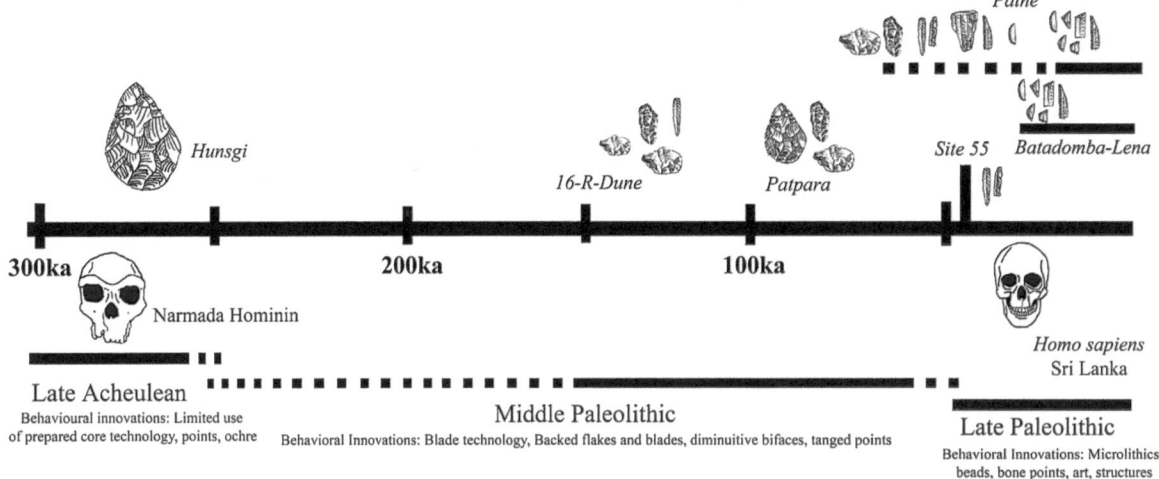

Figure 10. Chronology of behavioral innovations within the Indian subcontinent during the Later Pleistocene. Dashed lines represent possible relative dates, solid lines represent dates known through absolute dating techniques

though it is considerably smaller. Yet there is no sudden appearance of a modern behavioral package within South Asia that can be considered equivalent to the changes that signify the appearance of the Aurignacian within Europe. And while the sporadic and gradual appearance of explicitly symbolic artifacts and microlithic technology within the South Asian Late Paleolithic looks superficially similar to that within Middle Stone Age Africa, it occurs considerably later. The late, but gradual and sporadic appearance of modern human behavior within South Asia is a distinct phenomenon that may shed light on the processes behind the evolution of modern human behavior.

Conclusions

The discussion of the archaeological evidence from Africa, Western Europe and the Near East indicate that a number of questions remain to be answered regarding the evolution of modern human behavior. The origins of behavioral modernity were shown to be undoubtedly African. But the time depth of the functional and symbolic aspects of the modern behavioral package depended to some extent on

precisely how different the regional and temporal variation within assemblage composition and artifact form were from that seen in other regions of the world. Curiously, it was also apparent that the appearance of some of the best indicators of modern human behavior within the MSA was sporadic, with traits occurring in differing combinations and did not, at present, extend back to the origin of anatomically modern *H. sapiens* ca. 195–150 ka. The unique pattern of the emergence of modern human behavior within South Asia, may, if proved by further research, help answer some of these questions.

The use of blade technology and the spatial and chronological variation in assemblage composition within the South Asian Middle Paleolithic supports that seen within Africa and Western Europe. It is clear that the capability to produce a flexible technological package was shared between a number of hominin species. Interestingly, however, the South Asian Middle Paleolithic seems similar in its degree of variation to that seen within the MSA. Not only are there temporal and spatial variations in flake production and assemblage composition, there may also be differences within form within specific artifact types. This is not, however, a claim for stylistic variation within the South Asian Middle Paleolithic; instead the South Asian evidence suggests that more prosaic explanations for variation within the lithic record need to be dealt with first. Much of the variation within both the African and South Asian records is undoubtedly the result of different environments, functional needs and raw material. What remains may well be the result of cultural drift. This random accumulation of variation within the non-functional aspects of a particular type of artifact may lead to artifacts that look completely different to each other being produced by different populations. A difference in artifact form need not be deliberate or represent any portrayal of cultural identity. Crucially though, the effects of cultural drift on material culture vary with population size. In small populations, the effects of cultural drift are more pronounced than they are in larger ones (Shennan, 2000, 2001).

For any given society it is the relative proportions of cultural drift and "natural selection" that lead to the survival of a particular artifact within a material culture package. If an artifact is particularly useful in a given task that enables group survival then the group may be more likely to produce it. But in a small population, the effects of cultural drift may out way the effects of natural selection (Shennan, 2000, 2001). Even highly functional, "adaptive", artifacts may be lost from the archaeological record. The situation is only exasperated by the way in which material culture is bound by numerous social rules. Because part of the symbolism inherent in human culture is involved with the hierarchical partitioning of knowledge, including craft production, there are anthropologically noted cases where even "useful" artifacts have been lost from a societies cultural repertoire (e.g., Rivers, 1926). It is the effect of population demographics on cultural evolution that may provide the explanation

for the sporadic appearance of modern human behavioral traits in regions such as South Asia and Africa.

The fluctuating climates and environments within South Asia during the later Pleistocene must have impacted hominin populations. In periods of extreme aridity, marginal regions such as the Thar Desert, would have been increasingly difficult to survive. Population sizes are likely to have decreased, and population level extinctions may have occurred. In ameliorated climates, improved conditions may have lead to population expansion and dispersal. Instead of reflecting the accumulation of the necessary mental abilities, the sporadic appearance of explicitly symbolic artifacts instead perhaps reflects a mixture of societal pressures combined with the effects of population demographics on cultural evolution. These same micro-evolutionary processes of population expansion, dispersal, isolation and extinction, and their effect on genetic evolution, lead to the evolution of *H. sapiens* (Lahr and Foley, 1994, 1998) within sub-Saharan Africa. Arguably the variable and sporadic appearance of complex behavioral traits within the MSA is the archaeological reflection of demographic processes on cultural evolution. Under this model it is the particular demographic situation of Western Europe that leads to the Upper Paleolithic revolution. If the mental capabilities of anatomically modern humans are uncoupled from the archaeological record, then there can be a parsimonious explanation for why the archaeological record looks as it does.

The effects of cultural evolutionary processes, and their variable impact according to particular demographic situations, provide a parsimonious model for interpreting the evolution of modern human behavior. This is supported by the pattern of the emergence of modern human behavioral traits within South Asia. Given the initial stages of research into the appearance of behavioral modernity within South Asia, such conclusions can only be tentative. Further research is needed before the complex processes behind the evolution of modern human behavior can be understood. Hopefully, however, this initial study has indicated the potential implications of examining the archaeological record within South Asia for understanding the emergence of anatomically and behaviorally modern humans.

References

Ahmed, N. 1984. The Stone Age Cultures of the Upper Son Valley, Madhya Pradesh. Agam Kala Prakashan, Delhi.

Allchin, B. 1973. Blade and burin industries of West Pakistan and Western India. In: Hammond, N. (Ed.), South Asian Archaeology 1971. Duckworth, London, pp. 39–50.

Allchin, B., Goudie, A., Hegde, K. 1978. The Prehistory and Palaeogeography of the Great Indian Desert. Academic Press, London.

Andrews, J., Singhvi, A., Kuhn, R., Dennis, P., Tandon, S.K., Dhir, R. 1998. Do stable isotope data from calcrete record late Pleistocene monsoonal climate variation in the Thar Desert of India? Quaternary Research 50, 240–251.

Barham, L. 1998. Possible early pigment use in South-Central Africa. Current Anthropology 39, 703–710.

Barham, L. 2001. Central Africa and the emergence of regional identity in the Middle Pleistocene. In: Barham, L., Robson-Brown, K. (Eds.), Human Roots: Africa and Asia in the Middle Pleistocene. Western Academic and Specialist Press, Bristol, pp. 65–81.

Barham, L. 2002. Backed tools in Middle Pleistocene Central Africa and their evolutionary significance. Journal of Human Evolution 43, 585–603.

Bar-Yosef, O. 1998. The chronology of the Middle Palaeolithic of the Levant. In: Akasawa, T., Aoki, A., Bar-Yosef, O. (Eds.), Neanderthals and Modern Humans in Western Asia. Plenum Press, New York, pp. 39–57.

Bar-Yosef, O. 2002. The Upper Paleolithic revolution. Annual Review of Anthropology 31, 363–393.

Bar-Yosef, O., Kuhn, S. 1999. The big deal about blades: laminar technologies and human evolution. American Anthropology 101, 1–17.

Bar-Yosef, O., Vandermeersch, B. 1993. Modern humans in the Levant. Scientific American 4, 94–100.

Baskaran, M., Marathe, A. R., Rajaguru, S.N., Somayajulu, B.L.K. 1986. Geochronology of Palaeolithic cultures in the Hiran Valley, Saurashtra, India. Journal of Archaeological Science 13, 505–514.

Bednarik, R. 1990. An Acheulian haematite pebble with striations. Rock Art Research 7, 75.

Bednarik, R. 2003. The earliest evidence of palaeoart. Rock Art Research 20(2), 89–135.

Blumenschine, R., Brandt, S., Clark, J.D. 1983. Excavations and analysis of Middle Palaeolithic artifacts from Patpara, Madhya Pradesh. In: Sharma, G., Clark, J.D. (Eds.), Palaeoenvironments and Prehistory in the Middle Son Valley. Abinash Prakashan, Allahabad, pp. 39–114.

Brown, P., Sutikna, T., Morwood, M.J., Soejono, R.P., Jatmiko, Wayhu Saptama, E., Rokus Awe Due 2004. A new small bodied hominin from the Late Pleistocene of Flores, Indonesia. Nature 431, 1055–1061.

Bowler, J.M., Johstone, H., Olley, J., Prescott, J., Roberts, R., Shawcross, W., Spooner, N. 2003. New ages for human occupation and climate change at Lake Mungo, Australia. Nature 421, 837–840.

Cann, R.L. 2001. Genetic clues to dispersal in human populations: retracing the past from the present. Science 291, 1742–1748.

Carameli, D., Lalueza-Fox, C., Vernes, C., Lari, M., Casoli, A., Mallegni, F., Chiarelli, B., Dupanloup, I., Betranpetit, J., Barbujani, G., and Bertonelle, G. 2003. Evidence for a genetic discontinuity between Neanderthals and 24,000-year-old anatomically modern Europeans. Proceedings of the National Academy of Sciences, USA, 100, 6593–6597.

Clark, J.D. 1993. African and Asian perspectives on the origins of modern humans. In: Aitken, M.J., Stringer, C.B., Mellars, P.A. (Eds.), The Origin of Modern Humans and the Impact of Chronometric Dating. Princeton University Press, Princeton, pp. 148–178.

Clark, J.D., Brown, K.S. 2001. The Twin Rivers, Kopje, Zambia: stratigraphy, fauna and artifact assemblages from the 1954 and 1956 excavations. Journal of Archaeological Science 28, 305–330.

Close, A. 1978. The identification of style in lithic artefacts. World Archaeology 10(2), 223–237.

Conard, N.J. 1990. Laminar lithic assemblages from the Last Interglacial complex in Northwestern Europe. Journal of Anthropological Research 46, 243–262.

Deacon, H., Wurz, S. 2001. Middle Pleistocene populations of Southern Africa and the emergence of modern behavior. In: Barham, L., Robson Brown, K. (Eds.). Human Roots: Africa and Asia in the Middle Pleistocene. Western Academic Press, Bristol, pp. 55–65.

Dennell, R., Rendell, H., Halim, M., Moth, E. 1992. A 45,000-year-old open-air Paleolithic site at Riwat, Northern Pakistan. Journal of Field Archaeology 19, 17–33.

Deotare, B., Kajale, B. M., Rajaguru, S. and Basavaiah, N. 2004. Late Quaternary geomorphology, palynology and magnetic susceptibility of playas in western margin of the Indian Thar Desert. Indian Geophysical Union 8(1), 15–25.

Deraniyagala, S.U. 1992. The Prehistory of Sri Lanka: An Ecological Perspective. Department of the Archaeological Survey, Government of Sri Lanka, Colombo.

D'Errico, F. 2003. The invisible frontier. A multiple species model for the origin of behavioral modernity. Evolutionary Anthropology 12, 188–202.

D'Ericco, F., Zilhao, J., Julien, M., Baffier, D., Pelegrin J. 1998. Neanderthal acculturation in Western Europe? A critical review of the evidence and its interpretation. Current Anthropology 39, S1–S37.

Dibble, H.L., Rowland, N. 1992. On assemblage variability in the Middle Paleolithic of Western Europe. History, perspectives, and a new synthesis. In: Dibble, H.L., Mellars, P. (Eds.), The Middle Paleolithic: Adaptation, Behavior and Variability. University of Pennsylvania, Philadelphia.

Dobres, A.-M. 1995. Gender and prehistoric technology: On the social agency of technical strategies. World Archaeology 21(1), 25–49.

Gamble, C. 1986. The Palaeolithic Settlement of Europe. Cambridge University Press, Cambridge.

Gargett, R.H. 1999. Middle Palaeolithic burial is not a dead issue: the view from Qafzeh, Saint-Césaire, Kebara, Amud and Dederiyeh. Journal of Human Evolution 37, 27–90.

Henshilwood, C., Marean, C. 2003. The origin of modern human behavior: critique of the models and their test implications. Current Anthropology 44, 627–651.

Henshilwood, C.S., D'Errico, F., Vanhaeren, M., Van Niekerk, K., Jacobs, Z. 2004. Middle Stone Age shell beads from South Africa. Science 304, 404.

Henshilwood, C.S., D'Errico, F., Yates, R., Jacobs, Z., Tribolo, C., Duller, G., Mercier, N., Sealy, J., Valladas, H., Watts, I., Wintle. A. 2002. Emergence of modern human behavior: Middle Stone Age engravings from South Africa. Science 295, 1278–1280.

Henshilwood, C., Sealey, J. C. 1997. Bone artefacts from the Middle Stone Age at Blombos Cave, Southern Cape, South Africa. Current Anthropology 38, 890–895.

Hodder, I. 1982. Symbols in Action: Ethnoarchaeological Studies of Material Culture. Cambridge University Press, Cambridge.

Hovers, E., Ilani, S., Bar-Yosef, O., Vandermeersch, B. 2003. An early case of color symbolism: ochre use by modern humans in Qafzeh Cave. Current Anthropology 44, 491–522.

Ingman, M, Kaessmann, H., Paabo, S., Gyllensten, D. 2000. Mitochondrial genome variation and the origin of modern humans. Nature 408, 708–713.

James, H.V.A. 2003. Testing theories of modern human origins: the Middle Palaeolithic of South Asia. MPhil Thesis, University of Cambridge.

James, H.V.A., Petraglia, M.D. 2005. Modern human origins and the evolution of behavior in the later Pleistocene record of South Asia. Current Anthropology 46, S3–S27.

Joshi, R.V., Chitale, S.V., Rajaguru, S.N., Pappu, R.S., Badam, G.L. 1979–80. Archaeological studies in the Manjra Valley, Central Godavari Basin. Bulletin of the Deccan College Research Institute 40, 67–94.

Kar, A., Singhvi, S., Rajaguru, S., Juyal, N., Thomas, J., Banerjee, D., Dhir, R. 2001. Reconstruction of the late Quaternary environment of the lower Luni plains, Thar Desert, India. Journal of Quaternary Science 16(1), 61–68.

Kenoyer, J., Clark, J., Pal, J., Sharma, G. 1983. An Upper Palaeolithic shrine in India? Antiquity LVII, 88–94.

Kennedy, K. 2000. God Apes and Fossil Men: Paleoanthropology in South Asia. University of Michigan Press, Ann Arbor.

Kivisild, T., Papiha, S., Rootsi, S., Parik, J., Kaldma, K., Reidla, M., Laos, S., Metspalu, M., Pielberg, G., Adojaan, M., Metspalu, E., Mastana, S., Wang, Y., Gölge, M., Demirtas, H., Schnakenberg, E., De Stefano, G., Geberhiwot, T., Claustres, M., Villems, R. 2000. An Indian ancestry: a key for understanding human diversity in Europe and beyond.

In: Renfrew, C., Boyle, K. (Eds.), Archaeogenetics: DNA and the Population Prehistory of Europe. McDonald Institute for Archaeological Research, Cambridge, pp. 267–275.

Kivisild, T., Rootsi, S., Metspalu, M., Mastana, S., Kaldma, K., Parik, J., Mestspalu, E., Adojaan, M., Tolk, H., Stepanov, V., Gölgi, M., Usanga, E., Papiha, S., Cinnioğlu, C., King, R., Cavalli-Sforza, L., Underhill, P. 2003. The genetic heritage of the earliest settlers persists both in Indian tribal and caste populations. American Journal of Human Genetics 72, 313–332.

Kivsild, T., Bamshad, M., Kaldma, K., Metspalu, M., Metspalu, E., Reidla, M., Laos, S., Parik J., Watkins, W., Dixon, M., Papiha, S., Mastana, S., Mir, M., Ferak, V., Villems, R. 1999a. Deep common ancestry of Indian and western-Eurasian mitochondrial DNA lineages. Current Biology 9, 1331–1334.

Kivisild, T., Kaldma, K., Metspalu, M., Parik, J., Papiha, S., Villems, R. 1999b. The place of the Indian mtDNA variants in the global network of maternal lineages and the peopling of the Old World. In: Deka, R., Papiha, S. (Eds.), Genomic Diversity. Plenum Publishers, New York, pp. 135–152.

Klein, R. 1995. Anatomy, behavior and modern human origins. Journal of World Prehistory 9, 167–198.

Klein, R. 1999. The Human Career. University of Chicago Press, Chicago.

Klein, R. 2000. Archeology and the evolution of human behavior. Evolutionary Anthropology 9, 17–36.

Korisettar, R. 2004. Geoarchaeology of the Purana and Gondwana Basins of peninsular India: peripheral or paramount. Presidential Address, sixty-fifth session of the Indian History Congress, 28–30, December.

Korisettar, R., 2007. Toward developing a basin model for Paleolithic settlement of the Indian subcontinent. In: Petraglia, M.D., Allchin, B. (Eds.), The Evolution and History of Human Populations in South Asia: Inter-disciplinary Studies in Archaeology, Biological Anthropology, Linguistics and Genetics. Springer, Netherlands, pp. 69–96.

Kuman, K., Inbar, M., Clarke, R. J. 1999. Palaeoenvironments and cultural sequence of the Florisbad Middle Stone Age hominid site, South Africa. Journal of Archaeological Science 26, 1409–1426.

Kumar, G. 1996. Daraki-Chattan: a Palaeolithic cupule site in India. Rock Art Research 13, 38–46.

Lahr, M.M., Foley, R. 1994. Multiple dispersals and modern human origins. Evolutionary Anthropology 3, 48–60.

Lahr, M.M., Foley, R. 1998. Towards a theory of modern human origins: geography, demography and diversity in recent human evolution. Yearbook of Physical Anthropology 4, 137–176.

Le Monnier, P. 1991. Bark capes, arrowheads and concorde: on social representations of technology. In: Hodder, I. (Ed.), The Meaning of Things: Material Culture and Symbolic Expression. Harpercollins Academic, London, pp. 156–170.

Lombard, M. 2005. Evidence of hunting and hafting during the Middle Stone Age at Sibidu Cave, KwaZulu-Natal, South Africa: a multianalytical approach. Journal of Human Evolution 48, 279–300.

Malik, S.C. 1959. Stone Age Industries of the Bombay and Satara Districts. Faculty of Arts, Maharaja Sayajirao University of Baroda, Baroda.

Malinowski, B. 1922. Argonauts of the Western Pacific. Routledge and Kegan Paul, London.

Marathe, A.R. 1981. Geoarchaeology of the Hiran Valley, Saurashtra, India. Deccan College Postgraduate and Research Institute, Poona.

Mauss, M. 1970 The Gift, trans I. Cunnison. Cohen and West, London.

McBrearty, S., Brooks, A. 2000. The revolution that wasn't: a new interpretation of the origin of modern human behavior. Journal of Human Evolution 39, 453–563.

McDougall, I., Brown, F., Fleagle, J. 2005. Stratigraphic placement and age of modern humans from Kibish, Ethiopia. Nature 433, 733–736.

Mellars, P. 1989. Major Issues in the emergence of modern humans. Current Anthropology 30, 349–385.

Mellars, P. 1991.Cognitive changes and the emergence of modern humans in Europe. Cambridge Archaeological Journal 1, 63–76.

Mellars, P. 1996. The Neanderthal Legacy: An Archaeological Perspective from Western Europe. Princeton University Press, Princeton.

Mellars, P. 2002. Archaeology and the origins of modern humans: European and African perspectives. In: T. Crow (Ed.), The Speciation of Modern *Homo sapiens*. Proceedings of the British Academy 106, Oxford University Press, Oxford, pp. 31–47.

Mellars, P. 2005. The impossible coincidence: a single-species model for the origins of modern human behavior in Europe. Evolutionary Anthropology 14(1), 12–27.

Mercier, N., Valladas, H., Valladas, G., Reyss, L., Jelinek, A., Meignen, L., Joron, J.L. 1995. TL dates of burnt flints from Jelinek's excavations at

Tabun and their implications. Journal of Archaeological Science 22(4), 495–510.

Milo, R. 1998. Evidence for hominid predation at Klasies River Mouth, South Africa, and its implications for the behavior of early modern humans. Journal of Archaeological Science 25, 99–133.

Mishra, S. n.d.. Mehtakheri Excavations 1992. Unpublished report, Deccan College, Pune.

Mishra, S. 1995. Chronology of the Indian Stone Age: the impact of recent absolute and relative dating attempts. Man and Environment XX (2), 11–17.

Mishra, S., Jain, M., Tandon, S.K., Singhvi, A.K., Joglekar, P.P., Bhatt, S. C., Kshirsagar, S., Maik, S., Deshpande-Muhkerje, A. 1999. Prehistoric cultures and Late Quaternary environments in the Luni Basin around Balotra. Man and Environment XXIV (1), 39–49.

Mishra, S., Ota, S.B., Naik, S. 2004. Late Pleistocene ostrich egg shell bead manufacture at Khaparkhera, District Dhar, Madhaya Pradesh. In: Abstracts of Academic Symposia, Rock Art Society of India, International Rock Art Congress, Agra- 28 November to 02 December 2004, pp. 22.

Misra, V.D. 1977. Some Aspects of Indian Archaeology. Prabhat Prakashan, Allahabad.

Misra, V.N. 1967. Pre-and Proto- History of the Berach Basin South Rajasthan. University of Poona, Poona.

Misra, V.N. 1968. Middle Stone Age in Rajasthan. In: Bordes, F., de Sonneville Bordes, D. (Ed.), La Préhistoire, Problèmes et Tendancies. Centre National de la Recherché Scientifique, Paris, pp. 295–305.

Misra, V.N. 1985. The Acheulean succession at Bhimbetka, Central India. In: Misra, V. N., Bellwood, P. (Eds.), Recent Advances in Indo-Pacific Prehistory. Oxford I-B-H, New Delhi, pp. 35–47.

Misra, V.N. 1989. Stone Age India: an ecological perspective. Man and Environment XIV (1), 17–64.

Misra, V.N. 1995a. Geoarchaeology of the Thar Desert, North West India. In: Wadia, S., Korisettar, R., Kale, V. S. (Eds.), Quaternary Environments and Geoarchaeology of India. Memoir 32. Geological Society of India, Bangalore, pp. 210–230.

Misra, V.N. 1995b. The evolution of environment and culture in the Rajasthan desert during the Late Quaternary. In: Johnson, E. (Ed.), Ancient Peoples and Landscapes. Texas Tech University Press, Texas, pp. 77–103.

Misra, V.N. 2001. Prehistoric human colonization of India. Journal of Bioscience 26 (4), 491–531.

Misra, V.N., Rajaguru, S.N., Ganjoo, R.K., Korisettar, R. 1990. Geoarchaeology of the Palaeolithic site at Samnapur in the Central Narmada Valley. Man and Environment XV (1), 108–116.

Misra, V.N., Rajaguru, S.N., Raju, D.R., Ragnavan, H., Gaillard, C. 1982. Acheulian occupation and evolving landscape around Didwana in the Thar Desert, India. Man and Environment VI, 72–86.

Mitchell, P. 2002. The Archaeology of Southern Africa. Cambridge University Press, Cambridge.

Murty, M.L.K. 1979. Recent research on the Upper Palaeolithic Phase in India. Journal of Field Archaeology 6 (3), 301–320.

Morwood, M.J., Soejono, R.P., Roberts, R.G., Sutikna, T., Turney, C.S.M., Westaway, K.E., Rink, W.J., Zhao, J-X., Van den Bergh, G.D., Rokus Awe Due, Hobbs, D.R., Moore, M.W., Bird, M.I., Fifield, L.K. 2004. Archaeology and age of a new hominin from Flores in eastern Indonesia. Nature 431, 1087–1091.

Neiman, F.D. 1995. Stylistic variation in evolutionary perspective: inferences from decorative diversity and interassemblage distance in Illinois Woodland ceramic assemblages. American Antiquity 60, 7–36.

Okely, J. 1983. The Traveller Gypsies. Cambridge University Press, Cambridge.

O'Connell, J.F., Allen, J. 2004. Dating the colonization of Sahul (Pleistocene Australia-New Guinea): a review of recent research. Journal of Archaeological Science 31, 835–853.

Odess, D. 1998. The Archaeology of interaction: views from artifact style and material exchange in Dorset Society. American Antiquity 63, 417–435.

Ovchinnikov, I., Gotherstrom, A., Romanova, G., Kharitonov, V., Lidden, K., Goodwin, W. 2000. Molecular analysis of Neanderthal DNA from the northern Caucasus. Nature 404, 490–493.

Paddayya, K. 1982-83. Stone Age sites near Bhagi Mohari, Nagpur District, Maharashtra. Bulletin of the Deccan College Research Institute 43, 91–98.

Paddayya, K. 1982. Acheulian Culture of Hunsgi Valley. Deccan College, Pune.

Paddayya, K., 1984. Stone Age India. In: Mueller-Karpe, H. (Ed.), Neue Forschungen zur Altsteinzeit. C.H. Beck Verlag, Munich, pp. 345–403.

Pant, P.C. 1982. Prehistoric Uttar Pradesh (A Study of Old Stone Age). Agam Kala Prakashan, Delhi.

Pappu, R.S., Deo, S.G. 1994. Man-Land Relationships During Palaeolithic Times in the Kaladgi Basin, Karnataka. Deccan College, Pune.

Pappu, S. 2001. A Re-examination of the Palaeolithic Archaeological Record of Northern Tamil Nadu, South India: BAR International Series, Oxford, 1003.

Pappu, S., Gunnell, Y., Taieb, M., Brugal, J.-P., Touchard, Y. 2003. Excavations at the Palaeolithic site of Attirampakkam, South India: preliminary findings. Current Anthropology 44, 591–598.

Petraglia, M.D., Schuldenrein, J., Korisettar, R. 2003. Landscapes, activity, and the Acheulean to Middle Paleolithic transition in the Kaladgi Basin, India. Eurasian Prehistory 1 (2), 3–24.

Quintana-Murci, L., Semino, O., Bandelt, H., Passarino, G., McElreavey, K., Santachiara-Benerecetti, S.A. 1999. Genetic evidence of an early exit of Homo sapiens sapiens from Africa through eastern Africa. Nature Genetics 23, 437–441.

Raju, D.R. 1988. Stone Age Hunter-gatherers: An Ethnoarchaeology of Cuddapah Region, Southeast India. Ravish Publishers, Pune.

Rath, K., Thimma Reddy, K., Vijaya Prakash, P. 1997. Middle Paleolithic assemblage from Ramayogi Agraharam in the red sediments in the Visakhapatnam Coast. Man and Environment XXII (1), 31–38.

Rightmire, G.P. 1989. The Middle Stone Age humans from eastern and southern Africa. In: Mellars, P., Stringer, C. (Eds.) The Human Revolution: Behavioural and Biological Perspectives on the Origins of Modern Humans. Edinburgh University Press, Edinburgh.

Rivers, W.H.R. 1926. Psychology and Ethnology. Kegan, Paul, Trench, Trubner, London.

Sackett, J.R. 1977. The meaning of style in archaeology: a general model. American Antiquity 42, 369–380.

Sackett, J.R. 1982. Approaches to style in lithic archaeology. Journal of Anthropological Archaeology 1, 59–112.

Sackett, J.R. 1986. Style, function and assemblage variability: a reply to Binford. American Antiquity 51, 628–634.

Sali, S.A. 1989. The Upper Palaeolithic and Mesolithic Cultures of Maharashtra. Deccan College Post-Graduate and Research Institute, Pune.

Schiffer, M.B., Skibo, J.M. 1997. The explanation of artifact variability. American Antiquity 62, 27–50.

Shea, J.J. 2003. Neanderthals, competition and the origin of modern human behavior in the Levant. Evolutionary Anthropology 12, 173–187.

Shennan, S. 2000. Population, culture history, and change. Current Anthroology 41, 811–835

Shennan, S. 2001. Demography and cultural innovation: a model and its implications for the emergence of modern human cultures. Cambridge Archaeological Journal 11 (1), 5–16.

Shennan, S. 2002. Genes, Memes and Human History: Darwinian Archaeology and Cultural Evolution. Thames and Hudson, London.

Sinclair, A. 1995. The technique as symbol in Late Glacial Europe. World Archaeology 27 (1), 50–62.

Solecki, R.S. 1975. Shanidar IV, a Neanderthal flower burial in northern Iraq. Science 190, 880–881.

Solecki, R.S. 1989. On the evidence for Neanderthal burial. Current Anthropology 30, 324.

Sonawane, V.H. 1984. Prehistoric cultures of the Panchmahals, Gujarat. Man and Environment VIII, 20–30.

Stiner, M.C., Kuhn, S.L. 1992. Subsistence, technology, and adaptive variation in Middle Paleolithic Italy. American Anthropologist 94, 306–339.

Stringer, C. 2001. The morphological and behavioral origins of modern humans. In: Crow, T. (Ed.), The Speciation of Modern Homo sapiens. Proceedings of the British Academy, 106, Oxford University Press, Oxford, pp. 23–30.

Vandermeersch, B. 1981. Les Hommes Fossiles de Qafzeh (Israel). CNRS, Paris.

Van Peer, P. 1991. Interassemblage variability and Levallois styles: the case of the Northern African Middle Palaeolithic. Journal of Anthropological Archaeology 10, 107–151.

Vijaya Prakash, P., Rath, A., Krishna Rao, S. 1995. Prehistoric cultural evidence from Borra limestone Caves, Eastern Ghats. Man in India 75 (2), 163–173.

Walter, R.C., Buffler, R.T., Henrich Bruggemann, J., Guillaume, M., Berhe, S., Negassi, B., Libsekal, Y., Cheng, H., Lawrence Edwards, R., Von Cosel, R., Neraudeau, D., Gagnon, M. 2000. Early human occupation of the Red Sea coast of Eritrea during the last interglacial. Nature 405, 65–69.

White, R. 1997. Substantial acts: from materials to meaning in Upper Palaeolithic representation. In: Conkey, M, Soffer, O., Jablonski, N. (Eds.), Beyond Art: Pleistocene Image and Symbol. Memoirs of the California Academy of Sciences 23, 93–121.

White, T., Asfaw, B., Degusta, D., Gilbert, H., Richards, G., Suwa, G., F. Clark Howell. 2003. Pleistocene *Homo sapiens* from Middle Awash, Ethiopia. Nature 423, 742–747

Williams, M.A.J. and Clarke, M.F. 1995. Quaternary geology and prehistoric environments in the Son and Belan Valleys, North Central India. In: Wadia, S., Korisettar, R.. Kale, V.S. (Eds.), Quaternary Environments and Geoarchaeology of India. Memoir 32. Geological Societ of India, Bangalore, pp. 282–309.

Wiessner, P. 1991. Style and changing relations between the individual and society. In: Hodder, I. (Ed.), The Meaning of Things: Material Culture and Symbolic Expression. Routledge, London, pp. 56–62.

Wiessner, P. 1983. Style and social information in Kalahari San projectile points. American Antiquity 48, 253–76.

Wood, B., Strait, D. 2004. Patterns of resource use in early Homo and Paranthropus. Journal of Human Evolution 46, 119–162.

Wurz, S. 2002. Variability in the Middle Stone Age lithic sequence, 115,000-60,000 years ago at Klasies River, South Africa. Journal of Archaeological Science 29, 1001–1015.

Yellen, J.E., Brooks, A.S., Cornelissen, E., Mehlman, M.H., Stewart, K. 1995. A Middle Stone Age worked bone industry from Katanda, Upper Semliki Valley, Zaire. Science 268, 553–556.

10. Genetic evidence on modern human dispersals in South Asia: Y chromosome and mitochondrial DNA perspectives: *The world through the eyes of two haploid genomes*

PHILLIP ENDICOTT

Henry Wellcome Ancient Biomolecules Centre
Department of Zoology
University of Oxford
Oxford, OX1 4AU
England
phillip.endicott@zoology.oxford.ac.uk

MAIT METSPALU

Department of Evolutionary Biology
Estonian Biocenter
University of Tartu
Riia 23b
51010, Tartu, Estonia
mait@ebc.ee

TOOMAS KIVISILD

Leverhulme Centre for Human Evolutionary Studies
Henry Wellcome Building
Fitzwilliam Street
University of Cambridge
Cambridge CB2 1QH
England
t.kivisild@human-evol.cam.ac.uk

Introduction

Evidence of ancient human dispersals and settlement in South Asia is preserved in the genomes of its inhabitants, in the form of randomly accumulating mutations, which are passed down through the generations. Those mutations that occur on the uni-parental inherited Y chromosome and mitochondrial DNA permit the tracing of both paternal and maternal genealogies, respectively. Provided that there has been limited long range

M.D. Petraglia and B. Allchin (eds.), The Evolution and History of Human Populations
in South Asia, 229–244.

movement of people since the initial settlement of distinct geographic regions it is possible to differentiate, genetically, between the populations of these regions by the arrays of mutations arising locally. Besides the ongoing differentiation driven by mutation there is a process known as random genetic drift, whereby population expansions and contractions will tend to catalyze the random sorting of allele frequencies in populations, inducing and exaggerating further differences between them. Individuals sharing identical combinations of mutations at any single locus are grouped under the same haplotype (from the Greek Haplo for single). In bi-parentally inherited autosomal genes haplotypes have to be inferred from diplotypes i.e. information drawn from pairs of chromosomes that due to recombination display a more complicated mode of inheritance. When haplotypes are grouped together (into haplogroups) by their likeliest descent order they form a tree because successive generations of mutations can be traced back to root types, which due to random chance have been preserved in the current population. Using a high enough number of informative molecular markers it is possible to find a geographic separation between different branches of trees constructed in this way. This 'phylogeographic' approach (Avise, 2000) can in principle provide clues regarding both the peopling of a region of interest and about subsequent population movements into it, as well as bearing witness to demographic processes of the past, such as bottlenecks and expansions.

By examining the genomes of individuals representing different positions on the tree, inferences can be made about the movements of women and men in both space and time. The resolution obtained is dependent upon the amount of informative markers available and here the two loci under consideration vary enormously in their potential. The human mitochondrial genome is short (16,569 bases long) and can be sequenced for its entire length with contemporary technology (Anderson et al., 1981; Andrews et al., 1999). This allows trees of predetermined possible resolution to be drawn for the locus. Given the possibility of mutations arising in parallel and the relatively short length of the molecule, certain phylogenetic uncertainties may persist that can never be solved. Though the human Y chromosome is the smallest of its kind, the scale of variation within its sequence, and the different forms present within its length, theoretically allows one to distinguish between the Y chromosome of a father and his son. The difference is put in perspective when considering that the current state of the art Y phylogeny is largely based on around 50,000 nucleotides (approximately 0.1% of the chromosome length) sequenced in parallel in fifty three individuals (Underhill et al., 2000). This tree is continuously improving in its power of resolution as more parts of the Y chromosome are being sequenced (YCC, 2002; Jobling and Tyler-Smith, 2003).

As a 'daughter' population occupies a new territory local region-specific variations arise on the background of those gene alleles that were randomly taken by the migrating unit. More technically speaking, haplogroups that have reached different world regions randomly accumulate additional mutations, which identify new region specific sub-clades. These sub-clades are intrinsically concentrated in the specific core regions, which appear thus quite distinct from each other in their phylogenetic composition. Unless the geographical space is interrupted by strong barriers, there is often significant borderline overlap of haplogroup distributions between such regions. For example, lineages typical of both African and European populations exist in populations of North Africa and the Middle East (Rando et al., 1998; Richards et al., 2000; Bosch et al., 2001; Arredi et al., 2004; Cruciani et al., 2004). Depending on the level of subsequent admixture between neighboring populations such regional haplogroups may present themselves across space by clinal frequency gradients (one almost panmictic population unstructured in space), or in the form of highly distinguished regional patterns, with each population having specific characteristics. Applying a phylogeographic approach, it is then possible, with certain limitations, to reconstruct common ancestors between populations to reveal information about their origin and spread. In the case of high levels of gene flow between populations these phylogeographic inferences are more complicated, as is often the case with genetic variation within a single continental region. For example; mtDNA lineages at the basic haplogroup level of resolution do not present a clear geographic structuring within Europe compared to the intercontinental scale of variation of mtDNA.

In the case of population replacement, the study of mtDNA or Y chromosomal lineages in extant populations is incapable of revealing information about a population history before such an event. In principle, it is only possible then to estimate, from the level of variation within the extant population, the approximate time back to the replacement, and, in the case of geographic structure between the populations, to infer the likeliest source region of the intruders. Theoretically, the study of ancient DNA would facilitate the study of population histories diachronically but as far as South Asia is concerned the conditions of preservation (with the exception of the mountain regions) indicate there is little prospect for recovering evidence from prehistoric material (Kumar et al., 2000).

Evidence for the Out of Africa Hypothesis

The global phylogenetic pattern of extant mtDNA and Y chromosome lineages, as revealed by high resolution genotyping, is made up of haplogroups whose spread is highly continent or region specific (Figure 1). The measure of molecular resolution here is the number of informative polymorphic positions that are used as the basis for tree construction. It is the case with both haploid loci that African populations display the highest degree of variation in terms of the number of polymorphisms witnessed on average between individuals, and that the deepest splitting branches within both the mtDNA (Figure 1A) and Y chromosome trees (Figure 1B) are found exclusively in African populations. In contrast to the diversity we observe within Africa, all non-African mtDNA and Y chromosome lineages trace their ancestry back, in each case, to just three founder lineages: M, N, and R for mtDNA, and C, D, and F for the Y chromosome (Underhill et al., 2000; Ke et al., 2001; Endicott et al., 2003a, b; Kivisild et al., 2003; Oppenheimer, 2003; Forster, 2004; Kivisild et al., 2005). Viewed from the general perspective of these two uni-parental loci this suggests that all extant modern humans originate from Africa.

Establishing the Southern Route

The continental distributions of non-African mtDNA and Y chromosome haplogroups are not uniform. All three mtDNA founder haplogroups (M, N and R) can be found in present day populations from South Asia to Australia, while in West Eurasian populations only haplogroups derived from the N and R founders are present. Similarly, all West Eurasian Y chromosome variability can be seen as being derived from a single haplogroup, F. Within Asia the picture for the Y chromosome mirrors that of mtDNA quite closely with all three founder haplogroups being present. Y chromosome haplogroup F has a broad spread from Europe to Australia while founder group C is restricted, like mtDNA haplogroup M to the area East of and including South Asia. The third Y chromosome haplogroup, D, today is found at a noticeable frequency only in Tibet, Japan,

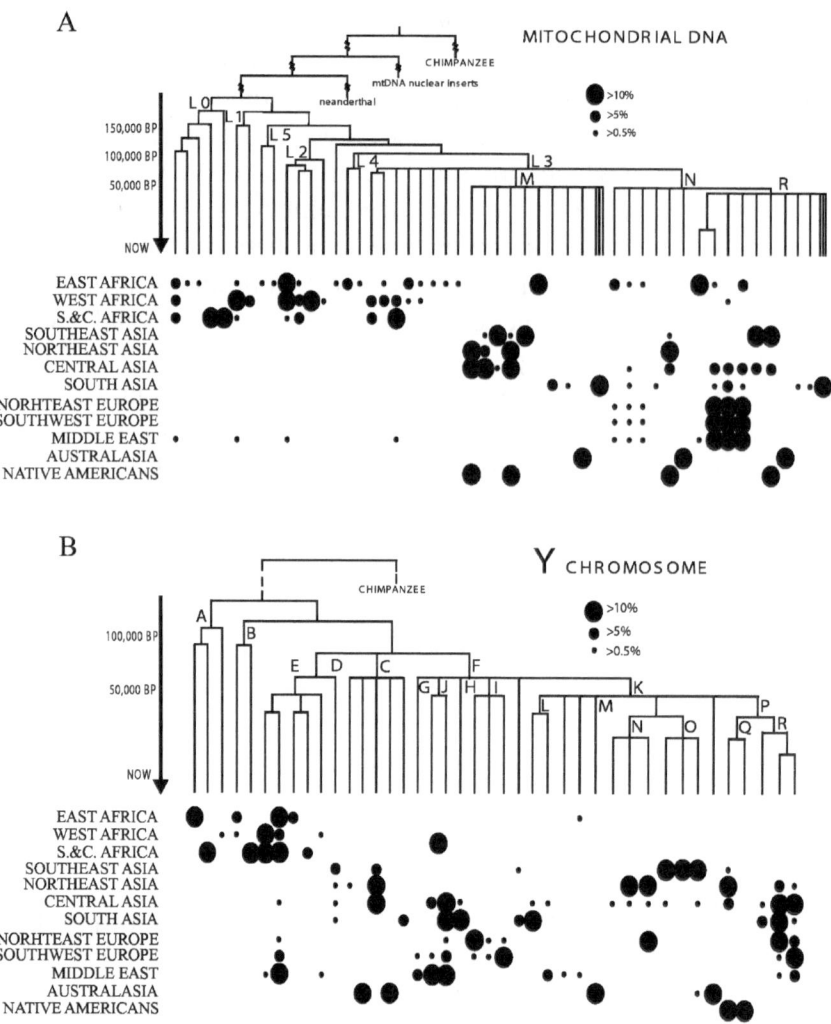

Figure 1. Phylogenetic trees of human mtDNA (1A) and Y chromosome (1B) lineages, showing their regional specificity. Circle sizes reflect the frequencies of the respective types in each region

Southwest China, and the Andaman Islands (Su et al., 1999; Su et al., 2000; Underhill et al., 2000; Thangaraj et al., 2003) and does not have any obvious parallel in the mtDNA phylogeography. Another exception in the Y chromosome phylogeny to this general pattern is haplogroup E3b, which can be considered as a relatively young actor on the scene of West Eurasian genetic diversity, appearing likely from East Africa only after the Last Glacial Maximum (Cruciani et al., 2004; Semino et al., 2004). Its absence in India suggests that the E3b transfer from East Africa to West Asia had to occur later than the transfer of those mtDNA (e.g., H, J, T, U, W, I) and Y

chromosome (e.g., J2) haplogroups in India considered to have a recent origin in West Asia. The distribution of such supposedly "recent" haplogroups (Kivisild et al., 1999a; Metspalu et al., 2004; Quintana-Murci et al., 2004) is largely concentrated in the northwest of Indian subcontinent.

What then are the sources of different regional haplogroup packages that can be determined in different non-African populations? Can these be traced back to a single or multiple dispersals from Africa? Leaving aside the various versions of the multiregional model of human evolution (Wolpoff, 1989; Templeton, 2002) there exist quite a wide range of opinions, even among the proponents of the recent African replacement scenario, concerning the routes and the number of migrations out of Africa. By the "classical" view a "northern route" over Sinai (Figure 2A) explains the appearance of all modern human groups outside Africa. According to Klein (Klein, 1999) this spread was triggered by the revolutionary transfer from Middle to Upper Paleolithic (UP) technology, which together with other indications, such as long distance communication networks and symbolic art, forms a package referred to as modern human behavior. Yet the presence of anatomically modern human bones in association with Late Pleistocene tools from Australia (Bowler et al., 2003), which pre-date the earliest UP tools from the Middle East cannot be easily

assimilated into this model, suggesting that this tendency of using UP technology as a proxy for modern humans outside of Africa needs to be abandoned (Endicott et al., 2003b). This conclusion is further underlined by the intrinsic differences between UP technologies in Europe and the Indian subcontinent, which has recently led to a call to distinguish the latter as Late Palaolithic (James and Petraglia, 2005).

To overcome this inconsistency Lahr and Foley (1994) proposed an additional, earlier, southerly migration route taken by the ancestors of early Australians, exiting from the Horn of Africa (Figure 2B). In the light of the discoveries of sea shell middens, together with Middle Paleolithic industries in the Red Sea, Stringer (2000) suggested that

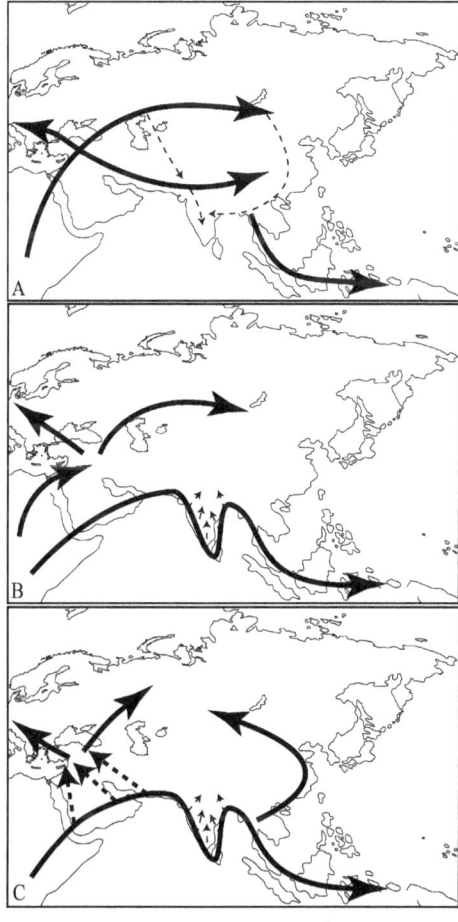

Figure 2. Three possible 'out-of-Africa' migration scenarios; (2A) the northern route, (2B) the northern route with a separate, earlier, southern exodus, (2C) the southern (coastal) route

the colonization of Australia proceeded by a coastal route that preceded a secondary wave, via the northern route, which finally brought forth the European Upper Paleolithic. The concept of a coastal migration was already envisioned in 1962 by the evolutionary geographer Carl Sauer (Sauer, 1962) who speculated that a coastal littoral would provide consistent access to familiar food resources for colonizing human groups. The phylogeography of mtDNA haplogroup M concords with the southern route dispersal (Kivisild et al., 1999a; Quintana-Murci et al., 1999, 2004; Metspalu et al., 2004) because it is absent among the populations that would immediately derive from the northern route migration over Sinai. The spread of the other mtDNA founder, N, can however be explained by either of the routes. A proposition of compromise from genetics has hence been put forward, supporting the conclusions of Lahr and Foley drawn from archaeology and anthropology, whereby mtDNA haplogroup M proceeded south while haplogroup N went north (Maca-Meyer et al., 2001; Tanaka et al., 2004).

South Asia at the Crossroads

The two route migration scenario becomes complicated as one studies more closely the extant genetic variation amongst human populations outside Africa. The concordant presence of all three mtDNA and Y chromosome founder haplogroups together in the Indian subcontinent is indicative of an early settlement of the region from Africa. Because neither the phylogeographies of mtDNA haplogroup M, nor Y chromosome haplogroups C and D, could be explained by the northern route dispersal, this settlement likely occurred along the southern route (Endicott et al., 2003b; Kivisild et al., 2003; Forster, 2004; Forster and Matsumura, 2005; Macaulay et al., 2005; Thangaraj et al., 2005b). It is hard to envisage a separate exodus to populate the Middle East and Europe because under a multiple migration scenario, unless the source population was identical, the daughter populations would differ from each other by the basic founder haplogroup composition (Figure 3A). Yet, this is not the case (Figure 3B); indeed, the two founder mtDNA lineages (M and N while R is nested within N) constituting the human mtDNA phylogeny outside Africa are just two of the myriad of lineages present in the source area – eastern Africa. The chance that two different migrations would randomly pick the same set of lineages would be extremely small. In fact, Western Eurasians are descended from two of the three mtDNA founder lineages present amongst the Australians. The exact time and birth place of M and N is even not so important here – whether it occurred in East Africa right before the exodus or on the early stages of the migration. Importantly, though, it is more parsimonious to see West Eurasia as having been settled from a common founding group that differentiated somewhere

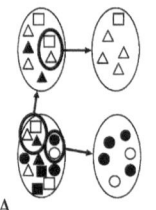
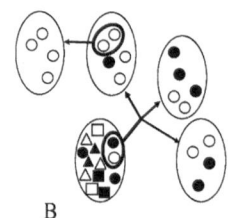

Figure 3. (A) The result of multiple migrations out of Africa sampling, by chance, different mtDNA haplogroups, and (B) the common origin of all mtDNA haplogroups outside Africa in one late Pleistocene migration drawing on the same set of lineages

between, most generally, East or Northeast Africa and South Asia. As haplogroup M, except for the African sub-clade M1, is not notably present in regions west of the Indian subcontinent, while it covers the majority of Indian mtDNA variation, it seems plausible that the split into daughter populations may have occurred somewhere within this zone (Endicott et al., 2003b; Kivisild et al., 2003; Metspalu et al., 2004).

A single southern route range expansion of the three mitochondrial and three Y chromosomal founder lineages can explain all extant variation observed in these loci not only in South Asia but also elsewhere in Eurasia and Oceania without invoking the time factor as a line of evidence (Figure 2C). Yet, we already distinguished between the set of founder haplogroups related with the "early" initial dispersal and the signatures of more "recent" ones, evinced, for example, by the Y chromosome haplogroup E3b distribution. Therefore, in the strictest sense a definition of all the potential number of Late Pleistocene migrations from Africa would be about "multiple dispersals". But when focusing our quest more specifically to the time frame relevant to the initial settlement of Eurasia and Australia the current evidence from mtDNA and Y loci can be explained by a single migration only.

The coalescence times of mtDNA haplogroups M, N and R are remarkably similar and ancient, $\sim 65,000$ years (Mishmar et al., 2003; Macaulay et al., 2005) and are more concordant with a single migration hypothesis. Because of the distant calibration point defined by the human and chimpanzee split, together with the generally low number of deeply splitting lineages in the trees constructed from human data, the coalescence times of the ancient haplogroups come all too often with broad error margins. However, direct comparisons of founder haplogroups M and R diversities (independent from the clock calibration) accumulated in India have

given highly similar estimates (Kivisild et al., 1999b; Metspalu et al., 2004; Quintana-Murci et al., 2004). The mtDNA based estimates are roughly consistent with the dating of the expansion of anatomically modern humans out of Africa (89-35 thousand years) using Y chromosome sequence information (Underhill et al., 2000). It seems most likely, therefore, given the existing evidence, that three Y chromosome founder lineages, accompanying mtDNA haplogroups M and N, were taken by the same wave of dispersal from Africa to the Indian subcontinent approximately 70–50 thousand years ago.

The absence of modern human remains in South Asia that match the genetic dates could be seen as a weakness of genetics, yet the early dates in Australia and the trail of deep rooting mtDNA lineages across South and Southeast Asia suggest that humans were present here in the same timeframe despite the current dearth of direct paleontological evidence. So far the dates of ~ 30 ka in Sri Lanka are the earliest in South Asia for modern human remains (Deraniyagala, 1984) and these are associated with some of the earliest deposits of microlithic technology in Eurasia (Kennedy, 2000).

Garden(s) of Eden, or How Fast was the Dispersal?

The question of whether the modern human expansion out of Africa was followed by a single and relatively fast range expansion (the so-called Strong Garden of Eden scenario, SGE) or by a gradual series of population expansions outside Africa (the Weak Garden of Eden, WGE, Harpending et al., 1993) has been approached recently through the analyses of complete mtDNA sequences (Forster and Matsumura, 2005; Macaulay et al., 2005; Thangaraj et al., 2005a). This has greatly augmented the information from the hypervariable region 1 (HVR1), where one mutation occurs on average in every 20 thousand years (Forster et al., 1996), which previously was the most common tool in use for analyzing mtDNA variation. The now widely available coding region information provides researchers with a matrix that is significantly more robust (less parallel events), yet four times as accurate (more mutations expected to occur in a given time frame).

According to the Weak Garden of Eden hypothesis one would expect to observe nested phylogenetic patterns (e.g., as the Amerind D1 clade is nested within East Asian haplogroup D phylogeny; Figure 4), whereby, for example, Southeast Asian populations would represent a phylogenetically nested subset of mtDNA lineages found in South Asia. This nested structure would be expressed in the number of shared phylogenetically deep mutations within South and Southeast Asians, which would be proportional to the time that the range expansion was delayed in South Asia before proceeding eastwards. Similarly, Papuan and Melanesian lineages, except for those derived from more recent migrations, would appear as a nested subset of Southeast Asians, and Australians, in turn, would appear to have a nested subset of Papuan mtDNA lineages. Yet, the available evidence from mtDNA variation directs us to investigate the alternative hypothesis.

Given the molecular clock calibrated over human and chimpanzee divergence at six million years ago, on average, one derived mutation in mtDNA coding region from any common founder would be expected to be shared on average between lineages from distinct geographic regions if their joint ancestral population existed for 5,138 years (Mishmar et al., 2003). However, the analysis of complete mtDNA genomic sequences has provided no evidence for such nested structuring of region-specific haplogroups: South Asian, Malaysian, Papuan, Australian, and East Asian lineages coalesce among themselves to a number of clades which do not share any common inner branches within the tree (Ingman and Gyllensten, 2003; Palanichamy et al., 2004; Tanaka et al., 2004; Macaulay et al., 2005; Rajkumar et al., 2005; Thangaraj et al., 2005a). One apparent exception to this pattern is the fact that the R founder is nested within haplogroup N, but both show similar coalescence times and do not show any sign of different migration routes – it may very well be that haplogroup R appeared early on the migration route out of Africa. A second deviation, haplogroup P (Forster et al., 2001), which is restricted to Oceania in its spread, is defined by a single coding region mutation that ties

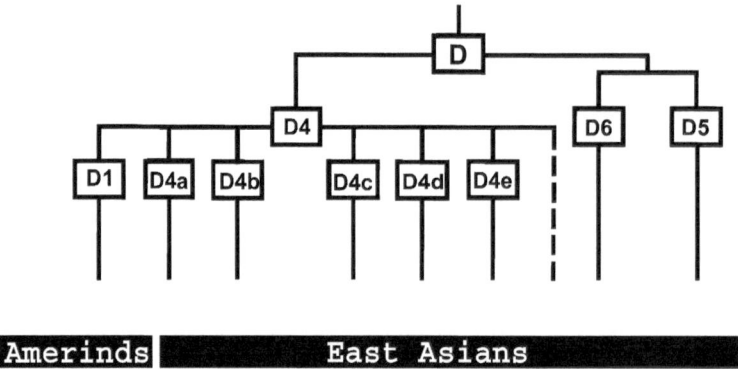

Figure 4. The nested position of the American-specific clade D1 within the East Asian haplogroup D. This pattern, typical of later colonizations, where the new founders are drawn from a pool of derived lineages, is not seen along the initial southern route, where autochthonous lineages arising immediately from the original founders are present throughout. This suggests a relatively rapid settlement of Eurasia and Near Oceania

together Papuans and Australians (Ingman and Gyllensten, 2003). So, the evidence from mtDNA is very much in favor of a rapid continuous dispersal occurring during a timeframe of thousands rather than tens of thousands of years.

The Role of Central Asia

Discussing the routes of the initial modern humans dispersals in Asia one cannot leave Central Asia without attention. According to the "pincer" model (Ding et al., 2000) this region was populated very early and therefore could have provided a substrate for further settlement of East and South Asia. In contrast, a single southern route migration model implies only a secondary settlement of Central Asia. It is true that mtDNA and Y lineages typical of East, West and South Eurasia are found in populations of Central Asia (Wells et al., 2001; Comas et al., 2004; Quintana-Murci et al., 2004). But, because a cocktail of lineages derived from different sources significantly increases the diversity present in the region it is not always easy to decide whether populations showing high levels of genetic heterogeneity have gone through recent admixture, receiving gene flow from multiple distinct sources, or if they have evolved in long-term separation from other populations, while retaining a high effective population size. If the latter scenario would be the case then one should observe deep rooting autochthonous lineages that, if found in other regions, would be present there in the form of more recent founder types. In contrast to South Asia, East Asia, Southeast Asia, Oceania and western Eurasia (and similar to the Americas), in Central Asia one cannot find autochthonous M, N or R lineages arising from the roots of these founder haplogroups. In fact, none of the prevalent haplogroups, except for some younger sub-haplogroups, in Central Asia show signs of radiation from their core. Indeed, virtually all of the extant

variation in Central Asia can be explained by an admixture scenario implicating movements from West, East, and South Eurasian sources (Comas et al., 2004; Quintana-Murci et al., 2004). The complex nature of Central Asian Y chromosome lineages can be explained not only as the region appearing as a "heartland" of genetic variation (Wells et al., 2001) but also, more parsimoniously, as a meeting point of paternal lineages coming from the east, west and south; thereby, leaving South Asia as the crossroads of early human migration.

Signals of Later Population Expansions in South Asia

A number of South Asian specific sub-clades of the three founder mtDNA haplogroups (M, N and R) show coalescence times equal to those of the founders themselves. This indicates that the first population expansion in South Asia took place shortly after the initial colonization (Kivisild et al., 2003; Metspalu et al., 2004). Yet, a secondary expansion of haplogroup N and R derived lineages (W, U7 and R2), in the region spanning from the Near East to western India, can be inferred. Judging from the coalescence times of these haplogroups this expansion probably took place around 30–20 ka. The coalescence times of the region-specific sub-clades of U7 indicate that the geographic segregation of this expansion between India and West Asia occurred during the onset of the Last Glacial Maximum (Kivisild et al., 1999a, b; Metspalu et al., 2004). At their root level both haplogroups U and W are more divergent in the Middle East and Europe than in South Asia. Haplogroup W is a sister clade to N2a (Derbeneva et al., 2002), and only W has been reported from South Asia. As for haplogroup U – most of its many equally deep rooted subclades are specific to western Eurasia while only three subclades of U2 are specific to South Asia (Kivisild et al.,

1999a; Palanichamy et al., 2004; Quintana-Murci et al., 2004). This pattern would suggest that the presence of haplogroups U and W in India might be due to an import rather than local origin. It has to be kept in mind though that the Indian varieties of haplogroup U2 are absent outside India and show coalescence times similar to the general founder haplogroups M, N, and R suggesting their recent import during the Holocene is unlikely.

Indian specific branches of haplogroup U7, as well as several local varieties of haplogroup M (e.g., M3a, M4a, M6 and M25) and R (e.g., R6), coalesce around 30–20 ka (Metspalu et al., 2004). The star-like radiation of these clusters is indicative of population expansions during the relatively favorable climate period before the onset of the Last Glacial Maximum around 18 ka. The phylogeographies of some of these expanding haplogroups show noticeable clines (e.g., M3a from northwest India) while for others the picture is less distinct (Metspalu et al., 2004). We note that, the timeframe of this expansion phase overlaps with that of the appearance of geometric microlithic technology in South Asia, which in the case of Sri Lanka occurs from ~28 ka (Deraniyagala, 1984).

As well as these relatively old additions to the initial late Pleistocene settlement patterns there have been more recent demographic events, although the effects of these have left little effect outside of certain key areas, usually acting as zones of interaction. In Pakistan and the western-most states of India the share of western Eurasian specific haplogroups (e.g., H, J, T, U4) climbs to nearly one half of the maternal gene pool (Kivisild et al., 1999a; Metspalu et al., 2004; Quintana-Murci et al., 2004). There is also mtDNA evidence of migrations through the northeastern region of present-day India from East Asia, which seem to be closely linked to Tibeto-Burman speakers in particular and likely date from the Holocene (Cordaux et al., 2003; Metspalu et al., 2004). This recent migratory phase through the northeast is supported by Y chromosomal haplogroup distributions, as those characterizing Tibeto-Burman speakers (O3e) appear to originate east of India (Cordaux et al., 2003). This corridor in the northeast of India may previously have seen the migration of Austro-Asiatic speaking groups who are found today mostly in the East Indian states of Orissa, Jharkhand, Bihar, and West Bengal, but who have a high percentage of O2a, a Y haplogroup thought to be associated with a SE Asian source (Su et al., 2000).

Caste, Language, and Agriculture: Cultural and Genetic Correlations?

The Austro-Asiatic and Tibeto-Burman language groups may retain a distinctive genetic signature due to their relatively recent introduction and limited subsequent male gene flow. However, consistent divisions between populations speaking Dravidian and Indo-Aryan languages are harder to define with reliability. The complex and intertwined history of changes in language, subsistence patterns, demography and political intervention, makes it difficult to relate genetic patterns to these widespread linguistic categories. The evidence from mtDNA argues against any strong differentiation between these (and other) major language groups (Kivisild et al., 1999a; Metspalu et al., 2004), and therefore nullifies attempts to trace, maternally, the large-scale population movements once speculated to have accompanied the arrival of Indo-Aryan languages (Barnabas et al., 1996; Passarino et al., 1996; Thapar and Rahman, 1996). Despite the presence of West Eurasian specific lineages in northwest India there has been quite limited diffusion southwards. In a very similar fashion the same region witnesses the spread of the West Eurasian Y haplogroup J2, but whether this is associated with the population expansions of the late Pleistocene or those of the more recent agricultural transition is, with present knowledge of the Y chromosome, less certain.

Similarly, the origins of the caste system are not at all clear and opinions about the trajectory and nature of its evolution range between polar opposites of recent versus ancient, and local versus external (Boivin, 2005; James and Petraglia, 2005). Very little attention has been paid to searching for caste in archaeology although this theoretically has potential. Genetic studies addressing the distinction between Indian castes and tribal communities also arrive at opposing conclusions; those based on phylogenetic analyses have generally provided support for the common origin of Indian tribes and caste groups irrespective of their social hierarchy (Roychoudhury et al., 2000; Kivisild et al., 2003; Metspalu et al., 2004) while those focusing on genetic distances have brought out the differences between tribal and caste groups, as a signal for the castes clustering with either West Asian (Cordaux et al., 2004b) or with Central Asian populations (Cordaux et al., 2004a).

A major obstacle in the analysis of Indian populations is the highly variable population size differences. Tribal groups often have extremely small effective population sizes, resulting in increased differentiation due to random genetic drift. This effect is exemplified most dramatically in island populations, such as the Jarawa and Onge groups of Andaman – where only two mitochondrial and one Y chromosomal lineages persist (Thangaraj et al., 2003). On the Indian mainland, where there are many more tribal populations, theoretically at least, it would seem helpful to pool together a number of these (to minimize the effect of drift in any of them taken individually) to arrive at more reliable estimates of genetic distances between tribal and caste populations (Cordaux et al., 2004a). However, this approach can still be problematic if the newly combined tribal groups come from the same narrow

geographic region and are likely to share a common history (Cordaux et al., 2004b).

On the basis of such approaches, claims for the external origin of major Y haplogroups of India (other than J2) have been made (R1a, R2, L) (Wells et al., 2001; Cordaux et al., 2004a). Yet their distributions can more parsimoniously be seen as local developments with movements taking place in the opposite direction. For example, R1a has been shown to have lower diversity in Central Asia and Eastern Europe (Kivisild et al., 2003) and, as such, these are unlikely sources for the Indian variants. Interestingly, R1a also displays high concentrations in the northwest of India suggesting a possible source of expansion in this region. As there is general agreement that the maternal heritage of India displays no recent widespread intrusion of mtDNA, the search for a haploid marker for the spread of the caste system, Indo-Aryan languages, or agriculture, lies with these particular Y haplogroups (Cordaux et al., 2004a). As yet the evidence is equivocal and there is no strong genetic signal for a major genetic component accompanying either the spread of Indo-Aryan languages or the caste system within India. Attempts to take this association of genetics with cultural continuums into the realms of subsistence categories (Cordaux et al., 2004b) are unlikely to be more successful (Sahoo et al., 2006).

The simplistic division between castes and tribes, or between subsistence categories, is most unlikely to be reflected in separate genetic histories, and might lead to misunderstanding amongst those not familiar with the limitations and biases of different types of analysis. Cultural change operates within a very different time frame to genetics and is likely to cross-cut and overlay the underlying genetic diversity of India, which appears, in the main, to derive from the late Pleistocene founder populations. Only the intrusion of Tibeto-Burman speakers appears to be clearly identifiable via distinct mtDNAs and Y chromosomes but this is due to the event having occurred relatively recently. For the origins of agriculture there is growing evidence of independent origins in South Asia (Misra, 2001; Fuller, 2003; Fuller et al., 2004; Fuller, this volume). Future developments in our knowledge of the Y chromosome will provide much more detail of both time depth (at the moment accuracy is limited to the Holocene due to uncertainty in our knowledge concerning the mutational process) and regional sub-structure. Until then, the temptation to explain the complex patterns of the general South Asian mitochondrial and Y chromosomal pools with reductive solutions, based on inadequate datasets, should be resisted.

Concluding Remarks

Evidence from both mtDNA and the Y chromosome argues for a rapid dispersal of modern humans from eastern Africa and subsequent settlement of South Asia. A single exodus along a southern, possibly coastal, route is a parsimonious conclusion to draw from contemporary patterns of haploid genetic distribution and diversity. The picture from the mtDNA perspective is clear, a predominantly late Pleistocene heritage with subsequent gene flow being limited and circumscribed in both space and time. In other words, it is not possible to detect any major population replacement events. To the contrary, it is still possible to detect demographic expansions from before the Last Glacial Maximum within the original gene pool. The Y chromosome tells a similar story of ancient inheritance, and like mtDNA, bears witness to multiple minor intrusions still visible due to their young age.

The population movements of the Holocene, together with the appearance of West Eurasian mtDNA lineages in the period 40–20 ka, indicate that South Asia has indeed been 'at the crossroads' for much of modern human prehistory, but that the autochthonous elements of its genetic heritage have not been dominated by these later comings and goings. Rather, only along the boundaries do we see significant changes in frequencies of different haplogroup distributions. As the sampling of extant populations will be increased and more detailed phylogenetic structure is recovered from the Y chromosome, together with information from diploid loci, our understanding of the complex history of this region, crucial to human prehistory, will become more refined. However, such developments will continue to emphasize the genetically complex patterns present, and are increasingly unlikely to support reductionist explanations of simplistic demographic and cultural scenarios. Rather, they should put weight behind the suggestion that West and South Asia, as conduits for the settlement of the rest of the world, are central to comprehending modern human evolution outside of Africa.

References

Anderson, S., Bankier, A.T., Barrell, B.G., de Bruijn, M.H., Coulson, A.R., Drouin, J., Eperon, I.C., Nierlich, D.P., Roe, B.A., Sanger, F., Schreier, P.H., Smith, A.J., Staden, R., Young, I G., 1981. Sequence and organization of the human mitochondrial genome. Nature 290, 457–465.

Andrews, R.M., Kubacka, I., Chinnery, P.F., Lightowlers, R.N., Turnbull, D.M., Howell, N., 1999. Reanalysis and revision of the Cambridge reference sequence for human mitochondrial DNA. Nature Genetics 23, 147.

Arredi, B., Poloni, E. S., Paracchini, S., Zerjal, T., Fathallah, D.M., Makrelouf, M., Pascali, V.L., Novelletto, A., Tyler-Smith, C., 2004. A predominantly neolithic origin for Y-chromosomal DNA variation in North Africa. American Journal of Human Genetics 75, 338–345.

Avise, J.C., 2000. Phylogeography: The History and Formation of Species. Harvard University Press, Cambridge, Massachusetts.

Barnabas, S., Apte, R.V., Suresh, C.G., 1996. Ancestry and interrelationships of the Indians and their relationship with other world populations: a study based on mitochondrial DNA polymorphisms. Annals of Human Genetics 60, 409–422.

Bosch, E., Calafell, F., Comas, D., Oefner, P. J., Underhill, P.A., Bertranpetit, J., 2001. High-resolution analysis of human Y-chromosome variation shows a sharp discontinuity and limited gene flow between northwestern Africa and the Iberian Peninsula. American Journal of Human Genetics 68, 1019–1029.

Boivin, N., 2005. Orientalism, ideology and identity: examining caste in South Asian archaeology. Journal of Social Archaeology 5, 225–252.

Bowler, J., Johnston, H., Olley, J., Prescott, J., Roberts, R., Shawcross, W., Spooner, N., 2003. New ages for human occupation and climatic change at Lake Mungo, Australia. Nature 421, 837–840.

Comas, D., Plaza, S., Wells, R. S., Yuldaseva, N., Lao, O., Calafell, F., Bertranpetit, J., 2004. Admixture, migrations, and dispersals in Central Asia: evidence from maternal DNA lineages. European Journal Human Genetics 12, 495–504.

Cordaux, R., Aunger, R., Bentley, G., Nasidze, I., Sirajuddin, S.M., Stoneking, M., 2004a. Independent origins of Indian caste and tribal paternal lineages. Current Biology 14, 231–235.

Cordaux, R., Deepa, E., Vishwanathan, H., Stoneking, M., 2004b. Genetic evidence for the demic diffusion of agriculture to India. Science 304, 1125.

Cordaux, R., Saha, N., Bentley, G., Aunger, R., Sirajuddin, S., Stoneking, M., 2003. Mitochondrial DNA analysis reveals diverse histories of tribal populations from India. European Journal Human Genetics 3, 253–264.

Cruciani, F., La Fratta, R., Santolamazza, P., Sellitto, D., Pascone, R., Moral, P., Watson, E., Guida, V., Colomb, E. B., Zaharova, B., Lavinha, J., Vona, G., Aman, R., Cali, F., Akar, N., Richards, M., Torroni, A., Novelletto, A., Scozzari, R., 2004. Phylogeographic analysis of haplogroup E3b (E-M215) Y chromosomes reveals multiple migratory events within and out of Africa. American Journal of Human Genetics 74, 1014–1022.

Deraniyagala, S.U., 1984. Mesolithic stone tool technology at 28,000 B.P. in Sri Lanka. Ancient Ceylon 5, 105–108.

Derbeneva, O.A., Starikovskaia, E.B., Volod'ko, N.V., Wallace, D.C., Sukernik, R.I., 2002. [Mitochondrial DNA variation in Kets and Nganasans and the early peoples of Northern Eurasia]. Genetika 38, 1554–1560.

Ding, Y.C., Wooding, S., Harpending, H.C., Chi, H.C., Li, H. P., Fu, Y.X., Pang, J.F., Yao, Y.G., Yu, J.G., Moyzis, R., Zhang, Y., 2000. Population structure and history in East Asia. Proceedings of the National Academy of Sciences USA 97, 14003–14006.

Endicott, P., Gilbert, M., Stringer, C., Lalueza-Fox, C., Willerslev, E., Hansen, A., Cooper, A., 2003a. The genetic origins of the Andaman Islanders. American Journal of Human Genetics 72, 178–184.

Endicott, P., Macaulay, V., Kivisild, T., Stringer, C., Cooper, A., 2003b. Reply to Cordaux and Stoneking. American Journal of Human Genetics 72, 1590–1593.

Forster, P., 2004. Ice Ages and the mitochondrial DNA chronology of human dispersals: a review. Proceedings of the Royal Society of London. Series B 359, 255–264.

Forster, P., Harding, R., Torroni, A., Bandelt, H.-J., 1996. Origin and evolution of Native American mtDNA variation: a reappraisal. American Journal of Human Genetics 59, 935–945.

Forster, P., Matsumura, S., 2005. Did early humans go north or south? Science 308, 965–966.

Forster, P., Torroni, A., Renfrew, C., Röhl, A., 2001. Phylogenetic star contraction applied to Asian and Papuan mtDNA evolution. Molecular Biology and Evolution 18, 1864–1881.

Fuller, D., 2003. An agricultural perspective on Dravidian historical linguistics: archaeological crop packages, livestock and Dravidian crop vocabulary. In: Bellwood, P., Renfrew, C. (Eds.), Examining the Farming/Language Dispersal Hypothesis. The McDonald Institute for Archaeological Research, Cambridge, pp. 191–213.

Fuller, D., 2007. Non-human genetics, agricultural origins and historical linguistics in South Asia. In: Petraglia, M.D., Allchin, B. (Eds.), The Evolution and History of Human Populations in South Asia: Inter-disciplinary Studies in Archaeology, Biological Anthropology, Linguistics and Genetics. Springer, Netherlands, pp. 393–443.

Fuller, D., Korisettar, R., Vankatasubbaiah, P., Jones, M., 2004. Early plant domestications in southern India: some preliminary archaeobotanical results. Vegetation History Archaeobotany 13, 115–129.

Harpending, H., Sherry, S., Rogers, A., Stoneking, M., 1993. The genetic structure of ancient human populations. Current Anthropology 34, 483–496.

Ingman, M., Gyllensten, U., 2003. Mitochondrial genome variation and evolutionary history of Australian and New Guinean aborigines. Genome Research 13, 1600–1606.

James, H., Petraglia, M., 2005. Modern human origins and the evolution of behavior in the Later Pleistocene record of South Asia. Current Anthropology 46, S3–S27.

Jobling, M.A., Tyler-Smith, C., 2003. The human Y chromosome: an evolutionary marker comes of age. Nature Reviews Genetics 4, 598–612.

Ke, Y., Su, B., Song, X., Lu, D., Chen, L., Li, H., Qi, C., Marzuki, S., Deka, R., Underhill, P., Xiao, C., Shriver, M., Lell, J., Wallace, D., Wells, R. S., Seielstad, M., Oefner, P., Zhu, D., Jin, J., Huang, W., Chakraborty, R., Chen, Z., Jin, L., 2001. African origin of modern humans in East Asia: a tale of 12,000 Y chromosomes. Science 292, 1151–1153.

Kennedy, K.A.R., 2000. God Apes and Fossil Men: Palaeoanthropology in South Asia. University of Michigan Press, Ann Arbor.

Kivisild, T., Bamshad, M. J., Kaldma, K., Metspalu, M., Metspalu, E., Reidla, M., Laos, S., Parik, J., Watkins, W. S., Dixon, M. E., Papiha, S. S., Mastana, S. S., Mir, M. R., Ferak, V., Villems, R., 1999a. Deep common ancestry of Indian and western-Eurasian mitochondrial DNA lineages. Current Biology 9, 1331–1334.

Kivisild, T., Kaldma, K., Metspalu, M., Parik, J., Papiha, S.S., Villems, R., 1999b. The place of the Indian mitochondrial DNA variants in the global network of maternal lineages and the peopling of the Old World. In: Papiha, S. S., Deka, R., Chakraborty, R. (Eds.), Genomic Diversity. Kluwer Academic/Plenum Publishers, pp. 135–152.

Kivisild, T., Rootsi, S., Metspalu, M., Mastana, S., Kaldma, K., Parik, J., Metspalu, E., Adojaan, M., Tolk, H.-V., Stepanov, V., Gölge, M., Usanga, E., Papiha, S. S., Cinnioglu, C., King, R., Cavalli-Sforza, L., Underhill, P. A., Villems, R., 2003. The genetic heritage of the earliest settlers persists both in Indian tribal and caste populations. American Journal of Human Genetics 72, 313–332.

Kivisild, T., Shen, P., Wall, D. P., Do, B., Sung, R., Davis, K. K., Passarino, G., Underhill, P. A., Scharfe, C., Torroni, A., Scozzari, R., Modiano, D., Coppa, A., de Knjiff, P., Feldman, M. W., Cavalli-Sforza, L. L., Oefner, P. J., 2005. The role of selection in the evolution of human mitochondrial genomes. Genetics.

Klein, R., 1999. The Human Career. University of Chicago Press, Chicago.

Kumar, S.S., Nasidze, I., Walimbe, S.R., Stoneking, M., 2000. Discouraging prospects for ancient DNA from India. American Journal of Physical Anthropology 113, 129–133.

Lahr, M., Foley, R., 1994. Multiple dispersals and modern human origins. Evolutionary Anthropology 3, 48–60.

Maca-Meyer, N., González, A. M., Larruga, J.M., Flores, C., Cabrera, V.M., 2001. Major genomic mitochondrial lineages delineate early human expansions. BMC Genetics 2, 13.

Macaulay, V., Hill, C., Achilli, A., Rengo, C., Clarke, D., Meehan, W., Blackburn, J., Semino, O., Scozzari, R., Cruciani, F., Taha, A., Shaari, N. K., Raja, J. M., Ismail, P., Zainuddin, Z., Goodwin, W., Bulbeck, D., Bandelt, H. J., Oppenheimer, S., Torroni, A., Richards, M., 2005. Single, rapid coastal settlement of Asia revealed by analysis of complete mitochondrial genomes. Science 308, 1034–1036.

Metspalu, M., Kivisild, T., Metspalu, E., Parik, J., Hudjashov, G., Kaldma, K., Serk, P., Karmin, M., Behar, D. M., Gilbert, M. T. P., Endicott, P., Mastana, S., Papiha, S. S., Skorecki, K., Torroni, A., Villems, R., 2004. Most of the extant mtDNA boundaries in South and Southwest Asia were likely shaped during the initial settlement of Eurasia by anatomically modern humans. BMC Genetics 5, 26.

Mishmar, D., Ruiz-Pesini, E., Golik, P., Macaulay, V., Clark, A. G., Hosseini, S., Brandon, M., Easley, K., Chen, E., Brown, M. D., Sukernik, R. I., Olckers, A., Wallace, D. C., 2003. Natural selection shaped regional mtDNA variation in humans. Proceedings of the National Academy of Sciences USA 100, 171–176.

Misra, V.N., 2001. Prehistoric human colonization of India. Journal of Biosciences 26, 491–531.

Oppenheimer, S., 2003. Out of Eden: The Peopling of the World. Constable, London.

Palanichamy, M., Sun, C., Agrawal, S., Bandelt, H.-J., Kong, Q.-P., Khan, F., Wang, C.-Y., Chaudhuri, T., Palla, V., Zhang, Y.-P., 2004. Phylogeny of mtDNA macrohaplogroup N in India based on complete sequencing: implications for the peopling of South Asia. American Journal of Human Genetics 75, 966–978.

Passarino, G., Semino, O., Bernini, L. F., Santachiara-Benerecetti, A. S., 1996. Pre-Caucasoid and Caucasoid genetic features of the Indian population, revealed by mtDNA polymorphisms.

American Journal of Human Genetics 59, 927–934.

Quintana-Murci, L., Chaix, R., Wells, S., Behar, D., Sayar, H., Scozzari, R., Rengo, C., Al-Zahery, N., Semino, O., Santachiara-Benerecetti, A., Coppa, A., Ayub, Q., Mohyuddin, A., Tyler-Smith, C., Mehdi, Q., Torroni, A., McElreavey, K., 2004. Where West meets East: the complex mtDNA landscape of the Southwest and Central Asian corridor. American Journal of Human Genetics 74, 827–845.

Quintana-Murci, L., Semino, O., Bandelt, H.-J., Passarino, G., McElreavey, K., Santachiara-Benerecetti, A.S., 1999. Genetic evidence of an early exit of *Homo sapiens sapiens* from Africa through eastern Africa. Nature Genetics 23, 437–441.

Rajkumar, R., Banerjee, J., Gunturi, H.B., Trivedi, R., Kashyap, V.K., 2005. Phylogeny and antiquity of M macrohaplogroup inferred from complete mt DNA sequence of Indian specific lineages. BMC Evolutionary Biology 5, 26.

Rando, J.C., Pinto, F., Gonzalez, A.M., Hernandez, M., Larruga, J. M., Cabrera, V.M., Bandelt, H.J., 1998. Mitochondrial DNA analysis of northwest African populations reveals genetic exchanges with European, near-eastern, and sub-Saharan populations. Annals of Human Genetics 62, 531–550.

Richards, M., Macaulay, V., Hickey, E., Vega, E., Sykes, B., Guida, V., Rengo, C., Sellitto, D., Cruciani, F., Kivisild, T., Villems, R., Thomas, M., Rychkov, S., Rychkov, O., Rychkov, Y., Golge, M., Dimitrov, D., Hill, E., Bradley, D., Romano, V., Cali, F., Vona, G., Demaine, A., Papiha, S., Triantaphyllidis, C., Stefanescu, G., 2000. Tracing European founder lineages in the Near Eastern mtDNA pool. American Journal of Human Genetics 67, 1251–1276.

Roychoudhury, S., Roy, S., Dey, B., Chakraborty, M., Roy, M., Roy, B., Ramesh, A., Prabhakaran, N., Rani, M.V.U., Vishwanathan, H.M., Mitra, M., Sil, S.K., Majumder, P.P., 2000. Fundamental genomic unity of ethnic India is revealed by analysis of mitochondrial DNA. Current Science 79, 1182–1192.

Sahoo, S., Singh, A., Himabindu, G., Banerjee, J., Sitalaximi, T., Gaikwad, S., Trivedi, R., Endicott, P., Kivisild, T., Metspalu, M., Villems, R., Kashyap, V. K., 2006. A prehistory of Indian Y chromosomes: evaluating demic diffusion scenarios. Proc Natl Acad Sci USA 103(4), 843–848.

Sauer, C., 1962. Seashore–primitive home of man? Proceedings of the American Philosophical Society 106, 41–47.

Semino, O., Magri, C., Benuzzi, G., Lin, A.A., Al-Zahery, N., Battaglia, V., Maccioni, L., Triantaphyllidis, C., Shen, P., Oefner, P. J., Zhivotovsky, L.A., King, R., Torroni, A., Cavalli-Sforza, L.L., Underhill, P.A., Santachiara-Benerecetti, A.S., 2004. Origin, diffusion, and differentiation of Y-chromosome haplogroups E and J: inferences on the neolithization of Europe and later migratory events in the Mediterranean area. American Journal of Human Genetics 74, 1023–1034.

Stringer, C., 2000. Coasting out of Africa. Nature 405, 24–27.

Su, B., Xiao, C., Deka, R., Seielstad, M. T., Kangwanpong, D., Xiao, J., Lu, D., Underhill, P., Cavalli-Sforza, L., Chakraborty, R., Jin, L., 2000. Y chromosome haplotypes reveal prehistorical migrations to the Himalayas. Human Genetics 107, 582–590.

Su, B., Xiao, J., Underhill, P., Deka, R., Zhang, W., Akey, J., Huang, W., Shen, D., Lu, D., Luo, J., Chu, J., Tan, J., Shen, P., Davis, R., Cavalli-Sforza, L., Chakraborty, R., Xiong, M., Du, R., Oefner, P., Chen, Z., Jin, L., 1999. Y-Chromosome evidence for a northward migration of modern humans into Eastern Asia during the last ice age. American Journal of Human Genetics 65, 1718–1724.

Tanaka, M., Cabrera, V. M., Gonzalez, A. M., Larruga, J. M., Takeyasu, T., Fuku, N., Guo, L.-J., Hirose, R., Fujita, Y., Kurata, M., Shinoda, K., Umetsu, K., Yamada, Y., Oshida, Y., Sato, Y., Hattori, N., Mizuno, Y., Arai, Y., Hirose, N., Ohta, S., Ogawa, O., Tanaka, Y., Kawamori, R., Shamoto-Nagai, M., Maruyama, W., Shimokata, H., Suzuki, R., Shimodaira, H., 2004. Mitochondrial genome variation in eastern Asia and the peopling of Japan. Genome Research 14, 1832–1850.

Templeton, A., 2002. Out of Africa again and again. Nature 416, 45–51.

Thangaraj, K., Chaubey, G., Kivisild, T., Reddy, A. G., Singh, V. K., Rasalkar, A. A., Singh, L., 2005a. Reconstructing the origin of Andaman Islanders. Science 308, 996.

Thangaraj, K., Singh, L., Reddy, A., Rao, V., Sehgal, S., Underhill, P., Pierson, M., Frame, I., Hagelberg, E., 2003. Genetic affinities of the andaman islanders, a vanishing human population. Current Biology 13, 86–93.

Thangaraj, K., Sridhar, V., Kivisild, T., Reddy, A.G., Chaubey, G., Singh, V. K., Kaur, S., Agarawal, P., Rai, A., Gupta, J., Mallick, C. B., Kumar, N., Velavan, T. P., Suganthan, R.,

Udaykumar, D., Kumar, R., Mishra, R., Khan, A., Annapurna, C., Singh, L., 2005b. Different population histories of the Mundari- and Mon-Khmer-speaking Austro-Asiatic tribes inferred from the mtDNA 9-bp deletion/insertion polymorphism in Indian populations. Human Genetics 116, 507–517.

Thapar, B.K., Rahman, A., 1996. The Post-Indus Cultures. In: Dani, A.H., Mohen, J.-P., (Eds.), History of Humanity. Clays Ltd., St. Ives plc., UK, pp. 266–279.

Underhill, P.A., Shen, P., Lin, A. A., Jin, L., Passarino, G., Yang, W. H., Kauffman, E., Bonne-tamir, B., Bertranpetit, J., Francalacci, P., Ibrahim, M., Jenkins, T., Kidd, J.R., Mehdi, S.Q., Seielstad, M.T., Wells, R.S., Piazza, A., Davis, R.W., Feldman, M.W., Cavalli-Sforza, L.L., Oefner, P.J., 2000. Y chromosome sequence variation and the history of human populations. Nature Genetics 26, 358–361.

Wells, R.S., Yuldasheva, N., Ruzibakiev, R., Underhill, P.A., Evseeva, I., Blue-Smith, J., Jin, L., Su, B., Pitchappan, R., Shanmuga-lakshmi, S., Balakrishnan, K., Read, M., Pearson, N.M., Zerjal, T., Webster, M.T., Zholoshvili, I., Jamarjashvili, E., Gambarov, S., Nikbin, B., Dostiev, A., Aknazarov, O., Zalloua, P., Tsoy, I., Kitaev, M., Mirrakhimov, M., Chariev, A., Bodmer, W.F., 2001. The Eurasian heartland: a continental perspective on Y-chromosome diversity. Proceedings of the National Academy of Sciences USA 98, 10244–10249.

Wolpoff, M., 1989. Multiregional evolution: the fossil alternative to Eden. In: Mellars, P., Stringer, C. (Eds.), The Human Revolution: Behavioral and Biological Perspectives on the Origins of Modern Humans. Edinburgh University Press, Edinburgh.

YCC, 2002. A nomenclature system for the tree of human Y-chromosomal binary haplogroups. Genome Research 12, 339–348.

11. Cranial diversity in South Asia relative to modern human dispersals and global patterns of human variation

JAY T. STOCK AND MARTA MIRAZÓN LAHR

Leverhulme Centre for Human Evolutionary Studies
The Henry Wellcome Building
Fitzwilliam Street
University of Cambridge
Cambridge, CB2 1QH
England
j.stock@human-evol.cam.ac.uk
m.mirazon-lahr@human-evol.cam.ac.uk

SAMANTI KULATILAKE

Department of Earth Sciences
Faculty of Science and Technology
Mount Royal College
4825 Mount Royal Gate S.W.
Calgary, Alberta, T3E 6K6
Canada
skulatilake@mtroyal.ca

Introduction

In 1840, William Whewell argued that hypotheses in the inductive sciences are either accepted or rejected through a 'consilience of inductions' (Whewell, 1840), under which the wealth of accumulated evidence from different perspectives supports a given hypothesis. A consilience of genetic, fossil and archaeological evidence now points to a recent African origin of modern humans. (Howells, 1976; Day and Stringer, 1982; Stringer and Hublin, 1984; Lahr 1994; Lahr and Foley, 1994, 1998). Studies of modern human genetic diversity suggest that all living humans share a common African ancestor who lived approximately 200-150 ka (Cann et al., 1987; Tishkoff et al., 1996; Harpending and Rogers, 2000; Ingman et al., 2000; Ke et al., 2001; Kivisild et al., 2001; Excoffier, 2002). Fossil crania from Omo Kibish, and Herto, dated between 200 and 120 thousand years ago, are the earliest crania with clear evidence for morphological features associated with anatomical modernity (Day, 1969; Day and Stringer, 1982, 1991; Lahr, 1996; White et al., 2003; McDougall et al., 2005).

M.D. Petraglia and B. Allchin (eds.), The Evolution and History of Human Populations in South Asia, 245–268.
© 2007 *Springer.*

Current fossil and archaeological evidence suggests that the first dispersal of anatomically modern humans out of Africa took the form of a temporary occupation of the Levant at approximately 100–90 ka, based upon remains found at the sites of Skhūl and Qafzeh (Bar-Yosef et al., 1986; Stringer et al., 1989; Stringer, 1992; Mercier et al., 1993; Turbon et al., 1997; Holliday, 2000). Fossil evidence for subsequent dispersals of modern humans is sparse. Following the early occupation of the Levant at Skhūl and Qafzeh, the subsequent modern human remains from this area date to 35 ka or more in Lebanon (Bergman and Stringer, 1989) and approximately 24 ka in Israel (Arensburg, 1977, 1981). Skeletal remains provide evidence for somewhat earlier occupation of modern humans elsewhere. There is little secure evidence for modern humans in Europe prior to 39 ka (Zilhão and d'Errico, 1999). Early modern human remains are relatively poorly represented in Asian archaeological deposits outside the Levant. Although there have been claims of remains with an antiquity of 67 ka, the earliest remains likely date to 33 ka (Etler, 1996). In Southeast Asia, early modern human remains have been identified as early as 44 ka at Niah Cave (Barker et al., 2002). Although two hominin teeth have been discovered at Liang Bua on Flores, in deposits dating to earlier than 74 ka, these are non-diagnostic to the level of species (Morwood et al., 2004). Human occupation in Australia may have been similarly early. While skeletal remains from Lake Mungo may date to as early as 62 ka (Thorne et al., 1999), they appear to be more conservatively dated to approximately 40 ka, with an earliest occupation at Lake Mungo dated to 50 ka (Bowler et al., 2003). The latter date brings the remains from Lake Mungo in line with the earliest reliable dates for occupation elsewhere in Australia, and provides a confident estimate of the earliest occupation of Australia to approximately 50 ka (Roberts et al., 1990;

Roberts et al., 1994; Turney et al.,2001; Gillespie, 2002).

Despite the evidence that modern humans were established in Australia by 50,000 years before present, evidence for early modern human occupation in South Asia is sparse. The earliest modern human skeletal remains come from the sites of Fa Hien and Batadomba-Lena in Sri Lanka, which yield dates of 30 ka and 28 ka respectively, for human remains associated with microlithic tool assemblages (Deraniyagala, 1992; Kennedy, 1999, 2000). While microlithic tool assemblages are found elsewhere in the subcontinent, their chronology is poorly understood and these assemblages have not yet been directly linked to modern human populations through fossil remains (James and Petraglia, 2005). The earliest modern human remains in India derive from terminal Pleistocene and Holocene contexts throughout much of the subcontinent (Kennedy et al., 1984; Lukacs and Pal, 1993, 2003: Kennedy, 2000). Middle Paleolithic technologies are abundant at many archaeological sites in India, and have been dated to various ages within the last 150,000 years (James and Petraglia, 2005). The earliest dates for Upper Paleolithic assemblages are ca. 45 ka at Site 55 in Pakistan (Dennell et al., 1992), or 26,210 BP at the 16R dune in Rajasthan (Misra, 1995). These sites provide the earliest clear evidence for modern human behavior in the Indian subcontinent. Since microlithic technologies are associated with anatomically modern humans, there is considerable indirect evidence for an earlier occupation of modern humans in the Indian subcontinent. The lack of early modern human fossil remains in this area may be due to a number of factors, including preservation bias or the relative lack of archaeological exploration compared to Upper Paleolithic contexts elsewhere.

Interpretations of possible human dispersal routes have been somewhat constrained due to archaeological chronologies and genetic distances between contemporary populations. With solid evidence for an early modern human occupation of Australia and Southeast Asia, any plausible interpretation of human migrations must allow for humans to reach these regions by 50 ka at the latest. Given the wealth of evidence that now supports a recent African origin of modern humans and their subsequent dispersal to other regions of the world, Asia must have acted as a corridor of migration between Africa and Australia. A commonly hypothesized route of modern human dispersal is a northern route out of Africa, across the Sinai peninsula into the Levant (Bar-Yosef, 1992; Lahr and Foley, 1994; Klein, 1999). Fluctuating environmental conditions opened this corridor during interglacials, and allowed for the migration of African fauna into Eurasia (Tchernov, 1992a, 1992b). In this context, human populations would have followed the movement of fauna across the Sinai Peninsula. However, the fluctuating environmental conditions also suggest that this route would also have been a considerable barrier during periods of glaciation (Foley and Lahr, 1997).

An alternative model involves a southern dispersal of modern humans originating in the Horn of Africa across the Bab al Mandab Strait, and following the Southern coast of Arabia into South Asia and Australia (Foley and Lahr, 1992; Lahr and Foley, 1994; Stringer, 2000). Proponents of the southern dispersal model have generally assumed that human populations who exploited marine resources would have been able to move very rapidly along the coastline towards Southeast Asia. A rapid movement of early modern humans would help to explain the similarity of times of genetic coalescence for the African exodus and the earliest dates of occupation in Australia. A recent geographic simulation of a southern dispersal route has demonstrated the plausibility of this model, but it suggests that a southern dispersal between Africa and Australia could

not have taken an entirely coastal route, and would have encountered both geographic and environmental barriers which required sustained reliance upon terrestrial resources in some regions (Field and Lahr, 2006). It is likely that these geographic and environmental constraints placed upon human populations dispersing through South Asia had considerable impact on the population history of these regions.

Genetics and the Origin of South Asian Populations

The recent African origin of all living humans is supported by considerable genetic evidence (Cann et al., 1987; Tishkoff et al., 1996; Harpending and Rogers, 2000; Ingman et al., 2000; Ke et al., 2001; Kivisild et al., 2001; Excoffier, 2002). One of the most critical aspects of this evidence is the fact that populations outside of Africa contain only a subset of the genetic diversity encompassed by populations within Africa (Tishkoff et al., 1996; Harpending and Rogers, 2000). Such findings not only suggest that all living humans share a recent African origin, but also indicate that subsequent to human dispersals out of Africa, there was little gene flow or migration back (Mitchell et al., 1999). Studies of Y-chromosome haplotype frequencies indicate that all non-African lineages coalesce to between 81 and 56 ka (Hammer and Zegura, 2002), and suggest that all non-African populations descend from one or a small number of migration events.

Recent analyses of mtDNA variation among isolated populations in Southeast Asia have identified unique lineages that clarify our understanding of South Asian migrations. Studies of complete mtDNA genomes of indigenous populations of Malaysia, Papua New Guinea and Australia have identified the presence of unique derived lineages of the M, N and R haplogroups (Macaulay et al., 2005). These haplogroups appear to date to between

63 and 60 ka, a time at which the northern dispersal route was likely inhospitable (van Andel and Tzedakis, 1996). Further evidence for unique M haplogroup lineages among indigenous Andaman Islanders, which coalesce between 70 and 50 ka, provide additional support for a rapid southern dispersal of modern humans followed by and extended period of geographic and genetic isolation of some populations (Thangaraj et al., 2005).

Genetic studies of Y-chromosome lineages and mtDNA variation (haplogroups and deletions) among human populations of the Indian subcontinent provide evidence for the population history of this region, and point towards a recent African origin of contemporary South Asians (Kivisild et al., 1999a, 1999b; Watkins et al., 1999; Underhill et al., 2001). Different researchers have estimated coalescence dates for the African origin of different lineages that converge on a similar chronology for dispersals into South Asia. MtDNA haplogroup frequencies place a potential East African origin of mtDNA lineages in India between 73 and 55 ka (Kivisild et al., 2000), and approximately 60 ka (Quintana-Murci et al., 1999). Both estimates fit a plausible chronology of the southern dispersal route and subsequent colonization of Southeast Asia and Australia. Furthermore, M haplogroups are common in South and East Asia, while N haplogroups are found in low frequencies in these regions. The inverse pattern is found in European and West Asian populations. Such marked differences in M and N haplogroup frequencies may indicate that these populations owe their origins to separate migration events from Africa. Indeed, Jobling and coworkers (Jobling et al., 2004) assert that the two major mtDNA haplogroups found outside of Africa, M and N, represent separate migrations out of Africa along southern coastal and northern migration routes, respectively.

The origin of agriculture in the Middle East, and its subsequent spread into the Indian subcontinent features prominently in the prehistory of this region, having been associated with the dispersal of the Indo-Iranian and Dravidian languages in the area (Renfrew, 1987, 1996; Cavalli-Sforza et al., 1994). However, this scenario remains controversial. Despite evidence from the distribution of languages and subsistence technology in recent history, recent genetic evidence has supported arguments in favour of interpreting the spread of agriculture into South Asia as the result of cultural diffusion (Whittle, 1996 – which suggests a mechanism of cultural transmission rather than population movement and gene flow), rather than demic diffusion (Ammerman and Cavalli-Sforza, 1984), which implies large-scale populations movements and gene flow. The latter view has been the dominant paradigm, a component of which is the assumption that tribal populations in India represent descendents of the original inhabitants, while Indian caste populations are thought to be derived from subsequent Indo-European migrations.

The genetic evidence for variation within South Asia is complex. Two recent studies of mtDNA (Cordaux et al., 2003) and autosomal markers (Vishwanathan et al., 2004) have found considerable genetic diversity between, and low diversity within, tribal groups, suggesting that these represent long-standing separate and relatively endogamic genetic pools. Furthermore, y-chromosome haplogroup diversity among traditional and recent agriculturalists and traditional hunter-gatherers has been used to argue in support of the demic diffusion model on genetic grounds (Cordaux et al., 2004), an interpretation also supported by an earlier study (Quintana-Murci et al., 2001).

However, other studies of frequencies of Indian-specific mtDNA haplogroup M sub-clades provide somewhat contradictory results. These data suggest that after the initial human colonization of India, gene flow in and out of this region has been surprisingly limited (Bamshad et al., 2001; Kivisild et al., 2003; Metspalu et al., 2004; Sahoo et al., 2006), with complex patterns that may reflect specific and potentially internal demographic events (Quintana-Murci et al., 2004). The issue of the timing of demographic events is a complex one. Genetic evidence has been used to suggest that there were only two significant demographic expansions in the history of modern humans in South Asia, the most recent of which occurred some 30–20,000 years ago, long after the initial colonization and expansion of modern humans into the subcontinent ca. 60–50 ka (Kivisild et al., 1999b). If this were the case, both demographic events would have occurred prior to, or concurrent with, the first anatomically modern skeletal remains in the region. This has been interpreted as evidence for population continuity from the Paleolithic to the present, and thus evidence of a small impact of external populations at the establishment of farming communities. While some mtDNA haplogroups are shared with Iranian populations, other M and R haplogroups that are specific to India are very rare elsewhere (Metspalu et al., 2004), further highlighting the apparent long-term trajectory of microevolution of populations within the subcontinent.

The above brief discussion stresses two key points regarding prehistoric human dispersals in and out of South Asia – first, that South Asia was an important migration route between western Asians and/or Africans and Southeast Asians and Australians (Roychoudhury et al., 2000; Majumder, 2001; Redd et al., 2002; Macaulay et al., 2005; Thangaraj et al., 2005); second, that the original Upper Pleistocene colonists of the region were the main, although not the only (Quintana-Murci et al., 2004), contributors to the living genetic pool.

The Biogeography of Modern Human Dispersals

The evidence for the relative genetic isolation of the Indian subcontinent from surrounding populations may be explained by geographic boundaries identified through computer simulation of a southern dispersal route (Field and Lahr, 2006). Figure 1 provides an adapted representation of this dispersal simulation, highlighting potential geographic barriers to the dispersal of modern humans during Oxygen Isotopic Stage 4. The Himalayas form a natural barrier between India and northern Asia, while the Indian subcontinent is bounded on the west by both the Sulaimān Mountain Range and the Indus river delta, which was likely comprized of salt flats and marshes during OIS4 (Field and Lahr, 2006). On the eastern edge of the Indian subcontinent, the Ganges-Brahmaputra River delta may have restricted populations movement out of Southern Asia due to a large expanse of mangroves (Agarwal and Mitra, 1991). Population flow to and from the east may have also been complicated by the Ragaing Yoma Mountains and the mouth of the Irawadi river (Field and Lahr, 2006). In contrast to the relatively inhospitable regions surrounding it, the Indian subcontinent would have been predominantly savannahs and dry woodlands, and as such was likely a hospitable environment for hunter-gatherers with economies that were based in part upon terrestrial resources. The combination of favorable conditions within and geographic barriers surrounding South Asia suggests that once humans were established in this area populations continued to develop *in situ*. The geographic expanse and natural resources of South Asia would have likely resulted in demographic expansion, a hypothesis which seems to be supported by the current genetic evidence from contemporary South Asian populations.

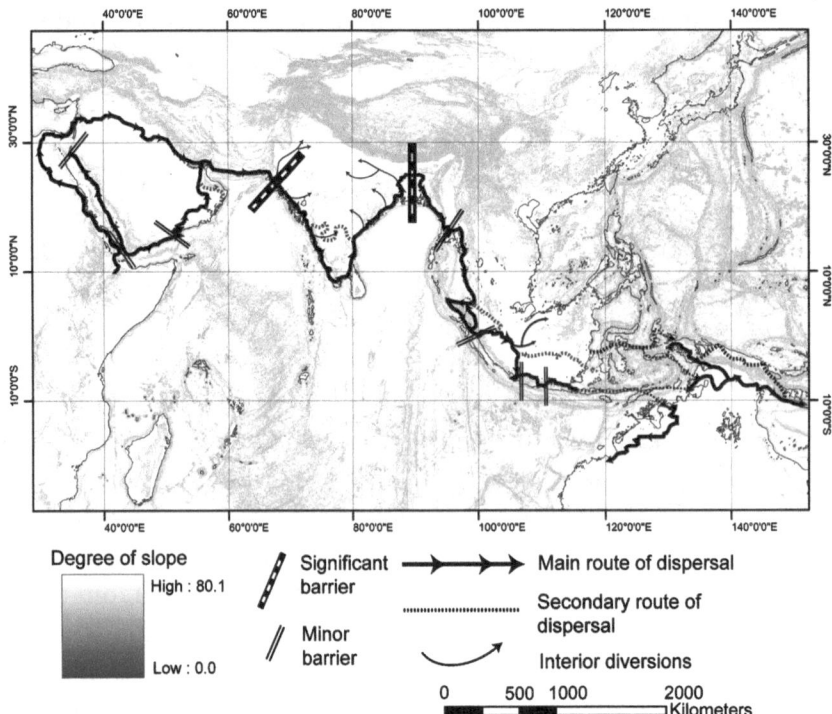

Figure 1. Possible routes of the southern dispersal of modern humans through South Asia, based upon sea levels in Oxygen Isotopic Stage 4 and a simulation model by Field and Lahr (2006). Major barriers to human movement are formed by the Indus River Delta and the Sulaimān Range, the Himalayas and the mangroves of the Ganges river delta

The evidence for long-term genetic and geographic isolation of populations in the Indian subcontinent has a number of implications for human diversity in India. Once populations reached the Indian subcontinent, a combination of abundant resources and geographic barriers would have led to a major demographic expansion in this area. As populations expanded in this region, a small number of groups would continue to migrate through Southern Asia, eventually leading to another demographic expansion within Australia.

The model of an early demographic expansion and subsequent *in situ* development of populations that is predicted by geography and supported by contemporary genetics has implications for modern human diversity in South Asia. If tribal populations represent remnants of Paleolithic societies then we would expect to see some similarities

between the morphology of Mesolithic hunter-gatherers and contemporary tribal groups, and considerable morphological homogeneity among contemporary populations. However, if contemporary Indian caste populations were structured on the basis of Neolithic expansions from Western Europe or Eurasia, a model of demic diffusion, we would expect to see evidence for morphological similarity between the contemporary populations of these regions. Conversely, if contemporary populations in India represent *in situ* development from the Paleolithic, we would expect to see greater homogeneity in the morphology of Indian peoples relative to populations elsewhere. Under this scenario, the Neolithic in India would correspond with diffusion from western Asia or Europe that was primarily cultural in nature.

The analysis of human skeletal diversity has been brought to bear on issues of population history, adaptation and genetic relationships. Studies of cranial morphology suggest that there are considerable differences between European and Australo-Melanesian crania, and that the latter show greater morphological similarities to African crania than to Upper Paleolithic or European crania (Lahr, 1996). The analysis of human skeletal remains from South Asia provides an important perspective on global human diversity. There is a long history of anthropometry of living and past populations in the Indian subcontinent. The majority of early work focused on the classification of populations into racial types, of which a great diversity was noted for the region (Guha, 1935; Risley, 1908; Sarkar, 1954, 1972; Majumdar, 1961). More recent work has focused on the variation within southern Asia, suggesting that despite the diversity of Indian populations, they do not easily fit racial typologies previously used to describe morphology elsewhere (Kennedy et al., 1984). Further research suggests that statistical differentiation of populations on the basis of physical characters is not particularly successful, and that geographic parameters explain more variability than population history (Majumder et al., 1990).

There have been many studies of skeletal morphology in South Asia, which predominantly focus on regional variation within the subcontinent. Mesolithic crania tend to be both large in size and relatively heterogeneous in morphology (Kennedy et al., 1984). Skeletal analyses of Mesolithic foragers in South Asia have emphasized likely interaction between Holocene hunter-gatherers and surrounding pastoralists and agricultural populations at sites in the Gangetic Plain (Kennedy, 1999), and in Rajasthan and Guajarat (Lukacs, 1990; Lukacs and Pal, 1993). Indeed, it appears as though interactions between foragers and pastoralists have been a common feature of social systems throughout much of the Holocene (Lukacs, 2002; Morrison, 2002; Possehl, 2002).

Analyses of the skeletal remains from the sites of Damdama, Mahadaha and Sarai Nahar Rai in the Gangetic Plain suggest that these populations were characterized by very large body and cranial sizes and considerable robusticity (Kennedy, 1999; Lukacs and Pal, 2003). However it is argued that similarities between this morphological pattern and Upper Paleolithic Europeans are due to morphological convergence on the basis of behavior and environmental conditions rather than close genetic relationships (Kennedy et al., 1984; Kennedy, 1999). While it has often been assumed that the contemporary tribal populations of South Asia are direct descendents of Mesolithic hunter-gatherers, it has been argued that tribal groups had considerable interaction and gene flow with surrounding populations (Kennedy, 1992), although phenotypic links may exist between the Upper Paleolithic skeletal remains from Bellanbandi Palassa in Sri Lanka and the contemporary Vedda (Kennedy, 1965).

Several studies of skeletal morphology in Harappan and subsequent periods also underscore the considerable heterogeneity of skeletal morphology in South Asia, which has commonly been interpreted within a framework of significant adaptive and genetic variation and considerable gene flow (Ehrhardt, 1963; Kennedy and Levisky, 1984; Hemphill et al., 1991; Hemphill and Lukacs, 1993). Some of this work has suggested that there is specific morphological evidence for genetic interaction with populations in western Asia (Hemphill et al., 1996; Rathburn, 1982), and that considerable variation developed within prehistoric societies (Bartel, 1979). The morphological variation documented in these skeletal series would seemingly lend support to demic diffusion models of the origins of agriculture in the Southern Asia.

Morphometric heterogeneity within the Indian subcontinent is a theme that is recurrently stressed in studies of skeletal morphology, however it remains a challenge to interpret population movement in South Asia on the basis of cranial morphology (Warusawithana-Kulatilake, 1996). In some ways, the level of morphological heterogeneity in south Asia is to be expected given the long history and social complexity of its people, and the disparate spatial and temporal scales from which well preserved archaeological remains are derived. To date only a few studies have investigated the cranial morphology of South Asian populations in detail across temporal and spatial axes, and fewer still have attempted to place south Asian diversity in the context of larger patterns of human diversity (Warusawithana-Kulatilake, 1996). The goal of this chapter is to place south Asian cranial diversity within the context of global patterns of human variation, and current theories of the modern human dispersals. In this respect, the scale of the questions posed is different to previous studies. Given limitations of extant skeletal series and archaeological assemblages, a comparison of this type will not reevaluate previous studies of population relationships with fine temporal and spatial resolution. In contrast, the current study looks from South Asia outwards rather than inwards, to provide a preliminary comparison of variation within South Asia to variability observed within our species more broadly.

Based upon genetic and archaeological evidence, alternative models of Indian population history involve either, relative isolation and *in situ* development of contemporary populations, or significant migration and gene flow from outside the subcontinent. Within this context, our understanding of morphological variation in South Asia would benefit from the placement of this morphology within a global framework. Using this approach, this study addresses the following questions:

1. Is human cranial morphology in South Asia relatively homogenous or heterogeneous?
2. Do contemporary South Asians show morphological affinities with populations in surrounding regions, and in particular, contemporary groups from western Asia?
3. Does the cranial morphology of South Asian populations reflect the long-term genetic and geographic isolation relative to global patterns of human diversity suggested by genetic evidence?
4. Do archaeological crania from the Mesolithic and mid-Holocene Neolithic of India show morphological affinities to contemporary South Asian populations?

Materials and Methods

The current study investigates these questions using a dataset representing global craniometric variation. The overall sample consists of measurements of 4463 adult crania representing most areas of the world (Table 1). The sample consists of a combination of craniometric data recorded by W.W. Howells, T. Hanihara, M. Mirazón Lahr and S. Kulatilake. The recent

Table 1. Recent human craniometric data[†] (n = 4463)

South Asian Regions	Other Asian Regions	Other Regions
Indus Valley (n = 103)	Andaman Islanders (n = 62)	Europe (n = 761)
Ganga Valley (n = 110)	N. Gulf (n = 23) – Afghanistan, Iran	Africa (n = 843)
South India/Deccan (n = 122)	Himalayan (n = 83) – Nepal, Bhutan, Tibet	Australia (n = 177)
NE India, Assam Tribal (n = 12)	S.E. Asia (n = 470) – Thailand, Vietnam, Philippines	Melanesia (n = 359)
Sinhalese, Sri Lanka (n = 4)	E. Asia (n = 430) – China, Japan, Taiwan, Korea	Polynesia (n = 403)
Vedda, Sri Lanka (n = 20)	W. Asia (n = 145) – Turkey, Arabia, Israel, Iraq	Micronesia (n = 37)
	N.E. Asia (n = 60) – Mongolia, Siberia	Americas (n = 274)

[†]Database includes measurements by Hanihara, Howells, Mirazón Lahr, Kulatilake

crania from Southern Asia were divided into several regions for subsequent analyses, based on geographic criteria. The major groupings include the Indus Valley, the Ganga Valley, Southern India/Deccan Plateau, Sri Lankan Sinhalese, Vedda, and N.E. Indian tribal groups, which include crania from the Mishmi, Naga, Thado-Kuki, Singho and other tribes. In order to test for morphological similarities between South Asian crania and populations from surrounding regions, the global dataset was grouped into the following seven surrounding Asian regions, examples of which are provided in brackets: the Northern Gulf (Afghanistan, Iran); Himalayan (Nepal, Bhutan, Tibet); Southeast Asia (Thailand, Vietnam, Philippines); East Asian (China, Japan, Taiwan, Korea); Western Asia (Arabia, Israel, Iraq); and N.E. Asia (Mongolia, Siberia). To place southern Asian cranial diversity into a global context, the remaining crania were divided into a further seven regions representing population groupings based on major geographic criteria. Groupings at this level include: Europe, Africa, Australia, Melanesia, Polynesia, Micronesia and the Americas.

Analyses of the global sample are based upon 23 standard cranial measurements common to all of the datasets (Table 2). The data were analyzed using canonical variates analysis to investigate the underlying patterns of cranial variability and morphometric relationships between population groups. In order to control for differences in overall cranial size, the data were size standardized using the geometric mean of maximum cranial length, breadth and height (GOL, XCB and BBH). This method of standardization was chosen as it accounts for the three main components of cranial size, and minimizes the problem of missing data. To calculate a geometric mean for all variables would involve either: a) the estimation of a considerable proportion of the dataset to eliminate missing variables, or, b) the exclusion of a considerable portion of the dataset from the analysis. Since the primary goal of this chapter is to place a general pattern of South Asian cranial diversity within global patterns of variation, the chosen approach maximizes the dataset, while still allowing the investigation of differences in cranial shape between populations.

Separate discriminant analyses were used to determine whether the cranial morphology of prehistoric individuals from archaeological sites in India show morphological affinities to any contemporary populations. The archaeological series derive from mid-Holocene (Bronze Age and Chalcolithic) contexts primarily in the Indus Valley region, with the exception of one individual from the site of Pandu Raja Dhibi in northeastern India (Figure 2). These sites date to between

Table 2. Craniometric variables (n = 23)

GOL – Glaello-Occipital Length*	OBB – Orbital Breadth*
NOL – Nasion-Occipital Length*	OBH – Orbital Height*
BNL – Basion-Nasion Length	MAB – Maxillary Breadth*
XCB – Maximum Cranial Breadth*	MDH – Mastoid Height*
BBH – Basion-Bregma Height*	MDB – Mastoid Breadth*
XFB – Maximum Frontal Breadth	BPL – Basion-Prosthion Length
ASB – Bi-Asterionic Breadth	ZMB – Bi-Zygomaxillary Breadth*
AUB – Bi-Auricular Breadth*	DKB – Interorbital Breadth*
ZYB – Bi-Zygomatic Breath	FRC – Frontal Chord*
NPH – Nasion-Prosthion Height*	PAC – Parietal Chord
NLH – Nasal Height*	OCC – Occipital Chord
NLB – Nasal Breadth*	

*Variables used in the analysis of Mesolithic, Neolithic and Iron Age cranial series

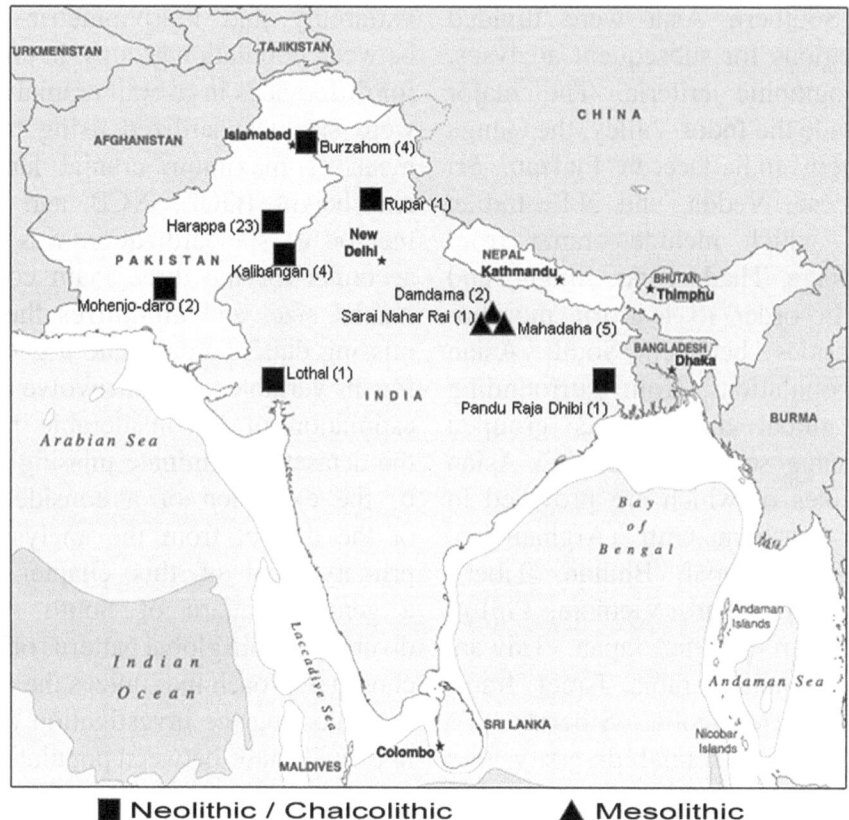

Figure 2. Location of archaeological sites yielding crania used in this study, with sample sizes in brackets

3,000 and 5,000 years ago. Early Holocene crania derive from the Mesolithic sites of Mahadaha, Sarai Nahar Rai and Damdama in the Ganga river basin, which date to between 15- 10 ka. Crania from these sites were sufficiently complete to provide sixteen common craniometric variables. While there are many more fragmentary crania of similar antiquity from archaeological contexts in India, the majority are too fragmentary to provide sufficient morphometric data.

Limitations on a comparison of this nature are imposed by an uneven spatial and temporal distribution of archaeological series. While the majority of Mesolithic crania are derived from the Gangetic plain, many of the mid Holocene crania are derived from the Indus valley. While this limits our ability to interpret temporal change throughout the Holocene, it should not deter us from comparing these

groups to contemporary series to investigate morphological similarities and differences.

Despite the relative completeness of the archaeological crania used in this study, some missing variables were estimated using multiple linear regressions to create a subset of the available Mesolithic crania from which a reasonable dataset could be derived. By estimating data for the fragmentary Mesolithic cranial series, the utility of available craniometric data is maximized. Missing data were not estimated for the global dataset, as the sample sizes were sufficient to provide global representation without potentially biasing or magnifying results through variable estimation. Discriminant analyses were used to determine whether the archaeological crania showed morphological affinities to contemporary populations in the Indian subcontinent. In order to maximize the sample size while minimizing the number of estimated variables, sixteen craniometric variables were used in this analysis (Table 2).

Results

Initially, canonical variates analysis was used to investigate the relative cranial variation within South Asia. The first factor produced by this analysis provides some discrimination between populations of the Indus Valley, the Ganga Valley and southern India (Figure 3). The crania from the Indus Valley tend to be longer, while those from Southern India tend to be broader (not only in terms of the head and face, but also the orbits) (Table 3). The NE Indian tribal crania are not clearly differentiated from those of other regions, and appear to be most similar to crania from non-tribal groups in the Ganga valley (Table 4). The discrimination of the Vedda and Sinhalese Sri Lankan crania on the second factor is marked; however, the small Sri Lankan sample limits the extent to which this contrast may be interpreted. This factor is weighted by constrasting crania with broad

mastoid processes, long frontals and short faces and noses, with those of opposite dimensions. The Indus, Ganga and South Indian crania are also characterized by high biasterionic breadths and mastoid heights. While these results highlight some cranial diversity between regions of South Asia, they also demonstrate that there are no clear regional distinctions between the groups from continental India.

Discriminant function analyses were used to classify South Asian cranial samples in relation to crania from the Andaman Islands, the Northern Gulf, Himalayan, Southeast, East, Northeast and West Asian regions, in order to determine whether crania from the Indian Subcontinent share morphological affinities with populations of surrounding areas. In this analysis, the crania of each of the South Asian regions were treated as unknown, and the discriminant functions derived from the analysis of populations from adjacent regions used (Table 5) to determine to which of the adjacent regions different South Asian crania were most similar to.

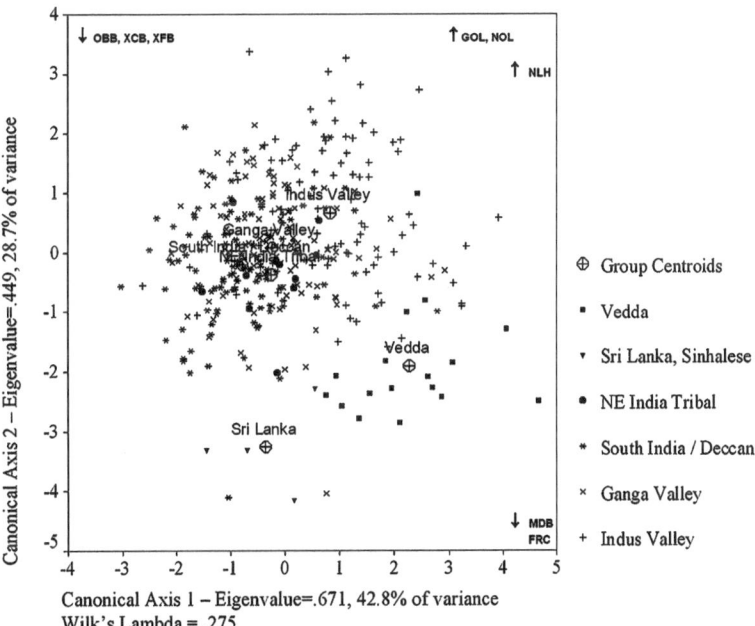

Figure 3. Scatterplot of CVA scores on the first two axes for South Asian skeletal series. Arrows indicate that high values of the listed variables are weighing the corresponding axis. Upward arrows indicate a positive weighting on the axis, downward arrows indicate a negative weighting

Table 3. Discriminant function structure matrix for the first three of five extracted functions, using the six cranial series from the Indian subcontinent

Variable*	Function 1 Eigenvalue=.671 42.8% of Variance	Function 2 Eigenvalue=.449 28.7% of Variance	Function 3 Eigenvalue=.246 15.7% of Variance
NOL	.491	.236	.380
GOL	.436	.241	.429
OBB	−.368	.336	.071
XCB	−.351	−.069	−.345
XFB	−.263	.075	−.112
ZMB	−.248	−.045	−.236
PAC	.226	−.013	.084
MAB	.491	.096	−.198
NLH	.119	.487	−.097
NPH	.065	.436	−.037
ASB	−.057	.409	−.234
MDH	−.011	.315	.039
MDB	.215	−.230	.085
BNL	.259	.232	.428
AUB	−.069	.215	−.380
ZYB	.018	.249	−.377
BPL	−.061	−.159	.354
FRC	−.006	−.235	.308
OBH	.239	.145	−.292
NLB	−.001	−.032	−.169
BBH	−.044	−.158	−.044
DKB	−.204	.143	.083
OCC	.097	−.051	−.035
NOL	.491	.236	.380
Wilk's Lambda for functions 1 through 5 = .275			

*All variables have been standardized to cranial size using the Geometric mean of GOL, BBH and XCB

Table 4. Classification results for the South Asian cranial series

Original Classification	Predicted Group Membership, % of cases					
	Indus Valley	Ganga Valley	South India /Deccan	NE India Tribal (Assam)	Sri Lanka	Vedda
Indus Valley	**65.3**	18.9	11.6	0	0	4.2
Ganga Valley	10.3	**57.0**	29.0	.9	0	2.8
South India/Deccan	9.7	22.1	**64.6**	1.8	.9	.9
NE India Tribal (Assam)	0	27.3	18.2	**54.5**	0	0
Sri Lanka	0	0	0	0	**100**	0
Vedda	5.9	0	0	0	0	**94.1**

The discriminant equations derived from these samples differentiate groups on the basis of facial height and cranial length on the first axis, and biasterionic and nasal breadths on the second. When these equations are applied to the South Asian samples, classifications consistently point towards a strong relationship between the populations of the Himalayas and the crania from the Ganges and N.E. tribal groups, with the South Deccan and Indus crania showing more general morphological similarities (Table 6). This reinforces the perception

Table 5. Discriminant function structure matrix for the first three of five extracted functions, using crania from the Andaman Islands, Western Asia, the Northern Gulf, Eastern Asia, Southeast Asia and the Himalayan regions

Variable*	Function 1 Eigenvalue=.819 45.9% of Variance	Function 2 Eigenvalue=.504 28.2% of Variance	Function 3 Eigenvalue=.296 16.6% of Variance
Nasion-prosthion height	.496	.060	−.068
Nasio-occipital length	.415	−.188	−.400
Occipital chord	.367	.149	.120
Basion-bregma height	.315	.258	−.084
Interorbital breadth	−.181	.032	−.055
Biasterionic breadth	.244	−.506	.343
Maximum cranial breadth	.158	.224	−.047
Glabello-occipital length	.386	−.158	−.438
Maximum frontal breadth	.131	.077	−.333
Frontal chord	.112	.085	−.330
Biauricular breadth	.288	.309	−.076
Orbit breadth-left	−.183	.107	−.222
Palate breadth external	.162	.217	.106
Nasal breadth	.069	.380	.053
Bizygomatic breadth	.316	.256	−.035
Nasal height	.262	.129	−.104
Orbit height-left	.201	.013	.224
Basion-prosthion length	.135	.130	−.182
Basion-nasion length	.307	.036	−.332
Parietal chord	.212	−.076	−.255
Wilk's Lambda for functions 1 through 5 = .241			

*All variables have been standardized to cranial size using the Geometric mean of GOL, BBH and XCB

Table 6. Discriminant function classification results for South Asian groups in relation to surrounding areas

| Original Classification | % grouped with adjacent region | | | | | |
	Andaman Is.	N Gulf	Himalayan	E Asia	SE Asia	W Asia
Indus Valley	1.1	24.2	**42.1**	4.2	2.1	2.6
Ganga Valley	4.7	10.3	**68.2**	3.7	8.4	4.7
South/Deccan	5.3	13.3	**54.0**	.9	9.7	16.8
NE Tribal	9.1	0	**63.6**	9.1	18.2	0
Sri Lankan	0	0	2.1	17.2	9.7	**71.0**
Vedda	0	0	17.6	5.9	35.3	**41.2**

of the Himalayas as a considerable barrier to the movement of people and genes. Although sample sizes are small, the Sri Lankan and Vedda crania classify most consistently with West Asian crania. The different morphological affinities of the continental south Asian groups and the Sri Lankan and Vedda crania to those from surrounding regions underscore the morphological differences and the geographic and genetic barriers between these groups.

The above analysis points towards a general level of homogeneity among the South Asian samples, but do these cranial series show any greater affinity to other human populations, whether by common ancestry or adaptive convergence? An analysis of South Asian cranial morphology in the context of the global pattern of human variation suggests that populations of the Indian subcontinent have a relatively unique and homogenous pattern of cranial morphology (Figure 4). In

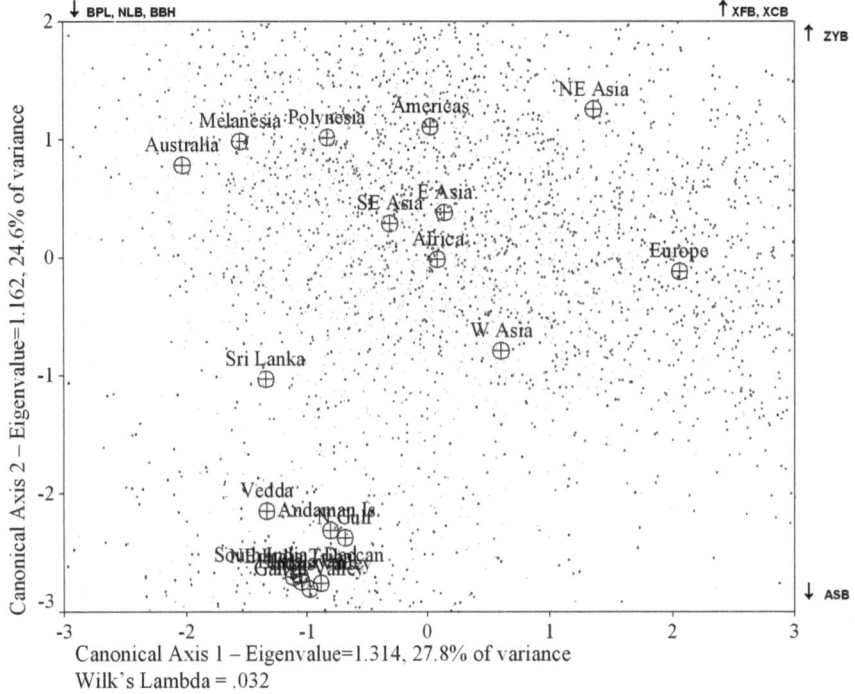

Figure 4. Scatterplot of CVA scores on the first two axes for the globally representative sample, emphasizing centroid locations rather than regional distributions. Arrows indicate that high values of the listed variables are weighing the corresponding axis. Upward arrows indicate a positive weighting on the axis, downward arrows indicate a negative weighting

this analysis (Table 7), factor one contrasts two morphologies – broad crania, with non-prognathic faces, narrow noses and relative low total height (characterizing Europeans, NE and Western Asians) with those of opposite shape (characterizing Australians and Melanesians). It is interesting to note the relative proximity of sample centroids on this axis, which demonstrates that while reasonable discrimination is observed, there is considerable overlap in sample distributions. In the distribution of regional variation on this axis, Europeans are morphologically distinct from Austalo-Melanesian groups, while cranial diversity within Africa falls intermediate between these morphological extremes (Figure 5). On the second factor, the majority of South Asian cranial series are clearly discriminated, with the exception of the small Sri Lankan Sinhalese sample, primarily on the basis of relative differences in zygomatic breadth and biasterionic breadth.

The centroids of the South Asian groups for the first two factors are tightly clustered. An expanded view of centroid locations is provided in Figure 5. Here it is clearly illustrated that crania from the Himalayan region are similar to the other series from within continental India, while the crania from the Andaman Islands and the Northern Gulf countries show some similarity to one another. The tight clustering of the continental Indian series suggests that crania in these regions are relatively homogenous when considered relative to global variation in cranial morphology. The homogeneity of these cranial series suggests that there may have been considerable barriers to gene flow into the Indian subcontinent from adjacent regions throughout much of human prehistory in this area.

Despite the evidence for relatively low variability in the cranial morphology of contemporary and recent South Asians, it remains possible that this homogenization

Table 7. Discriminant function structure matrix for the first four of eighteen extracted functions, incorporating all samples from the global dataset

Variable*	Function 1 Eigenvalue=1.314 27.8% of Variance	Function 2 Eigenvalue=1.162 24.6% of Variance	Function 3 Eigenvalue=.899 19.0% of Variance	Function 4 Eigenvalue=.509 10.8% of Variance
XFB	.580	−.228	−.021	.025
XCB	.504	−.017	.263	−.027
ASB	−.131	−.771	−.039	.253
AUB	.115	.175	.484	.212
GOL	−.200	−.002	−.475	.304
NOL	−.136	−.042	−.468	.164
ZMB	−.295	.042	.363	−.286
BNL	−.229	.026	−.136	.110
MAB	−.338	.060	.018	−.055
NPH	.095	.152	.202	.065
OBB	−.256	.089	−.207	.176
NLB	−.368	.016	−.180	−.377
MDB	−.289	.089	.022	.125
FRC	−.156	−.069	−.124	.016
BPL	−.458	.195	−.267	.044
OBH	−.088	.071	.151	.070
DKB	−.027	−.094	−.391	−.269
ZYB	−.108	.249	.262	.164
NLH	−.009	.127	.268	.097
BBH	−.365	.024	.180	−.263
OCC	−.144	.097	.088	−.107
PAC	−.204	−.142	−.284	−.113
MDH	−.261	.029	.076	.044
Wilk's Lambda for functions 1 through 18 = .032				

*All variables have been standardized to cranial size using the Geometric mean of GOL, BBH and XCB

was a relatively recent occurrence, as a result of either: a) gene flow within the Indian subcontinent, or, b) long-term adaptation to similar environmental and cultural conditions. It has been previously shown that Mesolithic crania were considerably larger than subsequent and recent human populations (Kennedy et al., 1984), a conclusion supported by examination of measures of cranial size in the current dataset. Discriminant analyses of size standardized craniometric data from Mesolithic crania suggest that they share characteristics of morphological shape with the recent Vedda and Sinhalese groups (Figure 6). While one Mesolithic cranium is quite similar in metric dimensions to contemporary crania from the Indus and Ganga Valleys, the rest show morphological similarity to the Vedda series on the basis of greater cranial lengths and

smaller orbital and maximum cranial breadths. A similar pattern of classification is found among the later Neolithic and Chalcolithic crania dating to between 5,000 and 3,000 years ago. In this case (Figure 7), the majority of crania cluster with either the Vedda or Sri Lankan series. In contrast, the Andamanese crania, while falling within the general pattern of South Asian cranial morphology, appear distinct in discriminant analyses. This could relate to a shared common ancestry followed by a founder effect and subsequent isolation among the Andamanese. Overall, these results suggest that while south Asian morphological diversity may be relatively limited today, the greater diversity observed in the Mesolithic and early Neolithic may be indicative of a process of homogenization of cranial shapes

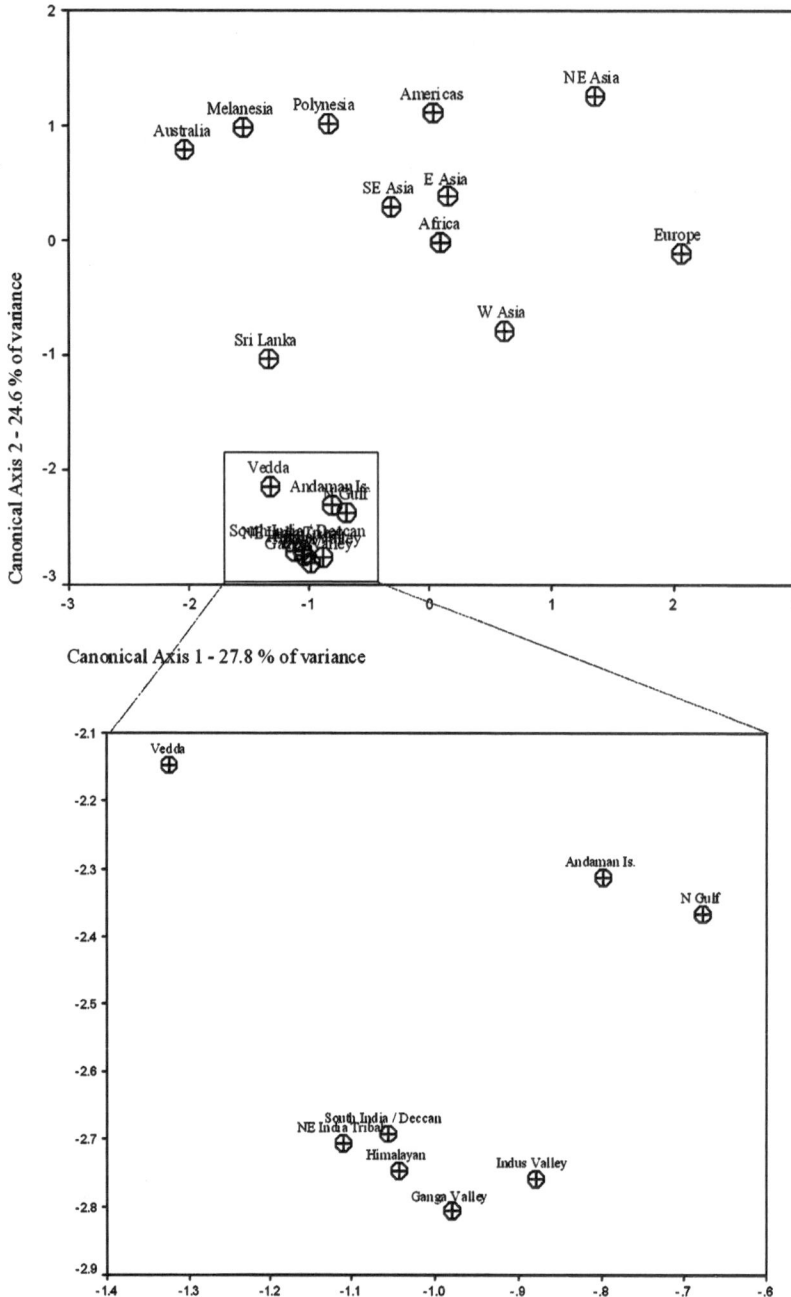

Figure 5. Global human sample, centroid coordinates on the first two Canonical Axes, with an enlarged representation of the centroids among the South Asian cranial series

in the mainland Indian subcontinent over the last few millennia.

Discussion

Several important aspects of the diversity of South Asian populations emerge from the analyses above. These will be discussed in order of scale.

First, when considering cranial variation within South Asia, populations appear to be quite heterogeneous. However, when this diversity is considered within the context of global patterns of human variation, South

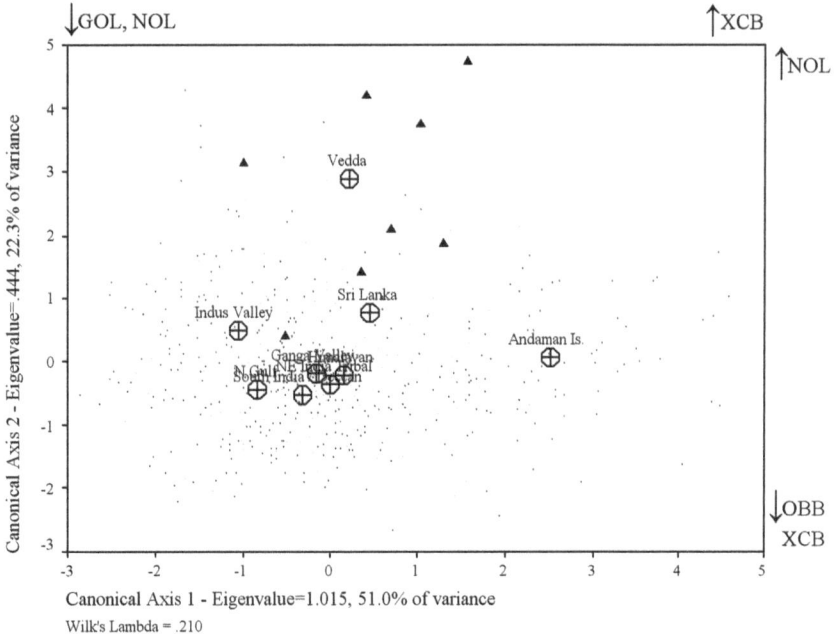

Figure 6. Scatterplot of CVA scores on the first two axes, placing the Mesolithic cranial series into the context of variation in South Asia, including the Andaman Islands and the N. Gulf and Himalayan regions. The Mesolithic crania (▲) cluster most consistently with the Sri Lankan and Vedda series, with 4/8 and 3/8 crania classified in these categories respectively, with the remaining individual being classified among the Indus Valley series

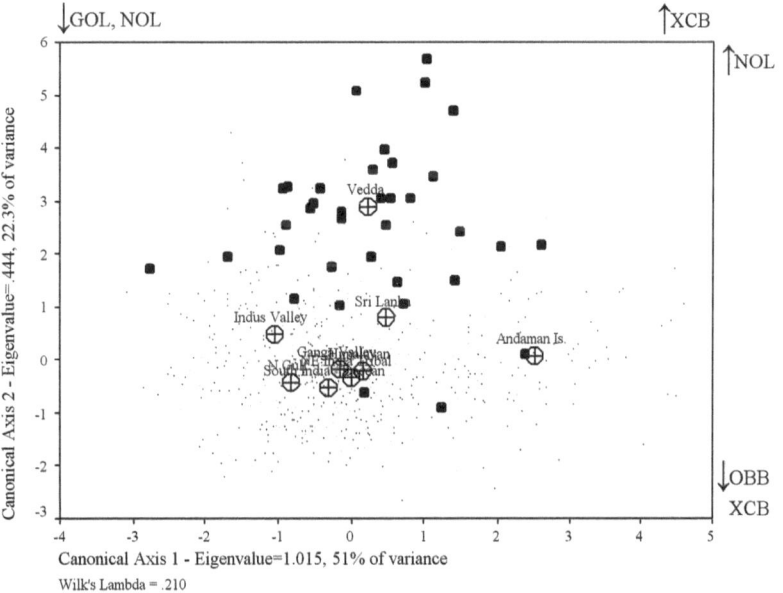

Figure 7. Scatterplot of CVA scores on the first two axes, placing the mid Holocene crania into the context of variation in South Asia, including the Andaman Islands and the N. Gulf and Himalayan regions. The mid Holocene crania (■) cluster most consistently with the Sri Lankan and Vedda series, with 17/36 and 15/36 crania classified in these categories respectively. Of the remaining four crania, one was classified in each of the Indus, Ganga, N. Gulf and Himalayan series

Asians appear to be relatively homogenous in cranial morphology and tend to form a cluster that is distinct from other groups. In other words, cranial morphology suggests that recent South Asian peoples are relatively closely related to each other.

Second, the socio-cultural structure of South Asian populations appears to have had a limited influence on diversity, as shown by the fact that the tribal populations of Assam used in this analysis are not morphologically distinct from other South Asians. Clearly, similar comparisons of other tribal groups are needed, but from the above it appears that either the population history of these individuals was similar to that of neighboring populations and not representative of tribal groups more generally, or the tribal/non-tribal boundary has been genetically more fluid than generally thought. This finding further strengthens the view that tribal populations, at least of the Assam, cannot be seen as remnants of earlier Mesolithic hunter-gatherers (Kennedy, 1999).

Third, the population history of the subcontinent extends to the North, as Himalayan populations clearly fit within the range of morphological diversity found within South Asia, but not to the South. There appears to be greater diversity between Sri Lankan populations and the continental groups within the geographic range of current-day India, than between Indian and Himalayan populations.

The morphological similarities among crania from the Indus and Ganga Valleys, Southern India, the Northeast Tribal populations and the Himalayas could be a reflection of the relative isolation of populations in the Indian Subcontinent subsequent to the earliest human occupation, as is suggested by genetic evidence (Kivisild et al., 2003; Metspalu et al., 2004). Alternatively, these similarities could be due to a combination of convergent evolution on the basis of similar environmental stressors and/or genetic admixture in the relatively recent past.

The former interpretation implies that geographic barriers may have had a significant impact on the current patterns of human diversity in the area. This supports the model of southern dispersal in which the barriers of the Sulaimān and Himalayan and Ragaing Yoma mountain ranges, in combination with the deltas of the Indus and Ganges-Brahmaputra rivers may have restricted population movement in and out of Southern Asia and led to a demographic expansion followed by relative isolation.

The second interpretation would argue for a recent process of homogenization within the subcontinent. The morphological differences between Mesolithic and more recent South Asian crania observed in this study (which had been observed and described before [see Kennedy, 2000]), do not rule out the possibility of gene flow during the spread of agriculture, particularly in terms of gene flow across southern Asia from a population source within the subcontinent. The relative pattern of homogeneity observed here suggests that, independent of whether or not there was an important demographic event in the mid-Holocene, there was relatively little gene flow from outside of south Asia afterwards.

The analysis of global patterns of craniometric variation in this study has further implications for our understanding of human dispersals and diversity. It is interesting to note that in the discriminant analyses carried out, cranial diversity within Africa falls in the centre of craniometric variation globally. This underscores the range of morphological variability found within Africa. European and Australo-Melanesian crania are clearly divergent in morphology, are separated by the first discriminant function, and are differentiated on either side of the African centroid. These results are consistent with the interpretation of an African origin for all non-African humans, but they highlight patterns of morphological divergence that may be reflective of multiple dispersal events out of Africa. A model of dispersals that could explain this variation is the presence of an initial, early dispersal that led to South Asian and Australo-Melanesian populations, and a subsequent dispersal that led to Europeans. While this is a plausible explanation for global variation in mitochondrial haplogroups M and N, the accuracy of these interpretations will require further testing through focused paleoanthropological and genetic research.

The current study places aspects of the cranial morphology of prehistoric and contemporary populations of the Indian subcontinent within a wider geographic context. In the process, it stresses that when human morphology in these regions is viewed on a very broad and global scale, South Asian people appear homogenous. This could be explained by either a relatively private evolutionary history characterized by little genetic exchange with neighbouring populations to the East, West and North, or by a relatively recent demographic expansion of a population from within South Asia accompanied by some admixture with the existing groups throughout the subcontinent. Both historical scenarios can be consistent with a growing body of inductive evidence that supports hypotheses of predominantly *in situ* development of contemporary populations within the Indian subcontinent following the initial occupation of the region from an African source with relatively minor genetic input from outside the subcontinent. However, the second model (that of a recent demographic event largely responsible for the homogenization of South Asian populations) can accommodate earlier gene flow from outside the region. The morphological differences observed among the archaeological south Asian crania are in line with previous studies, and do not rule out the possibility of some genetic input as suggested by demic diffusion models. More detailed analyses of skeletal material from archaeological contexts will provide important opportunities to test these relationships in the future, particularly when combined with the emerging genetic picture of the population history of India. Through further research we may gain a better understanding of the relationships between archaeological and more recent populations, the process of morphological adaptation among South Asian populations, and the relationship between morphology and phylogenetic relationships between populations.

Acknowledgments

The authors wish to thank T Hanihara and WW Howells for providing craniometric data used in this study. This research was partially funded by an NERC-EFCHED (Natural Environment Research Council, U.K) grant to one of the authors (MML). We further wish to thank three anonymous reviewers for comments and suggestions that improved this chapter.

References

Agarwal, R.P., Mitra, D.S., 1991. Paleogeographic reconstructions of Bengal Delta during Quaternary period. In: Vaidyanadhan, R. (Ed.), Quaternary Deltas of India: Memoirs of the Geological Society of India, pp. 13–24.

Ammerman, A.J., Cavalli-Sforza, L.L., 1984. The Neolithic Transition and the Genetics of Populations in Europe. Princeton University Press, Princeton.

Arensburg, B., 1977. New Upper Paleolithic human remains from Israel. Eretz-Israel 13, 208–215.

Arensburg, B., 1981. Recent evolution in Israël. Colloques internationaux du C.N.R.S. N. 599 - Les processus d l'hominisation, pp. 195–201.

Bamshad, M., Kivisild, T., Watkins, W.S., Dixon, M.E., Ricker, C.E., Rao, B.B., Naidu, J.M., Prasad, B.V.R., Reddy, P.G., Rasanayagam, A., Papiha, S.S., Villems, R., Redd, A.J., Hammer, M.F., Nguyen, S.V., Carroll, M.L., Batzer, M.A., Jorde, L.B., 2001. Genetic evidence on the origins of Indian caste populations. Genome Research 11(6), 994–1004.

Bar-Yosef, O., 1992. The role of western Asia in modern human origins. Philosophical Transactions of the Royal Society of London Series B-Biological Sciences 337(1280), 193–200.

Bar-Yosef, O., Vandermeersch, B., Arensburg, B., Goldberg, P., Laville, H., Meignen, L., Rak, Y., Tchernov, E., Tillier, A.M., 1986. New data on the origin of modern man in the Levant. Current Anthropology 27(1), 63–65.

Barker, G., Barton, H., Beavitt, P., Bird, M., Daly, P., Doherty, C., Gilberson, D., Hunt, C., Krigbaum, J., Lewis, H., Manser, J., McClaren, S., Paz, V., Piper, P., Pyatt, P., Rabett, R., Reynolds, T., Rushworth, G., Stephens, M., 2002. Prehistoric foragers and farmers in Southeast Asia: renewed investigations at Niah Cave, Sarawak. Proceedings of the Prehistoric Society 68, 147–164.

Bartel, B., 1979. A discriminant analysis of Harappan civilization human populations. Journal of Archaeological Science 6, 49–61.

Bergman, C.A., Stringer, C., 1989. Fifty years after: Egbert, an early Upper Palaeolithic juvenile from Ksar Akil, Lebanon. Paleorient 15, 99–111.

Bowler, J.M., Johnston, H., Olley, J.M., Prescott, J.R., Roberts, R.G., Shawcross, W., Spooner, N.A., 2003. New ages for human occupation and climatic change at Lake Mungo, Australia. Nature 421(6925), 837–840.

Cann, R., Stoneking, M., Wilson, A., 1987. Mitochondrial DNA and human evolution. Nature 325, 31–36.

Cavalli-Sforza, L.L., Menozzi, P., Piazza, A., 1994. The History and Geography of Human Genes. Princeton University Press, Princeton.

Cordaux, R., Deepa, E., Vishwanathan, H., Stoneking, M., 2004. Genetic evidence for the demic diffusion of agriculture to India. Science 304, 1125.

Cordaux, R., Saha, N., Bentley, G.R., Aunger, R., Sirajuddin, S.M., Stoneking, M., 2003. Mitochondrial DNA analysis reveals diverse histories of tribal populations from India. European Journal of Human Genetics 11, 253–264.

Day, M., Stringer, C., 1982. A reconsideration of the Omo Kibish remains. In: de Lumley, H. (Ed.), L'*Homo erectus* et la Place de L'homme de Tautavel parmi les Hominidés fossils. Centre National de la Recherche Scientifique, Nice, pp. 814–846.

Day, M.H., 1969. Omo human skeletal remains. Nature 222, 1135–1138.

Day, M.H., Stringer, C.B., 1991. Les restes crâniens d'Omo-Kibish et leur classification à l'intérieur du genre *Homo*. Anthropologie 95, 574–594.

Dennell, R., Rendell, H., Halim, M., Moth, E., 1992. A 45,000-year-old open-air Paleolithic site at Riwat, Northern Pakistan. Journal of Field Archaeology 19, 17–33.

Deraniyagala, S.U., 1992. The Prehistory of Sri Lanka: An Ecological Perspective. Department of the Archaeological Survey, Government of Sri Lanka, Colombo.

Ehrhardt, S., 1963. Frühmenschliche skelette aus Langhnaj in Gujarat. Zeitschrift für Morphologie und Anthropologie 5(2), 151–162.

Etler, D.A., 1996. The fossil evidence for human evolution in Asia. Annual Review of Anthropology 25, 275–301.

Excoffier, L., 2002. Human demographic history: refining the recent African origin model. Current Opinions in Genetic Developments 12(6), 675–682.

Field, J., Lahr, M.M., 2006. Assessment of the southern dispersal: GIS-based analyses of potential routes at Oxygen Isotopic Stage 4. Journal of World Prehistory 19(1):1–45.

Foley, R.A., Lahr, M.M., 1992. Beyond "out of Africa": reassessing the origins of *Homo sapiens*. Journal of Human Evolution 22, 523–529.

Foley, R.A., Lahr, M.M., 1997. Mode 3 technologies and the evolution of modern humans. Cambridge Archaeological Journal 7, 3–36.

Gillespie, R., 2002. Dating the first Australians. Radiocarbon 44(2), 455–472.

Guha, B.S., 1935. The racial affinities of the people of India. Census of India, 1931. Government of India Press, Delhi, pp. 2–22.

Hammer, M.F., Zegura, S.L., 2002. The human Y chromosome haplogroup tree: nomenclature and phylogeny of its major divisions. Annual Review of Anthropology 31, 303–321.

Harpending, H., Rogers, A., 2000. Genetic perspectives on human origins and differentiation. Annual Reviews Genomics – Human Genetics 1, 361–385.

Hemphill, B.E., Christensen, A.F., Mustafakulov, S.I., 1996. East meets West: a diachronic analysis of Bronze Age biological interactions across the Indo-Iranian borderlands. In: Allchin, B. (Ed.), South Asia Archaeology 1995. Oxford-IBH, New Delhi.

Hemphill, B.E., Lukacs, J.R., 1993. Hegelian logic and Harappan civilization: an investigation of Harappan biological affinities in the light of recent biological and archaeological research. In: Gail, A.J., Mevissen, G.J.R. (Eds.), South Asian Archaeology 1991. Franz Steiner Verlag, Stuttgart.

Hemphill, B.E., Lukacs, J.R., Kennedy, K.A.R., 1991. Biological adaptations and affinities of Bronze Age Harappans. In: Meadow, R.H. (Ed.), Harappan excavations 1986–1990: a multidisciplinary approach to third century urbanism. Prehistory Press, Madison, pp. 137–182.

Holliday, T.W., 2000. Evolution at the crossroads: modern human emergence in Western Asia. American Anthropologist 102(1), 54–68.

Howells, W.W., 1976. Explaining modern man. Journal of Human Evolution 4, 477–496.

Ingman, M., Kaessmann, H., Paabo, S., Gyllensten, U., 2000. Mitochondrial genome variation and the origin of modern humans. Nature 408(6813), 708–713.

James, H.V.A., Petraglia, M.D., 2005. Modern human origins and the evolution of behavior in the Later Pleistocene record of South Asia. Current Anthropology 46, S3–S27.

Jobling, M.A., Hurles, M.E., Tyler-Smith, C., 2004. Human Evolutionary Genetics: Origins, Peoples & Disease. Garland Science, New York.

Ke, Y.H., Su, B., Song, X.F., Lu, D.R., Chen, L.F., Li, H.Y., Qi, C.J., Marzuki, S., Deka, R., Underhill, P., Xiao, C.J., Shriver, M., Lell, J., Wallace, D., Wells, R.S., Seielstad, M., Oefner, P., Zhu, D.L., Jin, J.Z., Huang, W., Chakraborty, R., Chen, Z., Jin, L., 2001. African origin of modern humans in East Asia: a tale of 12,000 Y chromosomes. Science 292(5519), 1151–1153.

Kennedy, K.A.R., 1965. Human skeletal material from Ceylon, with an analysis of the Island's prehistoric and contemporary populations. Bulletin of the British Museum of Natural History (Geology) 11, 135–213.

Kennedy, K.A.R., 1992. Tooth size variations of the Vedda and prehistoric Sri Lankans. In: Lukacs, J.R. (Ed.), Culture, Ecology and Dental Anthropology. Journal of Human Ecology, Special Issue 2, pp. 171–182.

Kennedy, K.A.R., 1999. Paleoanthropology of South Asia. Evolutionary Anthropology 8(5), 165–185.

Kennedy, K.A.R., 2000. God-Apes and Fossil Men. The University of Michigan Press, Ann Arbor.

Kennedy, K.A.R., Chiment, J., Disotell, T., Meyers, D., 1984. Principal-components analysis of prehistoric South Asian Crania. American Journal of Physical Anthropology 64, 105–118.

Kennedy, K.A.R., Levisky, J., 1984. The element of racial biology in Indian megalithism: a multivariate analysis approach. Homo 35(3–4), 161–173.

Kivisild, T., Bamshad, M.J., Kaldma, K., Metspalu, M., Metspalu, E., Reidla, M., Laos, S., Parik, J., Watkins, W.S., Dixon, M.E., Papiha, S.S., Mastana, S.S., Mir, M.R., Ferak, V., Villems, R., 1999a. Deep common ancestry of Indian and western-Eurasian mitochondrial DNA lineages. Current Biology 9(22), 1331–1334.

Kivisild, T., Kaldma, K., Metspalu, E., Parik, J., Papiha, S., Villems, R., 1999b. The place of Indian mtDNA variants in the global network of maternal lineages and the peopling of the Old World. In: Deka, R., Papiha, S. (Eds.), Genomic Diversity. Plenum Publishers.

Kivisild, T., Papiha, S., Rootsi, S., Parik, J., Kaldma, K., Reidla, M., Laos, S., Metspalu, E., Pielberg, G., Adojaan, M., Metspalu, S., Mastana, S., Wang, Y., Gölge, M., Demirtas, H., Schnakenberg, E., DeStephano, G., Geberhiwot, T., Claustres, M., Villems, R., 2000. An Indian ancestry: a key for understanding human diversity in Europe and beyond. In: Renfrew, C., Boyle, K. (Eds.), Archaeogenetics: DNA and the Population Prehistory of Europe. McDonald Institute for Archaeological Research., Cambridge, pp. 267–275.

Kivisild, T., Reidla, M., Metspalu, E., Parik, J., Geberhiwot, T., Usanga, E., Chaventre, A., 2001. Eastern African origin of the human maternal lineage cluster, ancestral to people outside of Africa. American Journal of Human Genetics 69(4), 1386.

Kivisild, T., Rootsi, S., Metspalu, M., Mastana, S., Kaldma, K., Parik, J., Metspalu, E., Adojaan, M., Tolk, H.V., Stepanov, V., Golge, M., Usanga, E., Papiha, S.S., Cinnioglu, C., King, R., Cavalli-Sforza, L., Underhill, P.A., Villems, R., 2003. The genetic heritage of the earliest settlers persists both in Indian tribal and caste populations. American Journal of Human Genetics 72(2), 313–332.

Klein, R.G., 1999. The Human Career: Human Biological and Cultural Origins. University of Chicago Press, Chicago.

Lahr, M.M., 1994. The multiregional model of modern human origins: a reassessment of its morphological basis. Journal of Human Evolution 26, 23–56.

Lahr, M.M., 1996. The Evolution of Modern Human Diversity: A Study in Cranial Variation. Cambridge University Press, Cambridge.

Lahr, M.M., Foley, R.A., 1994. Multiple dispersals and modern human origins. Evolutionary Anthropology 3(2), 48–60.

Lahr, M.M., Foley, R.A., 1998. Towards a theory of modern human origins: geography, demography and diversity in recent human evolution. Yearbook of Physical Anthropology 41, 137–176.

Lukacs, J.R., 1990. On hunter-gatherers and their neighbors in prehistoric India: contact and pathology. Current Anthropology 31, 183–186.

Lukacs, J.R., 2002. Hunting and gathering strategies in prehistoric India: a biocultural perspective on trade and subsistence. In: Morrison, K.D., Junker, L.L. (Eds.), Forager-traders in South and Southeast Asia: Long-term Histories. Cambridge University Press, Cambridge, pp. 41–61.

Lukacs, J.R., Pal, J.N., 1993. Mesolithic subsistence in north India: inferences from dental attributes. Current Anthropology 34, 745–765.

Lukacs, J.R., Pal, J.N., 2003. Skeletal variation among Mesolithic people of the Ganga Plains: new evidence of habitual activity and adaptation to climate. Asian Perspectives 42(2), 329–351.

Macaulay, V., Hill, C., Achilli, A., Rengo, C., Clarke, D., Meehan, W., Blackburn, J., Semino, O., Scozzari, R., Cruciani, F., Taha, A., Shaari, N.K., Raja, J.M., Ismail, P., Zainuddin, Z., Goodwin, W., Bulbeck, D., Bandelt, H.J., Oppenheimer, S., Torroni, A., Richards, M., 2005. Single, rapid coastal settlement of Asia revealed by analysis of complete mitochondrial genomes. Science 308(5724), 1034–1036.

Majumdar, D.N., 1961. Races and Cultures of India. Asia Publishing House, Bombay.

Majumder, P., 2001. Ethnic populations of India as seen from an evolutionary perspective. Journal of Bioscience 26(4), 533–545.

Majumder, P.P., Shankar, B.U., Basu, A., Mahotra, K.C., Gupta, R., Mukhopadhyay, B., Vijayakumar, M., Roy, S.K., 1990. Anthropometric variation in India: a statistical appraisal. Current Anthropology 31, 94–103.

McDougall, I., Brown, F.H., Fleagle, J.G., 2005. Stratigraphic placement and age of modern humans from Kibish, Ethiopia. Nature 433(7027), 733–736.

Mercier, N., Valladas, H., Baryosef, O., Vandermeersch, B., Stringer, C., Joron, J.L., 1993. Thermoluminescence date for the Mousterian burial site of Es-Skhul, Mt-Carmel. Journal of Archaeolocial Science 20, 169–174.

Metspalu, M., Kivisild, T., Metspalu, E., Parik, J., Hudjashov, G., Kaldma, K., Serk, P., Karmin, M., Behar, D.M., Gilbert, M.T.P., Endicott, P., Mastana, S., Papiha, S.S., Skorecki, K., Torroni, A., Villems, R., 2004. Most of the extant mtDNA boundaries in South and Southwest Asia were likely shaped during the initial settlement of Eurasia by anatomically modern humans. BMC Genetics 5, article no. 26.

Misra, V.N., 1995. Geoarchaeology of the Thar Desert, North West India. In: Wadia, R., Korisettar, R., Kale, V.S. (Eds.), Quaternary Environments and Geoarchaeology of India. Geological Society of India, Bangalore, pp. 210–230.

Mitchell, R.J., Howlett, S., White, N.G., Federle, L., Papiha, S.S., Briceno, I., McComb, J., Schanfield, M.S., Tyler-Smith, C., Osipova, L., Livshits, G., Crawford, M.H., 1999. Deletion polymorphism in the human COL1A2 gene: genetic evidence of a non-African population whose descendants spread to all continents. Human Biology 71, 901–914.

Morrison, K.D., 2002. Historicizing adaptation, adapting to history: forager-traders in South and Southeast Asia. In: Morrison, K.D., Junker, L.L. (Eds.), Forager-traders in South and Southeast Asia: Long-term Histories. Cambridge University Press, Cambridge.

Morwood, M.J., Soejono, R.P., Roberts, R.G., Sutikna, T., Turney, C.S.M., Westaway, K.E., Rink, W.J., Zhao, J.X., van den Bergh, G.D., Due, R.A., Hobbs, D.R., Moore, M.W., Bird, M.I., Fifield, L.K., 2004. Archaeology and age of a new hominin from Flores in eastern Indonesia. Nature 431(7012), 1087–1091.

Possehl, G.L., 2002. Harappans and hunters: economic interaction and specialization in prehistoric India. In: Morrison, K.D., Junker, L.L. (Eds.), Forager-traders in South and Southeast Asia: Long-term Histories. Cambridge University Press, Cambridge, pp. 62–76.

Quintana-Murci, L., Chaix, R., Wells, R.S., Behar, D.M., Sayar, H., Scozzari, R., Rengo, C., Al-Zahery, N., Semino, O., Santachiara-Benerecetti, A.S., Coppa, A., Ayub, Q., Mohyuddin, A., Tyler-Smith, C., Mehdi, S.Q., Torroni, A., McElreavey, K., 2004. Where West meets East: the complex mtDNA landscape of the southwest and Central Asian corridor. American Journal of Human Genetics 74(5), 827–845.

Quintana-Murci, L., Krausz, C., Zerjal, T., Sayar, S.H., Hammer, M.F., Mehdi, S.Q., Ayub, Q., Qamar, R., Mohyuddin, A., Radhakrishna, U., Jobling, M.A., Tyler-Smith, C., McElreavey, K., 2001. Y-chromosome lineages trace diffusion of people and languages in southwestern Asia. American Journal of Human Genetics 68(2), 537–542.

Quintana-Murci, L., Seminol, O., Bandelt, H.J., Passarinol, G., McElreavey, K., Santachiara-Benerecetti, S., 1999. Genetic evidence for an early exit of Homo sapiens sapiens from Africa through eastern Africa. Nature Genetics 23(4), 437–441.

Rathburn, T.A., 1982. Morphological affinities and demography of Metal Age southwest Asian populations. American Journal of Physical Anthropology 59, 47–60.

Redd, A.J., Roberts-Thomsen, J., Karafet, T., Bamshad, M., Jorde, L., Naidu, J., Walsh, B., Hammer, M., 2002. Gene flow from the Indian subcontinent to Australia: evidence from the Y chromosome. Current Biology 12, 673–677.

Renfrew, C., 1987. Archaeology and Language: The Puzzle of Indo-European Origins. Jonathan Cape, London.

Renfrew, C., 1996. Language families and the spread of farming. In: Harris, D.R. (Ed.), The Origins and Spread of Agriculture and Pastoralism in Eurasia. Smithsonian Institution Press, Washington, DC, pp. 70–92.

Risley, H.H., 1908. People of India. Tacker Spink & Co., Calcutta.

Roberts, R.G., Jones, R., Smith, M.A., 1990. Thermoluminescence dating of a 50,000-year-old human occupation site in northern Australia. Nature 345(6271), 153–156.

Roberts, R.G., Jones, R., Spooner, N.A., Head, M.J., Murray, A.S., Smith, M.A., 1994. The human colonization of Australia – optical dates of 53,000 and 60,000 years bracket human arrival at Deaf-Adder Gorge, Northern-Territory. Quaternary Science Reviews 13(5–7), 575–583.

Roychoudhury, S., Roy, S., Dey, B., Chakraborty, M., Roy, M., Ramesh, A., Prabhakaran, N., Usha Rani, M., Vishwanathan, H., Mitra, M., Sil, S., Majumder, P., 2000. Fundamental genomic unity of ethnic India is revealed by analysis of mitochondrial DNA. Current Science 79(9), 1182–1192.

Sahoo, S., Singh, A., Himabindu, G., Banerjee, J., Sitalaximi, T., Gaikwad, S., Trivedi, R., Endicott, P., Kivisild, T., Metspalu, M., Villems, R., Kashyap, V.K., 2006. A prehistory of Indian Y chromosomes: evaluating demic diffusion scenarios. Proceedings of the National Academy of Sciences 103(4), 843–848.

Sarkar, S.S., 1954. The Aboriginal Races of India. Bookland, Calcutta.

Sarkar, S.S., 1972. Ancient Races of the Deccan. Munschiram Manoharlal, New Delhi.

Stringer, C., 2000. Palaeoanthropology – coasting out of Africa. Nature 405(6782), 24–27.

Stringer, C.B., 1992. Reconstructing recent human evolution. Philosophical Transactions of the Royal Society of London Series B-Biological Sciences 337(1280), 217–224.

Stringer, C.B., Grun, R., Schwarcz, H.P., Goldberg, P., 1989. ESR dates for the hominid burial site of Es Skhul in Israel. Nature 338(6218), 756–758.

Stringer, C.B., Hublin, J.J., 1984. The origin of anatomically modern humans in Western Europe. In: Smith, F.H., Spencer, F. (Eds.), The Origin of Modern Humans: A World Survey of the Fossil Evidence. Alan R. Liss, New York, pp. 51–135.

Tchernov, E., 1992a. The Afro-Arabian component in the Levantine mammalian fauna – a short biogeographical review. Israel Journal of Zoology 38, 155–192.

Tchernov, E., 1992b. Biochronology, paleoecology and disperal events of hominids in the southern Levant. In: Akazawa, T., Aoki, K., Kimura, S. (Eds.), The Evolution and Dispersal of Modern Humans in Asia. Hokusen-sha, Tokyo, pp. 149–188.

Thangaraj, K., Chaubey, G., Kivisild, T., Reddy, A.G., Singh, V.K., Rasalkar, A.A., Singh, L., 2005. Reconstructing the origin of Andaman Islanders. Science 308(5724), 996.

Thorne, A., Grun, R., Mortimer, G., Spooner, N.A., Simpson, J.J., McCulloch, M., Taylor, L., Curnoe, D., 1999. Australia's oldest human remains: age of the Lake Mungo 3 skeleton. Journal of Human Evolution 36(6), 591–612.

Tishkoff, S.A., Dietzsch, E., Speed, W., Pakstis, A.J., Kidd, J.R., Cheung, K., BonneTamir, B., Santachiara-Benerecetti, A.S., Moral, P., Krings, M., Paabo, S., Watson, E., Risch, N., Jenkins, T., Kidd, K.K., 1996. Global patterns of linkage disequilibrium at the CD4 locus and modern human origins. Science 271(5254), 1380–1387.

Turbon, D., PerezPerez, A., Stringer, C.B., 1997. A multivariate analysis of Pleistocene hominids: testing hypotheses of European origins. Journal of Human Evolution 32(5), 449–468.

Turney, C.S.M., Bird, M.I., Fifield, L.K., Roberts, R.G., Smith, M., Dortch, C.E., Grun, R., Lawson, E., Ayliffe, L.K., Miller, G.H., Dortch, J., Cresswell, R.G., 2001. Early human occupation at Devil's Lair, southwestern Australia 50,000 years ago. Quaternary Research 55(1), 3–13.

Underhill, P.A., Passarino, G., Lin, A.A., Shen, P., Lahr, M.M., Foley, R.A., Oefner, P.J., Cavalli-Sforza, L.L., 2001. The phylogeography of Y chromosome binary haplotypes and the origins of modern human populations. Annals of Human Genetics 65, 43–62.

van Andel, T.H., Tzedakis, P.C., 1996. Palaeolithic landscapes of Europe and Environs, 150,000 - 25,000 years ago. Quaternary Science Reviews 15, 481–500.

Vishwanathan, H., Deepa, E., Cordaux, R., Stoneking, M., Usha Rani, M.V., Majumder, P.P., 2004. Genetic structure and affinities among tribal populations of southern India: a study of 24 autosomal markers. Annals of Human Genetics 68, 128–138.

Warusawithana-Kulatilake, S., 1996. Cranial Variation and the Dispersal of Modern Humans in South Asia. Tharanjee Prints, Colombo.

Watkins, W.S., Bamshad, M., Dixon, M.E., Bhaskara Rao, B., Naidu, J.M., Reddy, P.G., Prasad, B.V.R., Das, P.K., Reddy, P.C., Gai, P.B., Bhanu, A., Kusuma, Y.S., Lum, J.K., Fischer, P., Jorde, L.B., 1999. Multiple origins of the mtDNA 9-bp deletion in populations of South India. American Journal of Physical Anthropology 109, 147–158.

Whewell, W., 1840. The Philosophy of the Inductive Sciences, Founded Upon Their History. John W Parker, London.

White, T.D., Asfaw, B., Degusta, D., Gilbert, H., Richards, G.D., Suwa, G., Howell, F.C., 2003. Pleistocene *Homo sapiens* from Middle Awash, Ethiopia. Nature 423 742–747.

Whittle, A., 1996. Europe in the Neolithic. Cambridge University Press, Cambridge.

Zilhão, J., d'Errico, F., 1999. The chronology and taphonomy of the earliest Aurignacian and its implications for the understanding of Neanderthal extinction. Journal of World Prehistory 13, 1–68.

PART III

NEW WORLDS IN THE HOLOCENE

12. Interpreting biological diversity in South Asian prehistory: *Early Holocene population affinities and subsistence adaptations*

JOHN R. LUKACS

Department of Anthropology
University of Oregon
Eugene, OR 97403-1218
jrlukacs@oregon.uoregon.edu

Introduction

The goal of this contribution is to provide an integrated synthesis of current research on prehistoric human skeletal remains from South Asia. Data for this study come from dental morphology, as well as from skeletal and dental pathology, and are derived from geographically and chronologically diverse samples. This approach to human biological affinities and subsistence patterns is conducted on the theoretical foundations of organic evolution (Mayr, 1982; Smith, 1993; Gould, 2002). An integrated yet multifaceted approach to population affinities and health in South Asian prehistory will reveal the array of influences that contribute to the complexity of human phenotypic diversity. Evolutionary genetics maintains that phenotypic diversity is the product of complex interactions among forces during ontogeny that include the combined influence of genotypic and environmental factors (Provine, 1971; Ridley, 2004). The deceptively simple yet elegant equation, which governs ontogenetic development generally is: phenotypic variance = genotypic variance + environmental variance. The simplicity of this equation is misleading. Its role in illuminating gene – environment interaction during ontogeny in real world situations is difficult, yet it provides a conceptual model for the dual approaches adopted in this interpretation of human skeletal and dental diversity in South Asia. In this analysis the more highly heritable and genetically influenced aspects of the dental phenotype, especially morphological variations, are employed in the estimation of biological affinities among prehistoric skeletal samples (Scott and Turner, 1997). Biological attributes whose variability is more directly influenced by the organism's interaction with the environment, such as pathological lesions of the oral cavity and the skeletal system, are analyzed to gain insight into activity patterns, subsistence systems, and dietary behaviors. The value of an integrated dualistic approach to human diversity in South Asian prehistory lies in the greater breadth and depth of understanding that it provides.

M.D. Petraglia and B. Allchin (eds.), The Evolution and History of Human Populations in South Asia, 271–296.

This chapter presents results of recent research on the population affinities and health status of early Holocene hunting and foraging peoples of the mid-Ganga plain. The questions and problems that drive the anthropological research reported here are of broad interest to a wide range of scholars. Fundamental issues involving the biological relationships of past human populations include such questions as: How are the early hunters and foragers of the Ganga Plain biologically related to Neolithic people of the greater Indus Plain? What are the biogenetic affinities between the 'Mesolithic' inhabitants of north India and skeletal samples in different regions and prehistoric periods? When interests shift to human – environment interaction in prehistory, a rather different set of research questions become relevant. How do local environments and subsistence systems impact dietary patterns and health status? Do subsistence and diet have discernibly different consequences on males and females?

Answers to these distinctive kinds of questions rely on different, yet complementary classes of data. One data set is comprised of over two dozen polymorphic genetically determined variations in dental morphology. The other consists of proliferative and degenerative pathological lesions that can be recognized in skeletal and dental tissues. These two data sets inform us about population affinities and health status, respectively, but they typically are not equally well preserved or manifested in single specimens. Normal patterns of tooth wear and pathological lesions, such as dental caries and antemortem tooth loss, are destructive in nature and gradually obliterate genetically informative morphological traits expressed on the tooth crown. A demographic effect is present here too, in which younger individuals yield more complete evidence of genetically influenced variables in the primary and permanent dentition. By contrast, older individuals in a skeletal series often do not preserve genetically informative biological traits well or in

abundance, because they are worn away or destroyed by pathological processes. Nevertheless, individuals from older age classes yield informative clues regarding environmental effects that yield insight into diet and health, ecology and disease.

Research Materials

The human remains on which this study is based come from four distinct chrono-cultural clusters. The focus of this analysis is referred to as the Mesolithic Lake Culture Complex. The remaining clusters are arranged in temporal sequence, from early to late, and provide comparative data for interpreting and evaluating biological attributes of the early Holocene, Mesolithic Lake Culture group (see Figure 1 for site locations). The five chrono-cultural clusters consist of the following:

1) Hunter/forager samples, from the Ganga Plain with 'Mesolithic' artifact associations (Uttar Pradesh, India). This sample consists of a total of 94 specimens from three main sites, including Sarai Nahar Rai (ca. 10,000 BP; n = 15), Mahadaha (ca. 4000 BP; n = 32), and Damdama (8500 BP; n = 47).

2) Early farming village samples from the site of Mehrgarh in the Kachi Plain (Baluchistan Province, Pakistan). These early farming sites produced a total sample of 163 specimens from two mortuary sites: an early Chalcolithic cemetery sample (ca. 6500 BP; n = 70) and a Neolithic cemetery sample (ca. 8000 BP; n = 93).

3) Urban farmers of the Indus Valley Civilization (Punjab Province, Pakistan). This mature phase sample was excavated and analyzed in 1987–1988 and consists of the Harappa cemetery R-37 sample (ca. 4500 BP; n = 90) and a subsequent addition to the sample made in 1994, known as the pottery yard sample (n = 5) yielding a total 95 specimens.

4) Northwest frontier sites with evidence of iron. Human remains from Sarai Khola

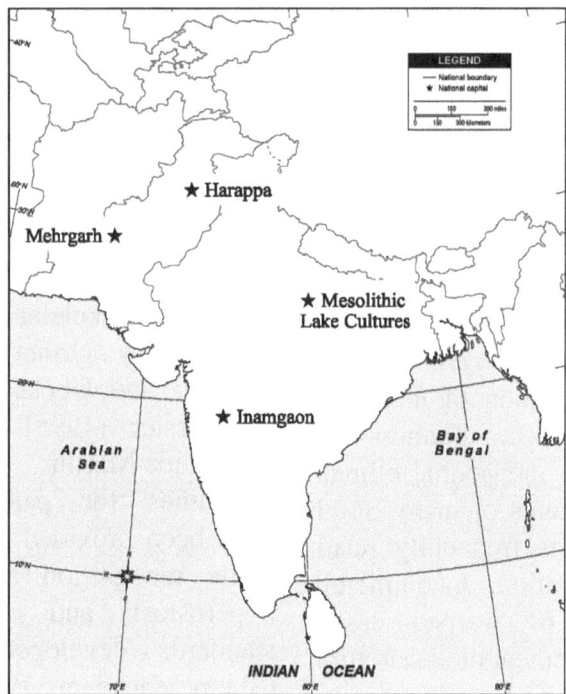

Figure 1. Map of site locations (modified from DeBlij and Muller, 2002)

(ca. 2000 BP; n = 46) are better preserved, but fewer in number than skeletons from Timargarha (ca. 3000 BP; n = 80). These sites provide a combined early Iron Age sample of 126 specimens.

5) Post-Harappan mixed subsistence early farmers of the Deccan Plateau with Chalcolithic material culture (Maharashtra, India). The thoroughly analyzed, primary site is Inamgaon (ca. 3350-2650 BP; n = 185). While other Deccan Chalcolithic sites have yielded human remains, the samples from Chandoli, Daimabad, and Nevasa are neither as abundant or well studied as the sample from Inamgaon.

Collectively, these skeletal samples total 663 individuals. Data from numerous additional sites are available, and have been analyzed by the author, yet are not included here due to small sample sizes, fragmentary condition or unique features of demography, geography or chronology that preclude their usefulness in this study. While one sample (from Harappa) was excavated and prepared for study by the author and colleagues, other samples were analyzed in either museum or field repositories. One important component of the research reported here is that all dental attributes, including morphological variations and pathological lesions were systematically examined by the author employing a single set of uniform methods. Therefore, issues of inter-oberver variability and differing systems of classification or methodological technique are obviated – improving the reliability of inter-sample comparisons.

Analytical Methods

The synthetic approach to understanding skeletal and dental variation in living and prehistoric populations of South Asia requires an array of analytical methods. Two complementary sets of methods are used: one for classifying variation in tooth crown morphology, and another for recognizing and diagnosing the disease processes responsible for causing pathological lesions of the skeleton and dentition.

Dental Morphology

The estimation of biological affinities among populations requires observing traits whose variable expression is controlled primarily by genetic variation. Populations that are similar for many independently controlled traits with high heritability are inferred to have close genetic affinities. When skeletal preservation is excellent, assessment of biological relationships from phenotypic variation may involve systematic observation of metric or non-metric variations of the cranium and postcranial skeleton. The monsoonal climate and post-burial environments of many South Asian archaeological sites frequently result in poor skeletal preservation. Incompletely preserved, fragmented or warped crania preclude accurate measurement or observation of non-metric variations. These concerns are exemplified at Damdama where most skulls are damaged beyond measurement (Figure 2), leaving the osteologist with small study samples rendering estimates of biological distance that are biased or unreliable. My research focuses on dental remains for this and several other practical and theoretical reasons. Teeth preserve well in archaeological contexts, even when crania are severely damaged, and details of dental morphology are clearly preserved (Figure 5, below). They present an abundant number of independent polymorphic variations of the tooth crown and root that are easily recorded. Morphological dental traits have a highly variable range of phenotypic expression that is under moderate to high levels of genetic influence. In addition, dental morphology can be observed and trait frequencies tabulated for living and prehistoric populations and direct comparisons of many trait frequencies can be computed with different multivariate statistical techniques. Systematic observation and classification of 25 morphological traits of the permanent teeth were conducted according to the guidelines of the Arizona State University Dental Anthropology System (Turner et al., 1991; Scott and Turner, 1997). A sub-set of ten morphological traits are used in this study.

Paleopathology

The recognition and scoring of skeletal pathology follows standards established and adhered to by practioners in the field. The primary sources for identifying and diagnosing pathological skeletal lesions include reference volumes by Ortner and Putschar (1981), Buikstra and Ubelaker (1994), Roberts and Manchester (1995), and Aufderheide and Rodriguez-Martín (1998). Supplemental guidelines for paleopathological analysis have been provided by Lovell (1997, 2000), while recognition and scoring of porotic hyperostosis and cribra orbitalia follows standards developed by Stuart-Macadam (Stuart-Macadam, 1989, 1992). Recognition and characterization of muscle attachment sites, or entheses, follows standards described by Hawkey and Merbs (1995) with attention to the concerns expressed by Robb (1998; Robb et al., 2001). Recognition of degenerative joint diseases, or osteoarthritis, rely on protocols established by Jurmain (1999), in which osteophytic lipping, porosity and eburnation were separately recorded.

Bioarchaeological Foundations and Archaeological Context of Research

This study of skeletal and dental variation in early Holocene north India is part of a larger and long term research agenda devoted to understanding the dynamics of biocultural transformation in ancient India. In the 1980s, the primary focus of bioarchaeological study was human skeletal and dental remains from Pakistan, including the Neolithic and Chalcolithic levels at Mehrgarh (Baluchistan Province), and from the Bronze Age cemetery at Harappa (Punjab Province). This prior research was devoted to descriptive and analytic aspects of variation in tooth size and morphology (Lukacs, 1986;

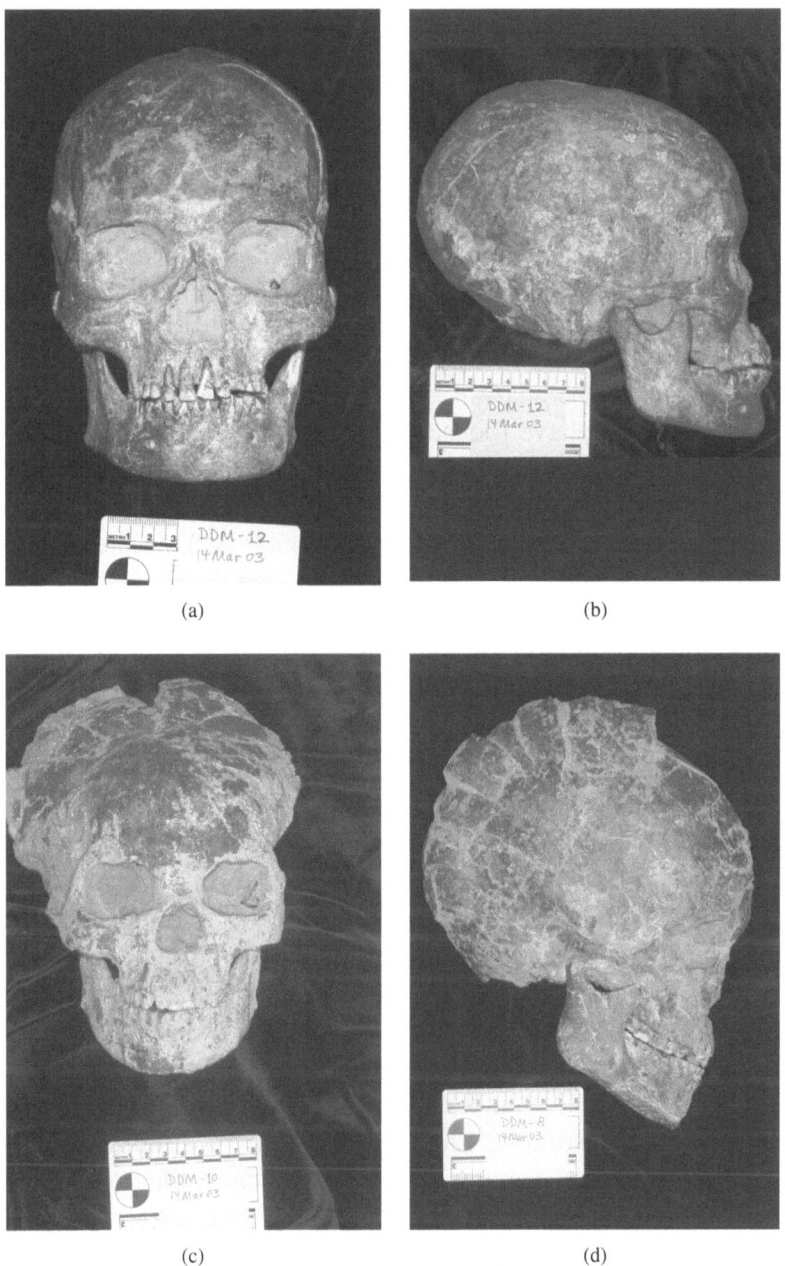

Figure 2. Variation in cranial preservation at Damdama, including well preserved skulls (a, b) and skulls affected by postmortem distortion (c, d). a) frontal view, and, b) right lateral view of DDM 12, a well-preserved, middle-aged adult female. c) Postmortem vertical compression (occipital-parietal) of DDM 10, and adult female. d) Postmortem lateral compression (occipital-parietal) of DDM 8, an adult male

Lukacs and Hemphill, 1991), dental wear patterns (Lukacs and Pastor, 1988, 1990), as well as multiple components of dental paleopathology (Lukacs, 1985; Lukacs et al., 1985; Lukacs and Minderman, 1992). Dental paleopathology at Harappa and a diachronic assessment of dental diseases yielded evidence for a decline in oral health with agricultural intensification in the greater Indus Valley (Lukacs, 1992).

Research on the bioarchaeology of human remains from Mesolithic Lake Culture sites of the Ganga Plain is extensive and on-going. Comprehensive analyses have been published on the skeletons from Sarai Nahar Rai (Kennedy et al., 1986) and Mahadaha (Kennedy et al., 1992). Preliminary reports on the dentition and stature of the Damdama series (Lukacs and Pal, 1993) confirm a phenotypic pattern similar to that described for Sarai Nahar Rai and Mahadaha. More recently, the analysis of human remains from Damdama have focused on aspects of skeletal pathology and subsistence patterns (Lukacs and Pal, 2007) and on issues of variation in stature and skeletal robusticity in response to activity levels and adaptation to climate (Lukacs and Pal, 2003). The research results reported in this study are built on this foundation of prior research and comprise the first inter-regional comparison of variation in dental morphology and skeletal and dental pathology between pene-contemporaneous early Holocene populations of the greater Indus Valley and middle Ganga Plains.

The site of Damdama is located in the middle Ganga Plain, 72 km NNE of the modern city of Allahadad (lat. $26^0 10' N$; long. $82^0 10' E$). Sister-sites are located nearby: Mahadaha – 5 km SSE and Sarai Nahar Rai – 40 km SW of Damdama. Mesolithic Lake Culture sites are similar in several features:

1) in geo-ecological setting they are situated on locally elevated ground adjacent to remnant ox-bow lakes; 2) in lithic technology all belong to the geometric microlithic tradition; and, 3) in site inventories they are all aceramic, and yield representative samples of fauna, stone and bone tools, and human burials. The antiquity of Mesolithic Lake Culture sites is problematic for several reasons, but recent AMS ^{14}C dates derived from human bone samples suggest an early Holocene antiquity for Damdama (8865 and 8640 BP; Lukacs et al., 1996). Absolute dates for Sarai Nahar Rai and Mahadaha are more variable and controversial,

yet all show a level of relative homogeneity in cultural inventory and biological attributes that suggests these sites are sampling one regional population.

The Mehrgarh site complex is situated in the Kachi Plain on flat alluvium, 15 m above the active Bolan River (lat.$29^0 28' N$; long. $67^0 39' E$). At the western-most margin of the greater Indus Valley, Mehrgarh is adjacent to the Bolan Pass, about 8 km from the front of the Harboi Hills, and 30 km SW of the modern town of Sibi. A multi-phase site, culture history at Mehrgarh begins in the Neolithic (ca. 8000 BP), and continues through early Chalcolithic (6500 BP), to late Chacolithic (4500 BP; Jarrige and Lechavellier, 1980). Cemeteries are similarly located within the living area at both sites, though the depth of burial features at Neolithic Mehrgarh (maximum = 11.0 m) is substantially greater than the depth of deposition at Mesolithic Lake Culture sites: Damdama (1.50 m), Mahadaha (0.60 m), Sarai Nahar Rai (0.06).

Burials exhibit significant differences in structural features, position of the deceased, and funerary accoutrements. Detailed description of Neolithic excavations and funerary practices at Mehrgarh includes an account of pit and wall construction, placement of the skeleton, and inferences derived from the sequential postmortem decomposition of joints (Lechevallier and Quivron, 1981, 1985; Sellier, 1990, 1992). A general description of Mesolithic Lake Culture settlements and burials (Pal 1986, 1992, 1994) document the close association of hearths, artifacts, and burials at Damdama and Mahadaha. Site-specific burial descriptions, line drawings and photographs are available for Sarai Nahar Rai (Sharma, 1973) and Mahadaha (Pal, 1985). Particular attention has been given to variation in the orientation and position of skeletons in five double burials and one triple burial at Damdama (Pal, 1988). Table 1 provides a comparison of the main features of human burials at Neolithic Mehrgarh and

Table 1. Comparison of burial features in early Holocene Indus and Ganga samples

Burial Feature	Neolithic Mehrgarh – Indus	Mesolithic Lake Culture – Ganga (Damdama, Mahadaha, Sarai Nahar Rai)
Structures	Mud brick wall	Ovoid pit
Orientation of body	East-west	East-west
Placement of body	Flexed, on left side, facing south head to east	extended, supine head to the west
Grave goods	Common:	
	body ornaments (turquoise)	few:
	bitumen lined basket	body ornaments (antler, bone)
	polished stone adze	lithics
	lithics: blades and cores	fauna
Example	MR 3–153 (Figure 3b)	DDM–38 (Figure 3a)

among the Lake Culture complex of the Ganga Plain. Representative burials from Neolithic Mehrgarh (MR 3-153) and Mesolithic Damdama (DDM-38) are provided in Figure 3. Note the absence of structural features and burial goods in the Mesolithic Lake Culture burial from Damdama. The overall impression derived from this comparison of burial features is one of greater cultural complexity and more developed material culture at Neolithic Mehrgarh than among the Mesolithic Lake Culture sites.

Research Results

Dental Morphology and Biological Distance

Morphological features of the Mesolithic Lake Culture dentition are documented and trait frequencies compared with data from Neolithic Mehrgarh. The degree of phenetic similarity in dental morphology is the basis for inferring the degree of biological affinity between these skeletal samples and between the Mesolithic Lake Culture sample and other prehistoric samples in South Asia. Homogeneity in the bio-cultural attributes of Mesolithic Lake Culture sites constitutes the basis for pooling dental morphological observations from Damdama, Mahadaha, and Sarai Nahar Rai for this analysis. In addition, the

high level of dental attrition documented in these series limits the number of individuals from whom morphological data could be attained (Lukacs and Pal, 1993). Dental data from Mahadaha and Sarai Nahar Rai alone, were inadequate to accurately characterize dental trait frequencies. The new morphological data from Damdama permit the first accurate descriptive and comparative evaluation of Mesolithic Lake Culture dental morphology.

Schematic comparison of dental crown trait frequencies for maxillary and mandibular teeth from Mesolithic Lake Culture sites and Neolithic Mehrgarh are presented in Figures 4a and 4b, respectively. Note the similarity in overall pattern of trait frequencies. Significant differences in dental morphology were found in 2 of 13 maxillary dental attributes and 3 of 14 mandibular traits. In maxillary traits, Neolithic Mehrgarh tends to have slightly higher frequencies of shovelling, interruption grooves (I2), and molar traits (hypocone, Carabelli's trait and metaconule), yet significant differences occur in only two traits: *tuberculum dentale* (I2) and the metaconule (M2). In mandibular dental traits, Mesolithic Lake Culture sites tend to have greater frequencies of the Y occlusal groove pattern and accessory molar cusps (C-6 and C-7), while 5 cusped molars are more frequent at Mehrgarh (M1 and M3).

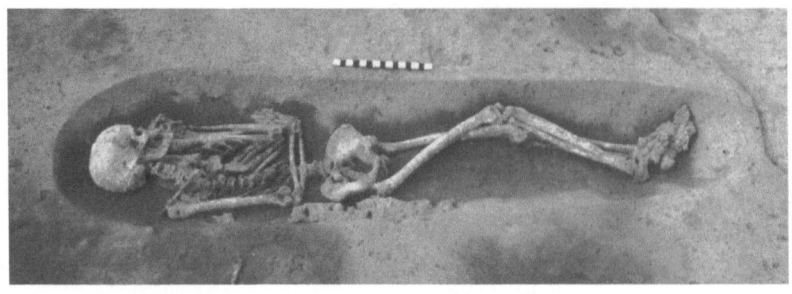

(a)

(b)

Figure 3. Comparison of burials: 'Mesolithic' Damdama and 'Neolithic' Mehrgarh. a) Damdama
burial 38 (DDM 38). Oriented east-west in a supine position, this specimen has legs slightly flexed,
the left arm tightly flexed and the right arm flexed 90^0. Note the absence of funerary goods. b)
Neolithic Mehrgarh burial 153. Oriented east-west, this individual was placed on their left side,
facing south, against a hand-made mud brick wall. Arms and legs are flexed. Burial goods include a
bitumen-lined basket (on right, adjacent to tibiae); a ground and polished stone axe, and body
adornments (small turquoise pieces) are present, but not visible in the photograph

Three lower molar traits are significantly different between groups: five cusped M3, Y groove M3, and the entoconulid (C-6, M3). Examples of dental preservation at Damdama, as well as representative morphological traits of the maxillary and mandibular dentition are provided in Figures 5 and 6, respectively.

A comparison of dental trait frequencies at Mesolithic Lake Culture sites and Neolithic Mehrgarh with five key South Asian prehistoric samples is designed to reveal similarities and differences in trait expression, and by inference biological affinity. Morphological trait frequencies for all seven groups are presented in Table 2. A chi-square test reveals that 54.5% (6/11) of maxillary dental

traits and 30% (3/10) of mandibular traits compared are significantly different among groups. In order to facilitate inter-group comparison of trait expression, standardized frequencies with mean $= 0$ and standard deviation $= 1$ were computed using the STANDARD procedure of SAS-PC (ver 8.0). Standardized frequencies of selected tooth crown traits are presented for four maxillary (Figure 7) and four mandibular (Figure 8) traits across all seven samples (see Table 3 for data). Arranged in approximate chronological order from left (earliest) to right (latest) these groups are: Mesolithic Lake Cultures (Damdama, Mahadaha, Sarai Nahar Rai; Ganga sample); Neolithic Mehrgarh (Indus

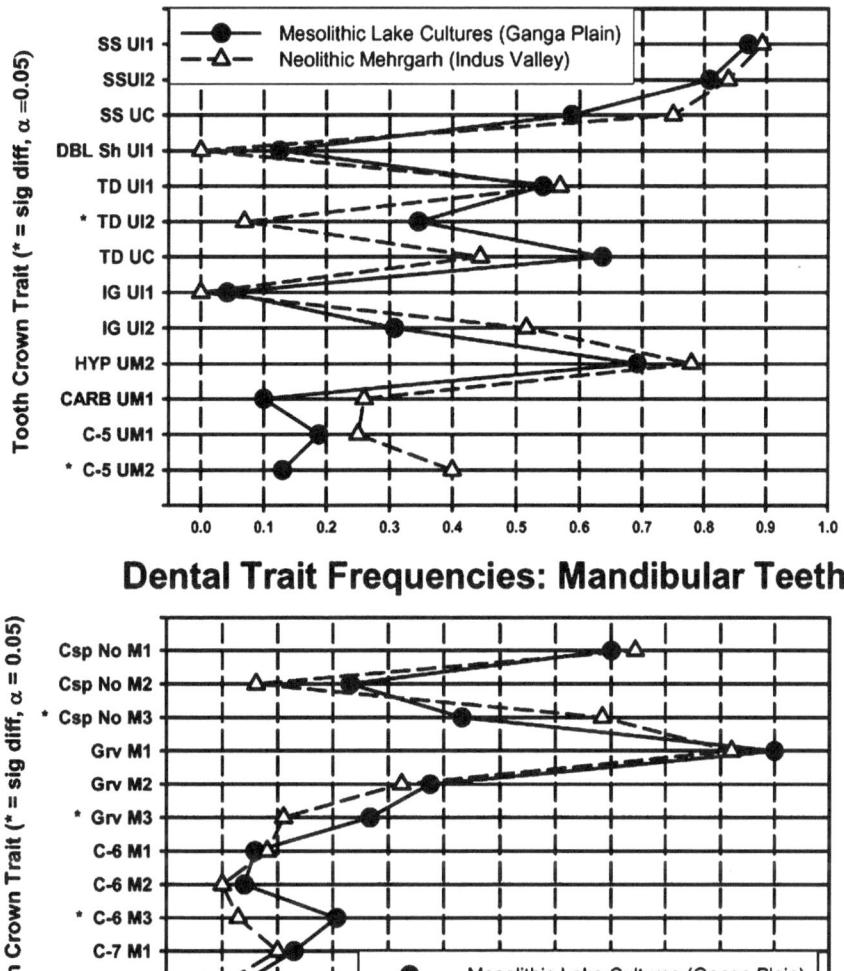

Figure 4. Dental morphology trait frequencies: Mesolithic Lake Cultures (Ganga Plain) and Neolithic Mehrgarh (Indus Valley) compared

sample); Chalcolithic Mehrgarh; Harappa; Inamgaon; Timargarha; and Sarai Khola.

In maxillary incisor traits (Figure 7), Mesolithic Lake Cultures and Neolithic Mehrgarh – Indus show strong positive deviations for I1 shovelling, but express divergent deviations in marginal interruptions for I2, with Mesolithic Lake Culture sites exhibiting a negative and Neolithic Mehrgarh a positive deviation. In maxillary molar morphology the Ganga Mesolithic Lake Culture and Indus samples share negative deviations for Carabelli's trait (M1), though the Lake Culture sample is more strongly negative than the Indus sample. Expression of the metaconule (M1) is divergent as well, with

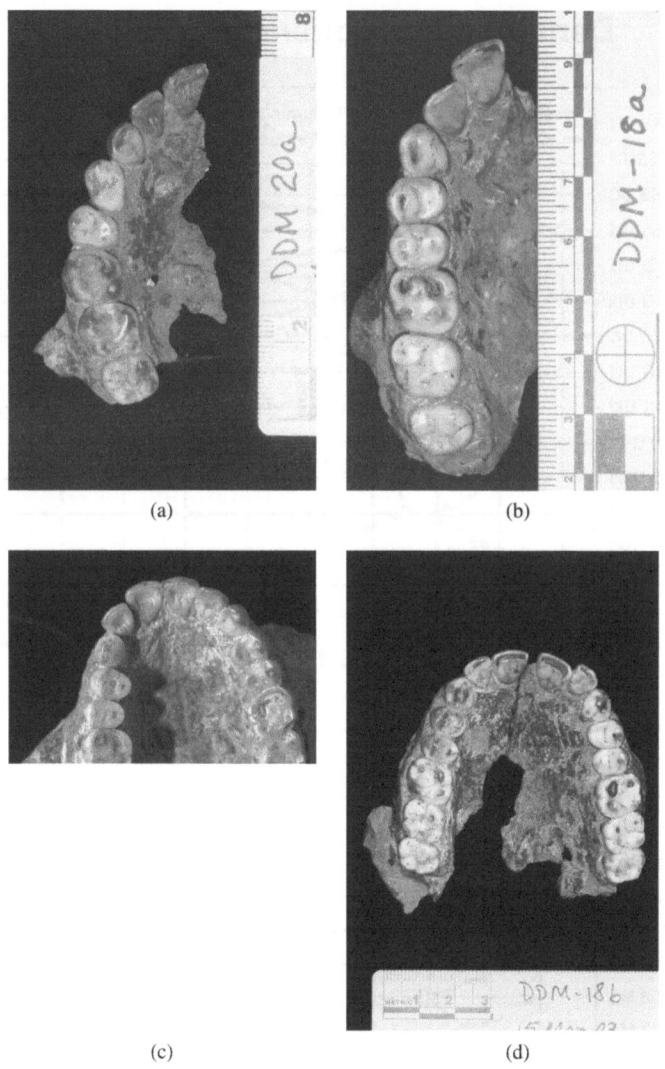

(a)   (b)

(c)   (d)

Figure 5. Morphological variation in the maxillary dentition of early Holocene foragers from Damdama, Mesolithic Lake Cultures. a) DDM 20a. Right maxillary dentition (young adult female). Note rotated premolar and C-5 (metaconulid) lower arrow. b) DDM 18a. Right maxillary dentition (young adult male). Note simple incisors (no shovel, no tuberculum dentale). c) DDM 36a. Anterior maxillary dentition (young adult male). Note shoveling of incisor teeth and well-developed tuberculum dentale (arrows). d) DDM 18b. Maxillary dentition (young adult male). Note simple incisors and well-developed hypocones (arrows)

moderately negative values for Mesolithic Lake Culture sites and a slightly negative deviation for Neolithic Mehrgarh, the Indus sample. Mandiblar molar traits (Figure 8) display different patterns as well, with 5 cusped second molars exhibiting a strong positive deviation in the Ganga sample and a modest negative deviation in the Indus sample, however both groups express moder-

ately positive deviation for the Y occlusal groove pattern in LM2. Variation in the expression of accessory lower molar cusps is greater for cusp 7 (the metaconulid), for which Mesolithic Lake Culture sites exhibit a positive deviation while the Indus sample from Neolithic Mehrgarh shows a moderately negative deviation. By contrast, the Mesolithic Lake Culture and Neolithic Mehrgarh samples

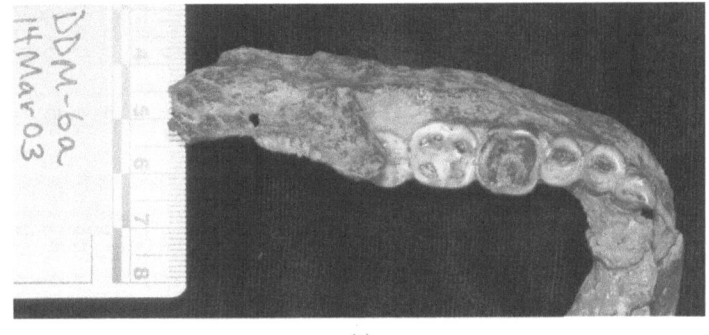

(a)

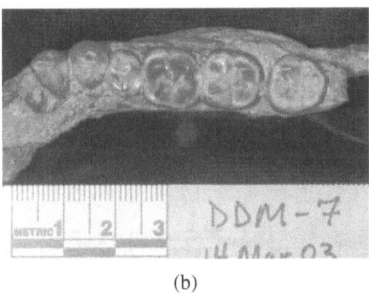

(b)

Figure 6. a) The right mandible of DDM 6a, an adult female. Note the heavy wear on M1 and the five cusped M2. b) The right mandibular dentition of DDM 7, an adult male. Note the Y groove pattern of M1 and M2, and the large size and presence of five cusps on M3

are more similar in their expression of accessory cusp 6 (the entoconulid), which exhibits modest and small negative deviations, respectively. Overall the Mesolithic Lake Culture dentition exhibits frequent but low grades of incisor and canine shoveling, low frequencies of double shovel and interruption grooves, while maxillary molars display low frequencies for Carabelli's trait, hypocone reduction, and the metaconule (Cusp 5). Mandibular dental traits common among Mesolithic Lake Culture samples include low levels of incisor shoveling and the distal canine accessory ridge, molarized second premolar (P4), and the presence large second lower molar teeth with the Y occlusal groove pattern and five cusps (Lukacs and Hemphill, 1992).

The relationship among South Asian prehistoric skeletal samples was preliminarily assessed using Ward's minimum variance cluster analysis (Ward, 1963). Ward's cluster technique is widely used in anthropological bio-distance studies. It is hierarchical, agglomerative and reliably yields clusters that accurately represent known group relationships. Clustering is based on standardized dental trait frequencies and the results are presented in Figure 9. Note the relatively close biological relationship between Neolithic Mehrgarh (8000 BP; Baluchistan) and Inamgaon (3350–2650 BP; Deccan Plateau). These groups are more closely related to one another than either is to the Mesolithic Lake Culture group. This triad – Neolithic Mehrgarh, Inamgaon, and the Mesolithic Lake Culture – is collectively distinct from, and distantly related to, other South Asian prehistoric dental samples. While close affinities exist between Harappa and Chalcolithic Mehrgarh, and between Sarai Khola and Timargarha, all four of theses groups are only distantly linked with the Mesolithic Lake Culture, Neolithic Mehrgarh and the Inamgaon cluster. The cluster pattern derived from standardized trait frequencies

Table 2. Dental morphology trait frequencies – prehistoric South Asia

Dental Trait	Tooth	Mesolithic Lake Culture 8700			Neolithic Mehrgarh 8000			Chalcolithic Mehrgarh 6500			Harappa 4500			Inamgaon 3350–2650			Timargarha 3000			Sarai Khola 2000		
Antiquity (BP)		P	N	f	P	N	f	P	N	f	P	N	f	P	N	f	P	N	f	P	N	f
Shovel-shape	UI1	20	23	0.870	25	28	0.893	21	25	0.840	10	18	0.556	22	24	0.917	1	7	0.143	3	9	0.333
	UI2	21	26	0.808	31	37	0.838	21	24	0.875	10	16	0.625	13	19	0.684	2	7	0.286	2	9	0.222
Tuberculum dentale	UI1	13	24	0.542	15	26	0.577	14	25	0.560	8	12	0.667	14	25	0.560	3	8	0.375	2	9	0.222
	UI2	9	26	0.346	2	29	0.069	7	24	0.292	6	13	0.462	1	20	0.050	0	7	0.000	0	9	0.000
Interruption	UI1	1	24	0.042	0	27	0.000	2	21	0.095	1	16	0.063	0	25	0.000	1	8	0.125	1	10	0.100
Groove	UI2	8	26	0.308	16	31	0.516	10	22	0.455	6	15	0.400	7	20	0.350	4	7	0.571	1	10	0.100
Hypocone Size	UM1	25	25	1.000	35	42	0.833	22	22	1.000	16	16	1.000	27	41	0.659	17	22	0.773	11	14	0.786
	UM2	18	26	0.692	2	41	0.049	10	18	0.556	2	18	0.111	0	20	0.000	0	13	0.000	2	13	0.154
Carabelli's	UM1	2	20	0.100	7	27	0.259	11	18	0.611	4	9	0.444	13	40	0.325	9	18	0.500	2	9	0.222
Metaconule	UM1	3	16	0.188	7	28	0.250	5	19	0.263	6	13	0.462	6	41	0.146	4	19	0.211	3	9	0.333
(C-5)	UM2	3	23	0.130	10	25	0.400	6	18	0.333	4	16	0.250	3	20	0.150	0	13	0.000	2	14	0.143
Cusp Number	LM1	14	20	0.700	39	43	0.907	20	23	0.870	17	20	0.850	32	39	0.821	19	25	0.760	9	15	0.600
	LM2	6	26	0.231	3	49	0.061	2	24	0.083	0	33	0.000	4	24	0.167	3	17	0.176	1	15	0.067
Y-Groove	LM1	17	17	1.000	23	25	0.920	15	21	0.714	15	17	0.882	32	35	0.914	12	17	0.706	5	7	0.714
Pattern	LM2	9	24	0.375	12	37	0.324	6	22	0.273	3	31	0.097	7	24	0.292	3	18	0.167	5	14	0.357
Entoconulid	LM1	1	17	0.059	3	37	0.081	5	23	0.217	1	20	0.050	4	37	0.108	0	22	0.000	1	14	0.071
(C-6)	LM2	1	25	0.040	0	44	0.000	1	23	0.043	0	28	0.000	0	24	0.000	1	18	0.056	0	15	0.000
Metaconulid	LM1	3	23	0.130	4	40	0.100	3	25	0.120	1	22	0.045	2	36	0.056	2	24	0.083	1	15	0.067
(C-7)	LM2	1	32	0.031	0	43	0.000	0	24	0.000	0	28	0.000	1	25	0.040	2	20	0.100	0	15	0.000

P = present; N = sample size; f = frequency

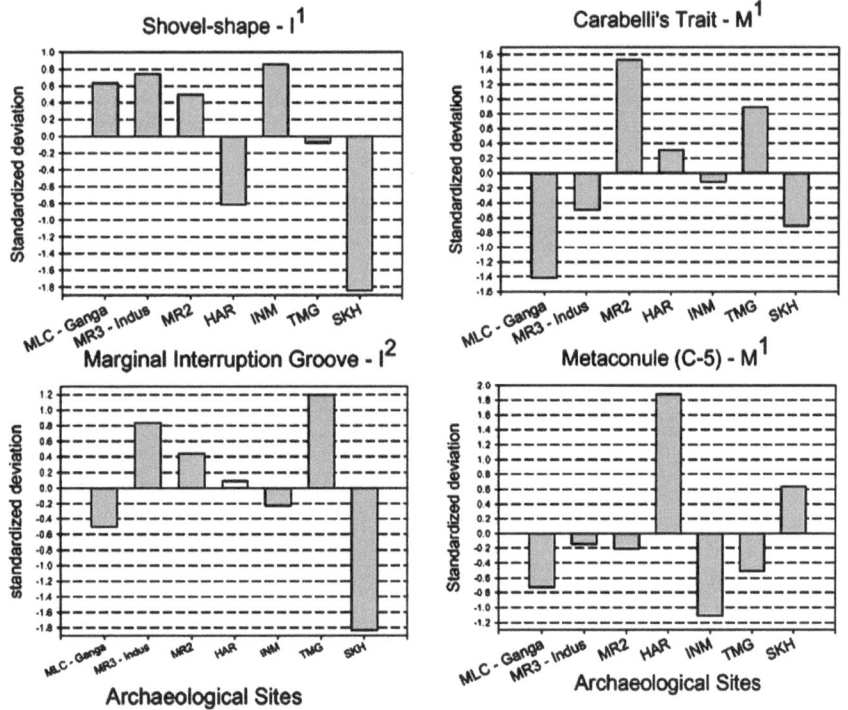

Figure 7. Standardized frequencies of selected polymorphic traits: maxillary teeth

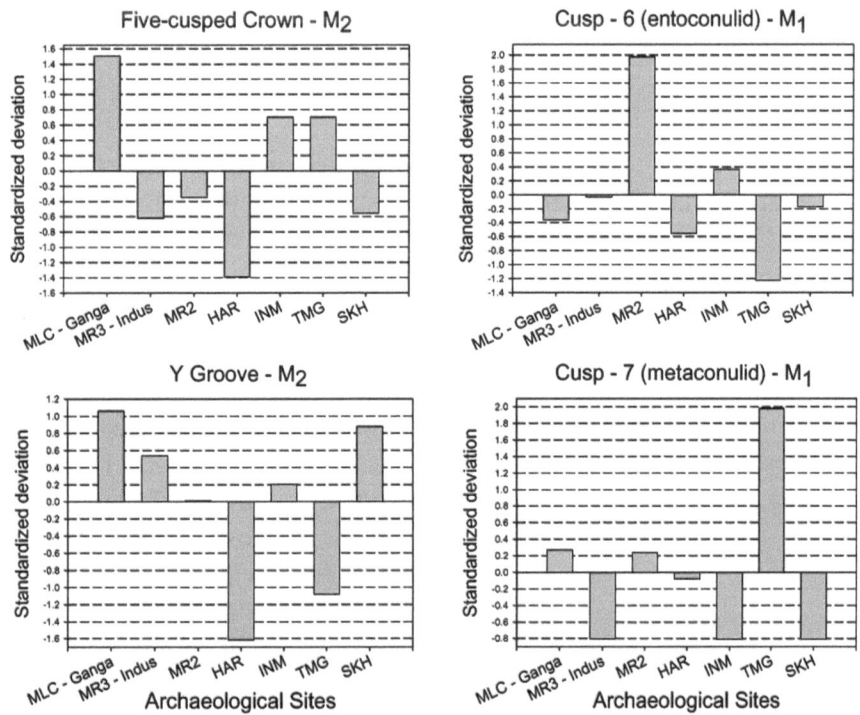

Figure 8. Standardized frequencies of selected polymorphic traits: mandibular teeth

Table 3. Standardized deviations for selected morphological dental traits

| | Standardized Frequencies: Maxillary dental traits | | | |
Site	SS I1	IG I2	CAR M1	C5 M1
MLC	0.63688	0.50095	1.41384	0.72806
MR3	0.74452	0.83785	0.49618	0.14364
MR2	0.50025	0.44229	1.53122	0.02060
HAR	0.81426	0.09195	0.31478	1.88238
INM	0.85455	0.22921	0.11738	1.11293
TMG	0.08072	1.19304	0.89099	0.51275
SKH	1.84121	1.83497	0.70959	0.63560

| | Standardized Frequencies: Mandibular dental traits | | | |
Site	CNO5M2	YGR M2	ENC6 M1	MTC7 M1
MLC	1.50656	1.060410	0.35845	0.27316
MR3	0.62187	0.54027	0.03094	0.80468
MR2	0.34432	0.01067	1.97478	0.23947
HAR	1.39048	1.61559	0.55516	0.07831
INM	0.70183	0.20507	0.36674	0.80468
TMG	0.70183	1.07795	1.22400	1.97974
SKH	0.55355	0.87712	0.17297	0.80468

Key to site abbreviations: MLC = Mesolithic Lake Cultures (Damdama, Mahadaha, Sarai Nahar Rai); MR 3 = Neolithic Mehrgarh; HAR = Harappa; MR 2 = Chalcolithic Mehrgarh; INM = Inamgaon; TMG = Timargarha; SKH = Sarai Khola.
Key to dental trait abbreviations: SS I1 – shovel-shape upper central incisor; IG I2 – Interruption groove upper lateral incisor; CAR M1 – Carabelli's trait upper first molar; C5 M1 – cusp 5 (metaconule) upper first molar; CNO5 M2 – cusp number lower second molar; YGR M2 – Y groove pattern lower second molar; ENC6 M1 – cusp 6 (entoconulid) lower first molar; MTC7 M1 – cusp 7 (metaconulid) lower first molar.

are in close agreement with previous studies (Lukacs, 1986) that found close affinities between Neolithic Mehrgarh and Inamgaon, a Chalcolithic site in peninsular west India. This analysis differs somewhat from prior studies (Lukacs and Hemphill, 1991) in displaying a closer linkage between Mesolithic Ganga Plain sites (Damdama, Mahadaha, Sarai Nahar Rai) and the cluster comprised of Neolithic Mehrgarh and Inamgaon.

Paleopathology: Clues to Health and Nutrition

Were semi-nomadic hunters and foragers of Damdama and the Lake Culture Complex healthy and well nourished? Is there evidence of environmentally induced physiological stress? Skeletal and dental indicators of health and nutrition were observed to answer these questions. Teeth and gnathic elements were examined for evidence of developmental or degenerative pathological lesions. Crania were examined for evidence of cribra orbitalia, and porotic hyperostosis (iron deficiency anemia), while postcranial remains were inspected for evidence of osteoarthritis (degenerative joint disease), and periostitis (a proliferative bone response, evidence of non-specific infection). Cranial and postcranial elements were scrutinized for evidence of trauma, including fractures, dislocations, wounds, and cutmarks. Vascular impressions of the tibia and hypertrophy of entheses were also systematically analyzed. Sample sizes fluctuate dramatically for skeletal and dental afflictions due to the role diagenetic and taphonomic factors

Ward's Cluster - Standardized Frequencies

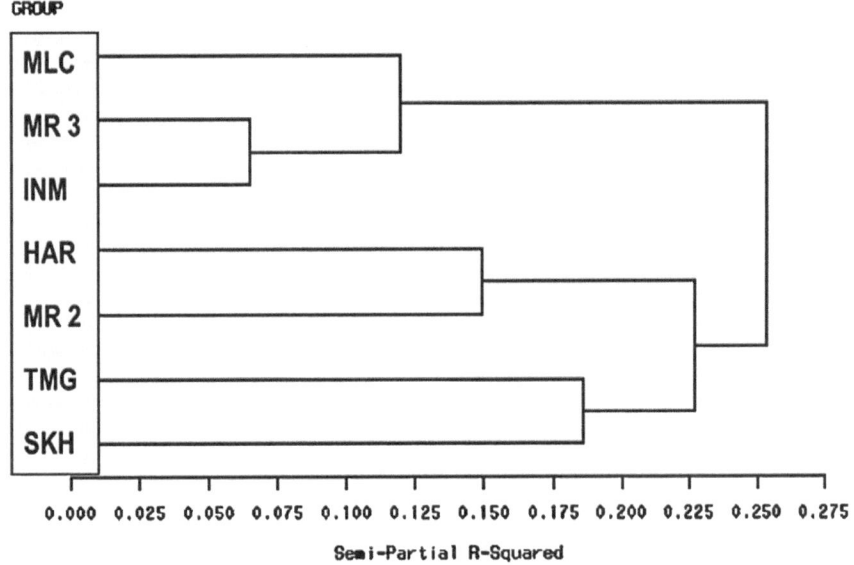

Figure 9. Ward's cluster based on standardized frequencies of dental morphological traits with significant inter-group differences. Abbreviations: MLC = Mesolithic Lake Culture; MR 3 = Neolithic Mehrgarh; MR 2 = Chalcolithic Mehrgarh; HAR = Harappa; INM = Inamgoan; TMG = Timargarha; SKH = Sarai Khola

play in influencing differential preservation of dental and skeletal elements.

Meaningful comparative analysis of the frequency of skeletal and dental lesions requires close attention to demographic variables of the groups under comparison. While some investigators make age-specific comparisons of lesion frequency, a viable alternative for small samples is testing for significant differences in demographic profiles. Although the Mesolithic Lake Culture sites analyzed here differ slightly in mean age, no significant inter-site differences in mean age at death were found between Damdama, Mahadaha and Sarai Nahar Rai (p < 0.05; Lukacs and Pal, 1993:751).

Pathological Lesions of Skeletal and Dental Remains
The prevalence of pathological dental lesions in the Damdama skeletal series was documented by prior research, but is included here to provide a holistic and comparative picture of health (Lukacs and Pal, 1993). The dental pathology profile of the Mesolithic Lake Culture and

Neolithic Mehrgarh are graphically compared in Figure 10; the supporting data are provided in Table 4. Dental caries rates are low, exposure of the pulp chamber is also rather low but caused by heavy dental wear, and evidence of systemic developmental growth disruption – as indicated by linear enamel hypoplasia – is common. A chi-square test of the prevalence of dental lesions at Mesolithic Lake Culture sites and Neolithic Mehrgarh reveals significant differences in five of the seven traits compared. Some of the difference may result from the more fragmentary nature of the Mehrgarh gnathic remains, which has a higher representation of isolated teeth. Consequently small sample size may contribute to the low rate of alveolar recession and dental abscesses at Mehrgarh. Lower rates of dental caries at Neolithic Mehrgarh are attributable to naturally fluoridated water in the Kachi Plain (Lukacs et al., 1985). The higher frequency of pulp exposure among Mesolithic Lake Culture samples results from greater rates of

Pathological Dental Lesions: 'Mesolithic' Lake Cultures (including DDM), and Neolithic Mehrgarh (MR 3).

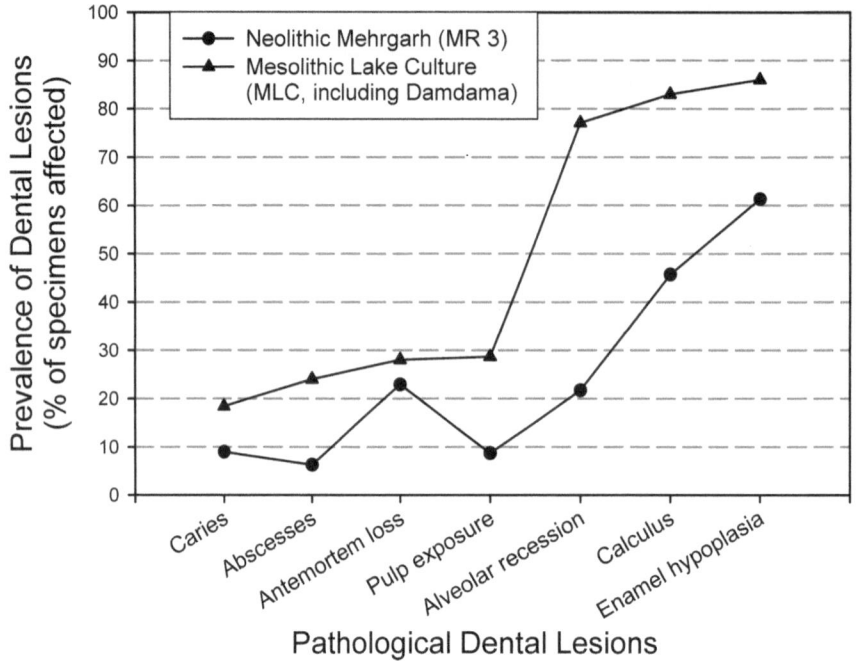

Figure 10. Dental pathology profile at Damdama

dental wear from the coarse diet associated with a hunter and gatherer subsistence.

The most frequent pathological dental lesion in both samples is enamel hypoplasia. Observed as transverse lines on the outer enamel surface, hypoplasia is evidence of deficient enamel formation and suggests that physiological disruptions in growth among

Table 4. Dental pathology of Ganga Mesolithic and Neolithic Mehrgarh compared

| Dental Lesion | Ganga Plain Mesolithic | | | | Neolithic Mehrgarh | |
| | DDM | | MLC | | MR 3 | |
	+/n	%	+/n	%	+/n	%
Caries	5/35	14.3	9/49	18.4	8/89	9.0
Abscesses	9/35	25.7	12/50	24.0	3/48	6.3
Antemortem loss	10/35	28.6	14/50	28.0	11/48	22.9
pulp exposure	11/35	31.4	14/49	28.6	8/92	8.7
Alveolar recession	27/33	81.8	37/48	77.1	10/46	21.7
Calculus	30/35	85.7	39/47	83.0	42/92	45.7
Enamel hypoplasia	30/32	93.6	37/43	86.0	57/93	61.3

MLC (Mesolithic Lake Cultures) = frequencies for Damdama (DDM) and Mahadaha (MDH) combined and arranged in ascending frequency.

children were common among early farmers of the Kachi Plain and semi-nomadic hunters and foragers of the Ganga Plain. Despite frequent disruptions to growth in childhood, the tall stature attained by Mesolithic Lake Culture men and women shows that periodic stress was balanced by flush periods that allowed for significant amounts of catch-up growth (Lukacs and Pal, 2003, 2007).

Results for skeletal pathology, trauma and other skeletal variations are presented in Tables 5 and 6, respectively. A significant finding of this study is the absence of pathological lesions for several morbid conditions indicative of nutritional and infectious disease. Lesions of the cranial vault, including: cribra orbitalia and porotic hyperostosis, and lesions of the postcranial long bones, such as periostitis were not observed in the Damdama skeletal series. None of these key indicators of nutritional and infectious disease have been reported for either Mahadaha (Kennedy et al., 1992) or Sarai Nahar Rai (Kennedy et al., 1986). Evidence of osteoarthritis was present, but rare. Osteoarthritis of the temporomandibular joint was lacking from the glenoid fossa of the temporal bone, but present on the mandibular condyle of one specimen (DDM 1, an adult female). This condition has also been reported in two specimens from Mahadaha (MDH 19 and 21, both older adult females) indicating that masticatory stress at the temporomandibular joint affected some individuals of Mesolithic Lake Culture (Lukacs and Hemphill, 1992). These observations are consistent with high rates of dental wear, abscesses due to heavy masticatory forces and high rates of pulp exposure. In general, vertebrae were poorly preserved and under-represented in the sample, however osteophytes were observed in three of seven specimens. Osteoarthritis of the appendicular skeleton is low in frequency and was observed bilaterally in the elbows of one individual (DDM-1, an adult female) and in the hand of another (metacarpals, DDM-12, adult female).

The skeletal analysis, including cranial and postcranial trauma, vascular impressions, and enthesial hypertrophy are presented in Table 6, and briefly discussed below.

Traumatic lesions were absent from all observable cranial remains, but three individuals preserved evidence of postcranial fractures: a) a compression fracture of the ninth thoracic vertebra (DDM 1); b) a simple fracture of a left ulna (DDM 24); and, c) a simple fracture of the right fibula (DDM 23). The latter injury was associated with multiple markers of dysfunction in the right leg including a prominent exostosis of

Table 5. Prevalence of pathological skeletal lesions at Damdama

Lesion Name	$N_{affected}$	$N_{observed}$	% affected
Cribra orbitalia	0	15	0.0
Porotic hyperostosis	0	30	0.0
Periostitis	0	40	0.0
Osteoarthritis			
Temporomandibular joint (glenoid fossa)	0	14	0.0
Temporomandibular joint (mandibular condyle)	1	12	8.3
Vertebral osteophytosis	3	7	42.9
Appendicular osteoarthritis	3	33	9.1

Key: $N_{affected}$ = number of specimens in which lesion was present;
$N_{observed}$ = total number of specimens observed.

Table 6. Prevalence of pathological lesions and skeletal variations at Damdama

Variation	$N_{affected}$	$N_{observed}$	% affected
Cranial Trauma	0	20	0.0
Post-Cranial Trauma	3	36	8.3
Vascular Impressions	5	21	23.8
Enthesial Hypertrophy	18	33	54.6

Key: $N_{affected}$ = number of specimens in which lesion was present; $N_{observed}$ = total number of specimens observed.

Figure 11. Vascular channels, features also known as vascular impressions or vessel tracks, on lateral aspect of the diaphysis of the left (lower) and right (top) tibiae of DDM 2. Four channels are easily viewed on each tibia

the posterior superior aspect of the femoral diaphysis.

Vascular impressions (channels, cortical grooves, vessel tracks) are shallow grooves on the cortical surface of long bone diaphyses, analogous to surficial grooves on the frontal bone (see Hauser and DeStefano, 1989). In humans they are most commonly found on the lateral (sub-periosteal) surface of the tibial diaphysis. DDM 2 presents numerous well demarcated vascular channels on right and left tibiae (Figure 11). Five of 21 specimens (23.8%) exhibited one or more vascular impressions on either the right, left or both tibia. While the functional significance of

such grooves remains unclear, the Damdama sample shows no significant difference by side, but suggests significant differences by sex – 5 of 10 females exhibit vascular impressions (7/15 tibiae; 46.7% of female tibiae), while none of 11 males exhibit the trait (0/14 tibiae). In a rare study of vascular impressions, Wells (1963a, 1963b) reports a frequency of 52.6% (n = 300) of tibiae affected in his Anglo-Saxon sample and reports no significant inter-sex difference. Like frontal grooves, vascular channels may indicate disharmony in the growth rate of different tissues. Exuberant bone growth, combined with slower development of neural or vascular tissues, may impose localized restriction in proliferation of sub-periosteal osseous tissues thereby creating sinuous or linear channels on the cortical surface. Vascular grooves, may mimic peri- or postmortem cutmarks or toothmarks, and have been observed on the lateral ramus of large bovid humeri (wildebeest) by Shipman and Rose (1984). Accurate description and diagnosis is required if mis-identification of these surface features of cortical bone is to be avoided (see Buikstra and Ubelaker, 1994:108, and Figure 77).

Entheses are sites throughout the skeleton where muscles attach to bone by tendon. Enthesial hypertrophy was systematically evaluated as an indicator of skeletal function and activity, and as one factor contributing to the perception of 'skeletal robusticity'. Enthesial hypertrophy was observed at multiple loci throughout the skeleton and 54% of specimens displayed enthesial hypertrophy at one or more sites. Common loci of enthesial hypertrophy include the forearm: proximally near the elbow at the supinator crest and radial tuberosity, and distally at the pronator quadratus insertion. The most prominent site of enthesial hypertrophy is on the posterior proximal surface of the tibia, where the soleal line is developed into a rugose welt-like swelling or in more extreme instances into a ridge-like crest from which the soleus muscle originates. Hypertrophy of the soleal line is a marker of plantar flexion related to high mobility, and is equally well developed in the Damdama series among males and females and among young and old individuals (Lukacs and Pal, 2003).

Discussion

This synthetic review documented aspects of human skeletal and dental diversity among semi-nomadic hunters and foragers of the Ganga Plain. These data and their interpretation bear upon two critical features of current anthropological debate: early human migrations and subsistence and health. Perspectives on these issues that can be derived from South Asian prehistory have the potential to contribute to global discussion of the biological affinities and the origin of modern humans and of how cultural shifts in subsistence and food production impact human health and nutritional status.

Dental Morphology, Biological Affinities, and Dispersal

The biological role of Mesolithic Lake Culture populations in the peopling of South Asia is illuminated by the pattern and frequency of dental morphological traits. The Mesolithic Lake Culture sample exhibits a dental morphological phenotype that is characterized by: a) simple anterior dentition in terms of low frequencies of incisor and canine shoveling, double shovel, *tuberculum dentale*, and marginal interruptions, b) infrequent accessory cusps in the maxillary (low frequency of Carabelli's trait and metaconule) and mandibular molar teeth (low frequency of entoconulid and metaconulid), combined with c) large second lower molars, featuring five cusps and Y occlusal groove patterns. This trait association shares some features in common with Turner's (1990) more generalized Sundadont dental pattern, and displays affinities with both Neolithic Mehrgarh (Baluchistan Province) and Chalcolithic Inamgaon (Deccan Plateau, western Maharashtra). Standardized dental trait frequencies and the cluster analysis derived from them show significant differences between Mesolithic Lake Culture populations and the sub-cluster composed of Harappa and Chalcolithic Mehrgarh. The Mesolithic Lake Culture and Neolithic Mehrgarh samples are most distantly removed from the sub-group comprised of later samples from northern Pakistan, including Sarai Khola and Timargarha.

The Mesolithic Lake Culture dental pattern can also benefit from comparison with an early South Asian pattern of dental variation described by Hawkey (1998, 1999) and given the label "Indodont". In contrast to the early world average for dental trait frequencies, the "Indodont" dental complex is characterized by Hawkey as comprised of a suite of high frequency (shovel shape I^1, hypocone M^2, Y-groove M_2, and four-cusped M_2), average frequency (double shovel I^1, interruption groove I^2, cusp 5 – M^1, parastyle M^3, six-cusped M_1, deflecting wrinkle M_1, protostylid M_1, and cusp 7 – M_1), and low frequency dental traits (winging I^1, *tuberculum dentale* I^2, distal accessory ridge C, Carabelli's trait M^1; Hawkey 1998:298).

Dental trait frequencies for three groups provide the basis for the "Indodont" or early South Asian dental complex: a) a North-west farming and herding group (2 sites; n = 3–76 individuals; includes Mehrgarh); b) a Sri Lanka hunter/gatherer group (4 sites; n = 3–18 individuals); and, c) northern Indian hunter/gatherer group (6 sites; n = 4–27 individuals; includes Mahadaha and Sarai Nahar Rai). Several dental trait frequencies described here for Mesolithic Lake Culture sites (Mahadaha and Sarai Nahar Rai, augmented by new data for Damdama) are comparable with Hawkey's characteri-zation of the India hunter/gatherer group. For example, maxillary dental traits such as *tuberculum dentale* I^2 (0.308 vs. 0.333), cusp 5 – M^1 (0.188 vs. 0.182), Carabelli's trait M^1 (0.100 vs. 0.067), and mandibular traits including cusp 7 – M^1 (0.130 vs. 0.118) and the protostylid (0.177 vs. 0.111) show close similarities. This result may not be surprising since Mahadaha and Sarai Nahar Rai were included along with data for four other geographically dispersed sites in Hawkey's India hunter/gatherer group. Other important traits are more difficult to compare reliably.

In a univariate analysis of 27 dental trait frequencies, Hawkey compared early South Asia with the early world average, yet for 12 of the 27 dental traits, or 44.4% of the comparisons, data for the India hunter/gatherer sample were either unavailable or consisted of sample sizes less than 10. This observation highlights the signif-icance of the present study, since dental trait frequencies for the Mesolithic Lake Culture sites are now based on an average sample size of 23 individuals (n ranges from 16 to 32) per dental trait. This study constitutes a more accurate and meaningful characterization of early hunter/forager dental variation because it is less affected by the bias of small sample size and is based on local breeding populations that are geographically circumscribed and culturally and biologically adapted to a similar riparian environment.

The archaeological and genetic evidence for human dispersal and the origin of modern human behavior in later Pleistocene South Asia was recently reviewed (James and Petraglia, 2005). Data from human biological diversity does not currently play a major role in this discussion (Lukacs and Pal, 2007), nor can dental morphology directly contribute to discussions regarding the early southern dispersal for lack of an adequate later Pleistocene hominin dental record. Nevertheless, research now in progress seeks to construct the dental morphological attributes of a hypothetical early southern dispersing population. Comparative investi-gation of later Pleistocene African (Irish and Guatelli-Steinberg, 2003) and early Holocene South Asian dental variations will enable a tentative morphological model to be proposed that will permit the biological identity of early dispersers to be recognized and classified as archaic or modern humans. As human skeletal and dental remains with Middle and Late Paleolithic associations are discovered in South Asia their dental attributes can

Table 7. Idealized model of subsistence systems correlates with demography and health

	Subsistence Strategy	
Population Attributes	Hunter-Gatherer/Forager	Agriculture
Size	Small	Large
Density	Low	High
Mobility	High	Low
Dietary Diversity	High	Low
Categories of Disease		
Contagious	Low	High
Dental	Low – moderate	Moderate – high
Infectious	Low	High
Nutritional	Low	High
Parasitic	Variable	Variable
Biological Adaptations and Variations		
Tooth size	Large	Small
Cranial/facial robusticity	High	Low
Post-cranial robusticity	Variable	Variable

be compared with the hypothetical dental morphotype and thereby make a more meaningful and direct contribution to current debates over dispersal and modern human origins in ancient India.

Paleopathology: Health Status and Biological Adaptation

The skeletal and dental evidence for health and nutritional status of early Holocene hunters and foragers of the Ganga Plain make a direct contribution to debates regarding the impact of subsistence shifts on human health and nutrition. Perspectives on the biological consequences of subsistence transitions were initially popularized by Mark Cohen (1989; Cohen and Bennett, 1993) and elaborated by Clark Larsen (Larsen, 1995, 1998). Transition theory contends that the subsistence shift from nomadic hunting and foraging to sedentary agriculture involved several biological costs. The biological toll of changing modes of food acquisition and preparation has been documented for specific geographic regions of the world (Cohen and Armelagos, 1984). A sequel to this influential volume, with greater emphasis on health and subsistence in Asian prehistory will soon be available

(Cohen and Crane-Kramer, 2007; Lukacs, 2007). The stresses of changing subsistence patterns are typically associated with a shift in population attributes (size, density, mobility) that tax a population's adaptive potential and stimulate an increase in contagious, infectious and nutritional diseases, and changes in stature and skeletal robusticity (Table 7). As research on paleopathology and paleodemography across the transition to agriculture becomes better documented, greater variation is evident in how populations adapt to agricultural subsistence (Cohen and Crane-Kramer, 2007).

For example, iron deficiency anemia, an affliction of sedentary agricultural populations, was unknown to humankind during our long history as hunters and foragers prior to about 10,000 years ago (Stuart-Macadam, 1998). Paleopathologists recognize skeletal indications of iron deficiency anemia as thinning and porosity of the external surface of cranial vault (porotic hyperostosis) and as porosity of the bony roof of the eye socket (cribra orbitalia). Dental diseases increase dramatically with the adoption and reliance on processed grains and cereals in the diet (Larsen, 1995). The frequency of dental caries in particular, increases with the adoption and intensification of agricultural subsistence and with technologically sophisticated food processing procedures (Lukacs, 1992). This decline in dental health with the adoption of agriculture has a greater impact on women, globally (Larsen, 1998) and in prehistoric South Asia, in particular (Lukacs, 1996).

Damdama and the Mesolithic Lake Culture samples described in this study fit the expectation of theoretical models of subsistence and health. Attributes of dental pathology are consistent with a hunting and foraging diet. Dental lesions that are common among Mesolithic Lake Culture sites, such as alveolar recession, abscesses, exposure of the pulp chamber, and high rates of attrition indicate heavy masticatory stresses. These conditions are consistent with robust structural features of the jaws, large tooth size and degenerative osteoarthritis of the temporomandibular joint (Lukacs and Pal, 1993; Lukacs, 2004). Low dental caries rates are consistent with expectations based on the generally coarse, unrefined, and abrasive diet of hunters and gatherers. Women among Mesolithic Lake Culture sites have higher caries rates than men and though this finding is not uncommon the explanation of this phenomenon is currently under re-evaluation (Lukacs and Largaespada, 2006). In addition to sex differences in diet, differences in hormone levels associated with reproductive events over the female lifespan may have a more significant role to play in explaining sex differences in caries rates than previously believed.

Osteological evidence of nutritional and infectious diseases is absent. The absence of pathological skeletal lesions from cranial and postcranial remains strongly suggests that the people of Damdama were free from nutritional deficiency and infectious diseases that produce lesions in skeletal tissue. In particular, the absence of porotic hyperostosis and cribra orbitalia suggests the absence of iron deficiency anemia, while the lack of periostitis is interpreted to indicate that infectious diseases were rare as well. In sum, the people of Damdama exhibit skeletal indicators of health commonly associated with mobile hunting and foraging societies. Trauma and markers of 'occupational' stress are rare. Bone fractures are uncommon suggesting that accidents, interpersonal violence, or intergroup conflict were not sufficiently common among the inhabitants of Damdama to be represented in their skeletal remains. Enthesial hypertrophy is evident in two main areas of the skeleton: the elbow (proximal ulna) and the leg (proximal tibia). These markers suggest repetitious use of the right forearm in throwing (spears, sling stone?) and habitual plantar flexion of the foot as in the propulsive phase of the human bipedal stride (distance walking possibly with loads) as typifies highly mobile people.

The people of Damdama were generally healthy and well adapted to their environment. Their skeletal pathology profile, pattern of trauma and activity markers, and tall stature are collectively consistent with a semi-nomadic foraging subsistence pattern and contrast dramatically with predictions for the expression of these variables among sedentary agriculturalists.

Acknowledgments

My long term investigation of the bioarchaeology and dental anthropology of South Asia over the past three decades has been supported by research grants from several agencies, including: the Alexander von Humboldt Foundation, the National Geographic Society, the National Science Foundation, the Smithsonian Institution (Foreign Currency Program), the Wenner Gren Foundation for Anthropological Research, and by fellowships from the American Institute of Indian Studies and the Council for International Exchange of Scholars (Indo-American

Fellowship Program). Access to museum or field collections of prehistoric human skeletal remains has been provided by University of Allahabad (V.D. Misra and J.N. Pal); Deccan College (G.L. Badam, S.R. Walimbe, and institute directors past and present); the French Archaeological Mission to Pakistan (J.-F. and C. Jarrige); University of Mainz (Wolfram Berhnard); University of California Berkeley Harappa Project (G.F. Dales). Dr Greg C. Nelson assisted with specimen preparation and skeletal inventory in Allahabad and offered comments on an earlier draft of the manuscript. The support and co-operation of agencies, administrators, and colleagues in research is deeply appreciated.

References

Aufderheide, A.C., Rodriguez-Martín, C., 1998. The Cambridge Encyclopedia of Human Paleopathology. Cambridge University Press, Cambridge.

Brace, C.L., 1980. Australian tooth-size clines and the death of a stereotype. Current Anthropology 21, 141–164.

Buikstra, J.E., Ubelaker, D.H., 1994. STANDARDS for Data Collection from Human Skeletal Remains. Arkansas Archaeological Survey, Fayetteville.

Cohen, M.N., 1989. Health and the Rise of Civilization. Yale University Press, New Haven.

Cohen, M.N., Armelagos, G.J., 1984. Paleopathology at the Origins of Agriculture. Academic Press, Orlando.

Cohen, M.N., Bennett S., 1993. Skeletal evidence for sex roles and gender hierarchies in prehistory. In: Miller, B.D. (Ed.), Sex Differences and Gender Hierarchies. Cambridge University Press, Cambridge, pp. 273–296.

Cohen, M.N., Crane-Kramer, G.M.M., 2007. Ancient Health: Skeletal Indicators of Agricultural and Economic Intensification. University Press of Florida, Gainesville.

De Blij, H.J., Muller, P.O., 2002. Geography: Realms, Regions, and Concepts. J. Wiley & Sons, New York.

Gould, S.J., 2002. The Structure of Evolutionary Theory. Cambridge: The Belknap Press/Harvard University Press.

Hawkey, D.E., 1998. Out of Asia: dental evidence for affinities and microevolution of early populations from India/Sri Lanka. Ph.D. Dissertation, Arizona State University.

Hawkey, D.E., 1999. The Indodont dental pattern of prehistoric South Asia and early world affinities. American Journal of Physical Anthropology, Supplement 28, 146–147.

Hawkey, D.E., Merbs, C.F., 1995 Activity-induced musculoskeletal stress markers (MSM) and subsistence strategy changes among ancient Hudson Bay Eskimos. International Journal of Osteoarchaeology 5(4), 324–338.

Hauser, G., De Stefano, G.F., 1989. Epigenetic variants of the human skull. Stuttgart: Schweizerbart.

Irish, J.D., Guatelli-Steinberg, D., 2003. Ancient teeth and modern human origins: an expanded comparison of African Plio-Pleistocene and recent world dental samples. Journal of Human Evolution 45, 113–144.

James, H.V.A., Petraglia, M.D., 2005. Modern human origins and the evolution of behavior in the later Pleistocene record of South Asia. Current Anthropology 46, S3–S27.

Jarrige, J.-F., Lechevallier, M., 1980. Les fouilles de Mehrgarh, Pakistan: problemes chronologique. Paleorient 6, 53–58.

Jarrige, J.-F., Meadow, R.H., 1980. Antecedents of civilization in the Indus Valley. Scientific American 243, 122–33.

Jurmain, R., 1999. Stories from the Skeleton: Behavioral Reconstruction in Human Osteology. Gordon and Breach Publ., Amsterdam.

Kennedy, K.A.R., Lovell, N.C., Burrow, C.B., 1986. Mesolithic Human Remains from the Gangetic Plains: Sarai Nahar Rai. South Asia Occasional Papers and Theses, No. 10. Cornell University, Ithaca.

Kennedy, K.A.R., Lukacs, J.R., Pastor, R.F., Johnston, T.L., Lovell, N.C., Pal, J.N., Burrow, C.B., 1992. Human Skeletal Remains from Mahadaha: A Gangetic Mesolithic Site. South Asia Occasional Papers and Theses, No. 11. Cornell University, Ithaca.

Lechevallier, M., Quivron, G., 1981. The Neolithic in Baluchistan: New evidences from Mehrgarh. In: Härtel, H. (Ed.), South Asian Archaeology 1979. Deitrich Reimer Verlag, Berlin, pp. 71–82.

Lechevallier, M., Quivron, G., 1985. Results of recent excavations at the Neolithic site of Mehrgarh, Pakistan. In: Schotsmans, J., Taddei, M. (Eds.), South Asian Archaeology 1983. Instituto Universitario Orientale, Naples, pp. 69–90.

Larsen, C.S., 1995. Biological changes in human populations with agriculture. Annual Reviews in Anthropology 24, 185–213.

Larsen, C.S., 1998. Gender, health and activity in foragers and farmers in the American southeast: implications for social organization in the Georgia Bight. In: Grauer, A.L., Stuart-Macadam P. (Eds.), Sex and Gender in Paleopathological Perspective. Cambridge University Press, Cambridge, pp. 165–187.

Lovell, N.C., 1997. Trauma analysis in paleopathology. Yearbook Physical Anthropology 34, 139–170.

Lovell, N.C., 2000. Paleopathological description and diagnosis. In: Katzenberg, M.A., Saunders, S.R. (Eds.), The Biological Anthropology of the Human Skeleton. Wiley-Liss, Inc., New York, pp. 217–258.

Lukacs, J.R., 1985. Dental pathology and tooth size at early Neolithic Mehrgarh: an anthropological assessment. In: Schotsmans, J., Taddei, M. (Eds.), South Asian Archaeology 1983. Instituto Universitario Orientale, Naples, pp. 121–50.

Lukacs, J.R., 1986. Dental morphology and odontometrics of early agriculturalists from Neolithic Mehrgarh, Pakistan. In: Russell, D.E., Santoro, J.-P., Sigogneau-Russell, D. (Eds.), Teeth Revisited: Proceedings of the VIIth International Symposium on Dental Morphology. Mem. Mus. Nat'l. Hist. Nat., Paris (Serie C), pp. 285–303.

Lukacs, J.R., 1990. On hunter-gatherers and their neighbors in prehistoric India: contact and pathology. Current Anthropology 31, 183–86.

Lukacs, J.R., 1992. Dental paleopathology and agricultural intensification in South Asia: new evidence from Bronze Age Harappa. American Journal of Physical Anthropology 87(1), 133–150.

Lukacs, J.R., 1996. Sex differences in dental caries rates with the origin of agriculture in South Asia. Current Anthropology 37, 147–153.

Lukacs, J.R., 2002. Hunter-gatherers, pastoralists, and agriculturalists in prehistoric India: a biocultural perspective on trade and subsistence. In: Morrison, K., Junker, L. (Eds.), Forager Traders in South and Southeast Asia. Cambridge University Press, Cambridge, pp. 41–61.

Lukacs, J.R., 2004. Human biological diversity in early Holocene India: a comparison of tooth size and dental morphology among inhabitants of the Indus and Ganga Plains. In: Srivastava, V.K., Singh, M.K. (Eds.), Issues and Themes in Anthropology: A Festschrift in Honor of Professor D.K. Bhattacharya. Palaka Prakashan, Delhi, pp. 389–414.

Lukacs, J.R., 2005. Comment on 'modern human origins and the evolution of behavior in the later Pleistocene record of South Asia' by H.V.A. James and M.D. Petraglia. Current Anthropology 46, S20–S21.

Lukacs, J.R., 2007. Climate, subsistence and health in prehistoric India: the biological impact of a short-term subsistence shift. In Cohen, M.N., Crane-Kramer, G.M.M. (Eds.), Ancient Health. University Press of Florida, Gainesville.

Lukacs, J.R., Hemphill, B.E., 1991. Dental anthropology of prehistoric Baluchistan: a morphometric approach to the peopling of South Asia. In: Kelley, M.A., Larsen, C.S. (Eds.), Advances in Dental Anthropology. Wiley-Liss, Inc., New York, pp. 77–119.

Lukacs, J.R., Hemphill, B.E., 1992. Dental anthropology. In: Kennedy, K.A.R., Lukacs, J.R., Pastor, R.F. et al. (Eds.), Human Skeletal Remains form Mahadaha: A Gangetic Mesolithic Site. South Asia Occasional Papers and Theses, No. 11. Cornell University, South Asia Program: Ithaca, pp. 157–270.

Lukacs, J.R., Largaespada, L., 2006. Explaining sex differences in dental caries: new data and the role of hormones and reproductive life history. American Journal of Human Biology 17(6), in press.

Lukacs, J.R., Minderman, L.L., 1992. Dental pathology and agricultural intensification from Neolithic to Chalcolithic periods at Mehrgarh (Balichistan, Pakistan). In: Jarrige, C., Meadow, R. (Eds.), South Asian Archaeology 1989. Prehistory Press, Madison, pp. 167–179.

Lukacs, J.R., Pal, J.N., 1993. Mesolithic subsistence in north India: inferences from dental pathology and odontometry. Current Anthropology 34, 745–765.

Lukacs, J.R., Pal, J.N., 2003. Skeletal variation among Mesolithic people of the Ganga Plains: new evidence of habitual activity and adaptation to climate. Asian Perspectives 42, 329–351.

Lukacs, J.R., Pal, J.N., 2007. Paleopathology and subsistence transition theory: new evidence from Mesolithic Damdama. In: Misra, V.D., Pandey, J.N., Pal, J.N. (Eds.), R.K. Varma Felicitation Volume. University of Allahabad, Allahabad, in press.

Lukacs, J.R., Pal, J.N., Misra, V.D. 1996. Chronology and diet in Mesolithic north India: a preliminary report of new AMS ^{14}C dates, ^{13}C isotope values, and their significance. In: Afanas'ev, G.,

Cleuziou, S., Lukacs, J.R., Tosi, M. (Eds.), The Prehistory of Asia and Oceania. ABAXO Edizioni, Forlì, pp. 301–311.

Lukacs, J.R., Pastor, R.F. 1988. Activity-induced patterns of dental abrasion in prehistoric Pakistan: evidence from Mehrgarh and Harappa. American Journal of Physical Anthropology 76(3), 377–398.

Lukacs, J.R., Pastor, R.F. 1990. Activity induced patterns of dental abrasion in prehistoric Pakistan. In: Taddei, M. (Ed.), South Asian Archaeology 1987. Instituto Italiano per il Medio ed Estremo Oriente, Rome, pp. 79–110.

Lukacs, J.R., Retief, D.H., Jarrige, J.F., 1985. Dental disease in prehistoric Baluchistan. National Geographic Research 1(2), 184–197.

Maynard Smith, J., 1993. The theory of evolution. Cambridge [England], New York: Cambridge University Press.

Mayr, E., 1982. The Growth of Biological Thought: Diversity, Evolution and Inheritance. Cambridge: Belknap Press of Harvard University Press.

Ortner, D.J., Putschar, W., 1981. Identification of Pathological Conditions in Human Skeletal Remains. Smithsonian Contributions to Anthropology, No. 28. Smithsonian Institution, Washington, D.C.

Pal, J.N., 1985. Some new light on the Mesolithic burial practices of the Ganga Valley: evidence from Mahadaha, Pratapgarh, Uttar Pradesh. Man and Environment 9, 28–37.

Pal, J.N., 1986. Microlithic industry of Damdama – Pratapgarh, U.P.: a preliminary analysis. Puratattva 16, 1–5.

Pal, J.N., 1988. Mesolithic double burials from recent excavations at Damdama. Man and Environment 12, 115–122.

Pal, J.N., 1992. Mesolithic human burials from the Ganga Plain, North India. Man and Environment 17, 35–44.

Pal, J.N., 1994. Mesolithic settlements in the Ganga Plain. Man and Environment 19, 91–101.

Provine, W., 1971. The Origins of Theoretical Populations Genetics. Chicago: University of Chicago Press.

Ridley, M., 2004. Evolution. Blackwell Publishing, Malden.

Roberts, C., Manchester, K., 1995. The Archaeology of Disease. Cornell University Press, Ithaca.

Robb, J.E., 1998. The interpretation of skeletal muscle sites: a statistical approach. International Journal of Osteoarchaeology 8(5), 363–377.

Robb, J.E., Bigazzi, R., Lazzarini, L., Scarsini, C., Sonego, F., 2001. Social status and biological status: a comparison of grave goods and skeletal indicators from Pontecagnano. American Journal of Physical Anthropology 115(3), 213–222.

Scott, G.R., Turner, C.G. II, 1997. The Anthropology of Modern Human Teeth: Dental Morphology and Its Variation in Recent Human Populations. Cambridge University Press, Cambridge.

Sellier, P., 1990. Anthropologie de terrain et gestes funéraires: le cimetière néolithique de Mehargarh (Pakistan). Les Nouvelles de l'Archéologie 40, 19–21.

Sellier, P., 1992. The contribution of paleoanthropology to the investigation of a functional funerary structure: the graves from Neolithic Mehrgarh, Period IB. In: Jarrige, C., Meadow, R. (Eds.), South Asian Archaeology 1989. Prehistory Press, Madison, pp. 253–266.

Sharma, G.R., 1973. Mesolithic lake cultures in the Ganga Valley, India. Proceedings Prehistoric Society 39, 129–146.

Shipman, P., Rose J.J., 1984. Cutmark mimics on modern and fossil bovid bones. Current Anthropology 25, 116–117.

Stuart-Macadam, P., 1989. Porotic hyperostosis: relationship between orbital and vault lesions. American Journal of Physical Anthropology 74, 511–520.

Stuart-Macadam, P., 1992. Porotic hyperostosis: a new perspective. American Journal of Physical Anthropology 87, 39–47.

Stuart-Macadam, P., 1998. Iron deficiency anemia: exploring the difference. In: Grauer, A.L., Stuart-Macadam, P. (Eds.), Sex and Gender in Paleopathological Perspective. Cambridge University Press, Cambridge.

Turner, C.G., 1990. Major features of Sundadonty and Sinodonty, including suggestions about East-Asian microevolution, population history, and Late Pleistocene relationships with Australian Aboriginals. American Journal of Physical Anthropology 82:295–317.

Turner, C.G. II, Nichol, C.R., Scott, G.R., 1991. Scoring procedures for key morphological traits of the permanent dentition: the Arizona State University dental anthropology system. In: Kelley, M.A., Clark, S.L. (Eds.), Advances in Dental Anthropology. Wiley-Liss, Inc., New York, pp. 13–31.

Varma, R.K., 1981–83. The Mesolithic cultures of India. Puratattva 13/14, 27–36.

Varma, R.K., 1986. The Mesolithic Age in Mirzapur. Paramjyoti Prakashan, Allahabad.

Varma, R.K., 1989. Pre-agricultural Mesolithic society of the Ganga Valley. In: Kenoyer, J.M. (Ed.), Old Problems and New Perspectives in the Archaeology of South Asia. Wisconsin Archaeological Reports, 2.

Varma, R.K., Misra, V.D., Pandey, J.N., Pal, J.N., 1985. A preliminary report of the excavations at Damdama (1982–84). Man and Environment 9, 45–65.

Ward, J., 1963. Hierarchial groupings to optimize an objective function. Journal of the American Statistical Association 58, 236–244.

Wells, C., 1963a. Cortical grooves on the tibia. Man (July), 112–114.

Wells, C., 1963b. Cortical grooves on the tibia. Man (November), 180.

13. Population movements in the Indian subcontinent during the protohistoric period: *Physical anthropological assessment*

S.R. WALIMBE

Department of Archaeology
Deccan College
Pune, 411 006
India
walimbes@vsnl.com

Introduction

An enormous amount of biological and cultural variability is present in prehistoric and contemporary populations in the Indian subcontinent, making it an important region to study population history. Various attempts have been made to describe and explain population diversity (e.g., Guha, 1935; Sarkar, 1954; Majumdar, 1961; Singh and Manoharan, 1993). Over the last five decades scholars from diverse disciplines, including social and biological anthropology, archaeology, and molecular biology have attempted to explain the patterns of population movements in the subcontinent. These studies aimed to identify the origins of populations in the region, evaluate chronological patterns in cultural behavior of past populations, and trace the origins of cultural diversity between contemporary groups. Some researchers attempted to correlate the phenotypic and genetic diversity of a group with its cultural and linguistic histories (Karve and Dandekar, 1951; Barnabas et al., 1996; Majumdar, 1998; Roychoudhury et al., 2001).

According to studies dealing with linguistic diversity, there are around 325 spoken languages in India, which are divided into four 'language families', i.e., Austric (Austrasiatic), Dravidian, Indo-European and Sino-Tibetan (Pattanayak, 1998). It is assumed that the speakers of the four language families represent at least four lineages. The problem of major concern among linguists is whether these language families developed within the region, or were introduced with migrations of people from outside the subcontinent. The geographical range of distribution of Austric, Indo-European and Sino-Tibetan speakers is extensive, while languages of Dravidian family are restricted largely to India. Therefore it is argued that Dravidian languages might have developed within India while migrating populations brought other languages to India.

297

M.D. Petraglia and B. Allchin (eds.), The Evolution and History of Human Populations in South Asia, 297–319.

Gadgil et al. (1998) claim that the Indian subcontinent has been populated by a series of migrations, propelled by significant techno-logical innovations outside India since the first major expansion of non-African *H. sapiens* probably around 65 ka before present. The likely migrations include: 1) Austric language speakers soon after 65 ka, probably from the northeast; 2) Dravidian language speakers around 6 ka from the Middle-East. Such populations are thought to have introduced knowledge about the cultivation of crops such as wheat, and domestication of animals, such as cattle, sheep or goats in the subcon-tinent; 3) Indo-European speakers in several waves after 4 ka. Such populations had control over horses and were familiar with iron technology; and, 4) Sino-Tibetan speakers in several waves after 6 ka, possibly introducing domesticated plant and animals. Since most of the presumed human migrations began in the protohistoric period, it is important to evaluate whether such information can be discerned in human skeletons. This is the subject matter of the present chapter.

Nature of the Human Skeletal Evidence

The Indian subcontinent provides an excellent array of human skeletal evidence belonging to various prehistoric periods (e.g., Kennedy, 2000; Walimbe and Tavares, 2002). These skeletal populations are part of a rich spectrum of cultural adaptations, including hunting and gathering in the Mesolithic, agro-pastoralism in the Neolithic-Chalcolithic, Iron-Age economy in the Megalithic and urbanization in the Harappan. With modern physical anthropological techniques, it is possible to examine the biological diversity and lifeways of past human populations.

There are obvious limitations, however, for using skeletal remains to assess biological distance. Firstly, research on human skeletal remains from archaeological sites is almost negligible in India. The discoveries of human

skeletons have figured merely as appen-dices to the main reports of excavated sites. Though more attention is now being paid to human remains, scientific recording procedures are far from satisfactory. While excavation reports for more than 300 sites record the recovery of human burials, detailed anthropological reports are available only for about 40 sites (Kennedy and Caldwell, 1984; Kennedy, 2000; Walimbe and Tavares, 2002). Secondly, barring a few exceptions like the Harappan or the Deccan Chalcol-ithic series, most of the skeletal collections have insufficient sample sizes. Consequently, the 'representativeness' of the skeletal series often remains doubtful. Further, the differ-ential preservation of skeletal elements makes the situation more difficult, limiting the number of observable parameters. Thirdly, only cranial elements have been taken into account, differential methodologies have been followed, and sexes are often 'pooled' while deriving statistical estimates. As a result, the validity of using human skeletons for inferring population distances and migrations remains difficult.

Until the early 1980's human skeletal studies were primarily focused to answer specific questions pertaining to establishing the 'ethnic' or 'racial' identity of the concerned population (see Kennedy, 2000). The inferences were based on the basis of phenotypic variations, primarily metric features of physical characterizations, like face or head shape, e.g., whether they were dolichocephalic (long-headed) or brachy-cephalic (broad-headed). The conjectures on 'racial' similarities or differences in two skeletal populations were used in archaeo-logical context to emphasize either cultural contact or changes in material culture (e.g., Sewell and Guha, 1931; Gupta et al., 1962; Sarkar, 1964, 1972). This approach was obviously in agreement with the trend of the early 20th century when theories of invasion, migration and 'mixing of blood' were the answers to diversities or 'discrepancies' noted in the skeletal record. In other words, anthropological research was guided by archaeological needs and the anthropological evidence was used primarily to complement archaeological hypotheses of cultural migration or diffusion.

Skeletal variations are now regarded as the net result of a highly complex process of genetic and non-genetic factors. Physical anthropologists recognize that human morphology and phenotypic traits are often adaptations to the environment. Adaptations may also be responsive to cultural adaptations including food procurement strategies and social behaviors. This conceptual change necessitates re-assessment of skeletal data and earlier hypotheses needed to be reframed. Such a distinct shift is noticeable in recent studies which used the paleodemographic approach for interpreting data (for a detailed bibliography see Walimbe, 1998; Walimbe and Tavares, 1996, 2002). These fresh studies led to a better understanding of the ability of humans to adapt to new environments (by developing appropriate cultural strategies) and responding biologically within a phenotypic range (as determined by the genotype).

Given this background it is necessary to evaluate the merits and weaknesses of using metric and morphological data for taxonomic usage, and thereby for assessing prehistoric population movements. Observations on the skeletal record can broadly be categorized into two research areas: 'continuous' metric traits and 'discrete' non-metric (morphological) traits.

'Continuous' Metric Data

Most researchers involved in skeletal analyses traditionally relied on cranial dimensions for inferring on population distances. The dimensions/indices most commonly used include cranial index, facial index, nasal index and facial-nasal perspective in profile.

The samples often include widely separated geographical, chronological and cultural periods, levels, and include various age-sex groups, obviously diminishing its value for taxonomic inferences. On the other hand, 'continuous' craniometric data, representing the broad temporal span of the last ten thousand years and well supported by archaeological evidence, is more important for understanding human evolutionary history. This is especially so for the recent past as it is a period of major cultural innovations, the most important of which is the domestication of plants and animals, enabling changes in settlement patterns and technological innovations. For conducting morphometric comparisons in an evolutionary perspective, human skeletal populations can be broadly divided into pre-agricultural and agricultural populations. As used here, the 'pre-agriculture' category includes the Mesolithic period, while the Harappan and Neolithic-Chalcolithic phases can be pooled together into the 'agriculture' category.

The Mesolithic represents a brief span of few thousand years between the Paleolithic and Neolithic stages and can be taken as a transitory phase from a food-gathering to a food-producing society. Widespread use of microliths and regular use of the bow and arrow indicate more intensive exploitation of natural resources of localized ecological niches, a subsistence based on hunting, gathering and fishing. Human skeletal evidence from the Mesolithic levels of Sri Lanka and India (e.g., Batadomba lena, Beli lena, Belan Bandi Palassa, Langhnaj, Mahadaha, Sarai Nahar Rai), though not large in number, yield important information relating to the course of human biological evolution in the Indian subcontinent (Ehrhardt and Kennedy, 1965; Kennedy, 1965; Kennedy et al., 1986a, 1986b, 1992).

With respect to cranial architecture, Mesolithic hunter-gatherers, in general, had a robust skull. The walls of cranial vault are usually thick and heavy. Skull shape varies in the range of 'medium' to 'very long' category (low mesocrany to hyperdolichocrany), with a spheroid or rambdoid vault. The forehead is usually exceedingly low with the frontal bone inclining gradually from glabella to bregma. The glabellary region is prominent with a heavy and divided frontal torus. The mastoid processes are moderately large to very large with prominent supramastoid crests. Sharp temporal lines are evident. Parietal bosses are pronounced, so also the occipital curvature. The occipital torus is usually remarkable for its large size and robusticity. Development of nuchal lines ranges from slight to pronounced.

The Harappan phase, ca. 3000-1500 BC, includes an extensive human skeletal series and has a well-documented record of cultural identity. The Harappan culture was urbanized and self-sufficient; it supported a large population that differentiated into a variety of occupations. The Harappans had well planned cities with public buildings, fortification walls, granaries and standardi-zation of weights and measures. Such traits reflect a highly developed socio-political and religious society depending on farming and domestication of cattle as well as other animals. The public buildings and granaries signify a well ordered and disciplined society. Besides Harappa itself, there are other sites yielding human skeletal remains; some of them are located outside the confines of the Indus valley proper. These represent the Mature Harappan phase, like, Rupar (Punjab), Rakhi Garhi (Haryana), Kalibangan and Tarkanwala Dera (Rajasthan), Lothal and Randal Dawa (Gujarat), Mohenjo-Daro and Chanhu-Daro (Sind). Not all of these series are studied, but skeletal information is available for around 400 individuals (Sewell and Guha, 1929, 1931; Gupta et al., 1962; Chatterjee and Kumar, 1963; Dutta et al., 1987; Hemphill et al., 1991).

Contemporary with the Harappan Civilization is the Neolithic phase which covers a wide time span from 3,000 to 1,000 BC Sites yielding skeletal material from this cultural phase cover a wide geographic area spreading from Burzahom (Kashmir) in the north to Assam and other sites in the northeast, right up to the south at sites such as Utnur, Nagarjunkonda, Palavoy, Hulikallu and Ieej in Andhra Pradesh, Tekkalakota, T. Narasipur and Budihal in Karnataka, and Paiyampalli in Tamil Nadu. Human skeletal collections from some of these sites have been studied (Ayer, 1960; Malhotra, 1965b; Basu and Pal, 1980; Walimbe and Paddayya, 1999; Walimbe, 2006a, 2006b).

Though a number of Chalcolithic sites have been excavated in northern and central India, information on the skeletal populations comes primarily from the Deccan plateau. These cultures flourished in this region between 2000-700 BC, and the period marks the beginning of sedentism. Chalcolithic subsistence was primarily on agriculture, supplemented by stock-raising and hunting-fishing. The food economy was based on a combination of agricultural products and animal food including both domestic and wild. Excavations at the sites like Inamgaon, Nevasa, Daimabad, Kaothe, Chandoli, and Walki have provided rich evidence of ceremonial human burial (Malhotra, 1965a; Kennedy and Malhotra, 1966; Lukacs and Walimbe, 1986; Walimbe, 1986, 1990; Mushrif and Walimbe, 2000; Mushrif, 2001).

The skeletal series representing the early agro-pastoral Neolithic-Chalcolithic populations is one of the largest human skeletal series in the Indian subcontinent which includes more than 400 individuals available for study. The adult segment of this collection is small, however.

Significant differences in cranial morphology become evident when the adult specimens from the 'pre-agricultural' groups are compared with the 'agricultural' populations. The adult specimens in the later populations are characterized by 'slightly long' to 'medium' cranium (mesocrany) with a tendency towards brachycrany. Body musculature is weak resulting in a gracile appearance. In general, early agricultural populations are characterized by a receding to vertical forehead with a faintly developed glabello-superciliary region, square to horizontal orbits, a broad nose with a depressed root, medium to low upper facial height, moderate sized cheekbones, and slight alveolar prognathism.

Craniometric data for some pre-agricultural and agricultural populations are summarized in Table 1. To avoid bias only male specimens are considered for the comparison. Figure 1 shows the approximate location of the sites mentioned in Table 1.

As seen in a few other incipient agricultural populations (Larson, 1984), the cross-cultural comparisons in the Indian context show differences in the cranial features of the pre-agricultural and early agro-pastoral populations, revealed by two significant changes in cranial morphometry. There is a gradual reduction in robusticity, and there are signficant changes in skull shape. The mean cranial index of the hunting-gathering Mesolithic populations is dolichocranial, 70.0 (long-headed). The pooled values for Harappan and Neo-Chalcolithic specimens indicate a cranial index that falls in the long- or medium-headed category (73.93, dolichocrany or mesocrany), with a tendency towards broad-head category (brachycrany) in the later levels. Other noticeable changes are a rotation of the facial region to a position more inferior to the cranium and a decrease in cranial length (Figure 2).

The question that needs to be addressed is whether the metric differences reflect a new or modified genetic composition, or reflect the trends of micro-evolutionary processes operative during the agricultural transition. This exercise is necessary to evaluate the theories of population identity and movements during the protohistoric period.

In the bioanthropological literature skull shape has often been used to establish population identity. For example, the appearance of brachycrany (broad-head) had often been interpreted as evidence for an 'intruder population', replacing earlier dolichocranial (long-headed) inhabitants. This evidence was also used to establish biological contacts of the broad-headed group with the indigenous long-headed group. For example, the abrupt decline of the civilizations at Harappa and Mohenjo-daro was attributed to a 'foreign racial element' (Sewell and Guha, 1931; Gupta et al., 1962). However, the brachy-cranial element was observed in different cultures and such a treatment is recognizable from widely separated geographical regions in India. While genetic makeup has a role in determining the phenotype of the population, alternative explanations can be offered for the population differences due to mechanisms of adaptation. While genetic influences in determining a regional phenotype cannot be ignored, changes in cranio-facial morphology can better be explained non-genetically, and appear to be primarily due to subsistence changes. The differential functional demands on the body with early farming societies (inclusive of more sophisticated food preparation techniques) could be the main factors influencing changes in cranio-facial morphology.

Robust body size and larger dentition in individuals are interpreted as successful biological adaptations in Mesolithic populations, essential for the exploitation of new ecological settings and a hunting-gathering way of life associated with the consumption of coarse-fibre food (Kennedy, 1984a). The overall gracile appearance of the later population, in comparison with their hunting-gathering predecessors, can mainly be attributed to two factors: decreased mechanical stress, and increased nutritional stress (Walimbe and Tavares, 1996; Walimbe,

Table 1. Comparative craniometric data for selected pre-agricultural and agricultural adult male populations

Population	Culture	Sp. No./ Sample Size	Measurements (mm) (MS Code)						Indices (MS Code)					ref
			1	8	17	9	45	66	I 1	I 2	I 3	I 73	I 40	
Bellan Bandi Palassa	Mesolithic	1	200	147	135	117	–	96	73.5	67.5	91.8	–	–	1
Mahadaha	Mesolithic	23	205	131	147	100	119	96	63.9	71.7	112.2	84.0	80.7	2
Mahadaha	Mesolithic	24	196	141	149	96	150	–	71.9	76.0	105.7	64.0	–	2
Mahadaha	Mesolithic	26	200	127	148	89	102	80	63.5	74.0	116.5	87.3	78.4	2
Sarai Nahar Rai	Mesolithic	72-III	198	135	153	97	–	90	68.2	77.3	113.3	–	–	3
Sarai Nahar Rai	Mesolithic	70-IV	192	146	124	107	145	118	76.0	64.6	84.9	73.8	–	3
Langhnaj	Mesolithic	V	187	137	130	103	138	107	73.3	69.5	94.9	74.6	77.5	4
Mean, nomadic/semi-settled pre-agricultural populations			**197**	**138**	**141**	**101**	**130**	**98**	**70.0**	**71.5**	**102.8**	**76.7**	**78.9**	**-**
Harappa R-37	Harappan	Mean (13)	188	133	134	95	131	89	71.1	71.4	100.3	71.3	69.7	5
Harappa Area G	Harappan	Mean (7)	181	138	134	99	128	91	76.1	74.0	96.9	77.9	75.6	5
Harappa H-I	Harappan	Mean (3)	187	138	135	97	134	87	73.8	70.8	97.9	72.5	65.6	5
Harappa H-II	Harappan	Mean (5)	189	145	135	96	136	79	76.7	71.5	93.7	72.4	55.5	5
Mohenjo-daro	Harappan	Mean (3)	197	130	139	95	127	102	66.0	70.6	107.2	74.8	80.3	6
Rupar	Harappan	1	181	138	–	89	–	94	76.2	–	–	–	–	7
Lothal	Harappan	Mean (7)	183	146	–	–	134	–	79.7	–	–	–	–	8
Nal	Harappan	1	189	132	146	93	120	–	69.8	77.2	110.6	77.5	–	9
Timargarh	Neolithic	Mean (9)	190	132	136	94	133	103	69.4	71.6	103.2	70.7	77.5	10
Burzahom	Neolithic	1	190	133	140	97	134	84	70.0	73.7	105.3	72.4	62.7	11
Piklihal	Neolithic	VIII-1	172	139	141	95	126	93	80.8	82.0	101.4	75.4	73.8	12
Ieej	Neolithic	1	179	145	–	97	–	112	81.0	–	–	–	–	13
Tekkalakota	Neolithic	5	185	134	146	91	–	95	72.4	78.9	109.0	–	–	14
Hullikallu	Chalcolithic	1	172	144	142	94	124	95	83.7	82.6	98.6	75.8	76.6	15
Agiripalli	Chalcolithic	1	182	138	132	95	130	–	75.8	72.5	95.7	73.1	–	16
Chinnamarur	Chalcolithic	1	187	134	–	98	–	100	71.7	–	–	–	–	17
Nevasa	Chalcolithic	1	186	125	–	89	122	90	67.2	–	–	73.0	73.8	18
Kaothe	Chalcolithic	4	175	129	129	96	135	99	73.7	73.7	100.0	71.1	73.3	19
Daimabad	Chalcolithic	18	194	133	145	89	121	96	68.6	74.7	109.0	73.6	79.3	20
Inamgaon	Chalcolithic	59	176	137	–	101	128	99	77.8	–	–	78.9	77.3	21
Inamgaon	Chalcolithic	146B	181	138	137	87	122	90	76.2	75.7	99.3	71.3	73.8	21
Chandoli	Chalcolithic	16	197	128	–	81	–	92	65.0	–	–	–	–	22
Balathal	Chalcolithic	1999-1	(179)	(131)	(122)	90	(127)	94	73.18	68.15	93.12	70.86	74.01	23
Mean, Early agro-pastoral rural/urban populations			**184**	**136**	**138**	**94**	**128**	**94**	**73.9**	**74.7**	**101.9**	**73.9**	**72.5**	**-**

Note: MS Codes used (measurements and indices):

(1) Maximum cranial length	(I 1) Cranial index
(8) Maximum cranial breadth	(I 2) Cranial length height index
(17) Cranial (basion-bregma) height	(I 3) Cranial breadth height index
(9) Minimum frontal height	(I 73) Jugo-frontal index
(45) Bizygomatic diameter	(I 40) Jugo-mandibular index
(66) Bigonial diameter	

References: 1. Kennedy, 1965; 2. Kennedy et al., 1992; 3. Kennedy et al., 1986b; 4. Ehrhardt and Kennedy, 1965; 5. Gupta et al., 1962; 6. Sewell and Guha, 1931; 7. Dutta et al., 1987; 8. Chatterjee and Kumar, 1963; 9. Sewell and Guha, 1929; 10. quoted, Kennedy et al., 1992; 11. Basu and Pal, 1980; 12. Ayer, 1960; 13. Walimbe, 1993a; 14. Malhotra, 1965a; 15. Walimbe, 1993b; 16. Walimbe, in prep-a; 17. Walimbe, in prep-b; 18. Kennedy and Malhotra, 1966; 19. Walimbe, 1990; 20. Walimbe, 1986; 21. Lukacs and Walimbe, 1986; 22. Malhotra, 1965b. 23. Robbins and Mushrif, in prep.

Figure 1. Approximate locations of studied sites. 1. Bellan Bandi Palassa; 2. Mahadaha; 3. Sarai Nahar Rai; 4. Langnhaj; 5. Harappa; 6. Mohenjodaro; 7. Rupar; 8. Lothal; 9. Nal; 10. Timargarh; 11. Burzahom; 12. Piklihal; 13. Ieej; 14. Tekkalakota; 15. Hullukallu; 16. Agripalli; 17. Chinnamarur; 18. Nevasa; 19. Kaothe; 20. Daimadab; 21. Inamgaoan; 22. Chandoli; 23. Balathal

1998). In addition, higher morbidity in the settled early farming communities might also have contributed for the comparatively delicate built. These three stress factors are described below.

Decreased Mechanical Stress

It has been postulated that, in comparison with hunting-gathering groups, agriculturally settled populations experienced reduction in the functional demand placed on the skeleton (Cohen and Armelagos, 1984). The advent of agriculture brought a major change in the food economy. Hunter-gatherers were required to be in search of food, to exerting themselves physically for long periods of time and in regular succession. On the contrary, the agricultural economy introduced the concept of food storage, with agricultural fields adjacent to their settlements. Assurance

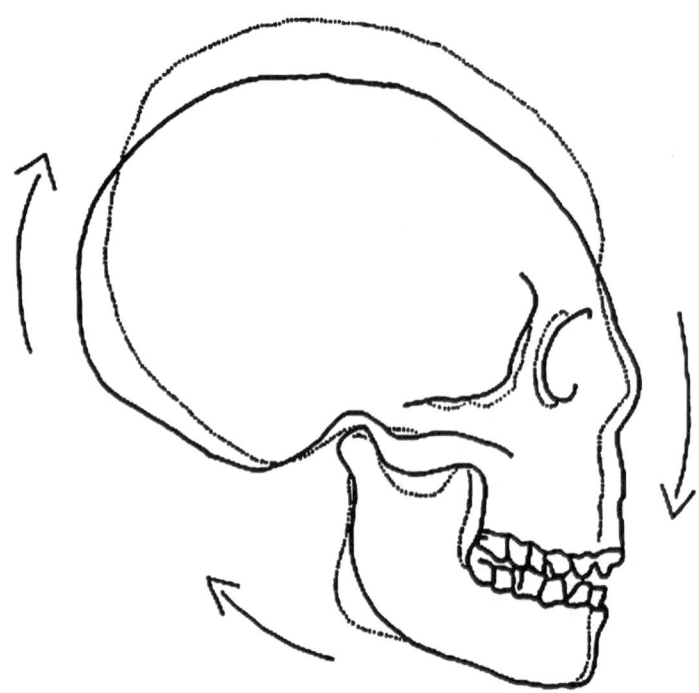

Figure 2. Cranial changes during the transition from the pre-agricultural (solid line) to the agricultural (hatched line) stage (after Martin et al., 1984)

of food supplies, with relatively less labor input, signifies a decrease in mechanical stress during the farming phase.

The argument gains support if the progress of dental crown reduction is traced from pre-agricultural to agricultural periods. The progress of dental reduction over time has been closely linked to the rate of techno-logical development and increased efficiency of food producing techniques (Brace, 1963, 1964). The Mesolithic populations have large teeth which are thought to be adaptive for a pre-agricultural, pre-pastoral society, whereas the subsequent Neolithic-Chalcolithic skeletal populations exhibit smaller dentition which is suited to the diet of an agricultural society. Crown size differences are evident when the total crown area (TCA), total cross-sectional area available for mastication, is compared for the two sets of populations (Table 2). The pooled TCA value for the Mesolithic popula-tions is 1258 mm^2, while the agricultural-pastoral Neolithic and Chalcolithic commu-nities have the TCA values of 1203 mm^2

and 1218 mm^2, respectively (Walimbe and Kulkarni, 1993). In addition to the soft carbo-hydrate diet, the sophistication of pottery use, quern processing, and other pounding activities enabled the settled agriculturalists to change their food preparation methods (i.e., from raw or roasted food to finely-ground and cooked food) eventually releasing the masticatory demand placed on the dental apparatus.

Dental size is related to the space available in the jaw bones during its development. The smaller dentition and smaller jaws in farming populations resulted in reduction of masti-catory muscles, thereby causing gracility in the gonial, zygomatic, glabellar and temporal region as reflected in the metric data. All of these changes contributed in the progressive tendency towards a vertical facial angle (orthognathus) seen in these populations. In the prognathus facial appearance (as in the Mesolithic population), to counterbalance weight of the skull posteriorly, neck muscles of the occipital region are more pronounced.

Table 2. Total crown area (TCA) and molar crown area (MCA) for Indian populations

Culture	TCA	MCA
Mesolithic	1258	673
Neolithic	1203	668
Bronze Age	1234	645
Chalcolithic	1218	699
Megalithic	1167	639
Modern	1101	597

Source: Walimbe and Kulkarni (1993)

Gracility of the occipital bone (nuchal area) in later farming communities is thus the outcome of the modifications in the facial region. These cranio-facial modifications resulted in a decrease in cranial length, thereby increasing the value of the cranial index.

Hence the change from dolichocrany to mesocrany (and further to brachycrany) in populations experiencing the agricultural transition is a consequence of decrease in head length and cannot be attributed to an increase in head breadth. As shown in Table 1, the average cranial length in hunting-gathering populations is as high as 197 mm, while in the later agricultural/pastoral communities the measurement is only 184 mm. On the other hand, the cranial breadth measurement remains more or less the same (138 mm and 136 mm, respectively). In spite of the higher value of head height the Mesolithic crania appear 'low' in profile. On the other hand, though there is no significant change in the height value in the later populations, the skull appears 'high', and, when compared with the length, the index (I 2) is considerably higher (basion-bregma height: 141 mm and 138 mm, and, index 71.5 and 74.7, respectively for pre-agricultural and agricultural groups). The relaxation of masticatory pressure is also reflected in lesser values of minimum frontal, bizygomatic and bigonial breadths (ft-ft diameter: 101 mm and 94 mm; zy-zy diameter: 130 mm and 128 mm; go-go diameter: 98 mm and 94 mm, respectively for pre-agricultural and agricultural groups).

The gracility of settled farmers is manifested in postcranial features. The robusticity of the Mesolithic people is displayed in the form of prominently developed processes and tubercles, indicating pronounced musculature. Mean statures for Mesolithic Sarai Nahar Rai and Mahadaha males was 181.47 cm and 181.69 cm, respectively, while the Chalcolithic populations were relatively short-statured, with a mean stature of 172.48 cm (Lukacs and Walimbe, 1984a; Kennedy et al., 1986b, 1992).

Increased Nutritional Stress

Other contributory factors for the overall reduction in robusticity with the advent of agriculture might have been greater nutritional stress. The diet in the hunter-gathering stage was rich in minerals, proteins, vitamins and trace nutrients and relatively low in starch; on the contrary, in settled early farming communities there were greater dependence on carbohydrate rich food (Kajale, 1991; Thomas, 2000). It has been postulated that agriculture was practiced to increase the carrying capacity of the land in order to accommodate recurrent population growth. Farming is a destabilized system that permits people to raise production above the natural capacity of the land, and the most desirable cultigens (i.e., high energy yielders) typically have low nutrient densities (Cohen, 1977). In other words, the intensified use of a vegetable food source in the farming stage assured greater food stability, it led to a less nutritive diet. It has been documented that the populations experiencing an agricultural transition tend to grow smaller in size because of protein malnutrition (Armelagos, 1990). It is possible that the same phenomenon might have occurred with the Indian protohistoric populations.

It appears that in the incipient agriculture stage, especially in the Deccan, the problem concerned not only quality of diet, but also quantity. Hunter-gatherers probably ate both meat and forest products whereas agriculturists had a narrow choice of food with little to supplement their diet when their crops failed. From the skeletal series of Inamgaon, Daimabad and Kaothe, markers of periodic stress have been noticed on the long bones and teeth (Harris lines and enamel hypoplasia, respectively) which probably indicate periodic famines, food shortages and high infection rates (Walimbe and Lukacs, 1992; Walimbe and Gambhir, 1994), supported by archaeological evidence for ecological stress (Dhavalikar, 1988).

Thus, both under-nutrition and mal-nutrition contributed to skeletal gracility in early farming protohistoric communities.

Higher Infection Rates

Sedentism and agriculture led to adverse effects on health (Cohen and Armelagos, 1984). Agriculturally dependent communities, who could not improve food production to meet the demands of an increasing population, experienced nutritional stress. Nutrition and disease are integrally and synergistically related with one another. Thus, a low quality diet and an increased population density contributed to deterioration in health of early farming communities. Hunter-gatherers moved about more often and hence could hardly be prone to epidemics; disease infection was thus not high, whereas in the more settled lifestyle incidence of infection

may have a greater chance of spreading. Domestication of animals might have introduced new pathogens in the pastoral populations.

Protohistoric skeletal data in the subcontinent have provided the primary evidence of physiological stress, an expected result of severe population pressure, and under- or malnourishment and infections. The important pathologies observed include evidences of general or episodic stress, specific or non-specific infectious lesions, traumatic lesions, degenerative pathologies and nutritional deficiencies (e.g., Lukacs et al., 1986; Walimbe and Lukacs, 1992; Mushrif, 2001). Higher morbidity levels in settled early agropastoral populations, which affected growth rate and metabolism in general, appears to be a yet another contributory factor for their delicate body built.

The interplay of non-genetic factors for influencing cranial shape is shown in Figure 3. Assessment of the quantum of the genetic and non-genetic influences on phenotypes is desired. However, an accurate picture cannot be drawn at present with the skeletal data available at hand due to two obvious limitations. First, the sample is statistically inadequate. Most of the archaeological sites yield remains of only a few individuals and preservation conditions are often far from satisfactory. Secondly, the archaeological populations come from a very large temporal span and occupy a wide geographic area. Forces of natural selection are always in operation and the population is expected to respond to local climatic situations, within the genotypically permissible limits. The validity of any statement on comparative morphology is therefore dependent on whether the skeletal data is 'representative' of the entire population under discussion. With the limited data available at hand, nevertheless, it can be concluded with fair certainty that life style and food habits appear to be more important than genetic inheritance in the expression

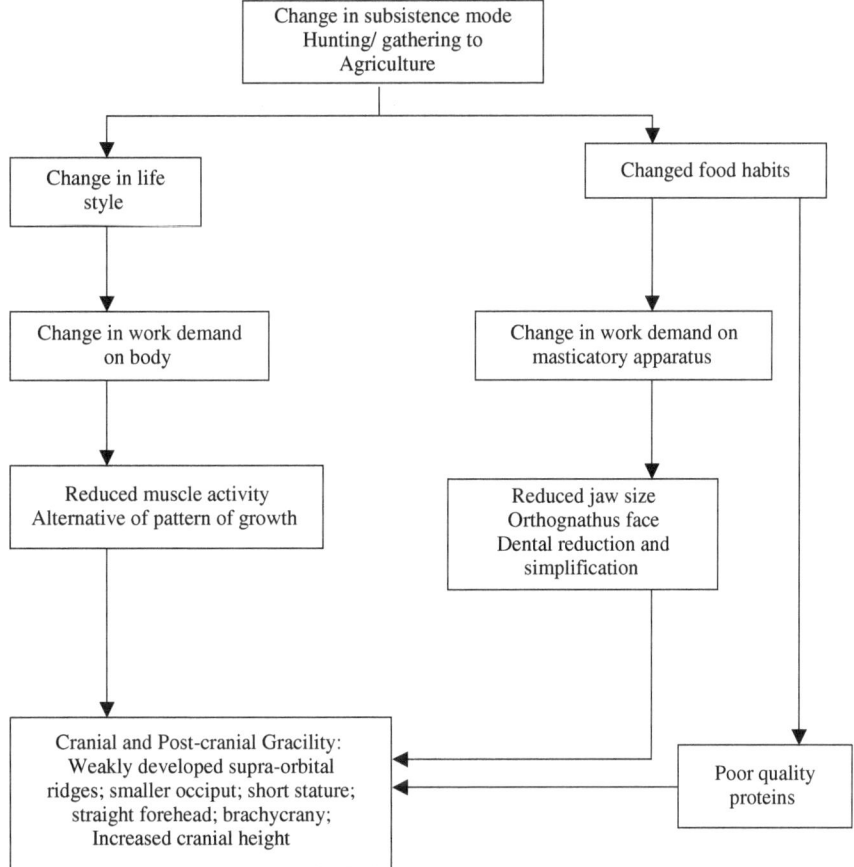

Figure 3. Model explaining cranio-facial morphological changes during the agricultural transition
(after Larsen, 1984; Walimbe, 1998)

of 'continuous' traits like craniometry. It seems therefore essential to evaluate the inferences based only on craniometric data cautiously while forwarding any hypothesis of entry of diverse genetic elements into India.

In sum, craniometric changes evidenced during the agricultural transition can more convincingly be interpreted in terms non-genetic factors rather than using them to suggest a new or modified genetic composition.

'Discrete' or 'Non-Metric' Data

Morphological traits with little to no sexual dimorphism, having low susceptibility to environmental change, and lacking age-related morphological changes have been given increased attention in recent years in Indian anthropology (Lukacs and Walimbe, 1984b; Hemphil et al., 1991, 1997; Walimbe and Kulkarni, 1994; Hawkey, 2002). In comparison with the studies based on continuous traits, non-metric cranial discrete traits and dental occlusal morphological features appear to be more important and relevant for understanding population distances, and thereby to infer past population relationships and movements. Dental morphological features are particularly important since they remain unchanged during the development process. The proto-historic skeletal series, especially in the Neolithic-Chalcolithic of India, is characterized by the over-representation of sub-adults, accounting for almost 70% of the collection. This population segment is of little use for calculating population distances in the conventional metric approach. Earlier bioanthropological research, therefore, focused only on complete adult crania, discarding infant, sub-adult and fragmentary bones. The immediate impact of the changed research perspective is the inclusion of immature and fragmentary bones and loose teeth, thereby radically increasing sample size and justifying 'pooling' of the male and female sample (Walimbe and Tavares, 1996, 2002).

A number of works have been undertaken with such a changed study design. Among the noteworthy institutions and scholars who included discrete traits in their skeletal studies are Kennedy from Cornell University, Lukacs from the University of Oregon (including Lukacs, Hemphill, Gwendolyn Robbins), Hawkey from Arizona University, and Walimbe from Deccan College (including Walimbe, Mushrif, Tavares and Kulkarni). These researchers have described the morphometric details of the skeletal series and have attempted to interpret the evidence from a biocultural perspective.

A recent study by Hawkey (2002) on 29 dental morphological features using a large sample size of 4,198 individuals is the most comprehensive account in this line of research. Major findings of this research are noteworthy. The major conclusions drawn from this study are as follows:

1) Both the Indus and the Deccan farming/herding communities share similarities with Indian Mesolithic hunter-gatherers, reflecting a common origin for the protohistoric communities. According to this study, the inhabitants of the Indus Civilization appear most likely to have been descendants of the indigenous hunter-gather populations of South Asia, rather than intrusive (and genetically distinct) populations from the West. Formation of the Thar desert (Rajasthan) during the Pleistocene was taken as the major geographic barrier preventing population movements from the

west. In other words, this inference implies genetic continuity in the subcontinent for at least during the last 20,000 years suggesting an indigenous and common origin for the people of the Indus and the Deccan Early agro-pastoral communities.

2) Dental parameters, which are less subject to environmental effects and more conservative to evolutionary changes, reflect the underlying genetic homogeneity of the Indus people. Importantly, the study concluded that there is no biological support to the argument that the Indus civilization belongs exclusively to the Dravidians. "The confusing of a linguistically-defined population ('Dravidian') with a biological population (Indus) has plagued Indian population affinity research for many years" (Hawkey, 2002:194). The peoples of the Indus Civilization are dentally similar not only to modern Dravidian-speakers (of South India, Sri Lankan Tamil), but also to the Indo-European (Afghanistan/Pakistan, Bengal, Sri Lankan Sinhalese), and Austro-Asiatic (East India) groups as well. The dental data supports the hypothesis that the Indus population was multilinguistic, and probably ethnically diverse (Hawkey, 2002:155). Even on the basis of craniometry it has now been appreciated that the Harappans do not exhibit any significant phenotypic type (Kennedy, 1984b). Though the skeletal material at Harappa is coming from three different deposits, the population belongs to a single morphometrically homogenous series. The variations in cranial size and form are very much in the acceptable range and can be understood as a normal range of physical variability present in most urban populations, past and present (Kennedy, 1982).

3) There is no substantial amount of gene flow between the Indus and the Deccan farming/herding communities. In other words, the Deccan Neolithic-Chalcolithic groups have evolved directly from

the hunting-gathering Mesolithic communities. In this case, few similarities seen between the Indus sample and contemporaneous peoples of the Deccan farming-herding cultures are due to cultural contacts (influencing food habits) rather than biological in nature. It may be noted that evidence for contact between the two groups, based on economic/trade exchange has been observed in the archaeological record (Possehl, 1976; Kennedy and Possehl, 1979). In Hawkey's (2002) opinion, the Deccan region maintains a relatively stable dental pattern from the prehistoric times to the present-day. While there was intra-regional gene flow in the populations to the north of the Vindhyan mountain range (e.g., between the Indo-Gangetic and northwestern regions) there is little dental evidence to support inter-regional gene flow between these northerly populations and the peoples of the Deccan. The Vindhyan Range is considered to be effective in restricting the Indo-European entrance into the Deccan during the proto-historic times, which may have further separated populations into the two regions. She further opines that, although the narrow coastal strip of the southwestern peninsula connects to the north, the Western Ghats may have also served to lessen contact with the somewhat inaccessible hill regions of the Deccan plateau.

4) The dental morphological data further indicates that there is little evidence to support an external origin for the Iron Age/Megalithic populations. The data rather suggest the origins of the Iron Age populations within central and southern peninsular India, and not from north-western regions. The Iron Age and the Early Historic populations of the Deccan are dissimilar to the contemporary populations from both Northwest and the Indo-Gangetic regions. The populations, however, maintain affinity with the farming-herding groups of the Deccan. The lack of a closer relationship between the Iron Age/Early Historic populations of the north and the Deccan suggests that gene flow between the two regions was disrupted in some manner by this time. Hawkey (2002) suggests that the adoption of the Vedic caste system, and the resultant marriage prohibitions especially after the urbanization process, may have helped to produce distinctive regional dental patterns. It seems clear that in comparison with the 'continuous' metric traits, the non-metric discrete traits, especially those related to dental occlusal morphological features, give far more reliable picture of the genetic affinities of the past populations. These parameters, being more genetically controlled, can more appropriately be used to hypothe-size about protohistoric human movements in the subcontinent. The use of the whole range of skeletal collections, including immature and fragmentary elements, which were earlier routinely discarded, has been shown to be of great advantage.

Discussion

When attempting to hypothesize about population dispersals and migrations, the limitations of the archaeological and anthropological data become obvious. Archaeological inferences are, of course, based on a sample of the currently available evidence recovered. More importantly, concepts and theories change as new discoveries are made. Traditional methods of biological anthropology, including those employed in skeletal analyses, are inadequate to trace human phylogenetic history and evolutionary change beyond a few generations. Therefore, additional skeletal and archaeological evidence, in conjunction with linguistic data and other biological tools (e.g., mitochondrial DNA, Y-chromosome DNA), is required to explain the plausible scenarios in the process of the peopling of the subcontinent. Study of human population migrations necessarily demands a re-evaluation of linguistic and archaeological concepts. In light of such problems, a number of observations may be made about the available archaeological and physical anthropological data (see Figure 4).

The subsistence economy of the Mesolithic period (10–4 ka) was based on hunting and gathering (Misra, 1976, 1989). There appears to be marked growth in human population size as is attested by the significant increase in number of sites (Misra, 2001). The explanation for the dramatic increase in archaeological sites lies in the increased rainfall and consequently availability of increased food resources. Introduction of microlithic tools led to enhanced efficiency in hunting, collection and processing of wild plant foods. Increased food security during this period led to reduction in nomadism and to seasonally sedentary settlement. This is reflected in the large size of Mesolithic sites, and the thickness of habitation deposits both in open-air and rock shelter sites (Misra, 2001).

The hunting-gathering way of life was slowly replaced by food production from about 8 ka. Mesolithic inhabitants eventually bred selected wild animals and cultivated selected wild grasses. The earliest evidence of agriculture in the subcontinent is reported from the site of Mehrgarh (ca. 8.5 ka). Evidence at the site of Bagor provides a date of ca. 6.6 ka for the domesti-cation of cattle and sheep/goats (Thomas, 2000). Recent genetic reports that take the antiquity of goat domestication back to 35 to 10 ka (Joshi et al., 2004) can only be supported when relevant archaeological evidence is available.

The evolution and spread of the Dravidian language is associated with early domesti-cation (Gadgil et al., 1998). The Dravidian populations may have initiated the cultivation of wheat, and domesticated cattle, sheep/goats. The origin of the Dravidian language family cannot be convincingly traced on the basis of archaeological data. The language may have evolved locally as the general morphological homogeneity seen in

the skeletal data from the Mesolithic to later stages does not support the idea of a new population spread.

The transition from food gathering to food production does not appear to have happened in a uniform pace across the subcontinent. At around 4ka, farming became the primary mode of food procurement, especially in the North where land is fertile (Misra 2001). In the peninsular region, the change was slower because of the semi-arid climate and the topography, which was composed of hilly and rocky areas.

Food production had a profound impact on the evolution of human society in the subcontinent. Farming demanded a sedentary life style and thus permanent villages were established (Smith, 1995). A surplus economy encouraged craft specialization. Knowledge of copper and bronze led to the invention of wheel which revolutionized transport and pottery production. By 6 ka, the urbanization process began to appear in and around the Indus valley. Despite the increasing dependence on farming in the south, the Chalcolithic culture of the Deccan remained less developed in socio-cultural complexity. Studies on dental morphological features (Hawkey, 2002) show that both the Indus and the Deccan farming/herding communities share similarities with Indian Mesolithic hunter-gatherers, reflecting their common origin. Yet, there is no substantial amount of gene flow between the Indus and the Deccan farming/herding communities. These observations, in general, suggest an independent evolution of Harappan and Neolithic-Chalcolithic populations from the Mesolithic populations inhabiting their respective regions. The variation in cultural attainment of the Indus and Deccan populations has to be understood in terms of their local ecology.

There are many hypotheses regarding the origin, development and extinction of the Indus Valley civilizations (Kenoyer, 1998). The cities of Mohenjodaro and Harappa were

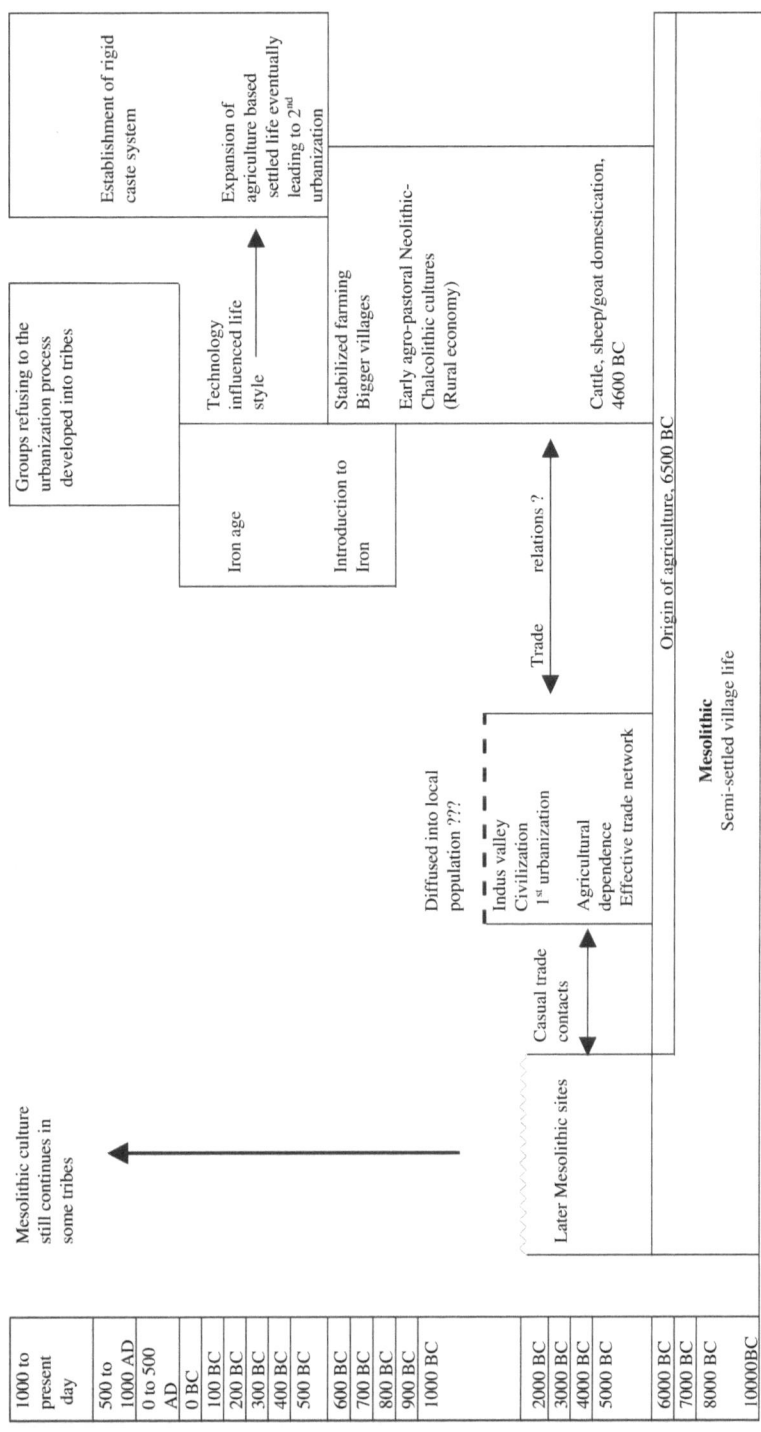

Figure 4. Assessment of cultural evolutionary trends using physical anthropological data

usually characterized as fully developed urban centres from their beginning. The inspiration for these settlements, if not the actual founding populations, was supposed to have come from Mesopotamia (Ratnagar, 1981; Misra, 2001). However, recent excavations at many sites, particularly at Harappa, Kalibangan, Mehrgarh, and Dholavira, highlight the process of the growth of urban settlements from small rural societies (Possehl, 1999). The Harappan civilization is now viewed as a uniquely South Asian development (Misra, 2001). This interpretation corroborates the idea that the Indus population gradually evolved from hunting and gathering societies and rudimentary farming groups.

The decline and disappearance of the Indus civilization has been attributed to many factors, including climatic decline, environmental degradation due to excessive use of soil and plant resources, tectonic movements, and foreign invasions (Mughal, 1990; Ratnagar, 2000; Dhavalikar, 2002; Agrawal and Kharakwal, 2003). The presumed invasion of Indo-European language speakers, who were thought to have destroyed the major urban centres of Harappa and Mohenjodaro, may be examined.

Physical anthropological studies do not support an a movement of Aryan speakers into the Indus Valley around 3.5ka (Hemphill et al., 1991, 1997). According to these investigators, gene flow from Bactria is an event of much later date, not having any impact on Indus Valley gene pools until around 2 ka. Kennedy (1984b) examined 300 skeletons from the Indus Valley Civilization and concluded that the ancient Harappans are not markedly different in their skeletal biology from the present-day inhabi-tants of North-western India and Pakistan. Kennedy (1995) also remarks that if an Aryan invasion had taken place, obvious discontinuities in the skeletal record should be found. Hemphill et al. (1991) and Kennedy (1995) suggest that there existed two phases of biological

discontinuity within the Indus Valley from the Neolithic times to around 2 ka. The first is said to occur between 8 and 4.5 ka, which is reflected in the strong differences irrespective of the occupational continuity between the Neolithic and Chalcoli-thic inhabitants of Mehrgarh. The second discontinuity exists between the inhabitants of Harappa, Chalcol-ithic Mehrgarh and post-Harappan Timar-garha on the one hand, and the Early Iron Age (better known as the Gandhara Grave culture) inhabitants Sarai Khola, on the other, between 2.8 and 2.2 ka. Kennedy (1995:53) concludes that, "if Vedic Aryans were a biological entity represented by the skeletons from Timargarha, then their biological features of cranial and dental anatomy were not distinct to a marked degree from what we encountered in the ancient Harappans." Comparing the Harppan and the Gandhara Grave cultures, Kennedy (1995:54) remarks, "our multivariate approach does not define the biological identity of an ancient Aryan population, but it does indicate that the Indus Valley and Gandhara peoples shared a number of craniometric, odontometric and discrete traits that point to a high degree of biological affinity."

The hypothesis of 'Aryan invasion', can also be questioned in the light of paleopathological evidence. Besides the 'foreign phenotypic element' (broad-headed individuals) in the later phases of at Harappa, the 'massacre' evidence at Mohenjodaro was also used as a convincing proof to plead the theory of 'Aryan invasion' (Gupta et al., 1962). At Mohenjodaro, a number of skeletons, unlike those found in other Harappan sites, were collected from the residential area or streets of this ancient city. There is no evidence of a cemetery so far found at the site. The evidence comprised of 37 skeletons, or parts thereof. Some of these skeletons were found in contorted positions and groupings that can suggest anything but ordinary and unceremonial burials. Many skeletons are either disarticulated or incomplete. Earlier anthropological works (Sewell and Guha, 1931; Gupta et al., 1962) claimed that the 'fractures' on these skulls were diagnostic of wound marks and can be directly related to the cause of death. The disorderly disposal of dead bodies indicated that the 'massacre' was a specific event at Mohenjodaro when armed foreign invaders destroyed the city and liquidated its citizens. Archaeological reassessment more than two decades later (Dales, 1964, 1965) disproved the massacre theory. As Kennedy (1984c) puts it, any conclusion to traumatic stress should rest on indisputable skeletal evidence. His re-study of Mohenjodaro skeletal collection in the light of new methodological tools provides a critical judgment in this regard. Of the 24 skulls studied, only one specimen had convincing and irrefutable signs of trauma which may be the cause of death. All other reputed wound marks appear to be erosional in origin or are cases of successfully healed lesions unrelated to circumstances and places of burial.

If the hypothesis of an 'Aryan invasion' cannot be supported using physical anthropological data, then the spread of Indo-European languages in the subcontinent needs to be explained on non-biological grounds. There is no doubt that surplus agricultural economy of Harappans induced increased trade contacts with others (especially to the West). It seems much more likely that multiple waves of Indo-European migration, in small numbers, are possible causing a mingling of the immigrants and local populations. There may have been significant exchange and assimilation of culture and language on both sides. The immigrants may have travelled back and forth to their original lands taking language and culture to other Indo-European peoples. Human skeletal remains excavated from sites of Harappa and Mohenjodaro show a mixed ethnic composition similar to the present (Kennedy, 1984b, 1995), showing support for migration rather than an invasion.

In recent years, human population genetics data corroborates some physical anthropological inferences, concluding that there is no material evidence for any large scale migrations into India over the period of 4500 to 800 BC. Though DNA-based studies on Indian populations began during the early 1990s, some of the initial studies dealt with populations which were neither anthropologically well defined nor were really representative of Indian populations (Barnabas et al., 1996; Roychoudhury et al., 2001). Studies were undertaken by molecular biologists themselves, and anthropologists were not involved in the research designs. The anthropological value of these works is limited owing to the defective sampling procedures. A significant change is seen in recent years where studies are designed by anthropologists. A broader range of populations is being examined following precise sampling procedures.

Basu et al. (2003) examine genetic variation in 44 geographically, linguistically, and socially disparate ethnic populations of India and use U2 frequencies to infer the existence of 'Aryan' movements. U2 comprises two sub-lineages, U2e (European-specific sub-lineage) and U2i (Indian-specific sub-lineage). Basu et al. (2003) have shown that U2e is not present in the Indian tribal groups, but only among castes. The U2e frequency is therefore more important in estimating the number of Aryan-speaking people entering India. Such evidence showed a much smaller estimate of migrants, though the actual number is difficult to estimate. Aryan speakers possibly came into India in small bands over a long period of time, as opposed to in a single wave of migration (Majumdar, personal communication).

The main issue is that the genetic data can only infer the migration of people and not the culture of the dispersing populations. The conjectures about language migrations can be substantiated only by linguistic evidence, texts and archaeological data. If language transmission takes place through contact and spread of farming (not a spread of the farmers), it will leave little signature in the genetic record. Similarly, if a small group of speakers become the dominant elite through military conquest or by economic supremacy they can impose their language on the general population, again without a significant sign in the genetic record. Is it then logical to assume that increasing trade relations probably persuaded people to adopt the Indo-European language for economic purposes?

One may wonder whether the 'dominance' of Indo-Aryan speakers in the north explains the large-scale recedence of 'indigenous' Dravidians towards the south. If the Dravidians speakers were widespread throughout India before the arrival entry of Indo-Europeans arriving (in small numbers and in multiple waves) their 'total retreat' to southern India to avoid dominance of the later appears to be sociologically unrealistic. It is quite logical to believe captivity of indigenous peoples (especially women) for labor by the dominant group. However, there is no 'Dravidian racial element' in the North. Therefore, the incorporation of an Indo-European language as a result of increasing trade is a more appealing explanation. The change to agriculture was slow in the South, but animal domestication was proving profitable, as seen in the rural-based Neolithic-Chalcolithic cultures. Usage of the Dravidian languages continued effectively in the South since their contacts with the north remained limited because of the difficult terrain. As Hawkey (2002) points out, the Deccan region has been a relatively stable dental pattern from the prehistoric times to the present-day. In her opinion, there is little dental evidence to support inter-regional gene flow between the northerly populations and the peoples of the Deccan for which probably the Vindhyan Range was an effective barrier.

The settled village life of early agro-pastorals beyond the domain of the Indus civilization can be divided into two culture groups, namely Neolithic and Chalcolithic, which flourished simultaneously during 6–4 ka. These early village cultures constitute the basis of present-day Indian rural society, which has changed slowly, as indicated by the replacement of iron to copper technology and the transformation of occupational groups into castes (Misra, 2001). According to Hawkey (2002), the dental morphological data suggests the origins of the Iron Age culture within central and southern peninsular India, and not from north-western regions. Iron Age populations, however, maintain affinity with the farming-herding groups of the Deccan.

The introduction of iron technology was of crucial importance to the expansion of an agriculture-based settled life. Copper and bronze were scarce and precious and accessible only to influential members of the society (Agrawal, 2000). However, once iron technology was mastered, more individuals had access to this material for tools, weapons and vessels. Agricultural surpluses, influenced by fertile soils, perennial availability of water, iron technology, and human enterprise, led to the emergence of a second urbanization in the Ganga valley around 2.6 ka, and a slow spread to peninsular India. The Early Historic phase began with neatly planned cities, increasing trade relations, efficient network of trade routes, defined social stratification, social complexity, refined knowledge of arts, literature, science, and medicine (Dhavalikar, 1999). There is archaeological evidence for the development of an early historic trade network from the 6th century BC onwards (Majumdar, 1968; Lahiri, 1992). There is every reason to believe that small-scale migrations of various bands of traders from the west, north and east were interacting. The well-established 'silk route', from the 2.3 ka onwards, and the establishment of port cities, on the eastern and western coast, provide testimony of maritime trade that was flourishing in and through India (Begley and Puma, 1992; Ray and Salles, 1996; Francis, 2002; Leena, 2004). Various groups continued coming into India in later phases as well, including during 7–11 century AD (Majumdar, 1993).

The introduction of iron technology greatly accelerated the expansion of agriculture, which in turn led to the dramatic increase in human population sizes. One of the consequences of these developments was increased deforestation and loss of wildlife depriving hunter-gatherers of the resources of their livelihood. Hunter-gatherers were forced to adopt agriculture or occupations associated with agriculture and get assimilated into the expanding farming society (Misra, 2001). The origin of caste structure in India may be correlated with the expansion of agriculture. Some populations, however, maintained former lifestyles and remained isolated, continuing to maintain their tribal identity, even up to the present day.

Concluding Remarks

Hypotheses regarding human migrations during the protohistoric time should primarily rely more on archaeological human skeletal evidence than mere cultural evidence. Previous works in human skeletal biology dealt with craniometric data, esp. cranial index, to hypothesize on population movements. Owing to limited sample size and few comparable metric variables these inferences remained doubtful. Recent studies based on morphological features of broadened the sample give contradicting results.

In this chapter, an attempt has been made to evaluate the merits and demerits of using metric features as against morphological characteristics of skeletal data in the Indian context. The change from dolichocephaly to brachycephaly head form appears more influenced by non-genetic factors, and cannot

be ascribed to 'foreign' gene flow. Relaxation of mechanical stress, increased use of starchy food and higher rates of infections in the early farming communities might have resulted in a delicate built and changes in craniofacial morphology. Cranial and dental morphological data clearly indicate genetic continuity from the Mesolithic era.

The hypotheses regarding massive population movements during the protohistoric period cannot be supported on available skeletal data. However, it must be noted that very few skeletal series have so far been studied. For better understanding of the process of peopling of the region it is necessary to generate more anthropological data on skeletal populations of various geographical and cultural brackets. A variety of metric and non-metric parameters should be observed and statistically analyzed. Moreover, skeletal data cannot be used in isolation to explain the process of the peopling of the subcontinent, but needs to be viewed in conjunction with archaeological and ethnographic data. Molecular information, based on reliable samples, will be valuable addition to the body of knowledge.

References

Agrawal, D.P., 2000. Ancient Metal Technology and Archaeology of South Asia. Aryan Books International, New Delhi.

Agrawal, D.P., Kharakwal, J.S., 2003. Bronze and Iron Ages in South Asia. Aryan Books International, New Delhi.

Armelagos, G.J., 1990. Health and disease in prehistoric populations in transition. In: Swedlund, A.C., Armelagos, G.J. (Eds.), Disease in Populations in Transition: Anthropological and Epidemiological Perspectives. Gergin and Garvey, New York, pp. 127–144.

Ayer, A.A., 1960. Report on human skeletal remains excavated at Piklihal near Mudgal. In: Allchin, F.R. (Ed.), Piklihal Excavations. Andhra Pradesh Government Archaeological Series 1, Hyderabad, pp. 143–154.

Barnabas, S., Apte, R.V., Suresh, C.G., 1996. Ancestry and interrelationship of the Indians and their relationship with other world populations: a study based on mitochondrial DNA polymorphisms. Annals of Human Genetics 60, 409–422.

Basu, A., Mukherjee, N., Roy, S., Sengupta, S., Banerjee, S., Chakraborty, M., Dey, B., Roy, M., Roy, B., Bhattacharyya, N.P., Roychoudhury, S., Majumder, P.P., 2003. Ethnic India: a genomic view, with special reference to peopling and structure. Genome Research, 2277–2290.

Basu, A., Pal, A., 1980. Human Remains from Burzahom. Anthropological Survey of India, Calcutta.

Begley, V., De Puma, R.D. (Eds.), 1992. Rome and India – The Ancient Sea Trade. University of Wisconsin Press, Madison.

Brace, C.L., 1963. Structural reduction in evolution. American Naturalist 97, 39–49.

Brace, C.L., 1964. The probable mutation effect. American Naturalist 98, 453–455.

Chatterjee, B.K., Kumar, G.D., 1963. Racial elements on Post Harappan skeletal remains at Lothal. In: Ratnam, B. (Ed.), Anthropology on the March. Social Science Association, Madras, pp. 104–110.

Cohen, M.N., 1977. The Food Crisis in Prehistory: Over Population and the Origin of Agriculture. Yale University Press, New Haven.

Cohen, M.N., Armelagos, J.G. (Eds.), 1984. Palaeoanthropology at the Origins of Agriculture. Academic Press, Orlando, Florida.

Dales, G.F., 1964. The mythical massacre at Mohenjodaro. Expedition 6(3), 36–43.

Dales, G.F., 1965. New investigations at Mohenjodaro. Archaeology 18, 2.

Dhavalikar, M.K., 1988. The First Farmers of the Deccan. Ravish Publishers, Pune.

Dhavalikar, M.K., 1999. Historical Archaeology of India. Books and Books, New Delhi.

Dhavalikar, M.K., 2002. Environment and Culture: A Historical Perspective. Bhandarkar Oriental Research Institute, Pune.

Dutta, P.C., Pal, A., Gupta, P., Dutta, B.C., 1987. Ancient Human Remains from Rupar. Anthropological Survey of India, Calcutta.

Ehrhardt, S., Kennedy, K.A.R., 1965. Excavations at Langhnaj: 1944–63, Part III: The Human Remains. Deccan College, Pune.

Francis, Jr. P., 2002. Early historic South India and the international maritime trade. Man and Environment XXVII (1), 153–160.

Gadgil, M., Joshi, N.V., Shambu Prasad, U.V., Manoharan, S., Patil, S., 1998. Peopling of India.

In: Balasubramanian D., Appaji Rao, N. (Eds.), The Indian Human Heritage. University Press, Hyderabad, pp. 100–129.

Guha, B.S., 1935. The Racial Affinities of the People of India. Census of India, 1931, New Delhi.

Gupta, P., Dutta, P.C., Basu, A., 1962. Human Skeletal Remains from Harappa. Anthropological Survey of India, Calcutta.

Hawkey, D.E., 2002. The peopling of South Asia: Evidence for Affinities and Microevolution of Prehistoric Populations from India/Sri Lanka. Spolia Zeylanica 39, 1–300.

Hemphill, B.E., Lukacs, J.R., Kennedy, K.A.R., 1991. Biological adaptations and affinities of Bronze Age Harappans. In: Meadow, R.H. (Ed.), Harappan Excavation 1986–1990: A Multidisciplinary Approach to Third Millennium Urbanism. Prehistory Press, Madison, pp. 137–182.

Hemphill, B.E., Christensen, A.F., Mustafakulov, S.I., 1997. Trade or travel: an assessment of interpopulational dynamics among Bronze Age Indo-Iranian populations. In: Allchin R., Allchin B. (Eds.), South Asian Archaeology 1995. Vol. 2. Oxford and IBH, New Delhi, pp. 855–871.

Joshi, M.B., Rout, P.K., Mandal, A.J., Tyler-Smith, C., Singh, L., Thangaraj, K., 2004. Phylogeography and origin of Indian domestic goats. Molecular Biology and Evolution 21–3, 454–462.

Kajale, M.D., 1991. Current status of Indian Palaeoethnobotany: Introduced and indigenous food plants with a discussion of the historical and evolutionary development of Indian agriculture and agricultural systems in general. In: Renfrew, C. (Ed.), New Light on Early Farming: Recent Developments in Palaeoethnobotany. Edinburgh University Press, Edinburgh, pp. 155–190.

Karve, I., Dandekar, V.M., 1951. Anthropometric Measurements of Maharashtra. Deccan College, Pune.

Kennedy, K.A.R., 1965. Human skeletal material from Ceylon with an analysis of the Island's prehistoric and contemporary populations. Bulletin of British Museum of Natural History and Geography 11, 135–213.

Kennedy, K.A.R., 1982. Palaeodemographic perspectives of social structural change in Harappan society. In: Pastner S., Flam, L. (Eds.), Archaeology of Pakistan: Recent Socio-cultural and Archaeological Perspective. Cornell University South Asia Program, Ithaca, pp. 211–18.

Kennedy, K.A.R., 1984a. Biological adaptation and affinities of Mesolithic South Asians. In: Lukacs, J.R. (Ed.), The People of South Asia: The Biological Anthropology of India, Pakistan and Nepal. Plenum Press, New York, pp. 29–55.

Kennedy, K.A.R., 1984b. A reassessment of the theories of racial origins of the people of the Indus Valley civilization from recent anthropological data. In: Kennedy, K.A.R., Possehl, G.L. (Eds.), Studies in the Archaeology and Palaeoanthropology of South Asia. Oxford-IBH Publishing, New Delhi, pp. 99–107.

Kennedy, K.A.R., 1984c. Trauma and disease in the Harappans. In: Lal, B.B., Gupta, S.P. (Eds.), Frontiers of the Indus Civilization. Books and Books, New Delhi, pp. 425–436.

Kennedy, K.A.R., 1995. Have Aryans been identified in the prehistoric skeletal record from South Asia? Biological anthropology and concepts of ancient races. In: Erdosy, G. (Ed.), The Indo-Aryans of Ancient South Asia: Language, Material Culture and Ethnicity. Walter de Gruyter, Berlin, pp. 32–66.

Kennedy, K.A.R., 2000. God-Apes and Fossil Men-Paleoanthropology of South Asia. University of Michigan Press, Ann Arbor.

Kennedy, K.A.R., Caldwell, P., 1984. South Asian prehistoric human skeletal remains and burial practices. In: Lukacs, J.R. (Ed.), The People of South Asia: The Biological Anthropology of India, Pakistan and Nepal. Plenum Press, New York, pp. 159–97.

Kennedy, K.A.R., Disotell, T., Roertgen, W.J., Sherry, J., 1986a. Biological Anthropology of Upper Pleistocene Hominids from Sri Lanka: Batadomba Lena and Belilena Caves. Ancient Ceylon, No. 6.

Kennedy, K.A.R., Lovell, N.C., Burrow, C.B., 1986b. Mesolithic Human Remains from the Gangetic Plain: Sarai Nahar Rai. South Asian Occasional Papers and Theses, No.10, South Asian Program. Cornell University, Cornell, NY.

Kennedy, K.A.R., Lukacs, J.R., Pastor, R.F., Johnston, T.L., Lovell, N.C., Pal, J.N., Hemphill, B.E., Burrow, C.B., 1992. Human Skeletal Remains from Mahadaha: A Gangetic Mesolithic Site. South Asian Occasional Papers and Theses, No. 11, South Asian Program. Cornell University, Cornell, NY.

Kennedy, K.A.R., Malhotra, K.C., 1966. Human Skeletal Remains from Chalcolithic and Indo-Roman Levels from Nevasa: Anthropometric and Comparative Analysis. Deccan College, Pune.

Kennedy, K.A.R., Possehl, G.L., 1979. Hunter-gatherer/agriculturalist exchange in prehistory: an indian example. Current Anthropology 20, 592–593.

Kenoyer, J.M., 1998. Ancient Cities of the Indus Valley Civilization. Oxford University Press, Oxford.

Lahiri, N., 1992. Archaeology of Indian Trade Route Up To 200 BC. Oxford University Press, Delhi.

Larsen, C.S., 1984. Health and disease in prehistoric Georgia: The transition to agriculture. In: Cohen, M. N., Armelagos, G. J. (Eds.), Paleopathology at the Origins of Agriculture. Academic Press. New York, pp. 367–392.

Leena, H.J. S., 2004. Sailing across seven seas (a study of maritime trade in Bengal). Ph.D. Dissertation, Deccan College, Pune.

Lukacs, J.R., Bogorad, R.K., Walimbe, S.R., Dunbar, D.C., 1986. Palaeopathology at Inamgaon: A Post-Harappan agrarian village in Western India. American Philosophical Society Proceedings 130(3), 289–311.

Lukacs, J.R., Walimbe, S.R., 1984a. Palaeodemography at Inamgaon: An early farming village in Western India. In: Lukacs, J.R. (Ed.), The People of South Asia: The Biological Anthropology of India, Pakistan and Nepal. Plenum Press, New York, pp. 105–132.

Lukacs, J.R., Walimbe, S.R., 1984b. Deciduous dental morphology and the biological affinities of a late chalcolithic skeletal series from western India. American Journal of Physical Anthropology 65, 23–30.

Lukacs, J.R., Walimbe, S.R., 1986. Excavations at Inamgaon, Vol.II, Part I: Physical Anthropology of Human Skeletal Remains: An Osteobiographic Analysis. Deccan College, Pune.

Majumdar, D.N., 1961. Races and Cultures of India. London: British Museum.

Majumdar, P.P., 1998. People of India: biological diversity and affinities. Evolutionary Anthropology 6, 100–110.

Majumdar, R.C. (Ed.), 1968. The Age of Imperial Unity. Bharatiya Vidya Bhavan, Bombay.

Majumdar, R.C. (Ed.), 1993. The Age of Imperial Kanauj. Bharatiya Vidya Bhavan, Bombay.

Malhotra, K.C., 1965a. Human skeletal remains from Neolithic Tekkalakota. In: Nagaraja Rao, M.S., Malhotra, K.C. (Ed.), The Stone Age Hill Dwellers of Tekkalakota. Deccan College, Pune, pp. 109–162.

Malhotra, K.C., 1965b. Human skeletal remains from Chandoli. In: Deo, S.B., Ansari, Z.D. (Eds.), Chalcolithic Chandoli. Deccan College, Pune, pp.143–184.

Martin, D.L., Armelagos G.L., Goodman A.H., Van Gerven D.P., 1984. The effects of Socioeconomic change in prehistoric Africa: Sudanese Nubia as a case study. In: Cohen, M.N., Armelagos, G.J. (Eds.), Paleopathology at the Origins of Agriculture. Academic Press, Orlando, pp. 193–216.

Misra, V.N., 1976. Ecological adaptations during the terminal stone age in western and central India. In: Kennedy, K. A. R., Possehl, G L. (Eds.), Ecological Backgrounds of South Asian Prehistory. Cornell University, Ithaca, pp. 28–51.

Misra, V.N., 1989. Stone age India: an ecological perspective. Man and Environment, 14, 17–64.

Misra, V.N., 2001. Prehistoric human colonization of India. Journal of Bioscience 26(4), 491–531.

Mughal, M.R., 1990. The decline of the Indus civilization and the Late Harappan period in the Indus valley. Lahore Museum Bulletin 3(2), 13–14.

Mushrif, V., 2001. Biological anthropology of the Deccan Chalcolithic populations: a case study on Nevasa human skeletal series. Ph.D. Dissertation, Deccan College, Pune.

Mushrif, V., Walimbe, S.R., 2000. A fresh look at the human skeletal remains from Chalcolithic Walki (Maharashtra). Man and Environment XXX (2), 19–34.

Pattanayak, D.P., 1998. The language heritage of India. In: Balasubramanian, D., Appaji Rao, N. (Eds.), The Indian Human Heritage, University Press, Hyderabad, pp. 95–99.

Possehl, G. L., 1976. Lothal: A gateway settlement of the Harappan civilization. In: Kennedy, K.A.R., Possehl, G.L. (Eds.), Ecological Backgrounds of Asian Prehistory. Cornell University South Asia Program, Ithaca, pp. 118–131.

Possehl, G.L., 1999. Indus Age: The Beginning. Oxford and IBH, New Delhi.

Ratnagar, S., 1981 Encounters: The Westerly Trade of the Harappan Civilization. Oxford University Press, Delhi.

Ratnagar, S., 2000. The End of the Great Harappan Tradition. Manohar Publisher, Delhi.

Ray, H.P., Salles, J. (Eds.), 1996. Tradition and Archaeology (Early Maritime Contacts in the Indian Ocean). Proceedings of the International Seminar 'Techno-archaeological Perspectives of Seafaring in the Indian Ocean 4th Century BC -15th century AD'. Manohar, New Delhi.

Roychoudhury, S., Roy, S., Basu, A., Banerjee, R., Vishwanathan, H., Usha Rani, M.V., Sil, S.K., Mitra, M., Majumdar, P.P., 2001. Genomic structures and population histories of linguistically distant tribal groups of India. Human Genetics 109, 339–350.

Sarkar, S.S., 1954. The Aboriginal Races of India. Asiatic Society, Calcutta.

Sarkar, S.S., 1964. Ancient Races of Baluchistan, Panjab and Sind. Bookland, Calcutta.

Sarkar, S.S., 1972. Ancient Races of Deccan. Munshiram Manoharlal, New Delhi.

Sewell, R.B.S., Guha, B.S., 1931. Human remains. In: Marshall, J. (Ed.), Mohenjo-Daro and the Indus Civilization. Volume 2. Arthur Probsthan, London, pp. 599–648.

Sewell, R.B.S., Guha, B.S., 1929. Report on the bones excavated at Nal. Memoirs of Archaeological Survey of India, 35, 56–89.

Singh, K.S., Manoharan, S., 1993. People of India: Languages and Scripts. Oxford University Press, Delhi.

Smith, B.D., 1995. The Emergence of Agriculture. Scientific American Library, New York.

Thangaraj, K., Chaubey, G., Kivisild, T., Reddy, A.G., Singh, V.K., Rasalkar, A.A., Singh, L., 2005 Reconstructing the origin of Andaman Islanders. Science 308, 996.

Thomas, P. K., 2000. Contribution of Deccan College to archaeozoological research. Bulletin of Deccan College Research Institute 60–61, 77–96.

Walimbe, S.R., 1986. Palaeodemography of protohistoric Daimabad. In: Sali, S.A. (Ed.), Daimabad 1976–79. Archaeological Survey of India, New Delhi, pp. 641–740.

Walimbe, S.R., 1990. Human skeletal remains. In: Dhavalikar, M.K., Shinde, V., Atre, S. (Eds.), Excavations at Kaothe. Deccan College, Pune, pp. 111–231.

Walimbe, S.R., 1998. Bio-cultural adaptations in cranial morphology among the early farming Chalcolithic populations of the Deccan plateau. In: Bhasin, M.K., Malik, S.L. (Eds.), Contemporary Studies in Human Ecology: Human Factor, Resource Management and Development. Indian Society for Human Ecology, Delhi, pp. 25–40.

Walimbe, S.R., 2006a. A report on adult human skeleton from Neolithic Ieej (A.P.). Bulletin of Andhra Pradesh Archaeology Department.

Walimbe, S.R., 2006b. An anthropometric and comparative analysis of the human skeletal remains from the Chalcolithic levels of Hullikallu

(A.P.). Bulletin of Andhra Pradesh Archaeology Department.

Walimbe, S.R., Gambhir, P.B., 1994. Long Bone Growth in Infants and Children: Assessment of Nutritional Status. Mujumdar Publications, Mangalore.

Walimbe, S.R., Kulkarni, S.S., 1993. Biological Adaptations in Human Dentition: An Odontometric Study on Living and Archaeological Populations in India. Deccan College, Pune.

Walimbe, S.R. and Kulkarni, S.S., 1994. Dental anthropology in India: a review. Man and Environment XIX (1–2), 205–216.

Walimbe, S.R., Lukacs, J.R., 1992. Dental pathology at the origins of agriculture: evidence from Chalcolithic population of the Deccan Plateau. In: Lukacs, J.R. (Ed.), Culture, Ecology and Dental Anthropology. Kamala Raj, Delhi, pp. 117–132.

Walimbe, S. R., Paddayya, K., 1999. Human skeletal remains from the Neolithic ashmound site at Budihal, Gulbarga District, Karnataka. Bulletin of Deccan College Research Institute 58–59, 11–47.

Walimbe, S.R., Tavares, A., 1996. Evolving trends in skeletal biology in the Indian sub-continent: a case study on the incipient agricultural populations of the Deccan plateau. In: Wadia, S., Korisettar, Kale, V.S. (Eds.), Quaternary Environments and Geoarchaeology of India: Essays in Honour of Prof. S.N. Rajaguru. Geological Society of India, Bangalore, pp. 515–529.

Walimbe, S.R., Tavares, A., 2002. Human skeletal biology: scope, development and present status of research in India. In: Paddayya, K. (Ed.), Recent Studies in Indian Archaeology. Indian Council of Historical Research, Monograph Series 6. Munshiram Manoharlal, New Delhi, pp. 367–402.

14. Foragers and forager-traders in South Asian worlds: *Some thoughts from the last 10,000 years*

KATHLEEN D. MORRISON

Department of Anthropology
University of Chicago
1126 East 59th Street
Chicago
Illinois, 60637
USA
k-morrison@uchicago.edu

Introduction

South Asia is sometimes seen as a particularly good place to study the earliest periods of human history because of the presence of contemporary peoples who gather and hunt, people usually referred to as "tribals," a label first affixed by the British but which has had considerable staying power (Singh, 1997:33; Morrison, 2002a:35). These groups are often viewed as natural analogues for the past, providing us with information about environmental adaptations, social organization, and the like. Much (but of course not all) ethnoarchaeology in South Asia is predicated on simple parallels between contemporary tribals and Paleolithic or Mesolithic peoples, a strategy which tends to erase the actual histories of these groups, and an assumption of timelessness which recent research (e.g., Stiles, 1993; Murthy, 1994; Morrison and Junker, 2002) has shown is not well-founded. Even where the complex histories of the Holocene are acknowledged, many still feel that "tribals" offer the

best analogues for the past, based on a notion of some essential tribal cultural (sometimes even "racial") identity as the link which joins the past and the present. In this chapter, I review a few examples of long-term histories of foraging groups in South Asia and, in so doing, suggest an alternative framework in which hunting and gathering, past and present, is seen as a set of *strategic practices* rather than as an essential identity, a distinction which allows us to view the complex, entangled histories of South Asian foragers and forager-traders not as masks obscuring "real" economic, ecological, and social relations but instead as resources for learning about the life-possibilities of those who lived in the distant past.

Essentialized Categories, Erased Histories

The sharp distinction between "tribe" and "caste," a classification which has considerable political and economic significance in contemporary Indian society and is thus

M.D. Petraglia and B. Allchin (eds.), The Evolution and History of Human Populations in South Asia, 321–339.

unlikely to disappear any time soon, is clearly a rather arbitrary one. The category of caste itself is highly contested, with some scholars seeing considerable changes in both the overall importance of caste as well as changes in specific caste boundaries and rankings as recently as the nineteenth century. There is broader agreement that caste categories, boundaries, and rankings were in considerable flux as late as the seventeenth century (Bayly 1999; Dirks 2001; and see Boivin, 2005, Boivin, this volume). In South Asia, the notion of a common "tribal" identity makes sense only as an index of exteriority to caste; there are a vast number of linguistically, ethnically, and religiously distinct "tribal" groups across the subcontinent whose histories have little apparent connection except that they all, in some way, pose a problem of identity to caste orders. While this sense of externality is important, marking what in many cases seems to be a real history of resistance to assimilation and rejection of dominant modes of being in the world, it also does not constitute "tribe" as an analytically robust category. Further, because what is meant by "caste" has changed radically over the last three thousand and even the last three hundred years, definitions of group membership constructed as anti-caste are subject to much of the same sense of uncertainty and flux. Thus, there is little justification for extending a clear caste/tribe boundary back in time.

"Tribal" designation, in the dual tribe/caste mode, is also often associated with a host of (usually derogatory) assumptions including primitivity, simplicity, lack of progress, and so on. More positively, tribal identities may also carry an aura of ecological wisdom, supernatural power, and – what is politically most important to the aspirations of contemporary people – an assumption of originality or indigeneity (Béteille, 1998; Karlsson, 2001; Morrison, 2002a, 2002c). Whether positive or negative, however, the essentialist mode which casts all "tribals" and especially those

who gather and hunt, into a common category provides a powerful incentive for archaeologists to deploy studies of these groups as simple analogues for the past. As I discuss below, however, the inadvisability of drawing simple analogies between contemporary foragers and the distant past does not imply that ethnographic and historical data have nothing to teach us.

Actual histories of foraging groups in the subcontinent show an interesting diversity of form throughout the Holocene (and no doubt before as well). Although it is difficult, and, at some point in time, eventually impossible to trace the specific histories of named groups known from the historic record deep into the past, both the overall record of Holocene economic diversity and the group-specific trajectories that can be reconstructed reveal: 1) that foraging continued to be important in South Asia throughout the Holocene; and, 2) that there was considerable mobility in group composition and flux in economic strategies. For some people, hunting and gathering were consistent core subsistence strategies to which animal husbandry, small-scale agriculture, trading, military service, wage labor, and/or raiding were added opportunistically. For others, there may have been an oscillation (cf. Gardener, 1985) between foraging and other modes of existence; still others seem to have adopted gathering and hunting only after years as agriculturalists (Stiles, 1993; Ratnagar, 1997). One feature of this economic flexibility has been the existence of longstanding connections between foraging groups and others in what is usually considered "caste society." Thus, there is little reason to believe that contemporary people who hunt and gather can be accurately described as remnants of or throwbacks to the Paleolithic, or even necessarily as descendants of an unbroken lineage of hunter-gatherers (Morrison, 2002a, 2002b).

By now the "revisionist debate" of the 1980s and 1990s (e.g., Schrire, 1980, 1984; Wilmsen, 1983, 1989, 1993; Denbow, 1984;

Wilmsen and Denbow, 1990; Lee and Gunther, 1991, 1993, 1995) has achieved the status of disciplinary history, but the terms of this debate are still important, especially since the discussion found little resonance in South Asia at the time. What was critical about this debate was not so much the specific assertions about the complex and entangled histories of particular foraging groups, but the conclusion drawn by the revisionists that the recuperation of hunter-gatherer history made ethnographic analogy impossible and, thus much of our understanding of early humans untenable (see Shott, 1992, Morrison, 2002a). The "recognition of history" that caused shock waves across the Kalahari, was, however, old news in South Asia where hunter-gatherers have long been recognized as participants in the worlds of differently-organized others (states, empires, agriculturalists, traders, etc.). As noted, in a few cases the evidence even suggests that some contemporary foraging groups assumed this line of work relatively recently, within the last three or four hundred years (Guha, 1999; Morrison, 2001, 2002c). Thus, it hardly seems controversial to note that the use of ethnographic observations of South Asian hunter-gatherers to retrodict past human behavior is going to be problematic.

That said, both sides in the revisionist debate appeared to share a common bias toward viewing foragers as an ideal type– and at the extreme edges of the debate as an essentialized category – that was either still pure (and hence useful for archaeology) or already tainted (and hence useless as an historical model). However, we do not learn from the ethnographic and historic record because it tells us about *foragers*, but because it tells us about *foraging*. Some people in the contemporary world hunt and gather and this is interesting and instructive. Just because there are no simple parallels between past and present does not mean there are no parallels. Even if we abandon the problematic caste/tribe dichotomy and

acknowledge the entangled histories of all South Asian foragers, this does not mean that archaeology is on its own. On the contrary, the rich store of information we possess about hunting and gathering activities constitutes a valuable storehouse for archaeological inference. Contemporary and historically-known foragers have much to teach us about the organizational possibilities for human action; their evident strategic flexibility seems quite germane when we consider that, as should become clear, much of the Mesolithic archaeological record of India is characterized by multiple economic strategies–hunting, gathering, agriculture, animal husbandry– as well as by presumed temporal/cultural markers which do not always neatly delineate specific groups as agriculturalists, hunter-gatherers, or pastoralists.

The Archaeology of Holocene Hunting and Gathering: Current Challenges

Although Paleolithic peoples and their forebears lived in a world of hunter-gatherers, by some time around the seventh millennium BC in the northwestern part of the subcontinent, agriculture and husbandry of domestic animals became part of the subsistence mix of this region (Constantini, 1984). Early agriculture in South Asia was a mosaic process, with some regions such as the far south not adopting cultivation until around 3000 BC (see overviews by Kennedy, 2000; Chakrabarti, 2001; Fuller, 2006, Fuller, this volume). Plant and animal domestications, too, took place in multiple loci and both indigenous and imported cultigens eventually became part of South Asian diets. In spite of this, hunting and gathering have persisted to the present, occasionally even replacing agriculture as a primary economic strategy. Elsewhere (Morrison, 2005), I have addressed the impact of this persistence on the broader histories of South and Southeast Asian states; here, however, I am concerned primarily with why our understanding of Holocene hunting and gathering in India is so diffuse. Three basic challenges can be identified; problems of systematics, problems of data, and problems of training.

Problems of Systematics

We can begin with the culture-historical term "Mesolithic" which, in South Asia, is generally used quite loosely to refer to a specific time period, tool technology, or a way of life–and sometimes to all three. Given the dominance of cultural-evolutionary labels, the term Mesolithic is often treated as a Post-Paleolithic, Pre-Neolithic phenomenon regardless of its actual location in time, a perspective which clearly elides significant contextual variability. There is, at present, no specific term or terms for foraging sites that fall in the late Holocene. Possehl and Rissman (1992) have suggested a "interactive trade and barter" sub-category of the Mesolithic and more recently, Miracle (2005) has deployed neologisms such as "Microlithic Iron Age," a label with much to recommend it. I would, however, suggest certain modifications based on the specific problem of microliths (see below and Morrison, 1999). Although I am aware of the terminological difficulties surrounding the word forager (Morrison, 2002b:12–13), perhaps a simple label such as "Iron Age[1] foragers" might suffice, although coming to this kind of conclusion may certainly require chronometric dates and such we may be forced to make preliminary assessments as well.

As is well known, the term Mesolithic is not a solid chronological designation. Chronometric dates from "Mesolithic" sites extend all the way from the early Holocene to the late Middle Periods, that is, to the sixteenth century or so. The term, insofar as it is useful for global comparisons, is really only meaningful for the early Holocene. There is really no reason that sites where we find evidence for gathering and hunting need always to be labeled Mesolithic, not least because we often do not know how hunting and gathering fit into the larger economic strategies of those who created the location–gathering and hunting are practiced by a wide range of people just as "hunter-gatherers" often deployed other strategies such as trade, wage labor, etc. Nevertheless, the tradition of using the term Mesolithic very broadly is well-established in the literature and if we cannot always reform its usage, we can at least practice a healthy skepticism about undated and cursorily-studied locales.

Problems of systematics also include the fact that the term "Microlithic" is sometimes used as a synonym for Mesolithic, assuming an association between microlithic blade technology and this culture-historical period. Microliths were actually made quite early in South Asia, as recent evidence of Paleolithic-era microliths from Sri Lanka has shown (Deraniyagala, 1998; Kennedy, 2000). However, as is again quite well-known, microliths as well as other blade-based lithic technologies were also used quite late and are not necessarily always markers of foraging (Misra, 1976:45). Kennedy (2000:212) notes, for example, that microliths are associated with Harappan sites, citing an association between microliths and Harappan assemblages at Allahdino. A systematic study of notices in *Indian Archaeology – a Review* to 1989 also noted the systematic co-occurrence of "microliths[2]" with Early Historic sites, among others; a few notices even mention their placement atop historic walls (Lycett and Morrison, 1989). Indeed, chipped stone tools, including microliths, made on moulded bottle glass from the seventeenth and eighteenth centuries have also been reported (Malik, 1959:50; Cooper, 2002:93). Although microliths were used by many foraging groups, then, they are neither unambiguous indicators of hunting and gathering nor are they chronologically restricted enough that the term "Microlithic" represents much of an improvement over Mesolithic.

Problems of Data

Our understandings of Holocene hunting and gathering have improved tremendously in the last several years, but we still face significant problems of data. There are still few well-excavated contexts, especially those with both chronometric dates and analysis of plant and animal remains. Few open-air sites of later foragers have been studied, in part because of preservation issues, but perhaps also because the archaeology of later historic sites is often organized quite differently from that of prehistoric sites, with excavators being less likely to screen deposits or, in some cases, to even systematically collect and record artifacts (as opposed to structures). It is quite difficult, further, to diagnose forager sites from surface remains alone, especially given the problems with systematics noted above. Misra (2002:112-3) notes, for example, that although more than 2,000 Mesolithic sites have been reported, only 29 have been excavated, some of those in very cursory fashion. Finally, very few systematic regional surveys have been carried out, so our sense of larger-scale connections is still much poorer than it should be.

Problems of Training

Because of the strict separation in South Asia between the study of prehistory and the archaeology of later, historic periods (usually allied with art, architecture, and history), the study of recent hunting and gathering has tended to be neglected. Site selection for excavation of more recent material tends toward large urban contexts while prehistorians often avoid (or do not report on) locations with obvious remnants of later periods (Medieval pottery, coins etc.). This reluctance on both sides is somewhat understandable, given differences in expertise. It is difficult to understand later contexts without an appreciation of the larger worlds in which foragers lived and here scholars need to engage with history as well as with the archaeology of cities and villages. For specialists in later periods, lithic analysis may present a challenge; certainly the range of expertise required, from numismatics to geoarchaeology is rarely possessed by a single person. However, this is true of all archaeology and is a problem that can certainly be overcome.

Frameworks for Understanding Diversity

Given, then, that dominant assumptions about deep continuity between contemporary foraging groups and prehistoric hunter-gatherers may not always be warranted, it is still possible to deploy the complex and entangled histories of South Asian foragers and forager-traders (terms defined below) to help understand the distant past. As I have already suggested, an approach that views gathering, hunting, and other activities as strategies, rather than as attributes of groups essentialized as "foragers" by nature, disentangles the analysis of cultural continuity and change from economic strategies per se. That is, members of self-identified groups may change subsistence strategies while still maintaining group identities. Conversely, some groups which have historically been defined primarily on the basis of subsistence practices (or presumed "primitivity") appear to have incorporated people from very diverse linguistic, social, and religious backgrounds as well as having had complex political histories (S. Guha, 1999; Skaria, 1999; Morrison, 2002c, 2005).

The Chenchus, for example, are a Central Indian "tribal" group about whom the well-known ethnographer von Furer-Haimendorf wrote (1982:4–5), "Until two or three generations ago, the Jungle Chenchus seem to have persisted in a life-style similar to that of the most archaic Indian tribal populations, and their traditional economy can hardly have been very different from that of forest-dwellers of earlier ages." He went on to make other, racially-based comparisons which will not concern us here (Morrison, 2002a:36–7). Furer-Haimendorf, whose fieldwork in India covered nearly sixty years of the early twentieth century, was hardly alone in seeing the Chenchus as timeless and traditional. A longer-term analysis, however, shows a very different picture. Furer-Haimendorf's Chenchus are the very same group studied by Murthy (1994) who noted their great wealth, the existence of Chenchu royalty, and their long record of service to various governments in eastern India from the fifth century AD. Sumit Guha (1999), in a similar historical analysis of western India between the seventeenth and nineteenth centuries, documents the rise of "tribal kingdoms" and the changing military alliances between "tribal" leaders and larger polities. If these are living stone age peoples, then our ideas about deep prehistory certainly need major revision.

Because we can no longer draw simple parallels between the ethnographic and historical record of Indian foragers and the archaeological past does not, however, make this record irrelevant for archaeology. On the contrary, it proffers some potentially exciting new avenues for historic archaeology as well as providing useful analogues for the potential structure and organization of hunting and gathering activities in South Asian environments and cultural worlds. Here I focus on the former, using the record of Holocene hunting and gathering to make an argument about the complementarity of foraging forms in Asia.

We can begin with a somewhat contrarian view that inverts the dominant culture-evolutionary paradigm of archaeology. This view asserts that South (and Southeast) Asian hunter-gatherers have actually been instrumental in the development of complex polities in the region. Much of the famed Indian Ocean trade, in particular, as well as the coastal polities and European colonies that lived off of the trade would not have existed without the active engagement of foraging

peoples. Thus, far from representing the peripheral activity of a series of marginal peoples, gathering in particular has played a far greater role in South Asian history than is usually acknowledged. This kind of active (sometimes coerced) engagement in trade relations characterizes what I have called a "forager-trader" form of foraging which, it is possible to argue, has actually *underwritten* the continued existence of "classic" subsistence foragers in South and Southeast Asia. These two forms are not, of course, mutually exclusive, with fluidity being a key part of the issue–given the often oppressive power relations involved in forager-trader situations, flight and avoidance become key strategies of resistance. Bird-David (1990, 1992a, 1992b) has stressed the cultural dimension of South Asian foraging, especially the maintenance of separate identities and practices (including mobility, foraging, and behavior seen as transgressive by the larger society [cf. R. Guha 1999]) that constitute forms of resistance to lowland society.

A history of Holocene hunting and gathering in India is, then, to a larger extent than one might think, a history of India in general. This is clearly not the venue for such a history; however, I would like to isolate here some of the kinds of conditions favorable to subsistence foraging as opposed to forager-trader forms, an exercise that may provide a provisional framework for such a history. We can begin with three basic and related factors that influenced the degree to which foragers were able to (or forced to) participate in outside worlds; occupational history and power relations, geography, and resources[3].

Occupational Histories and Power Relations

The long trajectory of anthropogenic landscape change in South Asia has meant that in certain areas of high agricultural potential, including most of the large river deltas, hunting and gathering, both as strategy and identity, have tended to disappear through time. Areas with the longest histories of agriculture, especially intensive agriculture, have the fewest foragers. It is certainly no accident that contemporary "tribal" groups who practice at least some hunting and gathering are differentially concentrated in the tropical northeast, the semi-tropical southwest (especially the uplands), and the hilly regions of central India where dense deciduous monsoonal forests still exist (Bird-David, 1999). The trajectory of agricultural adoption in South Asia begins in the far northwest, on the margins of and then onto the Indus plain, with later adoption of cultivation in Central India and even later in the south and northeast. Contrary to notions that South Asia has always been densely occupied, many areas actually had quite low population densities into the Early Modern period, making competition for land less fierce than in the last few centuries. Here a comparison between regions is useful; some of the radical upland deforestation experienced in South Asia in the late pre-colonial and colonial periods took place much later in Southeast Asia (e.g., Boomgaard et al., 1997) and in these places foragers, too, have had relatively more success in maintaining themselves at least partially independently. For the maintenance of subsistence hunting and gathering, it is perhaps sufficient that landscapes amenable to foraging and hostile to agriculture have few others competing for space. Thus, specific occupational histories, as well as raw economic potentialities are at in issue in understanding the organization of Holocene foraging.

For foraging forms adjusted to exchange, however, it is also necessary that locations available and suitable[4] (see below) for foraging contain resources desired outside local communities, as discussed below. Thus, both competition and cooperation with differently-organized others are at issue. The occupational histories that impinged on the lives of foragers were not only their own, but also those of expanding agriculturalists, miners, prospectors, royal courts, foresters, and others. Further, these others have had differential abilities to coax, coerce, entice, and influence the activities of foragers. Some of this potential relates to demographic density and everyday interaction; but differential technology (military, transportation) is also at issue as is the ability to produce and distribute materials desired by foragers. In cases where interactions were relatively intensive, such as in southern India, these objects of desire included metal tools, textiles, and food grains, perhaps pointing to the importance, again, of local wild resources. That is, where carbohydrate-rich foods were more rare, the allure of rice and other grains may have been proportionally greater[5]. Thus, the continuing importance of hunting and gathering is related to a fine balance between proximity and distance to others; too far away, and regular relationships cannot be maintained, too close and foragers are threatened by landscape transformation and competition from other would-be gatherers.

Geography

While subsistence foraging appears to have been possible, and practiced, wherever both resource distributions and challenges from others permitted (making remoteness from others a virtue), many forager-trader groups in South Asia live either on islands or on narrow mountain chains surrounded by agricultur-alists and state societies, marking the impor-tance of transportation links but also of the balance between access and remoteness. The spaces of gathering, further, tend also to be spaces of expert knowledge (Morrison, 2002c), where resources are dangerous to collect, or difficult to find, capture, and/or process. As I have argued elsewhere (2001, 2002a, 2002b; cf. Fox, 1969) forager-traders may be thought of as economic and cultural specialists, especially insofar as many of the products they gather and trade are not foodstuffs and indeed are not used by them at all except as a medium of exchange. Further, historical analysis of some groups in southern India (Morrison, 2002c) has shown a significant flow of goods, including food grains and metal tools, into foraging areas, suggesting that relations of exchange were often essential rather than incidental. In other cases, connections were clearly much more casual. This variability may be linked to, among other things, the transportation technologies and coercive powers of trade partners. For example, hunter-gatherers interacting with Harappan traders and craftspeople appear to have had much more bargaining power, as it were, than did upland foragers of island Southeast Asia confronted with Dutch colonial demands and desires for such goods as sappan wood, bird of paradise feathers, resins, dyes, and other "exotic" commodities (Boomgaard et al., 1997; Cribb, 1997; Boomgaard, 1998). The scale of extraction in the latter instance dwarfs that the former, not only because of the large capacity of European ships, but also because of the effective use of firepower and political coercion of a kind not possible in third-millennium Gujarat.

Resources

The co-occurrence of subsistence foraging and forager-traders forms in Holocene South and Southeast Asia highlights the salience of resource distributions for hunter-gatherer life ways. Despite assertions to the contrary[5], subsistence foragers have, in some cases, managed to exist in tropical forest settings (e.g., Bellwood 1997). At the same time, however, the small number of such instances points to the difficulty of foraging-based subsistence in these environments. In southern India, for example, the Western Ghat

mountains with their tropical and semi-tropical rainforests contain no Paleolithic or Mesolithic sites (Misra, 2002:112–4; Raju and Venkatasubbaiah, 2002:102); even Neolithic remains are rare (Korisettar et al., 2002:174). Archaeological and paleobotanical evidence suggest a relatively late occupation coincident with the Early Historic explosion of long-distance trade involving forest products (Morrison, 2002c). Highly modified anthropogenic environments, too, posed challenges to subsistence foragers.

At the same time, many of the same upland, forested areas where pre-agricultural sites are rare were also home to plant, animal, and mineral resources desired by outsiders. Chief among these were the so-called "minor forest products;" spices, resins, gums, dyes, saps, scented and other kinds of non-timber wood taxa, honey, wax, and wild animal products. Many of these are high-value and relatively compact items, often highly dispersed, difficult of access and requiring substantial local knowledge. Questions of occupational history, geography, and resources clearly structure, but do not by any means determine, the possibilities for foraging groups. Once agriculturalists, pastoralists, states, and empires came into the picture they were almost always part of the finely-tuned balancing act practiced by South Asian hunter-gatherers, but their impact was highly variable, as the examples sketched out below illustrate.

Holocene Hunting and Gathering in South Asia: Some Examples

Western India: Forager-Trader and Forager-Pastoralist Interaction with Harappans

As is well-known, the Indus Valley was host to one of the world's first urban societies, a polity or more likely series of linked polities, spread over a vast area of the floodplain of the Indus and its tributaries as well as adjacent upland areas. The different "domains" (Possehl, 2002) of the Indus civilization, or the Harappan, as it is also called, each contained one very large city as well as numerous smaller towns and villages. The Harappan world was one supported by an intensive agropastoral economy focused on winter cultivation of wheat, barley and other Near Eastern crops. Harappan faunal assemblages are dominated by domesticates, including cattle, buffalo, sheep, goats, and pigs. Trade was extensive, with connections across the Indus world marked by a common script, the use of seals, and a uniform system of weights and measures. By the mid-third millennium BC, the so-called Mature Period, the Indus floodplain and its surrounding hills had experienced a long history of environmental transformation, dense occupation, and close interconnection of cultural groups (Kenoyer, 1997; Possehl, 2002; Wright, in press). However, the Indus sphere also extended to the southeast, along the coast and into the salty flats of what is now the state of Gujarat, in India. This area, part of what Possehl (2002) calls the "Sorath Harappan," differs from the heartland not only in environmental features–differences in monsoon patterns meant that cropping patterns were more suited to summer cultivation of millets and other indigenous[6] crops – but also, at least in part, in cultural practices. Archaeological sites include both very "local" kinds of assemblages as well as a few places with full-fledged Harappan material culture inventories. One of the latter, Lothal, was posited by Possehl (1976, 2002) to be a "gateway community" where Harappan traders interacted with local hunter-gatherers, exchanging valuable local raw materials (such as semiprecious stones for bead and seal-making) for ceramics, metal artifacts, and domesticates. Dhavalikar et al. (1995) suggest a similar role for the small port-town of Kuntasi.

The evidence from nearby "Mesolithic" sites seems to support this general model (see discussions by Kennedy, 2000; Lukacs, 2002; Possehl, 2002; Morrison, 2005), with significant finds of domestic fauna, pottery, and metal tools in sites otherwise apparently used by semi-mobile hunter-gatherers. One of these sites is Langhnaj (Sankalia, 1965), where recent re-excavation by Shinde and colleagues has shown two cultural levels (Kashyap and Shinde, 2005), the lower an "aceramic Mesolithic" dating from 5600–4800 BC and the upper a "ceramic Mesolithic" dating from 4800 to 3500 BC. Critically, microwear analysis of quartz microliths from *both* levels showed that copper tools may have been used to retouch lithic blades. Sankalia had earlier obtained a date of 2440–2160 BC (Possehl and Rissman, 1992:462) from deposits containing a Harappan-style copper knife and disk beads, dates coincident with the Mature Harappan settlement at Lothal. Other "Mesolithic" sites in the area also contain small quantities of pottery and, more often, bones of domestic animals (Kennedy, 2000:200–212).

In Chalcolithic Western India, foragers, pastoralists, and other mobile groups had the advantage of occupational histories and local environments that left them room to expand and, if necessary, flee the attentions of urban-dwellers. There is, further, no evidence for coercion or unequal power relations between foragers and others, though the nature of the evidence is probably not robust enough to reveal such relationships were they present. Significant evidence for the use of animal domesticates, and apparently for animal husbandry as well, complicates our image of subsistence foragers interacting with traders and agriculturalists, inviting us to think more creatively about the possible roles of animal husbandry in local economic strategies. Further, dental evidence (Lukacs and Pal, 1993; Lukacs, 2002) shows a high rate of caries (cavities) at Langhnaj, suggesting a high carbohydrate diet and thus also perhaps some reliance on traded grain.

Despite the apparent "mixing" of categories, with hunting and gathering people clearly also experimenting with livestock and at least occasionally using ceramics and metal, there is still a good case to be made here for cultural and social diversity. Here we seem to see substantial interaction between local foraging groups and others, but not yet of the intensity that characterized later forager-traders. Settled farming populations at this time were apparently still sufficiently small so as not to seriously threaten the habitat of foraging peoples, allowing for a wide range of strategies from subsistence foraging to forager-trader and forager-pastoralist forms. On the other hand, the substantial biological evidence for exchange of (human) genetic material suggests connections went well beyond simple trade relations (Lukacs, 2002; Kennedy et al., 1984).

As in other cases, the kinds of resources desired by Harappan urban-dwellers seem to have primarily been high-value (precious and semi-precious stones, shell, ivory, and wood including teak from the Western Ghats; Possehl, 2002:73) and (with the exception of wood) relatively portable materials available in only limited contexts and whose exploitation required some local knowledge. One common factor in many instances of forager-based trade, relatively inexpensive transportation by sea, is also present here. Both technology and distance thus play a role in maintaining relationships between disparate groups that are in some ways close, in others quite distant. Water transport allowed goods collected by hunter-gatherers (and local farmers) to move to distant consumers; this technology as well as the social organization required to exploit regional differences in resources seems to be critical to the maintenance of such relationships.

Northern India: The Vindhya Hills and the Gangetic Plain

Although the earliest agriculture in South Asia was identified on the western upland edge of the Indus Valley at the site of Mehrgarh, dating to the seventh millennium BC (Jarrige, 1984), North-Central India's so-called Vindhyan Neolithic was the next to follow, with cultivation beginning around the fifth millennium BC, in an era marked, at least for a time, by relatively small-scale agricultural settlements. Here there are none of the sharp contrasts presented by the trade between urban Harappans and the mobile foragers, forager-pastoralists, and (perhaps) forager-agriculturalists of Gujarat. In this region, and on the Gangetic plain to the north, hunting and gathering clearly continued alongside agriculture, and many have debated whether or not the same level of interaction evident in Gujarat also obtained here (see Kennedy, 2000 for a review).

In this region there are many more excavated sites and, critically, a large sample of human skeletal material as well. Evidence from these sites (Baghai Khor, Damdama, Lekhahia, Mahadaha, and Sarai Nahar Rai) clearly shows temporal overlap between agriculturalists and foragers, with some sites, such as Sarai Nahar Rai, occupied quite late. The Mesolithic site of Sarai Nahar Rai was occupied into the first millennium BC (Agrawal, 1971; Possehl and Rissman, 1992; Kennedy, 2000; Lukacs, 2002), the beginning of the "second urbanization" of South Asia, when cities were built on the Gangetic plain. Many of these Vindhyan and Gangetic sites also show some evidence of interaction between hunter-gatherers and others, primarily in the form of ceramics.

Here, however, the rich biological record of the region has proven critical for determining the degree of interaction between farmers and others. Based both on some new radiocarbon dates from Damdama and Lekhahia which place them in the seventh millennium BC (and thus prior to the advent of agriculture in this region), as well as on skeletal indicators of both diet and post-cranial structure, Lukacs (2002:52–3) argues that the Vindhya Hills/Gangetic Plains "Mesolithic" sites do not support an interactive trade model, noting:

The biological evidence reveals similar dental and skeletal pathology profiles for "Mesolithic" sites (Damdama, Lekhahia, Mahadaha, Sarai Nahar Rai) that are fully consistent with a diet that is coarse and unrefined in texture and diverse enough to adequately satisfy nutritional requirements (Lukacs and Pal, 1993; Lukacs and Misra, 1996). In short, the pattern of variability in pathological lesions for these four Mesolithic sites is typical of people who practice a hunting-gathering mode of subsistence. The skeletal and dental pathology profile of these people is inconsistent with consumption of refined or agriculturally produced foods and with a sedentary lifestyle (Lukacs and Pal, in prep [2003]). Furthermore, biological variations in dental and skeletal structure, such as tooth size (large) and craniofacial structure (robust), of these groups is similar and consistent with expectations for groups adapted to an eclectic diet obtained through hunting and foraging (Lukacs and Pal, 1993; Lukacs and Misra, 1996).

Obviously, there is much work to be done in this region, particularly in clarifying chronological issues. However, even if Lukacs' new dates move Damdama and Lekhahia into the Early Holocene, what we might call the Mesolithic proper, it seems probable that other sites for which later dates have been reported (Possehl and Rissman, 1992; Kennedy, 2000) will indeed turn out to be contemporaneous with places occupied by agriculturalists. The fact that, as Lukacs has demonstrated, these later Holocene foragers did *not* appear to have very close interactions with others, is certainly important. Competition for land was almost certainly not a major issue at this time, when population levels were still very low. Although we still have little evidence for this, landscape transformation was probably not yet extensive. Here, too, there is nothing approaching a forager-trader

mode; existing information about resource procurement, mobility, and exchange (Sharma et al., 1980; Varma, 1981/83) involves lithic raw materials from the Vindhyan Hills as well as seasonal distributions of water, plants, and game. Demand for the kind of highly-valued resources that fuel forager-trader interactions may, perhaps, only be generated by highly stratified urban societies that can afford to pursue them.

The Andaman Islanders in the Indian Ocean World

Both the Andaman and Nicobar Islands consist of long, north-south oriented archipelagos lying along ancient sea routes between the Indian mainland and island Southeast Asia. Although both are part of the Indian Republic, they actually lie closer to Southeast Asia than to the Indian peninsula. Textual sources from as early as the second century AD describe Andaman Islanders as hostile, even as cannibals, a characterization that has continued to appear in the literature even into the present. This image of the fierce and reclusive Andamanese (see discussion in Mukerjee, 2003) is quite different from those of the neighboring Nicobarese, major suppliers of ambergris to the sixteenth century port city of Melaka, held at that time by the Portuguese. Indeed, the commercial lingua franca of the Nicobarese in the sixteenth century was Portuguese. The Nicobarese were, then, clearly deeply involved in the Indian Ocean trade, a vast circuit of goods and people that streched from East Asia to the Mediterranean and which was well-established by the first few centuries AD. Archaeological data, too, trace these connections, with finds of Chinese ceramics reported for the Nicobars, but not the Andamans where an (undated) excavated shell midden containing several seventeenth-century Sumatran gold coins is the primary material evidence for long-distance exchange (Cooper, 2002:2).

The Andaman islands, famous in anthropological circles for E.R. Radcliffe-Brown's early ethnography (1922) and in colonial history for a notorious British penal colony, have often been depicted as the home of a very "ancient" and "primitive" people. For example, the private Andaman Association's web site (*http://andaman.org*) calls them "a very ancient pygmy people living (until very recently and in some numbers today) a paleolithic hunting and gathering life, just as the ancestors of us all did until around the end of the ice age", an image consistent with the last eight hundred years of press on this region. Certainly hunting and gathering have been major components of island life ways from the date of first colonization to the present, but it also seems clear that the model of the "primitive isolate" that has been drawn for the Andamanese is somewhat exaggerated (Cooper, 2002, Morrison, 2005).

Local resources, clearly, did draw outside attention. Malay, Burmese and Chinese ships regularly visited caves and rockshelters containing nests of the white-nest swiftlet (*Collocalia fuciphaga inexpectata*), a delicacy in China and parts of Southeast Asia. Similarly, sea slugs caught in coastal waters made their way to eastern consumers. Thus, the Andamans, like the Ghats of southern India, held resources desired by wealthy foreigners which were difficult of access and which exerted little to no appeal to potential local consumers. Both remote and yet, by water, accessible for bulk transport, these islands were well-suited for the development of forager-trader strategies. What was missing, however, was the ability of outsiders to coerce local foragers into greater participation and it seems that, until the nineteenth century, Andaman islanders were able to set their own terms for outside interaction.

To what extent might islanders have been involved in this trade? Until recently, little archaeological work had been conducted in the Andamans, but with the publication of Cooper's sustained program of survey and excavation, we now have some clues to the occupational history and interactions of the Andaman islanders. Although iron objects are consistently found in both lower and upper levels of excavated midden sites, Cooper argues (2002:22–3) that this highly coveted material could have come entirely from shipwrecks. Cooper's recent excavations provide the only absolute dates for island history; her *earliest* radiocarbon assessment dates to between 162 and 290 AD (see Morrison, 2005 for information on calibration procedures). Interestingly, this is precisely the time of a great expansion of trade in the Indian Ocean (Ray, 1994). Clearly, more research is needed and there may well be sites with older deposits waiting to be studied, but the present lack of any very long-term occupational record is quite interesting. Further, the presence of both pig (*Sus scrofa*) bones and a small number of ceramics from the earliest levels of this site suggest that the Andaman islanders brought some of the accouterments of settled agricultural life with them at the time of initial colonization (Cooper, 2002:7, 83–93). Even if there are earlier sites waiting to be studied and it turns out that human occupation of the Andamans has lasted more than a few thousand years, the presence of pigs and pottery does indicate at least some contact with the outside world. At the same time, the material record does not suggest a high intensity of interaction with outsiders, consistent with the historical notices.

The archaeological data are, at least apparently, in direct opposition to genetic studies which suggest a great deal of genetic isolation which, in some views (Thangraj et al., 2002), would take more than the archaeologically-documented 2,000-year long occupation of the islands to produce. Of course, there is nothing to prevent this small population from having a long biological history in multiple locations, with their sojourn on the islands taking up only about the last 2,000 years.

After incorporation into the British Colonial Empire in 1858, and especially after the founding of the British penal colony, islanders were forced to interact closely with others. The fact that the Andamanese were very hard-hit by introduced diseases around this time seems to support the finding of their earlier isolation or semi-isolation. Still, in spite of the significant problems of the colonial era, the Andamanese showed a remarkable resiliency, maintaining a commitment to gathering and hunting as well as to indigenous technologies, even transferring chipped stone technology to molded bottle glass (Cooper, 2002). In this, they can be seen as akin to other foraging peoples in the region, for whom the values associated with their way of life were worth maintaining even in the face of great outside pressure.

Southern India: the Western Ghats

The Western Ghat mountains of southwest India rise steeply from the narrow coastal plain but fall more gently on the east down to the eastward-sloping plateaus of the southern peninsula. The orographic barrier presented by these mountains captures a significant amount of the summer southwest monsoon rains, creating a rainshadow over the peninsular interior. The Ghats support a graded series of vegetation associations, from humid tropical rainforests on the west to drier deciduous forest forms on the east. In this way the Ghats can be seen as forming a narrow island of elevation-graded resources unique to the region. These mountains and their foothills are home to a number of plants, in particular, that became highly prized throughout the world, a list that includes gums and resins, dyes, woods, and especially spices. Of the latter, the most important was certainly pepper (*Piper nigrum*), indigenous to the Malabar region. Pepper and other forest products grew wild

in Ghat forests though pepper and some other spices were also widely cultivated by the sixteenth century (Morrison, 2002a). Pepper circulated broadly across the Indian Ocean world as early as the first century AD, a circulation attested by both archaeological finds of pepper in Red Sea ports and by textual notices of this "black gold" in the Roman world (e.g., Tomber, 2000). Much later, Portuguese colonial efforts in India, reputedly fueled by the quest for "Christians and spices" (Subrahmanyam, 1993), also focused on the acquisition of forest products, most of which they obtained via intermediaries to local "tribal" groups (Morrison, 2002a).

The Western Ghats were (and are) home to a number of "tribal" groups whose occupations now include agriculture, wage labor, trade, pastoralism, hunting, and gathering (enumerated in Bird-David, 1999). Certain groups are historically and ethnographically known as foragers though it seems clear that subsistence strategies have been quite variable and flexible over the last several thousand years. In this region we can see the development of full-fledged forager-trader strategies, complex accommodations to the historically-specific ecological, demographic, and political realities of the region (Morrison, 2002a, 2002c). Although there are number of groups in the Ghats known as hunter-gatherers (Bird-David, 1999, and see papers in Lee and Daly, 1999), all of them have long histories of interaction with others, some involving such consistent exchanges of forest products for cultivated grains, cloth, metal objects, and other items that Fox (1969) dubbed them "professional primitives." I have discussed some aspects of the histories of these groups elsewhere (Morrison, 2001, 2002a, 2005), but it is important to note that although historical, archaeological, and paleoenvironmental evidence all points to the existence of a trade in forest products from the Early Historic period, it is also clear that the level of interference and the degree of powerlessness of forest peoples increased significantly after the fifteenth and sixteenth century, when the international spice trade took off and when European colonial domination of South Asia began.

The Ghats are passable primarily through well-defined passes or ghats ("steps"), though some of the many small rivers originating in the mountains and running down to the west coast are also navigable by small craft. The collection of many of the forest products involved in exchange required significant local knowledge (particularly given the high taxonomic diversity and low concentrations of individuals of any given taxon in this tropical setting). Some also involved hazardous conditions for collection and/or tedious preliminary processing. It was also the case that the land base of local foragers was increasingly squeezed by expanding agriculture and deforestation, especially into the Early Modern period, and many groups may have been pushed into higher elevations by competing groups. This area, then, possessed all the conditions for the development of forager-trader forms; unequal power relations, objects of desire that attracted both foragers and outsiders, a distant but not too-distant geography, and finally a complex occupational history of settlement and land use that both put pressure on hunting and gathering but also provided new opportunities. As I have suggested elsewhere (Morrison, 2002a, 2005), it was the development of forager-trader forms in this region that permitted particular groups to maintain their distance from lowland society and to retain values and beliefs, including the valorization of mobility and hunting, as well as a host of social and cultural practices that set them apart from the larger society and from the state (cf. Bird-David, 1990, 1992a, 1992b). In this way, then, forager-trader accommodations worked to underwrite the continued existence of hunting and gathering in this region, even to the present day.

Conclusion

The role of South Asia in hunter-gatherer studies has always been rather ambiguous (Morrison, 2002b), in part because of the low level of information about South Asian groups disseminated outside the region. However, it is also clear that South (and Southeast) Asian foragers have also been regarded as somewhat "compromised" by their interactions with agriculturalists, making them something like the alter-egos of such paradigmatic foragers as the San-speakers of southern Africa, once seen as untouched, timeless foragers, a vision rather rudely dissected in the acrimonious "revisionist debate" of the late twentieth century. Given the demise of that model of living stone age people, it is clear that we have to contend with the complex Holocene histories of all hunting and gathering groups, not only those of South Asia. It is worth revisiting the words of B.J. Williams, ethnographer of the Birhor of Bihar (northern India), made at the famous *Man the Hunter* symposium (1968:128, emphasis added):

In some important ways the Birhor do not meet the conditions assumed in the model of hunting-gathering society. They are neither politically autonomous nor are they economically autonomous. They live in an area that has been inhabited by tribal agriculturalists for a very long period of time. During the past 100-plus years the area has seen a large influx and growth of Hindu and Muslim agriculturalists that now far outnumber the tribal population. The Birhor traders hunted and collected items to the villagers in exchange for rice...The Birhors also spend some time making rope from the inner-bark fiber of certain vines. These they also trade for rice...Not only do the Birhor live a form of economic parabiosis with agriculturalists, but also they are in some ways a politically subjugated minority...These conditions which are the result of intensive interaction with dominant groups makes the Birhor less than ideal as a basis for inferences about possible forms of social organization in hunting groups living only among hunters. *On the other hand, they have the great advantage of being hunters now.*

South Asia has, indeed, a large number of people who have the advantage of being foragers now. The advantage, in the case, is obviously to scholars who wish to understand the possibilities presented by South Asian environments and conditions for life in the distant past.

South Asian foragers, across the long history of the Holocene, and over the diverse ecological and cultural landscapes of the subcontinent, have shown a remarkable array of economic and social strategies not only for subsistence but also for the continuation of ways of ways of life valued within specific groups but often denigrated outside them. Throughout the Holocene, it seems, the practice of hunting and gathering for subsistence continued while trade-based foraging was also important (and became more so over time). Forager-trader strategies may have, in fact, underwritten the continued existence of subsistence foraging in some times and places. Maintaining mobility patterns, religious practices, and social forms not shared by the majority population was clearly a long-term priority for some groups who were willing (or forced) to be flexible in subsistence practice, integrating agriculture, wage labor, military service, and trade into foraging life styles. Not all of this change was a process away from hunting and gathering; on the contrary, it seems that some groups historically known as foragers actually moved into hunting and gathering from agriculture. Some at least partially-foraging groups in South Asia established independent kingdoms, and most were integrated within lowland states, even if peripherally (Murthy, 1994; S. Guha, 1999; Skaria, 1999). It is even possible to argue (Boomgaard, 1989; Morrison, 2005) that hunting and gathering partially underwrote formation of coastal polities in South and Southeast Asia, including colonial enclaves, suggesting a need to integrate hunter-gatherer studies into larger histories of the region.

We must, then, squarely face the complex, long-term histories of foraging in Asia, including the Holocene record of integration of foraging strategies with agriculture, wage labor, trade and tribute relations, and pastoralism. These histories, far from representing a problem for archaeologists, should be seen as a rich record of possibilities for action and a source of new understandings of the past.

Notes

1. Here, of course, we must then use labels as indicators of *time periods* rather than as assessments of cultural "progress," or shorthand for ceramic distributions. The Iron Age, for example, thus needs to designate a specific chronological range and not, as I have argued elsewhere (Morrison in press, Morrison and Lycett 1998) be conflated with the occurrence of megaliths (i.e. using Megalithic as synonym for Iron Age), a problem exactly parallel to the microlith problem, given the long time-span of megalith construction and use.

2. Unfortunately, small chipped stone tools are sometimes referred to as microliths whether or not they employ blade technology and it is not always possible to say how lithic analyses were structured from these brief reports. It is worth exercising caution regarding reports of microliths.

3. This discussion largely follows Morrison (2005).

4. However, I do not wish to discount the considerable ideological apparatus surrounding choice around staple foods. It is clear that rice, in particular, packed a powerful symbolic punch so that the transition to rice consumption may be seen as akin to Srinivas' (1989) famous concept of "sanskritization" a mode of elite emulation (Morrison 2001, in press). Knapen (2001) notes a similar process at work in colonial-era Borneo.

5. This well-known controversy was initially fueled by assertions (Headland 1987; Headland and Reid 1989; Bailey et al. 1989) that the resources and resource structures of humid tropical forest environments prevented "pure" hunting and gathering forms of adaptation, an hypothesis meant in part to make sense of the scanty archaeological record of those regions. Since then, many authors have noted exceptions to this pattern (e.g. Brosius 1991; papers in Headland and Bailey eds. 1991; see Junker 2002 for a detailed review) both in terms of resource distributions and archaeological remains. It should be noted that the original argument was developed for humid tropical forests and not

for tropical forests with significant dry periods. The Ghat forests include both environmental types.

6.   African millets were also introduced to South Asia during this period.

References

Agrawal, D.P., 1971. The Copper Bronze Age in India. Munshiram Manoharlal, Delhi.

Bailey, R.C., Head, G., Jenike, M., Owen, B., Rechtman, R., Zechenter E. 1989. Hunting and gathering the tropical rainforest: is it possible? American Anthropologist 91, 59–82.

Bayly, S., 1999. Caste, Society and Politics in India from the Eighteenth Century to the Modern Age. Cambridge University Press, Cambridge.

Bellwood, P., 1997. The Prehistory of the Indo-Malaysian Archipelago. University of Hawai'i Press, Honolulu.

Béteille, A., 1998. The idea of indigenous people. Current Anthropology 39(2), 187–91.

Bird-David, N., 1992a. Beyond the 'hunting and gathering mode of subsistence': culture-sensitive observations on the Nayaka and other modern hunter-gatherers. Man (n.s.) 27, 19–44.

Bird-David, N., 1992b. Beyond the original affluent society: a culturalist reformulation. Current Anthropology 33(1), 25–47.

Bird-David, N., 1990. The giving environment: another perspective on the economic system of gatherer-hunters. Current Anthropology 31, 189–196.

Bird-David, N., 1999. Introduction: South Asia. In: Lee, R.B., Daly, R. (Eds.), The Cambridge Encyclopedia of Hunters and Gatherers. Cambridge University Press, Cambridge, pp. 231–237.

Boivin, N., 2005. Orientalism, ideology and identity: examining caste in South Asian archaeology. Journal of Social Archaeology 5, 225–252.

Boivin, N., 2007. Anthropological, historical, archaeological and genetic perspectives on the origin of caste in South Asia. In: Petraglia, M.D., Allchin, B. (Eds.), The Evolution and History of Human Populations in South Asia: Inter-disciplinary Studies in Archaeology, Biological Anthropology, Linguistics and Genetics. Springer, Netherlands, pp. 341–361.

Boomgaard, P., 1989. The VOC trade in forest products. In: Grove, R.H. Damodaran, V., Sangwan, S. (Eds.), Nature and the Orient: The Environmental History of South and Southeast Asia. Oxford University Press, Delhi.

Boomgaard, P., Columbijn, F., Henley, D. (Eds.), 1997. Paper Landscapes: Explorations in the Environmental History of Indonesia. KITLV Press, Leiden.

Brosius, P. 1991. Foraging in tropical rain forests: the case of the Penan of Sarawak, East Malaysia (Borneo). Human Ecology 19, 123–150.

Chakrabarti, D.K., 2001. India: An Archaeological History. Oxford University Press, Delhi.

Costantini, L., 1984. The beginnings of agriculture in the Kachi plain: the evidence of Mehrgarh. In: Allchin, B. (Ed.), South Asian Archaeology 1981. Cambridge University Press, Cambridge, pp. 29–33.

Cooper, Z., 2002. Archaeology and History: Early Settlements in the Andaman Islands. Oxford University Press, Delhi.

Cribb, R., 1997. Birds of paradise and environmental politics in colonial Indonesia, 1890–1931. In: Boomgaard, P., Columbijn, F., Henley, D. (Eds.), Paper Landscapes: Explorations in the Environmental History of Indonesia. KITLV Press, Leiden, pp. 379–408.

Denbow, J.R., 1984. Prehistoric herders and foragers of the Kalahari: the evidence of 1500 years of interaction. In: Schrire, C. (Ed.), Past and Present in Hunter Gatherer Studies. Academic Press, Orlando, pp. 175–193.

Deraniyagala, S.U., 1998. Pre- and Protohistoric settlement in Sri Lanka. In: XIII U.I.S.P. Congress Proceedings, Forli, 8–14 September, 1996, Volume 5: The Prehistory of Asia and Oceania. A.B.A.C.O., Forli.

Dhavalikar, M.K., Raval, M.R., Chitalwala, Y.M., 1995. Kuntasi: a Harappan Emporium on West Coast. Deccan College Postgraduate and Research Institute, Pune.

Dirks, N.B., 2001. Castes of Mind: Colonialism and the Making of Modern India. Princeton University Press, Princeton.

Fox, R.G., 1969. 'Professional primitives': hunters and gatherers of nuclear South Asia. Man in India 49(2), 139–160.

Fuller, D.Q., 2006. The Emergence of Agriculture in South India. University College London Press, London.

Fuller, D.Q., 2007. Non-human genetics, agricultural origins and historical linguistics in South Asia. In: Petraglia, M.D., Allchin, B. (Eds.), The Evolution and History of Human Populations in South Asia: Inter-disciplinary Studies in Archaeology, Biological Anthropology, Linguistics and Genetics. Springer, Netherlands, pp. 393–443.

Guha, R., 1999. Savaging the Civilized: Verrier Elwin, his Tribals, and India. University of Chicago Press, Chicago.

Guha, S., 1999. Environment and Ethnicity in India: 1200–1991. Cambridge University Press, Cambridge.

Headland, T.N. 1987. The wild yam question: how well could independent hunter-gatherers live in a tropical rainforest environment? Human Ecology 15, 463–491.

Headland, T.N., Bailey, R.C. 1991. Introduction: have hunter-gatherers ever lived in tropical rain forest independently of agriculture? Human Ecology 19, 115–122.

Headland, T.N., Reid, L. 1989. Hunter-gatherers and their neighbors from prehistory to the present. Current Anthropology 30, 43–66.

Jarrige, J.F., 1984. Continuity and change in the north Kachi Plain (Baluchistan, Pakistan) at the beginning of the second millennium BC. In: Schotsmans, J., Taddei, M. (Eds.), South Asian Archaeology 1983. Instituto Universitario Orientale, Dipartmento di Studi Asiatici, Naples, pp. 35–68.

Junker, L. L. 2002. Southeast Asia: introduction. In: Morrison, K.D., Junker, L.L. (Eds.), Forager-Traders in South and Southeast Asia: Long-Term Histories. Cambridge University Press, Cambridge, pp. 131–166.

Karlsson, B.G., 2001. Indigenous politics: community formation and indigenous peoples' struggle for self-determination in northeast India. Identities: Global Studies in Culture and Power 8(1), 7–45.

Kashyap, A., Shinde, V., 2005. Significance of copper residue on microliths from the Mesolithic site of Bagor, Rajasthan. Paper presented at the Biannual Conference of the European Association of South Asian Archaeologists, London.

Kennedy, K.A.R., 2000. God-Apes and Fossil Men: Paleoanthropology in South Asia. University of Michigan Press, Ann Arbor.

Kennedy, K.A.R., Chimet, J., Disotell, T., Meyers, D., 1984. Principal-components analysis of prehistoric South Asian crania. American Journal of Physical Anthropology 64(2), 105–118.

Kenoyer, J.M., 1997. Trade and technology of the Indus Valley: new insights from Harappa, Pakistan. World Archaeology 29(2), 262–280.

Knapen, H. 2001. Forests of Fortune? The Environmental History of Southeast Borneo, 1600–1880. KITLV Press, Leiden.

Korisettar, R., Venkatasubbaiah, P.C., Fuller, D.Q., 2002. Brahmagiri and beyond: the archaeology of the southern Neolithic. In: Settar, S, Korisettar, R. (Eds.), Indian Archaeology in Retrospect, Volume I, Prehistory: Archaeology of South Asia. Manohar and Indian Council of Historical Research, Delhi, pp. 151–238.

Lee, R.B., Daly, R. (Eds.), 1999. Cambridge Encyclopedia of Hunters and Gatherers. Cambridge University Press, Cambridge.

Lee, R.B., Gunther, M., 1991. Oxen or onions? The search for trade (and truth) in the Kalahari. Current Anthropology 32, 592–601.

Lee, R.B., Gunther, M., 1993. Problems in Kalahari historical ethnography and the tolerance of error. History in Africa 20, 185–235.

Lee, R.B., Gunther, M., 1995. Errors corrected or compounded? A reply to Wilmsen. Current Anthropology 36, 298–305.

Lukacs J.R., 2002. Hunting and gathering strategies in prehistoric India: a biocultural perspective on trade and subsistence. In: Morrison, K.D., Junker, L.L. (Eds.), Forager-Traders in South and Southeast Asia: Long-Term Histories. Cambridge University Press, Cambridge, pp. 41–61.

Lukacs, J.R., Misra, V.D., 1996. The people of Lekhahia: a bio-cultural portrait of Mesolithic hunter-foragers of north India. In: Allchin, B., Allchin, F.R. (Eds.), South Asian Archaeology 1995. Oxford and IBH, New Delhi.

Lukacs, J.R., Pal, J.N., 1993. Mesolithic subsistence in north India: inferences from dental attributes. Current Anthropology 34(5), 745–765.

Lukacs, J.R., Pal, J.N., 2003. Skeletal variation among Mesolithic people of the Ganga plains: new evidence of habitual activity and adaptation to climate. Asian Perspectives 42(2), 329–351.

Lycett, M.T., Morrison, K.D., 1989. Persistent lithics: post Iron Age lithic technology in South India. Paper presented at the Annual Conference on South Asia, Madison, Wisconsin, Nov. 3–5.

Malik, S.C., 1959. Stone Age Industries of the Bombay and Satara Districts. Maharaja Sayajirao University of Baroda, Baroda.

Miracle, P. 2005. The stability of hunting and gathering in the prehistory of south India. Paper presented at the Biannual Conference of the European Association of South Asian Archaeologists, London.

Misra, V.N., 1976. Ecological adaptations during the terminal Stone Age in Western and Central India. In: Kennedy, K.A.R., Possehl, G.L. (Eds.), Ecological Backgrounds of South Asian Prehistory. South Asia Occasional Papers, Cornell University, Ithaca, pp. 28–51.

Misra, V.N., 2002. The Mesolithic Age in India. In: Settar, S., Korisettar, R. (Eds.), Indian Archaeology in Retrospect, Volume I, Prehistory: Archaeology of South Asia. Manohar and Indian Council of Historical Research, Delhi, pp. 111–126.

Morrison, K.D., 1999. South Asia: prehistory. In: Lee, R.B., Daly, R. (Eds.), Cambridge Encyclopedia of Hunters and Gatherers. Cambridge University Press, Cambridge, pp. 238–242.

Morrison, K.D., 2001. Coercion, resistance, and hierarchy: local processes and imperial strategies in the Vijayanagara Empire. In: Alcock, S., D'Altroy, T., Morrison, K.,, Sinopoli, C. (Eds.), Empires: Perspectives from Archaeology and History. Cambridge University Press, Cambridge, pp. 253–78.

Morrison, K.D., 2002a. Introduction. In: Morrison, K.D., Junker, L.L. (Eds.), Forager-Traders in South and Southeast Asia: Long-Term Histories. Cambridge University Press, Cambridge, pp. 21–40.

Morrison, K.D., 2002b. Historicizing adaptation, adapting to history: forager-traders in South and Southeast Asia. In: Morrison, K.D., Junker, L.L. (Eds.), Forager-Traders in South and Southeast Asia: Long-Term Histories. Cambridge University Press, Cambridge, pp. 1–17.

Morrison, K.D., 2002c. Pepper in the hills: upland-lowland exchange and the intensification of the spice trade. In: Morrison, K.D., Junker, L.L. (Eds.), Forager-Traders in South and Southeast Asia: Long-Term Histories. Cambridge University Press, Cambridge, pp. 105–128.

Morrison, K.D., 2005. Historicizing foraging in Asia: power, history, and ecology of Holocene hunting and gathering. In: Stark, M. (Ed.), An Archaeology of Asia. Basil Blackwell, New York, pp. 279–302.

Morrison, K.D., in prep. Oceans of Dharma: Landscapes, Power, and Place in Southern India. University of Washington Press, Seattle.

Morrison, K.D., Junker, L.L. (Eds.), 2002. Forager-Traders in South and Southeast Asia: Long-Term Histories. Cambridge University Press, Cambridge.

Mukerjee, M., 2003. The Land of the Naked People: Encounters with Stone Age Islanders. Houghton Mifflin, Boston.

Murthy, M.L.K., 1994. Forest peoples and historical traditions in the Eastern Ghats, south India. In: Allchin, B. (Ed.), Living Traditions: Studies in the Ethnoarchaeology of South Asia. Oxford and IBH, New Delhi, pp. 205–218.

Possehl, G.L., 2002. The Indus Civilization: A Contemporary Perspective. AltaMira Press, Walnut Creek.

Possehl. G.L., 1976. Lothal: a gateway settlement of the Harappan civilization. In: Kennedy, K.A.R., Possehl, G. (Eds.), Ecological Backgrounds of South Asian Prehistory. Occasional Papers and Theses No. 4, Cornell South Asia Program, Ithaca, pp. 118–131.

Possehl, G.L., Rissman, P.C., 1992. The chronology of prehistoric India: from earliest times to the Iron Age. In: Ehrich, R.W. (Ed.), Chronologies in Old World Archaeology. University of Chicago Press, Chicago, Vol. I. pp. 465–490, Vol. II, pp. 447–474.

Radcliffe-Brown, A.R., 1922. The Andaman Islanders. Cambridge University Press, Cambridge.

Raju, D.R., Venkatasubbaiah, P.C., 2002. The archaeology of the Upper Palaeolithic phase in India. In: Settar, S., Korisettar, R. (Eds.), Indian Archaeology in Retrospect, Volume I, Prehistory: Archaeology of South Asia. Manohar and Indian Council of Historical Research, Delhi, pp. 85–110.

Ratnagar, S., 1997. Hunter-gatherer and early agriculturalist: archaeological evidence for contact. In: Nathan, D. (Ed.), From Tribe to Caste. Indian Institute of Advanced Study, Shimla, pp. 117–135.

Ray, H.P., 1994. The Winds of Change: Buddhism and the Maritime Links of Early South Asia. Oxford University Press, Delhi.

Sankalia, H.D., 1965. Archaeological Excavations at Langhnaj: 1944–63, Pt. 1. Deccan College Postgraduate and Research Institute, Pune.

Sharma, G.R., Misra, V.D., Mandal, D., Misra, B.B., Pal, J.N., 1980. Beginnings of Agriculture: Studies in History, Culture and Archaeology. Abinash Prakashan, Allahabad.

Schrire, C., 1980. An inquiry into the evolutionary status and apparent identity of San hunter-gatherers. Human Ecology 8(1), 9–32.

Schrire, C., 1984. Wild surmises on savage thoughts. In: Schire, C. (Ed.), Past and Present in Hunter-Gatherer Studies. Academic Press, Orlando, pp. 1–26.

Singh, K.S., 1997. Tribe into caste: a colonial paradigm(?). In: Nathan, D. (Ed.), From Tribe to Caste. Indian Institute of Advanced Study, Shimla, pp. 31–44.

Skaria, A. 1999. Hybrid Histories: Forests, Frontiers, and Wilderness in Western India. Oxford University Press, Delhi.

Stiles, D., 1993. Hunter-gatherer trade in wild forest products in the early centuries AD with the Port of Broach, India. Asian Perspectives 32(2), 153–167.

Subrahmanyam, S., 1993. The Portuguese Empire in Asia, 1500–1700: A Political and Economic History. Longmans, London.

Thangraj, K., Singh, L, Reddy, A.G., Raghavendra Rao, V., Seghal, S.C., Underhill, P.A., Pierson, M., Frame, I.G., Hagelberg, E., 2002. Genetic affinities of the Andaman Islanders: a vanishing human population. Current Biology.

Tomber, R., 2000. Indo-Roman trade: the ceramic evidence from Egypt. Antiquity 74(285), 624–631.

Varma, R.K., 1981–83. The Mesolithic cultures of India. Puratattva 13/14, 27–36.

von Fürer-Haimendorf, C., 1982. Tribes of India: The Struggle for Survival. University of California Press, Berkeley.

Williams, B.J., 1968. The Birhor of India and some considerations of band organization. In: Lee, R., DeVore, I. (Eds.), Man the Hunter. Aldine, Chicago, pp. 126–132.

Wilmsen, E.N., 1983. The ecology of illusion: anthropological foraging in the Kalahari. Reviews in Anthropology 10(1), 9–20.

Wilmsen, E.N., 1989. Land Filled With Flies. University of Chicago Press, Chicago.

Wilmsen, E.N., 1993. On the search for (truth) and authority: a reply to Lee and Gunther. Current Anthropology 34, 715–721.

Wilmsen, E.N., Denbow, J.R., 1990. Paradigmatic history of San-speaking peoples and current attempts at revision. Current Anthropology 31, 489–523.

Wright, R., in press. The Indus Civilization. Cambridge University Press, Cambridge.

15. Anthropological, historical, archaeological and genetic perspectives on the origins of caste in South Asia

NICOLE BOIVIN

Leverhulme Centre for Human Evolutionary Studies
The Henry Wellcome Building
Fitzwilliam Street
University of Cambridge
Cambridge, CB2 1QH
England
nlb20@cam.ac.uk

Introduction

A volume that addresses the population diversity, both past and present, of South Asia must obviously include a discussion of caste. The institution of caste undoubtedly plays, and may well long have played, a crucial role in the biological and social ordering of South Asian society. In addition, the analysis of caste has been central to many studies of South Asia, and in particular those undertaken from social anthropological, biological anthropological and historical perspectives. Indeed, it is entirely possible that more has been written about caste than any other single aspect of South Asian society.

In spite of this fact, however, or perhaps because of it, the topic of caste remains mired in controversy. This is perhaps nowhere more evident than in the controversy that surrounds questions concerning the origins of the caste system in South Asia. Given the centrality of caste in many reconstructions of South Asian population history, however, it is nonetheless crucial to grapple with the problem of caste origins, particularly in a volume such as this. Inspired by the interdisciplinary character of the present volume and the conference from which it stemmed, this chapter will therefore attempt to examine the complex question of caste origins from a multiplicity of disciplinary perspectives. In particular, it will examine ideas about caste origins that have emerged within the context of the disciplines of anthropology, history, archaeology and genetics.

My aim here is not to resolve the question of caste origins, a task that, at least on current evidence, is a long way from being possible, and is, in addition, very unlikely to lead to simple answers. Rather, I wish to take a comparative approach, and examine how the question of caste origins has been addressed in each of these disciplines, as

M.D. Petraglia and B. Allchin (eds.), The Evolution and History of Human Populations in South Asia, 341–361.
© 2007 *Springer.*

well as the sorts of insights have been acquired through such analyses. This is argued to be a crucial first step in pursuing a more critical and holistic approach to the study of caste as a historically-constituted (and variable) institution in South Asia. Inter-disciplinary communication regards the origins of caste in South Asia has been uneven and in many cases superficial, even between related disciplines like history and archaeology. Between more distantly related disciplines on separate sides of the social and natural sciences divide (Morell, 1993), like anthropology and genetics, isolationalist tendencies have been even more acute, and have often led to approaches and findings that are not only incompatible, but actively preclude inter-disciplinary discussion.

My view is that the reconstruction of the population histories of South Asia, and of the historical trajectories of social and biological features like caste, are highly complex issues that are unlikely to be resolved through the efforts of any one discipline. Instead, it will be necessary for disciplines to join hands and work collaboratively, a task that will undoubtedly call for all researchers studying caste to achieve a better awareness of the histories, approaches and findings of a variety of disciplines. The comparative, inter-disciplinary study undertaken in this chapter is an attempt to facilitate this task, and has been undertaken by an archaeologist with interests in social anthropology, history and genetics. The general nature of the comparison that can be undertaken in such a limited space and my own limitations in terms of disciplinary background and knowledge preclude a highly detailed investigation of the study of caste origins in each of the fields I will address. Nonetheless, the absence to date of any broadly comparative approach to caste, as far as I am aware, makes the enterprise a useful and indeed a necessary one, if only because it creates the starting point for further inter-disciplinary discussion and debate.

In the account that follows, I will look in turn at each of the disciplines of anthropology, history, archaeology and genetics, and assess both how they have looked at caste, and, most significantly, what they have concluded concerning its origins. Key debates surrounding caste within the individual disciplines will be highlighted. Given the fact that the study of caste origins through the use of genetics is still very much in its infancy, and debates still underdeveloped, the final section on genetics will focus on outlining some of the issues that appear necessary for this fledgling approach to address, based on the findings of the other disciplines discussed in the chapter. After examining the ideas about caste origins that have been generated within individual disciplines, I will conclude by outlining the lessons the can be drawn from this comparative exercise, and by sketching the outline of a developmental trajectory for caste that requires testing through detailed inter-disciplinary studies. Before beginning any of this, however, I will start by providing a definition of caste as the term applies to the South Asian context.

Defining Caste

In light of the broad comparative and inter-disciplinary approach of both the chapter and the volume, it is useful to start with a relatively standard and widely agreed upon, if rather traditional and uncritical, definition of caste. The definition offered here is therefore that castes are the hereditary, endogamous, hierarchically-ranked and sometimes occupationally specific groups to which all Hindus belong. What this means is that individuals inherit their caste at birth, and marry persons of the same caste. Castes themselves are ranked into hierarchies, so that some castes have higher status than other castes. This status is generally related to a traditional caste occupation and its perceived ritual purity, even if this occupation is not, as is generally the case in contemporary, particularly urban South Asian society, practiced by the members of a caste. As already indicated, this is a traditional definition and it is not without its addendums and caveats, some of which will become clear during the course of the chapter. Nor is it one that all scholars would agree with. Nevertheless, as it continues to be a widely cited definition, and has cross-disciplinary utility, it is a useful starting point, particularly for those with little formal background on the topic.

In India today, we find thousands of castes, though any one specific region will only contain a minute proportion of this overall number. In addition, while some castes do have a wide distribution, many are distinctive to particular regions. A village in north India, for example, may contain individuals of a dozen or half-dozen castes, and these castes will represent a completely different assortment from those found in a south Indian village. It should be pointed out that while caste is most closely associated with Hinduism, it is also found amongst non-Hindu communities not only in India, but in other parts of South Asia.

The Hindi term that is generally translated as 'caste' is 'jati', which actually means something akin to genus (Daniel, 1984; Mayer, 1997) and is applied not just to human beings but also to animals, plants and even inorganic materials (Daniel, 1984). Caste or jati as applied to human beings refers, as indicated, to a very specific, inter-marrying group. The term is to be differentiated from that of 'varna', which refers to a much wider, and highly abstract ordering of the whole of Indian society into four groups. The four varna are also hierarchically ranked, and consist of the Brahman (priest), Ksatriya (warrior and king), Vaisya (variably merchant or peasant) and Sudra (variably farmer or serf) varnas. The untouchables, those who work with impure substances such as hides, meat and human and animal refuse, are not included in this four-fold division.

Anthropology and the 'Invention' of Caste

Of all the disciplines I will address in this chapter, anthropology is certainly the one that demonstrates the longest-standing and most intensive interest in caste. Indeed, until fairly recently, anthropology focused a large proportion of its resources on the study of caste. This interest in caste dates back to the colonial era, and in particular the period that followed the 1857 "Mutiny", an event that accentuated the need for better understanding by the British of their Indian subjects (Dirks, 2001). Early anthropologists and administrators spent a great deal of time compiling catalogues and descriptions of individual castes in the Indian subcontinent, and this cataloguing of castes took off in earnest in 1872, when the first *Census of India* was produced. Each of the census reports, which were conducted on a decennial basis, would consist of piles of volumes containing statistical and descriptive records of castes for all parts of India (Dirks, 2001). This tradition reached its apogee with the publication in 1908 of H.H. Risley's classic text, *The People of India*, which served as the model for a series of volumes organized around the "encyclopaedic delineation of the customs, manners, and measurements of the castes and tribes of the different regions of India" (Dirks, 2001:50).

Subsequently, however, anthropologists have devoted energy less to describing and cataloguing, and more to understanding caste, through detailed ethnographic studies in particular regions or places. The 'village studies' ethnographies of the 1950s and 1960s, by McKim Marriott and others, exemplify this approach (see studies in Marriott, 1955, and many of those in Singer and Cohn, 1968, for example). Such studies often focused on trying to understand the inner workings and structure of the caste system, and how it related to other aspects of society. At a more general level, theoreticians were interested in determining precisely how the caste system differed from other systems of social organization, and often drew on textual sources and Indological writings in addition to ethnographic observations in elucidating models of caste and Indian society (e.g., Marriott and Inden, 1977; Marriott, 1989).

Probably the most influential work on caste has been Dumont's *Homo hierarchicus* (Dumont, 1980). This work drew heavily on ancient textual sources, although Dumont also relied on ethnographic data, and its primary argument was that the core of the caste system was the opposition between the pure and the impure. Thus, according to Dumont, what marked caste off from other forms of social organization was the way it gave precedence to the religious over the political and economic domains of social life. Not only were power and status to be understood as separate within Indian society, but they also existed in hierarchical relationship with one another: the priest is superior to the king, since the latter represents only the profane political world. Dumont's formulation has been strongly critiqued on numerous grounds (see discussion in Fuller, 1997b), in particular for taking a 'book' view of Indian society, in relying largely on ancient texts and the ethnographic writings of others (Béteille, 1990, 1991). It has also been noted that Dumont's structuralist approach led to a holistic and highly ordered perspective on Indian society that sits at best uneasily with the view from the ground.

Perhaps the most problematic aspect of the work of Dumont and many others of his and previous eras, however, is its striking ahistoricism (Fuller, 1997b). This is inherent in Dumont's structuralist approach, but was also a product of the reliance by him and others on textual sources, according to which it was assumed that contemporary practices could be interpreted in light of ancient meanings (as discussed critically by Cohn as early as 1968). While textual information was generally used in conjunction with ethnographic data, the assumption that Indian villages were static, traditional entities (as noted in Cohn 1968, 1990; and Fuller, 1997b) did little to relieve the overriding ahistoricism of most anthropological accounts of caste and society. Given the ahistorical nature of many anthropological approaches, therefore, it is probably not surprising that anthropology has not traditionally devoted much attention to the question of caste origins. Nonetheless, some of the most radical statements concerning the origins of the caste system come from more recent anthropological and historical approaches that have begun to grapple with issues of historical change and the impact of colonialism in South Asia.

One of the central tenets of much recent work in this area has been that caste has been over-emphasized in anthropological accounts of India. It has been argued that caste is an orientalist preoccupation that has often functioned as a 'foil to build up the West's image of itself' (Inden, 1990:83; Appadurai, 1986, 1988; Kaviraj, 1997). By focusing on caste, and its assumed correlates of irrationality, injustice and inequality, anthropologists have helped to uphold and reinforce a vision of Western superiority. In response to this quite justified critique, anthropologists have in recent years moved increasingly away from a focus on caste, in order to address such topics as politics, gender, colonialism, globalization, and communalism. It is now recognized that other aspects of social organization, including gender, age, class, ethnicity, and religion are at least as salient as caste, if not more so, in many realms of South Asian life (e.g., Chakravarti, 1975; Wadley, 1980; Srinivas, 1987; Raheja, 1988; Béteille, 1991, 2002; Kaviraj, 1997; Lambert, 1997; Dirks, 2001). Even practitioners who continue to study caste have now learned to accord it with less over-riding significance than it often previously held for Western anthropologists (see papers in Fuller 1997a), although this general turn is not without its dissenters (e.g., Fitzgerald, 1996; Gupta, 2004).

Closely linked to such insights into the imbalanced view resulting from an over-emphasis on caste in anthropological studies of South Asia has been the radical and more controversial argument that caste can also in some senses be seen as an 'invention' of the West. Dirks in particularly has made this argument quite forcefully on the basis of a detailed analysis of colonial period writings on India (Dirks, 1989, 1992, 2001). These demonstrate that colonial views about Indian society in general, and caste in particular, were transformed as the colonial role and degree of intervention in India changed. During the early colonial period, caste drew surprisingly little attention from colonial commentators. It was only later, in the mid- to late-nineteenth century that colonial accounts, drawing in many cases on textual evidence and Brahman commentary rather than ethnographic observation (the authors of a several very influential texts had never even visited India), began to describe caste as an important force in Indian society. The ancient *Laws of Manu* (discussed below) were an important source for many – as Dirks states, they have been "trotted out for the last two hundred years as the classical statement of the caste system" (2001:35) – and assessments of the caste system like these, assembled from several paragraphs of a longer work by the Reverend M.A. Sherring, became increasingly the norm:

"Caste is sworn enemy to human happiness"; "Caste is opposed to intellectual freedom"; "Caste sets its face sternly against progress"; "Caste makes no compromises"; "The ties of caste are stronger than those of religion"; and "Caste is intensely selfish." (cited in Dirks, 2001:47)

Caste, thus construed, justified British intervention by demonstrating the narrow-mindedness and chauvinism of the Hindu, and part of its increasing play in accounts of India in the late nineteenth century needs to be understood within the context of changing attitudes to Britain's role in the subcontinent. Caste also provided a means of enumerating and hence knowing and controlling Indian populations. Enumeration devices like the decennial census, in which caste featured as a principal category, were a key modality of colonial knowledge and rule (Dirks, 2001; also Kaviraj 1992, 1997; Appadurai, 1993; Cohn, 1996). They also transformed Indian society. As Kaviraj notes:

In the colonial period, because of the double process of statistical counting and spatial mapping, the template of identities underwent an historic change. From indeterminate, plural, context-dependent identities, these became enumerated. (Kaviraj, 1997:327)

Enumeration it is argued, not only accentuated caste consciousness, but also crystallized identities that were previously much more fluid, variable and diverse. "Caste" became a single term capable of expressing, organizing and systematizing India's diverse forms of social identity, community and organization (Dirks, 2001:5). Two hundred years of colonial intervention, conquest and rule did not leave caste as it had been found, and understandings that assume some sort of traditional caste society are increasingly seen by anthropologists today as problematic.

Essentially then, this is where anthropology stands on caste at the moment. Scholars may disagree about the extent to which caste was altered by the colonial project, but almost all recognize that caste as we know it today in India was in many important ways shaped by colonialism. If we are to search for the origins of caste, it is necessary to look, at least partly, in the colonial period.

Texts, Historical Change and Ideology in the Study of Caste Origins

Ancient Indian texts, as we have seen, were frequently drawn upon by anthropologists to understand contemporary Indian society during the colonial period and even afterwards. The ancient texts have also been used, perhaps with greater justification, to attempt to understand ancient Indian society and accordingly, the putative origins of the caste system. For many interested in the origins of the caste system, particularly in the colonial period, the ancient texts of South Asia have been seen as a goldmine of information. Some of these texts, like the *Rg Veda*, are argued to predate even the origins of writing in South Asia, and to thus shed light on a prehistoric (or as some describe it, protohistoric) period. The very ancient date of many of the texts suggests the possibility of examining very early forms of society in South Asia.

The subcontinent's earliest texts, the *Vedas*, have been argued by Max Muller and subsequent scholars to have been compiled between 1500–1000 BC (Deshpande, 1995), though they were preserved in oral form and not written down until ca. 400 BC. The *Vedas* are essentially a large body of sacred writings, the generally earliest group of which make up the *Rg Veda*. The origins of caste in South Asia have traditionally been traced by many scholars to the encounter between native, dark-skinned Dasu and immigrant Aryas described in the *Rg Vedic* texts. The Aryas are argued to have conquered the Dasas, and imposed upon them their hierarchical status system, which placed the Dasas at the bottom. The emergence of something akin to modern-day caste is traced to the description of four ranked varnas that are compared in a verse of the *Rg Veda* to the original body (*Purusha*) that fell to the earth in a primordial sacrifice:

> When they divided the Man,
> into how many parts did they divide him?
> What was his mouth, what were his arms,
> what were his thighs and feet called?
> The Brahman was his mouth,
> of his arms was made the warrior
> His thighs became the vaisya
> of his feet the sudra was born.
> *Rg Veda* 2.2.1.1

Caste and varna (often seen as closely related and even interchangeable by writers up to the present day) are also described in later texts like the *Bhagavad Gita* and the *Laws of Manu* (Bailey, 1999; Dirks, 2001). The *Bhagavad Gita* is a spiritual text contained within the *Mahabharata*, an epic composed over a long period between the last few centuries BC and AD 500 (Witzel, 1995a). As Bailey discusses, it includes a section warning against the evils that can lead to and result from a confusion of castes or varnas (Bailey, 1999:13–14). The *Laws of Manu*, argued to have been composed around the first century AD (Bailey, 1999:14), have been even more influential in traditional accounts of caste origins, having taken on canonical legal and anthropological significance under early British rule (Dirks, 2001:34). The *Manu* text provides an account of the origins of the four varnas, as well as an explanation for the generation of the myriad of actual caste groups, described as resulting from processes of marriage and miscegenation (Dirks, 2001:35).

While many scholars of the colonial period took these mentions of caste or varna in the ancient Indian texts to indicate an ancient origin for the caste system, however, contemporary historians generally treat references to caste and varna in a much more critical fashion. And just as historians have critiqued anthropological interpretations, anthropologists have also critiqued historians. Indeed, one of the most vehement critiques of the naiveté of textual interpretation in the South Asian context has come from the anthropologist Edmund Leach, who has commented disparagingly on how the *Rg Veda* has traditionally been used by scholars studying India (Leach, 1990). Leach argues that it is highly problematic to treat religious documents like the *Rg Veda* as historical records. He regards the *Rg Veda* as a myth, and points out that while all history has mythical elements, myth and history are not the same:

From this point of view all history is myth. But the converse is not the case. Although some texts can function either as history or as myth, history and myth are, in a fundamental sense, categories of quite different kinds. History is anchored in the past; it is time-bound; it cannot be repeated. Myth is timeless; it is constantly re-enacted in ritual performance. ... I ... wish to insist that when any of us who are anthropologists are presented with stories that purport to be history, we should be sceptical. We always need to ask: In whose interest is it that the past should be presented to us in this way? (Leach, 1990:229).

While Leach's critique probably underestimates the potential value of the *Rg Veda*, and the ability of historians and others to discern information about ancient Indian society from it, his attack of the superficial interpretations attached to the text by many scholars is nonetheless relevant, particularly in the context of discussions of caste. And indeed many scholars who draw on the texts today do recognize that they are mythical rather than historical texts, that they likely reflect the attempts of a priestly class to create or legitimate a certain position for themselves within society, and that they can thus not be taken at face value (Erdosy, 1995b; Witzel, 1995b). Attempts to link groups described in the *Rg Veda* to linguistic, racial and caste groups have increasingly given way to discussions of 'ethnicity' and at least at times, of identity as something constructed by, rather than reflected in, the ancient texts (Allchin, 1995; Deshpande, 1995; Erdosy, 1995a, 1995b). Related to all of this is a recognition that the varna model that is frequently the focus of early textual discussions reflects an idealized scheme whose link to the thousands of castes that exist in India today is far from clear (Bayly, 1999; Maisels, 1999).

Overall then, what we find is an attempt to problematize these ancient texts rather than just take them at face value. This shift has also been closely associated with specific attempts to move beyond a notion of caste as traditional and unchanging from time immemorial. Olivelle, for example, argues that a concern for ritual purity – understood by Dumont

to be central to the notion of caste – arose much later in Indian society (Olivelle, 1998). His analysis of the texts on Dharma suggests that the notion of purity primarily served a means of strengthening and sustaining existing social boundaries. Susan Bayly, based on an analysis of much later historical texts, has similarly argued that hierarchical and purity-conscious social codes become popularized and more widespread only in the post-Moghul period, just prior to colonialism (Bayly, 1999). She sees evidence for a two-stage transformation between 1650 and 1850, leading to the creation of "the elaborately ritualized schemes of social stratification which have come to be thought of as the basis of 'traditional' faith and culture in the subcontinent" (Bayly, 1999:26). The first stage is marked by the "rise of the royal man of prowess", and attempts by ruling elites and the priests and ascetics with whom they were able to associate their power "to establish firm social boundaries between themselves and the non-elite tillers and arms-bearers to whom their forebears had often been closely afflilated" (Bayly, 1999:26). Here we see the emergence of more caste-like forms of social order. In the second stage, which reached its culmination in the colonial period, Brahmans become ever more widely deferred to, achieving a quasi-independent status in such areas as legal codes and colonial administrative practice (Bayly, 1999:27). Bayly argues that it is during "this period that the power of the pollution barrier became for many Indians the defining feature of everyday caste experience" (Bayly, 1999:27). Thus, Olivelle and Bayly are in essential agreement concerning the late emergence of both ritual purity and more ritualized forms of caste practice. They share with Dirks an understanding that caste does not constitute an ancient, unchanging presence in Indian society, though their accounts point to a somewhat more gradual emergence and give more priority to indigenous developments and forms of agency (see also Deshpande,

2005) than does that of Dirks. In any case, the key point is that scholars now very much see caste as it exists today in India as something that cannot clearly be traced back into the ancient past in the ways that scholars used to expect.

Materiality, Migrations and Modern Politics: Caste and Archaeology

While history and particularly anthropology have devoted much attention to caste, archaeologists have only rarely engaged with the topic (Boivin, 2005; examples of archaeologists who have addressed caste include Erdosy, 1986; Fairservis, 1995; and Coningham and Young, 1999). I have argued elsewhere that this likely has much to do with dominant political and archaeological paradigms at the time of archaeology's establishment as a scholarly discipline in South Asia (Boivin, 2005). Colonial politics and migrationist tendencies in archaeology did much to encourage firstly the importation of models from the West to interpret South Asia's ancient remains (Boivin and Fuller, 2002; Morrison, 1994), and secondly the interpretation of change not as coming from within a dynamic and evolving South Asian society, but as resulting from the migration of new peoples, inevitably from the west (Cohn, 1996; Fuller and Boivin, 2002). Accordingly, when caste has been considered by archaeologists in South Asia, it has been almost entirely within the context of migrationist paradigms. In particular, considerations of caste have frequently been linked to the notion of an Indo-Aryan invasion of the subcontinent in later prehistory.

The Indo-Aryan invasion model traces its roots to the discovery, announced in 1786 by Sir William Jones, of a close relationship between Latin, Greek, Sanskrit, Germanic and Celtic languages. These languages became known as the Indo-European languages, and their relatedness raised the puzzling question

of how they had come to be spoken in such geographically distant and culturally distinctive places. It was postulated that the speakers must have originally been one people, sharing a homeland. The challenge therefore became one of locating the whereabouts of this Indo-European homeland, and tracing the subsequent spread of the Indo-European languages as its people dispersed. The translation of the *Rg Veda* by Muller in the mid-1800s provided some key clues. Muller assigned a date for the compilation of the *Rg Veda* of 1200–1500 BC at least, and the text was interpreted as reflecting the arrival of an off-shoot of the original Indo-European group, the Indo-Aryans, into the Indian subcontinent around this time. The Indo-Aryans were then supposed to have, much like the subsequent British, invaded and conquered India. It has been argued that they imposed caste on the indigenous peoples of India, who they placed at the bottom of the caste hierarchy.

This interpretation of the linguistic and cultural evidence, provided by linguists and others, was taken on board by archaeologists, whose role was to look for material evidence of the proposed invasion. Finding such evidence was relatively unproblematic in the early days of the discipline, when archaeology operated under a culture-history paradigm that equated material culture groups with ethnic groups. If archaeologists observed a change in material culture, this was generally interpreted as reflecting the invasion of a new group of people, or at the very least the diffusion of new ideas from elsewhere. The challenge was thus one of finding a major material culture change in South Asia around the proposed date of the Indo-Aryan migrations based on linguistic evidence. Such a transition was eventually located in the demise of the Indus Valley civilization, which had been discovered by archaeologists in the 1920s. Two influential British archaeologists in particular, Stuart Piggott and Sir Mortimer

Wheeler, argued that impetus for both the origins and decline of the ancient urbanized civilization came from the west: the former from a diffusion of ideas from Mesopotamia, and the latter from an invasion of Indo-Aryan peoples (Piggott, 1950; Wheeler, 1968). These invading conquerors were argued to be responsible for the new burial practices and pottery styles seen in Cemetery H at Harappa (Childe, 1934; Vats, 1940), as well as what were argued to be disorganized piles of bodies at Mohenjo-daro, indicative of a massacre of the city's inhabitants (Wheeler, 1947).

Over the course of the past half century, the model of an Indo-Aryan population invasion have been thoroughly problematized, and largely discredited within archaeology (Shaffer, 1984; Erdosy, 1995a, 1995b; Kennedy, 1995, 2000; Possehl, 2002; Shaffer and Lichtenstein, 1995; Kenoyer, 2005). The demise of the Indus Valley Civilization is now understood largely in terms of much more gradually unfolding, localized processes that have been convincingly linked to major hydrological and environmental changes in the north-western part of the subcontinent (Dales, 1966; Raikes, 1968; Mughal, 1982). Other attempts to link major material culture transformations in South Asian archaeology, including the introduction of Painted Grey Ware in the north, and the beginnings of megalith-creation in south India, to the proposed Indo-Aryan invasions have proven similarly unconvincing (Shaffer, 1984; Erdosy, 1995b). What the accumulation of archaeological evidence over the course of the twentieth century has inevitably demonstrated is that the major transitions in South Asian pre- and proto-history are gradual and often show little evidence for any outside origin (Shaffer and Lichtenstein, 2005). Even where potential external material culture links are found, as in the Late Harappan period, they can in no sense be taken to indicate any large-scale influx of people. Archaeologists in particular have thus very much moved away from migrationist models, including the idea

of Indo-Aryan invasions, as an explanation for cultural change in South Asia. And those scholars, both archaeological and otherwise, who continue to embrace an Indo-Aryan migration paradigm now generally present a very different model that sees the language change as resulting more from social processes than any substantial population movements (Allchin, 1995; Witzel, 1995a; Thapar, 1999).

What these developments imply in terms of the study of caste origins is that attempts to trace it to an Indo-Aryan invasion or Vedic culture are problematic from an archaeological point of view. The idea that a new and hierarchical social system was imposed by Indo-Aryan invaders in the second millennium BC is, like other aspects of the Indo-Aryan invasion model, simply not supported by any archaeological evidence. Nor, as some scholars have attempted to argue (Kenoyer, 1989, 1995; Maisels, 1994), is there adequate evidence to support the existence of caste during either the Harappan period or its immediate aftermath. As has been pointed out, these and other attempts to locate caste in the archaeological record suffer from a tendency to infer caste from otherwise ambiguous data strictly on the basis of a South Asian context (Sinopoli, 1991; Boivin, 2005). The paucity of research into the material correlates of caste (though see Khare, 1976; Miller, 1985; Moore, 1989; Kramer and Douglas, 1992; Noble, 1997; and Abeyaratne, 1999), as well as the limited quantity of detailed, high-resolution archaeological data for most South Asian contexts (the site of Harappa providing a clear exception) compared to other parts of the world, mean that present-day archaeology offers at best a very crude instrument for looking at the origins of caste. These problems, as well as perhaps more importantly the tendency for discussions relating to past ethnic and social identities to become politicized within contemporary South Asian archaeology (elements of this are discussed in Mandal, 1993; Rao, 1994, 1999; Hassan, 1995; Lamberg-Karlovsky, 1997; Coningham and Lewer, 2000; Ratnagar, 2004; and Fosse, 2005), also help explain why archaeologists have generally steered clear of discussing caste. As I have argued elsewhere, however, to avoid caste can be to actually play into the hands of communalist forces (Boivin, 2005). Caste and other aspects of identity must at the very least be discussed and theorized by archaeologists, so that they do not become the preserve of reactionary forces.

In conclusion then, it is clear that in archaeology as well, the idea that caste can be traced back into the deep past lacks support from most scholars. In archaeology's case, however, it is limitations of method and theory more than a lack of data that have precluded recognition of caste in the past. This is not to say that archaeologists will find caste in the archaeological record, but only to point out until they have linked more systematically-collected, detailed archaeological databases with more developed interpretive frameworks for examining issues of social, political and cosmological organization in South Asian archaeology, there is very little that they will be able to say with any authority about aspects of identity like caste in the past.

The Newcomer: Genetics and the Challenges Posed by Caste

If the disciplines of the social sciences have demonstrated that ideas about caste and its origins are often shaped by contemporary social, political and economic agendas, then perhaps the answer to questions about caste origins in South Asia is best sought in the more objective data offered by the natural sciences. As is now widely recognized, the field of molecular genetics offers a powerful new means of examining population histories, including processes of population movement, interaction, isolation, and transformation (e.g., Cann et al., 1987). The relevance of these methods to the South Asian setting has not escaped geneticists, who are

the most recent in a long line of scientists to take an interest in South Asian populations, and the issue of caste. However, while colonial period research generally pursued now discredited attempts to link caste and race (e.g., Risley, 1915), more recent investigations of caste using genetics have taken a historical approach, and have focused on tracing its origins through the analysis of the contemporary South Asian gene pool. Much of this work has been closely linked to the analysis of tribes. These groups, however, are generally seen as the result of much earlier population events such as the colonization of South Asia by modern humans during the Pleistocene (e.g., Basu et al., 2003; Cordaux et al., 2003, 2004b; Endicott et al., 2003; Kumar and Reddy, 2003; Vishwanathan et al., 2004; Thangaraj et al., 2005).

The genetic study of caste has been underway for little more than a decade, and this formative research period has been characterized by significant optimism and ambition, despite recognition of the fact that South Asia's remarkable genetic, cultural and linguistic diversity (Majumder, 1998; Kivisild, 2000) suggests the operation of complex underlying processes (Majumder, 2001a). This optimism is particularly marked in a recent series of papers that claim to have traced the origins of the caste system to an ancient, male-mediated, Indo-Aryan invasion that pushed indigenous Dravidian speaking populations southwards, and established Indo-Aryans at the top of the caste hierarchy (e.g., Bamshad et al., 2001; Quintana-Murci et al.,2001; Basu et al.,2003; Cordaux et al., 2004a). This is supported, for example, by analyses of maternally-inherited mitochondrial (mt) DNA that show that Indian caste groups, regardless of rank, are more closely related to Asians and most dissimilar from Africans (Bamshad et al., 2001; see Table 1). In contrast, the paternally-inherited Y-chromosome tells a different story: while lower castes are still more similar to Asians,

Table 1. Genetic distances between caste groups from Andhra Pradesh and continental populations, based on analysis of MtDNA (HVR1 sequence). From Bamshad et al., 2001 (Table 1); see original article for data and methodology

Caste Group	Africans	Asians	Europeans
Upper	.179	.037	.100
Middle	.182	.025	.086
Lower	.163	.023	.113
All castes	.196	.026	.077

upper castes are more similar to Europeans, and middle castes are equidistant between the two (Bamshad et al., 2001; Table 2). Caste-dependent links with Europeans are confirmed by analyses of mtDNA haplogroups and autosomal *Alu* elements (Bamshad et al., 2001). The findings of this study by Bamshad and colleagues have been echoed by others. Basu et al, for example, argue that Central Asian populations are genetically closer to upper-caste populations, particularly in the north of the subcontinent, than to middle- or lower-caste populations (Basu et al., 2003). This and other findings by Basu et al. are again taken to confirm the historical reality of the Indo-Aryan invasion, which is argued to have pushed Dravidian-speaking indigenous peoples southwards (Basu et al., 2003:2287). Similar arguments have been made on the basis of comparisons of caste and tribal DNA sequences, in some cases with tribals seen as the indigenous inhabitants overwhelmed by invading Indo-Aryan populations (e.g., Cordaux et al., 2004a).

The substantial number of genetic studies confirming the migration into South Asia in late prehistory of Indo-Aryans is noteworthy,

particularly in light of the fact that the theory of substantive Indo-Aryan population displacements into South Asia, not to mention the whole notion of caste having extremely ancient origins, have come under such heavy critique in the disciplines addressed previously. Nonetheless, as I will argue, the findings of these studies need to be treated with substantial caution, if not outright scepticism, based on problems concerning both the genetic patterns and their interpretation.

Problems concerning the genetic patterns themselves are illustrated by several important studies that contradict these findings. A number of studies, for example, argue for much more limited Holocene-period gene flow into South Asia than the studies outlined above. The work of Kivisild and colleagues, in particular, has argued on the basis of an analysis of mitochondrial, Y-chromosome and autosomal DNA sequences that Indian tribal and caste populations have received limited gene flow from external regions since the beginning of the Holocene (Kivisild et al., 1999a, 1999b, 2000, 2003a, 2003b). As Endicott et al. note in the present volume,

Table 2. Genetic distances between caste groups from Andhra Pradesh and continental populations, based on analysis of Y chromosome (STRs). From Bamshad et al., 2001 (Table 3); see original article for data and methodology

Caste Group	Africans	Asians	Europeans
Upper	.0166	.0104	.0092
Middle	.0156	.0110	.0108
Lower	.0131	.0088	.0108
All castes	.0151	.0101	.0102

the contradictory findings seem to some degree at least to be the result of the different types of analyses undertaken: while studies based on phylogenetic analyses have generally provided support for the idea of a *common* origin of Indian tribal and caste groups, irrespective of their status in the social hierarchy, those focusing on genetic distances have resulted in an emphasis on the *differences* between tribes and castes. Other problems exist as well. It has been pointed out by Kivisild et al., for example, that larger sample sizes, more popula- tions, and increased molecular resolution are required to adequately resolve current genetics debates about South Asian population histories (Kivisild et al., 2003a:329). These various problems simply highlight the fact that interpretation is involved at even the early levels of genetic analysis, a fact that is overlooked when we characterize, as I have above, the natural sciences as objective.

While methodological, analytical and inter- pretive issues of a genetic nature clearly exist with respect to the nascent molecular genetics studies of South Asian populations, and of caste in particular, I am not adequately equipped to deal with these, and am keen instead, in light of the foregoing discussion, to examine some of the anthropological, historical and archaeological problems with this work at the broader interpretive level. This analysis is not intended to take geneti- cists to task for not having a sufficient grasp of relative issues in the social sciences (although, admittedly, a little more interest in and knowledge of such issues might reasonably be anticipated in light of the relevance of social factors to topics like caste, tribe and South Asian prehistory that are frequently addressed by geneticists). Instead my primary aim is to encourage greater discussion and interaction between all of the four fields, and in particular those undertaking the analysis of South Asian populations from biological perspectives on

the one hand, and social perspectives on the other.

In reading the genetics literature on South Asia, it is very clear that many of the studies actually start out with some assump- tions that are clearly problematic, if not in some cases completely untenable. Perhaps the single most serious problem concerns the assumption, which many studies actually start with as a basic premise (e.g., Watkins, 1999; Roychoudhury et al., 2000; Bamshad et al., 2001; Quintana-Murci et al., 2001), that the Indo-Aryan invasions are a well-established (pre)historical reality. The studies confirm such invasions in large part because they actually assume them to begin with. Thus Bamshad and colleagues state as a premise of their study that:

... Indo-European-speaking people from West Eurasia entered India from the Northwest and diffused throughout the subcontinent. They purportedly admixed with or displaced Indigenous Dravidic-speaking populations. Subsequently they may have established the Hindu caste system and placed themselves primarily in castes of higher rank. (Bamshad et al., 2001:1).

The fact that Bamshad et al. then go on to establish an Indo-Aryan migration, as outlined above, is thus less surprising. One clear reason that the assumption of Indo-Aryan invasions is repeatedly made in the literature undoubtedly relates to the fact that little archaeological literature is referenced, and that which is referred to is generally at least some twenty if not thirty or more years out of date. Thus we find statements like:

The origins of the nearly one billion people inhab- iting the Indian subcontinent and following the customs of the Hindu caste system are controversial: are they largely derived from Indian local populations (i.e. tribal groups) or from recent immigrants to India? *Archae- ological* and linguistic *evidence support the latter hypothesis*, whereas recent genetic data seem to favor the former hypothesis. (Cordaux et al., 2004a:231; emphasis mine).

Leaving   aside   the   also   problematic identification   of   tribal   groups   as   the original inhabitants of India (see Morrison, this volume), another standard theme in the genetics literature, I wish to highlight rather that the statement concerning archaeological evidence is supported by a single, 30-year-old reference (Fairservis' 1975 *The Roots of Ancient India*). There seems to be an assumption amongst many of the geneticists that archaeological knowledge (like, as we shall see, caste itself) is static and non-evolving. Clearly, however, as outlined previously, archaeological understandings of Indo-Aryan invasions have changed, and indeed have changed dramatically, over the course of the last three decades, and it is simply not accurate to state any longer that archaeological evidence supports the idea that modern Indian caste populations derive from recent (i.e., Indo-Aryan) immigrants. Similar limited familiarity with the archaeological literature is also highlighted in the work of Basu et al., who cite another 30-year-old reference (along with, strangely, another geneticist) to support the idea of population invasions along an old-style culture-history migrationist model – in this case one that links megaliths in north India with those in Iran and Baluchistan, and megaliths in south India with those in Europe (Basu et al., 2004:2).

The assumption of Indo-Aryan invasions into the subcontinent is so entrenched that when genetics findings contradict the notion of invasions, geneticists generally see it as problematic, and often try to interpret it away. Thus, Cordaux and colleagues are so wedded to the idea of a full-fledged Indo-Aryan invasion that they explain the contradictory mitochondrial evidence, indicative of a high indigenous contribution, as resulting from combined practices over the centuries since invasion of hypergyny and preferential female infanticide, the latter apparently at rates as high as 30–80% (this range is based, it should be pointed out, on the general sociobiological literature rather

than any South Asian historical or ethnographic sources, undoubtedly since they are completely lacking) (Cordaux et al., 2004a). Given the claimed preference in the natural sciences for parsimonious explanations, this hardly seems the most likely explanation of the observed patterns, which really do seem to indicate at the very least something less than a full-fledged population movement (Bamshad et al., 2001; Kivisild et al., 2003a; Metspalu et al., 2004).

Part of the reason that many geneticists prove Indo-Aryan invasions so frequently is that they give little if any consideration to other populations that have or may have entered South Asia in prehistoric and historic times. Another problematic assumption that therefore needs to be highlighted is that in much of the genetics literature, the only (or only significant) possible post-Holocene source of non-indigenous genetic material is Indo-Aryans. This is remarkable in light of the fact that while we have little (if any) evidence in the archaeological record of Indo-Aryans, we do possess clear archaeological and/or historical evidence of contacts with regions west (and east) of South Asia from the Neolithic onwards. This evidence is found in levels dating to as early as the 7–8th millennium BC at Baluchistani sites like Mehrgarh (Jarrige, 1981, 1982), where domesticated crops and animals from the Near East first enter South Asia, one would assume attended by at least some degree of human genetic interaction (and see Cordaux et al., 2004b). Genetic introductions from the west then continue through to the colonial period, and indeed up to the present day. For example, to take a less well-known instance of such admixture, substantial inter-breeding between European men and Indian women has been historically documented for the early colonial period (Dalrymple, 2002). Speaking specifically about the relationship between Europeans and Indians, Dalrymple notes that contrary to popular expectations of racial and cultural segregation in colonial period India: "[a]t all times up to the nineteenth century, but perhaps especially so during the period 1770 to 1830, there was wholesale interracial sexual exploration and surprisingly widespread cultural assimilation and hybridity" (Dalrymple, 2002:10).

The probability of long-term, and to some degree continuous, genetic introductions to the South Asian gene pool from populations to the west is rarely acknowledged by geneticists. It is therefore encouraging to find in the genetics literature recent acknowledgement of diverse population influences, "by the Huns, Greeks, Kushans, Moghuls, Muslims, English and others" (Kivisild et al., 2003a:329), and the probability of "*multiple* waves of immigration of Caucasoid peoples, originating from a large geographical area, into India over a relatively long period of historical time" (Roychoudhury et al., 2000; emphasis mine). To what degree these other migrations and interactions have affected the South Asian gene pool remains to be seen. However, this will remain a mystery, and hence unquantifiable, unless non-Indo-Aryan invasions are entertained, and geneticists stop sloppily attributing non-indigenous genetic signatures to population invasions that they have not adequately researched and critically evaluated. Essentially, geneticists need to work with prehistorians, historians, linguists and other scholars specializing in the South Asian past.

Another problematic assumption in the genetics literature is that caste is unchanging. Of particular relevance in light of the subject matter is the fact that in much of the genetics literature it is generally taken for granted that castes have been endogamous since they originated (e.g., Rajkumar and Kashyap, 2004:2; though see, for example, Bamshad et al., 1998). And yet, one of the questions that should really be *asked* using the genetics data is to what degree endogamy has actually been practiced in South Asia, and over what time period. Interestingly, some studies show a surprising degree of genetic similarity, both between castes and between tribes and castes. Cordaux et al., for example, provide evidence of gene flow between agriculturalists and hunter-gatherers who have recently adopted agriculture (although the conclusions they reach concerning the relevance of this data to the issue of demic diffusion with agriculture are problematic; see Cordaux et al., 2004b). Another interesting study, of Y-chromosome SNP haplotypes in caste and tribal populations in Andhra Pradesh, showed substantial admixture of non-African haplotypes in the male gene pool of the Siddis, a group known to have migrated from Africa to India as recently as AD 1100 (Ramana et al., 2001). This study indicated that "at least 56% of the male genes of the Siddis could be of Indian origin" (Ramana et al., 2001:699), indicating that neither the Siddis nor their caste neighbors have practiced endogamy to the degree expected. Whether this is because caste practices have changed over the past millennium, because theory has not met with practice, or, as likely, both, remains unclear, but whatever the reason, such data strongly imply that the issue of caste endogamy deserves detailed investigation.

One final problem I would like to raise concerning the genetics literature on caste origins in South Asia is the failure to consider relevant anthropological issues, processes and patterns. For example, at a general level, the fact that marriage fields vary across India, and indeed between castes, seems relevant (Sopher, 1980). In the south, for example, marriage is frequently between members of a kin group from the same village, while in the north, women are generally married to non-kin from distant villages. These kinds of sociological patterns seem potentially relevant to the interpretation of genetically observed patterns, which often show differences between north and south India. Micro-level processes also need to be considered. It is clear, for example, that caste histories are exceedingly diverse, and that these should be studied in detail before castes are sampled for genetics studies. Some castes are of much more recent origins than others, while certain groups have raised their status through processes of what is commonly termed 'Sanskritization' (Srinivas, 1952; Charsley, 1998). This particular issue has been raised in a recent assessment of genetics studies by Majumder, who points out that:

... a caste bearing the same name may have very different origins in different geographical regions. There are examples in which a tribe dispersed over a large geographical region, took up different occupations in different sub-regions, and 'fitted' itself into the caste hierarchy at different rungs. Karve's work has also indicated that each of the different Brahmin castes ... in Maharashtra probably has a different origin. (Majumder, 2001b:932).

Ramana and colleagues have genetically documented this kind of process in Andhra Pradesh, where they note that the Vizag Brahmins and Peruru Brahmins are genetically distant from each other despite belonging to the same social rank (Ramana et al., 2001). In another example of the same kind of discrepancy, Roychoudhury and colleagues note that while Watkins et al. (1999) documented high frequencies of the intergenic COII/tRNALys 9-bp deletion amongst Irula tribes, their own testing of Irula mtDNA did not reveal the 9-bp deletion at all. They state:

We are unable to offer any clear explanation of this discrepancy of estimated frequencies, except to state that we have sampled the Irulas from their original habitat, Nilgiri Hills, while the samples of Watkins et al., were drawn from the coastal areas of Andhra Pradesh, where the Irulas are presumably migrants. (Roychoudhury et al., 2000:1190)

Such findings demonstrate very emphatically that shared caste names cannot be mistaken for shared caste histories. Anthropological, sociological, geographical and historical information is crucial to the genetics enterprise.

Conclusions

Caste then, it is clear, is a complicated matter, and these complexities are compounded when what is at stake is not caste in its contemporary form, but the history and origins of caste in South Asian society. This has posed a challenge to each of the disciplines addressed in this chapter. What appeared at first to be relatively straightforward and in many cases even obvious solutions to the question of caste origins have proven, as a result of more sustained research, and the introduction of new methodologies and interpretive frameworks, far less convincing answers than earlier investigators had believed. Rather than moving closer to the 'truth' about caste origins, anthropology, history, archaeology and genetics have all had to grapple with fundamental issues concerning not just caste origins but the very nature of caste itself, and the challenge of understanding social categories that have been and continue to be implicated in political ideologies, both national and global. As apparent with even the most supposedly objective of the scientific approaches addressed here, understandings of caste and its origins are undeniably shaped by the assumptions that are brought to their study.

Beyond illustrating the complexity of caste and its origins in South Asia, the comparison undertaken in this chapter also suggests a number of conclusions concerning the history of caste (see Table 3). For one thing, they strongly indicate that the search for a nice, neat, straightforward 'single origins' model of caste development (see Boivin, 2005) is unlikely to yield results. Rather than some sort of simple event to be traced back to the migration into the subcontinent of a group of Indo-Aryans and/or the conquering and domination of one group by others, the beginnings of caste are much more likely to be the result of long-term processes whose understanding requires an appreciation of social dynamics and developments that took

*Table 3. Hypotheses concerning the origins of caste, on the basis of anthropological,
historical, archaeological and genetics findings*

- Caste origins **complex**
- **No** single origin
- **Gradual** emergence of caste
- **Significant** regional variation
- **Variable** integration into caste fold of different groups
- Differences in what people do perhaps an **early** feature of caste
- Ritual purity and perhaps endogamy **later** features
- Substantial **recent** change

place in various parts of South Asia during various time periods. Archaeology, history, and more recently genetics, have all failed to yield clear, undisputed patterns suggestive of an unchanging monolithic system that has reigned in the subcontinent for millennia. Anthropology challenges even the wisdom of looking for one. Despite many years of searching, and many models claiming to draw on it, the evidence for a long-term caste system of the type found in India today remains lacking. It is time to accept that whatever the origins of caste, the specific constellation of traits we associate with caste today have complex and varied histories that require the application of more subtle and sophisticated approaches than have often been undertaken by anthropologists, archaeologists, historians, geneticists and others.

It is also clear that the importance of caste has varied regionally and contextually. Both archaeological and historical data reveal significant subcontinental diversity, even within the framework of increasingly long-distance exchange networks and cultural links through time. This diversity operated at various levels, creating differences between north and south at one end of the spectrum, and between various communities living in one region at the other. To speak of an ancient caste system becomes extremely problematic in light of these highly apparent social, cultural and political discontinuities. Historical and contextual variation are also important. Patterns of authority and the means to negotiate, challenge and transform them

were clearly not constant, and must certainly have led to changing understandings of the relationships between different groups within society. It is likely that even when caste has been a factor in social life, its importance has been shaped by context, and the particular needs of specific groups.

So what can we say about the origins of caste in South Asia in light of these complexities? Well, the evidence for regional and even local variability indicates that different groups have been integrated into the caste fold in different ways and at different times, perhaps over a long period. Moreover, the various features of the caste system do not seem to have appeared all at once, and indeed some of them may have arisen relatively late in Indian history. The archaeological record, which demonstrates a mosaic of economically differentiated groups from a very early date, suggests that differences in what people do was likely an early feature of social differentiation in the subcontinent, and one that may perhaps have been linked to the eventual emergence of something like caste. Historical studies suggest that caste differences only became ritualized during the later development of caste, however, and that their crystallization as context-independent, essentialist markers of identity took place as late as the colonial period. It therefore seems probable that while early economic and social differentiation may have led to some degree of endogamy, narrowly defined and strictly prescribed endogamy of the type found today in the subcontinent was also likely a late development, linked to the ritualization of caste and the crystallization of caste identities.

These various assertions concerning the origins of caste, made on the basis of the diverse findings of the disciplines addressed in this chapter, remain to be investigated through more detailed and systematic studies. That such studies need to be marked by a greater degree of inter-disciplinary inter-action and informed discussion than has occurred to date is clear. It is also apparent that scholars from all disciplines need to continue to develop a greater theoretical and methodological appreciation of the complex social, economic, political and cultural trajec-tories that have led to what is today referred to as the "caste system". Furthermore, as the internal struggles and debates of many of the disciplines highlight, colonial period and contemporary political agendas need to be taken into account. However, recog-nition of their role in the story of caste and its origins should not frighten scholars away from addressing caste at all. Simple, often politically motivated stories about caste origins need to be challenged by detailed inter-disciplinary research that demonstrates the complexity of both caste and its devel-opmental trajectory. Rather than shying away from the investigation of caste in light of these complexities or caste's political implica-tions, scholars need to work together to create and maintain an informed discussion about the origins of caste, its role in the structuring of biological diversity, and its place in South Asian society.

Acknowledgments

I am very grateful to Michael Petraglia, Bridget Allchin, Mark Kenoyer and Peter Forster for helpful comments on an earlier draft of this paper. I would emphasize, however, that any errors or omissions remain entirely my own responsibility.

References

Abeyaratne, N., 1999. The role of caste hierarchy in the spatial organisation of a village landscape in the Dry Zone of Sri Lanka. In: Ucko, P.J., Layton, R. (Eds.), The Archaeology and Anthropology of Landscape: Shaping Your Landscape. Routledge, London, pp. 135–45.

Allchin, F.R., 1995. The Archaeology of Early Historic South Asia: The Emergence of Cities and States. Cambridge University Press, Cambridge.

Appadurai, A., 1986. Is homo hierarchicus? American Ethnologist 13(4), 745–61.

Appadurai, A., 1988. Putting hierarchy in its place. Cultural Anthropology 3, 36–49.

Appadurai, A., 1993. Number in the colonial imagi-nation. In: Breckenridge, C.A., van der Veer, P. (Eds.). Orientalism and the Postcolonial Predicament. University of Pennsylvania Press, Philadelphia, pp. 314–39.

Bamshad, M., Kivisild, T., Watkins, W.S., Dixon, M.E., Ricker, C.E., Rao, B.B., Naidu, J.M., Prasad, R. B.V., Reddy, P.G., Rasanayagam, A., Papiha, S.S., Villems, R., Redd, A.J., Hammer, M.F., Nguyen, S.V., Carroll, M., Batzer, M.A., Jorde, L.B., 2001. Genetic evidence on the origins of Indian caste populations. Genome Research 11.

Bamshad, M.J., Watkins, W.S., Dixon, M.E., Jorde, L.B., Rao, B.B., Naidu, J.M., Prasad, B.V.R., Rasanayagam, A., Hammer, M.F., 1998. Female gene flow stratifies Hindu castes. Nature 395, 651–652.

Basu, A., Mukherjee, N., Roy, S., Sengupta, S., Banerjee, S., Chakraborty, M., Dey, B., Roy, M., Roy, B., Bhattacharyya, N.P., Roychoudhury, S., Majumder, P.P., 2003. Ethnic India: A genomic view, with special reference to peopling and structure. Genome Research 13, 2277–2290.

Bayly, S., 1999. Caste, Society and Politics in India from the Eighteenth Century to the Modern Age. Cambridge University Press, Cambridge.

Béteille, A., 1990. Race, caste and gender. Man (n.s.) 25, 489–504.

Béteille, A., 1991. Society and Politics in India: Essays in Comparative Perspective. Athalone Press, London.

Béteille, A., 2002. Caste, Class and Power: Changing Patterns of Stratification in a Tanjore Village. Oxford University Press, Delhi.

Boivin, N., 2005. Orientalism, ideology and identity: Examining caste in South Asian archaeology. Journal of Social Archaeology 5(2), 225–252.

Boivin, N., Fuller, D., 2002. Looking for post-processual theory in South Asian archaeology. In: Settar, S., Korisettar, R. (Eds.), Indian Archaeology in Retrospect, Volume IV: Archaeology and Historiography. Manohar, New Delhi, pp.191–215.

Cann, R.L., Stoneking, M., Wilson, A.C., 1987, Mitochondrial DNA and human evolution. Nature 325, 31–36.

Cavalli-Sforza, L.L., Menozzi, P., Piazza, A., 1994. The History and Geography of Human Genes. Princeton University Press, Princeton.

Chakravarti, A., 1975. Contradiction and Change: Emerging Patterns of Authority in a Rajasthan Village. Oxford University Press, Delhi.

Charsley, S., 1998. Sanskritization: The career of an anthropological theory. Contributions to Indian Sociology (n.s.) 32(2), 527–49.

Childe, V.G., 1934. New Light on the Most Ancient East. Routledge and Kegan Paul, London.

Cohn, B.S., 1968. Notes on the history and study of Indian society and culture. In: Singer, M., Cohn, B.S. (Eds.), Structure and Change in Indian Society. Wenner-Gren Foundation, New York.

Cohn, B.S., 1990. An Anthropologist Amongst the Historians and Other Essays. Oxford University Press, Delhi.

Cohn, B.S., 1996. Colonialism and Its Forms of Knowledge. Princeton University Press, Princeton.

Coningham, R., Young, R., 1999. The archaeological visibility of caste: an introduction. In: Insoll, T. (Ed.), Case Studies in Archaeology and World Religion: The Proceedings of the Cambridge Conference. BAR International Series 755, Archaeopress, Oxford, pp. 84–93.

Coningham, R., Lewer, N., 2000. Archaeology and identity in south Asia – interpretations and consequences. Antiquity 74, 664–7.

Cordaux, R., Aunger, R., Bentley, G., Nasidze, I., Sirajuddin, S.M., Stoneking, M., 2004a. Independent origins of Indian caste and tribal paternal lineages. Current Biology 14, 231–235.

Cordaux, R., Deepa, E., Vishwanathan, H., Stoneking, M., 2004b. Genetic evidence for the demic diffusion of agriculture to India. Science 304, 1125.

Cordaux, R., Saha, N., Bentley, G.R., Aunger, R., Sirajuddin, S.M., Stoneking, M., 2003. Mitochondrial DNA analysis reveals diverse histories of tribal populations from India. European Journal of Human Genetics 11, 253–264.

Dales, G.F., 1966. The Decline of the Harappans. Scientific American 241(5), 92–100.

Dalrymple, W., 2002. White Mughals. Flamingo, London.

Daniel, E.V., 1984. Fluid Signs: Being a Person the Tamil Way. University of California Press, London.

Deshpande, M.M., 1995. Vedic Aryans, non-Vedic Aryans, and non-Aryans: Judging the linguistic evidence of the Veda. In: Erdosy, G. (Ed.), The Indo-Aryans of Ancient South Asia. Munshiram Manoharlal, New Delhi, pp. 67–84.

Dirks, N.B., 1989. The original caste: power, history and hierarchy in South Asia. Contributions to Indian Sociology n.s. 23(1), 59–77.

Dirks, N.B., 1992. Castes of mind. Representations 37, 56–78.

Dirks, N.B., 2001. Castes of Mind: Colonialism and the Making of Modern India. Princeton University Press, Princeton.

Dumont, L., 1980. Homo Hierarchicus: The Caste System and its Implications. University of Chicago Press, Chicago.

Endicott, P., Thomas, M., Gilbert, P., Stringer, C., Lalueza-Fox, C., Willerslev, E., Hansen, A.J., Cooper, A., 2003. The genetic origins of Andaman islanders. American Journal of Human Genetics 72, 178–184.

Erdosy, G., 1986. Social ranking and spatial structure: examples from India. Archaeological Review from Cambridge 5(2), 154–66.

Erdosy, G., 1995a. Language, material culture and ethnicity: theoretical perspectives. In: Erdosy, G. (Ed.), The Indo-Aryans of Ancient South Asia. Munshiram Manoharlal, New Delhi, pp. 1–31.

Erdosy, G., 1995b. The prelude to urbanization: Ethnicity and the rise of Late Vedic Chiefdoms. In: Allchin, F.R. (Ed.), The Archaeology of Early Historic South Asia: The Emergence of Cities and States. Cambridge University Press, Cambridge.

Fairservis Jr., W.A., 1995. Central Asia and the Rgveda: the archaeological evidence. In: Erdosy, G. (Ed.), The Indo-Aryans of Ancient South Asia. Munshiram Manoharlal, New Delhi, pp. 206–12.

Fitzgerald, T., 1996. From structure to substance: Ambedkar, Dumont and orientalism. Contributions to Indian Sociology (n.s.) 30(2), 273–88.

Fosse, L.M., 2005. Aryan past and post-colonial present: the polemics and politics of indigenous Aryanism. In: Bryant, E.F., Patton, L.L. (Eds.), The Indo-Aryan Controversy: Evidence and Inference in Indian History. Routledge, London, pp. 434–467.

Fuller, C.J. (Ed.), 1997a. Caste Today. Oxford University Press, New Delhi.

Fuller, C.J., 1997b. Introduction: Caste today. In: C.J. Fuller (Ed.), Caste Today. Oxford University Press, New Delhi, pp. 1–31.

Fuller, D., Boivin, N., 2002. Beyond description and diffusion: A history of processual theory in the Archaeology of South Asia. In: S. Settar, Korisettar, R. (Eds.), Indian Archaeology in Retrospect, Volume IV: Archaeology and Historiography. Manohar, New Delhi, pp. 161–90.

Gupta, D., 2004. The certitudes of caste: When identity trumps hierarchy. Contributions to Indian Sociology (n.s.) 38(1/2), v–xv.

Hassan, F.A., 1995. The World Archaeological Congress in India: politicizing the past. Antiquity 69, 874–7.

Inden, R., 1990. Imagining India. Basil Blackwell, Oxford.

Jarrige, J.-F., 1981. Economy and society in the early Chalcolithic/Bronze Age of Baluchistan. In: Härtel, H. (Ed.), South Asian Archaeology 1979. Dietrich Verlag, Berlin, pp. 93–113.

Jarrige, J.-F., 1982. Excavations at Mehrgarh: Their significance for understanding the background of the Harappan Civilization. In: Possehl, G. (Ed.), Harappan Civilization. Oxford & IBH, New Delhi, pp. 79–84.

Kaviraj, S., 1992. The imaginary institution in India. In: Chatterjee, P., Pandey, G. (Eds.), Subaltern Studies, volume VII. Oxford University Press, Delhi, pp. 1–39.

Kaviraj. S., 1997. Religion and identity in India, Ethnic and Racial Studies 20(2), 325–44.

Kennedy, K.A.R., 1995. Have Aryans been identified in the prehistoric skeletal record from South Asia? Biological anthropology and concepts of ancient races. In: G. Erdosy (Ed.), The Indo-Aryans of Ancient South Asia. Munshiram Manoharlal, New Delhi, pp. 32–66.

Kennedy, K.A.R., 1999. God-Apes and Fossil Men: Palaeoanthropology in South Asia. University of Michigan Press, Ann Arbor, MI.

Kenoyer, J.M., 1989. Socio-economic structures of the Indus Civilization as reflected in specialized crafts and the question of ritual segregation. In: Kenoyer, J.M. (Ed.), Old Problems and New Perspectives in the Archaeology of South Asia. University of Wisconsin, Madison, pp. 183–192.

Kenoyer, J.M., 1995. Ideology and legitimation in the Indus State as revealed through symbolic objects. The Archaeological Review 4(1/2), 87–131.

Kenoyer, J.M., 2005. Culture change during the Late Harappan period at Harappa: New insights on Vedic Aryan issues. In: Bryant, E.F., Patton, L.L. (Eds.), The Indo-Aryan Controversy: Evidence and Inference in Indian History. London: Routledge, pp. 21–49.

Khare, R.S., 1976. The Hindu Hearth and Home. Carolina Academic Press, Durham, N.C.

Kivisild, T., Bamshad, M.J., Kaldma, K., Metspalu, M., Metspalu, E., Reidla, M., Laos, S., Parik, J., Watkins, W.S., Dixon, M.E., Papiha, S.S., Mastana, S.S., Mir, M.R., Ferak, V., Villems, R., 1999a. Deep common ancestry of Indian and western-Eurasian mitochondrial DNA lineages. Current Biology 9, 1331–34.

Kivisild, T., Kaldma, K., Metspalu, M., Parik, J., Papiha, S., Villems, R., 1999b. The place of the Indian mtDNA variants in the global network of maternal lineages and the peopling of the Old World. In: Deka, R., Papiha, S.S. (Eds.), Genomic Diversity. Kluwer Academic/Plenum Publishers, New York, pp. 135–52.

Kivisild, T., Papiha, S.S., Rootsi, S., Parik, J., Kaldma, K., Reidla, M., Laos, S., Metspalu, M., Pielberg, G., Adojaan, M., Metspalu, E., Mastana, S.S., Wang, Y., Gölge, M., Demirtas, H., Schankenberg, E., Franco de Stefano, G., Geberhiwot, T., Claustres, M., Villems, R., 2000. An Indian Ancestry: A Key for Understanding Human Diversity in Europe and Beyond. In: Renfrew, C., Boyle, K. (Eds.), Archaeogenetics: DNA and the Population History of Europe. McDonald Institute for Archaeological Research, Cambridge, pp. 267–75.

Kivisild, T., Rootsi, S., Metspalu, M., Mastana, S., Kaldma, K., Parik, J., Metspalu, E., Adojaan, M., Tolk, H.-V., Stepanov, V., Gölge, M., Usanga, E., Papiha, S.S., Cinnioğlu, C., King, R., Cavalli-Sforza, L., Underhill, P.A., Villems, R., 2003a. The genetic heritage of the earliest settlers persists both in Indian tribal and caste populations. American Journal of Human Genetics. 72, 313–332.

Kivisild, T., Rootsi, S., Metspalu, M., Metspalu, E., Parik, J., Kaldma, K., Usanga, E., Mastana, S.S., Papiha, S.S., Villems, R. 2003b. The genetics of language and farming spread in India. In: Bellwood, P., Renfrew, C. (Eds.), Examining the Farming/Language Dispersal Hypothesis. McDonald Institute Monographs, Cambridge, pp. 215–222.

Kramer, C., Douglas, J.E., 1992. Ceramics, caste, and kin: spatial relations in Rajasthan, India. Journal of Anthropological Archaeology 11, 187–201.

Kumar, V., Reddy, B.M., 2003. Status of Austro-Asiatic groups in the peopling of India: An exploratory study based on the available prehistoric, linguistic and biological evidences. Journal of Bioscience 28(4), 507–522.

Lamberg-Karlovsky, C.C. 1997. Colonialism, nationalism, ethnicity and archaeology, Part 1. The Review of Archaeology 18(2), 1–14.

Lambert, H., 1997. Caste, gender and locality in rural Rajasthan. In: C.J. Fuller (Ed.), Caste Today. Oxford University Press, New Delhi, pp. 93–123.

Leach, E., 1990. Aryan invasions over four millennia. In: E. Ohnuki-Tierney (Ed.), Culture Through Time: Anthropological Approaches. Stanford University Press, Stanford, pp. 227–45.

Maisels, C.K., 1999. Early Civilizations of the Old World. Routledge, London.

Majumder, P.P., 1998. People of India: Biological diversity and affinities. Evolutionary Anthropology 6, 100–110.

Majumder, P.P., 2001a. Ethnic populations of India as seen from an evolutionary perspective. Journal of Bioscience 26(4), 533–545.

Majumder, P.P., 2001b. Indian caste origins: Genomic insights and future outlook. Genome Research 11, 931–32.

Mandal, D. F., 1993. Ayodhya: Archaeology After Demolition. Orient Longman, New Delhi.

Marriott, M. (Ed.), 1955. Village India: Studies in the Little Community. University of Chicago Press, Chicago.

Marriott, M., 1989. Constructing an Indian ethnosociology. Contributions to Indian Sociology 23(1), 1–40.

Marriott, M., Inden, R.B., 1977. Toward an ethnosociology of South Asian caste systems. In: David, K. (Ed.), The New Wind: Changing Identities in South Asia. Mouton, Paris, pp. 227–38.

Mayer, A., 1997. Caste in an Indian village: change and continuity 1954–1992. In: Fuller, C.J. (Ed.), Caste Today. Oxford University Press, Delhi, pp. 32–64.

Metspalu, M., Kivisild, T., Metspalu, E., Parik, J., Hudjashov, G., Kaldma, K., Serk, P., Karmin, M., Behar, D.M., Gilbert, M.T.P., Endicott, P., Mastana, S., Papiha, S.S., Skorecki, K., Torroni, A., Villems, R., 2004. Most of the extant mtDNA boundaries in south and southwest Asia were likely shaped during the initial settlement of Eurasia by anatomically modern humans. BMC Genetics 5, 26.

Miller, D., 1985. Artefacts as Categories: A Study of Ceramic Variability in Central India. Cambridge University Press, Cambridge.

Moore, M.A., 1989. The Kerala House as Hindu cosmos. Contributions to Indian Sociology (n.s.) 23(1), 169–202.

Morell, V., 1993. Anthropology: nature-culture battleground. Science 261(5129), 1789–1802.

Morrison, K.D., 1994. States of theory and states of Asia: regional perspectives on states in Asia. Asian Perspectives 33(2), 183–86.

Morrison, K.D., 2007. Foragers and forger-traders in South Asian worlds: some thoughts from the last 10,000 years. In: Petraglia, M.D., Allchin, B. (Eds.), The Evolution and History of Human Populations in South Asia: Inter-disciplinary Studies in Archaeology, Biological Anthropology, Linguistics and Genetics. Springer, Netherlands, pp. 321–339.

Mughal, R., 1982. Recent Archaeological Research in the Cholistan Desert. In: Possehl, G. (Ed.), Harappan Civilization: A Contemporary Perspective. Oxford and IBH, New Delhi, pp. 85–95.

Noble, W.A., 1997. Toda huts and houses. In: P. Hockings (Ed.), Blue Mountains Revisited: Cultural Studies on the Nilgiri Hills. Oxford University Press, Delhi, pp. 192–230.

Olivelle, P. 1998. Caste and purity: a study in the language of the Dharma literature. Contributions to Indian Sociology (n.s.) 32(2), 189–216.

Piggott, S., 1950. Prehistoric India. Penguin, Harmondsworth.

Possehl, G. 2002. The Indus Civilization: A Contemporary Perspective. Walnut Creek: Altamira.

Quintana-Murci, L., Krausz, C., Zerjal, T., Sayar, S.H., Hammer, M.F., Mehdi, S.Q., Ayub, Q., Qamar, R., Mohyuddin, A., Radhakrishna, U., Jobling, M., Tyler-Smith, C., McElreavey, K., 2001. Y-chromosome lineages trace diffusion of people and languages in southwestern Asia. American Journal of Human Genetics 68, 537–42.

Raheja, G.G., 1988. The Poison in the Gift: Ritual, Prestation, and the Dominant Caste in a North Indian Village. University of Chicago Press, Chicago.

Raikes, R.L., 1968. Kalibangan: Death from Natural Causes. Antiquity, 42, 286–291. Rajkumar, R., Kashyap, V.K., 2004. Genetic structure of four socio-culturally diversified caste populations of southwest India and their affinity with related Indian and global groups. BMC Genetics 5, 23.

Ramana, G.V., Su, B., Jin, L., Singh, L., Wang, N., Underhill, P., Chakraborty, R., 2001. Y-chromosome SNP haplotypes suggest evidence of gene flow among caste, tribe, and the migrant Siddi populations of Andhra Pradesh, south India. European Journal of Human Genetics 9, 695–700.

Rao, N., 1994. Interpreting silences: symbol and history in the case of Ram Janmabhoomi/Babri Masjid. In: E.C. Bond, A. Gilliam (Eds.), Social Construction of the Past: Representation as Power. Routledge, London, pp. 154–64.

Rao, N., 1999. Ayodhya and the Ethics of Archaeology. In: T. Insoll (Ed.), Case Studies in Archaeology and World Religions. The Proceedings of the Cambridge Conference. BAR International Series 755, Archaeopress, Oxford, pp. 44–47.

Ratnagar, S. 2004. 'Archaeology at the heart of political confrontation: the case of Ayodhya', Current Anthropology 45, 239–59.

Risley, H.H., 1915. The People of India, 2nd ed. W. Thacker, London.

Roychoudhury, S., Roy, S., Dey, B., Chakraborty, M., Roy, M., Roy, B., Ramesh, A., Prabhakaran, N., Usha Rani, M.V., Vishwanathan, H., Mitra, M., Sil, S.K., Majumder, P.P., 2000. Fundamental genomic unity of ethnic India is revealed by analysis of mitochondrial DNA. Current Science 79(9), 1182–92.

Shaffer, J.G., 1984. The Indo-Aryan invasions: cultural myth and archaeological reality. In: Lukacs, J.R. (Ed.), The People of South Asia: The Biological Anthropology of India, Pakistan and Nepal. Plenum, London, pp. 77–90.

Shaffer, J.G., Lichtenstein, D.A., 1995. The concepts of "cultural tradition" and "palaeoethnicity" in South Asian archaeology. In: Erdosy, G. (Ed.), The Indo-Aryans of Ancient South Asia. Munshiram Manoharlal, New Delhi, pp. 126–54.

Shaffer, J.G., Lichtenstein, D.A., 2005. South Asian archaeology and the myth of Indo-Aryan invasions. In: Bryant, E.F., Patton, L.L. (Eds.), The Indo-Aryan Controversy: Evidence and Inference in Indian History. Routledge, London, pp. 75–104.

Singer, M., Cohn, B.S. (Eds.). 1968. Structure and Change in Indian Society. Wenner-Gren Foundation, New York.

Sinopoli, C.M., 1991. Seeking the past through the present: recent ethnoarchaeological research in South Asia. Asian Perspectives 30(2), 177–192.

Sopher, D.E. (Ed.), 1980. An Exploration of India: Geographical Perspectives on Society and Culture. Cornell University Press, Ithaca, NY.

Srinivas, M.N., 1952. Religion and Society among the Coorgs. Clarendon, Oxford.

Srinivas, M.N., 1987. Dominant Caste and Other Essays. Oxford University Press, Delhi.

Thangaraj, K., Chaubey, G., Kivisild, T., Reddy, A.G., Singh, V.K., Rasalkar, A.A., Lingh, L., 2005. Reconstructing the origin of Andaman islanders. Science 308, 996.

Thapar, R., 1999. The Aryan question revisited. Lecture delivered to the Academic Staff College, JNU, New Delhi.

Vats, M.S., 1940. Excavations at Harappa. Government of India Press, Delhi.

Viswanathan, H., Deepa, E., Cordaux, R., Stoneking, M., Usha Rani, M.V., Majumder, P.P., 2004. Genetic structure and affinities among tribal populations of southern India: a study of 24 autosomal DNA markers. Annals of Human Genetics 68, 128–138.

Wadley, S.S., 1980. The Powers of Tamil Women. Maxwell School of Citizenship and Public Affairs, Syracuse.

Watkins, W.S., Bamshad, M., Dixon, M.E., Bhaskara Rao, B., Naidu, J.M., Reddy, P.G., Prasad, B.V.R., Das, P.K., Reddy, P.C., Gai, P.B., Bhanu, A., Kusuma, Y.S., Lum, J.K., Fischer, P., Jorde, L.B., 1999. Multiple origins of the mt-DNA 9-bp deletion in populations of south India. American Journal of Physical Anthropology 109, 147–58.

Wheeler, M., 1947. Harappa 1946: The defences and cemetery R37. Ancient India 3, 58–130.

Wheeler, R.E.M., 1968. The Indus Civilisation. Cambridge University Press, Cambridge.

Witzel, M., 1995a. Early Indian history: Linguistic and textual parameters. In: Erdosy, G. (Ed.), The Indo-Aryans of Ancient South Asia: Language, Material Culture and Ethnicity. Munshiram Manoharlal, New Delhi, pp. 85–125.

Witzel, M., 1995b. Rgvedic history: poets, chieftains and polities. In: Erdosy, G. (Ed.), The Indo-Aryans of Ancient South Asia: Language, Material Culture and Ethnicity. Munshiram Manoharlal, New Delhi, pp. 307–352.

16. Language families and quantitative methods in South Asia and elsewhere

APRIL McMAHON

Linguistics and English Language
University of Edinburgh
14 Buccleuch Place
Edinburgh, EH8 9LN
Scotland
april.mcmahon@ed.ac.uk

ROBERT McMAHON

Molecular Genetics Laboratory
Western General Hospital
Crewe Road
Edinburgh
Scotland
rmcmahon@staffmail.ed.ac.uk

Introduction and Aims

This book sets out to generate a new understanding of South Asian prehistory; and as the title suggests, contributors (and readers) are interested in both evolution and diversity. Framing this in terms more familiar to linguists, the approaches adopted and tested in the chapters collected here focus on both the historical, diachronic perspective, and on synchronic distributions of features and groupings, both today and at particular points in the past.

Our own aim, in this chapter, is to contribute to these greater goals by suggesting ways in which our understanding of language families and connections between language families in South Asia could become deeper. We hope to show in particular how recent quantitative approaches to language comparisons could cast new light on South Asian language affiliations; but also how applications of such methods might use linguistic evidence to illuminate the histories of South Asian populations through correlations of linguistic and genetic evidence. We are not ourselves experts in any South Asian language, and will therefore conduct very little of this analysis ourselves, beyond a few illustrations of the scope of the methods we are proposing. Instead, we will focus on introducing and defining the different quantitative methods currently

363

available, outlining the data they require, the assumptions they make, and the results they have produced in other applications. We hope that South Asian specialists, once armed with this toolkit for dealing with linguistic data, will be motivated to explore the potential of quantitative approaches for themselves.

Why Quantitative Methods?

Two relatively recent changes of emphasis in linguistics have contributed to the current rise in popularity in quantitative approaches to language comparison and classification. First, the power and benefits of computational work have become very clear in, for instance, sociolinguistics and corpus linguistics. Without a means of identifying and counting particular linguistic features, and correlating these with non-linguistic factors like gender, social class, age, ethnicity and attitudes, the sociolinguistic paradigm initiated by William Labov (see Labov, 1994, 2001) could never have made discoveries about the meaning, use and speaker control of language; nor could we have confirmed aspects of what had been suspected from prior, qualitative research. Likewise, the development of substantial linguistic corpora allows linguists to search for examples of even low-frequency words and constructions which might rarely if ever be encountered by an individual linguist reading through texts; and again, this kind of approach has led to a much more nuanced understanding of the ways new elements enter and spread through languages.

Linguists interested in comparison and classification face many of the same problems which have been tackled successfully by quantitative means in sociolinguistics and corpus linguistics: we need replicable, robust and objective methods to develop new results and to confirm those reached by older methods based more on scholarly insight and judgement; and we face challenges of correlating linguistic data with evidence from related but more

strongly quantifying fields, notably archaeology and genetics. We might anticipate, then, that historical linguists would be keen to exploit the potential of quantitative approaches in identifying whether languages belong to the same family, and how closely related, or how structurally similar they might be. However, a further development in historical linguistics has added to the urgency of the case, and this involves language contact. There has, of course, been a long-standing awareness that speakers can be influenced by languages other than their own, and that 'borrowing' of words between languages can therefore happen. This awareness is reflected in the parallel development of two nineteenth-century models for language change, namely the family tree model (Schleicher, 1863), which depicts descent with differentiation of daughter languages from a common ancestor, and the wave model (Schmidt, 1872), which foregrounds instead the spread and transfer of innovations from one dialect or language to others.

However, recent thinking in historical linguistics has emphasized that virtually all languages are formed from an interaction of family relationships and contact events. The challenge is to identify which elements of vocabulary and structure come from which source; and in some circumstances this is harder than others. For instance, some languages are isolates (such as Basque or Burushaski), with no discernible family affiliations. Others have features which do not match those of their sister groups (Germanic is one such case, within Indo-European); but it may not be clear where those novel features have come from. Perhaps most importantly, historical linguists have had to adjust to the idea that the effects of language contact can be much more profound and far-reaching than once supposed (see Thomason, 2001 for an overview). Far from being a minor factor which might affect a few words, contact-induced change can influence every level of the grammar, to the extent that languages particularly strongly affected may become 'non-genetic' (Thomason and Kaufman, 1988), with unclear affiliations, or connections with their historical family effectively severed by contact events. There are, in addition, several categories of contact languages, which would not exist but for the effects of contact: these include pidgins (typically structurally simple systems used by speakers of a range of other languages for necessary interactions such as trade); the more complex creoles (which are structurally and functionally expanded, and have native speakers of their own); and a currently much-debated class of mixed languages, which often take their vocabulary from one language and their grammar from another.

Together, the recognition of the benefits quantitative and computational approaches have brought to other subdisciplines within linguistics, and the turn towards a recognition of the true impact of contact, have made the introduction of quantitative methods into historical linguistics at least timely, and perhaps inevitable. In particular, historical linguists can no longer assume that contact-induced change can safely be ignored; one of the most important goals is now to find methods which will distinguish the effects of descent with divergence from those of contact. Without a means of making this distinction, linguistic evidence may become effectively inadmissible in attempts to trace population histories, since the source of particular data is obviously key to establishing the meaning and significance of correlations between linguistics, archaeology and genetics.

There are few parts of the world where such concerns are more pressing than South Asia. The linguistic situation here is complex. Turning first to family membership, a number of South Asian languages are known to belong to the Indo-Iranian subgroup of Indo-European, the most extensively investigated language family in the world; and there has been fairly extensive structural work on the Dravidian languages of the south. However, there are also Sino-Tibetan languages in the north, and the less-researched Munda family, alongside at least one isolate, Burushaski; and affiliations of individual languages are not always completely clear.

One barrier to attaining clarity on family groups is precisely contact. The Indian subcontinent is a deeply multilingual area, with at least 15 languages having official status in different states; in 1951, 62 languages were listed as having more than 100,000 speakers (Thomason, 2001). Such circumstances will inevitably create sociolinguistic complexity, in that we are unlikely to be able to predict straightforwardly which languages will borrow from which, and which directions influence is likely to travel in. Loans from one South Asian language to another are also well documented. However, lexical borrowing is by no means the end of the story: when many words are borrowed, sounds and grammatical structures may also follow, and in addition South Asia is home to several potential examples of a further type of contact-induced change, namely convergence, or the formation of linguistic areas. Indeed, one of the earliest documented examples of convergence is in South Asia, with the investigation of languages in Kupwar by Gumperz and Wilson (1971). Convergence at its most extreme leaves the vocabulary of languages almost untouched, but they become more and more similar in their structures so that ultimately it is possible to have virtual word-by-word intertranslatability. Convergence seems to be motivated both socially and acquisitionally. From a social perspective, keeping vocabulary relatively distinct and intact provides an easy marker of the particular language being used; and this may have particular salience in a caste-based system, for example. Meanwhile, it may be more straightforward for children to learn several languages if they share a degree of structural parity.

Thomason (2001) discusses the South Asian Sprachbund, a convergence area involving Indic, Dravidian and Munda languages. The area is defined by various grammatical and phonological features (including retroflex consonants, agglutination in noun declensions, quotative constructions, certain particles, rigid SOV word order, and the absence of a verb 'to have'), which have probably spread across the area historically, though it is notoriously difficult in convergence situations to say exactly which features have spread from where; and the situation here is complicated further by the overlaying of a whole series of mini-convergence areas, with their own smaller sets of more geographically restricted overlapping features. Convergence itself is a relatively unclear topic in modern historical linguistics, and it is by no means easy to say where a linguistic area begins and 'normal' borrowing ends (see Campbell, 2006; Matras et al., 2006 for discussion); but the existence of such situations clearly adds to the complexity of the South Asian linguistic situation, and to the need for methods to disentangle common ancestral features from features spread through contact. We turn in the next sections to an outline of a selection of the currently available quantitative methods, and initially to the definition of the steps involved in applying them.

Stages in Applying Quantitative Methods

The goals of applying quantitative methods to linguistic data are to reveal similarities between languages, and ideally to help us ascertain the source of these similarities. In order to achieve these goals, we do not require a single step, but three.

First, it is necessary to **code** the data. Quantitative methods, by their nature (and indeed by definition), apply to quantities: so we must convert our linguistic data into numbers. As the second step, we **process**

these coded data, producing pictorial representations (such as family trees), and may also conduct further numerical or statistical tests. Finally, there is a stage of **interpretation**, which involves making hypotheses about the origin and meaning of the particular features and connections we have isolated and tested through our processing phase.

Of these three stages, the first and third crucially involve linguists, and dedicated linguistic methods. Coding, in our view, cannot be conducted in an automatic way, since it is essential to be sensitive to the structures and complexities characteristic of language; so any coding process will involve either a judgement-based system, or a direct translation into numerical terms, based on a programme constructed specifically for linguistic data. Likewise, the stage of interpretation requires both a knowledge of likely linguistic patterns, and expertise in the particular area and languages at issue, though clearly there will be situations where the latter is not available because we simply do not know much about the languages under comparison. For example, increasingly linguists find themselves collecting data from endangered languages (Nettle and Romaine, 2000), whose speakers may be few, elderly, and rather uncertain about a language they may not have spoken outside a few formulaic utterances for many years. In such situations, where there is no recorded history for a language and little reliable data available, interpretations are inevitably more difficult to reach and to support, and here we are likely to have to rely on conclusions reached in comparable cases where the same patterns have emerged from processing but the history is known. In addition, such patterns are increasingly checked against simulated data: if a particular signal emerges consistently from simulations of a certain 'known' history, we can more securely argue for that interpretation in unclear situations involving real but poorly attested languages (always bearing in mind the problem that in historical linguistics, the unclear and undocumented histories outnumber the clear and documented ones). The three stages of coding, processing and interpretation are therefore to some extent interdependent, though in the next two sections we discuss coding and processing separately for ease of exposition. Finally, we bring in questions of interpretation through an initial application of network methods to Indo-Iranian data in particular.

Coding the Data

The first step in applying any quantitative method, then, involves coding the data, or transforming linguistic evidence into numbers, thus making it eligible for further statistical or computational processing. There are three main issues to be confronted here: these involve deciding on the area of the grammar from which data are to be considered (phonetics/phonology, morphosyntax, vocabulary and lexical semantics, or more than one of these); whether data are then to be treated as characters or distances; and of course, how the data can be collected in the first place. In the next three subsections we shall consider these issues separately, although in fact they are interdependent and, as we shall see, decisions in one domain will inevitably have repercussions in the others.

Which Area of the Grammar?

The non-quantitative, traditional Comparative Method in historical linguistics, by which recognized language families and subfamilies have typically been established, relies on a mixture of evidence, most notably involving recurrent matches of sound and meaning throughout the vocabulary (see Harrison, 2003; Rankin, 2003). Where these regular correspondences can be found, we establish both the existence of cognates between sister languages, and a plausible reconstruction of the likely common ancestral form; where the resemblances suggest regular changes in the different sisters, rather than manifesting themselves as complete identity, we also safeguard against the effects of borrowing. Moreover, the findings of the Comparative Method are strengthened considerably when they include not only sound and meaning, but also regular correspondences in the morphology, including case or number systems, for instance. The Comparative Method itself cannot be quantified directly, since it relies on judgements of plausibility, at least in assessing the number and frequency of correspondences we require to rule out chance as their source; but we shall see that some quantitative methods get closer to its *modus operandi* than others.

Most quantitative methods to date have focused on the vocabulary and therefore on matches of sound and meaning; moreover, there has been a tendency to concentrate on the basic vocabulary, the supposedly universal and borrowing-resistant part of the lexicon featuring words for common human experience, body parts, simple kinship terms, small numerals, and so on. These basic vocabulary items are collected in standard 100- or 200-item lists, so-called Swadesh lists, after Morris Swadesh (1952, 1955), who proposed their use in lexicostatistics. Note that our discussion here relates only to lexicostatistics *per se*, and not to its extension into dating, glottochronology, which we reject. Lexicostatistics shares some common ground with the Comparative Method, and indeed to a considerable extent presupposes it, since it requires an analysis of whether the items filling the same 'slot' on the Swadesh list in two languages (say, the common, everyday words for 'sky' or 'hand') are cognate or not. Where detailed comparative work has not taken place between two languages prior to this lexicostatistical work, assessment of likely cognacy can be based only on superficial pattern-matching, which is susceptible to the many problems identified for

Greenberg's controversial method of mass comparison (Greenberg, 1987; Ruhlen, 1991), which would be rejected by most historical linguists.

In lexicostatistics, then, a score of 1 will be assigned where the items filling a given slot in the two languages being compared are cognate, and a 0 when they are not. The scores are summed for the full 100 or 200 items, to give a figure for the overall similarity between these languages. Different pairs of languages can then be compared to indicate relative closeness or distance. Many difficulties have been identified with this method (see, for instance, the papers in Renfrew et al., 2000): for example, how universal are the supposedly universal meanings? What do we do when we have no single translation equivalent for the slot meaning, or where we have two or more, apparently synonymous equivalents? Heggarty (n.d.; and see also McMahon and McMahon, 2005: Chapter 6) proposes extensions to the method to recognize partial matches across slots, and to code for different degrees of cognate plausibility in cases where the comparative method has not been carried out on the languages under comparison, or cannot be carried out fully for reasons of availability of data. Slaska (2006) further proposes that traditional lexicostatistics must urgently be revised to ensure that the basic vocabulary items for each language are collected from a range of speakers, not just from a single individual, or from the linguist's own knowledge of a language, or from a dictionary; she establishes that including multiple speakers leads to inclusion of a range of responses for each slot in the list. Word-list comparisons in general are discussed further in Embleton (1986) and Kessler (2001).

However, regardless of the refinements made to traditional lexicostatistics, the method retains the difficulty of requiring a two-step coding of data: first, we assess whether items are cognate or not, and then we assign a number. The numbers achieved in a lexicosta-

tistical comparison do not represent a direct numerical transformation of the data, but a secondary process mediated by judgements, which may not be wholly objective despite our best intentions, and which are themselves not quantitative. Moreover, lexicostatistics focuses on only a single domain of the grammar, and is predicated on the assumption that basic vocabulary is resistant to borrowing; yet it is clear that, as Embleton (1986), Kessler (2001) and McMahon and McMahon (2005) have recently shown in quantitative terms, this resistance is only relative, and a good deal of borrowing can go on in a basic meaning list, even if this is not so concentrated as in the lexicon at large. McMahon and McMahon (2003, 2005; see also McMahon et al., 2005) suggest that the effects of borrowing can be used to our advantage in diagnosing likely contact by contrasting more and less conservative sublists from a Swadesh-list, and we return to this possibility below. But even if this is the case, concentrating on the vocabulary and lexical semantics will remain seriously problematic for an area like South Asia, with its likely cases of convergence: if contact is affecting not only the vocabulary but also structural domains of the grammar (or indeed not the vocabulary at all), it is vital that we should be able to conduct quantitative explorations of these.

To some extent, work on non-lexical quantitative comparisons is already under way. For instance, Ringe et al. (2002) and their co-workers (see for example Nakhleh et al., 2005) have pioneered an approach using a set of characters for Indo-European drawn from the lexicon, morphology and phonology; as we shall see below, the drawback with this method resides in the fact that it is character-based, and therefore strictly applicable only to Indo-European, at least using the set of characters developed thus far. However, in its use of a mix of characters from different levels of the grammar this approach arguably comes closest to a quantification of the Comparative

Method. A more detailed illustration of this method is given in the next section.

When we turn to quantitative methods for other, individual areas of the grammar over different language families, the picture is currently less clear. The situation for morpho-syntactic comparison is at a parti- cularly early stage: our lack of detailed knowledge to date of the likely directionality of syntactic change, and consequent difficulties in reconstructing syntax, are contributory reasons for this, though there are some signs of initial design of methods in Heggarty (n.d.) and ongoing typological work by Longobardi and his co-workers. On the other hand, there is already a range of possible methods for comparing phonetics, and these are reviewed in Kessler (2005); but for the most part these have been designed for purposes outside language comparison and classification, for instance being intended to measure how far an L2 pronunciation or a child's pronunciation depart from an L1 or adult model. Heggarty (n.d.; see also Heggarty et al. 2005) expresses considerable concern about many of these phonetic similarity metrics on the basis that they are driven by considera-tions of computational simplicity rather than by a true awareness of the complexities of linguistic structure and variation. Heggarty's own method for phonetic comparison, itself under development, involves a computational instantiation of articulatory distinctive features, crucially weighted to recognize universal struc-tural biases of sound systems, and mediated by a node form derived from the common ancestral form through which the reflexes of daughter languages or dialects are compared directly, to allow for slot or segment matching.

Clearly, we do not anticipate that the most enlightening future applications of quantitative methods in South Asia will involve just a single one of these data-types: on the contrary, we anticipate that including more data-types should give more inte- resting and robust results. However, there are inevitable trade-offs to be taken into account, so that the availability of a straightforward method for comparison in the basic vocabulary can to an extent countermand the limitations inherent in using only a single data-type; while the complex-ities of phonetic comparison programming may make this unsuitable at present as an initial step in analysis. Furthermore, different types of comparison can reveal different histories because the similarities on which they work are motivated by different factors. So, lexico-statistical work depends on prior assessment of whether items are cognate; and because cognacy is an essentially historical notion, based on whether two items and therefore two systems have a single common ancestor or not, the resulting output will be phylo-genetic and will tell us about branching order, subgrouping and history. In contrast, comparisons based on phonetic similarity are measuring similarity, regardless of its origin, and we will have to develop further post-processing and interpretation techniques to assess whether those similarities have arisen from common origin and subsequent change, or from contact, or from parallel but independent innovation. Close similarity in phonetics but clear difference in lexicostatistics would not, then, constitute an inexplicably contradictory result: the former provides a phenogram, and the latter a phylogram, and a mismatch is intriguing and suggestive that more than one history has been involved. Any prior knowledge of languages and their embedding in complex systems of social, political and personal interactions is likely to ensure that this does not surprise us at all.

Character and Distance Data

A second decision to be made prior to any quantitative comparison is whether we are dealing with character or distance data. This is relevant at the level of coding, because it will determine the type and range of numbers we assign; but it is even more relevant at the subsequent stage of processing, since certain programs, as we shall see, are designed to handle character data, while others work on distance data.

The distinction between character and distance data interacts with the question of whether those data are categorial or scalar from the perspective of measurement. We can either be dealing with straightforward presence versus absence of an element or feature – a categorial question – or with more or less of something gradable. However, the distinction is not always so clear, since for example we can reinterpret a scalar distinction as categorial: this type of idealisation is quite common in linguistic analysis. Thus, the dimensions of tongue height and backness in the vowel space within the vocal tract (these dimensions themselves idealized conceptual-isations of many interacting articulatory and acoustic movements and signals) contribute in a gradient way to the realisation of vowels, which are on a continuum in terms of higher and lower, fronter and backer realisations. However, we typically divide the continuum at impressionistically determined points, assigning one realisation to the /ɪ/ vowel in *bit*, and another to the /ɛ/ in *bet*, though the range of realisations may in fact overlap for different speakers or different environments of utterance. We could take a gradient approach and measure the formant frequencies involved; or we could make categorial decisions and assign each realisation to a particular, abstract phonological vowel. Whether we take a scalar or a categorial perspective might depend, for example, on how much weight we assign to claims that speakers and hearers work essen-tially on a system of categorial perception, so that these decisions are often theoretically colored.

Here, we shall consider only one example of a character-based, and one of a distance-based approach, and these are drawn from the coding methods already introduced in the previous section. The computational cladistics approach of Don Ringe and his co-workers (Ringe et al., 2002; Nakhleh

et al., 2005) involves the prior identifi-cation and scoring of a set of lexical, morphological and phonological characters. These characters are selected to bear in particular on the first-order subgrouping of Indo-European languages (including the Indo-Iranian subgroup, for which Vedic Sanskrit, Avestan and Old Persian are included as representatives). Ringe et al. (2002) use a set of 333 lexical, 22 phonological, and 15 morphological characters; and they argue strongly that the morphological and phono-logical characters are both more salient and more reliable in establishing subgroups than the more numerous lexical cases. Each lexical character is handled essentially as it would be in traditional lexicostatistics, being assigned a character state of 1 or 0 depending on whether there is an etymological match across the cases compared or not. As for the non-lexical data, some examples are given in Table 1.

The use of characters of this kind is strongly advantageous because it is based on more than a century of intensive investigation of the subgroup structure of Indo-European, so that characters supporting particular subgroups or identifying them uniquely are already known. This does not mean that the work of Ringe et al. is condemned only to repeat conclu-sions which have been established through prior application of the Comparative Method: this leaves a number of inclarities which the computational cladistics approach can hope to resolve. Moreover, applying quantitative methods in this way allows the simultaneous consideration of more than 350 essentially independent characters to assess their contri-bution to the overall pattern of relationships, and this alone is well worthwhile.

Nonetheless, the character-based approach brings limitations as well as advantages. In order to achieve a close to perfect phylogeny, where virtually all characters are compatible with a single outcome in terms of subfamily relationships, Ringe et al. have rejected a considerable number of characters, for a range

Table 1. Examples of non-lexical characters (from Ringe et al., 2002)

---

a. Phonological characters

P5: medial $*k^w >* g^w$ unless $*s$ follows immediately

(Vedic Sanskrit, Avestan and Old Persian all share state 1, meaning they do not show this change and maintain the ancestral state).

P14: merger of voiceless aspirated stops and pre-consonantal voiceless stops as fricatives

(Vedic Sanskrit maintains the ancestral state here, coded 1; but Avestan and Old Persian have the derived state, showing the effects of this change. They are coded 2. This supports a subgroup of Avestan and Old Persian as against Vedic Sanskrit, within Indo-Iranian).

b. Morphological characters

M8: most archaic superlative suffix

(Vedic Sanskrit, Avestan and Old Persian all share state 1, indicating that the most archaic superlative suffix is $*$-*isto*)

M5: mediopassive primary marker

(Vedic Sanskrit, Avestan and Old Persian all share state 2, the non-ancestral $*-y$ marker).

---

of reasons. Furthermore, comparability across language families is impossible in any direct sense: the methodology is potentially generalizable (though note that it depends on rigorous and time-consuming construction of a set of relevant characters), but the set of characters selected will be completely family-dependent. This means that this approach is limited to questions of subgrouping within already established families. It cannot be used to assess whether languages are related or not where affiliations are unclear, or to distinguish the effects of descent with divergence from those of contact. It might be possible, then, to apply the computational cladistics method as it stands to a larger group of Indo-Iranian languages; but a wholly different set of characters would need to be developed for Dravidian, or for Munda; and comparison across these families to cast light on issues of convergence, for instance, must seek another method altogether.

Lexicostatistical methods, on the other hand, are typically distance-based. They start from the same point, of assigning values to particular states, but the crucial difference is that those values are then summed to create an overall score for each language or variety. In traditional lexicostatistics, the values available are typically only 0 (for non-

cognate) and 1 (for cognate); working over a 100-meaning Swadesh list will then produce a percentage similarity, or distance score for each comparison. In principle, however, there is no reason why different values cannot be assigned to reflect, for instance, absence of data, or likely borrowing, and this is precisely the approach adopted in the Dyen et al. (1992) database. This lexicostatistical database provides a 200-meaning list for each of 95 Indo-European languages and varieties, and is structured as shown in Table 2, which lists two examples for English and eight South Asian languages. This database is rapidly becoming the most widely used electronic resource for quantitative work; there are some inaccuracies in coding, but McMahon and McMahon (2005) demonstrate that quantitative methods can identify those miscodings as inconsistent.

When we move beyond individual pairwise comparisons of languages to simultaneous comparisons of a wide range of systems, simply assigning 1 for cognate and 0 for non-cognate is no longer sufficient. Take, for example, English: some items will be cognate between English and German; other English words find matches in North Germanic languages such as Swedish, or in Romance languages like French, and some of these may not be cognates, since the explanation for

Table 2. Two examples from the Dyen et al. (1992) database

Language	'Sky'	Cognate Group	'Child'	Cognate Group
English	sky	202	child	32
Lahnda	esman	005	becca	005
Wakhi	osmon	005	kuduk	202
Afgan	asman	005	tifl	34
Persian list	aseman	005	bachche	200
Nepali	akas	006	balakha	006
Singhalese	ahasa	006	lamaya	33
Gujarati	akas	006	chokro	100
Hindi	akas	006	becca	005

these similarities may well involve contact and borrowing. In some cases we might really not quite know what is going on, and might want to mark a unique state, or hedge our bets by using a code which will remove an item for further processing, at least for the time being. This is exactly what is allowed by the system of coding adopted by Dyen et al. For instance, in the entries for 'sky' in Table 2 above, there is a reasonably straightforward picture among the South Asian languages, with two different cognate classes, one shared by Lahnda, Wakhi, Afghan and Persian, and the other by Nepali, Singhalese, Gujarati and Hindi (which, however, also has an alternative *asman* form, and might alternatively or additionally be coded 005). When we turn to the 'child' entries, the picture is much more complex: only Lahnda and Hindi are coded as having items from the same cognate class. Elsewhere, items coded in the 30s (as for English, Afghan and Singhalese) are unique states, which signal either reflexes apparently unrelated to anything else in the database, or known borrowings. In a further stage of coding, each pairwise comparison giving a match between cognate classes for a particular meaning will be assigned a 1, and every discrepant pair will be assigned a 0. Composite similarity scores for each language pair can then be generated and combined into an overall distance matrix, subsequently being processed in the ways described below.

Collecting Data

In the future, we might hope that scholars wishing to apply particular quantitative methods to linguistic data could start by calling up an off-the-peg electronic database covering the languages, or geographical area, or part(s) of the grammar in which they are interested. These readily available data could then be coded in the desired way, and subsequently passed through whichever processing programme might be appropriate.

Unfortunately, we are not quite at this stage yet, and often the only solution to getting the right data for the right languages is to go out and collect it yourself, as Heggarty (n.d.) did for lexical and phonetic comparison work on Andean languages. Even then, there are obvious limitations to what any field-worker can achieve; but physically collecting data in this way goes some way towards ensuring that, for phonetic work, a consistent level of detail is achieved in transcriptions; and that a consistent and appropriate list of meanings is translated for lexical comparisons. Furthermore, at this relatively early stage of quantitative comparisons, the stages of data collection, coding and processing are not wholly independent: it was only in attempting to collect the words instantiating certain meanings, for instance, that Heggarty (n.d.) encountered a range of difficulties with Andean languages, for instance regularly finding more than one eligible item for each slot in the list. In response to such challenges to comparison,

Heggarty proposed emendations to the meaning list, developing the CALMA list (intended to be 'Culturally And Linguistically Meaningful for the Andes', and named by analogy with Matisoff's CALMSEA list [1978, 2000], in turn Culturally and Linguistically Meaningful for South East Asia). Heggarty also recognizes that such many-to-one and one-to-many matches between forms and meanings indicates that a more nuanced set of matching metrics is required: the advent of modern computing power means we can allow ourselves, where it is motivated, to progress beyond scores of 1 or 0 for each point of comparison.

Individual data collections of this kind can give us invaluable insights into particular languages and varieties. They are also typically carried out by individual scholars who know the languages concerned thoroughly, so that they have the additional benefits of consistency and reliability: more extensive and wide-ranging databases balance their reach against the fact that they are typically composed from a number of different and potentially discrepant sources. On the other hand, however, we face difficulties of comparability: the more closely tailored a particular set of features is to a given language group or geographical area, the more difficult it is to match it with ostensibly parallel datasets collected elsewhere. The subtly revised meaning lists developed for South East Asia, for instance, or for the Andean languages, or for Australia (O'Grady, 1960; Alpher and Nash, 1999), may provide superb resolution for those specific languages and dialects; but they are all discrepant to some extent with the standard Swadesh lists, and those discrepancies would multiply if we wanted to compare one non-standard list with another. Scholars collecting revised lists must therefore weigh up the advantages and disadvantages of standard versus revised lists, and might wish where possible to gather material for both.

One target for the future of quantitative methods in this area is the provision of databases covering an increasing range of languages both for meaning-list comparisons and for the development of methods in other areas of the grammar. Ringe, Warnow and their co-workers have led the way in this respect, with their character set being available on-line (at http://www.cs.rice.edu/~nakhleh/CPHL). We also, however, urgently need a central repository for 100- or 200-meaning lists for as many languages as possible; as we have seen, we already have such a resource for Indo-European, in the Dyen et al. (1992) database (http://www.ntu.edu.au/education/langs/ielex/HEADPAGE.html).

An exciting but preliminary parallel database, again based on cognacy judgements over 200-meaning lists, is under development for Austronesian languages (http://language.psy.auckland.ac.nz/index.php). If such lists were also transcribed both phonemically and in considerably narrower phonetic terms, this would provide an invaluable resource for the development and testing of phonetic comparison methods. Interestingly, it is in the domain of morphosyntactic comparison, where quantitative methods are least developed, that a particularly promising and extensive database is under construction: this is the World Atlas of Linguistic Structures, based at the Max Planck Institute for Evolutionary Anthropology in Leipzig, but including contributions from more than 40 typologists and descriptive linguists worldwise (see http://emeld.org/workshop/2004/bibiko/bibiko-original.html, and Haspelmath et al., 2005). Although this database will in due course include specifications of a whole range of structures for more than 2,500 languages, including a selection from South Asia, it is inevitably a long-term undertaking and is currently still heavily underpopulated, with more than 90% of cells currently unfilled. This will in the future be an invaluable resource for comparison and historical investigation, though again it must be remembered that the database shows which features match, without prejudice to the source and meaning of these features in particular cases. It follows that here as elsewhere, we must pay particular attention to coding, processing and interpreting these raw data in order to understand them and to construct and distinguish between phylograms and phenograms based on them.

Processing the Data

Tree-Drawing and Tree-Selection Programs

Processing the coded data involves the use of computer programs to display tree or network representations, as well as statistical testing of results, though we shall not be pursuing the latter further here. Clearly, this stage is not independent of the coding stage, since as we shall see certain programs are limited to character data while others are designed for distances (and likewise, some statistical tests require continuous, scalar data while others are incompatible with results of this kind). What all these processing tools and techniques have in common is their objectivity and repeatability, offering the opportunity to run and rerun programs and thereby test their outputs for significance, as well as revealing any apparent inconsistencies in pattern which may be highly relevant at the next stage of interpretation.

Though there are exceptions (one being the perfect phylogeny software designed and used by Ringe, Warnow and co-workers), most of the tree-drawing programs used for linguistic comparison and classification are derived from biology, with the most commonly encountered being the PHYLIP suite of programs (Felsenstein, 2001). The most salient advantages of all tree-drawing programs is that they generate hundreds of thousands of trees and select, if not the best, at least one of those compatible with all or most of the data. Not all theoretically possible trees can be considered individually, since the number of available tree representations mounts up rapidly: five languages would have 15 possible unrooted trees and 105 possible rooted ones; but just 10 languages would have 2,027,025 unrooted and 34,459,425 rooted possible trees (Page and Holmes, 1998:18). Programs therefore operate with heuristics to remove large numbers of implausible candidates early in the process, and we cannot in principle rule out the possibility that the true tree has been rejected at this point. This can, however, be checked by bootstrapping or other resampling techniques, to check how often a particular tree or range of similar trees emerges from repeated runs.

We shall not spend much time on tree-drawing programs here, since in our view at least they have been virtually supplanted by programs allowing for networks. However, it is worth pointing out that the clustering techniques involved in these programs are in fact quite varied: some operate by Neighbour Joining; others by Maximum Parsimony, or UPGMA (Unweighted Pair Grouping Method of Agglomeration). In a test of these various clustering techniques (and others) against their Indo-European character-set, Nakhleh et al. (2005) found that all operated reasonably well, with the exception of UPGMA, which was clearly inferior in all cases. However, they do note that UPGMA incorporates an assumption that all the systems under comparison are at the same time-depth, and this is clearly violated for the group of languages Nakhleh et al. selected, which include Hittite, Latin and Old English, for instance, with clearly different periods of attestation. It should also be noted that some tree-drawing programs generate meaningful branch-lengths, such that a longer branch indicates more change since the common ancestor: within the PHYLIP package, the Fitch program (based on a maximum likelihood clustering process) produces representations with meaningful branch-lengths, while its 'sister' program Kitch does not.

Tree-drawing programs, then, have clear advantages over the single linguistic scholar making potentially subjective decisions on the best tree. They have been shown to produce (with the possible exception of UPGMA, as discussed above) representations which do include the expected first-order subgroups of Indo-European, for example in Nakhleh et al. (2005). McMahon and McMahon (2003, 2005) also show that such programs can provide both rooted and unrooted trees which closely resemble the subgrouping structure conventionally proposed for Indo-European, when applied to the Dyen et al. database. However, these applications involve lexical data already coded for cognacy, or in the case of Nakhleh et al., a mixture of lexical data and morphological and phonological characters explicitly selected for their salience in establishing first-order branching for an intensively-studied family; in both cases, the lexical data are also from the basic vocabulary. These factors should minimize the disruption caused to the programs by the effects of contact-induced change, but they cannot be expected to bypass it altogether: Embleton (1986) and Kessler (2001) have shown that even in the most basic vocabulary borrowing is possible and attested within Indo-European.

McMahon et al. (2005) summarize a process for diagnosing and using such lexical borrowing: this involves subdividing the conventional 200-meaning Swadesh list into two sublists, one more and the other less susceptible to borrowing. Although the criteria for inclusion in these sublists cannot be discussed fully here, they are based on work by Lohr (1999), and incorporate measures of relative reconstructibility and relative retentiveness. The items least susceptible to borrowing tend to be reconstructible for more protolanguages (and therefore have higher universality), and are also more retentive, showing fewer substitutions of form for the same meaning over a particular time period. Though such substitutions can take place for a number of reasons, borrowing is clearly one of the major reasons why a lexical item is displaced from its earlier semantic slot. McMahon and McMahon (2003) showed that trees drawn by the same program for the same languages, using the more conservative and more changeable lexical data, may differ in the positions of certain languages: notably, English tends to shift towards North Germanic in runs based on the more changeable data due to borrowing from Old Norse, while Frisian likewise gravitates towards Dutch, and Romanian towards the margins of Romance (because of loans from Slavic). Such applications are promising, but inevitably compromised by the fact that only by rerunning tree-drawing programs for different data can we hope to isolate the signs of borrowing. This means that extending such programs to other language groups in order to establish possible relatedness or subgrouping, and to guard against confusing the effects of contact with signals of common origin, can never be more than a hit-or-miss affair. It may scarcely be fair to blame tree-drawing programs for the fact that they are designed only to draw trees; but it is nonetheless true that there would be considerable advantages in working with programs which can recognize the effects of contact without requiring multiple runs on partial data-sets.

Network Programs

Network programs and approaches are now quite commonplace for biologists, who have for some considerable time been developing methods which accept that different features may have different histories. Linguists, however, find it almost too good to be true when they are told that programs are available which will depict relationships between languages (or any systems) as tree-like where they involve only descent with differentiation from a common ancestor, but as diagrammatically more complex when the histories are multiple or unclear. This is precisely what network programs do, constructing reticulations or additional connections within trees to reflect parallel innovation or horizontal transfer. Since these are precisely the issues historical and comparative linguists also have to confront, it is no wonder they have been particularly struck by the potential of network approaches, and that such models figure prominently in the papers in Forster and Renfrew (2006) and McMahon (2005).

Network programs in general, and their application to linguistic data in particular, are outlined in McMahon and McMahon (2005: Chapter 6). In brief, the original Network program itself (Bandelt et al., 1995; Bandelt et al., 1999; Forster et al., 2001, http://www.fluxus-engineering.com) operates by split decomposition (Bandelt and Dress 1992), a technique for grouping data according to shared characters. First, each system being compared (whether this is a molecule of genetic material, or a lexical list) is assigned a certain state at each location – say, a 1 for cognate at slot position 57, a 0 at position 35, and so on down the 200-item basic vocabulary list. Splits will be generated at points where one language or subgroup has a certain state, while all others have another. Network draws branch lengths depending on the number of state changes between nodes, and then inserts reticulations in cases where splits are incompatible with a single tree. Such incompatible splits suggest that one node, or language has similarities with more than one other node; these cannot be displayed simultaneously on a single straightforward tree, but can be shown if we use the options of reticulations to collapse the different incompatible trees into a single network graph.

Network has two additional features of note. First, the volume of reticulations generated can be controlled by setting an internal parameter, epsilon: this can be extremely useful in cases where contact between languages has been particularly extensive and therefore creates so many reticulations that the overall pattern can be difficult to discern. Resetting epsilon to a slightly less sensitive level can allow us to zoom out so we can see the wood for the trees. Secondly, Network does not only produce reticulations where they are appropriate, but also provides a list of the specific characters which are behaving in a non-tree-like way. In the later stage of interpretation, we can therefore examine those specific characters to assess the possible significance of the reticulations we see. When we are dealing with languages whose histories have been extensively researched, we can check the etymologies of these discrepant characters; in cases where histories are less clear, we can nonetheless inspect the non-tree-like characters to see how they are out of line with majority patterns.

However, Network is not the only relevant program to be used by biologists and linguists. In biological terms, Network is intrinsically limited in its use of character data: it can produce a detailed and informative diagram for a comparison of individual molecules, but is not generalizable to the population level, since it cannot provide an easily-read composite picture of all the relevant molecules. If we return to the coding of the Dyen et al., 1992 database illustrated in Table 2 above, we can apply Network to the individual state codings for a particular meaning (like the 005 and 006 for the 'sky' meaning). It can also be used for a relatively small set of characters – up to 500, though fewer if any are multistate characters, as virtually all the lexical meanings are. But Network cannot be used for data which have been coded into percentage cognacy scores or composite similarity matrices, or for data which are intrinsically scalar.

A number of alternative network programs for distance data have therefore now been developed: these include Splitstree (Huson, 1998) and NeighbourNet (Bryant and Moulton, 2004), though we shall focus only on the latter, since here the resolution seems particularly well-suited to linguistic data and we are able to produce reticulations in cases where they seem appropriate and explicable, without being swamped by them. NeighbourNet is based on a neighbour-joining type of algorithm, and its suitability for distance data means it can be used for a wide range of linguistic evidence – not just standard Swadesh lists, but also modified lists allowing one-to-many or many-to-one matches within slots, or weighted comparisons of phonetic data, as proposed by Heggarty (n.d.; and see McMahon and McMahon, 2005: Chapter 8). On the other hand, we must remember that NeighbourNet is, as a result of these character-istics, a phenetic method: it produces diagram-matic representations on the basis of similar-ities, but there is no undertaking as to the meaning or origin of those similarities. The order of branching in the tree-like portion of a NeighbourNet cannot be expected necessarily to mirror the order of historical branching for a language family; similarities may well arise from not only borrowing but also independent, parallel changes. This is particularly likely in the case of phonetic comparisons, since certain sound changes in certain contexts are quite strongly recurrent regardless of language family, while globally rare changes may be quite frequent in individual families. The later stage of interpretation is therefore particularly important for NeighbourNets, which cannot be 'read' directly as historical representations: and nor can we depend here on the generation of a set of problem characters as we can for Network, since NeighbourNet is operating on a distance matrix, and cannot disentangle this bigger, composite picture to find the specific features contributing to reticulations. All these investigations, especially at this early stage in their development for linguistic data, involve compromises: so for NeighbourNet we must weigh the need for further work at the inter-pretation stage against the helpful composite diagrams we can derive for overall distance comparisons.

Some South Asian Illustrations

In the previous sections we have provided an outline of the coding and processing stages of the application of quantitative methods, though in the processing section we focused on tree- and network-drawing programs rather than dealing with the statistical options we might include subsequently in, for example, assigning confidence intervals to particular branches in trees. In this final section, we shall provide some brief illustrations of these approaches for a small subset of South Asian languages. This subset necessarily includes only the languages included in the Dyen et al. (1992) database, and is therefore restricted both to Indo-European, and to lexical data. The idea is to provide a flavor of what quanti-tative methods might contribute to linguistic work in South Asia, and to subsequent attempts to correlate such linguistic data with genetic and archaeological evidence, or to explain linguistic patterns with reference to those other fields. But the essential first steps of data selection, collection and coding for a wide range of other South Asian languages are necessary precursors to further and more extensive work of this kind.

As discussed in earlier sections, there seems great potential in applying NeighbourNet to language data, so long as this is distance-based. The resolution of the programme seems ideal, and of course as a network programme, NeighbourNet is crucially not limited to drawing trees, but can introduce reticulations where there are incompatibilities with a simple tree configuration. In turn, these can then be investigated further.

Figure 1 shows a NeighbourNet for the South Asian languages in the Dyen et al. lexical database, for the 30 most conservative items in the 200-meaning list: remember that these are the meanings which seem highest in universality and retentiveness, and which therefore seem most resistant to borrowing.

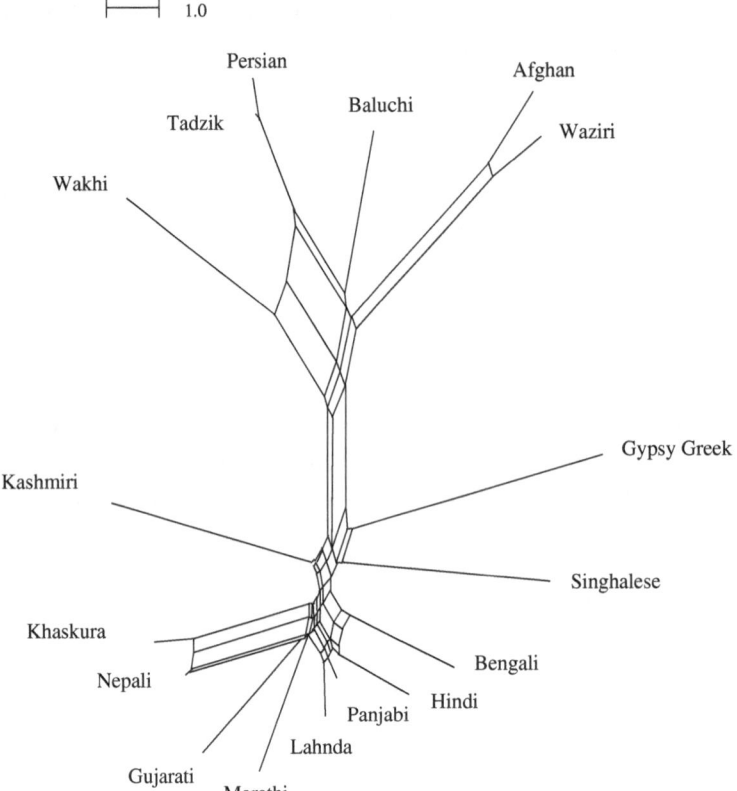

Figure 1. NeighbourNet of 30 highly conserved meanings from the Dyen et al. (1992) database for their South Asian languages

For comparison, Figure 2 shows a parallel NeighbourNet for the same languages and the same database, but this time picking out 24 of the least conservative meanings: these are less reconstructible in universal terms; tend to have a quicker turnover of forms historically to instantiate each meaning; and are typically more prone to borrowing.

In fact, even this very preliminary exploration produces one issue worthy of further investigation. In previous investigations of Indo-European and Andean languages, McMahon and McMahon (2005) have shown that networks drawn on the basis of the most conservative data are typically strongly more tree-like than those drawn from less conservative meanings. Typically, there will be fewer reticulations in the diagrams based on the more conservative data, and branch lengths here will also be shorter, since less change will have

taken place in the most conservative data, which by their nature are more likely to maintain the ancestral state or something close to it.

However, for these South Asian languages we see a strikingly different pattern. If anything, Figure 2, from the least conservative data, is more tree-like than Figure 1; and moreover, the branch lengths in both cases are very similar in length. Figure 2 does show reticulations, for instance between Punjabi, Hindi, Bengali and Marathi; but equally, reticulations are plentiful and highly visible in Figure 1.

There are several possible reasons for this intriguing pattern: as we shall show, we can go some way towards investigating this further using the quantitative methods at our disposal, but a full understanding of the situation would necessarily rely on the expertise of linguists familiar with the South Asian languages and

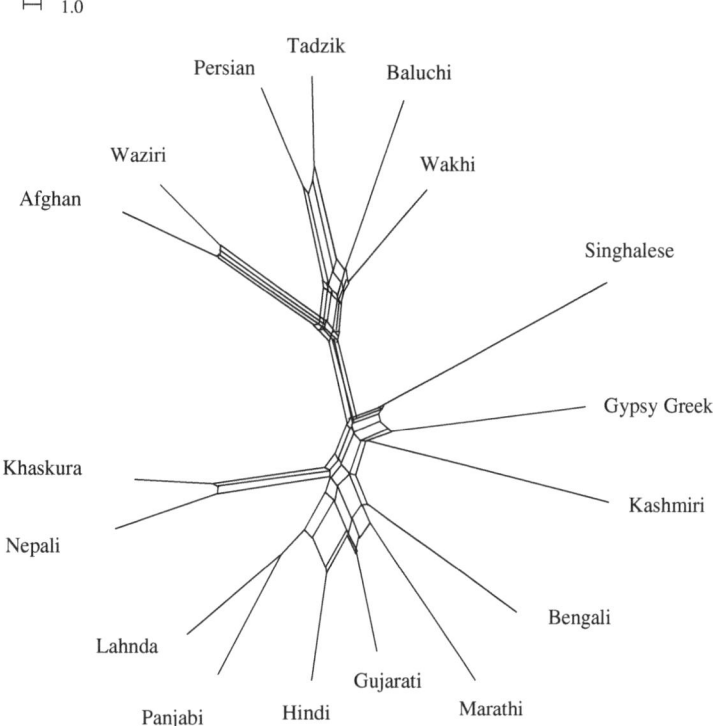

Figure 2. NeighbourNet for 24 least conserved meanings from the Dyen et al. (1992) database for their South Asian languages

contexts. We might suggest, however, as a possibility for future investigation, that the longer than usual branches in Figure 1 may reflect the speed of the initial split of some of these languages. If the period of early diversification was particularly intense and swift, then we might well see greater divergence over a shorter than usual period for the more conservative database. Furthermore, the unusually high volume of reticulations in Figure 1 may indicate a higher than usual volume of shared retentions from the common ancestor, or indeed a period of selective, differential 'sampling' from the ancestral system by the different daughters in a dialect continuum situation. Indeed, we know that in the case of Hindi, for example, there has been borrowing from the close-to-direct ancestor, Sanskrit, as is also attested for Romance languages and Latin, where the ancestor was similarly maintained at least as a literary and religious language during

the historical period: such borrowing from an ancestor is also relatively rare, but in addition is extremely hard to spot linguistically. The reticulations we observe in Figure 1 may also, however, have the same significance as those we see, and would expect to see, for the less conservative data in Figure 2: they may show simply the density of inter-borrowing among these South Asian languages. In that case, the patterns we find might suggest that borrowing of even the most basic vocabulary is possible at relatively high frequencies in convergence situations or in dialect continua with parti- cularly strong multilingualism.

All these possibilities require, and deserve, further consideration by specialists. However, we can take one step towards a clearer answer, by comparing our NeighbourNets for the South Asian languages for those derived from simulated data over known and controlled histories, incorporating borrowing between

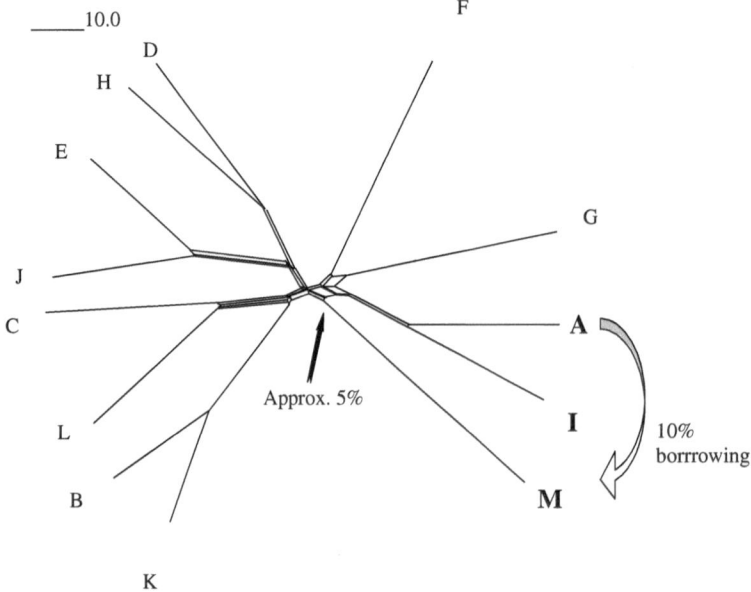

Figure 3. NeighbourNet for simulated data, with 10% borrowing from A to M

languages. The NeighbourNet in Figure 3 is for a simulated equivalent of our 200-meaning list, allowing 10% borrowing from Language A into Language M.

This NeighbourNet can be compared with Figure 4, where the same languages and data are included, but no borrowing is allowed in the simulation.

A comparison of Figures 3 and 4 then shows that the 10% borrowing effectively pulls A and its sister language I towards M and away from the other related languages in the network, most notably F and G. There

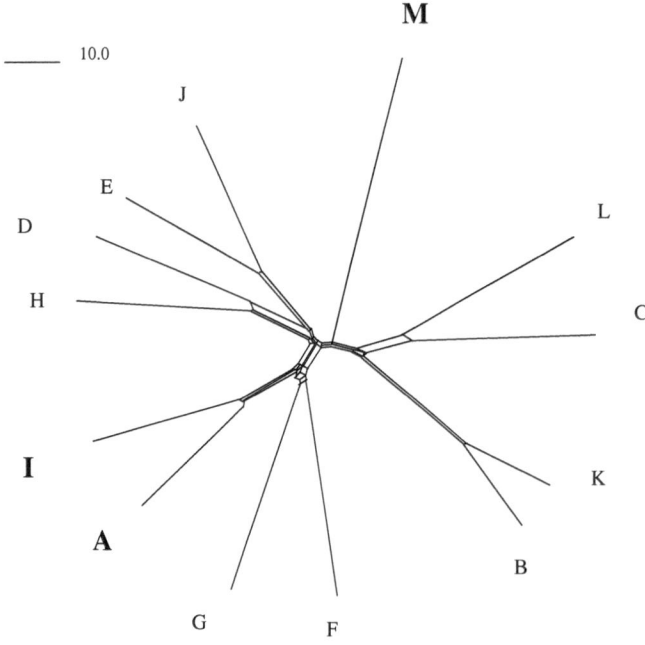

Figure 4. NeighbourNet for simulated data with no borrowing

are also several reticulations connecting these languages in Figure 3; but then again, there are also reticulations in Figure 4. Further investigation reveals that the reticulations in Figure 3 do indeed relate to the items borrowed from A into M, while those in Figure 4 in fact show shared retentions from the common ancestor. The point is that such further investigation is required: it is not simply the presence of reticulations that matters, though of course this does help us isolate where we should focus in further work. We also crucially need ways of interrogating those reticulations to show what is causing them.

We return, then, to our non-simulated Dyen et al. data and to the NeighbourNets derived from subsets of these data for South Asian languages in Figures 1 and 2. The question is what the reticulations in Figure 1 mean: why are we seeing links between these languages for even the most retentive and conservative data? Could this mean interborrowing, or is there some other explanation? The best way of homing in on these reticulations is to run

Network, which as we have seen generates a list of the specific characters which are behaving in a non-tree-like way and therefore contributing to the reticulations. Network can be used in this case because the most restrictive sublist contains only 30 meanings, and the number of languages involved is also small, so that we do not exceed the upper limit on the characters Network can cover.

Figure 5 shows a Network for these most conservative meanings, again for the South Asian languages in the Dyen, Kruskal and Black database, and constructed using a median joining algorithm so that the volume of reticulations shown is relatively restricted. The location of these reticulations, however, clearly matches the pattern in Figure 1, the parallel NeighbourNet.

We can take a further step towards interpreting these reticulations by isolating the features which are leading to the construction of incompatible and cross-cutting clusters. A selection of languages and three relevant meanings is given in Table 3.

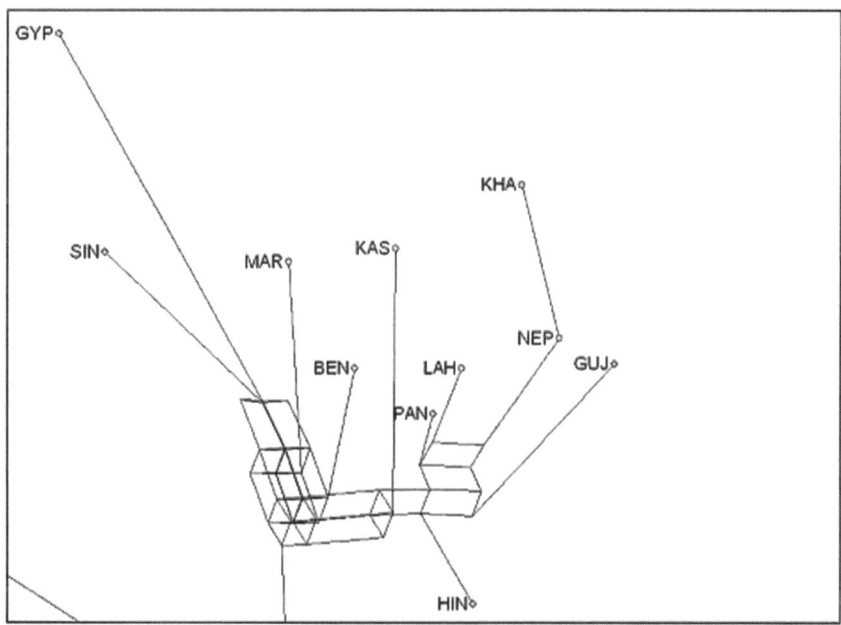

Figure 5. Network for the most conservative meanings, for South Asian languages in the Dyen et al. database. (GYP = Gypsy Greek; SIN = Singhalese; MAR = Marathi; BEN = Bengali; KAS = Kashmiri; PAN = Panjabi; LAH = Lahnda; HIN = Hindi; NEP = Nepali; KHA = Khaskura; GUJ = Gujarati)

Table 3. Three meanings contributing to tree incompatibilities for South Asian languages

Language	'long'	'salt'	'to sleep'
Bengali	lamba 003	nun 202	nid jaoa 006
Lahnda	lemba 003	lun 202	semmen 200
Gujarati	lambu 003	mithu 200	suwu 200
Marathi	lamb 003	mith 200	nijne 006
Hindi	lemba 003	nemek 32	sona 200
Singhalese	diga 200	lunu 202	nidiyagona 006
Wakhi	deroz 200	kimuk 003	ruxp- 201
Persian	deraz 200	namuk 003	khabidan 002

Clearly, the forms listed here fall into at least two cognate groups for each of the three meanings (with a few other items marked as unique states, whether because they are known borrowings or for other reasons). However, the configurations of these cognate groups are not consistent between meanings: for 'long', we find a grouping of Bengali, Lahnda, Gujarati, Marathi and Hindi, while Singhalese, Wakhi and Persian share the alternative state; but for 'salt', Singhalese groups with Bengali and Lahnda. This processing method provides a list of items which should then as the next step be considered by linguists familiar with the histories and structures of these languages.

Our aim in this chapter has not been to contribute new insights directly to current knowledge of language groupings and histories in South Asia, but to introduce the idea that some of the quantitative methods currently under development might usefully be employed in this area. Such quantitative methods clearly require the participation of specialists on South Asian languages if they are to realize their potential. We hope that our outline of the different stages in application of quantitative methods, and of the available methods themselves, will help to make colleagues aware of the possibilities these new approaches bring, and will encourage them to test such methods for themselves. We look forward to hearing in due course what contribution such quantitative investigations may make to our greater understanding of the complex patterns of interaction among South Asian languages, and between linguistic and other data in South Asian populations.

References

Alpher, B., Nash, D., 1999. Lexical replacement and cognate equilibrium in Australia. Australian Journal of Linguistics 19, 5–56.

Bandelt, H-J., Dress, A.W.M., 1992. Split decomposition: a new and useful approach to phylogenetic analysis of distance data. Molecular Phylogenetics and Evolution 1, 242–52.

Bandelt, H-J., Forster, P., Sykes, B.C., Richards, M.B., 1995. Mitochondrial portraits of human populations using median networks. Genetics 141, 743–753.

Bandelt, H-J., Forster, P., Röhl, A., 1999. Median-joining networks for inferring intraspecific phylogenies. Molecular Biology and Evolution 16, 37–48.

Bryant, D., Moulton, V. 2004. Neighbornet: an agglomerative algorithm for the construction of planar phylogenetic networks. Molecular Biology and Evolution 21, 255–265.

Dyen, I., Kruskal, J.B., Black, P., 1992. An Indoeuropean classification: a lexicostatistical experiment. Transactions of the American Philosophical Society 82, Part 5.

Embleton, S.M., 1986. Statistics in Historical Linguistics. Brockmeyer, Bochum.

Felsenstein, J., 2001. PHYLIP: Phylogeny Inference Package. Version 3.6. Department of Genetics, University of Washington.

Forster, P., Renfrew, C. (Eds.), 2006. Phylogenetic Methods in Historical Linguistics.

McDonald Institute for Archaeological Research, Cambridge.

Forster, P, Torroni, A., Renfrew, C., Röhl, A., 2001. Phylogenetic star contraction applied to Asian and Papuan mitochondrial DNA. Molecular Biology and Evolution 18, 1864–81.

Greenberg, J.H., 1987. Language in the Americas. Stanford University Press, Stanford.

Gumperz, J. J., Wilson, R., 1971. Convergence and creolization: a case from the Indo-Aryan/Dravidian border. In: Hymes, D. (Ed.), Pidginization and Creolization of Languages. Cambridge University Press, Cambridge, pp. 151–67.

Harrison, S.P., 2003. On the limits of the Comparative Method. In: Joseph, B., Janda, R. (Eds.) The Handbook of Historical Linguistics. Blackwell, Oxford, pp. 213–243.

Haspelmath, M., Dryer, M., Gil, D., Comrie, B. (Eds.), 2005. World Atlas of Language Structures. Oxford University Press, Oxford.

Heggarty, P., n.d. Measured Language. Blackwell, Oxford. (Publications of the Philological Society).

Heggarty P., McMahon, A., McMahon, R., 2005. Dialect classification by phonetic similarity: a computational approach. In: Delbecque, N, van der Auwera, J., Geeraerts, D. (Eds.) Perspectives on Variation. Mouton De Gruyter, Amsterdam, pp. 43–91.

Huson, D. H., 1998. Splitstree: a program for analysing and visualizing evolutionary data. Bioinformatics 14, 68–73.

Kessler, B., 2001. The Significance of Word Lists. CSLI Publications, Stanford.

Labov, W., 1994. Principles of Linguistic Change, I: Internal Factors. Blackwell, Oxford.

Labov, W., 2001. Principles of Linguistic Change, II: Social Factors. Blackwell, Oxford.

Lohr, M., 1999. Methods for the genetic classification of languages. Ph.D. Dissertation, University of Cambridge.

Matisoff, J., 1978. Variational Semantics in Tibeto-Burman. Institute for the Study of Human Issues, Philadelphia.

Matisoff, J., 2000. On the uselessness of glottochronology for the subgrouping of Tibeto-Burman. In: Renfrew, C., McMahon, A., Trask, R.L. (Eds.), Time Depth in Historical Linguistics. McDonald Institute for Archaeological Research, Cambridge, pp. 333–72.

Matras, Y., McMahon, A., Vincent, N. (Eds.), 2006. Linguistic Areas, Convergence and Language Change. Palgrave, London.

McMahon, A. (Ed.), 2005. Quantitative Methods in Language Comparison. Special issue of Transactions of the Philological Society, 103.2.

McMahon, A., Heggarty, P., McMahon, R., Slaska, N., 2005. Swadesh sublists and the benefits of borrowing: An Andean case-study. In: McMahon, A. (Ed.), Quantitative Methods in Language Comparison. Special issue of Transactions of the Philological Society, 103.2, 147–170.

McMahon, A., McMahon, R., 2003. Finding families: quantitative methods in language classification. Transactions of the Philological Society 101, 7–55.

McMahon, A., McMahon, R., 2005. Language Classification by Numbers. Oxford University Press, Oxford.

Nakhleh, L., Warnow, T., Ringe, D., Evans, S., 2005. A comparison of phylogenetic reconstruction methods on an Indo-European dataset. In: McMahon, A. (Ed.), Quantitative Methods in Language Comparison. Special issue of Transactions of the Philological Society, 103.2, 171–192.

Nettle, D., Romaine, S., 2000. Vanishing Voices: The extinction of the world's languages. Oxford University Press, Oxford.

O'Grady, G.N., 1960. More on lexicostatistics. Current Anthropology 1, 338–339.

Page, R., Holmes, E., 1998. Molecular Evolution: A phylogenetic approach. Blackwell, Oxford.

Rankin, R.L., 2003. The Comparative Method. In: Joseph, B., Janda, R. (Eds.), The Handbook of Historical Linguistics. Blackwell, Oxford, pp. 183–212.

Renfrew C., McMahon, A., Trask, L. (Eds.), 2000. Time Depth in Historical Linguistics. 2 Vols. McDonald Institute for Archaeological Research, Cambridge.

Ringe, D., Warnow, T., Taylor, A., 2002. Indo-European and computational cladistics. Transactions of the Philological Society 100, 59–129.

Ruhlen, M., 1991. A Guide to the World's Languages. Volume 1: Classification. Edward Arnold, London.

Schleicher, A., 1863. Die Darwinische Theorie und die Sprachwissenschaft. Offenes Sendschreiben an Herrn Dr. Ernst Haeckel, o. Professor der Zoologie und Direktor des zoologischen Museums an der Universität Jena. Böhlau, Weimar.

Schmidt, J., 1872. Die Verwantschaftsverhältnisse der indogermanischen Sprachen. Böhlau, Weimar.

Slaska, N., 2006. Meaning lists in historical linguistics – a critical appraisal and case study. Ph.D. Dissertation, University of Sheffield.

Swadesh, M., 1952. Lexico-statistic dating of prehistoric ethnic contacts. Proceedings of the American Philosophical Society 96, 453–46.

Swadesh, M., 1955. Towards greater accuracy in lexicostatistic dating. International Journal of American Linguistics 21, 121–137.

Thomason, S.G., 2001. An Introduction to Language Contact. Edinburgh University Press, Edinburgh.

Thomason, S.G., Kaufman, T., 1988. Language Contact, Creolization, and Genetic Linguistics. University of California Press, Berkeley.

17. Duality in *Bos indicus* mtDNA diversity: *Support for geographical complexity in zebu domestication*

DAVID A. MAGEE

Smurfit Institute of Genetics
Department of Genetics
Trinity College Dublin 2
Ireland
mageeaa@tcd.ie

HIDEYUKI MANNEN

Laboratory of Animal Breeding and Genetics
Faculty of Agriculture
Kobe University
1-1 Rokkoudai
Kobe, 657
Japan
mannen@kobe-u.ac.jp

DANIEL G. BRADLEY

Smurfit Institute of Genetics
Department of Genetics
Trinity College Dublin 2
Ireland
dbradley@tcd.ie

Introduction

For the past decade mitochondrial DNA diversity (mtDNA) has been the primary genetic focus for researchers investigating domestic livestock origins (MacHugh and Bradley, 2001). This is attributable to a combination of genetic characteristics. Firstly, mtDNA exhibits a rapid rate (particularly the mtDNA control region) of evolution. This ensures that mtDNA sequences will have accumulated sufficient variation over the shallow 10,000-year timeframe of animal domestication to allow the phylogenetic relationships of sequences sampled from different geographic regions to be

M.D. Petraglia and B. Allchin (eds.), The Evolution and History of Human Populations
in South Asia, 385–391.

reconstructed. Secondly, mtDNA is almost exclusively maternally inherited and is (at least predominantly) exempt from recombination. Hence the origin of any modern domestic mtDNA lineage can be traced unambiguously to a single ancestral lineage which must, at some stage in its history, have entered the domestic pool via the physical capture, taming and ultimate domestication of a wild female animal. Furthermore, in the absence of recombination it is possible to identify ancestral mtDNA lineages recruited into the domestic pool from different captures from the wild despite the millennia of interbreeding that has occurred since the initial domestication events.

A series of phylogenetic surveys of mtDNA sequence variation have highlighted the complexity of livestock domestication (reviewed by Bruford et al., 2003). Undoubtedly, the most striking feature to emerge from these studies is the existence, typically of two or several distinct and divergent mtDNA sequence clusters within the mtDNA pool of many domestic livestock species; for example, pig, sheep, cattle, goat and water buffalo. These sequence clusters presumably each represent a group of modern descendants of ancestral founder mtDNA lineages adopted into the domestic pool during discrete domestication processes. Furthermore, for many species these clusters tend to be geographically distributed, primarily along an East-West axis, suggesting that different mtDNA lineages were recruited from the wild as a result of geographically and possibly temporally separate domestication events. Thus, it has become increasingly clear that multiple domestication is a recurring theme in domestic animal history.

In cattle, a duality in domestic mtDNA sequence diversity has been identified, in which samples of sequences invariably divide into one of two divergent groups, where their difference in sequence suggests a last common ancestor hundreds of thousands of years ago (Loftus et al., 1994). This separation of bovine mtDNA sequences follows the well-established taxonomic distinction between the two domestic cattle forms, namely the humpless variety of Europe, the Middle East and West Africa (*Bos taurus* or taurine) and the humped cattle of South Asia (*Bos indicus* or zebu). These phylogeographic patterns strongly support the domestication of both *Bos taurus* and *Bos indicus* from at least two genetically- and geographically-differentiated wild progenitor (aurochs) populations. Indeed, archaeological evidence argues that *Bos taurus* had its domestic origins in the early agrarian societies of the Near East and possibly North Africa some 10,000 years ago, while *Bos indicus* originated in the Indian subcontinent some 2,000 years later (Zeuner, 1963; Meadow, 1996; Bökönyi, 1997).

MtDNA Diversity Patterns within *Bos taurus*

Recently, more extensive surveys of mtDNA sequence diversity have afforded a higher resolution of ancestry within *Bos taurus* (Troy et al., 2001). In these, taurine mtDNA sequences fall into one of five phylogenetically distinct, star-like haplotypic clusters or haplogroups. Each haplogroup consists of a numerically predominant centrally-positioned sequence (presumed to be ancestral), while the surrounding haplotypes are thought to be derived from the ancestral sequence due to the accretion of variation over time. Another noteworthy feature of *Bos taurus* mtDNA sequence diversity is the spatial distribution of the major haplogroups. In particular, one haplogrouop (T3) predominates in Europe (Figure 1) and is one of three detected at appreciable frequencies in the Near East, while symmetrically, African variation is almost exclusively composed of members of a separate haplogroup (T1), which is encountered rarely elsewhere. The strong geographic partitioning of these

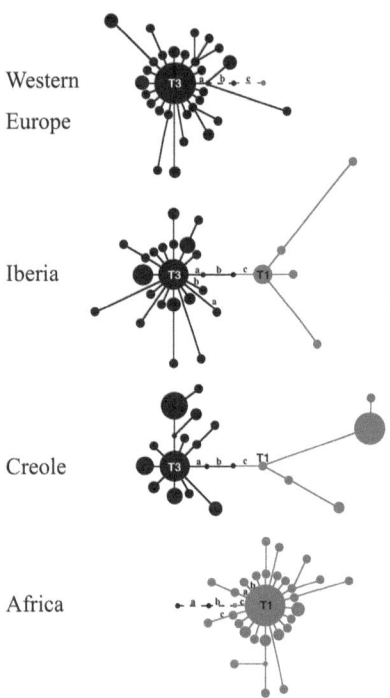

Figure 1. Reduced median networks featuring *Bos taurus* mtDNA lineages (240bp of D-loop) sampled in 67 Western European, 60 Iberian, 59 Caribbean Creole and 61 African cattle (from Magee et al., 2002). Within these networks, sequences are represented by circles and base pair differences between haplotypes are denoted by connecting lines; longer lines indicate multiple changes. Both the central European haplotype (T3) and African haplotype (T1) are shown. The relationship between the European (T3) and African (T1) central haplotypes are defined by transitions at nucleotide positions 16255, 16113 and 16050 (denoted a, b and c, respectively); dotted lines and smaller circles show this axis of relationship when either sequence is absent. Circle areas are proportional to the frequency of each haplotype and shading indicates which of the two major haplotypes they root to. Strikingly, T1 and its derivatives predominates in Africa, T3 predominates in Western Europe and the mix between the two in Iberian and Creole cattle is an indication of the mixed continental heritage of both these regions

major *Bos taurus* haplogroups has permitted geographical exceptions to be interpreted as signatures of secondary migratory events, such as the occurrence of African mtDNA sequences in Iberian and American populations which may be suggestive of historical importations of African cattle to these regions (Cymbron et al., 1999; Magee et al., 2002).

The starlike topology produced by both the T1 and T3 haplogroups in Africa and Europe, respectively, is representative of the classic phylogeny expected from a population sample which has undergone a demographic expansion from a relatively small base (Jorde et al., 1998). The obvious candidate process for such an expansion is domestication itself, where it may be envisaged that a relatively small number of wild female animals were incorporated into early domestic flocks, which then gave rise to increasing numbers of offspring under improving systems of animal husbandry; a population expansion which has continued to the present.

An application of the molecular clock allows some test of this scenario. If, in a haplogroup, the central haplotype, e.g., T1, is an ancestral sequence and the surrounding minor haplotypes represent sequences which have accumulated mutations since the expansion of T1, then the calibrated sequence diversity within the T1 haplogroup can be used to estimate the age of the T1 expansion (Forster et al., 1996). Estimates for both T1 (8.5–12 ka), and T3 (9–11.6 ka) are consistent with the time-depth of cattle domestication (Troy et al., 2001). Furthermore, these phylogenetic patterns together with the disjunct geographies of the T1 and T3 *Bos taurus* haplogroups have been argued as support for separate domestic origins for the cattle of Europe and Africa (Wendorf and Schild, 1994; Bradley et al., 1996).

Two mtDNA Clusters within *Bos indicus*

Detailed examination of *Bos indicus* mtDNA sequence diversity also reveals a separation of sequences into discrete haplogroups. Analyzing the most variable 240bp fragment of the bovine mtDNA control region, Baig et al. (2005) and Magee (2002) have presented zebu sequences, which clearly coalesce to two moderately divergent predominant haplotypes, termed Z1 and Z2. *Bos indicus* mtDNA sequence variation is summarized in Figure 2. This reduced median network, comprized of

samples of Indian provenance (Bradley et al., 1996; Baig et al., 2005) is constructed from the same mtDNA control region as those *Bos taurus* sequences shown in Figure 1 and is therefore directly comparable. The dual star cluster motif is strikingly analogous to that seen in African and European cattle and we therefore suggest that they are each signatures of domestication-induced population expansion. When the same mtDNA calibration is applied to these data, the estimated ages are 2.6–7.6 ka for Z1 and 8.7–14.9 ka for Z2. These dates are compatible with bovine domestic history, however, it must be noted that these ranges are subject to further uncertainty due to inherent difficulties associated with estimating the substitution rate within hypervariable regions of the mitochondrial genome (Macaulay et al., 1997).

Geographic Distribution of *Bos indicus* mtDNA Clusters

Whereas the two major *Bos taurus* mtDNA clusters show striking phylogeographical distribution which aids the inference that they represent alternate incorporations from the wild, both the zebu haplogroups are found in the purported center of zebu diversity, the Indian subcontinent (Magee, 2002; Baig

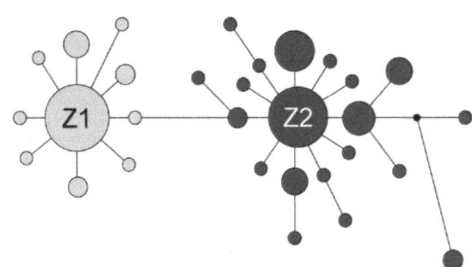

Figure 2. Reduced median networks featuring *Bos indicus* mtDNA lineages from Indian cattle breeds (Bradley et al., 1994; Ritz et al., 2000; Troy et al., 2001; Magee, 2002; Baig et al., 2005). The major feature is that these sequences group into two clusters centered around predominant haplotypes, Z1 and Z2. The numerical and topological dominance of these haplotypes suggests these as alternate ancestral sequences from which the outlying sequences derive. The two major haplotypes are separated by four mutations within the 240bp region surveyed, at positions 16084, 16085, 16141 and 16232

et al., 2005). However, Magee (2002) has noted some tendency for an east-west trend in frequencies of these two haplogroups. Therefore, to establish if any significant geographical trend was apparent we examined our unpublished and GENBANK-deposited *Bos indicus* sequences which are unambiguously East and West of the Indian subcontinent in provenance.

Previously, Troy et al. (2001) and Magee (2002) described a total of 14 zebu mtDNA sequences sampled from morphologically taurine animals of the Near East. Their presence attest to a process of admixture which may be ancient in origin and which is likely to have been predominantly influenced by animals from the more Western regions of the Indian subcontinent (Loftus et al., 1999; Kumar et al., 2001). Of these 14 zebu introgressor sequences, 10 belonged to the Z1 clade. In addition, we examined GENBANK sequences which could be identified by homology search as being *Bos indicus* in type and which could be identified as being sampled from regions east of the subcontinent; including substantial samples from China. There were 149 of these, of which the great majority, 134, were from the Z2 clade. Thus it seems that a majority of western zebu mtDNAs are of the Z1 type, but emphatically, a great majority of eastern zebu are of the Z2 type – this difference is highly significant when tested by Chi-squared or Fishers Exact tests ($p < 0.001$).

Origins of the Alternate *Bos indicus* mtDNA Clusters

Domestic *Bos indicus* cattle first appear within the archaeological record of Baluchistan, some 8,000–7,000 years ago and excavations from Mehrgarh, the principal site of Baluchistan, suggest that *Bos indicus* cattle were derived from a morphologically different wild progenitor to *Bos taurus* cattle (Jarrige and Meadow, 1980; Meadow,

1996). By the 3rd millennium BC zebu cattle were prominent components of Indus Valley civilization sites, such as those at Harappa and Mohenjo-daro (Allchin and Allchin, 1968; Meadow, 1996). Overall, archaeological evidence for zebu domestication in the northwest of the subcontinent is strong.

It seems likely that both the moderately divergent ancestral Z1 and Z2 haplotypes were recruited into the domestic pool and the resulting star-like phylogenies surrounding these represent the legacy of domestication-induced expansions. It could be argued that both of these could have been incorporated as part of a single process, such as that in Baluchistan. Indeed, both Z1- and Z2-type sequences have been sampled within populations of the Indian subcontinent and limited analysis of diversity does not reveal any major differences within the populations of this region. However, the most striking feature of the *Bos indicus* mtDNA phylogeography, namely the high incidence of the Z2 haplogroup in sampled regions to the east of the subcontinent, and predominance of Z1 in the west does not point to a single origin.

Alternately, these data may provide some credence for secondary domestication(s) of the *Bos indicus* wild progenitor in separate areas throughout the Indian subcontinent. For example, it is possible that the increased incidence of the Z2 haplogroup sequences in East Asian zebu may reflect the legacy of a domestication of the *Bos indicus* ancestor somewhere to the East of Baluchistan. While the archaeology of Mehrgarh strongly supports the candidature of Baluchistan as a primary center of *Bos indicus* domestication, some authors argue that zebu cattle may have also been domesticated within Peninsular India (Allchin and Allchin, 1982). The Neolithic culture of Deccan Plain in South India features distinctive ashmounds thought to be have been produced by the burning of cattle dung (Misra, 2001). These are regarded as potential ancient cattle pens which may have served as enclosures for the capturing, taming and ultimate domestication of wild aurochsen ca. 5 ka. An assertion in this argument is that earliest representations of cattle within southern Neolithic sites depict cattle with delicate limbs and humps and long horns – in contrast to the heavier-bodied representations on Harappan seals (Allchin and Allchin, 1974). However, in opposition to this is the recorded variability in morphology of Harappan faunal remains (Meadow, 1991). Posited morphological differences are far from conclusive and may simply reflect the artistic preference of seal cutters. That said, faunal evidence suggests that the wild ancestor of *Bos indicus* was present in South India possibly as late as 4 ka and was therefore available for domestication (Grigson, 1985; Joglekar and Thomas, 1992).

A third possibility is that members of the Z2 haplogroup represent the descendants of those haplotypes adopted into the domestic lineage during the initial phase of aurochs capture and taming in Baluchistan, ca. 8–7 ka, while the increased occurrence of the Z1 haplogroup in regions west of the Indian subcontinent points towards a more western origin. In support of this assertion, metric analysis of bone assemblages uncovered from Shahr-I Sokhta (Sistan, Iran) has led Bökönyi (1997) to argue for a potential Iranian domestication of zebu cattle during the 4th–3rd millennia BC, although the basis for this assertion is not strong.

Conclusion

Thus we have discovered, as has Baig et al. (2005) that there are two clusters of mtDNA sequence types within *Bos indicus*. By analogy with *Bos taurus* results and from general population genetic theory, these may credibly be posited as the products of the domestication of two alternate types of indicine mtDNA. The patterns and quantifiable extent of diversity around each cluster are consistent with a domestication-induced population expansion. It is not clear that they necessarily come from culturally or geographically separate captures of *Bos primigenius namadicus*, but some inference may be drawn from their present-day distribution. Whereas both types of mtDNA are found within Indian samples there is a clear difference in their frequencies when samples of origin east and west of the peninsula are considered. In a limited sample of *Bos indicus* mtDNA haplotypes from Near Eastern cattle, 71% were of Z1 type whereas in a large sample of indicine haplotypes sampled in regions east of India, 90% were Z2 in character. It is tempting to posit that this current geographical disparity may represent a geographical distinction in the origins of these; with the Z1 cluster representing the products of domestication in Baluchistan and Z2 haplotypes being the legacy of one or more events to the east of this complex. However, making specific assertions about candidate regions is beyond the scope of the data discussed here. India possesses the largest national herd of cattle in the world, including much morphological diversity. Clearly, a detailed study of the geography of bovine mtDNA and other genetic variants within the subcontinent will be a priority in making further inference about the detail of domestication of zebu cattle.

References

Allchin, B., Allchin, R.A., 1968. The Birth of Indian Civilization: India and Pakistan Before 500 B.C. Penguin, London.

Allchin, B., Allchin, R.A., 1982. The Rise of Civilization in India and Pakistan. Cambridge University Press, Cambridge.

Baig, M., Beja-Pereira, A., Mohammad, R., Kulkarni, K., Farah, S., Luikart, G., 2005. Phylogeography and origin of Indian domestic cattle. Current Science 89, 38–40.

Bokonyi, S., 1997. Zebus and Indian wild cattle. Anthropozoologica 25–26, 647–654.

Bradley, D.G., MacHugh, D.E., Cunningham, P., Loftus, R.T., 1996. Mitochondrial diversity and the origins of African and European cattle. Proceedings of the National Academy Science of the United States of America 93, 5131–5135.

Bruford, M.W., Bradley, D.G., Luikart, G., 2003. DNA markers reveal the complexity of livestock domestication. Nature Reviews Genetics 4, 900–910.

Cymbron, T., Loftus, R.T., Malheiro, M.I., Bradley, D.G., 1999. Mitochondrial sequence variation suggests an African influence in Portuguese cattle. Proceedings of the Royal Society London Series B-Biological Science 266, 597–603.

Grigson, C., 1985. *Bos indicus* and *Bos namadicus* and the problem of autochthonous domestication in India. In: Misra, V.N., Bellwood, P. (Eds.), Recent Advances in Indo-Pacific Prehistory. Oxford and IBH, New Delhi, pp. 425–428.

Jarrige, J.-F., Meadow, R. H., 1980. The Antecedents of Civilization in the Indus Valley. Scientific American 243, 122–133.

Joglekar, P.P., Thomas, P.K., 1992. Ancestry of *Bos* species: myth and reality IV: the origins of humped cattle. Man and Environment 17, 51–54.

Jorde, L.B., Bamshad, M., Rogers, A.R., 1998. Using mitochondrial and nuclear DNA markers to reconstruct human evolution. Bioessays 20, 126–136.

Kumar, K., Freeman, A.R., Loftus, R.T., Gaillard, C., Fuller, D.Q., Bradley, D.G., 2003. Admixture analysis of South Asian cattle. Heredity 91, 43–50.

Loftus, R.T., MacHugh, D.E., Bradley, D.G., Sharp, P.M., Cunningham, P., 1994. Evidence for two independent domestications of cattle. Proceedings of the National Academy Science 91, 2757–2761.

Loftus, R.T., Ertugrul, O., Harba, A.H., El-Barody, M.A., MacHugh, D.E., Park, S.D., Bradley, D.G., 1999. A microsatellite survey of cattle from a centre of origin: the Near East. Molecular Ecology 8, 2015–2022.

Macaulay, V.A., Richards, M.B., Forster, P., Bendall, K.E., Watson, E., Sykes, B., Bandelt, H.J., 1997. MtDNA mutation rates–no need to panic. American Journal of Human Genetics 61, 983–990.

MacHugh D.E., Bradley D.G., 2001. Livestock genetic origins: goats buck the trend. Proceedings of the National Academy Science 98, 5382–5384.

Magee, D.A, Meghen, C., Harrison, S., Troy, C.S., Cymbron, T., Gaillard, C., Morrow, A., Maillard, J.C. Bradley, D.G., 2002. A partial African ancestry for the Creole cattle populations of the Caribbean. Journal of Heredity 93, 429–432

Magee, D.A., 2002. Molecular genetic investigation of the diversity and origins of Old and New world cattle populations. Ph.D. Dissertation, University of Dublin.

Misra, V.N., 2001. Prehistoric human colonization of India. Journal of Biosciences 26 (Supplemental), 491–531.

Meadow, R.H., 1991. Faunal remains and urbanism at Harappa. In: Meadow, R.H. (Ed.), Harappa Excavations 1986–1990: A Multidisciplinary Approach to Third Millennium Urbanism. Prehistory Press, Madison, pp. 89–106.

Meadow, R.H., 1996. The origins and spread of pastoralism in northwestern South Asia. In: Harris, D.R. (Ed.), The Origins and Spread of Agriculture and Pastoralism in Eurasia. University College London Press, London, pp. 25–50.

Ritz, L.R., Glowatzki-Mullis, M.L., MacHugh, D.E., Gaillard, C., 2000. Phylogenetic analysis of the tribe Bovini using microsatellites. Animal Genetics 31, 178–185.

Troy, C.S., MacHugh, D.E., Bailey, J.F., Magee, D.A., Loftus, R.T., Cunningham, P., Chamberlain, A.T., Sykes, B.C., Bradley, D.G., 2001 Genetic evidence for Near-Eastern origins of European cattle. Nature 410, 1088–1091.

Zeuner, F.E., 1963. The history of domestication in cattle. In: Mourant, A.E., Zeuner, F.E. (Eds.), Man and Cattle: Proceedings of a Symposium on Domestication. Royal Anthropological Society of Great Britain and Ireland, Occasional Paper 18, pp. 9–15.

Wendorf, F., Schild, R., 1994. Are the early Holocene cattle in the Eastern Saharan domestic or wild? Evolutionary Anthropology 3, 118–128.

18. Non-human genetics, agricultural origins and historical linguistics in South Asia

DORIAN Q FULLER

Institute of Archaeology
University College London
31–34 Gordon Square
London, WC1H 0PY
England
d.fuller@ucl.ac.uk

Introduction

In the histories of human populations, the origins of agriculture marks a major demographic watershed. In most cases, hunter-gatherer societies were mobile, or at least mobility was used strategically to cope with seasonal shortages in the surrounding environments. Agriculture made an important change from this situation because, even though it relies on a seasonal cycle of planting, growing and harvesting, it provides a storable surplus that can sustain populations through lean seasons. Other important changes usually associated with the beginnings of agriculture are those brought about by a reliable source of carbohydrate-rich staples such as cereals (or in some tropical regions, tubers). Starchy staples such as cereals, which can be cooked into soft gruel (or porridge), make a useful weaning food for infants. This allows babies to be weaned off mother's milk at an earlier age, and therefore agriculture increases the potential rate of population growth (Cohen, 1991). A related side effect of stored starchy agricultural produce is the increase in starch and sugars in the diet that tends to cause increased dental cavities, an effect usually detectable in the skeletal remains of early agricultural societies, in contrast to those of earlier hunter-gatherer societies (Larsen, 1997). Another side effect of agriculture often visible in skeletal remains is increased malnutrition brought about by the vitamin deficiencies of starch-rich diets, but poor in vegetable diversity, of many early agriculturalists. Thus, although agriculture was fundamental to later developments of civilizations, its beginnings may not have been advantageous to populations when measured in terms of health. This begs the question as to why hunter-gatherers who were successful in most environments ever resorted to cultivation and agriculture. But once agriculture was adopted those groups who employed it had potentially vast demographic advantages, in terms of rate of population growth, over their hunter-gatherer contemporaries.

M.D. Petraglia and B. Allchin (eds.), The Evolution and History of Human Populations
in South Asia, 393–443.

It is this demographic advantage of farming which is the fundamental premise of models of prehistory in which significant migration is supposed to have occurred in the Neolithic, with genetic consequences and language replacement. The basic extension of the demographic advantage to patterns of geographical spreads is the 'wave of advance' model of Ammerman and Cavalli-Sforza (1971), which was then related to the dispersal of major language families (or macro-families) by Renfrew (1987, 1996, 2000) and Bellwood (1996, 2001, 2005, Diamond and Bellwood, 2003). In the case of South Asia this premise has been used to propose a Neolithic influx of Indo-European speakers from southwest Asia into India (Renfrew, 1987; Bellwood, 2005), as well as the ancestors of the Dravidian speakers of South India from the same general direction, on the assumption of an Elamo-Dravidian macro-family (Renfrew, 1987; Bellwood, 1996, 2005:210–216). In addition, the Munda languages spoken by hill tribes in Eastern and parts of central India, which are clearly part of the larger Austro-Asiatic family in Southeast Asia, have been suggested to represent an agricultural Neolithic influx from the Northeast (Bellwood 1996, 2005:210–216; Glover and Higham, 1996:419; Higham, 2003). These models have offered alternative populational prehistories, especially in terms of dating, to conventional views in which all major populations coming into South Asia came from the northwest: first the Paleolithic ancestors of the Munda, then the agricultural ancestors of the Dravidians and finally the chariot- and horse-riding pastoralists who brought Indo-European (e.g., Fuchs, 1973; Gadgil et al., 1998; Kumar and Mohan Reddy, 2003, and for more recent linguistic and archaeological data see, e.g., Parpola, 1988; Witzel, 2005). The agriculture/language dispersal hypothesis also provides a clear explanatory framework: that of demographic growth of farmers with a long-term advantage over hunter-gatherers. Despite the potential attraction of a demographic prime-mover for simplifying patterns in prehistory, like all hypotheses, it requires testing against the empirical evidence for human prehistory. As evidence for human prehistory, we can turn to archaeology, historical linguistics and physical anthropology (including human genetics), as all of these sources preserve to varying degrees of precision information about past population histories (Rouse, 1986). The present contribution will attempt such an assessment of the role of agricultural dispersals in structuring the major cultural divisions and linguistic geography of South Asia, by assessing some of the empirical details available from archaeology and historical linguistics. In developing this subject, I will expand upon and update a recent effort to correlate archaeology (especially archaeobotany) and linguistics (Fuller, 2003a). One issue which requires further consideration, but will not be pursued in the present chapter, is the impact of an endogamous, cross-cousin marriage system, which can be inferred for early Dravidian speakers but not other language groups, on genetic patterns and demography.

This chapter will move from genetic and biogeographic evidence of non-human species, through archaeology, towards a revised tabulation of linguistic data with implications for South Asian prehistory. While the picture of human genetics and physical anthropology are best dealt with by others (e.g., see chapters by Endicott et al., Stock et al., Lukacs, this volume), I will start by looking at genetics of selected non-human species, in particular those key companion species of farmers, crops and livestock. The genetics, and, at a less precise level, the general phylogenetic inferences and biogeography of crops and livestock, encodes information about histories of movement, as people have acted as important agents in the dispersal of these species. While this dispersal may occur through exchanges between humans groups, and not necessarily through human population migration, the patterns of origins and dispersals in crops and livestock provides clear geographical and chronological parameters which must be accounted for in any model of human prehistory. Once we have set the scene, in terms of the non-human players and elements, I will turn to the archaeological evidence as it stands today. Archaeology provides the most clear, empirical and datable evidence for past economies and cultural practices, although it remains limited by gaps in the evidence. The patchiness of the archaeological record is particularly stark for the earliest agriculturalists in most parts of South Asia and their hunter-gatherer ancestors. Nevertheless, it is becoming increasingly clear that when farming groups began to settle permanently in villages, they were already agricultural in regionally distinctive ways, with at least three plausible indigenous South Asian foci of plant domestication (Fuller, 2003a, 2003b), plus an important northwestern agricultural tradition with its roots in Southwest Asia. I will then attempt to match this archaeological picture with that available from historical linguistics, in which increasing progress has been made at characterising not just cognates across existing, related languages, such as Dravidian languages of South India, but also in terms of inferring the past existence of now extinct substrate languages that have left their mark through loan words, especially relating to the Indian flora and agriculture.

Where and When: Biogeography and Genetics

Starting from the basics, we must ask what species served as the basis for early agricultural systems and where is it likely that hunter-gatherers regularly engaged and selected such plants as food sources. Biogeography and biological systematics provide essential information about how species known today in domesticated form developed. Through systematics, from traditional taxonomy to the increasingly powerful tools of molecular genetics, the closest free-growing or free-ranging relatives of crops and livestock can be identified, i.e. the wild progenitors. Comparisons between these provide a basis for identifying wild progenitors and how the domesticated forms differ from their wild relatives and may thus be identified archaeologically. Once identified, the ecology and geographical distribution of wild progenitors in the present day provides essential evidence from which to infer where these species would have been available to past human groups, and thus where they could have been first brought under human control. This information about modern distribution does, however, need to be considered in relation to past climate and environmental changes. In the case of southwest Asia there are a number of crops which occur wild in the transitional zone between the Mediterranean woodlands of oak and other trees, with open park woodland and the transition to grassland steppe, in a zone that averages 400–600 mm of annual rainfall, especially in the Levant, Anatolia and the parts of the Taurus Mountains (Moore et al., 2000:58; Zohary and Hopf, 2000). These are the founder crops of agriculture in the fertile crescent, most of which were also of importance to the agriculture of South Asia, especially in the northwest and the greater Indus region. The areas in which they were potentially domesticated have been inferred by combining their modern geography with information about paleoecology through the late Pleistocene and early Holocene (Hillman, 1996; Hillman, 2000:327–339; Willcox, 2005). The wild progenitors and ecologies of the most important seed crops of African origin were outlined by Harlan (1971, 1992), with only minimal refinements through more recent work (see Fuller 2003c; Neumann, 2004). The equivalent level of information is not available for crops originating in other regions, and for some South and Southeast Asian species we are still in the early stages of documenting the distribution and environmental tolerance of wild progenitors, let alone trying infer from paleoecological sources their distribution immediately prior to domestication. Nevertheless, a first attempt to synthesize information from agronomic and floristic sources for grain crops of Indian origins has been published (Fuller, 2002:292–296, for some vegetables and fruits, see Fuller and Madella, 2001, although some revision is now possible, see below). For crops originating in China and Southeast Asia, Simoons (1991) provides a useful overview.

Despite there being much to learn about the wild progenitors of many South Asian crops, there is much that is already known which has not been incorporated in the reasoning of many archaeological syntheses. This is notably the case in language macro-dispersal models of the last few years (e.g., Diamond and Bellwood, 2003; Bellwood, 2005), which are contradicted by clear indications for multiple domestications in key subsistence taxa of South and East Asia as well as many indications of indigenous Indian domestications. In the proposals of Bellwood (2005), agriculture came to India from the outside, primarily by human dispersals. This is not a new conclusion, as the earlier attempt by MacNeish (1992) to synthesize early agriculture worldwide suggested essentially the same thing for South Asia. Similarly, Harlan (1975, 1995) viewed South Asian agriculture as a derivative mix of Southwest and Southeast Asian origins. Agriculture is argued to derive from the well-documented early domestications in the Near Eastern 'fertile crescent' brought to South Asia by the ancestors of both Dravidian speakers and Indo-European speakers. Meanwhile, rice-focused agriculture is assumed to derive

from early domestication in the Yangzi river basin of China and spread to India from the northeast together with ancestors of the Munda language family (Glover and Higham, 1996; Higham, 2003; Bellwood, 2005). While I will return to the language issues later, I would like to start by examining evidence that indicates that species shared between South and East Asia suggest a recurrent pattern of multiple origins, with separate East Asian and South Asian domestications.

On the Origins and Spread of Rice

Rice (*Oryza sativa*) is one of the most utilized crops of the world today, but the complexities of its early history remains largely unraveled. Rice is now cultivated in a wide range of habitats from temperate northern China and Korea to the eutropical areas of Indonesia. It is grown as broadcast sown crops on hillsides, often as part of extensive slash-and-burn systems, and it is grown in highly labor intensive, flooded 'paddy' lands in which seedlings grown in one paddy are dug up and individually replanted into another field. The assumption, which is widespread in the literature, that all Asian rice derived from a single domestication, somewhere in the wild rice belt from eastern India across northern Indo-China or South China (e.g., Chang, 1995, 2000), has been based more on the presumption of single origins for crops in general, coupled with problematic archaeological inferences. Starting with the assumption that rice was domesticated once, there have been some rather extreme attempts to relate East Asian and South Asian archaeology, such as via comparisons between Neolithic China (sixth through fourth millennium BC) and Neolithic Kashmir (2500-1000 BC) (e.g., Van Driem, 1998), even though the latter had agriculture based on Near Eastern crops (wheat, barley, lentils and peas) and not rice! More recently, Kharakwal et al.'s (2004) attempt to link cord-impressed ceramics with rice agriculture suggests hyper-diffusionism based on superficial similarities in ceramics, including the Jomon of Japan (which is non-agricultural), parts of Neolithic China of the early to mid-Holocene, and much later 4th to 2nd millennium BC material from the Ganges. All such hyperdiffusionist studies are flawed, not only because they stretch archaeological logic by drawing comparisons across such vast areas and time-spans, but most importantly because they fail to take into account what we already know from botany about rice origins. Historical linguists have been mistaken in trying to make sense of a vast array of potential rice words on the assumption of a single centre of rice origin from which such words ought to originate (e.g., Mahdi 1998; Pejros and Snirelman, 1998; Witzel, 1999:30–33). Less explicitly reasoned attempts to link all of South and East Asian rice into a single story, are the grand narratives linking agriculture and language spread, in which the spread of rice from the middle Yangzi to India with demographically expanding and migrating farmers is argued largely on the basis of model assumptions rather than archaeological evidence (e.g., Bellwood, 1996, 2005; Higham and Glover, 1996). Any attempt to make a single narrative about Asian rice is already falsified by phylogenetic evidence from rice itself.

Asian rice, despite being lumped under the species name, *Oryza sativa* (a Linnaean convention in use since the 1750s), is composed of two distinct phylogenetic species, *indica* and *japonica*. This has long been suggested by plant breeding research, in which hybridization between these two cultivars is found to be difficult and imperfect, with the majority of crosses between *indica* and *japonica* cultivars being wholly or partly sterile (Wan and Ikehashi, 1997). As a result, the botanical literature has had a persistent debate between hypotheses of rapid divergence after a single origin or two domestications (Oka, 1988; Chang, 1989, 1995; White, 1989; Thompson, 1996),

although it is the single origin that has tended to be assumed in archaeological syntheses (e.g., Bellwood, 1996, 2005; Glover and Higham, 1996; Higham, 1998; Bellwood and Diamond, 2003), perhaps largely due to the influence of T. T. Chang (1989, 1995, 2000). There now is substantial evidence for genetic distinctions between *indica* and *japonica* from a range of data (Sato et al., 1990; Sano and Morishima, 1992; Chen et al., 1993a, 1993b; Sato, 2002; Cheng et al., 2003). Most significant is genetic evidence from the chloroplast (a plant organelle like the mitochondria inherited maternally) and nuclear DNA variants called SINEs. A sequence deletion in the chloropast DNA of *indica* cultivars links them with wild annual "*O. rufipogon*" (i.e., *O. nivara* in the taxonomy used here) (Chen et al., 1993a, 1993b; Cheng et al., 2003; for current rice taxonomy see Vaughan, 1989, 1994). Meanwhile, there are some seven SINEs that separate the *nivara-indica* group from the *rufipogon-japonica* (Cheng et al., 2003). Figure 1 shows the phylogenetic model produced by Cheng et al. (2003), in which the *japonica* cultivars form a very tight group in relation to the dispersed groupings of wild rufipogon types. By contrast, the grouping of *indica* is looser and more interspersed with wild *nivara*. This contrast might even suggest that *indica* is composed of more than one domestication event from wild *nivara* populations. On the basis of the modern geography of wild forms and cultivars at least one of these *indica* domestications in likely to have occurred in northern or eastern South Asia (Figure 2), while the *japonica* domestication can be placed in Southern China, probably the Yangzi basin.

The available archaeological evidence also suggests two distinct centres of early rice cultivation. In China, despite continuing controversies about the antiquity of rice use, cultivation, and domestication, it is widely accepted that rice cultivation was underway in

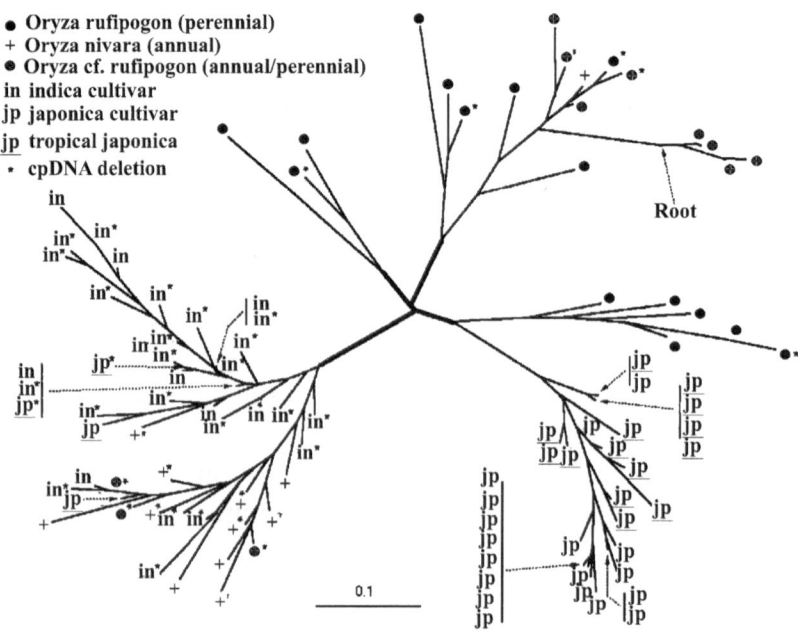

Figure 1. A phylogenetic representation of modern rice cultivars and wild populations based on SINE genetics (after Cheng et al., 2003; taxonomy revised to follow Vaughan, 1994). This shows the clearly distinct lineages of *japonica* (including most tropical forms, sometimes called *javanica)* and *indica* cultivars, which are interspersed with the annual wild populations *(Oryza nivara)*

the Middle Yangzi, and adjacent South China by the sixth millennium BC (e.g., Crawford and Shen, 1998; Lu, 1999, 2006; Cohen, 2002; Yan, 2002; Crawford, 2006). While rice spreads down the Yangzi river and northwards into parts of central China, and probably the Shandong peninsula during this early period, archaeological evidence from further north, south or the upper Yangzi post-dates 3000 BC (see Figure 2). In India, rice cultivation is quite widespread by ca. 2500 BC from the eastern Harappan zone in the upper Ganges

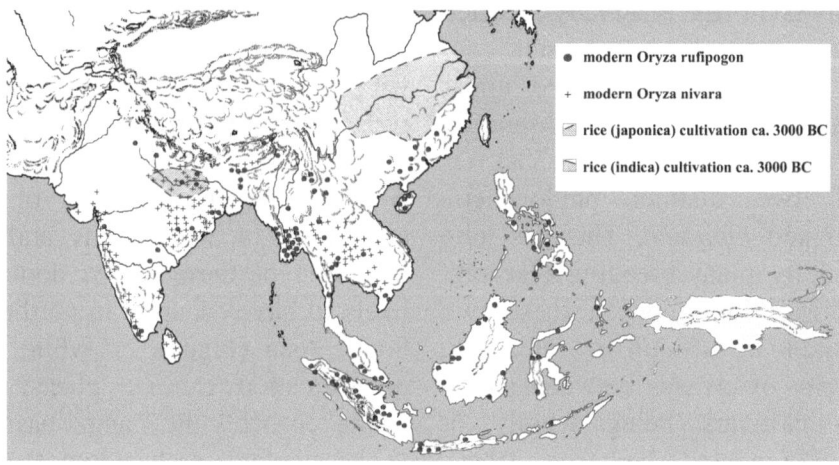

Figure 2. A map of wild rice distribution and likely zones of domestication. The distribution of the two wild progenitors of rice is plotted after Vaughan (1994). Some of these populations may be 'feral', e.g., along the Malabar coast. The extent of rice cultivation ca. 3000 BC indicated, based on archaeological evidence (for China, after Yan, 2002; for India, based on Fuller, 2002, with updated evidence discussed in text)

basin (e.g., at Kunal: Saraswat and Pokharia, 2003) and the Swat valley in northern Pakistan (Costantini, 1987) through the middle Ganges (see Fuller, 2002, 2003c; Saraswat, 2004a, 2005). A few sites with evidence for rice impressions in pottery (not necessarily domesticated) date back to the fourth millennium BC (Kunjhun II and Chopanimando), while recent excavations at Lahuradewa have been suggested to put rice cultivation back to as early as ca. 7000 BC, based on an AMS on a piece of a charred mass of rice (Tewari et al., 2003, 2005; Saraswat, 2004c, 2005; I. Singh, 2005). It must be cautioned, however, that criteria for recognizing domesticated rice as opposed to wild gathered rice remains weak and unsubstantiated, and the presence of cultivation practices is unclear. The sample size is very small, with less than a dozen grains recovered from the first season of work. While further research is needed, the recent evidence from Lahuradewa indicates at the very least that foragers were exploiting (wild) rice in the Ganges plain from ca. 7000 BC and perhaps already producing some ceramics at this date (and undoubtedly by the Fourth Millennium BC) (cf. Saraswat, 2005; I. Singh, 2005; Tewari et al., 2005). Sometime after this cultivation began and selection for domesticated rice, which may have taken one or two millennia, had taken place by 3000-2500 BC (see below, 'The Ganges Neolithic'). It is after this time when rice had spread towards the northwest in the first half of the third millennium BC, indicated by finds at Early Harappan Kunal and at Ghaleghay (see Figure 2). Whether early rice cultivation in Eastern India (e.g., Orissa) should be seen as dispersal from this same centre or a separate process, perhaps rather later, requires further archaeobotanical investigation (see below, 'The Eastern Neolithic').

East and South Again: Water Buffalo and Chicken

One of the major animal domesticates of Asia is the water buffalo. Its association with wet rice agriculture in China and Southeast Asia is well-known. Biological and archaeological evidence, however, suggest separate origins, which are unlikely to be tied directly to the centres of rice origins. Traditional taxonomy distinguishes between the swamp and river types of water buffalo, with the latter being prominent in the more semi-arid environments of South Asia and the former from the wetter lowlands of East Asia (Grove, 1985; Hoffpauir, 2000). Pleistocene or early Holocene fossil evidences include Pakistan and north-central China, as well as presumably most of the South and Southeast Asian mainland were in the wild buffalo range (see Figure 3). Traditional taxonomy suggested that distinctive swamp and river morphotypes might be distinguished, possibly with separate domestications (Zeuner, 1963). More recently mitochondrial DNA sequence data suggests at least two distinct clusters of phylogenetic diversity, suggesting two separate geographical sub-samples of the wild genetic diversity (Lau et al., 1998; Bruford et al., 2003:905). Based on modern distributions, this points again towards South Asia and East Asia. Archaeologically the challenge is to use bone evidence to distinguish wild from domesticated populations. Despite claims in the literature for a domestication in the Lower Yangzi (e.g., Chang, 1986; Bellwood, 2005), this has been based thus far on the assumption that finds of buffalo are necessarily domesticated, rather than on any morphometric data. Recently the study of water buffalo from the site of Kuahuqiao (ca. 6000-5400 BC) suggests no clear size reduction in relation to contemporary or early wild populations, and kill-off profiles are consistent with hunting, rather than specialized management (Liu and Chen, 2004). In China the first indication

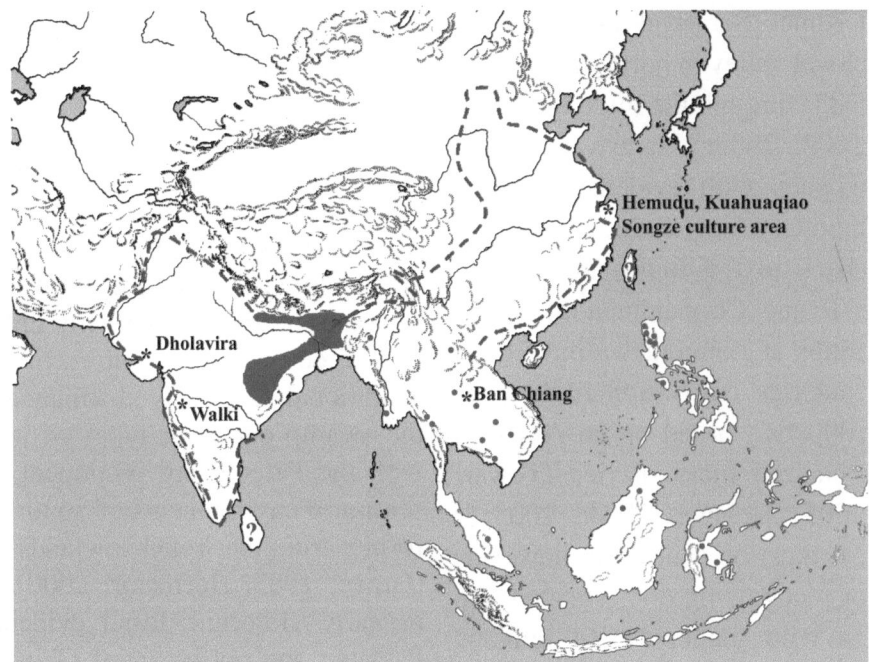

Figure 3. A map of probable Holocene distribution of wild water buffaloes, modern refugia of wild populations and important archaeological sites of buffalo remains. Modern wild distribution shown as grey areas and grey dots, while Early Holcene distribution based on Late Pleistocene/Early Holocene fossil evidence indicated by dashed line (after Hoffpauir, 2000). Question marks indicate islands where past presence of wild populations is uncertain. Note that some island populations could represent feral escapes from domestication. Selected archaeological sites, discussed in text, are indicated

for domesticated water buffalo is indirect and artifactual. The presence of large stone plough tips from the Songze Neolithic culture of the Lower Yangzi area occur for the first time by ca. 3500 BC (Shanghai Cultural Relics Protection Committee, 1962:465). These tools imply the use of animal traction, of which the water buffalo is the only indigenous candidate, and the traditional source of power. Assuming that western (or South Asian) cattle had not yet been introduced to China this date provides a minimum age for domestication of water buffalo in the Lower Yangzi. When water buffalo came into use, perhaps by dispersal, in Southeast Asia remains unclear. Water buffalo bones at Ban Chiang in Thailand date back to 1600 BC, although it is not clear whether these represent domestic animals (cf. Bellwood, 1997; Higham and Thosarat, 1998).

In South Asia by contrast, bone evidence comes from the Harappan site of Dholavira by ca. 2500 BC. Here smaller sized animals are present and make up a substantial proportion of the animal bone assemblage and present kill-off patterns that could indicate management (Patel, 1997; Patel and Meadow, 1998; Meadow and Patel, 2003). Water buffalo from Walki on the northern Peninsula from the mid-Second millennium BC have been argued to be domesticated (Joglekar, 1993).

The situation with chickens is similarly problematic in terms of determining domestic status and geographical origins (Blench and MacDonald, 2000). Wild *Gallus* sp. are well-known in South Asia, such as *G. sonnerati* in the peninsula, while the wild progenitors of domestic chickens are distributed across north and northeast India through mainland southeast Asia and Southern China.

In addition, there are several other gallinaceous birds native to South Asia, and clear comparative criteria for determining these are needed. If we give reported identifications the benefit of the doubt, then, in China, the widespread occurrence of *Gallus*-type bones by the fifth millennium BC would seems to argue for husbandry/domestication at the northern margin of the wild distribution in central China (West and Zhou, 1988; Blench and MacDonald, 2000). If we take a similar view of the numerous *Gallus* reports from South Asia, which are by and large restricted to agricultural periods (see Fuller, 2003a: Table 4), we can suggest the pattern of chicken dispersal. In western regions (Gujarat and the Indus Valley), where the wild progenitor is absent today (although this need not have been in the case in prehistory), several finds point to chicken-keeping by the Mature Harappan phase. Similarly, most early finds from north India also come from the second half of the third millennium BC. Amongst these are the quantities of 'chicken' bones from Damdama (Thomas et al., 1995a). This site is culturally Mesolithic in the sense of lacking pottery, but clearly incorporates material dating to the second half of the third millennium BC, including domesticated cereals (see discussion, below), but with an apparently wholly wild fauna (Chattopadyaya, 1996, 2002). This might suggest a particular cultural context in which chickens came to be managed in Northern India.

Thus chickens, water buffaloes and rice show essentially the same pattern, that of likely East and South Asian origins. While it is still possible, even likely, that varieties of these domesticates were introduced to South Asia from the northeast, these would only have been new forms that added to diversity already established in South Asia on the basis on indigenous domestication. Thus there is little basis to attribute agricultural origins in parts of India to demographic influx from the northeast, but we should investigate independent processes in India that paralleled those in China.

In the following section I will begin by addressing the other conventional source for diffusionist models of South Asian prehistory, population entry via the northwest. In this case archaeological, and archaeobotanical, evidence, can be considered. While domesticates of Southwest Asian origin are clearly important in South Asian agriculture, a significant early importance in subsistence is only found in northwestern South Asia. Meanwhile evidence for these Southwest Asian domesticates is limited or absent from the earliest food production in at least three parts of the subcontinent implying that local sources of food production were already established.

Indian Agricultural Traditions: Five Local Centers

In outlining the archaeology of early agricultural traditions in South Asia, I will simplify this into five key zones (Figure 4, building on Fuller, 2002, 2003b). First there is the northwest, including the greater Indus valley and its hilly flanks to the west and north. In these regions summer monsoon rains are limited or unreliable and much cultivation depends either on the limited regular winter rains or else river water, which rises in the spring and summer as Himalayan snow melts (Leshnik, 1973; Fuller and Madella, 2001). Second, there is the middle Ganges zone, an area with the benefits of both significant monsoon rains and numerous perennial river systems that are fed by the monsoons. This area incorporates significant cultural diversity in the archaeological record. Thirdly, it may be necessary to consider Neolithic traditions in Eastern India (Orissa and Jarkhand) as distinct from the Gangetic Neolithic, although the Neolithic there is still poorly documented and could relate to the Gangetic pattern (cf. Fuller, 2003a; Harvey et al., 2005). Fourthly, there is

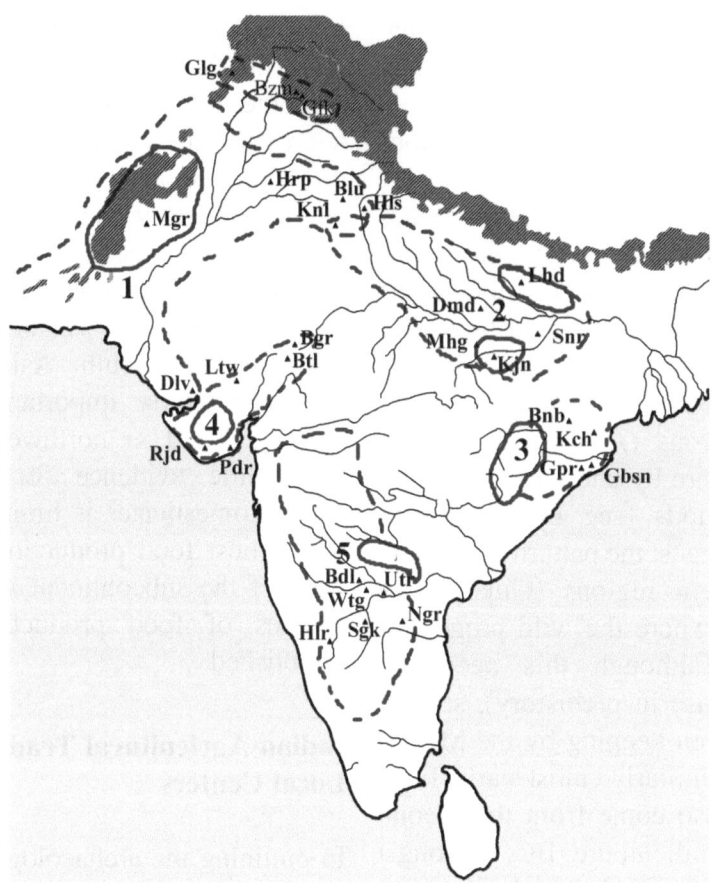

Figure 4. The major independent Neolithic zones of South Asia, with selected archaeological sites. For each zones the solid grey outline indicates best guess region(s) for indigenous domestication processes and/or earliest adoption of agriculture. The dashed lines indicates the expanded region of related/derivative traditions of agriculture; selected sites plotted. 1. The northwestern zone, with the disjunct area of the Northern Neolithic shown: Mgr. Mehrgarh; Glg. Ghaleghay; Bzm. Burzahom; Gfk. Gufkral; Hrp. Harappa; Knl. Kunal; Blu. Balu; Hls. Hulas 2. The middle Ganges zone with two possible rice domestication areas: Dmd. Damadama; Lhd. Lahuradewa; Mhg. Mahagara; Kjn. Kunjhun; Snr. Senuwar; 3. Eastern India/Orissan zone: Bnb. Banabasa; Kch. Kuchai; Gpr, Gopalpur; Gbsn. Golabai Sassan; 4. Gujarat and southern Aravalli zone: Ltw. Loteshwar; Rjd. Rojdi; Pdr. Padri; Btl. Balathal; Bgr. Bagor; 5. Southern Indian zone: Bdl. Budihal; Wtg. Watgal; Utr. Utnur; Sgk. Sanganakallu and Hiregudda; Hlr. Hallur; Ngr. Nagarajupalle

Western India, mainly evidence from Gujarat, especially the Saurashtra peninsula but possibly also parts of Southeast Rajasthan and the area around Mount Abu. This region also is favored by monsoons and represents the ecological transition from the dry Thar desert into the semi-arid monsoon tropics that support a mosaic of savannahs and deciduous woodlands. Fifthly, there is the Southern Neolithic zone in the semi-arid peninsular interior which has received increasing attention as a region of domestication of monsoon-adapted pulses and millets in the later middle Holocene (Fuller et al., 2001, 2004; Fuller and Korisettar, 2004; Asouti et al., 2005).

The Northwest and the Indus

In northwestern South Asia, the dominant crops from the time of earliest evidence derived from the Southwest Asian Neolithic Founder crops (Zohary, 1996; Zohary and Hopf, 2000). These crops, especially wheats and barley, but also lentils, peas, chickpeas, grasspea, flax and safflower, can now be placed in the Levantine zone and southeastern Anatolia. Cultivation of some of the cereals has now been postulated for the Late Pleistocene, after ca. 11,000 BC, while domesticates are clearly widespread in the region by the beginning of the Pre-Pottery Neolithic B (ca. 8800 BC) (Harris, 1998a; Willcox, 1999, 2002; Garrard, 2000; Moore et al., 2000; Hillman et al., 2001; Colledge and Conolly, 2002; Charles, 2006). Representatives of this crop package had spread to Central Asia by ca. 6000 BC, the time of the Djeitun Neolithic (Harris, 1998b) and to western Pakistan by the time of Neolithic Mehrgarh. The second ceramic phase at Mehrgarh begins ca. 6000 BC, as recent stratigraphic reassessment indicates (Jarrige et al., 2006). The earlier aceramic period at the site is estimated to have begun by ca. 7000 BC (Jarrige, 1987; Meadow, 1993; Possehl, 1999; Jarrige et al., 2006). Despite some arguments in favor of cereal domestication in Pakistan (e.g., Possehl, 1999), the lack of wild progenitors (for wheats, all the pulses, flax and safflower) and the late available dates by comparison to Southwest Asia, points towards the spread of crops, and this could have involved the spread of farmers, although diffusion of just the crops is possible too. This Southwest Asian agricultural package was well-established and widespread in the Indus region by the time of Harappan urbanism in the Third Millennium BC (Meadow, 1996, 1998; Fuller and Madella, 2001), although it is not yet clear whether all of the crops which were present by then had arrived already by the Neolithic.

While the staple crops were all introduced, livestock and other crops indicate a number local domestications. The best documented of these is the domestication of zebu cattle inferred from metric changes in bones through the Mehrgarh sequence as well as distinctive humped cattle figurines (Meadow, 1984, 1993). Phylogenetic evidence from DNA is also clear in indicating a separate domestication (or two) of humped zebu cattle from Near Eastern (and African) taurine cattle (MacHugh et al., 1997; Bradley et al., 1998; Bruford et al., 2003; Kumar et al., 2003; Magee et al., this volume). Goats appear domesticated from the earliest occupation at Mehrgarh, but recent genetics suggests one or two domestications of goats additional that of the Near East (probably Iran) (Luikart et al., 2001; Bruford et al., 2003:905). Genetic evidence for sheep is similar, with a plausible domestication in Central Asia or Baluchistan (Hiendleder et al., 2002; Bruford et al., 2003:905). Bone evidence from Mehrgarh could indicate a sheep domestication process in this region (Meadow, 1984, 1993). In addition the fibre crop cotton appears at Mehrgarh during the Neolithic, perhaps by 5000 BC, and is a likely domesticate of this region (Costantini and Biasini, 1985; Fuller, 2002; Moulherat et al., 2002). The native cotton, *Gossypium arboreum*, is a woody shrub and as such was likely to have been cultivated in perennial orchards like fruits. Mehrgarh also provides evidence for grapes and jujube that might have been cultivated or managed for fruit. The status of the large true date seeds from Mehrgarh is problematic as they are uncharred and undated, but at the Harappan site of Miri Qalat in Makran wild type date stones (probably *Phoenix sylvestris*) occur confirming date consumption (and probably cultivation) in this region (Tengberg, 1999), while true dates (*Phoenix dactylifera*) were certainly present in Iran (Tengberg, 2005). Sesame is also domesticated in this region although the earliest finds are from the Mature Harappan period (Fuller, 2003d; Bedigian, 2004). Another important domesticate of the Indus region is the water buffalo, which has

been well-documented as a domesticate at
the Harappan city of Dholavira in the great
Rann of Kutch, culturally and climatically an
outlier of the Sindh region (Patel and Meadow,
2003).

This Harappan agricultural system, with
a large component derivative from further
west, was constrained by a major climatic
frontier from spreading further east. The
greater Indus region and the Indo-Iranian
Borderlands lack reliable monsoon rainfall,
whereas in the eastern zones of the Harappan
civilization (such as eastern and northern
Punjab and Haryana), monsoon rains are
consistently more reliable. It is such a zone
where we would expect reliance on rainfed
summer crops to have been important, and
indeed Early Harappan and Mature Harappan
archaeobotanical evidence from this region
consistently shows the presence of native
Indian monsoon crops alongside the Harappan
(Near Eastern) winter crops (e.g., Saraswat,
1991, 1993, 2002; Willcox, 1992; Saraswat
and Pokharia, 2002, 2003). While many of
the monsoon crops may have spread to the
region from areas to the east, such as the
middle Ganges, hard evidence for this is yet
to be established for this origin. It is possible
that some indigenous domestication occurs
in the Himalayan foothills or the Ganges-
Yamuna Doab region. Of particular interest
in this regard is the presence of small, Indian
millets from Early Harappan levels at Harappa
(back to the Ravi Phase, ca. 3200 BC),
especially *Panicum sumatrense* (Weber, 2003)
as this hints at domestication of monsoonal
millet crops that is earlier than and perhaps
independent of those further south, in peninsular India, or in Gujarat. Further archaeological evidence is needed to document the
emergence of agricultural villages and pre-
Harappan sites in this eastern Harappan zone
and the upper Ganges as well as their cultural
relations to developments in the middle
Ganges.

The Northern Neolithic

Another but later Neolithic tradition is
documented from Kashmir and the far north of
Pakistan (the Swat Valley). Generally known
as the Northern Neolithic, this tradition is best
represented by sites in the Kashmir valley,
although related sites can be identified in
Swat (Northwest Pakistan). Here sites occupy
the milder valley bottoms and begin to be
occupied in the later Fourth Millennium BC in
an aceramic phase, known from recent exaca-
vations at Kanishpur (Mani, 2004) as well
as older work at Gufkral (Sharma, 1982).
Ceramic production has begun ca. 3000 BC
and sites appear to be significantly more
widespread by the end of the third Millennium
BC (e.g., Allchin and Allchin,1982:111–116;
Sharma, 1982, 1986; Mani, 2004). The earliest
phases are characterised by broad deep pits,
with bell-shaped profiles. While these have
conventionally been interpreted as pit houses,
recent debates have raised the likelihood that
they were large storage features (Conningham
and Sutherland, 1998). Whatever the case it
is clear from these sites that the dominant
crops were winter wheat (including free-
threshing and emmer), barley, peas and lentils
(Kajale, 1991; Lone et al., 1993; Pokharia
and Saraswat, 2004), and thus derive from
the same ultimate Near Eastern source. Faunal
evidence includes sheep, goat, and cattle,
while the status of buffalos and pigs requires
confirmation (see review by Kumar, 2004).
The plant evidence is therefore opposed to the
idea that the Kashmir Neolithic can be related
to a westward dispersal of millet-growing
Sino-Tibetan speakers as some have argued
(Parpola, 1994:142; Van Driem, 1998:76–84;
Possehl, 2002:39). The crops and livestock
species present are clearly not those of
Yangshao China. The presence of Chinese
like stone harvesting knives in Kashmir
remains curious but must be regarded as
a technological diffusion given the subsistence data, and these forms only occur in
later Neolithic phases such as Burzahom

II and Gufkral 1C (Allchin and Allchin, 1982:figure 5.9; Sharma, 1982; Kumar, 2004). These harvesters also appear around this time further south in Baluchistan in the Late Harappan era, as at Pirak (Jarrige, 1985,1997). The agricultural situation might therefore be congruent with the suggestion of a distinct linguistic substrate in Kashmir (Witzel, 1999:6–7). It is possible that the Near Eastern crops had diffused to local hunter-gatherers from the Indus region to the South or from Central Asia (the latter favored by Lone et al., 1993), together with domesticated animals. Although an immigration of farmers from these directions is also possible. It is tempting to suggest that the late arrival of agriculture here was due to an ecological barrier, as cultivation here requires winter tolerant, vernalizing forms of cereals and might therefore be compared to the processes involved in the delay of agricultural spread between Southeast Europe and the central European plains (cf. Bogaard, 2004:160–164).

Subsequently, early in the Second Millennium BC, during the Late Harappan transition, we can infer that the northern Pakistan/Kashmir region had developed contact with cultural groups to the north/east in the Chinese cultural sphere, indicating either long-distance trade or immigration into adjacent Himilayan zones of Sino-Tibetan speaking groups. At this time stone harvest knives appear in Kashmir, and similarly they appear further south in Baluchistan in the Late Harappan era, as at Pirak (Jarrige, 1985, 1997). As discussed by Jarrige (1985, 1997) this period sees important changes in cooking techniques as well. Impressions in pottery from Ghalegay, together with grains from Bir-Kot-Gwandhai, suggest some localized *indica* rice cultivation by 2500 BC (Constantini, 1987), which must have diffused from the Gangetic region to the Southeast. By contrast later Harappan rice from Pirak (after 1900 BC), has notably shorter, plumper grains, suggesting *japonica* type (Costantini, 1979), which is also supported by the form of bulliform phytoliths from the site that suggest *japonica* (Sato, 2005), which therefore supports the contention of diffusion from China by the early Second Millennium BC.

The Ganges Neolithic

Although there is much to be resolved in terms of dating and domestication status of remains from the middle Ganges, this region is a likely centre of domestication. The earliest well-sampled levels contain potentially native crops, including rice, millets and slightly later monsoon pulses, while later levels include introduced winter crops. This suggests that when wheat, barley and lentils diffused from the west they were adopted into already established systems of cultivation. At the site of Mahagara, south of Allahabad on the Belan river, the adoption of these winter crops occurs ca. 1800-1700 BC (Harvey et al., 2005; Harvey and Fuller, 2005, unpublished dating evidence), whereas further north and east at Senuwar this adoption occurred perhaps ca. 2200 BC (Saraswat, 2004a). Recently directly dated barley from Damadama is ca. 2400 BC (Saraswat, 2004b, 2005a), while from new research at Lahuradewa, it occurs in Phase 2, 2500-2000 BC, directly dated to ca. 2200 BC (Saraswat and Pokharia, 2004; Saraswat, 2005). The crop that is consistently present at all these sites from the earliest phases is rice, although small millets are also consistently reported. In the case of Mahagara these include the widespread *Brachiaria ramosa* and *Setaria vertcillata*, whereas *Setaria pumila* is reported from Senuwar and Lahuradewa. While there remains room for concern over consistency of millet identification criteria, as well as problems of intrusive millets from later periods, it is nevertheless clear that one or more small millets were part of the early cultivation systems of the Ganges. Native Indian pulses are also present, especially *Vigna radiata* and *Macrotyloma uniflorum*, but these are in no case present from the earliest levels of sites and might therefore be adopted from an adjacent region of India. While the mungbean has wild progenitor population in parts the Himalayan foothills and central Indian hill ranges, wild horsegram is not yet documented close to this zone, which therefore suggests dispersal of native pulses from further south, or perhaps west, by ca. 2000 BC, although extinct progenitor populations might conceivably have occurred in drier parts of central India or the southern Vindhyas. Although there are cucurbit (gourd) crops native to north India (Decker-Walters, 1999; Fuller, 2003a), hard archaeological evidence is still limited to ivy gourds (*Coccinia grandis*) from (early?) Harappan Kunal (Saraswat and Pokharia, 2003), Balu (Saraswat, 2002) and Late Harappan Hulas (Saraswat, 1993), and *Luffa cylindrica* after it had dispersed to South India (Neolithic Hallur) by the mid-Second millennium BC (Fuller et al., 2004).

Still to be clarified is whether there was one main trajectory towards agriculture or dispersed parallel trajectories in different local traditions, and what role interactions between early farmers and hunter-gatherers played. At present we might discern at least three contemporary cultural/economic traditions in the region. At present three distinct cultural traditions can be defined, each of which passed through two or three economic stages. First there is a tradition located in the eastern part of this region. Its earliest stage, represented by the site of Lahuradewa shows evidence for occupation on a lake edge back to the 7th millennium BC (Tewari et al., 2003, 2005;Saraswat, 2004c, 2005; I. Singh, 2005). Already in this period, or certainly by sometime in the the fifth millennium, ceramics had begun to be produced, and rice was part of the diet, and may even have been cultivated, although the very limited evidence available to date is inconclusive and is more suggestive of wild rice collecting. All the fauna thus far studied from that period were wild (Joglekar, 2004), and it is likely that occupation was intermittent (with hiatuses), or else highly seasonal to account for the long timespan of 3000–3500 years that relates to this lowest layer less than 50 cm thick). Intriguingly, the ceramic assemblage does not yet suggest much perceptible change during the period, although the third millennium levels include several new forms including some that suggest influence from the Harappan zone to the west. In the third millennium and certainly during the period 2500-2000 BC, settlement probably became more regular, evidence for cultivation is less ambiguous, and new species from external sources were adopted, in particular barley (Saraswat, 2004c, 2005), as well as pulse species that may also be non-local. In this period at least some domesticated sheep/goats are present (also adopted from the west). At this period agricultural village settlements are being founded over a wider region, such as Senuwar (Saraswat, 2004a, 2005), suggesting the filling in of the landscape with agriculturalists and the emergence of sedentary settlements. After 2000 BC a wider crop repertoire is present, including summer and winter pulses and the faunal assemblage is predominantly domesticated including cattle, sheep and goats. Clay lined storage bins suggest more investment in permanent facilities at the site. A second tradition that shows parallel economic developments, but possibly following regionally distinct timing is found in the northern Vindhyan hills and the Son and Belan river valleys. An earlier phase of seasonal settlement, ceramic production and some rice use (if not cultivation) is indicated by sites like Kunjhun II and Chopanimando, dating back to the fourth millennium BC, with earlier preceramic roots (Sharma et al., 1980; Clark and Khanna, 1989). It should be noted that the pottery from Chopanimando is a distinct cord-impressed style that does not match that from most other sites in the region, and suggests a local ceramic

'Mesolithic' tradition that developed amongst some Vihdyan hunter-gatherers. It is only in the early second millennium BC that sedentary village sites are widely founded in the region, including sites like Mahagara and Koldihwa (the latter possibly seasonal) (cf. Sharma et al., 1980; Harvey and Fuller, 2005; Harvey et al., 2005) and Tokwa (Misra et al., 2001, 2004). These sites have evidence for monsoonal crops, such as rice and millet from the earliest period and then at later levels the addition of Indian pulses, and winter crops like wheat, barley and lentils. By this period there is also clear evidence of animal herding, including sheep/goat and cattle, and features such as an animal pen with hoof impressions at Mahagara (Sharma et al., 1980).

The third tradition in the region is a persistent tradition of hunter-gatherer-fishers focused on oxbow ponds of the greater Ganges floodplain. Numerous Mesolithic sites are known in the region, especially in the region north of the modern Ganges river, such as Damadama (see Pandey, 1990; Lukacs and Pal, 1993; Chattopadyaya, 1996; V.D. Misra, 1999; Kennedy, 2000:200–205; Lukacs, 2002, see Lukacs, this volume). Although the available dates from these sites (Mahadaha, Sahar-Naha-Rai, and Damdama) range widely from the start of the Holocene (8000–10,000 BC) to 2000 BC, there are now clear grounds for assuming at least some overlap between this aceramic 'Mesolithic' cultural tradition and the ceramic 'Neolithic' food producers in adjacent regions to the South and East. This comes in the form of two direct AMS dates of the second half of the Third Millennium BC on barley (an introduced domestic) and rice (plausibly a domesticate, especially by this time) from Damdama (Saraswat, 2004b, 2005). Thus crop cultivation, or at least significant quantities of traded cereals, must have contributed to the economy of the hunter-fishers of the Ganges at least after 2500 BC; these groups remained hunting wild fauna and did not use pottery. The interrelationships between these traditions still need to be elucidated (cf. Lukacs, 2002) and the role of local domestications versus crop adoptions needs to be assessed. The presence of crops that plausibly originated in this zone, such as rice, by the early third millennium BC in the upper Ganges region, e.g. at Kunal (Saraswat and Pokharia, 2003) and further afield in Swat (Ghalegay, Costantini, 1987) suggest that agriculture was established in the middle Ganges by 3000 BC, but if so, the communities of these early farmers have remained largely undiscovered, and were presumably less sedentary than their late third millennium successors.

The Eastern Neolithic

Early agriculture in eastern India (Orissa) is still largely unknown. As has often been discussed this region has widespread populations of wild rice (*O. nivara* and *O. rufipogon*). The native millets and *Vigna* pulses could also be domesticated in this region, as could the north Indian cucurbits and the tuber crop taro (*Colocasia esculenta*). Uniquely wild in this region is the pigeonpea (*Cajanus cajan*). At present the main excavated sites are late Neolithic mounds from the coastal plains or the Mahanadi River valley, such as Golbai Sassan, Gopalpur and Khameswaripalli established by the end of the 3rd millennium BC or during the 2nd millennium BC (Sinha, 1993, 2000; Mohanty, 1994; Kar, 1995, 2000; Kar et al., 1998; Behera, 2002; Harvey et al., 2006) These sites probably relate to the settling down of already agricultural populations, and the earliest phases of agriculture in this region are yet to be documented archaeologically. Archaeobotanical evidence from the later and better established phases of Gopalpur and Golbai Sassan (after 1500 BC) indicates cultivation of rice and native pulses (mung, urd, horsegram and the local pigeonpea). Small millets are present (including *Panicum sumatrense, Setaria* sp. and *Paspalum sp.*) but these may occur as rice weeds or subsidiary crops (Harvey et al., 2006). A single winter crop, lentils, is present indicating a contrast from the Ganges where a wider range of winter crops is prominent. The available faunal data indicates domestic fauna (including bovines and caprines), while artifacts point to the importance of riverine fishing. Reconnaissance of upland Neolithic sites in the Orissa hills suggests a very different Neolithic tradition. Here, sites such as Banabasa (Harvey et al., 2006), appear to have been non-sedentary and largely non-ceramic, suggesting the likelihood of a pattern of shifting cultivation. An older excavation at the site of Kuchai, in the northern Orissa foothills, can probably be connected to this upland tradition, and showed a transition from microlithic technology to ceramics with ground stone axes (including the shouldered celts which are a typical component at these upland sites) (Thapar, 1978, *Indian Archaeology 1961–62-a Review*). Ceramics are reported to include rice husk impressions (Vishnu-Mittre, 1976), but there is no further basis for inferring a more complete subsistence system.

Pre-Harappan Western India: Gujarat and Adjacent Rajasthan

Gujarat is likely to have been a centre for the domestication of local, monsoon-adapted crops, after livestock was adopted into this area from the Indus region to the west. Archaeobotanical evidence for the beginnings of cultivation in this region is not yet available, and the earliest ceramic bearing sites, of the Padri and Anarta traditions (ca. 3500–2600 BC) have so far not yielded plant remains. Nevertheless, these sites have produced evidence for some domestic fauna, including directly dated cattle bones from the fourth millennium BC from Loteshwar (Patel, 1999; Meadow and Patel, 2003) and probable domestic fauna from Padri (Joglekar, 1997;

Shinde, 1998a) and Prabas Patan (P. Thomas, 2000). Other sites, such as Bagor, which are often cited as evidence for adoption of livestock by mid-Holocene hunter-gatherers (e.g., Possehl, 1999), need archaeozoological reassessment in light of a refined understanding of the difficulties of separating sheep and goat from blackbuck antelopes (cf. Meadow and Patel, 2003). While livestock are being adopted into this region, it is plausible that ceramic bearing sites in the wetter Saurashtra, as opposed to the desert fringe sites, were sites of communities of cultivators. In the Mature Harappan period (from 2600 BC), a period from which systematic archaeobotanical evidence is available, a stark contrast can be drawn between millet-dominated agriculture of Saurashtra and wheat-barley-winter pulse agriculture of the Indus valley and the Harappan core (Weber, 1991; Reddy, 2003). While there have been recent controversies over identification of millets in this region (Fuller et al., 2001, 2002, 2003b), it is clear that native Indian small millets were predominant.

The crop, little millet (*Panicum sumatrense*, which is native to monsoonal India), and a species (or two) of *Setaria*, were cultivated (probably those which are native such as *S. verticillata* and *S. pumila*). In addition, it is now apparent that *Brachiaria ramosa* was present at Rojdi (probably replacing the reported identifications of the introduced *Setaria italica*) (Weber, personal communication). It is possible that these species were domesticated in Saurashtra, although hard evidence for the process is lacking and other regions may also have witnessed domestication of these species (such as *Brachiaria ramosa* in South India and *Panicum sumatrense* in Punjab). By the latest period of Rojdi C (2000–1700 BC), crops from Africa were introduced, including sorghum, pearl millet and finger millet – the presence of some of the latter now seems clear on morphological grounds despite earlier concerns (Weber, personal communication, cf. Fuller, 2003c), although a full reassessment of contextual dates of these crops is needed. The pulse urd, *Vigna mungo*, which is native to the northern Peninsula or the southern Aravallis, is present from early Rojdi (ca. 2500 BC) and could represent a local domesticate, while horsegram (*Macrotyloma uniflorum*) and mungbean (*Vigna radiata*) are adopted by Rojdi C (2000-1700 BC). In general despite ties in trade and culture with the Harappan Indus valley, the archaeobotany of Gujarat is much more peninsular in character, suggesting a tradition of cultivation distinct from that of the Indus valley but plausibly from hunter-gathering roots similar to that of the Southern Neolithic. Recent research in Rajasthan on the Ahar/Banas culture region, indicates that agricultural villages were clearly established by ca. 3000 BC, as at Balathal (Shinde, 2002). What is less clear is whether this should be connected with Gujarat and indigenous domestications or agricultural dispersal from the Indus region (as postulated in Fuller, 2003b). The archaeobotanical evidence from the mid to late third millennium BC (Kajale, 1996), indicates predominance of the Near Eastern winter crops, a clear contrast with Gujarat.

The Southern Neolithic

The Southern Neolithic, of northern Karnataka and southwest Andhra Pradesh, provides the earliest evidence for pastoralism and agriculture in Peninsular India (Korisettar et al., 2001a, 2001b; Fuller, 2003b, 2006). A well-known site category of the Southern Neolithic is the ashmound, which has been shown (especially at Utnur and Budihal) to be an accumulation of animal dung at ancient penning sites that have been episodically burnt, sometimes to an ashy consistency, and sometimes to a scoriaceous state (Allchin, 1963; Paddayya, 1998, 2001). Animal bones (at all sampled sites) indicate the dominance of cattle in the animal economy, with a smaller presence of sheep and goat (Korisettar et al., 2001a, 2001b). Although Allchin and Allchin (1974, 1995) have made a case for local domestication of zebu varieties in the South, this suggestion is not yet corroborated by archaeological bone evidence. Their argument is based on the morphology of rock art depictions which contrast with contemporary Harappan depictions and suggest the kind of varietal differentiation between southern and northwestern zebus was already established. Recent archaeobotanical research has provided a picture of recurrent staples and occasional secondary crops of the Southern Neolithic (Fuller et al., 2001a, 2001b, 2004; Fuller, 2003b, 2006). The staples include two native species of millets (*Brachiaria ramosa* and *Setaria verticillata*) and two pulses (*Vigna radiata* and *Macrotyloma uniflorum*). What is known of the ecology of these species suggests that domestication occurred in a Dry Deciduous woodland zone that interfingered with savannah scrub (favored by *Macrotyloma uniflorum*) and moist deciduous woodland (favored by *Vigna radiata*). The millets would have occurred patchily throughout these zones. While this zone has been argued to be on the inside of the Western Ghats (Fuller and Korisettar, 2004), patches along the Eastern Ghats between the Krishna and the Godavari river are now favored on the basis of recently gathered data on wild progenitors of the *Vigna* pulses (Fuller and Harvey, 2006). The modern distribution of these ecological zones in the peninsular region is illustrated in Figure 5. When climatic conditions were wetter during much of the early and mid-Holocene we would expect the Moist Deciduous zones to have expanded (especially eastwards towards the central peninsula, and for the savannah/scrub zones to have been reduced by impinging dry deciduous woodlands (Fuller and Korisettar, 2004). Some of the areas that are today Dry Deciduous forests with a significant teak (*Tectona grandis*) element that occur in

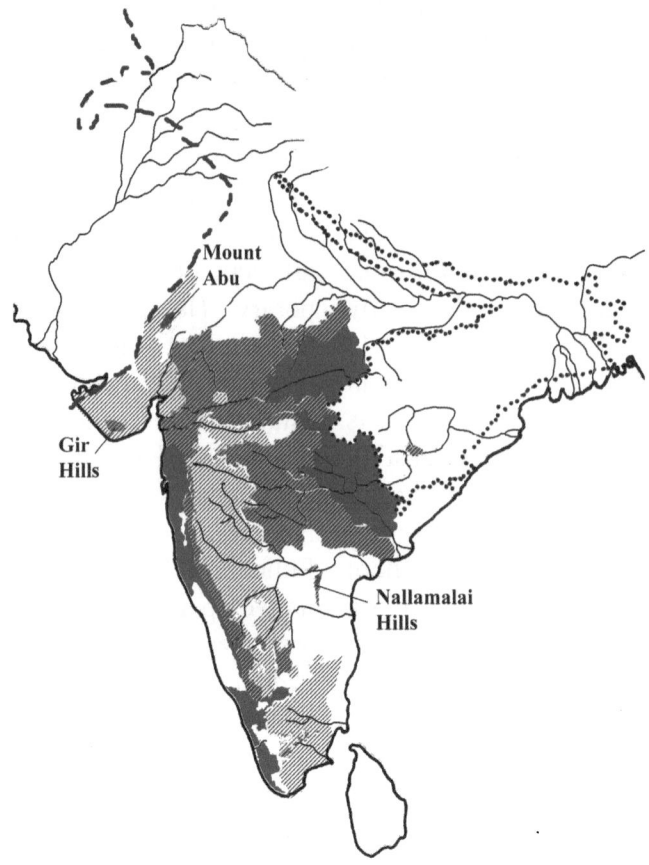

Figure 5. Map of important ecological zones of peninsular India relevant to understanding agricultural origins (after Asouti and Fuller, in press; based on Puri et al., 1983, 1989; Meher-Homji, 2001). The dark grey zone indicates Moist Deciduous forests with teak (*Tectona grandis*) as an important element, while the black dots indicates the western extent of the sal tree (*Shorea robusta*). The Dry Deciduous teak forests are darkly hatched (*Hardwickia* dominated dry deciduous forests have been excluded), while savannah-scrub areas are lightly hatched. The grey dashed line indicates the western boundary of the monsoon zone, east of this line summer rainfall averages more than 40 cm per year

the hills of the eastern peninsula (Eastern Ghats) would have been Moist Deciduous in character. It is such forests where we might expect former extensions of the wild mungbean, from which domestication could have occurred.

In addition there are data that non-native taxa were adopted into cultivation during the Southern Neolithic. These include wheat and barley by ca. 1900 BC (but only on a minority of sites), hyacinth bean (*Lablab purpureus*, probably a native of East Africa), African pearl millet (*Pennisetum glaucum*) and pigeonpea (*Cajanus cajan*, from

Orissa or adjacent parts of eastern India) and the vegetable *Luffa acutangula* (from North India), all of the latter by ca. 1500 BC. There is still no clear sequence from foraging to farming, and indeed archaeobotanical evidence to assess the earliest Southern Neolithic agriculture is still lacking from archaeological Phase I (3000–2200 BC). Nevertheless, the existing evidence indicates dependence on a group of species that are native to the peninsula, with non-native species being rare (on a minority of sites) or occurring only in the latest Neolithic period (Phase III), e.g., the African crops. Although there are a few grains of rice from Hallur, these are most likely grains from a wild form (Fuller, 2003b:378, n.2), which could have infested millet fields along the upper Tungabhadra as a weed. Evidence for cultivation and consumption of rice occurs only in the Iron Age (Kajale, 1989; Fuller, 2002). The archaeology of the Southern Neolithic suggests increasing sedentism over most of the region only after 2000 BC and especially during Phase III. This suggests that population densities began to fill in the landscape by comparison the earlier phases of the Neolithic, when we might expect forms of shifting cultivation (and perhaps shifting settlement) to have been practiced. This filling in of the landscape is reflected in the west coast pollen evidence for deforestation focused on ca. 1500 BC (Fuller and Korisettar, 2004).

It is only at this time that settled agricultural villages become widespread on the peninsula, consistent with a model of demographic expansion of early peninsular farmers. For example, the millet-pulse-livestock agriculture of the Ashmound Tradition dispersed southwards and eastwards to adjacent regions. Evidence from the Kunderu river basin, just beyond the eastern distribution of the ashmounds indicates that the same subsistence package was established between 1900 and 1700 BC (Fuller et al., 2001b; Fuller, 2006). There is now new evidence for contemporary hunter-gatherer groups living in caves of the Erramalai hills who were in interaction with the ceramic producing farmers of the Kunderu plains. The cultural differences, in terms of the lack of ashmounds and some distinctive aspects of ceramic style, might suggest that this represents cultural diffusion, it is equally likely that this represents an immigrant group with some cultural traits that set out from the core ashmound tradition into agriculturally virgin land where they could continue traditions of shifting cultivation rather than more intensive methods that would have been adopted in the

Southern Neolithic core. This is suggested for example by limited evidence for thin ashmound-like deposits at the base of the Nagarajupalle Neolithic site in the Kunderu river basin (author's data). It may also be the case that this Southern Neolithic agricultural tradition dispersed northwards, but if so it was of a less sedentary and less visible form of settlement than the later Malwa tradition, which became established ca. 1800 BC with well documented village sites on the middle Tapti river, the upper Godavari and the upper Bhima (Shinde, 1998b; Panja, 1999, 2001). At this later stage agriculture had a large component of Harappan elements, wheat, barley and the winter pulses, but also the native (or Southern Neolithic) pulses and small millets. Full identification details of Malwa/Jorwe millets is not available, but it is clear the *Brachiaria ramosa* is amongst them (Kajale, personal communication; for important published datasets see, e.g., Kajale, 1979, 1988, 1990, 1994), in addition to the urd bean which may have originated in this northern peninsular zone (or Gujarat).

A general process which can be perceived in the archaeological evidence is the replacement of older millet species by more productive millet types and in many cases by rice. This has clearly occurred in Peninsular India since Neolithic *Brachiaria ramosa* and *Setaria verticillata* have largely given way to Central Asian/Chinese *Setaria italica* and African *Pennisetum, Sorghum* and *Eleusine*, a process that can be perceived in Early Historic archaeobotanical samples and has finished by the colonial period. These later cereals are more productive and, in the case of the African cereals, generally free-threshing making them less labor intensive to prepare. In other areas millets have been replaced with rice, a process which began when rice first appeared at some sites in the first millennium BC, after 1000 BC. Dry rice cultivation is essentially equivalent in ecology to the wetter forms of millet cultivation, such as river bank cultivation of *Panicum sumatrense* or *Echinochloa* and in some areas such as the drier Bellary and Kurnool district has occurred in the past couple of decades with the expansion of irrigation canals. A significant implication of this process is that we might expect a semantic shift to have occurred from more ancient millets to more recent intro- duced millets or rice which came to take their place in agricultural and dietary impor- tance. We must therefore consider the possi- bility that linguistic evidence may prove to be biased towards these modern replacements and mask prehistoric semantic shifts which have occurred in parallel across separate language family branches.

Setting the Speech Scene: Languages Real and Inferred

Historical linguistics is doubtless a reflection of past population movements and interactions, as are genetics. Much recent research on integrating linguistics with archaeology (and genetics) has happened in the past two decades since the publication of Renfrew's (1987) *Archaeology and Language* (see also Blench and Spriggs, 1999; Renfrew, 2000; Blench, 2004). As physical anthro- pology cannot define races, neither can pure languages be defined. The process of language change *and mixing* is complex, as variants enter a pool in which selection takes place for a variety of social and cultural reasons (Mufwene, 2001). Variants from different speakers are pooled, recombined and selected for transmission to subsequent generations. In cases of general cultural homogeneity, without significant migration, the variants are all similar, thus most language lineages have traditionally remained stable through time, but in some contexts speakers of diverse origins may influence each other and thus transmit to future generations a mixed linguistic heritage. All historical linguists accept that substrate languages have left their mark on now dominant languages, implying considerable periods of interaction amongst different language speakers and bilingualism (Crowley, 1997:197; Witzel, 1999; Southworth, 2005a:98–125); this is perhaps difference in *degree*, but not in kind, to the kinds of processes of language transmission involved in creating historical creole languages, where the speakers contributing to a speech variant pool are from much more diverse backgrounds (see Mufwene, 2001). Thus while it is undoubtedly true that languages are carried with the movement of speakers (Bellwood, 2001, 2005:190–193), the number of speakers vis-à-vis pre-existing populations is a matter that is more difficult to infer (but for a model, see Ehret, 1988). In order to get at this we need to try to frame periods of language interaction in time and space so that we can consider the likely historical and social circumstances that were involved, which ultimately can be informed by archaeological evidence.

Our improving grasp of early agricul- tural traditions in South Asia (at least those that were becoming sedentary), and the biogeography of their cultivars as well as wild flora, means that there is a basis for assessing linguistic data. The assessment that follows improves upon and revises that of Fuller (2003a). This earlier study began with an assessment of the antiquity of different plants in the archaeology of South India and then looked at the distri- bution and probable antiquity of words for these selected species across the Dravidian languages (building on Southworth, 1988). Some initial comments were also formulated on possible north Indian domesticates and unknown substrate language(s) of Indo-Aryan (based on Masica, 1979) as well as Proto- Munda agricultural vocabulary (based on Zide and Zide, 1976). In addition to archaeob- otanical advances, there have been signif- icant linguistic advances in recent years. Of note are efforts to identify distinct substrata that have influenced Indo-Iranian and Indo-Aryan languages at different periods and a relative chronology of these substrates (Kuiper, 1991; Witzel, 1999, 2005, 2006; Southworth, 2005a, 2005b; Southworth and Witzel, 2006), and new efforts to reconstruct early Dravidian vocabulary (Krishnamurti, 2003; Southworth, 2005a; but see some reservations, below). Recent analysis that explains much of the evolutionary divergence of Austroasiatic into Munda and Mon-Khmer, which are opposite in many linguistic structures, also has significant historical implications (Donegan and Stampe, 2004). One clear indication of this work is that we need to break free of the present as a complete key to the past: there were languages spoken in the past that are not reflected directly in those known at present. There are dead language families. But these have nevertheless left their mark through loanwords and other substrate features.

In this consideration of South Asian linguistic prehistory I focus on the three major living language families: Dravidian, Austro-Asiatic and Indo-European. For the present consideration I will leave aside the complex Himalayan situation and the northwestern periphery of the subcontinent with its isolate Burushaski and the Dardic group of Indo-European languages (but see Witzel, 1999, 2005). There are thus three major families, plus the isolate of the upper Tapti river, Nahali. Indo-European languages are represented by the Indo-Aryan languages located today throughout northern and northwestern South Asia, with earlier linguistic forms preserved in Sanskrit literature such as the Rig-Veda (Southworth, 2005a). The peninsula is predominantly Dravidian. While Munda language groups are concentrated in the hills of Eastern India, where they often encapsulate smaller Dravidian languages, including the poorly documented North Dravidian Kurux (Oraon) and Malto. On the hills of northern

Maharashtra the isolated North Munda Korku language, occurs adjacent to the isolate Nahali (Figure 6). Nahali has been related by some authors to a hypothetical extinct Bhil language (Witzel, 1999:62–63; Southworth, 2005a). In addition, extinct substrate languages are clearly indicated for the Nilgiri hills (Emeneau, 1997; Witzel 1999:64) and the Veddas of Sri Lanka (Witzel, 1999:64; Southworth, 2005a). While most of these substrate languages are likely to have been of hunter-gatherers, two major extinct agricultural languages can be inferred for north and northwest South Asia (see below).

Although there has been archaeological discussion of an agriculturally-driven dispersal of Indo-European (specifically Indo-Aryan) into India (e.g., Renfrew, 1987; Bellwood, 2005), this hypothesis lacks support from specific linguistic or archaeological evidence (cf. Fuller, 2003a). Witzel (2005) provides the most recent, comprehensive attempt to infer the route and historical context of Indo-European entry into the subcontinent, including inferred substrate words from a lost Central Asian language, attributed to the Bractria-Margiana archaeological complex of the third millennium BC (e.g., wheat, hemp, sheaf, seed, Bactrian camels and donkeys), as well as words shared with northwestern substrates of the northwestern frontier (Burushaski) and Kashmir. The important evidence for an inferred Harappan substrate is taken up below. A model of two different branches of Indo-Aryan, an 'inner' branch focused on the central Ganges and an outer branch that extended from Sindh through the northern Peninsula and central India towards the east, will not be pursued below as these must relate to cultural processes that occurred after the establishment of agriculture in most regions but they may nevertheless be significant elements in Late Chalcolithic/Iron Age cultural processes in parts of India (for discussion, see Southworth, 2005a).

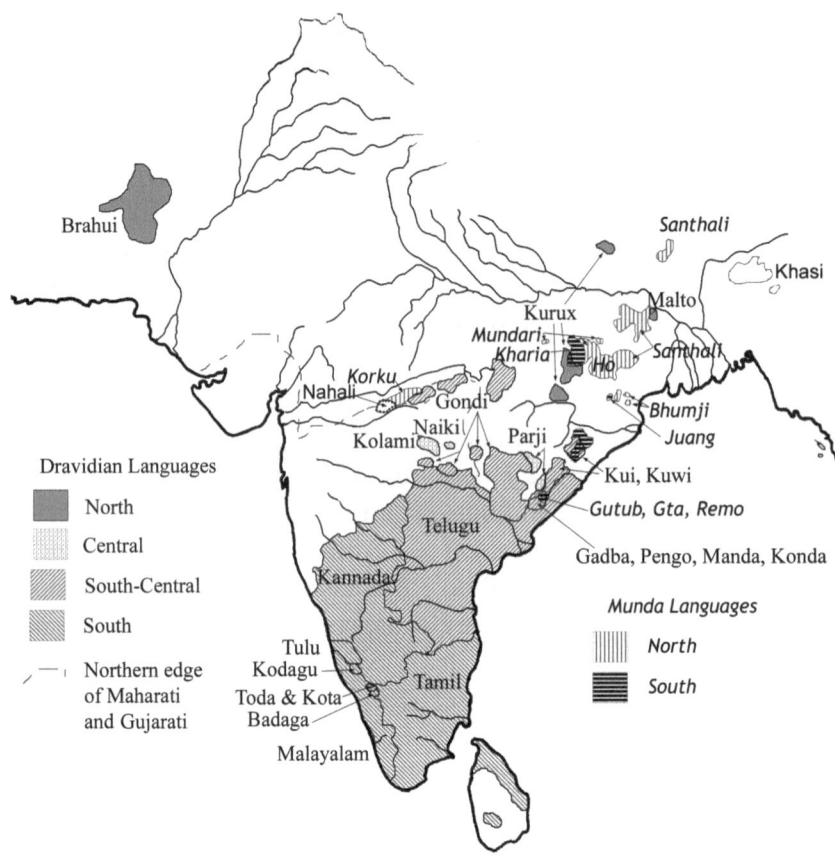

Figure 6. Map of non-Indo Aryan languages in South Asia (excluding Himalayan zone)

There is still room for some controversy with regards to how to represent Dravidian phylogenetically. Four major Dravidian subgroups are well-established (Figure 7), although recent controversy has arisen about how these should be grouped in a hierarchical, phylogenetic framework (Krishnamurti, 2003:figures 11.2A, B; Southworth,

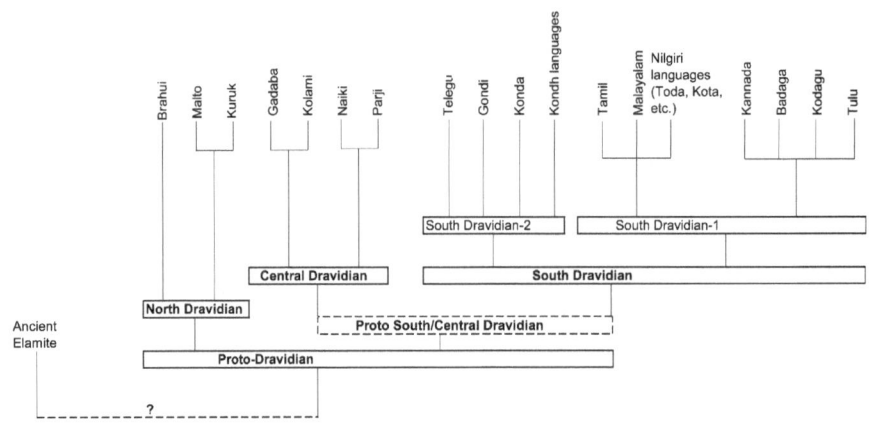

Figure 7. A phylogenetic representation of the Dravidian languages. Well-established groups are indicated by solid boxes (North, Central and South) (Krishnamurti, 2003). I have retained the hypothesis of a Proto-South/Central group, indicated by dashed box (after McAlpin, 1981; Southworth, 1988; Fuller, 2003d) for reasons offered in the text

2005a:233–236). A major issue concerns whether or not a nested hierarchy can be inferred between north, central and southern (including south-central) Dravidian subfamilies. I will continue to use the nested hierarchy of North, Central and South (Fuller, 2003a; following McAlpin, 1981, Southworth, 1988), as opposed to the more cautious but less historically informative three-branch polytomy of the most recent books. As the first botanical assessment of Fuller (2003a) revealed, there appears to be some archaeobotanical grounds for accepting this order of branching. Latecomer crops are generally only documented in South Dravidian, while native crops tend to be documented as cognates with Central Dravidian, while for the most part wild peninsular species may sometimes be documented for the North Dravidian languages as well (see below). As discussed by Southworth (2005a:234–5) there is evidence for a longer and more recent history of contact between the South and Central subfamilies, and there are cases of shared innovations in semantics in South and Central Dravidian as opposed to North Dravidian languages. Thus, even if clear shared phonological or morphological changes are absent, there are grounds for suggesting a phylogenetic hierarchy which groups Central and South Dravidian (Proto-South/Central Dravidian); the lack of clear phonological innovations may suggest that these branches diverged quite rapidly as we might associate with rapidly expanding and dispersing (Neolithic) populations.

Another issue has been the placement Brahui, spoken by pastoralists in Western Pakistan surrounded by Baluchi speakers (an Iranian language). This isolated location has often been taken to indicate a dispersal of early Dravidian speakers from the northwest, with a subsequent language shift to Indo-European languages. It seems to now be increasingly accepted that the ancestral Brahui, found today in Baluchistan, migrated within the past

millennium from a North Dravidian area in central India (Elfenbein, 1987, 1998; Witzel, 1999:30, 63; Southworth, 2005a; but for a dissenting view see Parpola, 1994:161). As noted by Witzel (1999:63), there is a lack of older loanwords from Iranian languages such as Avestan or Pashto, but only from modern Baluchi. In addition, it was the latter position, which implied an early divergence of Brahui, that has long been taken to support to dispersal of the early Dravidian speakers from the northwestern subcontinent, perhaps to be connected with a shared ancestral relationship to the ancient Elamite speakers of Iran (McAlpin, 1981; Fairservis and Southworth, 1989; Bellwood; 2005). As will be argued below, the evidence of lexical reconstructions relating to flora, as well as placenames, modern language geography and archaeological correlations all point to Proto-Dravidian located on the peninsula, and thus Brahui must be accounted for by a migration from the Peninsular region (possibly including Saurashtra or parts of Rajasthan) towards Iran.

The Munda language family includes a number of relatively small and often isolated languages in two main sub-groups (Bhattacharya, 1975; Zide and Zide, 1976; Donegan and Stampe, 2004; Southworth, 2005a): South Munda, including the Sora and Kharia languages, and North Munda, including Santali of northern Orissa and Bihar, and the grouping of Mundari, Ho and Bhumij, further south (Figure 8). The isolated Korku in Madhya Pradesh is also grouped more distantly with the Northern group. This disjunct location of the Korkus suggests that the Mundaric dispersal westward (or alternatively eastward) preceded the northward expansion of Gondi (central Dravidian) speakers, who presumably moved from the southeast. Nahali, further west still, includes many Munda elements but is now generally excluded from this group (Bhattacharya, 1975; Tikkanen, 1999), and has been suggested as

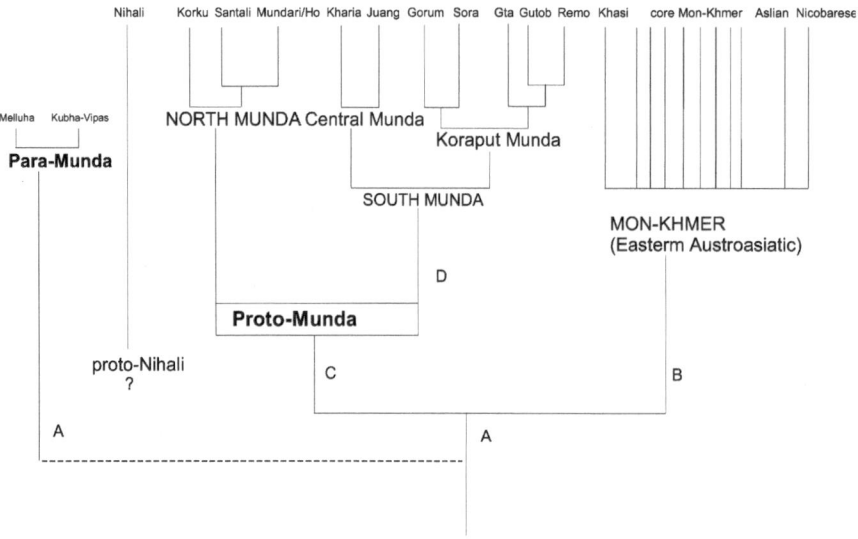

Figure 8. A phylogenetic representation of the Munda and Austroasiatic languages (top), with a hypothestical macro-phylogeny incorporting Witzel's 'Para-Munda' languages and the Mon-Khmer branches. Important cultural developments (derived traits) are indicated by letters (see text for discussion). It remains unclear whether Nihali should be incorporated in this phylogeny

a linguistic remnant of the earliest modern human dispersal out of Africa on the basis of possible distant relations with extinct Ainu (of north Japan) (Witzel, 1999:63). The entire Munda group is placed more as a distinct distant branch of the Austroasiatic family of languages, which is widely distributed in mainland Southeast Asia including the literary languages of Mon and Khmer in Burma and Cambodia (Blench, 1999:66; Diffloth, 2005). Of crucial significance to population history is how the Munda group is related to the rest of the Austro-Asiatic family, and whether the direction of spread should be seen as to or from India, an issue to which I return below. The centre of gravity of the Munda is clearly Eastern India, with the highest language diversity in Southern Orissa (the greater Koraput region), where the north and south Munda subfamilies overlap and where the highly diverse Koraput group of South Munda languages occur. One important lexical item, reconstructed by Zide and Zide (1976), which points also towards an Eastern India focus for Proto-Munda is the sal tree (*Shorea robusta*) since this species is confined to eastern India

and through the Central Ganges, but absent from the west, south and southeast Asia (although related species occur there).

Extinct North Indian Languages

Beyond the modern languages, there is possible evidence for at least two major extinct language groups (see especially Witzel, 1999, 2005). Of particular significance is the evidence for agricultural and botanical terminology borrowed into Indo-Aryan (Table 1), and to a lesser extent Dravidian, which appears to be neither Dravidian nor Munda (Mascia, 1979, 1991:42; Fairservis and Southworth, 1989:137; Kuiper, 1991:14–15; Fuller, 2003a). This includes a possibly earlier, and more upper Gangetic centred 'Language X' (Masica, 1979), which I have previously suggested might be linked to the Neolithic of the Ganges valley, or to be more precise the dispersal of the 'Language X' might be connected with the spread of rice, pulse, millet and cucurbit agriculture in northern India, from a possible epicentre in the hilly flanks of the

Table 1. Vegetation and agricultural loanwords from the Harappan substrate(s) in Indo-Aryan languages (Based on Masica, 1979; Witzel, 1999, 2005a, 2005b; cf. Fuller, 2003d: Table 16.8). Words marked with a 'kv' have been identified by Witzel as etyma of the Kubhā-Vipāś or "Para-Munda" language with phonological affinities to Munda/Austro-Asiatic. Witzel has divided those from Vedic sources into 'levels' in terms of probable relative chronology within the textual corpus, with 1.1 being earliest and 1.5 being latest. Some of Masica's substrate words are only attested in more recent languages (MIA = Middle Indo-Aryan, NIA = New Indo-Aryan)

Term/species	Sanskrit/OIA	Vedic Level	Origins/Archaeology	Linguistic Comments
Plough (ard)	*Lāngala*	1.1	Present in Early Harappan period (Kalibangan Ardmarks); Harappan models. Also Bronze Age Mesopotamia, Late Neolithic Europe	Also to Dr. and to PMunda. From a Sumerian original for 'sickle' (Witzel, 1999:16)?
Sow	*Vap-*	1.1		Possibly also in Indo-Iranian from Hittite?
Ploughman, two ploughmen	*Kinasa,*[kv] *Kinara*[kv]		See plough (above)	
Sow, furrow	*Sītù*	1.1		
Winnowing basket	*Śúrpa*	1.2		
Lentils, *Lens culinaris*	*Masura*	1.2/3	Domesticated in Near East probably by PPNB (8500 BC)	see Table 5.
Linseed (flax), *Lnium ussitatissimum*	*Atasī*	1.1	Domesticated in Near East probably by PPNB (8500 BC)	Similar source for PSDr word, see Table 5.
Date, *Phoenix* sp.	*Khajúra*[kv]	1.2/3	*P. sylvestris* wild in Sindh and through most of India; *P. dactylifera* possibly wild in Iranian plateau, or domesticated in Arabia/Mesopotamia	Distinct from PDr and PMunda words, see Table 2.
Cotton, *Gossypium arboreum*	*Karpasa*[kv]	1.5	Probably domesticated in Pakistan/Baluchistan. At Mehrgarh by c. 5000 BC	
Indian jambos, *Syzygium cumini*	*Jambu*	1.5	Moist and Dry Deciduous woodlands of South Asia	
Indian jujube, *Ziziphus mauretania*	*Badara-*	1.5	Wild throughout drier savanna and steppe zones of South Asia	
	Karkandu[kv]-	1.5		
Chaff, straw	*Busa*	1.5		From *busá* (Vedic 1.1)
Setaria italica	*Priyángu*	1.2/3	Domesticated in North China by 5000 BC. Also in Caucasus(?). Finds in South Asia in Late Harappan period. Related *Setaria* spp. Native to South Asia (see Fuller 2002; 2003b)	

(Continued)

Table 1. (Continued)

Term/species	Sanskrit/OIA	Vedic Level	Origins/Archaeology	Linguistic Comments
Panicum miliaceum	Ánu	1.2/3	Domesticated in North China by 5000 BC. Also in Caucasus(?). Finds in South Asia in Late Harappan period (see Fuller 2003b). Similar *Panicum sumatrense* native to South Asia, cultivated at Harappa by 3000 BC (Weber, personal communication).	
Vigna radiata	Khálva	1.2	Domestication(s) on peninsula (south/east) and northern India. Neolithic finds from Ganges and Southern Neolithic.	
Vigna mungo	Mása	1.2	Domestication on northern peninsula/S. Rajasthan. Early finds from Harappan Gujarat and Neolithic Ganges(?)	
Horsegram, *Macrotyloma uniflorum* (syn. *Dolichos biflorus* auct. pl.)	Khala-kula, [=Skt. kulattha]	1.4/5	Domestication(s) Indian savannah zones from Rajasthan through peninsula. Widespread Neolithic finds (Ganges, South India)	Ultimately from PDr, biogeographically less likely from PMunda.
Sesame, *Sesamum indicum*	Tila[kv?]		Domestication in southern Harappan zone(?)	Kv > Skt.;also > SDr1 *ellu*; > Sumer. *ili*; > Akkadian *ellu/alu*
Wild sesame, *Sesamum malabaricum*	Jar-tila[kv]		Wild in Sindh(?), Punjab, Malabar coast	
Sieve, filter	kārotara[kv]			
Silk-cotton tree, *Bombax ceiba* (syn. *Salmalia malabarica*)	śalmali[kv]		Native to Moist Deciduous forests and wetter variants of Dry Deciduous (e.g. teak zone)	
Papal tree, *Ficus religiosa*	Pippala		Wild throughout monsoonal South Asia, formerly in Baluchistan(?)	
Chickpea, *Cicer arietinum*	Canaka CDIAL 4579		Domesticated in Near East probably by PPNB (8500 BC)	See Table 5. Attested in Pali, Pkt.
Grasspea, *Lathyrus sativus*	K(h)ēsārī CDIAL 3925		Domesticated in Near East probably by PPNB (8500 BC)	
Pea, *Pisum sativum*	*mattara CDIAL 9724		Domesticated in Near East probably by PPNB (8500 BC)	Only in NIA

(Continued)

Table 1. (Continued)

Term/species	Sanskrit/OIA	Vedic Level	Origins/Archaeology	Linguistic Comments
Cucmber, *Cucumis sativus*	*Ksīraka* CDIAL 3667, 3698, 3703		Domesticated in northern India/Himalayan foothills	Only in NIA. Cf. Munda. Remo *Sarlay*, Kharia *kenra*, Santali *taher*
Bitter gourd, *Momordica charantia*	*Kāravella*[kv]		Domesticated in northern India/Himalayan foothills	MIA
Ivy gourd, *Coccinia grandis*	*Kunduru*		Domesticated in northern India/Himilayan foothills. Archaeological finds from Ganges plain by 1800 BC	Unconvincing Iranian and Austro-Asiatic etymologies have been suggested.
Sponge gourd, luffa, *Luffa acutangula*	*tori CDIAL 5977		Domesticated in northern India/Himalayan foothills. Southern Neolithic finds from mid-Second Millennium BC (Fuller et al., 2004)	
Okra, *Abelmoschus esculentus*	*Bhinda*		Domesticated hybrid of Gangetic *A. tuberculatus x A. ficulneus*, of semi-arid western/peninsular India. Could originally refer to cultivars or wild forms of either parent species.	
Grape, *Vitis vinifera*	*Drākshā*		Domesticated in Southwest Asia, also Indo-Iranian borderlands (?). Present in Pre-Harappan Baluchistan. Harappan fruit crop.	Southworth, 2005:107
Sheep, *Ovis aries*	*Bhedra*		Domestication in Near East by late PPNB; additional Asian domestication(s) may include Afghanistan/Baluchistan	MIA, NIA, <?> PMunda *medra*

middle Ganges zone (from Allahabad towards western Bihar). Early texts indicate that Indo-Aryan speakers picked up retroflexion as they moved into northwest India/Pakistan (Deshpande, 1995; Tikkanen, 1999), which might be connected with this extinct language. More recently, it has become increasingly clear that another, distinct substrate language or languages heavily influenced early Vedic Sanskrit, probably mainly in the greater Punjab region (Witzel, 1999, 2005). This has been inferred therefore to be substrate influence from the Harappan language, or the *Kubhā-Vipāś* language (to use Vedic terms) (Witzel, 1999:8–16, 2005:176–179). On the basis of prefixes and consonant clusters, Witzel suggests that this language shares phonological structure (especially prefixes) with Munna or the greater Austro-Asiatic family of languages, and thus refers to it as 'Para-Munda'. Witzel has further inferred a separate dialect or related language that seems to have been focused in the southern Indus or greater Sindh region, thus a southern Harappan language, or *Meluhhan*, to apply to an ancient Mesopotamia term for the region. Loanwords, and versions of the same word, from the Southern and Northern Harappan dialects can be shown to have regular phonological differences (Witzel,1999:30–37). Current archaeological orthodoxy implies that actual Proto-Munda was a relative latecomer to the subcontinent from the Northeast (e.g., Higham, 1998; Fuller, 2003c; Bellwood, 2005), a problem which requires reconsideration.

This range of substrate words clearly indicates indigenous agriculturalists at the time of the arrival of Indo-Aryan speakers in the subcontinent. The crop species represented point towards Indus agricultural traditions and/or that of the upper Ganges, including species of Southwest Asian origins as well as Indian species of northern origins. This also indicates that if 'Language X' is indeed to be related to a Gangetic Neolithic tradition, that this had already intermingled with the Harappan (*Kubhā-Vipāś*) tradition, presumably already by the period of urbanism. Indeed in the Eastern Harappan zone, including the upper Yamuna basin there is growing evidence for an Early Harappan tradition that incorporated the Southwest Asia crops with native rice, pulses and probably millets (cf. Saraswat, 2002, 2003, 2004), and became part of the Harappan civilization area in the later Mature period (from 2300–2200 BC). Vedic terms for singing, dancing and musical instruments also come from the *Kubhā-Vipāś* substrate source (Kuiper, 1991:19–20; Witzel; 1999:41). The loans from the *Kubhā-Vipāś* language, and probable 'Language X' is pronounced in the earliest parts of the Rig Veda, whereas plausible

Dravidian loans are few and later in the Rig Veda, or post-Rig Veda (and possibly indirect through an intermediary language) as are those of the *Meluhha* language (Witzel, 1999:18–23; cf. Southworth, 2005a). Witzel (1999:24), however, has continued to accept that early Dravidian must have entered the subcontinent via Sindh as non-agricultural farmers, a view which can be contrasted with either the Proto-Dravidian farming vocabulary suggested by Southworth (1976, 1988, 2005a) or the development of agriculture early within the divergent lineages of Proto-Dravidian hunter-gatherer-herders (Fuller, 2003a). Evidence for placing early Dravidian, and perhaps Proto-Dravidian speakers needs to be considered, both through reconstructible vocabulary as well as toponyms.

Early Dravidian Ecology and Agriculture

A challenge is to untangle reliable Proto-Dravidian cultural vocabulary and to relate this to archaeology and evidence for place-names (which may relate to later dispersal of subfamilies of Dravidians). Evidence for a more widespread distribution of Dravidian cultural groups (but not necessarily Proto-Dravidian) in the past, with subsequent conversion to Indo-European languages is clear (see Figure 4; Trautman, 1979; Fairservis and Southworth, 1989; Parpola, 1994; Southworth, 2005a; 2005b). Southworth, for example has traced village place-name endings typical of South India throughout Maharashtra and the Saurashtra peninsula, and with a few in Sindh and Rajasthan (Southworth, 2005a: Chapter 9). In these regions (specifically Gujarat and Maharashtra) cross-cousin marriages are either typical or practiced by some cultural/caste groups, as discussed by Trautman (1979, 1981). There appears to be no evidence that cross-cousin marriages were ever practiced in Gangetic India (Trautman,

1981). This implies that this characteristically Dravidian cultural practice has persisted in areas where Indo-Aryan languages are now spoken. This terminology is reconstructed for Proto-Dravidian by Krishnamurti (2003:10). The practice of cross-cousin marriages within the North Dravidian subfamily remains problematic, with the practice only recorded amongst the Kurukh but neither Malto or Brahui; the absence from the latter can be explained by cultural influences due to their encapsulation. The reconstruction of this practice has potential implications for archaeology and paleodemography, as it implies a particular kind of extended kin-network and endogamy that we might expect to influence aspects of settlement pattern and perhaps genetic structure within populations.

Two difficulties face historical linguistic reconstruction: incomplete recording and anachronistic definitions. As is well-known, the better recorded languages are the large and literary languages (Tamil, Telugu, Kannada, Malayalam), whereas the word lists available for other languages are more limited (e.g., absence of data for names of many crops in North Dravidian and often Central Dravidian in Fuller, 2003a). While it is undoubtedly true that absence of a cognate word in these incompletely recorded languages is not necessarily evidence for absence, it seems methodoligically flawed to reconstruct Proto-Dravidian from cognates just across the South (SDr1) and South-Central (SDr2) families, as Krishnamurti (2003) does. These larger and more widespread language subfamilies share a more recent common ancestry and as such can be expected to preserve later cultural-historical developments, such as greater social complexity. The fact that these are the most widespread and diverse subfamilies also suggests that they have expanded more recently and successfully, which may itself relate to demographic and cultural factors related to the emergence of more intensive agriculture and social complexity. There is also a danger in projecting into prehistory more modern definitions of words that have arisen metaphorically in parallel in the more recent past. Krishnamurti (2003) had reconstructed a Proto-Dravidian word for "write," but the cognates in all Central Dravidian and South-Central languages, as well as most South Dravidian languages is glossed as 'scratch' or make 'lines' and indeed only in Tamil has this meaning been extended to 'inscribe' or 'write' (Dravidian Etymologycal Dictionary [DEDR], entry number 1623, Burrow and Emeneau, 1984). He has also reconstructed 'king' from cognates found only in the four literary languages (DEDR 527), i.e. those languages which have been historically associated with states, and which derives from a compound word meaning "the high one," a fairly recurrent way to make terms for rulers (e.g., English, 'her highness'). Meanwhile his large state territory is a term (*natu* DEDR 3638) that has extended in Tamil from an original meaning of village or cultivated land (cf. Krishnamurti, 2003:7–8), and weaving (DEDR 3745) is widely glossed as 'to do matwork' or even 'thatch', and need not imply a textile industry! In other words he has inferred an essentially urban and Bronze Age (or even Iron Age, as he reconstructs iron, but from a word meaning 'black') for the Proto-Dravidians, and he cites their identification with the Harappans as possible (although the Harappans did not have iron). Nevertheless there are many things which have cognates across a large number of Dravidian languages, and many are to be found in terms of plants. While there remain gaps in recording, especially for the North Dravidian languages, these need to be filled by new linguistic field recording or use of sources beyond the Dravidian Etymological Dictionary (Burrow and Emeneau, 1984).

In terms of pinning down early Dravidians, an ecological assessment of tree names may be useful (compare with Figure 5). In Table 2 there is a selection of trees, that are found in the Dry Deciduous forests of the Peninsula and central India (Puri et al., 1989; Meher-Homji, 2001; Asouti and Fuller, 2006). Many of them also occur in Eastern India and in parts of the Himalayan foothills, but some do not, notably teak. They are entirely monsoonal, absent from the northwest, and also present in smaller patches in Saurashtra (Gir hills) and Rajasthan (Mount Abu). The fact that several of these species have good cognates across all Dravidian subfamilies strongly supports a Proto-Dravidian homeland somewhere in Peninsular India. Culturally, it is of interest that some of these species are ecological dominants in the Dry Deciduous woods of the peninsula, suggesting that this was a particularly salient environment to these people. In addition, a number of these species are useful, as sources of edible fruits, medicines or lac (used for lacquering and as dye). In the drier savannah zones, that in reality intergrade with the dry deciduous, two more fruit trees can be definitely reconstructed to proto-Dravidian, and another nearly so (Table 3). By comparison, Moist Deciduous trees in Table 4 in no cases are recorded to extend to North Dravidian, although they do consistently have cognates across the South and Central branches (absence from North Dravidian could be a limitation of recording). Of interest from this zone is the likely tuber food (perhaps cultivated), taro. Those wetter species present, both *Syzygium* and *Artocarpus* favor watercourses and along rivers extend their ranges into drier zones. Of the species on these lists, only *Ziziphus* and the date palm(s), might possibly have been known in Baluchistan/Iran, and only a few more species (toddy palm, the *Ficus* spp., *Terminalia* spp.) would have occurred in Sindh (and probably very patchily). Thus, taken together, the tree words and place-names point to a restricted peninsular zone for the early Dravidian speakers focused on the Dry Deciduous and savannah zones. If the Moist Deciduous elements are taken into account (assuming incomplete recording for North Dravidian) then even Saurashtra is less likely (although these species could be found as relicts on Mount Abu, Rajasthan). Thus the plant name evidence clearly contradicts Krishnamurti's (2003:15) claim that early Dravidians were throughout the subcontinent "even as far as Afghanistan."

From similar vegetation zones we find the wild progenitors of the crops that also have wide Dravidian cognates (included in Tables 2–4, also, Fuller, 2003a). It is not possible to know whether knowledge of these plants implies their cultivation (although that is often assumed, e.g., Southworth, 1988), if they might have been encountered wild in the environment. As previously argued (Fuller, 2003a) those species with the deepest Dravidian roots, based on recorded cognates, correspond to those with the oldest archaeological occurrences in South India, and suggest an identification with the Southern Neolithic (also concluded by Southworth, 2005a). Crops that are non-native and archaeologically turn up somewhat later, such as wheat, barley and African crops, tend to have recorded cognates only for Proto-South Dravidian, although in many cases these plants are poorly recorded in the DEDR (which calls for moving to further sources or new recording). There remain some unresolved issues. Crops such as urd and pigeonpea are not part of a widespread and early Southern Neolithic crop package. Pigeonpea arrived later, ca. 1500 BC, spreading from Orissa while urd has been found as a trace occurrence on a few sites, and is rather to be associated with cultures like the Deccan Chalcolithic and Late Harappan Gujarat. If we assume that some (like horsegram and mung) will prove to be cognate in Kurukh and Malto (once additional linguistic sources become available), while others (urd, pigeonpea) do

Table 2. Trees and shrubs of the Dry Deciduous zone cognate across Dravidian subfamilies, indicating those languages for which cognates are documented in their respective subfamilies. DEDR entry numbers indicated (Burrow and Emeneau, 1984). Protoform reconstructions from Southworth (2005). For comparison Indo-Aryan (after Turner 1966) and Munda languages (after Zide and Zide, 1976) are included

Species	Uses	SD1	SD2	CDr	ND	DEDR	CDIAL nos.	PMunda
Butea monosperma, flame of the forest	Lac host, resin: Bengal kino, 'holi powder' yellow pigment, medicinal uses	X	X	X	X	4981 *mur-ukk-	3149 su-kimśu-ka (from Witzel's K-V language)	
Pterocarpus marsupium, Malabar kino tree	resin: Malabar kino,	X	X	X		5520 Ta. venkai		
Moringa sp., Drumstick tree, horseradish tree	*M. oleifera* wild in W. Himalayan foothills, but similar *M. concanensis* in Nallamalais, Conkan, inner Western Ghats	X	X	X	X	4982 *murum-	> 10209 murangi (H., Or.). 12437 sigru	
Schleichera oleosa, Ceylon oak	Lac host (true shellac), edible leaves, fruits and seeds	X	X	X	X	4348 *puc-/*puy-		
Ficus religiosa, Pipal	One of the Sacred figs. Introduced to peninsula?	X	X			202 PSDr *ar-ac-al	8205 pippala	
		X	X	X		2697 *cuw-		
Ficus benghalensis, banyan	One of the sacred figs, introduced to peninsula??	X	X	X		382 *āl	7610 nyagrodha	
Phyllanthus emblica, emblic myrobalan	Edible fruit, medicinal	X	X	X		3755 *nelli-	1247 amalaka	
		X	X			574 Te. usirika		
Feronia limonia, wood apple	Edible fruit	X	X	X		(?) 5509 *wel-V-	2749 kapittha	
Bombax ceiba, silk-cotton tree	Source of fibre	X	X	X		495 & 5539	12351 Śalmali, Śimbala	
Gmelina arborea	Edible fruit, medicinal root and bark	X	X	X		1743 Ta. kumir	3082 karsmarya^{kv} 4030 gambhari	
Tectona grandis, teak	Medicinal uses	X	X	X		3452 *tēnkk-	? >12369 saka	
Terminalia tomentosa	Dominant peninsular deciduous tree	X	X	X		4718 *mar-Vt-	963 asana	
Terminalia bellerica	Medicinal uses	X	X	X		3198 *tānt-i	11817 vibhidakā	

(Continued)

Table 2. (Continued)

Species	Uses	SD1	SD2	CDr	ND	DEDR	CDIAL nos.	PMunda
Phoenix sylvestris/ dactylifera, wild forest date, domestic date	Edible fruit	X	X	X	X	2617 *cīnt(t)-*	*Khajúra*	*Vn-deñ, *raloXg
Borassus flabellifer, toddy palm (may also mean *Caryota urens*, the west coast's toddy palm)	Sweet fruit, edible, generally fermented	X	X	X	X	3180 *tāZ*	>Skt. *Tāla* CDIAL 5750	
Cordia myxa, sebestan plum	Edible fruit		X	X		3627 5408	1990 *uddala* 12610 *selu*	
Azadirachta indica, neem	Medicinal uses, sacred	X	X			5531 *wē-mpu*	7245 *nimba*	

Table 3. Trees and shrubs of the dry evergreen scrub zone cognate across Dravidian subfamilies, indicating those languages for which cognates are documented in their respective subfamilies. DEDR entry numbers indicated (Burrow and Emeneau, 1984). Protoform reconstructions from Southworth (2005). For comparison words of Indo-Aryan (after Turner, 1966) and Munda languages (after Zide and Zide, 1976) are included

Species	Uses	SD1	SD2	CDr	ND	DEDR	CDIAL nos.	PMunda
Diopsyros melanoxylon	Edible berry, a kind of ebony wood, used in tanning	X	X	X		3329	>5872 *tumburu-.* 3464 *kendu*	
Tamarindus indica, tamarind	Edible fruits, native(?) to India as well as Africa	X	X	X	X	2529 *cin-tta*	1280 *amla*	*R-tiXn also(?) *(ro)joXd
Ziziphus mauritania, Indian jujube	Edible fruit	X	X	X	X	475	Skt. *badara-*	
Macrotyloma uniflorum, horsegram	Edible pulse, crop	X	X	X		2153 *koL*	>Skt. *kulattha*, or from PM (?)	*kodaXj <?>Skt./PDr. Dr. source more likely

not, then we would have clear linguistic stratification that reflects that of archaeobotany, and implies that indigenous peninsular agriculture (perhaps focused on the Eastern Ghats Dry Deciduous zones north of the Krishna River) can be associated with Proto-Dravidians. The Southern Neolithic, as it is currently known, would then reflect one of the cultural offshoots as this early Dravidian

agriculture expanded. While the status of plant cultivation amongst Proto-Dravidians remains unresolved, the herding of animals seems clear with reconstructed words for cow *ām (DEDR 334), bull *erum- (DEDR 815), two probable sheep/goat terms (one for each species, or female and male?) *yātu- (DEDR 5153), *kat-ā- (DEDR 1123) (Southworth, 2005a: Chapter 8, Appendix A).

Table 4. Trees and shrubs of the Moist Deciduous zone cognate across Dravidian subfamilies, indicating those languages for which cognates are documented in their respective subfamilies. DEDR entry numbers indicated (Burrow and Emeneau, 1984). Protoform reconstructions from Southworth (2005). For comparison words from Indo-Aryan (after Turner, 1966) and Munda languages (after Zide and Zide, 1976) are included

Species	Uses	SD1	SD2	CDr	ND	DEDR	CDIAL nos.	PMunda
Artocarpus integrifolia	Edible fruit	X	X	X		3988 *pal-ac/ *pan-ac	7781	
Syzygium cumini, Indian jambos or java plum	Edible fruit	X	X	X		2917 Ga. Nendi *ñānṭ-Vl also SDr 2914 Ta. *naval*	*Jambu*	NM *koXda SM *ko?-deX
Vigna radiata, mung bean	Edible pulse, crop	X	X	X		3941 *payaru (S) *pac-Vṭ/*pac-Vl	10198 *mudgà*, *khálva*	
Vigna mungo, urd bean	Edible pulse, crop	X	X	X		690 *uZ-untu	>1693 *uddida 10097 *mása* >Skt. *malada*	*rVm
			X	X		4862 *minimu		
Cajanus cajan, pigeon pea	Edible pulse, crop	X	X	X		3353 *tu-var-	>Skt. *tubarika*	*sVr/d – u/aj *sVr/d – oXm
		X	X			1213 *kar-Vnti		
Colocasia esculentum, taro	Edible, tuber crop	X	X	X	X	2004 *kic-ampu	?> Skt. *Kemuka*, *kacu*, *kacvi*	
Sesame (wild?), *Sesamum indicum/ malabaricum*	Edible oil seed	X	X	X		3720 *nuv-	Skt. *tila*, *jar-tila* (wild sesame); cf *ellu* in SDr, and similar in ancient Sumer and Akkad.	

Some challenges for further investigation remain. First, it should be noted that tables used here have excluded the native millets and rice. As discussed in Fuller (2003a), millet terms that can be extracted from botanical sources are often unrepresented in the DEDR, and key millet species that occur archaeologically, especially *Brachiaria ramosa*, are not recorded at all. Between (some) millet species we might expect a substantial degree of semantic shift, as these species have many superficial similarities. Thus, it is of interest that Southworth (2005a) has reconstructed two millet terms to Proto-Dravidian, with another four added at the Proto-South Dravidian stage, and two more to the proto-language of Tamil and Kannada (Southworth, 2005a:247–248). From southern Neolithic sites there are two predominant millet crops (Fuller et al., 2001, 2004), whereas by the early historic period as documented on archaeological sites in Tamil Nadu seven millets have been identified archaeologically (but not including Sorghum) (Cooke et al., 2005). I have also omitted rice, for which Southworth (2005a: Chapter 7, B8) reconstructs 3 possible early Dravidian terms, although glosses in some languages suggest that these might originally have been more general terms for ears of grain, crops, cooked grain (and I would suggest perhaps some other crop, such as a millet). While South and North Munda each have a reconstructible term for rice, with apparent cognates in other Austroasiatic languages, there is not one coherent rice etymology for the whole family, and etymologies like those in Mahdi (1998) and Witzel (1999:30–33) also use proto-forms for millet terms. In general, I would regard such a semantic shift as more likely to have occurred in the other direction, from older millet terms to rice (which is everywhere a more productive and increasingly widespread crop in historical times). Horse terms, which include those for donkeys and probably wild hemiones, are also problematic with three possible terms reconstructed to Proto-Dravidian or Proto-S/C Dravidian (Southworth, 2005a, Chapter 8, Appendix A; cf. Witzel, 2005:103–104). One problem is that the archaeozoology of the equid species in peninsular India is still poorly documented and the actual semantic categories of the proto-words may not be clearly fixed. Southworth expresses the most confidence in a Proto-South/Central Dravidian term for donkeys (DEDR 1364, *kaz-ut-ay), which might plausibly have spread to South India by the third millennium BC (from ultimate origins in Egypt or the Sahara). Sesame also raises questions, as linguistic data suggest a reconstruction for one term back to Proto-South/Central Dravidian, although there is no archaeobotanical evidence yet for its early use, as early as native pulse (and millet) crops which we know were being cultivated in Neolithic South India. While I previously suggested that this species may have been encountered by early Dravidians in wild form (Fuller, 2003a), since it is native to South Asia, further consideration makes this less likely. The habitats on the peninsula where sesame occurs are restricted to the wet west coast near sea-level, including coastal sand dunes (personal botanical field observation), and such an ecology is incompatible with the deciduous woodland species that readily reconstruct to Proto-Dravidian or Proto-South/Central Dravidian. Sesame is likely to have been domesticated prior to the Mature Harappan period somewhere in the greater Indus region (Fuller, 2003d; Bedigian, 2004), in line with its Para-Munda etymology. There is no evidence to suggest dispersal to the peninsula prior to the Late Neolithic/Chalcolithic period, i.e. the same time horizon as wheat, barley and some African crops, which would be in line with the northwestern tila loanword in Proto-South Dravidian.

Archaeological evidence can make a significant contribution to dating the antiquity of languages. While one might suggest correlation between a reconstructed proto-language vocabulary and an archaeological culture horizon, it is easy for dating to be wrong, since technologies and crops will have continued in use. On the other hand, when different language sub-families have distinct words for items of culture, we may hypothesize that such technologies (or domesticates) entered the cultural repertoire independently in each of the language/culture sub-families, and archaeological evidence for the adoption of such technologies might be used to place a general minimal age for the separation of these branches. Evidence for a number of items which have distinct roots across the South, Central and North Dravidian language groups, suggest a mid-second millennium BC minimal divergence for the Central and South Dravidian languages on the basis of archaeological dates. This includes domesticates that have distinct etyma across these three language subfamilies, including several tree-fruit cultivars (mangoes, Citrus spp., bael fruits), as well as chickens (see Table 5). In addition, adopted tree crops from Southeast Asia can be reconstructed only for Proto-South Dravidian, Areca nuts, coconuts and sandalwood. Wood charcoal evidence for sandalwood indicates its establishment in South India by ca. 1300 BC, with probable Citrus tree cultivation from the same period (Asouti and Fuller, 2006). Bananas may have been introduced even later since the two South Dravidian branches have different roots. In the future we may expect archaeological phytolith evidence to be able to pin down the date of introduction of Bananas to this region; it now appears that some banana cultivar was established in the lower Indus region already in Harappan times (Madella, 2003). These data suggest therefore that Proto-South Dravidian might be identified with the latest phase of the Southern Neolithic and the transition to

Table 5. Selected plants and livestock with separate linguistic roots from different Dravidian subfamilies, indicating those languages for which cognates are documented in their respective subfamilies. DEDR entry numbers indicated (Burrow and Emeneau, 1984). Protoform reconstructions from Southworth (2005). This list includes introduced crops. For comparison words from Indo-Aryan (after Turner, 1966) and Munda languages (after Zide and Zide, 1976) are included

Species	Uses, comments on	Dravidian Languages (DEDR entry nos.)				Indo-Aryan	Munda
	Origins (in relation to South India)	PSDr [PDr.3]					
		SDr1 [SDr]	SDr2 [SCDr]	CDr	NDr		
Mango *Mangifera indica*	Edible fruit, wet Western Ghats forests and introduced cultivars from northeast India (Assam)	4782 PSDr *māmŭ*		4772 *mat-kāy* (>Go., Kui, Kon., Kuwi)	2943		*uXli/ *uXla SM *kaj'-er/ *kag'-er (green mango)
Bael *Aegle marmelos*	Edible fruit, introduced as cultivar from central/north India(?)	1910 Ta. Kuvilam, cf. Skt.	4821 Te. *maredu*	4821 SDr2>Nk	2072 Kur. *Xotta*	[p.457] Skt. *bailvam,* Pkt. *Billa-*	
Mast tree *Calophyllum inophyllum*	Restricted distribution: Western Ghats wet forests, west and east coast pockets	4343 PSDr *pun-ay*					
Coconut *Cocos nucifera*	Introduced from Malaysia/ Indonesia, via Sri Lanka(?)	3408 PSDr *ten-kāy* "southern-fruit" 1254 PSDr *kairu* (coconutfibre)				*Nārikela* Ramayana Skt.	
Citron, *Citrus medica*	Introduced to south by 1300 BCE from central-eastern Himalayas	4808 Ta. *Matalai,* PSDr. *Māt-al*				Cf. 10013 Skt. *Matu-lunga-*	
Orange, *Citrus aurantium*	Introduced from SE Asia via NE India(?)	552 PSDr *ize*					
Sandalwood, *Santalum album*	Introduced from Indonesia by 1300 BCE	2448 PSDr *cāntu*					
Banana, Musa *paradisiaca*	Introduced from Malaysia/ Indonesia, via Sri Lanka(?). In Sindhi Harappan Kot Diji by 2000 BCE	5373 PSDr1 *wāz-a-*	205 PSDr2 *ar-Vnṭṭi*	754 Pa., Ga.			
Areca nut, *Areca catechu*	Introduced from Southeast Asia	88 PSDr *at-ay-kkāy*					
Mustard, *Brassica* sp., probably *B. juncea*	In northwestern subcontinent by the Harappan civilization. Native there(?)	921 PSDr *ay-a-*					

(Continued)

Table 5. (Continued)

Species	Uses, comments on	Dravidian Languages (DEDR entry nos.)				Indo- Aryan	Munda
Barley *Hordeum vulgare*	Introduced crop (Near Eastern). In Southern Neolithic by 1900 BC	1106 PSDr1 **koc-/Kac-*					
Wheat *Triticum* spp. Mainly *T. aestivum* words	Introduced crop (Near Eastern). In Southern Neolithic by 1900 BC	PSDr **koo-tumpai* 1906 PDr. *kāl-i* ('rice/wheat')					
Pearl millet *Pennisetum glaucum*	Introduced crop (African). In Southern Neolithic by 1500 BC	1242 PSDr **kampu*				Skt. *kambu*	
Sorghum *Sorghum bicolor*	Introduced crop (African)	2896 PDr-2 **connel*				Pkt. *Gajja* >Dr. Skt. *yavanala* (from IE)	
Hyacinth bean *Lablab purpureus*	Introduced crop (African). In Southern Neolithic by 1500 BC	? 2496 PSDr **cikk-Vt*; Also, 262.					
Peas *Pisum sativum*	Introduced crop (Near Eastern)	Probable cognates with Guj, Mah.				9724 NIA **mattara*	
Lentils *Lens culinaris*	Introduced crop (Near Eastern)	Probable cognates, <Skt.				Skt. *masúra*	
Chickpea *Cicer arietinum*	Introduced crop (Near Eastern)	1120 PSDr1 **kapalai*	Te.'sana-galu'	Cf. Mah. 'harbara'		4579 *Caṇaka*	
Fenugreek *Trigonella fenugraecum*	Introduced crop (Near Eastern). Finds Harappan and Late Harappan Punjab/Haryana	5072 PSDr **mentt-i*				10313 **mētthī.* <? >PSDr	
Flax *Linum ussitatissimum*	Introduced crop (Near Eastern)	3 PSDr **ak-V-ce* (**akace*)				OIA **atasi-* <? >PSDr	
Cotton *Gossypium arboreum* (>*Gossypium* spp.)	Introduced crop (from Pakistan)	3393 PSDr **tuu* (but='feather') 3976 PSDr **par-utti* 3726 PSDr **nūl* (cotton thread)				5904 Skt. *tula-* <? >SDr Skt. *karpāsa-* (Table 1)	
Chicken *Gallus gallus*	Introduced domestic animal	2248 PSDr **kōz-i* (>Nk.)	2160 (>Go.)	2013		**kukhro,* Skt. *kukkutah*	**si(X)m*
Pig, domestic	Introduced/local ?	4039 PS/CDr **pan-ti*					
Water buffalo (female)	Introduced from NW(?)/local ?	816 PSDr **erum-*					Kharia *Bontel* Sant. *bitkel* **oreXj* ('draft animal')

the Megalithic period in South India, in the time horizon 1500–1300 BC, and certainly no earlier than 1800–1700 BC. Central Dravidian is likely to have diverged prior to this date (by ca. 2000 BC, before the introduction of wheat and barley), and North Dravidian even earlier (but further linguistic clarification is needed on native crop words before a date can be assigned). Further support comes from other technologies such as those of metal working. Terms for gold and smelting can be reconstructed from Proto-South Dravidian only (Southworth, 2005a). Archaeological evidence for metals is restricted to Phase III of the Southern Neolithic (i.e., 1800–1400 BC), including gold objects from Tekkalakota (1700–1400 BC) (Nagaraja Rao and Malhotra, 1965, Korisettar et al., 2001a). It is also at the Proto-South Dravidian level that a number of terms that suggest incipient social hierarchy (and political economy) are found (e.g., chiefs or lords, tribute, commodity/ware, 'money' [some standard of exchange value], battle/army, a range of buildings and settlement types) (Southworth, 2005a: Chapter 8, Appendix B), which is congruent with the evidence for the evolution towards social complexity from Neolithic Phase III towards the Megalithic (Fuller and Boivin, 2005; Fuller et al., 2007).

Early Munda Agriculture and Austroasiatic Dispersals

New linguistic research suggests that Munda ancestry, and the larger Austroasiatic family, should be placed in South Asia. In recent discussions archaeologists have assumed that Munda was a relative late-comer to the subcontinent, coming from Southeast Asia/Southwest China (e.g., Higham, 1998, 2003; Bellwood, 2001, 2005; Bellwood and Diamond, 2003; Fuller, 2003c). This has also tended to be the assumption of linguists, since the Southeast Asian Mon-Khmer languages

form the sister group to Munda languages (e.g., Zide and Zide, 1976; Diffloth, 2005; see also, Blench, 1999, 2005). Implicit in most of this literature is the assumption that rice has a single origin to be located in South China. For reasons already reviewed above, this assumption is in error. It is contradicted by genetic evidence from rice, and is inconsistent with currently available archaeobotanical evidence, which instead indicates that Chinese *japonica* rice domestication is distinct from *indica* rice domestication, probably in the Ganges and perhaps an additional locus. Since Mon-Khmer and Munda share (some) agricultural vocabulary (Zide and Zide, 1976; Blench, 2005), including terms for rice, but not a strongly rice-focused vocabulary (Fuller, 2003a; Blench, 2005) this was taken to imply dispersal from the Chinese centre of rice domestication. The archaeobotanical case negates this, leaving it an open question whether Mon-Khmer or Proto-Munda should be seen as dispersing.

The evidence of an Austroasiatic substrate in the Indus valley and new linguistic research on comparative phonology and syntax both support an indigenous development for Proto-Munda and a dispersal eastwards for Mon-Khmer. If the Austroasiatic affiliation of the inferred *Kubhā-Vipāś* and *Melluha* languages ('Para-Munda') are correct then this would imply a much earlier and more widespread distribution of pre-Munda/Austro-Asiatic. As already noted, the reconstructed vocabulary (e.g., Sal trees) and modern linguistic geography suggest an Eastern Indian (Orissan) homeland for Proto-Munda, which would suggest that these language substrates, as well as Munda-like placenames in the Gangetic zone (Witzel, 1999:15, 2005:179–180) come from an earlier pre-Proto-Munda branch of Austro-Asiatic. This is also suggested by the phonological structure of Para-Munda vis-à-vis modern Austro-Asiatic languages. As discussed by Witzel (2005:178–179), these substrate loanwords have active prefixing, a small number of possible infixes and no clear suffixes. This is typical of the eastern Austro-Asiatic languages of the Mon-Khmer family (Diffloth, 2005; Donegan and Stampe, 2004), whereas Munda tends to be suffixing (with other infixes). As explored in detail by Donegan and Stampe (2004:20) proto-Austroasiatic is inferred to have had a 'rising rhythm' with one or two syllable words stressed on the second syllable, prefixing and analytic grammar (i.e., without complex declensions and conjugations) based on subject-verb-object ordering. This rhythm has been retained in Mon-Khmer, whereas in Munda it has evolved in an opposite direction, to a 'falling rhythm' in which grammar became synthetic based on subject-object-verb ordering in which suffixes became necessary for marking gender, tense, etc. for subordinate clauses. While falling rhythm is typical across language families in South Asia, the Munda suffixes do not appear to be either borrowings or calques (translations) from Dravidian (Donegan and Stampe, 2004:19), but instead they evolved for reasons of simplifying speech rhythm (a 'trochaic bias') (ibid.:25–26). This falling rhythm is an important trait uniting Munda languages (*sensu stricto*), and thus the lack of clear suffixing in Witzel's 'Para-Munda' would place this language lineage prior to, or separate from, the Proto-Munda lineage. Donegan and Stampe (2004:27) conclude that the diversity of Munda structures and low level of Munda cognates, in contrast to Mon-Khmer, argues that this is the older branch of this language family, thus suggesting a South Asian Austroasiatic homeland. Similarly, acceptance of 'Para-Munda' as a branch prior to the diversification of Proto-Munda (and presumably Mon-Khmer) also argues for greater antiquity of Austroasiatic in South Asia than in Southeast Asia. This further implies that if the Austric hypothesis, which links Austronesian languages of island Southeast Asia with Austroasiatic, is accepted (cf. Blust,

1996b; Higham, 2003) then this divergence must be placed in deeply pre-agricultural times and related probably to a Pleistocene demographic process (see also, Blench, 1999, 2005).

In terms of agricultural history, we probably need to assume at least two origins (or adoptions) of agriculture within Austroasiatic, as indicated by the label "A" on Figure 8. In the history of the 'Para-Munda' lineage Near Eastern wheat-barley agriculture was adopted, as documented archaeologically in Baluchistan and the Indus valley. Note that neither of these cereals or the winter pulses or flax have 'Para-Munda' etymologies. Additional local domesticates were added, such as cotton, sesame and some fruits (*Phoenix sylvestris*, jujube and Indian jambos), all with 'Para-Munda' etymologies. Some species from the Gangetic basin were also adopted, carrying with them loanword names and perhaps accompanying some immigrant farmers (of Language X), such as rice, cucumbers (and other gourds) and native *Panicum* and *Setaria* millets (which would have been subsequently replaced by larger grained *P. miliaceum* and *S. italica*), and native Indian pulses (horsegram, mung and urd).

By contrast the (pre-)Proto-Munda lineage somewhere in Eastern India followed a different trajectory to agriculture. These people adopted (or domesticated) two or three small millets, rice, probably pigeon pea and mungbean, while adopting horsegram and perhaps a small millet from early Dravidian groups or some intermediary, extinct group. It may be that during this process of agricultural beginnings in Eastern India that demographic expansion and cultural differentiation led some offshoot group to move eastwards towards Southeast Asia retaining some tradition of shifting cultivation that involved rice and/or millets (ancestral to Mon-Khmer) (labelled 'B'). If this group had an economic emphasis on shifting cultivation in hilly zones then we might tentatively identify them with the Neolithic of the Orissa hills which produced some shouldered celts, which have long been taken to indicate connections with Southeast Asia (e.g., Wheeler, 1959), but the arrow of dispersal needs to now be reversed to an out-of-India dispersal. Proto-Munda agriculture should perhaps be placed in the Orissan lowlands. The reconstructed rice and millet terms in Proto-Munda all show evidence of having suffered semantic shift between species (including between rice and millets) and often plausible connections with other language families as loanwords in one direction or another (cf. Zide and Zide, 1976:1311; Mahdi, 1998; Witzel, 1999: 30–33). Words for goat, chicken, and draught cattle (zebu?) suggest that the Proto-Munda speech community existed at the time these taxa were dispersed as domesticates across northern India, i.e., in the mid to late third millennium BC. The reconstructed word for water buffalo is perhaps more likely to imply a separate domestication in eastern India, as there is no archaeological basis to infer that the domesticated water buffalos of the Sindhi Harappan (e.g., Dholavira) dispersed widely. It is of note that the water buffalo is symbolically significant amongst ethnographic Munda-speaking peoples (Zide and Zide, 1976:1319). It would be within the cultural context of these emergent agriculturalists of eastern India, that key linguistic changes occurred (marked as "C" in Figure 8, such as the rhythmic and word order changes). Then one cultural lineage (North Munda) must have been more prone to dispersal, perhaps with more of an ancestral emphasis on shifting cultivation (a second wave of hill cultivators), while the other (South Munda) was more prone to sedentarisation and increasing population density. It was within this more sedentary group that pigs were domesticated or adopted and became culturally salient (Figure 8, "D").

Conclusion: A Mosaic of Origins, Expansions and Interactions

Currently we are on the brink of being able to produce a new synthesis of early agriculture and later Holocene population history in South Asia. Both the archaeology of early agriculture and the historical linguistics of South Asia have undergone major advances in data collection and analysis in recent years. Nevertheless there remain major gaps in the evidence. In archaeology, there are major regional biases in Neolithic excavation and in systematic archaeobotany. Key regions such as central India (Madhya Pradesh) and Eastern India (Jarkhand, Chattisgarh, Orissa, northern Andhra) are still largely unknown and we are forced into speculative scenarios. In the Gangetic basin and South India we face the archaeological challenge that our better documented Neolithic sites are already fully agricultural and more or less sedentary. Their less sedentary, more archaeologically ephemeral predecessors await discovery, although the new research findings at Lahuradewa (Uttar Pradesh) hint at some of the insights such sites may soon yield. As some have long-maintained (e.g., Possehl and Rissman, 1992; Possehl, 1999) there may be a stage during which animal herding spread prior to the beginnings of plant cultivation, but which parts of South Asia and which cultural traditions participated in this remains to be clearly documented through archaeology, in which modern archaeozoology is critical. In the northwest of India and Pakistan a research focus on the Harappan civilization has left Neolithic developments poorly understood.

In terms of linguistics, further collection of data from small languages and relating to 'minor' crops is needed. As noted, millets are poorly represented in linguistic sources, both because the botany of linguistic sources is not always clear (and always poorly documented in botanical terms) and because these crops are often not of great subsistence significance in the modern day. Similar problems surround certain vegetable crops, such as the numerous indigenous gourd (cucurbitaceae) crops of northern India. In addition, a more realistic and botanically informed assessment of semantic shift between millets, rice and other cereal crops is needed. As recent research indicates (e.g., Witzel, 1999, 2005; Southworth, 2005a), there is much to gained by further assessment of substrate loanwords and ancient borrowing between languages. The integration of such linguistic findings with an archaeological framework of cultural complexes and chronology offers the greatest promise for an integrated long-term cultural history of South Asian populations. Some working hypotheses in this direction have been offered in the present chapter. Once such a framework is in place, historical linguistics potentially offers archaeologists access to less material aspects of culture, such as concepts of kinship and the supernatural.

The Neolithic revolution fuelled a major demographic expansion. While population density can be theorized to have promoted sedentism (e.g., Rosenberg, 1998), this in turn helped to accelerate population growth. Archaeology indicates a number of distinct Neolithic cultural traditions likely to be based on separate transitions from hunting-and-gathering that involved domestication. This is likely to have occurred at least in South India, Western India (Gujarat), the middle Ganges and probably the Orissan region, as well as distinctive developments in the Indus basin and hill regions to its west. These, and possibly other, Neolithic beginnings must have involved population expansions of culturally distinct groups, presumably with different languages. In addition, the spread of farming through the incorporation of hunter-gatherers might also be expected to have involved language shift to established farmer languages, presumably through high degrees of bilingualism that can account

for some of the varied substrates detectable in South Asian languages. As suggested above, the Neolithic languages that underwent expansion, and subsequent diversification, include Proto-Munda (in Eastern India), Proto-Dravidian (or an early derivative) in South India, 'Para-Munda' in the Greater Indus region, and perhaps 'Language X' in the Ganges basin. In Gujarat or south Rajasthan we might perhaps think in terms of a proto-Nahali agricultural language or a second early branch of Dravidian. All of this implies that a large degree of cultural (and linguistic) diversity was already established in South Asia prior to the Neolithic, and this must be accounted for by population expansions during an era of hunter-gatherers, such as during the Pleistocene.

The language history of South Asia extends back to the entry of modern humans, and must be complicated by processes of internal expansion and differentiation and further influxes. In general terms such population processes are indicated in the genetic diversity of modern populations in South Asia, which points to a substantial proportion of human biological diversity as developing within South Asia since the Pleistocene (e.g., Su et al., 1999; Kumar and Mohan Reddy, 2003; Kivisild et al., 2003; see Endicott et al., Stock et al., this volume). Technological innovations and climatic changes must have contributed to these processes (James and Petraglia, 2005). Oxygen Isotope Stage 3 saw the expansion of wet forests as well as grass-dominated savannas, especially after ca. 50,000 years ago (Prabhu et al., 2004), and this presumably promoted the expansion of human groups and facilitated migrations between South Asia and areas to the west. Subsequent dry climate of the last glaciation may have forced population distributions to adjust and separated lineages on either side of the greater Thar Desert. The wetter conditions of the terminal Pleistocene and early Holocene, provided a context that would have encouraged expansion and migration again. It is presumably to such processes, and numerous still imperceptible local processes, that language dispersals into South Asia and deep separations with related cultural lineages must be attributed. Linguistic macro-phyla hypotheses need to be considered against such a backdrop, including the proposed links between Nahali and Ainu (perhaps at the earliest stage), links between Austroasiatic and Austronesian (and perhaps Sumerian, see Witzel, 1999:15–16) or Dravidian and Elamite (and perhaps Afro-asiatic or Sumerian, see Blazek 1999) (at a later stage, but probably still Pleistocene). It is within these earlier stages in which Austroasiatic became widespread across northern South Asia, from the Para-Munda Indus region to the Proto-Munda Orissan region, and during which the ancestors of Proto-Dravidian became established on the Peninsula. The Neolithic revolution then provided a major demographic transition through which established languages expanded and diversified in parallel in several areas of the subcontinent. Subsequently language changes occurred through processes of social interactions that were political as much as demographic, reflected in the extensive evidence for substrates and loanwords (e.g., in Indo-Aryan), and contextualized by the increasing social complexity of the Chalcolithic and Iron Age societies of South Asia. Further research in linguistics, archaeology and their integration has much to reveal about the dynamics of these cultural histories.

Acknowledgments

I would like to acknowledge my debt to several friends and collaborators who have introduced me to the Neolithic archaeology of various parts of India, including Ravi Korisettar, P. C. Venkatasubbaiah, J. N. Pal, M. C. Gupta, Rabi Mohanty, Kishor Basa, Basanta Mohanta, and K. Rajan. My ideas have also developed through ongoing discussions with Nicole Boivin, Emma Harvey, Michael Petraglia, Eleni Asouti and Marco Madella. I have benefited from several discussions and arguments with my archaeobotanist colleagues, Mukund Kajale, K.S. Saraswat and Steve Weber in recent years, which have helped me to clarify my thinking on our present state of knowledge with regards to early plant cultivars, and I accept responsibility for my differences of opinion. My avocational thinking on Indian linguistics has benefited from recent discussions and correspondence with Frank Southworth, although any mistakes are likely my own. I must thank Qin Ling for introducing me to aspects of the Chinese Neolithic, including material published in the Chinese language. This paper has been improved by those who took time to read and comment on various drafts, including Emma Harvey, Edgar Samarasundara, Archana Verma, Mary Anne Murray and three anonymous peer-reviewers.

References

Allchin, B., Allchin, F.R., 1982. The Rise of Civilization in India and Pakistan. Cambridge University Press, Cambridge.

Allchin, B., Allchin, F.R., 1995. Rock art of north Karnataka. Bulletin of the Deccan College Post-Graduate and Research Institute 54–55, 313–339.

Allchin, F.R., 1963. Neolithic Cattle Keepers of South India. A Case Study of the Deccan Ashmounds. Cambridge University Press, Cambridge.

Allchin, F.R., Allchin, B., 1974. Some new thoughts on Indian cattle. In: van Lohuizen-de Leeuw, J.E., Ubaghs, J.N. (Eds.), South Asian Archaeology 1973. E.J. Brill, Leiden, pp. 71–77.

Ammerman, A.J., Cavalli-Sfroza, L., 1971. Measuring the rate of spread of early farming in Europe. Man (n.s.) 76, 674–688.

Asouti, E., Fuller, D.Q., 2006. Trees and Woodlands in South India: An Archaeological Perspective. UCL Press, London.

Asouti, E., Fuller, D.Q., Korisettar, R., 2005. Vegetation context and wood exploitation in the southern Neolithic: preliminary evidence from wood charcoals. In: Franke-Vogt, U., Weisshaar, J. (Eds.), South Asian Archaeology 2003. Proceedings of the European Association for South Asian Archaeology Conference, Bonn, Germany, 7th - 11th July 2003. Linden Soft, Aachen, pp. 336–340.

Bedigian, D., 2004. History and lore of sesame in southwest Asia. Economic Botany 58(3), 330–353.

Behera, P.K., 2002. Khameswaripali: a protohistoric site in the middle Mahanadi Valley, Orissa: results of first season (1996–97) excavation. In: Sengupta, G., Panja, S. (Eds.), Archaeology of Eastern India: New Perspectives. Jayasree Press, Kolkata, pp. 487–514.

Bellwood, P., 1996. The origins and spread of agriculture in the Indo-Pacific region: gradualism, diffusion or revolution and colonization. In: Harris, D.R. (Ed.), The Origins and Spread of Agriculture and Pastoralism in Eurasia. UCL Press, London, pp. 465–498.

Bellwood, P., 1997. Prehistory of the Indo-Malaysian archipelago. University of Hawaii Press, Honolulu.

Bellwood, P., 2001. Early agriculturalist population diasporas? Farming, languages and genes. Annual Review of Anthropology 30, 181–207.

Bellwood, P., 2005. First Farmers: The Origins of Agricultural Societies. Blackwell, Oxford.

Bhattacharya, S., 1975. Studies in Comparative Munda Linguistics. Indian Institute for Advanced Study, Simla.

Blazek, V., 1999. Elam: abridge between Ancient Near East and Dravidian India? In: Blench, R., Spriggs, M. (Eds.), Archaeology and Language IV. Routledge, London, pp. 48–78.

Blench, R., 1999. Language phyla of the Indo-Pacific region: recent research and classification. Bulletin of the Indo-Pacific Prehistory Association 18, 59–76.

Blench, R., MacDonald, K.C., 2000. Chickens (II.G.6). In: Kiple, K.F., Ornelas, K.C. (Eds.), The Cambridge World History of Food. Cambridge University Press, Cambridge, pp. 496–499.

Blench, R.M., 2004. Archaeology and language: methods and issues. In: Bintliff, J. (Ed.), Blackwell's Companion to Archaeology. Blackwell, Oxford, 52–74.

Blench, R.M., 2005. From the mountains to the valleys: understanding ethnoliguistic geography in Southeast Asia. In: Blench, R.M., Sagart, L., Sanchez-Mazas, A. (Eds.), Perspectives in the Phylogeny of East Asian Languages. Curzon Press, London, pp. 31–50.

Blench, R.M., Spriggs, M., 1999. General introduction. In: Blench, R.M., Spriggs, M. (Eds.), Archaeology and Language IV. Routledge, London, pp. 1–20.

Blust, R., 1996a. Austronesian culture history: the windows of language. In: Goodenough, W. H.(Ed.), Prehistoric Settlement of the Pacific. American Philosophical Society, Philadephia, pp. 28–35.

Blust, R., 1996b. Beyond the Austronesian homeland: the Austric hypothesis and its implications for archaeology. In: Goodenough, W.H. (Ed.), Prehistoric Settlement of the Pacific. American Philosophical Society, Philadelphia, pp. 117–160.

Bogaard, A., 2004. Neolithic Farming in Central Europe. Routledge, London.

Bradley, D.G., Loftus, R., Cunningham, P., MacHugh, D.E., 1998. Genetics and domestic cattle origins. Evolutionary Anthropology 6, 79–86.

Burrow, T., Emeneau, M.B., 1984. A Dravidian Etymological Dictionary. Clarendon Press, Oxford.

Chang, K.-C., 1986. The Archaeology of China. Yale University Press, New Haven.

Chang, T.T., 1989. Domestication and spread of the cultivated rices. In: Harris, D.R., Hillman, G.C. (Eds.), Foraging and Farming: The Evolution of Plant Exploitation. Unwin, London, pp. 408–417.

Chang, T.T., 1995. Rice. In: Smartt, J., Simmonds, N.W. (Eds), Evolution of Crop Plants. Longman Scientific, Essex, pp. 147–155.

Chang, T.T., 2000. Rice (II.A.7). In: Kiple, K.F., Ornelas, K.C. (Eds.), The Cambridge World History of Food. Cambridge University Press, Cambridge, pp. 132–149.

Charles, M.P., 2006. East of Eden? A consideration of the Neolithic crop spectra in the eastern Fertile Crescent and beyond. In: Colledge, S., Conolly, J. (Eds.), The Origins and Spread of Domestic Plants in Southwest Asia and Europe. UCL Press, London.

Chattopadyaya, U.C., 1996. Settlement pattern and the spatial organization of subsistence and mortuary practices in the Mesolithic Ganges Valley, North-Central India. World Archaeology 27, 461–476.

Chattopadyaya, U.C., 2002. Researches in archaeozoology of the Holocene period (including the Harappan Tradition in India and Pakistan). In: Settar, S., Korisettar, R. (Eds.), Indian Archaeology in Retrospect, Volume III. Archaeology and Interactive Disciplines. Manohar, New Delhi, pp. 365–422.

Chen, W.-B., Nakamura, I., Sato, Y.-I., Nakai, H., 1993a. Distribution of deletion type in cpDNA of cultivated and wild rice. Japanese Journal of Genetics 68, 597–603.

Chen, W.-B., Nakamura, I., Sato, Y.-I., Nakai, H., 1993b. Indica and Japonica differentiation in Chinese landraces. Euphytica 74(3), 195–201

Cheng, C., Motohashi, R., Tchuchimoto, S., Fukuta, Y., Ohtsubo, H., Ohtsubo, E., 2003. Polyphyletic origin of cultivated rice: based on the interspersion patterns of SINEs. Molecular Biology and Evolution 20, 67–75.

Clarke, G.D., Khanna, G.S., 1989. The site of Kunjhun II, Middle Son Valley, and its relevance for the Neolithic of Central India. In: Kenoyer, J.M. (Ed.), Old Problems and New Perspectives in the Archaeology of South Asia. Department of Anthropology, University of Wisconsin, Madison, pp. 29–46.

Cohen, D.J., 1998. The origins of domesticated cereals and the Pleistocene-Holocene transition in East Asia. The Review of Archaeology 19, 22–29.

Cohen, D.J., 2002. New perspectives on the transition to agriculture in China. In: Yasuda, Y. (Ed.), The Origins of Pottery and Agriculture. Lustre Press and Roli Books, New Delhi, pp. 217–227.

Cohen, M.N., 1991. Health and the Rise of Civilization. Yale University Press, New Haven.

Colledge, S., Conolly, J., 2002. Early Neolithic agriculture in Southwest Asia and Europe: re-examining the archaeobotanical evidence. Archaeology International 5, 44–46.

Committee, S., C.R.P. 1962. The test excavation of Guang Fu Lin Neolithic site at Songjia county of Shanghai. Kao Gu (archaeology), 9.

Conningham, R., Sutherland, 1998. Dwellings or granaries? The pit phenomenon of the Kashmir-Swat Neolithic. Man and Environment 22, 29–34.

Cooke, M., Fuller, D.Q., Rajan, K., 2005. Early Historic agriculture in southern Tamil Nadu: archaeobotanical research at Mangudi, Kodumanal and Perur. In: Franke-Vogt, U., Weisshaar, J. (Eds.), South Asian Archaeology 2003. Proceedings of the European Association for South Asian Archaeology Conference, Bonn, Germany, 7th – 11th July 2003. Linden Soft, Aachen, pp. 341–350.

Costantini, L., 1979. Plant remains at Pirak. In: Jarrige, J.-F., Saontoni, M. (Eds.), Fouilles de Pirak, Volume 1. Difussion de Boccard, Paris, pp. 326–333.

Costantini, L., 1983. The beginning of agriculture in the Kachi Plain: the evidence of Mehrgarh. In Allchin, B. (Ed.), South Asian Archaeology 1981. Cambridge University Press, Cambridge, pp. 29–33.

Costantini, L., 1987. Appendix B. Vegetal remains. In: Stacul, G. (Ed.), Prehistoric and Protohistoric Swat, Pakistan. Instituto Italiano per il Medio ed Estremo Orientale, Rome, pp. 155–165.

Costantini, L., Biasini, L.C., 1985. Agriculture in Baluchistan between the 7th and 3rd Millenium B.C. Newsletter of Baluchistan Studies 2, 16–37.

Crawford, G., Shen, C., 1998. The origins of rice agriculture: recent progress in East Asia, Antiquity 72, 858–866.

Crawford, G., 2006. East Asian plant domestication. In: Stark, M. (Ed.) Archaeology of Asia. Blackwell, Oxford, pp. 77–95.

Crowley, T., 1997. An Introduction to Historical Linguistics. Oxford University Press, Oxford.

Decker-Walters, D.S., 1999. Cucurbits, Sanskrit, and the Indo-Aryans. Economic Botany 53(1), 98–112.

Deshpande, M.M., 1995. Vedic aryans, non-Vedic aryans, and non-Aryans: judging the linguistics evidence of the Veda. In: Erdosy, G. (Ed.), The Indo-Aryans of Ancient South Asia. Language Material Culture and Ethnicity. Walter de Gruyte, Berlin, pp. 67–84.

Diamond, J., Bellwood, P., 2003. Farmers and their languages: the first expansions. Science 300, 597–603.

Diffloth, G., 2005. Austroasiatic languages. In Encyclopedia Britannica, 2005 online edition, retrieved from 10 July 2005 from http://www.britannica.com/eb/article-9109792.

Donegan, P., Stampe, D., 2004. Rhythm and the synthetic drift of Munda. In: Singh, R. (Ed.), The Yearbook of South Asian Languages and Linguistics 2004. Mouton de Gruyter, Berlin, pp. 3–36.

Ehret, C., 1988. Language change and the material correlates of language and ethnic shift. Antiquity 62, 564–74.

Elfenbein, J., 1987. A periplus of the 'Brahui Problem'. Studia Iranica 16, 215–233.

Elfenbein, J., 1998. Brahui. In: Steever, S.B. (Ed.), The Dravidian Languages. Routledge, London, pp. 388–414.

Emeneau, M.B., 1997. Linguistics and botany in the Nilgiris. In: Hoskins, P. (Ed.), Blue Mountains Revisited: Cultural Studies on the

Nilgiri Hills. Oxford University Press, New Delhi, pp. 74–105.

Endicott, P., Metspalu, M., Kivisild, T., 2007. Genetic evidence on modern human dispersals in South Asia: Y chromosome and mitochondrial DNA perspectives. In: Petraglia, M.D., Allchin, B. (Eds.), The Evolution and History of Human Populations in South Asia: Inter-disciplinary Studies in Archaeology, Biological Anthropology, Linguistics and Genetics. Springer, Netherlands, pp. 229–244.

Fairservis, W.A., Southworth, F., 1989. Linguistic archaeology and the Indus Valley Culture. In: Kenoyer, J.M. (Ed.), Old Problems and New Perspectives in the Archaeology of South Asia. Department of Anthropology, University of Wisconsin, Madison, pp. 133–141.

Fuchs, S., 1973. The Aboriginal Tribes of India. Macmillan Press, Madras/London.

Fuller, D.Q., 2001. Harappan seeds and agriculture: some considerations. Antiquity 75, 410–413.

Fuller, D.Q., 2002. Fifty years of archaeobotanical studies in India: laying a solid foundation, In: Settar, S., Korisettar, R. (Eds.), Indian Archaeology in Retrospect, Volume III. Archaeology and Interactive Disciplines. Manohar, Delhi, pp. 247–363.

Fuller, D.Q., 2003a. An agricultural perspective on Dravidian historical linguistics: archaeological crop packages, livestock and Dravidian crop vocabulary. In: Bellwood, P., Renfrew, C. (Eds.), Examining the Farming/Language Dispersal Hypothesis. McDonald Institute for Archaeological Research, Cambridge, pp. 191–213.

Fuller, D.Q., 2003b. Indus and non-Indus agricultural traditions: local developments and crop adoptions on the Indian peninsula. In: Weber, S.A., Belcher, W.R. (Eds.), Indus Ethnobiology. New Perspectives from the Field. Lexington Books, Lanham, pp. 343–396.

Fuller, D.Q., 2003c. African crops in prehistoric South Asia: a critical review. In: Neumann, K., Butler, A., Kahlheber, S. (Eds.), Food, Fuel and Fields: Progress in African Archaeobotany. Heinrich-Barth Institut, Köln, pp. 239–271.

Fuller, D.Q., 2003d. Further evidence on the prehistory of sesame. Asian Agri-History 7(2), 127–137.

Fuller, D.Q., 2006. Dung mounds and domesticators: early cultivation and pastoralism in Karnataka. In: Jarrige, C., Lefèvre, V. (Eds.), South Asian Archaeology 2001, Volume I. Prehistory. Éditions Recherche sur les Civilisations, Paris, pp. 117–127.

Fuller, D.Q., Boivin, N.L., 2005. From domestic economy to political economy: a framework for thinking about changes in artefacts and agriculture in prehistoric South India. Paper presented at the Bienneal Conference of the European Association of South Asian Archaeology, London, July, 2005.

Fuller, D.Q., Boivin, N.L., Korisettar, R., 2007. Dating the Neolithic of south India: new radiometirc evidence for key economic, social and ritual transformations. Antiquity, in press.

Fuller, D.Q., Harvey, E.L., 2006. The archaeobotany of Indian pulses: identification, processing and evidence for domestication. Environmental Archaeology, in press.

Fuller, D.Q., Korisettar, R., 2004. The vegetational context of early agriculture in South India. Man and Environment 29, 7–27.

Fuller, D.Q., Korisettar, R., Venkatasubbaiah, P.C., Jones, M.K., 2004. Early plant domestications in southern India: some preliminary archaeobotanical results. Vegetation History and Archaeobotany 13, 115–129.

Fuller, D.Q., Madella, M., 2001. Issues in Harappan archaeobotany: retrospect and prospect. In: Settar, S., Korisettar, R. (Eds.), Indian Archaeology in Retrospect, Vol. II. Protohistory. Manohar, New Delhi, 317–390.

Fuller, D.Q., Korisettar, R., Venkatasubbaiah, P.C., 2001a. Southern Neolithic cultivation systems: a reconstruction based on archaeobotanical evidence. South Asian Studies 17, 171–187.

Fuller, D.Q., Venkatasubbaiah, P.C., Korisettar, R., 2001b. The beginnings of agriculture in the Kunderu River Basin: evidence from archaeological survey and archaeobotany. Puratattva 31, 1–8.

Thompson, G.B.T., 1996. The Excavations of Khok Phanom Di: A Prehistoric Site in Central Thailand. Volume IV: Subsistence and Environment: The Botanical Evidence. Oxbow Books, Oxford.

Gadgil, M., Joshi, N.V., Shambu Prasad, U.V., Manoharan, S., Patil, S. 1998. Peopling of India. In: Balasubramanian, D., Appaji Rao, N. (Eds.), The Indian Human Heritage. Universities Press, Hyderabad, pp. 100–129.

Garrard, A., 2000. Charting the emergence of cereal and pulse domestication in South-West Asia. Environmental Archaeology 4, 67–86.

Glover, I.C., Higham, C.F.W., 1996. New evidence for early rice cultivation in South, Southeast and East Asia. In: Harris, D.R. (Ed.), The Origins and Spread of Agriculture and Pastoralism in Eurasia. UCL Press, London, pp. 413–441.

Grove, C.P., 1985. On the agriotypes of domestic cattle and pigs in the Indo-Pacific region. In: Misra, V.N., Bellwood, P. (Eds.) Recent Advances in Indo-Pacific Prehistory. Oxford and IBH, New Delhi, pp. 429–438.

Harlan, J.R., 1971. Agricultural origins: centers and noncenters. Science 174, 468–474. Harlan, J.R., 1992. Crops and Ancient Man. American Society for Agronomy, Madison.

Harris, D.R., 1998a. The origins of agriculture in Southwest Asia. The Review of Archaeology 19, 5–11.

Harris, D.R., 1998b. The spread of Neolithic agriculture from the Levant to Western Central Asia, In: Damania, A.D., Valkoun, J., Willcox, G., Qualset, C.O. (Eds.), The Origins of Agriculture and Crop Domestication. Proceedings of the Harlan Symposium 10–14 May 1997, Aleppo, Syria. International Center for Agricultural Research in the Dry Areas, Alllepo, pp. 65–82

Harvey, E., Fuller, D.Q., 2005. Investigating crop processing through phytolith analysis: the case of rice and millets. Journal of Archaeological Science 32, 739–752.

Harvey, E., Fuller, D.Q., Basa, K.K., Mohanty, R., Mohanta, B., 2006. Early agriculture in Orissa: some archaeobotanical results and field observations on the Neolithic. Man and Environment, 30, in press.

Harvey, E., Fuller, D.Q., Pal, J.N., Gupta, M.C., 2005. Early agriculture of Neolithic Vindyhas (North-Central India). In: Franke-Vogt, U., Weisshaar, J. (Eds.), South Asian Archaeology 2003. Proceedings of the European Association for South Asian Archaeology Conference, Bonn, Germany, 7th - 11th July 2003. Lindin Soft, Aachen, pp. 329–334.

Hiendleder, S., Kaupe, B., Wassmuth, R., Janke, A., 2002. Molecular analysis of wild and domestic sheep questions current nomenclature and provides evidence for domestication from two different subspecies. Proceedings of the Royal Society of London B 269, 893–904.

Higham, C.F.W., 1998. Archaeology, linguistics and the expansion of the Southeast Asian Neolithic. In: Blench, R., Spriggs, M. (Eds.), Archaeology and Language II. Routledge, London, pp. 103–114.

Higham, C.F.W., 2003. Languages and farming dispersals: Austroasiatic languages and rice cultivation. In: Renfrew, C., Bellwood, P. (Eds.) Examining the Farming/Language Dispersal Hypothesis. McDonald Institute for Archaeological Research, Cambridge, pp. 223–232.

Higham, C.F.W., Thosarat, R., 1998. Prehistoric Thailand: From early settlement to Sukothai. River Books, Bangkok.

Hillman, G.C., 2000. Abu Hireyra 1: The Epipalaeolithic. In: Moore, A.M.T., Hillman, G.C., Legge, A.J. (Eds.), Village on the Euphrates: From Foraging to Farming at Abu Hureyra. Oxford University Press, New York, pp. 327–398.

Hillman, G.C., Hedges, R., Moore, A.M.T., Colledge, S., Pettitt, P., 2001. New evidence of Late Glacial cereal cultivation at Abu Hureyra on the Euphrates. The Holocene 11, 383–393.

Hillman, G.C., Mason, S., de Moulins, D., Nesbitt, M., 1996. Identification of archaeological remains of wheat: the 1992 London Workshop. Circaea 12, 195–209.

Hoffpauir, R., 2000. Water Buffalo, (II.G.23). In: Kiple, K. F. and Ornelas, K. C. (Eds.), The Cambridge world history of food. Cambridge University Press, Cambridge, pp. 583–607.

James, H.V.A., Petraglia, M.D., 2005. Modern human origins and the evolution of behavior in the Later Pleistocene record of South Asia. Current Anthropology 46(S5), S3–S28.

Jarrige, J.-F., 1985. Continuity and change in the North Kachi Plain (Baluchistan, Pakistan) at the beginning of the second millennium BC. In: Schotmans, J., M. Taddei, M. (Eds.), South Asian Archaeology 1983. Instituto Universitario Orientale, Dipartimento di Studi Asicatici, Naples, pp. 35–68.

Jarrige, J.-F., 1987. Problèmes de datation du site néolithique de Mehrgarh, Baluchistan, Pakistan. In: Aurenche, O., Evin, J., Hours, F. (Eds.), Chronologies du Proche Orient/Chronologies in the Near East: Relative Chronologies and Absolute Chronology 16,000–4,000 B.P. British Archaeological Reports International Series 379, Oxford, pp. 381–386.

Jarrige, J.-F., 1997. From Nausharo to Pirak: continuity and change in the Kachi/Bolan region from 3rd to 2nd Millennium BC. In: Allchin, R., Allchin, B. (Eds.), South Asian Archaeology 1995. Oxford-IBH, New Delhi, pp. 35–68.

Jarrige, J.-F., Jarrige, C., Quivron, G., 2006. Mehrgarh Neolithic: the updated sequence. In: Jarrige, C., Lefèvre, V. (Eds.), South Asian Archaeology 2001. Éditions Recherche sur les Civilisations, Paris, pp. 129–142.

Joglekar, P.P., 2004. Animal economy at Lahuradewa, preliminary results. Paper presented at the Indian Archaeological Society and Indian Society for

Quaternary Science and Preshitoric Studies annual conference, Lucknow, December 2004.

Joglekar, P.P., Thomas, P.K., 1993. Faunal diversity at Walki: a small Chalcolithic settlement in western Maharashtra. Bulletin of the Deccan College Post-Graduate and Research Institute 53, 75–94.

Kajale, M.D., 1979. On the occurrence of ancient agricultural patterns during the Chalcolithic periods (c. 1600–1000 BC) at Apegaon, District Aurangabad in central Godavari valley, Maharashtra. In: Deo, S.B., Dhavalikar, M.K., Ansari, Z.D. (Eds.), Apegaon Excavations. Deccan College, Pune, pp. 50–56.

Kajale, M.D., 1988. Plant economy. In: Dhavalikar, M.K., Sankalia, H.D., Ansari, Z.D. (Eds.), Excavations at Inamgaon. Deccan College Postgraduate and Research Institute, Pune, pp. 727–821.

Kajale, M.D., 1989. Archaeobotanical investigation on Megalithic Bhagimohari, and its significance. Man and Environment 13, 87–96.

Kajale, M.D., 1990. Observations on the plant remains from excavation at Chalcolithic Kaothe, District Dhule, Maharashtra with cautionary remarks on their interpretations. In: Dhavalikar, M.K., Shinde, V.S., Atre, S.M. (Eds.), Excavations at Kaothe. Deccan College, Pune, pp. 265–280.

Kajale, M.D., 1991. Current status of Indian palaeoethnobotany: introduced and indigenous food plants with a discussion of the historical and evolutionary development of Indian agriculture and agricultural systems in general. In: Renfrew, C. (Ed.), New Light on Early Farming – Recent Developments in Palaeoethnobotany. Edinburgh University Press., Edinburgh, pp. 155–189.

Kajale, M.D., 1994. Archaeobotanical investigations on a multicultural site at Adam, Maharashtra, with special reference to the development of tropical agriculture in arts of India. In: Hather, J. (Ed.), Tropical Archaeobotany: Applications and New Developments. Routledge, London, pp. 34–50.

Kajale, M.D., 1996. Palaeobotanical investigations at Balathal: preliminary results. Man and Environment 21, 98–102.

Kar, S.K., 1995. Further exploration at Golpalpur, Orissa. Puratattva 26, 105–106.

Kar, S.K., 2000. Gopalpur; A Neolithic-Chalcolithic site in coastal Orissa. In: Basa, K.K., Mohanty, P. (Eds.), Archaeology of Orissa. Pratibha Prakashan, Delhi, pp. 368–391.

Kar, S. K., Basa, K.K., Joglekar, P.P., 1998. Explorations at Gopalpur, District Nayagarh, Coastal Orissa. Man and Environment 23, 107–114.

Kennedy, K.A.R., 2000. God-Apes and Fossil Men: Paleoanthropology in South Asia. University of Michigan Press, Ann Arbor.

Kharrakwal, J.S., Yano, A., Yasuda, Y., Shinde, V.S., Osada, T., 2004. Cord impressed ware and rice cultivation in South Asia, China and Japan: possibilities of inter-links, Quaternary International 123–125, 105–115.

Kivisild, T., Rootsi, S., Metspalu, M., Mastana, S., Kaldma, K., Parik, J., Metspalu, E., Adojaan, M., Tolk, H.-V., Stepanov, V., Goge, M., Usanga, E., Papiha, S.S., Cinniogu, C., King, R., Cavalli-Sforza, L., Underhill, P.A., Villems, R., 2003. The genetic heritage of the earliest settlers persists both in Indian tribal and caste populations. American Journal of Human Genetics 72, 313–332.

Korisettar, R., Joglekar, P.P., Fuller, D.Q., Venkatasubbaiah, P.C., 2001b. Archaeological re-investigation and archaeozoology of seven southern Neolithic sites in Karnataka and Andhra Pradesh. Man and Environment 26, 47–66.

Korisettar, R., Venkatasubbaiah, P.C., Fuller, D.Q., 2001a. Brahmagiri and beyond: the archaeology of the southern Neolithic. In: Korisettar, R., Settar, S. (Eds.), Indian Archaeology in Retrospect, Volume I. Prehistory. Manohar, New Delhi, pp. 151–238.

Krishnamurti, B., 2003. The Dravidian Languages. Cambridge University Press, Cambridge.

Kuiper, F.B.J., 1991. Aryans in the Rig Veda. Rodopi, Amsterdam/Atlanta.

Kumar, P., Freeman, A.R., Loftus, R.T., Gaillard, C., Fuller, D.Q., Bradley, D.G., 2003. Admixture analysis of South Asian cattle. Heredity 91, 43–50.

Kumar, V., Mohan Reddy, B., 2003. Status of Austro-Asiatic groups in the peopling of India: an exploratory study based on the available prehistoric, linguistic and biological evidence. Journal of Bioscience 28, 507–522.

Larsen, C.S., 1997. Biological changes in human populations with agriculture. Annual Review of Anthropology 24, 185–213.

Leshnik, L.S., 1973. Land use and ecological factors in prehistoric north-west India. In: Hammond, N. (Ed.), South Asian Archaeology. Duckworth, London, pp. 67–84.

Liu, L., Chen, X., 2004. The measurement and primary analysis of the buffalo bones from the Kua Hu Qiao site. In: Zhejian Provincial Institute of Archaeology and Cultural Relics, Kua Hu Qiao: A Neolithic Site Excavation Report. Wenwu Press, Beijing. [In Chinese]

Lone, Farooq A., Maqsooda Khan, Buth, G.M., 1993. Palaeoethnobotany – Plants and Ancient Man in Kashmir. A.A. Balkema, Rotterdam.

Lu, T.L.D., 1999. The Transition from Foraging to Farming and the Origin of Agriculture in China. British Archaeological Reports, Oxford.

Lu, T.L.D., 2006. The origin and dispersal of agriculture and human diaspora in East Asia. In: Sagart, L., Blench, R., Sanchez-Mazas, A. (Eds.), The Peopling of East Asia: Putting Together Archaeology, Linguistics and Genetics. Routledge Curzon, London, pp. 51–62.

Luikart, G., Gielly, L., Excoffier, L., Vigne, J.-D., Bouvet, J., Taberlet, P., 2001. Multiple maternal origins and weak phylogeographic structure in domestic goats. Proceedings of the National Academy of Sciences (USA) 98, 5927–5932.

Lukacs, J.R., 2002. Hunting and gathering strategies in prehistoric India: a biocultural perspective on trade and subsistence. In: Morrison, K.D., Junker, L.L. (Eds.), Forager-Traders in South and Southeast Asia. Cambridge University Press, Cambridge, pp. 41–61.

Lukacs, J.R., 2007. Interpreting biological diversity in South Asian prehistory: Early Holocene population affinities and subsistence adaptations. In: Petraglia, M.D., Allchin, B. (Eds.), The Evolution and History of Human Populations in South Asia: Inter-disciplinary Studies in Archaeology, Biological Anthropology, Linguistics and Genetics. Springer, Netherlands, pp. 271–296.

Lukacs, J.R., Pal, J.N., 1993. Mesolithic subsistence in north India: Inferences from dental attributes. Current Anthropology, 34(5), 745–765.

MacHugh, D.E., Shriver, M.D., Loftus, R.T., Cunningham, P., Bradley, D.G., 1997. Microsatellite DNA variation and the evolution, domestication and phylogeography of taurine and zebu cattle (Bos taurus and Bos indicus). Genetics 146, 1071–1086.

MacNeish, R.S., 1992. The Origins of Agriculture. University of Oklahoma Press, Norman, Oklahoma.

Madella, M., 2003. Investigating agriculture and environment in South Asia: present and future considerations of opal phytoliths. In: Weber, S.A., Belcher, W.R. (Eds.), Indus Ethnobiology: New Perspectives from the Field. Lexington Books, Lanham, pp. 199–250.

Magee, D.A., Mannen, H., Bradley, D., 2007. Duality in Bos indicus mtDNA diversity: support for geographical complexity in zebu domestication. In: Petraglia, M.D., Allchin, B.

(Eds.), The Evolution and History of Human Populations in South Asia: Inter-disciplinary Studies in Archaeology, Biological Anthropology, Linguistics and Genetics. Springer, Netherlands, pp. 385–391.

Mahdi, W., 1998. Transmission of southeast Asian cultigens to India and Sri Lanka. In: Blench, R., Spriggs, M. (Eds.), Archaeology and Language II: Archaeological Data and Linguistic Hypotheses. Routledge, London, pp. 390–415.

Mani, B.R., 2004. Further evidence on Kashmir Neolithic in light of recent excavations at Kanishkapura. Journal of inter-disciplinary Studies in History and Archaeology 1(1), 137–142.

Masica, C.P., 1979. Aryan and non-Aryan elements in north Indian agriculture. In: Deshpande, M.M., Hook, P.E. (Eds.), Aryan and Non-Aryan in India. Center for South and Southeast Asian Studies, University of Michigan, Ann Arbor, pp. 55–151.

Masica, C.P., 1991. The Indo-Aryan Languages. Cambridge University Press, Cambridge.

McAlpin, D.W., 1981. Proto-Elamo-Dravidian: The Evidence and its Implications. American Philosophical Society, Philadelphia.

Meadow, R., 1984. Animal Domestication in the Middle East: A View from the Eastern Margin. In: Clutton-Brock, J., Grigson, C. (Eds.), Animals in Archaeology 3. Early Herders and their Flocks. British Archaeological Reports, Oxford, pp. 309–337.

Meadow, R., 1993. Animal domestication in the Middle East: a revised view from the eastern Margin. In: Possehl, G.L. (Ed.), Harappan Civilization: A Recent Perspective. Oxford and IBH, New Delhi, pp. 295–320.

Meadow, R., 1996. The origins and spread of agriculture and pastoralism in northwestern South Asia In: Harris, D.R. (Ed.), The Origins and Spread of Agriculture and Pastoralism in Eurasia. UCL Press, London, pp. 390–412.

Meadow, R., 1998. Pre- and Proto-Historic agricultural and pastoral transformations in northwestern South Asia. The Review of Archaeology 19, 12–21.

Meadow, R., Patel, A.K., 2003. Prehistoric pastoralism in northwestern South Asia from the Neolithic through the Harappan Period. In: Weber, S.A., Belcher, W.R. (Eds.), Indus Ethnobiology: New Perspectives from the Field. Lexington Books, Lanham, pp. 65–94.

Meher-Homji, V.M., 2001. Bioclimatology and Plant Geography of Peninsular India. Scientific Publishers, Jodhpur.

Misra, V.D., 1999. Agriculture, domestication of animals and ceramic and other industries in prehistoric India: Mesolithic and Neolithic. In: Pande, G.C. (Ed.), The Dawn of Civilization up to 600 BC. Centre for Studies in Civilization, Delhi, pp. 233–266.

Misra, V.D., Pal, J.N., Gupta, M.C., 2001. Excavation at Tokwa: a Neolithic-Chalcolithic settlement. Pragdhara 11, 59–72.

Misra, V.D., Pal, J.N., Gupta, M.C., 2004. Significance of recent excavations at Tokwa in the Vindhyas and Jhusi in the Gangetic Plains. Journal of Inter-disciplinary Studies in History and Archaeology 1(1), 120–126.

Mohanty, B., 1994. Golbai: a new horizon in Orissan archaeology. Orissa Historical Research Journal 39, 30–32.

Moore, A.M.T., Hillman, G.C., Legge, A.T., 2000. Village on the Euphrates: From Foraging to Farming at Abu Hureyra. Oxford University Press, New York.

Moulherat, C., Tengberg, M., Haquet, J.-F., Mille, B., 2002. First evidence of cotton at Neolithic Mehrgarh, Pakistan: analysis of mineralized fibres from a copper bead. Journal of Archaeological Science 29, 1393–1401.

Mufwene, S.S., 2001. The Ecology of Language Evolution. Cambridge University Press, Cambridge.

Nagaraja Rao, M.S., Malhotra, K.C., 1965. Stone Age Hill Dwellers of Tekkalakota. Deccan College, Pune.

Neumann, K., 2004. The romance of farming: plant cultivation and domestication in Africa. In: Stahl, A.B. (Ed.), African Archaeology: A Critical Introduction. Blackwell, Oxford, pp. 249–275.

Oka, H. I., 1988. Origin of Cultivated Rice. Japan Science Society Press, Tokyo.

Paddayya, K., 1998. Evidence of Neolithic cattle-penning at Budihal, Gulbarga District, Karnataka. South Asian Studies 14, 141–153.

Paddayya, K., 2001. The problem of ashmounds of Southern Deccan in the light of the Budihal excavations, Karnataka. Bulletin of the Deccan College Post-Graduate and Research Institute 60–61, 189–225.

Pandey, J.N., 1990. Mesolithic in the Middle Ganga Valley. Bulletin of the Deccan College Post-Graduate and Research Institute 49, 311–316.

Panja, S., 1999. Mobility and subsistence strategies: a case study of Inamgaon, a Chalcolithic sites in western India. Asian Perspectives 38, 154–185.

Panja, S., 2001. Research on the Deccan, Chalcol-ithic. In: Settar, S., Korisettar, R. (Eds.), Indian Archaeology in Retrospect, Volume I. Prehistory. Manohar, New Delhi, pp. 263–276.

Parpola, A., 1994. Deciphering the Indus Script. Cambridge University Press, Cambridge.

Parpola, A., 1988. The coming of the Aryans to Iran and India and the cultural and ethnic identity of the Dasas. Studia Orientalia (Helsinki) 64, 195–302.

Patel, A.K., 1997. The pastoral economy of Dholavira: a first look at animals and urban life in third millennium Kutch. In: R. Allchin, Allchin, B. (Eds.), South Asian Archaeology 1995. Oxford-IBH, New Delhi, pp. 101–114.

Patel, A.K., 1999. Paper presented at Fifteenth Inter-national Conference on South Asian Archaeology, Leiden Univerisy, July 5–9, 1999.

Patel, A.K., Meadow, R., 1998. The exploitation of wild and domestic water buffalo in prehis-toric northwestern South Asia. In: Buitenhuis, H., Bartosiewicz, L., Choyke, A. M. (Eds.), Archaeo-zoology of the Near East III. Centre for Archae-ological Research and Consultancy, Rijksuniver-siteit Groningen, Groningen, pp. 180–199.

Pejros, I., Snirelman, V., 1998. Rice in Southeast Asia: a regional inter-disciplinary approach. In: Blench, R., Spriggs, M. (Eds.), Archaeology and Language II: Archaeological Data and Linguistic Hypotheses. Routledge, London, pp. 379–389.

Pokharia, A.K., Saraswat, K.S., 2004. Plant resources in the Neolithic Economy at Kanishpur, Kashmir. Paper presented at National Seminar on the Archaeology of the Gange Plain, Joint Annual Conference of the Indian Archaeological Society, Indian Society of Prehistoric and Quaternary Studies, Indian History and Culture Society, December 2004, Lucknow.

Possehl, G.L., 1999. Indus Age: The Beginnings. University of Pennsylvania Press, Philadephia.

Possehl, G.L., 2002. The Indus Civilization: A Contem-porary Perspective. Alta Mira, Walnut Creek.

Possehl, G.L., Rissman, P., 1992. The chronology of prehistoric India from earliest times to the Iron Age. In: Ehrich, R.W. (Ed.), Chronologies in Old World Archaeology. University of Chicago Press, Chicago, vol. 1, pp. 465–490, vol. 2, pp. 447–474.

Prabhu, C.N., Shankar, R., Anupama, A., Taieb, M., Bonnefille, R., Vidal, L., Prasad, S., 2004. A 200-ka pollen and oxygen-isotopic record from two sediment cores from the eastern Arabian Sea. Palaeogeography, Palaeoclimatology, Palaeoe-cology 214, 309–321.

Puri, G. S., Gupta, R.K., Meher-Homji, V.M., Puri, S., 1989. Forest Ecology (second edition), Volume II. Plant Form, Diversity, Communities and Succession. Oxford and IBH, New Delhi.

Puri, G.S., Meher-Homji, V.M., Gupta, R.K., Puri, S., 1983. Forest Ecology (second edition), Volume I. Phytogeography and Forest Conservation. Oxford and IBH, New Delhi.

Reddy, S.N., 2003. Discerning Palates of the Past: An Ethnoarchaeological Study of Crop Cultivation and Plant Usage in India. Prehistory Press, Ann Arbor.

Renfrew, C., 1987. Archaeology and Language: The Puzzle of Indo-European Origins. Cambridge University Press, Cambridge.

Renfrew, C., 1996. Language families and the spread of farming. In: Harris, D.R. (Ed.), The Origins and Spread of Agriculture and Pastoralism in Eurasia. UCL Press, London, pp. 70–92.

Renfrew, C., 2000. At the edge of knowability: towards a prehistory of languages. Cambridge Archaeological Journal 10, 7–34.

Rosenberg, M., 1998. Cheating at musical chairs: territoriality and sedentism in an evolutionary context. Current Anthropology 39, 653–681.

Rouse, I., 1986. Migrations in Prehistory. Yale University Press, New Haven.

Sano, R., Morishima, H., 1992. Indica-Japonica differentiation of rice cultivars viewed from variations in key characters of isozyme, with species reference to Himilayan hilly areas. Theoretical and Applied Genetics 84, 266–274.

Saraswat, K.S., 1991. Crop economy at ancient Mahorana, Punjab (c. 2100–1900 B.C.). Pragdhara 1, 83–88.

Saraswat, K.S., 1993. Plant economy of Late Harappans at Hulas. Purattatva 23, 1–12.

Saraswat, K.S., 2002. Banawali (29′37′5″N; 75′23′6″E), District Hissar. Indian Archaeology 1996–97- A Review, 203.

Saraswat, K.S., 2004a. Plant economy of early farming communities at Senuwar, Bihar. In: Singh, B.P. (Ed.), Senuwar Excavations. Banares Hindu University, Varanasi.

Saraswat, K.S., 2004b. Plant economy of Damdama. Paper presented at Indian Archaeological Society and Indian Society for Quaternary Science and Preshitoric Studies, Lucknow, December 2004.

Saraswat, K.S., 2005. Agricultural background of the early farming communities in the Middle Ganga Plain. Pragdhara 15, 145–178.

Saraswat, K.S., Chanchala, 1995. Palaeobotanical and pollen analytical investigations. Indian Archaeology 1990–91 - A Review, 103–104.

Saraswat, K.S., Pokharia, A.K., 2002. Harappan plant economy at ancient Balu, Haryana. Pragdhara 12, 153–172.

Saraswat, K.S., Pokharia, A.K., 2003. Palaeoethnobotanical investigations at Early Harappan Kunal. Pragdhara 13, 105–140.

Saraswat, K.S., Pokharia, A.K., 2004. Archaeological studies in the Lahuradewa Area 2. Plant economy at Lahuradewa: a preliminary contemplation. Paper presented at National Seminar on the Archaeology of the Ganga Plain, Joint Annual Conference of the Indian Archaeological Society, Indian Society of Prehistoric and Quaternary Studies, Indian History and Culture Society, December 2004, Lucknow.

Saraswat, K.S., Sharma, N.K., Saini, D C., 1994. Plant economy at ancient Narhan (Ca. 1,300 B.C. - 300/400 A.D.). In: Singh, P. (Ed.), Excavations at Narhan (1984–1989). Banaras Hindu University, Varanasi, pp. 255–346.

Sato, Y.-I., 2002. Origin of rice cultivation in the Yangtze River Basin. In: Yasuda, Y. (Ed.), The Origins of Pottery and Agriculture. Lustre Press and Roli Books, New Delhi, pp. 143–150.

Sato, Y.I., 2005. Rice and Indus civilization. In: Osada, T. (Ed.), Linguistics, Archaeology and Human Past. Research Institute for Humanity and Nature, Kyoto, pp. 213–214.

Sato, Y.-I., Ishikawa, R., Morishima, H. 1990. Nonrandom association of genes and characters found in indica x japonica hybrids of rice. Heredity 65, 75–79.

Shanghai Cultural Relics Protection Committee, 1962. The Test Excavation of Guang Fu Lin Neolithic Site at Songjia County of Shanghai, Kao Gu (archaeology), 9. [in Chinese]

Sharma, A.K., 1982 Excavations at Gufkral, 1981. Purattatva 11, 19–25.

Sharma, A.K., 1986 Neolithic Gufkral. In: Buth, G.M. (Ed.), Central Asia and Western Himalaya – A Forgotten Link. Scientific Publishers, Jodhpur, pp. 13–18.

Sharma, G.R., Misra, V.D., Mandal, D., Misra, B.B., Pal, J.N., 1980. Beginnings of Agriculture (Epi-Palaeolithic to Neolithic: Excavations at Chopani-Mando, Mahadaha, and Mahagara). Abinash Prakashan, Allahabad.

Shinde, V., 1998a. Pre-Harappan Padri culture in Saurashtra: the recent discovery. South Asian Studies 14, 173–182.

Shinde, V.S., 1998b. Early Settlements in the Central Tapi Basin. Munshiram Manoharlal, New Delhi.

Shinde, V.S., 2002. The emergence, development and spread of agricultural communities in South Asia. In: Yasuda, Y. (Ed.), The Origins of Pottery and Agriculture. Lustre Press and Roli Books, New Delhi, pp. 89–115.

Simoons, F.J., 1991. Food in China. A Cultural and Historical Inquiry. CRC Press, Boca Raton.

Singh, I.B., 2005. Landform development and palaeovegetation in Late Quaternary of the Ganga Plain: implications for anthropogenic activity. Pragdhara 15, 5–31.

Sinha, B.K., 1993. Excavations at Golbai Sasan, District Puri, Orissa. Puratattva 23, 48–50.

Sinha, B.K., 2000. Golbai: a protohistoric site on the coast of Orissa. In: Basa, K.K., Mohanty, P. (Eds.), Archaeology of Orissa. Pratibha Prakashan, Delhi, pp. 322–355.

Southworth, F.C., 1976. Cereals in South Asian prehistory: the linguistic evidence. In: Kennedy, K.A.R., Possehl, G.L. (Eds.), Ecological Backgrounds of South Asian Prehistory. South Asia Program, Cornell University, Ithaca, New York, pp. 52–75.

Southworth, F.C., 1979. Lexical evidence for early contacts between Indo-Aryan and Dravidian. In: Deshpande, M.M., Hook, P.E. (Eds.), Aryan and Non-Aryan in India. Center for South and Southeast Asian Studies, University of Michigan, Ann Arbor, pp. 191–233.

Southworth, F.C., 1988. Ancient economic plants of South Asia: linguistic archaeology and early agriculture. In: Jazayery, M.A., Winter, W. (Ed.), Languages and Cultures: Studies in Honor of Edgar C. Polome. Mouton de Gruyter, Amsetrdam, pp. 649–688.

Southworth, F.C., 1992. Linguistics and archaeology: prehistoric implications of some South Asian plant names. In: Possehl, G.L. (Ed.), South Asian Archaeology Studies. Oxford and IBH, New Delhi, pp. 81–85.

Southworth, F.C., 2005a. The Linguistic Archaeology of South Asia. Routledge, London.

Southworth, F. C. 2005b. Prehistoric implications of the Dravidian element in the NIA lexicon with special reference to Marathi. International Journal of Dravidian Linguistics 34(1), 17–28.

Southworth, F.C., Witzel, M., 2006. The SARVA (South Asia Residual Vocabulary Assemblage) Poject Website (http://www.aa.tufs.ac.jp/sarva/).

Stock, J., Lahr, M.M., Kulatilake, S., 2007. Human dispersals and cranial diversity in South Asia relative to global patterns of human variation. In: Petraglia, M.D., Allchin, B. (Eds.), The Evolution and History of Human Populations in South Asia: Inter-disciplinary Studies in Archaeology, Biological Anthropology, Linguistics and Genetics. Springer, Netherlands, pp. 245–268.

Su, B., Xiao, J., Underhill, P., Deka, R., Zhang, W., Akey, J., Huang, W., Shen, D., Lu, D., Luo, J., Chu, J., Tan, J., Shen, P., Davis, R., Cavalli-Sforza, L. L., Chakraborty, R., Xiong, M., Du, R., Oefner, P., Chen, Z., Jin, L., 1999. Y-Chromosome evidence for a northward migration of modern humans into Eastern Asia during the last Ice Age. American Journal of Human Genetics 65, 1718–1724.

Tengberg, M., 1999. Crop husbandry at Miri Qalat, Makran, SW Pakistan (4000–2000 B.C.). Vegetation History and Archaeobotany 8, 3–12.

Tengberg, M., 2005. Exploitation and use of plants in the Halil Valley during the Bronze Age: first results from the archaeobotanical analysis at Kunar Sandal A and B, Southeast Iran. Paper presented at the 15th Conference of the European Association of South Asian Archaeologists, London.

Tewari, R., Srivastava, R.K., Singh, K.K., Saraswat, K.S., Singh, I.B., 2003. Preliminary report of the excavation at Lahuradewa, District Sant Kabir Nagar, U.P. 2001–2002: wider archaeological implications. Pragdhara 13, 37–68.

Tewari, R., Srivastava, R.K., Singh, K.K., Vinay, R., Trivedi, R.K., Singh, G.C., 2005. Recently excavated sites in the Ganga Plain and North Vindhyas: some observations regarding the pre-urban context. Pragdhara 15, 39–49.

Thapar, B.K., 1978. Early farming communities in India. Journal of Human Evolution 7, 11–22.

Thomas, P.K., Joglekar, P.P., Mishra, V.D., Pandey, J.N., Pal, J.N., 1995. A preliminary report of the faunal remains from Damdama. Man and Environment 20, 29–36.

Tikkanen, B., 1999. Archaeological-linguistic correlations in the formation of retroflex typologies and correlating areal features in South Asia. In: Blench, R., Spriggs, M. (Eds.), Archaeology and Language IV: Language Change and Cultural Transformation. Routeledge, London, pp. 138–148.

Trautman, T.R., 1979. The study of Dravidian kinship. In: Deshpande, M.M., Hook, P.E. (Eds.), Aryan and Non-Aryan in India. Center for South and

Southeast Asian Studies, University of Michigan, Ann Arbor, pp. 153–173.

Trautman, T.R., 1981. Dravidian Kinship. Cambridge University Press, Cambridge.

Turner, R.L., 1966. A Comparative Dictionary of the Indo-Aryan Languages. Oxford University Press, Oxford.

Van Driem, G., 1998. Neolithic correlates of ancient Tibeto-Burman migrations. In: Blench, R., Spriggs, M. (Eds.), Archaeology and Language II: Archaeological Data and Linguistic Hypotheses. Routledge, London, pp. 67–102.

Vaughan, D.A., 1989. The Genus *Oryza L.*: Current Status of Taxonomy. International Rice Research Institute, Los Banos, Philippines.

Vaughan, D.A., 1994. The wild relatives of rice: a genetic resources handbook. International Rice Research Institute, Los Banos, Philippines.

Vishnu-Mittre, 1976. The archaeobotanical and palynological evidence for the early origin of agriculture in South and Southeast Asia. In: Arnott, M.I. (Ed.), Gastronomy. Mouton and Co., The Hague, pp. 13–21.

Wan, J., Ikehashi, H., 1997. Identification of two types of differentiation in cultivated rice (*Oryza sativa L.*) detected by polymorphism of isozymes and hybrid sterility. Euphytica 94, 151–161.z

Weber, S.A., 1991. Plants and Harappan Subsistence: An Example of Stability and Change from Rojdi. Oxford and IBH., New Delhi.

West, B., B.-X., Z., 1988. Did chickens go north? New evidence for domestication. Journal of Archaeological Science 15, 515–533.

Wheeler, R.E.M., 1959. Early India and Pakistan. Thames and Hudson, London.

Willcox, G., 1992. Some differences between crops of Near Eastern origin and those from the tropics. In: Jarrige, C. (Ed.), South Asian Archaeology 1989. Prehistory Press, Madison, pp. 291–299.

Willcox, G., 1999. Agrarian change and the beginnings of cultivation in the Near East: evidence from wild progenitors, experimental cultivation and archaeobotanical data. In: Gosden, C., Hather, J. (Eds.), The Prehistory of Food: Appetites for Change. Routledge, London, pp. 478–500.

Willcox, G., 2002. Geographical variation in major cereal components and evidence for independent domestication events in Western Asia. In: Cappers, R.T.J., Bottema, S. (Eds.), The Dawn of Farming in the Near East. Ex Oriente, Berlin, pp. 133–140.

Willcox, G., 2005. The distribution, natural habitats and availability of wild cereals in relation to their domestication in the Near East: multiple events, multiple centres. Vegetation History and Archaeobotany 14(4), 534–541.

Witzel, M., 1999. Early sources for South Asian substrate languages. Mother Tongue Special Issue, 1–76.

Witzel, M., 2005. Central Asian roots and acculturation in South Asia: linguistic and archaeological evidence from Western Central Asia, the Hindukush and northwestern South Asia for early Indo-Aryan language and religion. In: Osada, T. (Ed.), Liguistics, Archaeology and the Human Past. Research Institute for Humanity and Nature, Kyoto, pp. 87–211.

Witzel, M., 2006. South Asian agricultural terms in Indo-Aryan. In: Osada, T., Sato, Y.-I., Witzel, M. (Eds,), Ethnogenesis in South and Central Asia. Harvard-Kyoto Roundtable (7th ESCA), Research Institute for Humanities and Nature, Kyoto, pp. 96–120

Yan, W., 2002. The origins of rice agriculture, pottery and cities. In: Yasuda, Y. (Ed.), The Origins of Pottery and Agriculture. Lustre Press and Roli Books, New Delhi.

Zeuner, F.E., 1963. A History of Domesticated Animals. Hutchinson, London.

Zide, A.R.K., Zide, N.H., 1976. Proto-Munda cultural vocabulary: evidence for early agriculture. In: Jenner, P.N., Thompson, L.C., Starosta, S. (Eds.), Austroasiatic Studies, Part II. University of Hawaii Press, Honolulu, pp. 1295–1334.

Zohary, D., 1996. The mode of domestication of the founder crops of Southwest Asian agriculture, In: Harris, D.R. (Ed.), The Origins and Spread of Agriculture and Pastoralism in Eurasia. UCL Press, London, pp. 142–158.

Zohary, D., Hopf, M., 2000. Domestication of Plants in the Old World. Oxford University Press, Oxford.

PART IV
CONCLUDING REMARKS

19. Thoughts on *The Evolution and History of Human Populations in South Asia*

GREGORY L. POSSEHL

Department of Anthropology
325 University Museum
University of Pennsylvania
Philadelphia, Pennsylvania 19104-6398
USA
gpossehl@sas.upenn.edu

This book will probably be recorded as a significant moment for the study of early populations in the Indian subcontinent. While paleolithic remains have been known there since the 19th century, until recently the general state of knowledge was very restricted and not on par with other world regions. The reasons for this are not of concern to me now, but clearly this situation has changed, and will continue to do so. *The Evolution and History of Human Populations in South Asia* can be taken as marking a turning point. Archaeology, environmental studies, linguistics and genetics have been brought together to offer us the first modern, holistic view of early India. There is still much work to be done, but a new paradigm is in place, and many new, exciting scholars are engaged with it.

Much of what follows is a discussion of hypotheses and thoughts on how we might proceed with future research. Participating in the conference, and reviewing the chapters, has not changed my sense of how little we really know about the early history of the Indian subcontinent, especially about the Paleolithic; but how much promise there is to dramatically change this. In this situation, it is scientifically sound to propose hypotheses that are broad and multi-dimensional, casting a wide net, rather than a narrow one. As knowledge accumulates, as we learn more about the ancient subcontinent, we will find that some parts of these broad, multi-dimensional hypotheses are not as viable as others, or perhaps not even viable at all. Thus, over time, we can winnow down the alternatives, but at the moment, given the present state of knowledge a liberal breadth of possibilities would seem to be the most prudent approach. One of the research themes that has much promise deals with the earliest inhabitants, and the "out of Africa" theme.

The Siwalik Hills

The earliest indications of human habitation of the subcontinent come from the Upper Siwaliks of Pakistan in the Soan Valley and Pabbi Hills. This is reviewed in this volume by Dennell, who suggests that there is much promise for

447

M.D. Petraglia and B. Allchin (eds.), The Evolution and History of Human Populations in South Asia, 447–459.
© 2007 *Springer.*

further research in this region. Artifacts are clearly present, and the dates that are available indicate a Mode I presence. But most of the tools are in surface context. At the minimum, the Soan Valley and Pabbi Hills inform us that the Upper Siwaliks are an outstandingly good place to search for Mode I sites, and moreover, they are rich in fossils including the remains of hominoids. These facts combine to suggest that more basic exploration is needed to find in situ Mode I sites which are likely to be there. There are no certainties in archaeological exploration, but it may well be that it is only a matter of time, and much hard work, before such sites are known.

At the moment the Siwaliks of Pakistan are not accessible for scientific research; however the Indian Siwaliks are, and there is even research being conducted there directed toward the finding of Mode I sites (e.g., Chauhan, 2004). But, the Siwalik Hills are an immense territory, and there is much room of others to join the hunt.

The Route to and within the Subcontinent

The route into the subcontinent and beyond is discussed in this book, principally by Turner and O'Regan and again by Korisettar. The former authors put forth the hypothesis that early hominins would have used the sea coasts as their primary routes. On the other hand, Korisettar argues that the rich inland resources of the subcontinent would have attracted and kept substantial early populations there as well. This would seem to be the case for populations associated with the well-documented Indian Acheulean industry.

I think most would agree that the initial "out of Africa" was made by small populations over many millennia, or tens of millennia. The direct distance from Suez to the Yellow Sea is about 8,000 kilometers. If the Mode I travelers moved along this route averaging one kilometer per year in a generally eastward

direction, it would take them only 22 years to complete this journey. This is clearly not the way to look at this sort of problem in an analytical way, but for purposes of illustration it does show that if we make the assumption that the "out of Africa" travelers took millennia, or tens of millennia to get from Africa to the Yellow Sea, then average distances moved each year in that direction were very, very small. Moreover, the earliest of them did not know where they were going, because no one had been there before. And even the later travelers may have been in the same position. Given the low population densities, and the ambiguities of communication in Mode I times the movements and knowledge of the very first (and second, and third, etc.) pioneers may have been totally unknown to later travelers of the same route(s). Thus, Eurasia in Mode I times may have been discovered over and over again.

The greater Himalayan land mass was in place two million years ago and this is a formidable landscape. It could have captured the curious or "lost," but may not have been attractive as a route from west to east. So, most of the very few, Mode I hominins probably would have gone around it, on the whole taking lines of least resistance to movement. These "lines of least resistance" would have included factors such as the availability of the kinds of food that the Mode I travelers ate, a need that may not have been exactly the same for all Mode I traveling parties, the availability of shelter (either natural or built) and, if Dennell is correct, the availability of stone for the manufacture of tools.

Some travelers could have moved to the north, especially during warm periods there, and others to the south into the subcontinent, along its shores, and into the richest landscapes of the interior. It may well turn out that one of these rich inland landscapes were the Siwalik Hills, similar to what we see at a place like Dmanisi in Georgia. The northern route, around the Himalayas is a long one, but the time for travel was immense. During warm periods, the route to the Yellow Sea via the Altai is certainly a possibility. And our Mode I travelers may have gone even farther north, if the weather and their material culture (especially clothing) allowed it.

There are two interesting routes in a southern part of this immense territory. One takes the valley of the Syr Darya, with the Tien Shan to the north and the greater Pamirs to the south, up the narrow mountains separating Central Asia from what is under modern environmental conditions the Takla Makan Desert (Figure 1). This is attractive because, while a part of it is reasonably daunting territory, the distances involved are relatively short, about 200 kilometers in the mountains, and once crossed there is a virtually unobstructed route to the Yellow Sea. There is another route in this area up the Amu Darya Valley into the Pamirs themselves, and down onto the Takla Makan. It is more daunting than the Syr Darya, but possible in a warm season of hard walking and climbing.

What comes of this discourse? First, distances are relatively short, if measured against the immense time frame within which they were crossed. Second, we should be looking at many routes into the subcontinent, and around it, into China and Southeast Asia. The sea coasts are attractive, and even may have been the predominant route as Turner and O'Regan note, but they are not so attractive that this logic rules out other routes, at least until this entire matter has been put to a proper scientific test.

The Movius Line

One of the great boundaries of deep prehistory has come to be called the "Movius Line." Hallam Movius, the Harvard archaeologist, first made the observation that there was a boundary that could be drawn roughly down the northern edge of the Ganges Valley that separated the "Acheulean" industries of peninsular India, Africa and the west, from the "pebble" or chopper-chopping, tool industries of Eastern and Southeastern Asia (Movius,

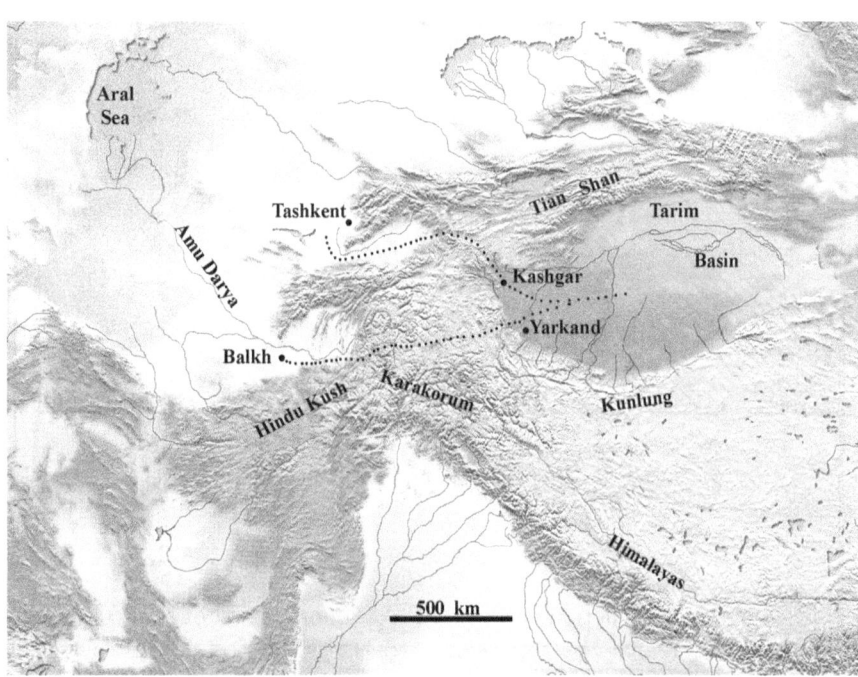

Figure 1. Potential routes in Central Asia

1944:103–104). I have reproduced his map showing the distribution as Figure 2. There is not a specific chapter in this book that deals with this issue, but it deserves some attention.

For example, the Movius Line may not be not a line at all, since there are many chopper-chopping assemblages in the subcontinent, some of them south of the Narmada River, something Movius was aware of (1944:104). In 1956, B.B. Lal published a map showing the relative percentages of chopper-chopping tools verses bifacial handaxes and cleavers in India and Pakistan, reproduced here as Figure 3. It should be noted that the Soan Industry, and presumably the other industries in the Siwaliks with high percentages of chopper-chopping tools, would be on the northeastern side of the Movius Line.

In the Irrawaddy River Valley of Myanmar there is a paleolithic industry, called the Anyathian, also studied by Movius (De Terra and Movius, 1943). It is a chopper-chopping tool industry that used fossilized wood and silicified tuffs as raw materials. The chronology of the Anyathian industry is based on river terrace observations, and is not at all certain. Moreover, all of these observations are old, of the 1940s and could stand a re-examination from a more contemporary perspective.

The fact that the Anyathian sits at the eastern frontier of the subcontinent, just across the mouth of the Ganges-Brahmaputra delta is interesting, since it may have some effect on our evaluation of the sea coast route, for example. The chronological and techno-typological position of the Anyathian industry vis-avis the other chopper-chopping tool industries of the subcontinent would also be worth investigating since such observations could inform us about the reality of the Movius Line:

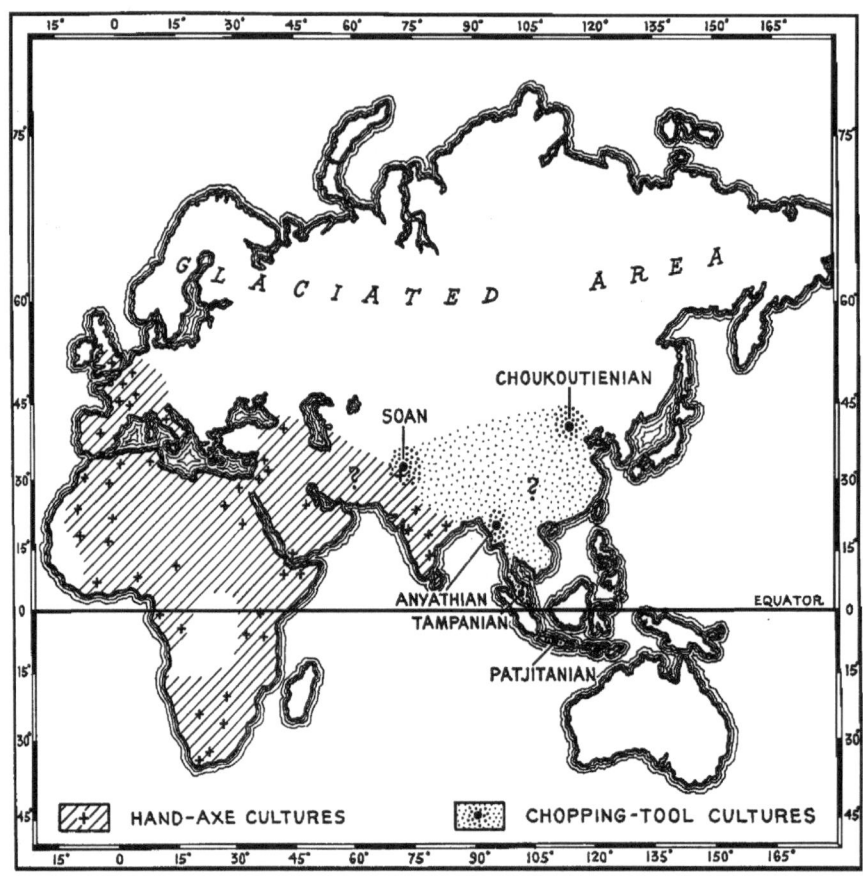

Figure 2. The Movius Line (after Movius, 1944:103)

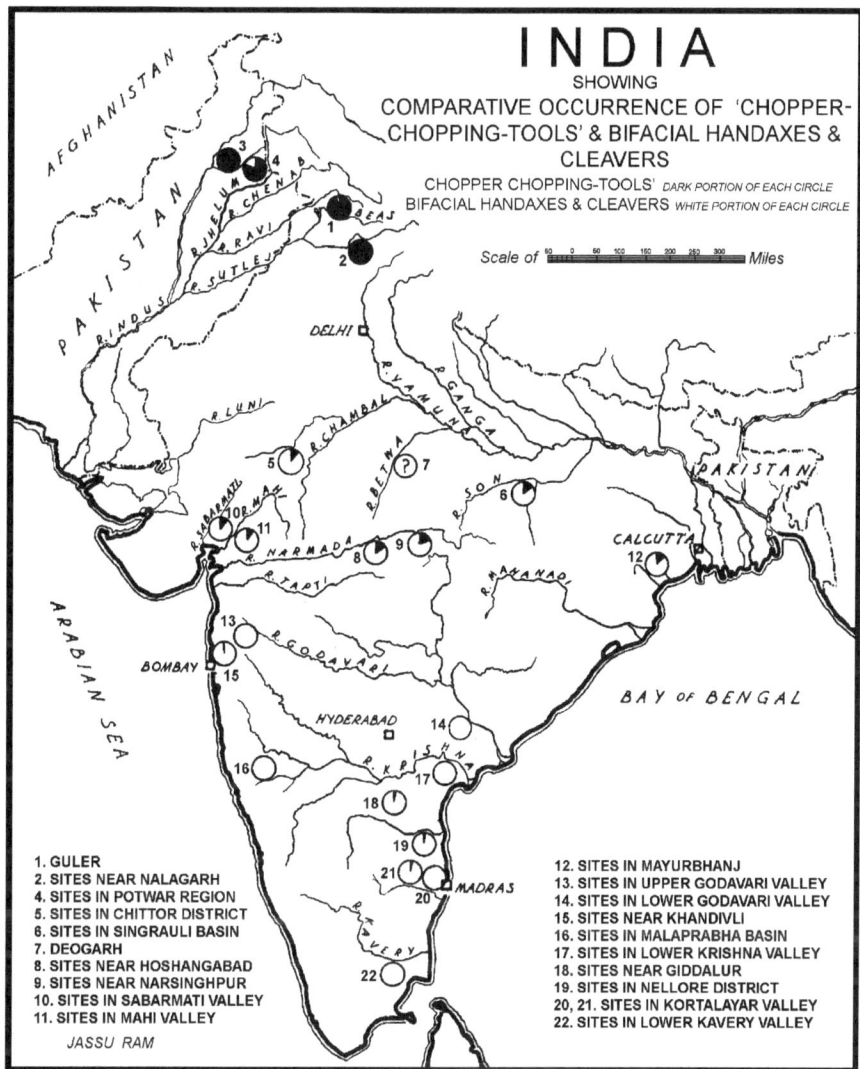

Figure 3. Comparative frequencies of care bifaces and Chopper chopping tools in the Subcontinent (after Lal, 1956:87)

does it actually separate two different cultural spheres, east and west, in paleolithic times? Is the Movius Line a "line" or more of a cline?

The Hoabinian is a long-lived (ca. 15–5 ka) industry found in Southeast Asia, both mainland and insular (Gorman, 1970). There is some fairly strong evidence for it at the Rongram Terrace site in Assam (Sharma, 1990). Since we have learned that not all simple stone tools are early, it would be worth knowing what the techno-typological relationship is between the Hoabinhian, the Anyathian and the chopper-chopping tool assemblages in India and Pakistan. While not nearly enough is

known of the Late Pleistocene-Early Holocene paleolithic industries of the subcontinent, the presence of the Hoabinhian in Assam suggests that if it is "real" the Movius Line may have persisted into the Holocene, thus characterizing much more than the Lower Paleolithic.

The Trajectory for the South Asian Paleolithic

There is the outline of a core biface/chopper-chopping tool transition to a flake-based technology using the Levallois technique in the subcontinent. More work needs to be done on this important transition, but it seems in general to conform to transitions in other parts of the Old World west of the Movius Line. But, the Upper Paleolithic is a weakly expressed in the subcontinent, so much so, that one could wonder if this is really a period there at all, as noted by James in this volume. But, microlithic technology is expressed quite robustly. In Sri Lanka there are microlithic tools, in association with *H. sapiens* that date to 31 ka at Fa Hien Cave and ca. 28.5 ka at Batadomba Lena (Deraniyagala, 1992:695–697). Sites with microliths are abundant in India, and there are a few sites with them in Pakistan, especially in the Karachi area, but we do not know the dates for most of them. Outside of Sri Lanka, none have been documented back beyond the Holocene boundary, ca. 10–8 ka.

It is a common-place in Indian archaeology to find sites where microlithic tools are associated with copper and iron artifacts and ceramics, even ceramics of the historical eras. Jacobson (1970:22) noted that there have been two reports from India of microliths made from bottle glass (see also Todd, 1932:42; Gordon and Gordon, 1943:95; Allchin, 1966:102), which would imply that they were made after the seventeenth century. Some of the peoples who used microlithic tools may also have made use of domesticated animals.

The microlithic tools are also techno-typologically astoundingly similar to their counterparts in the Near East, North Africa and Europe, at least to my eye. There are crescents, lunates, trapezoids and a wide variety of triangular forms, and beautiful polyhedral cores. Sound comparative work on this topic would serve to test this claim, and may shed light on inter-regional contact and population movements to and from the subcontinent and Sri Lanka.

A word should be said about the use of the terms "microlith" and "mesolithic" in South Asian archaeology. In many instances an author may use the term "microlith" in a literal way, "small stone tool." There are a number of such diminutive industries in the subcontinent. A good example of this is the Chalcolithic short blade industry associated with Sorath Harappan sites in Gujarat. But the same term is also used for the geometric and non-geometric microliths. South Asian archaeologists should be weaned from the habit of using the term "microlith" for anything other than the latter, because not doing this is so confusing. It is also true that all sites in the subcontinent with geometric or non-geometric microliths can, and are, called "mesolithic" even if they have pottery, domesticated animals and tools made of bottle glass. Is the Harappan village of Allahdino a mesolithic site because it has microlithic tools (Hoffman and Cleland, 1977)? In Europe, where the term "Mesolithic" was first used, I believe it means archaeological sites of the Holocene that were occupied by peoples who did not know the art of food production. In the end archaeologists working in the subcontinent should sit together and decide what "Mesolithic" means in their world area, and either abandon the term, or use it in a well defined manner.

As noted by Morrison in this volume not nearly enough is known about even the chronology of microlithic sites in the subcontinent and a concerted effort is needed to correct this fundamental deficiency. For the moment it is reasonable to hypothesize that microlithic industries were present in the subcontinent from at least 31,000 years ago and persisted in use until the 18th or 19th centuries of the common era. There may be a transition more-or-less directly from the flake industries of the Middle Paleolithic to microlithic technology, explaining the very weak expression of an Upper Paleolithic. But, these are hypotheses, in need of proper testing. Also in need of further testing is the notion that there were a number of different adaptations that peoples using microlithic tools in

India assumed. One of them seems to have involved interactive trade and barter between hunter-gatherers and agro-pastoralists.

Hunter-Gatherer/Agro-Pastoralist Interaction

There are at least three different cultural and historical contexts in the subcontinent for sites with a microlithic stone working technology: the Mesolithic Aspect, the Early Food Producing Aspect and the Interactive Trade and Barter Aspect (Possehl and Rissman, 1992), also recognized by Morrison in this volume. Sites of the Interactive Trade and Barter Aspect have the classic microlithic technology associated with variable faunal and floral assemblages. At times domesticated plants and animals seem to have been a part of the picture, although this is not exclusively the case, and it needs further testing. Of interest in sites of this aspect is the presence of technologically sophisticated materials such as copper/bronze, iron and glass. Other materials such as carnelian beads, sea shells and steatite also occur. Coins are known from some of these sites. Pottery is generally present and usually can be tied to the ceramics of surrounding village farming communities. Trade and interaction between these sites and surrounding communities, at times some distance away, can be inferred from these ceramics and the known source areas for the "exotic" materials found at these small settlements.

This can be illustrated by an examination of Langhnaj, one of the most important microlithic sites to have been excavated in Gujarat. It is situated in dunes and alluvial hillocks about 160 kilometers north of Lothal. Langhnaj is a site with an abundant microlithic industry found in three phases (Sankalia, 1965). There is one radiocarbon determination for Phase II at Langhnaj (TF-744) which calibrates to 2440-2160 BC, coincident with the Mature Harappan occupation of Lothal.

Bits of pottery came from all three levels of the site, along with stone tools. The ceramics were so poorly fired that they come in very small sherds, shapes being apparent only in the latest Phase III. But, those of Phase II are definitely a coarse black and red ware, with some typological similarity to the black and red wares of Lothal and it is perfectly possible that the inhabitants of Langhnaj learned the potter's art from the Harappans, most likely the Early Harappan pioneers who preceded the people of Lothal. Pottery and stone tools continue into Phase two. In addition, two ground stone artifacts were discovered: a pointed butt axe and a ring-stone or mace head. A knife of 98.12% pure copper and steatite disk beads were found in Phase II as well. These are all pieces of "advanced technology" in so far as the hunter-gatherers were concerned, somewhat out of place at Langhnaj, especially the copper knife. A very fine iron arrow head comes from Period III, with good Early Historic (ca. 300 AD) typological parallels. This date, along with the other radiocarbon dates and the stratigraphic evidence from Kanewal, with a microlithic camp stratigraphically situated between two Chalcolithic "villages" (Mehta et al., 1980), adds much strength to the case that there were hunters and gatherers in Gujarat at the time of the Sindhi Harappan occupation of Lothal and the other sites in Kutch.

These chronological considerations are important because they at least admit the possibility that the copper knife, steatite disk beads, ground stone tool making technology, possibly even the black and red ware pottery, came to Langhnaj, and doubtless other sites in North Gujarat as well, as items of barter with the Mature Harappans. Lothal emerges as a particularly important place because of its trading post character. It should also be reiterated for emphasis that Lothal was not a fortified site and this can be taken as a good indication that it enjoyed peaceful relations with its neighbors, some of whom should not have been included as a part of the Harappan Civilization.

There is more evidence for interaction between the Harappans of Gujarat and the hunting and gathering population of the region. This comes from physical anthropology and the analysis of the burials from Lothal. K.A.R. Kennedy has examined these remains, as well as those from Langhnaj, and other sites in this region. He and his colleagues (Kennedy et al., 1984) have noted that the individuals interred in the cemetery at Lothal fall within the range of variability for the Mature Harappan population as a whole, but is statistically somewhat to one side of the norm. Some of the metrical variables that seem to be "pushing" these individuals off the Harappan norm are features of facial robusticity (prognathasism, tooth size, skull thickness and the like) that are physical characteristics of the hunter-gatherers at Langhnaj and other sites of this type in the region. He proposes that we therefore have good reason to believe that more than economic intercourse took place between the Harappans in Gujarat and their hunter-gatherer neighbors (Kennedy et al., 1984:116).

J. Lukacs and his colleague J. N. Pal (1993) have also examined the Langhnaj skeletal series. Lukacs (in this volume) has additional information on Ganges Valley populations. They have noted that they a very high rate of dental caries. Other hunter-gatherer groups from the subcontinent, and other parts of the world as well, are characterized by low incidents of this malafliction, but it is generally high among food producing peoples, who consume large amounts of processed carbohydrates. The residue from these foods tends to stick on the teeth where the enzyme that causes tooth decay can do its work. The people of Langhnaj were not food producers. There were no domesticated animals found there, nor were harvesting tools. Thus, Lukacs and Pal believe that they may have been

getting a significant portion of their food from farmers in their region, through exchange. Lothal would be one of the prime candidates for participation in such an arrangement.

This evidence for trade and/or exchange and gene flow between the Harappans and hunter-gatherers in Gujarat supports the notion that the hunter-gatherers were people who procured raw materials for the factories and traders who lived at Lothal, and possibly other Sindhi Harappa sites in the region. This was probably only one way that the Harappans obtained such materials, but it would have been important for them since the hunter-gatherers would have been intimately acquainted with their own terrain and therefore could find the products in which the Harappans had shown an interest. These would have been materials like those found at Lothal: agate, carnelian, rock crystal, steatite, shell, ivory, as well as wood, like teak from the Western Ghats. Tin should also be mentioned because alluvial tin has been reported from North Gujarat. It does not seem likely that the hunter-gatherers of Gujarat played a role in the acquisition of copper, unless the Harappan smiths trained them to find the ores, mine and concentrate them. We do not know the answer to this question, but we should not rule out the possibility of quasi-formal training being needed in order for the Harappans to get what they wanted.

This symbiosis between hunter-gatherers and settled folk in the subcontinent is a characteristic of life there, that persists today. Since this lifeway has disappeared in Pakistan, we can focus on India, where there a few hunting and gathering groups do survive today, but were much more numerous in the nineteenth century. We learn from studies of these people that they were hunters and gatherers in the sense that they did not keep (many?) domesticated animals or engage in (much?) agriculture and earned most of their livelihood from the extraction of forest products. But, the key to their survival lies not in isolated self-reliance, but on a complex, symbiotic relationship with the cultivator, peasants around them. The forest people hunted wild animals and gathered forest products that were traded to their neighbors for agricultural products, metal implements, cloth and the like. Richard Fox has expressed this relationship in the following way:

Rather than being independent, primitive fossils, Indian hunter-and-gatherers represent occupationally specialized productive units similar to caste groups such as carpenters, shepherds or leather-workers. Their economic regimen is geared to trade and exchange with the more complex agricultural and caste communities within whose orbit they live. Hunting and gathering in the Indian context is not an economic response to a total undifferentiated environment. Rather it is a highly specialized and selective orientation to the natural situation: where forest goods are collected and valued primarily for external barter or trade, and where necessary subsistence or ceremonial items—such as iron tools, rice, arrow heads, etc.—are only obtainable this way. Far from depending wholly on the forest for their own direct subsistence, the Indian hunters-and-gatherers are highly specialized exploiters of a marginal terrain from which they supply the larger society with desirable, but otherwise unobtainable forest items such as honey, wax, rope and twine, baskets, and monkey and deer meat. Unlike the Australian aborigines or the Paiutes, their economic processes and well-being are dependent on the barter of these items for the crops and crafts of their more complexly organized plainsmen neighbors. The economic activity of Indian hunting-and-gathering groups is more akin to the specialization of caste hereditary occupation, than it is to the generalized environmental response of the Australians or Paiute (Fox, 1969:141–142).

The Origins of Food Production

Fuller (in this volume) notes the older notion that domesticated animals were introduced into the subcontinent through a process of diffusion from the Fertile Crescent. Some propose that they arrived with new human populations, and their languages (Renfrew, 1987; Bellwood and Renfrew, 2002). I have argued elsewhere that this set of ideas is in need of rethinking (Possehl, 2002:25–29).

The reasons for this are simple. Potential domesticates like the cattle, sheep and goats, as well as barley are as much at home in Pakistani Baluchistan as they are in the Fertile Crescent. Only wheat is missing from this package, but the early food producing peoples of the western subcontinent seem to have grown little of this crop. Moreover, the distribution of these plants and animals comes from nineteenth and twentieth century exploration, and may not accurately reflect their distribution 10,000 to 12,000 years ago. So, the fact that we do not find wild wheat in Baluchistan today does not mean that it was not there at 10,000 BC, or even earlier.

R. Meadow has noted that it is now clear that cattle, the humped South Asian zebu (*Bos indicus*) were locally domesticated at Mehrgarh (1998:12) (see Magee et al., this volume). Meadow goes on to note that:

1. Sheep are likely to have been domesticated from local wild stock during Period I (ca. 7000 BC).
2. Goats were kept from the time of the first occupation of the site.
3. Size diminution in goats was largely complete by late Period I, in cattle by Period II, and in sheep perhaps not until Period III.

The most parsimonious explanation for the domestication of these three key animals is that the process was local, using animals from their home ranges at the far eastern edge of the Iranian Plateau in Pakistani Baluchistan.

As to the key plants, barley and wheat, much research remains to be done. Period I at Mehrgarh is dominated by domesticated, naked six-row barley. There are two other varieties of domesticated barley as well. Domesticated wheat is present in very small amounts, seemingly less that 10% of the total food grains, in the form of einkorn, emmer, and a free threshing hard "durum." It is thus evident that Period I at Mehrgarh does not document the beginning of plant domestication, since there is already so much diversity. We need to find and excavate another site, slightly earlier than Mehrgarh, to document this part of the story of the development of food production on the western borderlands of the subcontinent.

Fuller's contribution to this volume also makes a very strong case for the indigenous development of the "Southern Neolithic" with its own suite of plants, but including cattle, sheep and goats. He further notes that it is possible that individual plants and animals were more-or-less independently domesticated in a number of regions of the subcontinent: Baluchistan, Gujarat, the mid-Ganges Valley, Orissa and South India. While there is much more interesting and important research to be done on the origins of agro-pastoralism in the subcontinent it is clear that this side of research is proceeding along in a vigorous way.

Human Biology, Language and the Remainder of Culture

In 1940, Franz Boas published an important book titled *Race, Language and Culture*. Using more contemporary terminology this translates to: human biology, language (a part of culture) and the remainder of culture. Implicit in many of the chapters in this volume are notions that there can be broad, underlying correlations between human biology, language and other features of culture, such is lifeways (e.g., peasants, tribals), ideology and subsistence practices (e.g., farming, pastoralism, hunting and gathering).

Boas addressed many themes in *Race, Language and Culture*. One of the most important is that, with few exceptions, human biology, language and the rest of culture are historically independent variables. They do not necessarily tend to "travel" together over long distances, or for long periods of time. There are many reasons for this, but they tend to fall into three categories: inherent cultural dynamism, interaction and environment,

which are certainly not independent of one another at a higher level of abstraction.

Cultural Dynamics

In biological terms, humans seem to change rather slowly, without interaction with others of our species. But, the same is not true for language and the rest of culture. These are in a constant dynamic state, never still, always changing. While culture change is difficult to quantify, there is a deep sense that this dynamic is not a constant, that sometimes it is rather faster than other times, giving us the sense of "ramps" and "steps" in long term historical trends (Braidwood and Willey, 1962:351; Adams, 1966:17–18).

Since the dynamic of culture change is always operative, peoples removed from one another, tend to drift apart, to become culturally different, without any other influences. In language, a good example of this can be found in systematic sound shifts, such as the initial 's' shift to 'h' which is one of the most common. This takes place not necessarily because people are in contact with others and gravitate to their way of speaking, but because of the dynamic nature of culture and factors influencing "linguistic drift" (see McMahon and McMahon, this volume, for interesting examples and measures).

Interaction

Interaction among peoples also leads to human biological and cultural change. It may well be true that any aspect of human culture can be borrowed from one community to another. "We know that myths, religious ideas, types of social organization, industrial devices, and other features of culture may spread from point to point, gradually making themselves at home in cultures to which they were at one time alien" (Sapir, 1921:205). And this list could be vastly increased. As for language, Dixon (1997:19–27) offers a very lucid discussion of what features of language tend to be borrowed, or shared, or lent and under what conditions this takes place.

After the initial peopling of Eurasia in all probability those who lived there were seldom, if ever, isolated for long, at least in terms of an archaeological perspective on time. The more-or-less sedentary people move about in the course of their day-to-day activities, and over the span of their lives. And they had other peoples moving among them: travelers, bards, tinkers, pastoralists and the like who served a host of useful purposes such as buying and selling commodities, trading goods and services, managing exchanges, fixing things that were broken, repairing things that needed attention, telling stories, giving performances, spreading news and gossip, lending "money," arranging marriages. More-or-less balanced, reciprocal interregional exchanges of goods took place between friends and allies.

There is every reason to believe, if we look to the documentation from history, and believe in notions of uniformatarianism, that the predominant theme of Eurasian history was human interaction, contact and sharing, not stagnation, isolation and closure. Ancient Eurasia would at times have been a seething cauldron of cultural dynamism. And, humans being humans, they shared their biological features along with their languages and other aspects of culture.

Environment

The natural environment also shapes humans, particularly their biology. Since the environment is also a dynamic, human adaptation, in cultural, linguistic and biological terms is never at a standstill. This is also a part of why human biological races have long been abandoned as a scientific proposition. There are good references to document this, but the ones I like best were written by Kennedy (1976, 1977, 1995). As people lived their lives in during prehistoric times in Eurasia the dynamics of human biology, language and the rest of culture

were all at work within a dynamic of culture, human interaction and an environmental setting. These are the forces which come into play to make "race, language and culture" independent historical variables.

Caste and its Origins

Caste is probably the most distinctively Indian social institution. This volume contains a contribution from N. Boivin that is a review of this topic from the perspectives of anthropology, history, archaeology and genetics. Along the way she also has useful comments on the problem of the identification of Indo-European speakers and the so-called "Aryan invasion" of the subcontinent.

The origins of caste have been the subject of some rather speculative writing, but there is some history. Some take it back to the Indus Civilization (Kenoyer, 1995). Megasthenes, who was in India during the Mauryan Empire of the third century BC, does not mention caste, but rather a seven endogamous occupational "classes": philosophers, peasants herdsmen, craftsmen, traders, soldiers government officials and councilors (McCrindle, 1877). The Laws of Manu note many occupations, within the system of four varnas and accounts for them by marriages across occupational boundaries. H.H. Risley believed that the system arose out of the racial context of blood purity, especially that posed by the Aryans and their dark skinned conquests (1915). Professor J.H. Hutton believed that the system arose at the village level when a dominant community of people raised themselves above their peers and enforced this social standing through the employment of taboos (Hutton, 1946).

It is clear that most discussions of the origins of caste are rife with racial overtones, historical bias, imprecise thought, bad science and various attitudes towards colonialism. But, the Boivin contribution is such a well researched and clear presentation of the topic that I can end my discussion of it with the following quote from the conclusions section of her chapter.

These various assertions concerning the origins of caste, made on the basis of the diverse findings of the disciplines addressed in this chapter, remain to be investigated through more detailed and systematic studies. That such studies need to be marked by a greater degree of inter-disciplinary interaction and informed discussion than has occurred to date is clear.

References

Adams, R. McC., 1966. The Evolution of Urban Society. Aldine, Chicago.

Allchin, B. 1966. The Stone-Tipped Arrow. Barnes and Noble, New York.

Boas, F., 1940. Race, Language and Culture. Macmillan Company, New York.

Braidwood, R.J., Willey, G.R. (Eds.), 1962. Courses Toward Urban Life: Archaeological Considerations of Some Cultural Alternates. Aldine Publishing Company, Chicago.

Deraniyagala, S.U. 1992. The Prehistory of Sri Lanka. Memoir Volume 8, Parts I and II. Department of Archaeological Survey, Government of Sri Lanka, Colombo.

Dixon, R.M.W., 1997. The Rise and Fall of Languages. Cambridge University Press., Cambridge.

Fox, R.G., 1969. Professional primitives: hunters and gatherers of nuclear South Asia. Man in India 49(2), 139–160.

Gordon, D.H., Gordon, M.E., 1943. The cultures of Maski and Madhavapur. Journal of the Royal Asiatic Society of Bengal, Letters 9, 83–96.

Gorman, C., 1970. The Hoabinhian and after: subsistence patterns in Southeast Asia during the late Pleistocene and early Recent periods. World Archaeology 2(3), 300–320.

Hoffman, M.A., Cleland, J.H., 1977. Excavations at the Harappan Site of Allahdino: The Lithic Industry at Allahdino. Papers of the Allahdino Expedition, No. 2, New York.

Hutton, J.H. 1946. Caste in India: Its Nature, Function and Origins. Cambridge University Press, Cambridge.

Jacobson, J., 1970. Microlithic Contexts in the Vindhyan Hills of Central India. Ph.D. Dissertation, Department of Anthropology, Columbia University.

Kennedy, K.A.R., 1976. Biological anthropology of prehistoric populations in South Asia. In: Kennedy, K.A.R., Possehl, G.L. (Eds.), Ecological Backgrounds of South Asian Prehistory. Cornell University South Asia Occasional Papers and Theses, No. 4, Ithaca, pp. 166–78.

Kennedy, K.A.R., 1977. A reassessment of the theories of the racial origins of the people of the Indus Valley Civilization from recent anthropological and archaeological data. Paper read at the 6th University of Wisconsin South Asia Conference, Madison, November 4–6.

Kennedy, K.A.R., Chiment, J., Drisotell, T., Meyers, D., 1984. Principal-components analysis of prehistoric South Asian crania. American Journal of Physical Anthropology 64(2), 105–118.

Kennedy, K.A.R., 1995. But Professor, why teach race identification if races don't exist? Journal of Forensic Science September, 797–800.

Lal, B.B., 1956. Palaeoliths from the Beas and Banganaga Valleys, Punjab. Ancient India 12, 58–92.

Lukacs, J.R., Pal, J.N., 1993. Mesolithic subsistence in North India: inferences from dental attributes. Current Anthropology 34(5), 745–765.

McCrindle, J.W., 1877. The fragments of the Indika of Megasthenes. The Indian Antiquary 6, 113–35, 236–50.

Meadow, R.H., 1998. Pre- and Proto-historic agricultural and pastoral transformations in northwestern South Asia. Review of Archaeology 19(2), 12–21.

Mehta, R.N., Momin, K.N., Shah, D.R., 1980. Excavation at Kanewal. Maharaja Sayajirao University, Archaeology Series, No. 17, Baroda.

Movius, H.L., 1944. Early Man and Pleistocene Stratigraphy in Southern and Eastern Asia. Papers of the Peabody Museum of American Archaeology and Ethnology, Harvard University, 19(3). Peabody Museum of American Archaeology and Ethnology, Harvard University, Cambridge.

Possehl, G.L., Rissman, P.C., 1992. The chronology of prehistoric India: from earliest times to the Iron Age. In: Ehrich, R.W. (Ed.), Chronologies in Old World Archaeology (2 Vols.). University of Chicago Press, Chicago, pp. 465–490, 447–474.

Renfrew, C., 1987. Archaeology and Language: The Puzzle of Indo-European Origins. Cambridge University Press, Cambridge.

Risley, H.H., 1915. The People of India. Thacker and Spink, Calcutta.

Sankalia, H.D., 1965. Excavations at Langhnaj: 1944–63, Part 1, Archaeology. Deccan College Postgraduate and Research Institute, Poona.

Sapir, E., 1921. Language. Harcourt Brace, New York.

Sharma, T.C., 1990. Discovery of Hoabinian cultural relics in north-east India. In: Ghosh, N.C., Chakrabarti, S. (Eds.), Adaptation and Other Essays: Proceedings of the Archaeological Conference, 1988. Visva-Bharati Research Publications, Santiniketan, pp. 136–139.

de Terra, H., Movius, H.L., 1943. Research on Early Man in Burma. Transactions of the American Philosophical Society, Philadelphia.

Todd, K.R.U., 1932. Prehistoric man round Bombay. Journal of the Prehistoric Society of East Anglia 7(1), 35–42.

Index

VERTEBRATE PALEOBIOLOGY
AND PALEOANTHROPOLOGY

PUBLISHED AND FORTHCOMING TITLES

Neanderthals Revisited: New Approaches and Perspectives
Edited by K. Harvati and T. Harrison
ISBN: 1-4020-5120-4, 2006

The Evolution and Diversity of Humans in South Asia
Edited by M. Petraglia
ISBN: 1-4020-5561-7, 2007

Hominin Environment in the East African Pliocene: An Assessment
of the Faunal Evidence
Edited by R. Bobe, Z. Alemseged and A.K. Behrensmeyer
ISBN: 1-4020-3097-5, *forthcoming 2007*